PROGRESS IN SOLAR PHYSICS

Detail of the golden ceiling in the Golden Temple of Amritsar, India, 18th century.

PROGRESS IN SOLAR PHYSICS

Review Papers Invited to Celebrate the Centennial Volume
of *Solar Physics*

Edited by

C. DE JAGER and Z. ŠVESTKA

Laboratory for Space Research, Utrecht, The Netherlands

Reprinted from Solar Physics, Vol. 100, Nos. 1–2

D. Reidel Publishing Company / Dordrecht / Boston

Library of Congress Cataloging-in-Publication Data
Main entry under title:

Progress in solar physics.

   "Reprinted from Solar physics, vol. 100, nos. 1-2."
   Bibliography: p.
   1. Sun--Addresses, essays, lectures.   2. Solar
physics.   I. Jager, C. de (Cornelis de), 1921-
II. Švestka, Zdenek.   III. Solar physics.
QB521.6.P76   1986   523.7   86-473
ISBN-13: 978-94-010-8538-0   e-ISBN-13: 978-94-009-4588-3
DOI: 10.1007/978-94-009-4588-3

Published by D. Reidel Publishing Company,
P.O. Box 17, 3300 AA Dordrecht, Holland.

Sold and distributed in the U.S.A. and Canada
by Kluwer Academic Publishers,
190 Old Derby Street, Hingham, MA 02043, U.S.A.

In all other countries, sold and distributed
by Kluwer Academic Publishers Group,
P.O. Box 322, 3300 AH Dordrecht, Holland.

TABLE OF CONTENTS

(Progress in Solar Physics)

PUBLISHER'S NOTE

We are very proud to present herewith the 100th Volume of *Solar Physics*, which is a landmark in the history of this journal.

This 100th Volume is unique in the sense that it does not contain regular research papers, but a collection of outstanding review papers by eminent solar physicists. Because of this it was felt appropriate to also publish this Volume in a hardback edition to make it more widely available, and to acquaint those, who are not regular subscribers to *Solar Physics*, with its high-quality contents.

It is only 18 years ago since this journal began publication. During these years over 50 000 pages of high-quality research have been published. We are honored to be associated with the many distinguished scientists who have made *Solar Physics* the prime journal in its field. We are in particular grateful to the Editors, Prof. C. de Jager and Dr Z. Švestka, who have been the stimulus behind this journal from its very inception. We also would like to extend our thanks to those who have been serving on the editorial board, and all who have been contributing to the journal.

We are looking forward to a long-lasting association with those working in the field that is embraced by the scope of *Solar Physics*, and we do hope that the journal will continue to receive the same excellent support from the community.

J. F. HATTINK

EDITORIAL

When we started this journal in 1967, we expected one or two volumes per year. Now, only 18 years later, we celebrate the centennial volume. Apparently, our colleagues have been much more prolific, and much more interested to publish in *Solar Physics*, than we ever dared to anticipate.

It has been a real pleasure to serve as the editors for the one hundred volumes that so far had appeared and we have greatly enjoyed the cooperation with all the colleagues who contributed to the journal or helped us in our editorial work.

In particular, we would like to use this opportunity for expressing our deep gratitude to the members of the *Solar Physics Editorial Board* for all the help we have received from them. Altogether, 89 scientists served in this function, some of them all the time since the first volume had appeared in 1967: R. G. Athay (Boulder), J. M. Beckers (Sac Peak/Tucson), V. Bumba (Ondřejov), R. F. Howard (Pasadena/Tucson), J. Kleczek (Ondřejov), A. B. Severny (Crimea), V. E. Stepanov (Irkutsk), and H. Zirin (Pasadena). The list of all those who were on the Board during the past 18 years is given below:

H. Alfvén	Sweden	1967–1971
R. G. Athay	U.S.A.	1967–
J. M. Beckers	U.S.A.	1967–
Chen Biao	China	1985–
L. Biermann	F.R.G.	1967–1974
D. E. Blackwell	U.K.	1967–1975
R. M. Bonnet	France	1982–
H. A. Brück	U.K.	1967–1974
A. Bruzek	F.R.G.	1967–1980
V. Bumba	Czechoslovakia	1967–
C. Chiuderi	Italy	1981–1984
F. L. Deubner	F.R.G.	1976–
A. Dollfus	France	1967–1975
A. R. Dupree	U.S.A.	1978–1981
J. W. Evans	U.S.A.	1967–1976
J. W. Firor	U.S.A.	1967–1974
A. D. Fokker	Netherlands	1967–1977
E. Fossat	France	1985–
H. Friedman	U.S.A.	1968–1975
A. H. Gabriel	U.K.	1978–
V. Gaizauskas	Canada	1984–
R. Giovanelli (†)	Australia	1967–1981
M. N. Gnevyshev	U.S.S.R.	1967–1976
L. Goldberg	U.S.A.	1967–1974
D. O. Gough	U.K.	1983–
J. W. Harvey	U.S.A.	1975–
T. Hirayama	Japan	1979–
J. Houtgast (†)	Netherlands	1967–1974
R. F. Howard	U.S.A.	1967–
H. S. Hudson	U.S.A.	1980–
A. Hundhausen	U.S.A.	1976–1983
F. W. Jäger	G.D.R.	1967–1979
J. T. Jefferies	U.S.A.	1967–1981

C. Jordan	U.K.	1975–
K. O. Kiepenheuer (†)	F.R.G.	1967–1975
J. Kleczek	Czechoslovakia	1967–
V. A. Krat (†)	U.S.S.R.	1967–1977
M. R. Kundu	India/U.S.A.	1968–
M. Kuperus	Netherlands	1975–
J. W. Leibacher	U.S.A.	1982–
J. Linsky	U.S.A.	1978–
W. C. Livingston	U.S.A.	1975–1978
M. E. Machado	Argentina	1982–
R. M. MacQueen	U.S.A.	1980–
P. Maltby	Norway	1981–
S. L. Mandelstam	U.S.S.R.	1976–
A. Maxwell	U.S.A.	1967–1974
F. Meyer	F.R.G.	1975–1982
R. Michard	France	1967–1977
D. M. Mihalas	U.S.A.	1981–
O. C. Mohler	U.S.A.	1967–1975
G. Newkirk	U.S.A.	1968–
J. C. Noyes	U.S.A.	1967–1974
R. W. Noyes	U.S.A.	1967–1981
Y. Öhman	Sweden	1967–1974
F. Q. Orrall	U.S.A.	1976–
E. N. Parker	U.S.A.	1968–
A. Keith Pierce	U.S.A.	1967–1976
S. B. Pikelner (†)	U.S.S.R.	1968–1977
E. R. Priest	U.K.	1977–
G. Righini (†)	Italy	1967–1975
J. Rösch	France	1967–1980
I. W. Roxburgh	U.K.	1978–
D. M. Rust	U.S.A.	1977–1984
H. U. Schmidt	F.R.G.	1983–
E. H. Schröter	F.R.G.	1975–
A. B. Severny	U.S.S.R.	1967–
S. F. Smerd (†)	Australia	1978–1979
B. V. Somov	U.S.S.R.	1982–
J. O. Stenflo	Sweden/Switzerland	1975–
V. E. Stepanov	U.S.S.R.	1967–
P. A. Sturrock	U.S.A.	1975–1978
Z. Suemoto	Japan	1967–1976
S. I. Syrovatsky (†)	U.S.S.R.	1976–1980
T. Takakura	Japan	1967–1977
E. A. Tandberg-Hanssen	U.S.A.	1980–
R. Tousey	U.S.A.	1967–1976
Y. Uchida	Japan	1976–
R. K. Ulrich	U.S.A.	1978–
W. Unno	Japan	1967–1975
M. Waldmeier	Switzerland	1968–1977
N. O. Weiss	U.K.	1982–
J. M. Wilcox (†)	U.S.A.	1976–1979
J.P. Wild	Australia	1967–1980
G. Withbroe	U.S.A.	1979–1982
V. V. Zheleznyakov	U.S.S.R.	1977–
H. Zirin	U.S.A.	1967–
J. B. Zirker	U.S.A.	1977–
C. Zwaan	Netherlands	1977–

We also want to thank the referees, many of which were not members of the Editorial Board. We realize the burden related to receiving for refereeing an unrequested paper, and therefore appreciate that most referees did their work well and reasonably fast. For those few who did not we have a polite request:

Please, if you cannot referee a paper within three or four weeks, send it back to our Editorial Office by return mail. If you leave for three or more weeks, do not let your mail pile on your desk, burying requests for referee reports under heaps of other matters. If you tend to forget it, kindly imagine that you yourself are the author and some other referee did receive your manuscript.

We have decided to celebrate the centennial volume of *Solar Physics* by a series of invited review papers over timely topics in solar research. Originally we had in mind just a few contributions in the leading part of the volume; but, after discussions with many members of the Editorial Board, and following their attractive suggestions, we have eventually invited altogether 33 distinguished colleagues to contribute. To our very pleasant surprise, only two of them have turned us down, and only two of those who agreed to contribute, failed to send us their papers. This has left us with 29 excellent contributions, greatly differing in style and contents, which, in our feeling, adds to the flavour of this volume. To that, we have taken the liberty to add one more paper, prepared by ourselves. The total extent is now far in excess of the usual 400 pages per volume, but we believe that the contents are worthy of this size.

Among the reviews are a few which discuss stars of the solar type, whatever this may mean. This reflects our long-lasting wish to extend the scope of *Solar Physics* to stellar problems, for which a comparison with solar data is essential. The fact that our Sun is the only star, of which the detailed structure can be seen, and that there are many stellar problems in which this knowledge can be very usefully applied, is well known, but rarely practically applied. We would like very much to use more pages of *Solar Physics* in the future for papers that compare the Sun and solar-type stars, solar and stellar magnetic fields, chromospheres, coronae, winds, and solar and stellar activity. Any help from the members of our Editorial Board and from any other reader of this volume will be highly appreciated.

CORNELIS DE JAGER and ZDENĚK ŠVESTKA
Solar Physics Editors

HIGH RESOLUTION SOLAR TELESCOPES

RICHARD B. DUNN

National Solar Observatory, National Optical Astronomy Observatories*,
Sunspot, NM 88349, U.S.A.

Abstract. The advantages and disadvantages of the configurations for high resolution solar telescopes are discussed within two broad groups: those with steerable mountings and those with fixed mountings. We then consider simple optical tests, stabilization of the internal optical path, windows, vibration, guiding and alignment systems, improving the observations, and solutions for large-aperture telescopes for Stokes polarimetry observations. This review does not address all the problems. It is not a compendium of solar telescopes, nor does it include any discussion of focal-plane instrumentation.

1. Introduction

There is a great deal of variety in solar telescopes. No two are alike. Each designer has a different idea of how to preserve and focus the wave front through the last 100 or so meters of air path that is more or less under his control. No one is content to copy an existing design, which suggests that the perfect high resolution solar telescope has not yet been conceived, or perhaps that each designer is favoring a different set of parameters or weighting them differently.

These parameters include the compactness and simplicity of the optical system, the number of large optical components that must be perfect over their entire aperture, central obscuration, the need for a window, the effect of image rotation, the requirements in alignment of the components to maintain high resolution, the amount of polarization introduced by inclined optical surfaces, the stabilization of the internal optical path, solar heating, and all the mechanical problems including treatment of the dome (interface between telescope and atmosphere), the height of the tower needed to be above ground turbulence, need for double towers to reduce the effects of the wind, reduction of vibration, size of the steerable structure, need for a heated observing room, drives, and so on. To discuss some of the possible tradeoffs we will divide solar telescopes into two broad classes: *Steerable Telescopes* that are pointed directly at the Sun and *Fixed Telescopes* that remain in one orientation and that use large mirrors to reflect the light to the telescope objective. (Some of the steerable telescopes send the light through the axes of the mounting to fixed instruments in an observing room, but we will still treat them as steerable.) We will then discuss some simple optical tests, stabilization of the internal optical path, windows, vibration, guiding and alignment systems, improving the observations, and solutions for a large-aperture relescope for Stokes observations. (Solar telescope parameters are also discussed by Engvold and Hefter (1982) and Giovanelli (1966).)

* Operated by the Association of Universities for Research in Astronomy. Inc., under contract with the National Science Foundation.

Solar Physics **100** (1985) 1–20.   0038–0938/85.15.

2. Steerable Telescopes

Steerable telescopes often look like conventional telescopes that are enclosed in tubes that form the structural members, or like a group of components attached to an equatorially-mounted optical bench or 'spar' (Evans, 1956; Carroll, 1970) or table (Cachon, 1984, p. 57) that is guided to track the Sun.

The simplest optical system for a steerable telescope consists of a single lens or mirror, perhaps with a magnifier at focus, pointed directly at the Sun. Examples of telescopes with a singlet simple lens as an objective include a 40-cm that used to be at Sac Peak (Dunn, 1965), and the twin 25-cm Hα telescopes at Big Bear (Zirin, 1970), which are evacuated to remove internal seeing. Because of the large amount of chromatism, these telescopes must always be used with narrow-band filters to obtain good focus.

Achromatic objective lenses are also used. The most successful of these is the 50-cm $f/13$ at Pic du Midi (Rösch, 1961; Mehltretter, 1979; Rösch, 1981; Cachon, 1984), with which the best granulation pictures have been taken. Recent instruments include the telescope for the Beijing Observatory (Li Ting, 1984) and the 30-cm at Culgoora (Loughead, 1968). The 60-cm achromat at Irkutsk also serves as a coronagraph and has the largest aperture of this type of telescope.

One concern with lenses is the change in spherical aberration with color. This variation is small on singlets, which must have an asphere to correct spherical, but can be very large in fast achromats whose curves and airspace are balanced to correct spherical aberations and coma simultaneously at only one wave length. A second concern is scattered light between the elements of the achromat, especially if the inner surfaces have the same radius. This should be small enough if the surfaces are coated, but needs to be checked in detail. (Albregtsen and Hansen (1974) and Mattig (1983) discuss scattered light in solar telescopes.) There is often condensation between the elements of the achromat. I evacuated the airspace between the elements of the 25-cm lens of the Solar Observing Optical Network (SOON). The airspace was considerable (1.5 cm) and was maintained uniformly by 5-mm diameter 'O' rings in a metal ring. The 'O' ring prevents the glass from touching the metal.

All coronagraphs have singlet simple objectives and are pointed directly at the Sun because additional optical elements in front of the 'Lyot stop' would scatter too much disc light into the corona. Examples include the 40-cm coronagraph at Sac Peak* (Evans, 1956) and the 40-cm High Altitude Observatory (HAO) instrument, which is no longer in use (Rush, 1963; Rush and Schnable, 1964). These instruments have different schemes for correcting the large amount of chromatism (45 cm for a 40-cm, $f/19$ singlet) introduced by a simple objective. The Sac Peak instrument corrects the color with a 20-cm diameter BK1-fluorite lens that is divided into seven elements to reduce the change of spherical abberation with color. The lens is immersed in oil (Dow Corning DC 200). The two simple lenses in the HAO instrument are almost perfectly corrected for color by a single back-aluminized Mangin mirror (Baker, 1954). The focus

---

* Formerly the Sacramento Peak Observatory. Now operating under the title National Solar Observatory/Sacramento Peak.

drifted in this telescope, perhaps because the mirror was formed on a corrector that was made from a high expansion crown glass (BK-7). With the addition of two more surfaces or a layer of oil that could expand, the mirror could have been made from a low expansion material. The corrections (figuring) for spherical aberration and the heating of the Mangin mirror are appreciably smaller if the corrector is larger than the usual $\frac{1}{3}$ to $\frac{1}{2}$ diameter of the objective. These coronagraphs may be operated with polarimeters at the focus of the simple objective lens, in order to avoid introducing spurious polarization from the inclined surfaces that follow. None of the coronagraphs have been evacuated to eliminate internal seeing, although one portion of an earlier design for the Sac Peak coronagraph did have an evacuated tube through the polar axis.

The simplest mirror system for a steerable telescope consists of a tilted or off-axis mirror that focusses the Sun to the side of the incoming beam where the heat can be removed (Zirin, 1970). This usually introduces seeing problems if operated in air (see Section 5), but should perhaps be pursued further, especially in conjunction with a large aperture telescope on an altazimuth mounting with the focussed sunlight placed above the incoming beam so that the heat can be easily removed. Examples of possible designs for tilted telescopes are the Schiefspiegler (Kutter, 1958, 1975a, b) and some telescopes proposed by Gelles (1975), Buchroeder (1976), Shafer (1978), and Hallam (1983). Correcting the aberrations with decenter and tilt has become possible in special telescopes, but has not been applied to solar telescopes.

An evacuated Newtonian telescope, for example the Kiepenheuer Institute 50-cm (Mattig, 1975), eliminates the airpath and is attractive for a smaller telescope whose window does not pose a manufacturing problem. A small flat mirror at 45° near focus deflects the light to a magnifying system. See also the solar balloon telescopes (Schwarzschild *et al.*, 1958; Kiepenheuer and Mehltretter, 1964). The Newtonian, which may have a small field of view due to coma, avoids the highly accurate internal alignment required by conventional Cassegrain and Gregorian telescopes that have fast primaries.

For larger apertures with long focal lengths and large fields a Gregorian telescope, again evacuated and sealed with a window, is attractive. Because of heating of the secondary, the Cassegrain configuration is largely avoided by solar astronomers even though it is the most popular stellar telescope and has been used in solar rocket and Shuttle telescopes. The Gregorian, which is longer and has some curvature of the field, can have a field stop that deflects or absorbs the heat from the Sun before it reaches the secondary. There are several examples of steerable compound telescopes, all evacuated, including the 60-cm (Cassegrain/Gregorian) now operated by Northridge College (Mayfield *et al.*, 1969), the 60-cm Gregorian at Big Bear (Prout, 1975), the 60-cm 'domeless' Gregorian telescope at Hida (Kühne, 1979; Shinoda, 1979; Nakai, 1981; Nakai and Hattori, 1985; Zeiss, 1981), shown in Figure 1, and the 60-cm Gregorian of the Kiepenheuer Institute (Wiehr and Duensing, 1979; Wiehr, 1984; Schröter, 1983). In the Gregorian, a flat third mirror can be placed near the image of the objective formed by the Gregorian to deflect the light coming from the Gregorian to the side of the tube. If the primary is fast the obscuration can be small and this mirror may be avoided. Baker

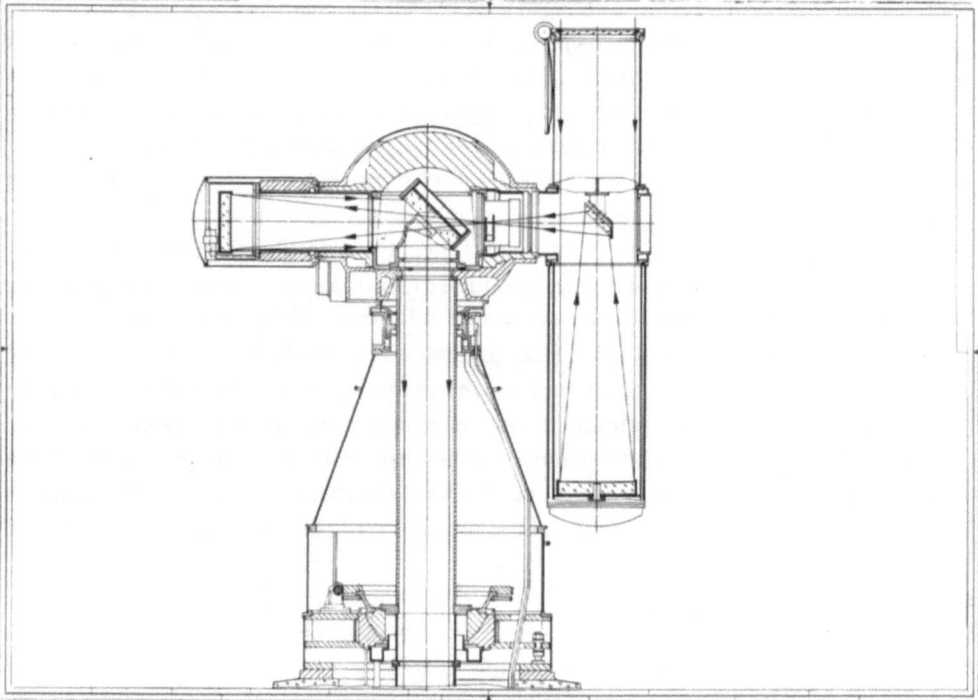

Fig. 1.   The mechanical and optical arrangement of the 60-cm Domeless Solar Telescope of the Hida
Observatory Japan. Courtesy of Zeiss, West Germany.

(1978) offered a wide-field version of a solar telescope with the secondary mirrors perpendicular to the main telescope.

The compound telescopes have considerable 'figure' on the conic mirrors and they must be accurately aligned to avoid coma across the entire field (Simmons *et al.*, 1980; Meinel and Meinel, 1984; Shack and Thompson, 1980; Fehniger, 1980). For small fields of view the tilt and decenter tolerances of the secondary are equivalent to specifying the degree of coincidence of the conic foci of the two conic mirrors. One may tolerate a change in direction of the axes of the conics, providing the conic foci themselves coincide. This concept is used to scan the primary of the Solar Optical Telescope (SOT) over the Sun and it can also be used to simplify the alignment system. For large fields this scheme introduces asymmetry at the edge of the field (Harned *et al.*, 1980). The tolerance of alignment of the two mirrors is extremely small (Weatherall and Rimmer, 1972) and amounts to a lack of coincidence in the lateral position of the conic foci of 170 microns and 50 microns of 'despace' for a $f/3.6$ primary mirror. In a stellar telescope the Serrurier truss is supposed to maintain these tolerances at all attitudes and most stellar astronomers believe it does, but I am not convinced. Resolution in stellar telescopes is usually much poorer than the diffraction limit, i.e., specified as '90% of the light within 1 arc sec', hardly acceptable for high resolution in solar work. To me the resolution of the compound telescopes in solar work has always been questionable.

Pictures taken with these telescopes, while very good, just do not look as if they have the very highest spatial frequencies that appear in the best Pic du Midi pictures. I suggest that the passive alignment systems and mirror supports may not be good enough or that they are not regularly aligned well enough to completely eliminate the coma. Perhaps an active alignment system is needed. The altazimuth mounting (for example, the Hida telescope) may be better for a passive system since the deflections are only a function of elevation and it should be possible to devise a simple lever and weight system on the secondary or primary to keep the conic foci in coincidence. Experience with the Multiple Mirror Telescope (MMT) (Beckers *et al.*, 1982) suggests that the hysteresis of telescope structures is small and that the deflections are reproducable, again suggesting that an active alignment system for an altazimuth telescope may not be necessary.

A spherical primary for these short telescopes also might be investigated because one thus avoids the aspheric figuring on the large surface, but spherical primaries usually result in a small field, which nevertheless may be acceptable (Paul, 1935; Jones, 1954; Strong, 1967; Brueggemann, 1968; Shafer, 1978; Meinel *et al.*, 1984). The corresponding secondaries are strongly figured.

3. Fixed Telescopes

Fixed Telescopes use large mirrors to reflect the light into the telescope thus avoiding the need to point the telescope itself. The telescope can be very long and arranged horizontally, vertically or along the polar axis and one can use a simple optical system consisting of a spherical primary mirror ($f/40$–$f/100$) or a long focal-length lens. One substitutes problems of manufacturing and supporting the large mirrors and the stabilization of the long optical path for the alignment, figuring and heating problems of the more compact telescopes. The choice between a fixed and steerable telescope is largely one of mechanical constraint in pointing a long truss or tube. The 'Hevelius' (King, 1979, Fig. 23), 'Lord Rosse' (Ibid., Figs. 88 and 90), 'Herschel' (Ibid., Fig. 59), 'Ramage' (Ibid., Fig. 85), and 'Cooper' (Ibid., Fig. 78) mountings offered solutions to this problem many years ago. I suggested extending the old designs to a solar telescope, as shown in Figure 2.

The problem with these coelostats, heliostats and turrets that redirect the light to the fixed telescope is that the mirrors are larger than the primary (up to twice the aperture for a heliostat) and, because of their high angle to the light beam, they must be very flat or they will introduce astigmatism (at 45° the mirror must be flat to 1/20 wave). Other problems are vignetting and polarization from the highly inclined surfaces and some of the mirror configurations cause the image to rotate.

Examples of fixed telescopes with coelostats that do not cause the image to rotate, are the Mount Wilson tower (Hale, 1912; Abetti, 1929, Fig. 11; Hale and Nicholson, 1938; King, 1979, Fig. 137), which feeds a 30-cm achromatic lens ($f/150$) as do the Rome Tower (Cimino, 1964), the Einstein tower (Abetti, 1929, Fig. 13), and the Arcetri Tower (Ibid., Fig. 8). Coelostats that feed vertical mirror telescopes are the Kitt Peak vacuum Telescope (Livingston *et al.*, 1976; Engvold, 1976) and the Kiepenheuer 60-cm

Fig. 2.   Model of the 'Rosse Telescope' concept for the LEST Foundation. A 2.5-m aperture $f/12$ sphere is pointed directly at the Sun by an altazimuth mount. Three additional mirrors stabilize the image and reflect the light to ground level in a manner similar to the Hida Telescope. The 3-m diameter, 30-m long, helium chamber is supported at its center of percussion by an elevation drive formed from cables. The entire three-legged structure rotates in azimuth on three rollers each supported by high pressure oil.

(Mehltretter, 1975, 1977, 1978; Soltau, 1979, 1983) (These two are very similar!). Horizontal telescopes include the Snow (King, 1979, Figs. 130, 131). A 228-cm diameter heliostat feeds the very long (244-m air path) McMath telescope (McMath and Pierce, 1960a, b; Pierce, 1964, 1969), which is mounted at the polar angle. The heliostat introduces uniform image rotation that is removed by rotating the spectrograph. A 'turret' that avoids vignetting and that can be evacuated consists of two mirrors at 45° that feed the vertical Vacuum Tower Telescope at Sac Peak (Dunn, 1964, 1969). This altazimuth design causes nonuniform image rotation that is removed by rotating all the instruments together. A similar design, but with an outer tower for a windscreen, was proposed for Ondřejov (Artus, 1976). The Stockholm Observatory has selected a turret with two 70-cm flats at 45° for their 50-cm telescope (Scharmer *et al.*, 1985).

As the aperture of the telescope becomes large (> 2 m?) the one or two large flats needed to feed a fixed telescope become impractical and one must return to a compact, steerable, compound (Gregorian) design.

4. Simple Optical Tests

If the image of the granulation taken with a short exposure shows sharp patches interspersed with blurred patches, then the light to the telescope is passing through two separate paths, and the troubles are most likely in the tropopause and not near to the aperture, since a defect there would spoil the entire image. If the image is always blurred then the chances are that the image is being destroyed by bad optics or by an 'upset' air path within the telescope itself.

There are a few simple tests that one can make to locate where in the optical path the image might be spoiled. A Foucault knife edge test using the limb of the Sun will show steady-state aberrations such as spherical aberration caused by imperfect optics or by heating of the edge of the mirror or lens (Mehltretter, 1979). It will also show pulses of heat (Schlieren) rising from the white-painted mirror cells of the coelostat. On the Sac Peak tower, it showed disturbances caused by the edge of the original cylindrically-shaped bezel on the turret (Dunn, 1969). This edge was subsequently eliminated. To perform the Foucault test, isolate a small part of the solar limb near focus and cut off the limb with a sharp edge parallel to the limb. Use a lens to image the objective on a white card. Shading over this image shows the aberrations (Porter, 1947; Wilson, 1975).

Often the telescope can be autocollimated by adjusting one of the mirrors in the coelostat or by tilting the mirror in the heliostat. Recently we studied the lower path of the McMath independently from that in the rest of the telescope by simply rotating the 157-cm diameter mirror that normally sends the light to the observing room so that it returned the light back to the primary.

If one has a window, the internal path in the telescope can be studied at the telescope focus by watching the change in the diffraction pattern, however poor, in the image reflected back from the window. One can also study the window itself by obtaining interference fringes between the two surfaces (Fizeau test) as follows: The telescope, which can be poor, collimates laser light on to the window. A lens at the focus of the telescope combines the beam reflected from both surfaces of the window to form an image of the objective covered by interference fringes. If the fringes are not staight there is distortion in the window and the number of fringes shows the wedge in the window (Dunn, 1972). If the two beams returning from the window are separated at the telescope focus, a shearing interferometer can compensate the wedge in the window in order to lower the fringe count and to place the fringes in any orientation.

5. Stabilization of Internal Optical Path

The light has travelled many thousands of meters in the air just above the entrance pupil, so one would think it possible to stabilize the path in the remaining 10–200 m within the telescope itself, but this may not be possible. The Pic di Midi 50-cm achromat has a short air path and produces a very fine image. With air in the Sac Peak telescope the lower 60 meters, which is below ground, is stable even with the Sun shining intermittently on the black cover on the main mirror at the bottom of the pit, but the top 40 m is convectively upset, presumably because the turret is always colder than the tower interior.

I tried once, with poor results, to use a 20-cm aperture, $f/10$ mirror telescope with an air path with air sucked through a plate at prime focus to cool the field stop. Similarly, Zirin reports that his 40-cm off-axis mirror telescope (referred to earlier) does not work without a vacuum. The 250-m long McMath air path is temperature controlled with some success and the auxiliary telescopes on the side of the heliostat have given images of high quality (Gillespie, 1976). Recently an IR thermograph (Beckers, 1984) has shown that there are hot spots that need to be insulated in the optical path of the McMath. Coulman (1969, 1974) has made studies of the optical path in the McMath. The Culgoora telescope has an airpath, and uses fans to suck air through perforated plates. Good internal seeing is claimed for it. One concludes that there is a better chance of stabilizing the air path if the path is short or if it is all below ground. If one can accept a lens or window, evacuation is a sure solution. Engvold has shown that helium is a real possibility for inhibiting convection (Engvold *et al.*, 1981, 1983, 1984). (With helium, one first pumps a vacuum and then vents in the helium.) In a preliminary test in the Sac Peak tower, helium at one half atmospheric pressure still showed convection in the upper part, suggesting that it might be better if the path were basically stable. Helium did work, however, in the Kitt Peak vacuum telescope.

Preferably the coelostat mirrors should also be enclosed in a vacuum. This necessity led to the 'turret' design at Sac Peak.

Part of the problem of using a mirror in air is the absorption of solar energy by the mirror's reflecting surface and the subsequent heating of the air near the mirror. Overcoated silver has been used to alleviate this source of heat.

Hammerschlag (1981a) is completing a 40-cm telescope with minimal metal structure ('everywhere are triangles') near the airpath. The mirrors are specially coated and the optical system is relatively short.

More experiments and calculations are needed on stabilizing internal air paths. Is it feasible to stabilize the Sac Peak tower when it is filled with air of helium? This would permit the use of a much thinner window with advantages in the reduction of polarization and more rapid response of the window to changes the temperature. What is the image quality of a short off-axis mirror telescope whose altazimuth mount places the prime focus always *above* the main telescope incoming beam where the heat generated at the focus can be collected by a blower system. This design would eliminate the window and vacuum and could be an inexpensive solution for a large-aperture telescope.

6. Windows

The window needed for supporting a vacuum or for containing the helium is not readily available in diameters larger than 1.6 m. The maximum working stress for glass is 80 bar, which leads to a diameter-to-thickness ratio of approximately 15 : 1 for a flat window that supports the vacuum load. A thin shell (1 cm thick with a 500-cm radius for a 2.5-m aperture) can also support the load and is optically acceptable, but it would be expensive to manufacture and it would have to be properly supported at its edge.

The variation of optical path with temperature is half as much with glass as with quartz because the larger dn/dT in quartz more than offsets the smaller change in thickness from its lower thermal expansion.

The variation of optical path with temperature is less if the window is thin (Mehltretter, 1979; Dunn and November, 1984). The edge of a thick window may have to be cooled (Dunn, 1972) or, at the least, its cell may have to be protected from the direct sun (Mehltretter, 1979).

What happens to the image when the lens (or window) supports a vacuum? Tests by the author on a 25-cm $f/20$ acromatic lens and also on the Sac Peak tower (Dunn, 1972) show that little happens to the image. The deflections and their optical effects from the two sides of the lens almost completely cancel each other. (Scharmer *et al.* (1985) has calculated the change for a 50-cm diameter achromat.) The same observation holds if the window is distorted by a slightly supporting surface.

Stresses due to annealing, bending, edge support, and thermal gradient can cause polarization in windows (Bernet, 1979; Dunn, 1984). The annealing introduces 'radial polarization' that can not be compensated by normal polarizers and retarders. In supporting the vacuum, stress birefringence (polarization) caused by the compression in the material above the neutral plane of the glass disk is cancelled by the birefringence in the opposite sense caused by the tension in the surface below the neutral plane. The stress birefringence caused by the load of the vacuum on the edge support of the window, however, is not cancelled. Its effect is less if the disk is thin. The temperature effect that causes stress birefringence is 12 times less for fused quartz then for glass because of the lower thermal expansion in fused quartz. Again, the effects are less if the window is thin.

An approximately 2 cm thick window made from fused quartz has been proposed for the 2.4 m LEST telescope to seal its helium-filled tube (Dunn, 1984). The effects of a change in temperature on the optical path would be 6 times less than for the current Sac Peak glass (K-50) window (10 cm thick and 86 cm aperture). The effects of temperature in causing polarization would be 120 times less. Polarization introduced at the edge support should be eliminated by the use of helium at near atmospheric pressure. The amount of material is well within current manufacturing capabilities, but the large diameter is a problem. Testing during the manufacture of this window might be done by monitoring the interference fringes between the two surfaces. To test the concept of a thin window, an experimental window (3–6 mm thick) is being manufactured by Zeiss for the 50-cm Newtonian at the Kiepenheuer Institute (Schröter, 1984).

7. Vibration

If the site in windy the telescope may vibrate and one must have a dome or windscreen that tracks the telescope and protects it. If the site is free from wind, there may still be a need for an enclosure to protect the telescope or coelostat from the weather. The enclosure may roll off or retract out of the way so that it does not disturb the air path and the telescope is then 'domeless' and operates in free air. Note the clever solutions on the Kitt Peak vacuum telescope and on the Northridge telescope. A conventional dome is not attractive for high resolution because there is bound to be a turbulent interface between the slot in the dome and the outside air. There are also problems with differential heating between the telescope, the dome floor and the outside air. Fans and cooling systems have been tried without much success in solving these problems. Although Big Bear observatory houses their telescope in a conventional dome with good results, I much prefer one of the Pic-du-Midi solutions (Rösch, 1961; Cachon, 1984) where the telescope tube is protected and the objective protrudes far from the dome itself. Wind tunnel tests (Engvold *et al.*, 1984; Courtés, 1962) show that this is a better solution for interfacing the window or lens to the windy atmosphere. A dome that could be used with such a 'snorkel' protects the 2-m stellar telescope at Pic du Midi (Cachon, 1984; Creusot-Loire, 1981) and is proposed for the Large European Solar Telescope (LEST). The 'domeless' solar telescope on Capri (Zeiss (no data available); Kiepen-heuer, 1964, 1966) is an extreme case of a telescope 'surrounded' with a dome.

In addition to those telescopes that are unprotected from the wind during observation there are some 'domeless' solar telescopes whose drives are stiff enough that the telescope can operate even when it is windy. Examples are the Sac Peak Vacuum tower and the Hida Gregorian telescope mentioned earlier. The Utrecht telescope designed by Hammerschlag is also domeless. The telescope has a small cross section that does not catch the wind and the gear drives are extremely stiff (Hammerschlag, 1984).

Usually the dome is supported by a separate tower. Sometimes these towers are concentric or enclose the telescope path, as on the McMath or surround the members of the structure that supports the telescope itself, as on the 150-ft Mount Wilson tower. Designers always worry about the coupling between the towers through the air that separates them, about the relative eigenfrequencies of the towers, and about the transmission of vibration through the foundations (Sjölund, 1984). Double towers are expensive and are not always certain to eliminate problems caused by the wind. I like concrete towers because the mass is greater than the steel towers so that the servos have something to react against. The damping in concrete is also higher than steel. At the other extreme, Hammerschlag (1981b) has designed and interferometrically tested a light steel tower that moves as a parallelogram without angular motion. It has a small cross section that reduces wind load as well as solar heating.

Concrete towers tend to become quite warm in the Sun (Giovanelli, 1966). Cooling was tried only once on the 90-cm thick, concrete Sac Peak tower because it was thought the wind would sweep away any thermals rising from the concrete surface. The Hida tower is covered with a cooled stainless steel skin (Ishiura, 1979) that has been shown to improve the seeing (Nakai, 1981).

As will be discussed subsequently, the vibration of the image, caused either by the telescope or the atmosphere, can be greatly reduced if one of the optical elements can be servoed rapidly to track the limb or a sunspot. If one plans to use such an 'active' mirror then the tower should be designed with a low frequency in combination with a low amplitude – a combination that is not easy to attain.

For these towers there is need for measurements of vibration that can be compared to the calculated values. Could one design a relatively flexible structure and then correct all the vibration with an 'active' mirror?

8. Guiding and Alignment

All solar telescopes should have a limb guider that centers the telescope on the center of the Sun. Blind pointing by means of encoders is unnecessary and usually inaccurate because the encoders are never perfectly aligned. The guiders on telescopes that rely on mechanical alignment of the optical beam to a mechanical axis are unsatisfactory (Sac Peak 40-cm coronagraph, McMath, and the original design for the Sac Peak Tower where the objective mirror at the bottom of the vacuum tube was not servoed). A 7.5-cm aperture lens or mirror is adequate for a guider. The null may be sensed by solar cells connected to a DC electronic system (AC balance chopping systems were originally used at Sac Peak, but are no longer necessary because DC amplifiers no longer drift). Since there is plenty of light, it is not necessary to resolve the image to obtain sub-arc sec guiding. An aperture 100 arc sec along the limb and 5 inside the limb gives a sufficient signal. Some means for matching the guider to the diameter of the Sun through the year is necessary. The focal length of the guider is of no consequence, but is chosen to match the scale of the mechanism for offset guiding. This mechanism can be an $X - Y$ table driven by stepping motors, tilt plates of optical glass, counter-rotating prisms before the guider objective, a tilt mirror, etc., but it should have a linear motion without backlash.

The guider must detect the errors on the final image after any image rotation. Alternatively, it can be referenced to an auxiliary internal alignment system. If the image is too large to guide on directly, and often it is (the 90-cm image on the McMath would require a offset motion of $\pm$ 90 cm), of if the field is small, one can rely on the laser alignment scheme developed at Sac Peak (Dunn *et al.*, 1981) and also used at the vacuum telescope at Kitt Peak. It locks the Sun to a laser whose beam is transmitted through the telescope. A quad cell detects the centering of the laser image in the final focal plane. The laser alignment problems due to the internal seeing near the telescope on the MMT (Beckers *et al.*, 1982) are not applicable here if most of the telescope is evacuated or filled with helium.

A very useful device that maintains the registration of the image even if it is locally stretched or displaced by the Earth's atmosphere or vibration is the high-speed 'sunspot tracker' ('agile' or 'active' mirror) demonstrated by Tarbell and Smithson (1981). Four photocells ('quadrant detector') generate a signal from the image of the sunspot that drives a gimbaled mirror to continually servo ('null' the signal). The results are impressive when used on the Sac Peak tower and 'every solar telescope should have one'. Lack of a sunspot can be inconvenient, but even then the residual uncorrected signal from the main guider can be sent to the servo in a 'feed forward' fashion that improves the stability providing the signal is coherent over the Sun. A 'correlation' detector that would process images of granulation would remove the need for a sunspot. Such a correlation tracker has been under development at Sac Peak and the Kiepenheuer Institute for some time.

In principal, using the laser alignment system discussed previously, the guider and laser could be mounted rigidly together on a mount that is floppy relative to the telescope and the laser image could be stabilized by the 'active' mirror to correct all sorts of telescope-caused pointing errors that are coherent over the image of the Sun. This 'active' mirror reduces the amplitude of the vibration in proportion to the gain of the open loop servo response (For example see Cushman, 1958). A servo that has 0 db gain at 100 Hz and that has a gain slope of 20 db/decade will reduce the amplitude of a 1 Hz vibration by a factor of 100, and a 10 Hz vibration by a factor of 10.

The driven mirror can be large if it is lightweight and mounted on a 'reactionless' gimbal such as the 'chopping secondary' of a large infrared stellar telescope. A 60-cm diameter light-weight reactionless mirror with voice coil drives can have peak bandwidths as high as 400 Hz.

The non-linear image rotation introduced by an altazimuth mount can best be calculated by computer. An error of 30 arc sec in rotation around the center of the Sun (the guider always constrains the image about its center) introduces an error of only 0.1 arc sec at the limb. The rotation can be corrected by an image rotator, or the entire focal plane instrument can be rotated.

A new telescope should have a guider with capability of raster scanning, an internal laser alignment system; some means for compensating image rotation; and an 'active' mirror whose input can be the image of a sunspot, the laser image in the internal guider, the uncorrected error signal from the guider (used as a 'feed forward' signal to the mirror), or the output of a correlation tracker. If the telescope is a Gregorian or Cassegrain with a fast primary it might also need an active alignment system to maintain perfect coincidence of the conic foci!

9. Improving the Observations

At those moments of the best seeing at a good site, for example at the Pic du Midi, the sharpness of the image is limited by the angular resolution of the aperture itself (50 cm). At that site, and presumably at others, we do not know how large the telescope aperture could be before the 'seeing' would spoil the images.

On other occasions, the scheme for overcoming the blurring, displacement, and stretching of image introduced by the Earth's atmosphere is to shorten the exposure as much as possible, wait for a moment of good seeing, which may take months if dedicated telescope time, and then trigger a rapid burst of pictures. One then selects the sharpest scene as judged by the eye and makes it into a Christmas card or postcard – which is by far the best way to ensure wide distribution and accurate reproduction. The next step is to assemble a movie using the same technique. Each burst can last 30 s before the feature of the Sun changes. In the past I have used this technique with good results. Recently on a filligree-granulation movie, I chose each frame for sharpness over a particular area of interest (Dunn and November, 1985a). (At Sac Peak the frames are rarely uniformly sharp over the entire frame.) The overall drift of the image was removed, and the guiding of the individual frames could have been improved with more patience. The amount of work in this process is not prohibitive and it is unfortunate that more high resolution movies made on film are not available.

With the advent of the computer, both the selection and alignment process can be automated. Dunn and November (1985b) have processed CCD data of images of granulation stored on magnetic tapes to bring them into registration ('correlate track'), adjust the position of the details smoothly over the scene to fit a running average ('destretch') and then select parts of them for sharpness. Clearly this computer program could also work on a CCD attached to the optical printer. A correlation tracker attached to the optical printer would make the alignment of local scenes less of a chore.

One of Dunn and November scheme's computes the contrast within a high spatial frequency band over each of 64 subrasters. The contrast within each of the subrasters is compared throughout the burst. The best subraster is selected and is then shifted to track the running average. This technique could be applied in real time so that the observer would see the best mozaic every 30 s. Dunn and November are extending the selection technique to CCD observations of magnetic and velocity fields, granulation and filigree, all made simultaneously (with beamsplitters and common shutter). They plan to determine the corrections from the successive granulation scenes and then apply the same degree of correction to the fainter, blurred, higher-noise pictures taken with the birefringent filter.

These techniques are not useful for spectra unless the spectrograph uses an image slicer to obtain an extended field. For polarization measurements, all Stokes parameters would have to be made simultaneously over an extended field through a common shutter.

Other post-observation techniques have been tried to 'restore' a sharp image from a sequence of blurred images taken as a burst. The 'Knox–Thompson' scheme (Knox and Thompson, 1974) shows promise and has been demonstrated by Stachnik *et al.* (1983) and von der Lühe (1985a). It could also be implemented in real time.

Observations that can not be made on the days of the best seeing or that require long exposures should be greatly improved at the telescope with the 'active' mirror discussed previously. At the very least, the active mirror guarantees that a highly magnified image is kept centered on the detector. In addition one can consider automatic focussing that may sharpen images made with both long and short exposures. When the seeing is good at Sac Peak and the images have a mix of sharp and blurred areas, the blurring is sometimes caused largely from defocus, suggesting that a high-speed focus detector detecting the contrast of the granulation image may improve the image over a small area (Dunn, 1985).

A higher order 'real time' correction device that goes beyond improving guiding and focus is the 'rubber' ('adaptive') mirror, that is made up of many small steerable optical elements (Smithson and Tarbell, 1981; von der Lühe, 1985b). It holds the promise of correcting the image for all the effects of the atmosphere over a small area, even with poor seeing. It would be very useful for spectra, where the exposures are long and the scenes can not be post-processed unless one has a large image slicer. A 19-element 'rubber' mirror is being tested at Sac Peak by Smithson. Like the 'active' mirror, it needs a sunspot to track. A more sophisticated version might use a correlation tracker that processes multiple images to track the image when the sunspot is not available.

In addition to the guider, laser alignment system, and an active mirror that guides on a variety of inputs, the 'complete' high resolution solar telescope might have an automatic high-speed focus detector, a scene selector, and a 'rubber' mirror! There is really very little work being done on this gadgetry. I believe that if it were available we would obtain many good sequences of the evolution of the fundamental, physical processes of the Sun.

10. Large Aperture Telescope for Stokes Polarization

There is interest in a high resolution (less than 0.5 arc sec), large aperture (1 to 2.4 m), polarization-free (to less than 0.1%) telescope for observations of Stokes polarization parameters in spectral lines operating on the ground (Engvold and Hefter, 1982). This telescope would support theoretical work and would lead to Stokes observations in space. It could also support the Solar Optical Telescope (SOT), which may fly on the shuttle in the mid 1990s. In SOT there are two inclined mirrors prior to the first focus. Title introduces a 1/2 wave plate between them that should compensate their polarization to some degree, but a precision polarimeter is not included in the current instrumentation.

The simplest polarization-free telescope is a steerable telescope with no inclined surfaces in front of the modulator. All currently-planned Stokes telescopes are of this form. Other attractive mechanical and optical configurations would be possible if for example, one 45° mirror, rotating with the telescope and never changing its aspect with respect to the line of sight, is permitted. This would require a fixed compensation that remains aligned with the 45° reflection. Astronomers worry about the stability of such a compensation and prefer to avoid it. *Tests are needed* that would guide the design of a polarization-free telescope without a window that had a tilted mirror as an objective, or a single 45° inclined surface prior to the modulator.

The next lower level of polarization-free telescope might be the Sac Peak tower with its turret, where the inclination of the mirrors remains at 45° and the illumination of the aperture remains the same, but each mirror rotates about an axis along the line of sight. (Currently the polarization in the Sac Peak tower is dominated by the residual polarization from the annealing of the window.) It may be possible to achieve the necessary polarization stability on this telescope but it remains to be shown.

It is much more difficult to compensate the fixed telescopes with their heliostats and coelostats whose illumination and mirror inclination and rotation along the line of sight

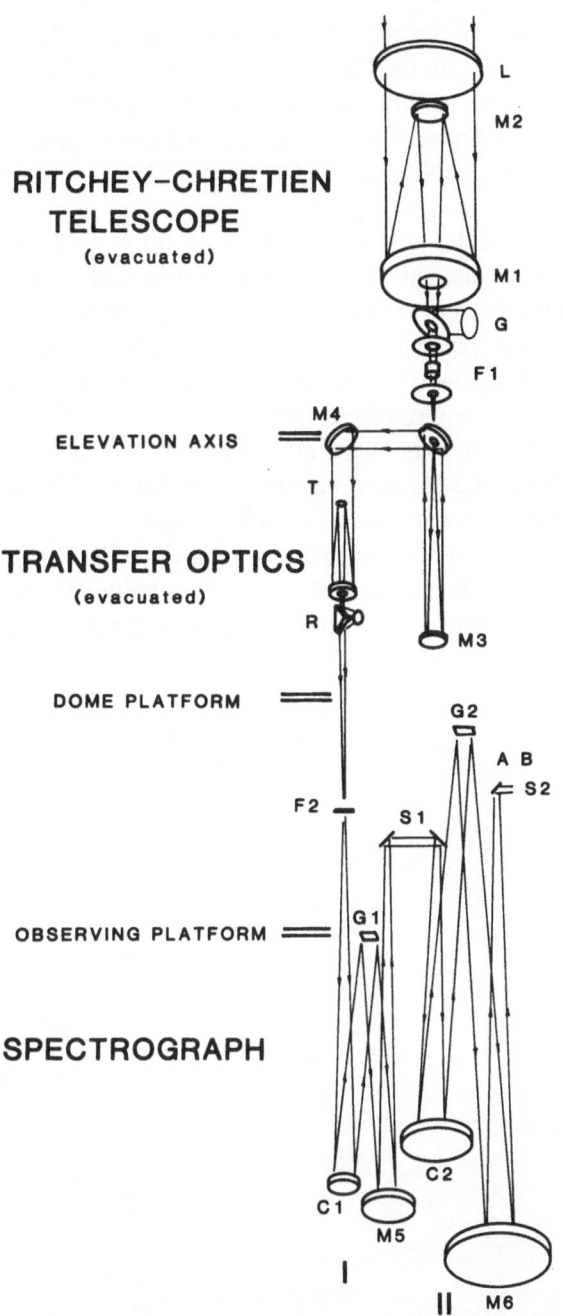

Fig. 3. Optical system for the THEMIS. L – Entrance window ($D = 1.1$ m); M2 – Secondary mirror ($f = 0.86$ m); M1 – Primary mirror ($D = 0.9$ m, $f = 3.15$ m); G – Photoelectric guider; F1 – Primary focus ($f = 15$ m), Polarization analyzer; M4 – Tilting mirror (100 Hz); T – Transfer optics ($\times 4$); R – Field rotator; M3 – Collimator mirror; F2 – Entrance slit ($f = 60$ m); I – Predisperser spectrograph; C1 – Collimator mirror ($f = 7.3$ m); G1 – Grating; M5 – Camera mirror; ($f = 6.3$ m); S1 – Focal plane; II – Echelle spectrograph; C2 – Collimator mirror ($f = 7.5$ m); G2 – Grating (79 l mm^{-1} blaze 63 degrees); M6 – Camera mirror ($f = 8.5$ m); S2 – Detector package; A – Line profiles; B – Monochromatic images.

are changing throughout the day. Nevertheless Harvey (1985) has a passive compensator for the heliostat on the McMath that works to 0.1% in the near infrared. He has also investigated the variation in polarization across the mirrors and finds that it is remarkably uniform. Others have analyzed coelostats in considerable detail (Macák, 1979; Makita *et al.*, 1982; Nikonov and Nikonova, 1982), but practical compensation looks difficult.

I compensate the circular polarization of the 25-cm SOON telescope throughout the day by placing a 4-cm diameter polaroid and quarter-wave plate on the objective and reimaging it onto the slit of the spectrograph. Two wave plates behind the slit are rotated automatically by a computer until the signal is maximized. Stress from the annealing of the glass in the achromatic lens limits the degree of compensation. It is never of the low magnitude as that required for Stokes.

The polarimeter itself must take advantage of the various image improvement techniques to achieve the required 0.5 arc sec resolution. Harvey (1985) and Stenflo (1984) have published recent reviews of polarimeters.

New large-aperture telescopes for Stokes include THEMIS (Telescope Heliographique pour l'Etude du Magnetisme et des Instabilities Solares) (Rayrole, 1981), a 1-meter aperture telescope which is funded and is being designed by J. Royrole in the

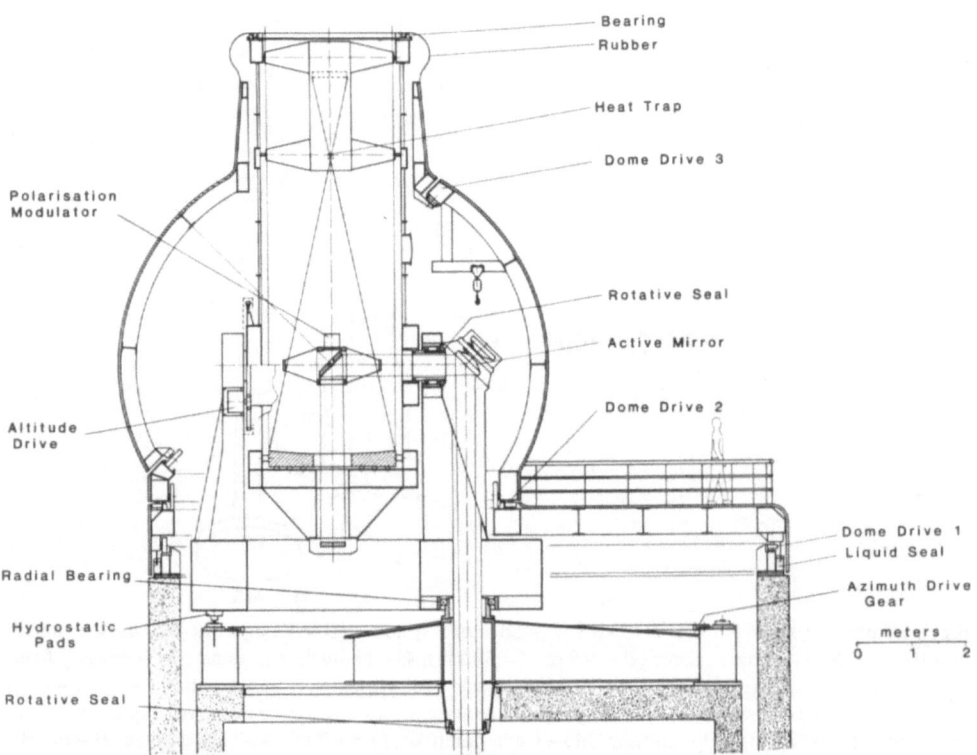

Fig. 4.   The current design for the Large European Solar Telescope (LEST).

Meudon group and A. Tournaire at the Institut National des Sciences de L'Univers, and the Large European Solar Telescope (LEST) (Engvold and Hefter, 1982; Andersen *et al.*, 1984), which has an aperture of 2.4 m. (A convincing case for the LEST as an ongoing ground-based telescope in competition with SOT was made by Beckers (1978). Focal plane instrumentation for LEST is specified by Wöhl *et al.* (1984).) These telescopes are shown in cross section in Figures 3 and 4. Both are pointed directly at the Sun and do not have any inclined optical components prior to the polarimeter. They represent a compromise among the various options and adopt many of the features discussed previously.

The Utrecht telescope, designed by Hammerschlag, will also be useful for Stokes, but its aperture is somewhat small (30 cm); however, even partial success of its many innovations will open up many exciting new design possibilities for high resolution solar telescopes.

References

Abetti, G.: 1929, *Handbuch Astrophys.* **4**.

Albregtsen, F. and Hansen, T.: 1974, 'The Theoretical Instrument Profile of the Combination Telescope-Pinhole Photometer and Its Effect upon Observations of Umbra Intensities', Report No. 42, Inst. of Theoretical Astrophysics, Blindern, Oslo.

Andersen, T., Dunn, R., and Engvold, O.: 1984, 'LEST Design Study', LEST Foundation Technical Report No. 7, September 1984.

Artus, H.: 1976, *Jena Rev.* **3**, 138.

Baker, J.: 1954, *Astron. J.* **59**, 74.

Baker, J.: 1978, *SPIE* **156**, 76, Fig. 4.

Beckers, J.: 1978, in G. Godoli, G. Noci, and A. Righini (eds.), *Future Solar Optical Observations Needs and Constraints*, Astrophysical Observatory of Arcetri.

Beckers, J.: 1984, 'IR Thermograph Study of CT10 Telescopes', Tech Report No. 15, Multiple Mirror Telescope Observatory, Univ. of Arizona, Smithsonian Institution.

Beckers, J., Poland, C., Ulrich, B., Angel, J., Low, F., and Wyatt, W.: 1982, *SPIE* **332**, 42.

Bernet, G.: 1979, 'Polarization of Glass Discs under Mechanical and Thermal Stress', JOSO Annual Report 1979.

Brueggemann, H.: 1968, *Conic Mirrors*, Focal Press Limited, p. 99.

Buchroeder, R.: 1976, *Tilted Component Optical Systems*, PhD dissertation, University of Arizona.

Cachon, A.: 1984, *The Pic-du-Midi Observatory*, Les Éditions Pyrénéennes, France.

Carroll, G.: 1970, *Sky Telesc.* **40**, 10.

Cimino, M.: 1964, 'Osservatorio Astronomico di Roma', Technical Final Report, January 1964, Grant AF EOAR 63–88, Air Force Cambridge Research Laboratory, OAR Document AFCRL–64–114.

Courtés, G.: 1962, in J. Rösch, G. Courtés and J. Dommanget (eds.), 'Site Testing', *IAU Symp.* **19**, 160.

Coulman, C.: 1969, *Solar Phys.* **7**, 122.

Coulman, C.: 1974, *Solar Phys.* **34**, 491.

Creusot-Loire: 1981, *Sky Telesc.* **62**, 333.

Cushman, P.: 1958, in J. Truxal (ed.), *Control Engineers Handbook*, McGraw Hill, Section 2.15–2.19.

Dunn, R.: 1964, *Appl. Opt.* **3**, 1353.

Dunn, R.: 1965, *Photometry of the Solar Chromosphere*, Air Force Cambridge Research Laboratory, AFCRL–65–398, Environmental Research Papers, No. 109.

Dunn, R.: 1969, *Sky Telesc.* **38**, 368.

Dunn, R.: 1972, 'High Resolution Solar Observations', *Space Res.* **XII**, Akademie-Verlag, Berlin, 1657–1669.

Dunn, R.: 1984, 'Window Considerations for LEST', LEST Foundation Technical Report No. 3, January 1984.

Dunn, R.: 1985, in preparation.

Dunn, R. and November, L.: 1985a, 'Filigree-Granulation Movie', shown at the May 12–15, 1985, Solar Physics Division meeting in Tucson, Arizona.

Dunn, R. and November, L.: 1985b, in R. Muller (ed.), *High Resolution in Solar Physics*, Lecture Notes in Physics, No. 233, Springer Verlag, p. 85.

Dunn, R., Spence, G., Stebbins, R., and Hill, H.: 1981, in R. Dunn (ed.), *Solar Instrumentation: What's Next?*, Sacramento Peak Observatory, p. 613.

Engvold, O.: 1976, *Astron. Tidsskr. (Sweden)* **9**, 145.

Engvold, O. and Hefter, M.: 1982, 'Phase A Feasibility Study on Principal Aspects on Design of a High Resolution Solar Telescope', JOSO Study Report, Institute of Theoretical Astrophysics, University of Oslo, and Arne Johnson Consulting Engineering, Stockholm.

Engvold, O., Brahde, R., and Fossum, B.: 1981, *Publ. Astron. Soc. Pacific* **93**, 526.

Engvold, O., Dunn, R., Livingston, W., and Smartt, R.: 1983, *Appl. Opt.* **22**, 10.

Engvold, O., Fossum, B., and Hold, E.: 1984, 'A Wind Tunnel Study of the LEST Structure', LEST Foundation Technical Report No. 6, August 1984.

Evans, J.: 1956, *Sky Telesc.* **15**, 436.

Fehniger, M.: 1980, *SPIE* **251**, 21.

Gelles, R.: 1975, *JOSA* **65**, 1141.

Gillespie, B.: 1976, *Sky Telesc.* **51**, 157.

Giovanelli, R.: 1966, *Galileo Celebrative Publication*, Proceeding of the Meeting on Sunspots, Firenze, Astrophysical Observatory of Arcetri, 68.

Hale, G.: 1912, *Publ. Astron. Soc. Pacific* **24**, 223.

Hale, G. and Nicholson, S.: 1938, 'Magnetic Observations of Sunspots', Volume 1, Carnegie Institute of Washington, No. 498, p. 9.

Hallam, K.: 1983, *SPIE* **445**, 295.

Hammerschlag, R.: 1981a, in R. Dunn (ed.), *Solar Instrumentation: What's Next?*, Sacramento Peak Observatory, p. 547.

Hammerschlag, R.: 1981b, in R. Dunn (ed.), *Solar Instrumentation: What's Next?*, Sacramento Peak Observatory, p. 583.

Hammerschlag, R.: 1984, 'A Telescope Drive with Emphasis on Stability', Astronomical Institute at Utrecht, preprint No. 444–22.

Harned, N., Harned, R., and Melugin, R.: 1980, *SPIE* **251**, 163.

Harvey, J.: 1985, 'Trends in Measurement of Solar Vector Magnetic Fields using the Zeeman Effect', Huntsville Symposium, 1984.

Ishiura, K.: 1979, Reito and Kūcho (Freezing and Air Conditioning) No. 226, p. 16 (in Japanese).

Jones, R.: 1954, *JOSA* **44**, 630.

Kiepenheuer, K.: 1964, *Appl. Opt.* **3**, 1363.

Kiepenheuer, K.: 1966, *Sky Telesc.* **31**, 256.

Kiepenheuer, K. and Mehltretter, J.: 1964, *Appl. Opt.* **3**, 1359.

King, H.: 1979, *The History of the Telescope*, Dover Publications, Inc., New York.

Kühne, C.: 1979, '60 cm .DST Domeless Solar Telescope', C. Zeiss (Oberkochen), Document No. 56–041–e.

Knox, K. and Thompson, B.: 1974, *Astrophys. J. Letters* **193**, L45ff.

Kutter, A.: 1958, *Sky Telesc.* **18**, 65.

Kutter, A.: 1975a, *Sky Telesc.* **49**, 46.

Kutter, A.: 1975b, *Sky Telesc.* **49**, 115.

Li Ting: 1984, 'Solar Telescope-Magnetograph', Report of Nanjing Astronomical Instruments Factory, Nanjing China.

Livingston, W., Harvey, J., Pierce, A., Schrage, D., Gillespie, B., Simons, J., and Slaughter, C.: 1976, *Appl. Opt.* **15**, 33.

Loughead, R.: 1968, *Solar Phys.* **4**, 185.

Macák, P.: 1979, *Bull. Astron. Inst. Czech.* **30**, 174.

Makita, M., Hamana, S., Kawakami, H., and Nishi, K.: 1982, *Ann. Tokyo Astron. Obs.* **19**, 24.

Mattig, W.: 1983, *Solar Phys.* **87**, 187.

Mattig, W. and Casanovas, J.: 1975, JOSO Annual Report 1975, p. 18.

Mayfield, E., Vrabec, D., Rogers, E., Janssens, T., and Becker, R.: 1969, *Sky Telesc.* **37**, 208.

McMath, R. and Pierce, A.: 1960a, *Sky Telesc.* **20**, 132.

McMath, R. and Pierce, A.: 1960b, *Sky Telesc.* **20**, 64.

Mehltretter, J.: 1975, 'The Fraunhofer Institute Vacuum Tower Telescope', JOSO Annual Report 1975, p. 36.

Mehltretter, J.: 1977, 'The Fraunhofer Institute Vacuum Tower Telescope for the Canary Islands', Progress Report, JOSO Annual Report 1977, p. 50.

Mehltretter, J.: 1978, 'Kiepenheuer-Institut 60 cm Vacuum Tower for the Canary Islands', Progress Report, JOSO Annual Report 1978, p. 33.

Mehltretter, J.: 1979, *J. Opt. (Paris)* **10**, 93.

Meinel, A. and Meinel, M.: 1984, *Opt. Engin.* **23**, 801.

Meinel, A., Meinel, M., Su, D., and Wang, Y.: 1984, *Appl. Opt.* **23**, 3020.

Nakai, Y.: 1981, in F. Moriyama and J. Henoux (eds.), *Proceedings of the Japan-France Seminar on Solar Physics*, p. 275.

Nakai, Y. and Hattori, A.: 1985, 'Mem. Coll. Ski.', *Kyoto Univ.* **36**, 385, also *Kwasan Obs. Kyoto*, No. 260.

Nikonov, O. and Nikonova, E.: 1982, *Izv. Gl. Astron. Obs. Pulkovo (USSR)* **199**, 103.

Paul, M.: 1935, *Revue Opt.* **14**, 13.

Pierce, A.: 1964, *Appl. Opt.* **3**, 1337.

Pierce, A.: 1969, *Solar Phys.* **6**, 498.

Porter, R.: 1947, in A. G. Ingalls (ed.), *Amateur Telescope Making*, Mann and Co, Inc., p. 1.

Prout, R.: 1975, 'ATM Photoheligraph', NASA contract NSR 05-002-071, Final Report.

Rayrole, J.: 1981, in F. Moriyama and J. Henoux (eds.), *Proceedings of the Japan-France Seminar on Solar Physics*, p. 258.

Rösch, J.: 1961, *Proceeding of the Symposium on Solar Seeing*, published by Consiglio Nazionale delle Ricerche, 1962, Roma, p. 35 (Fig. 11).

Rösch, J.: 1981, in F. Moriyama and J. Henoux (eds.), *Proceedings of the Japan-France Seminar on Solar Physics*, p. 269.

Rush, J.: 1963, *Sky Telesc.* **26**, 20.

Rush, J. and Schnable, G.: 1964, *Appl. Opt.* **3**, 1347.

Scharmer, G., Brown, D., Pettersson, L., and Rehn, J.: 1985, *Appl. Opt.* **24**, 2558.

Schröter, E.: 1983, 'The German Solar Telescope to be Installed in the Spanish Observatorio del Teide, Tenerife; Present Status', in *Proceedings of the International Workshop on Solar Physics and Interplanetary Traveling Phenomena*, Kunming, China, 21–25 November 1983, in press.

Schröter, E.: 1984, private communication.

Schwarzschild, M., Rogerson, J., Jr., and Evans, J.: 1958, *Astrophys. J.* **63**, 313.

Shack, R. and Thompson, K.: 1980, *SPIE* **251**, 146.

Shafer, D.: 1978, *Appl. Opt.* **17**, 1072.

Shinoda, M.: 1979, 'A Special Hydraulic Lifting System for the Hida Observatory', *Mitsubishi Denki Giho* **53**, No. 3, 192 (in Japanese).

Simmons, J., Schoening, W., Graham, D., and Ott, L.: 1980, *SPIE* **251**, 138.

Smithson, R. and Tarbell, T.: 1981, in R. Dunn (ed.), *Solar Instrumentation: What's Next?*, Sacramento Peak Observatory, p. 480.

Sjölund, J.: 1984, 'LEST Telescope Tower Study', Lest Foundation Technical Report No. 8.

Soltau, D.: 1979, 'Progress with the Kiepenheuer-Institute's 60 cm Vacuum Tower Telescope (VTT)', JOSO Annual Report 1979, p. 136.

Soltau, D.: 1983, 'The German 60 cm-Vacuum Tower Telescope and its post-focus facilities', JOSO Annual Report 1983, p. 43, see also *Proceedings of the International Workshop on Solar Physics and Interplanetary Traveling Phenomena*, Kunming, China, 21–25 November, 1983.

Stachnik, R., Nisensen, P., and Noyes, R.: 1983, *Astrophys. J. Letters* **271**, L37.

Stenflo, J.: 1984, 'Polarimeter Package for LEST', LEST Foundation Technical Report No. 4.

Strong, J.: 1967, *Appl. Opt.* **6**, 179.

Tarbell, T. and Smithson, R.: 1981, in R. Dunn (ed.), *Solar Instrumentation: What's Next?*, Sacramento Peak Observatory, p. 491.

von der Lühe, O.: 1983, 'Adaptive Optical Systems for LEST', LEST Foundation Technical Report No. 2, November 1983.

von der Lühe, O.: 1985a, in R. Muller (ed.), *High Resolution in Solar Physics*, Lecture Notes in Physics, No. 233, Springer Verlag, p. 96.

von der Lühe, O.: 1985b, in R. Muller (ed.), *High Resolution in Solar Physics*, Lecture Notes in Physics, No. 233, Springer Verlag, p. 62.

Wetherall, W. and Rimmer, M.: 1972, *Appl. Opt.* **11**, 2817.

Wiehr, E. and Duensing, K.: 1979, 'New Facilities at the Locarno Observatory; Improvement of the Telescope', JOSO Annual Report 1979, p. 138.

Wiehr, E.: 1984, 'The German 45 cm Vacuum-Gregory-Coudé Telescope', JOSO Annual Report 1984, p 36.

Wilson, R.: 1975, *Appl. Opt.* **14**, 2286.

Wöhl, H., Huber, M., Mein, P., and Smaldone, L.: 1984, 'Recommendations on Post Focus Instrumentation for LEST', LEST Technical Publication No. 5.

Zeiss, C. (Oberkochen): 1981, *Sky Telesc.* **62**, 127.

Zirin, H.: 1970, *Sky Telesc.* **39**, 215.

PRESENT PROBLEMS OF THE SOLAR INTERIOR

IAN W. ROXBURGH

Theoretical Astronomy Unit, School of Mathematical Sciences, Queen Mary College, University of London

Abstract. The standard model of solar evolution is reviewed and a number of problems highlighted. A fundamental question is whether there is any mixing of matter in the central regions, since such mixing could radically alter the model of the present Sun and modify our understanding of the evolution of other stars. Standard models of solar evolution become unstable to ^{3}He driven global oscillations at an age of 3×10^8 years and this may drive some mixing, even if this is not the case the finite amptitude limit of these oscillations is likely to produce modifications in the standard model. Convective overshooting at the bottom of the outer convective zone leads to an increased depth of this zone and small changes in the interior. It is pointed out that the young Sun had a ^{12}C driven convective core whose extent and duration depends on the extent of overshooting. Such a core is likely to produce a magnetic field which will affect the internal dynamics. The internal rotation of the Sun remains an enigma and absence of knowledge of any internal magnetic field makes it difficult to study the problem. Rotationally driven instabilities are ineffective in the central chemically inhomogenous regions but may contribute to the inward diffusion of lithium from the convective zone. These and other problems are considered, but few solutions are proposed.

1. Introduction

Until recently the observational constraints on the speculations of those who sought to determine the internal structure of the sun were very few: the observed mass, luminosity and radius; some incomplete knowledge of the composition of the surface layers and estimates of the age of the Sun – taken to be the same as estimates of the age of the solar system. Other parameters such as rotation, magnetic fields and mass loss in the solar wind were considered to be unimportant or at most a very small perturbation on the structure of a simple spherical model.

The situation has changed and is changing; the measured value of the flux of electron neutrinos at the Earth is less than theoretical predictions of 'standard models' by a factor of 3 (see Bahcall, 1985), and the expanding volume of data on solar oscillations is leading to empirical determinations both of the mean internal structure and of the interior dynamics. Already this has indicated some difficulties with standard models and contradicted the predictions of many non-standard models (see Gough, 1985b). The combination of more accurate and extensive oscillation measurements with an improvement of techniques of inversion may soon allow us to realise the hope expressed by Eddington in the concluding sentence of his pioneering book 'On the Internal Constitution of the Stars' (Eddington, 1926):

'... but it is reasonable to hope that in a not too distant future we shall be competent to understand so simple a thing as a star.'

Solar Physics **100** (1985) 21–51. 0038–0938/85.15.
© 1985 *by D. Reidel Publishing Company*

2. The 'Standard Model' of the Sun

The phrase 'standard model' is frequently used by many authors – there is no standard model as such and this phrase is taken to describe models that are built on the basis of a set of simple assumptions, the precise assumptions vary slightly from author to author, as do the numerical and mathematical approximations. Not surprisingly the results of such calculations do not agree in detail. There is, I believe, a need for model builders to compare these various assumptions and techniques to give at least a standard of reference for future work.

The basic features of a standard model are based on the following assumptions:

(1) The Sun is spherically symmetric and in hydrostatic equilibrium.

(2) Energy is transported by radiative transport except in regions which are convectively unstable.

(3) The energy of the Sun is produced by nuclear reactions converting hydrogen to helium.

(4) The Sun is in thermal equilibrium except for small changes in entropy due to slow change in pressure density and temperature during evolution.

(5) The chemical evolution is due to the nuclear reactions from the proton–proton chain and the carbon-nitrogen cycle.

(6) Matter is unmixed except in regions that are convectively unstable.

(7) The Sun was initially homogenous and has evolved without mass loss to have the present observed luminosity $L_\odot$ and radius $R_\odot$ at the present age.

Before a model can be constructed we need knowledge of the abundance of the elements, the opacity of stellar material, the equation of state and the reaction rates of different processes. For a precise determination of such models we also need a theory of convective energy transport – which we do not have – fortunately for modelling the interior this uncertainty is not too important but this is not true if we wish to use our models for calculating oscillation periods where the structure of the surface layers can be of importance. The opacity of stellar material is known only to low accuracy (approx $\pm 10\%$) and the abundance of helium is known, if at all, to even lower precision.

The 'standard model' is then computed by assuming some initial composition, for example that determined from observations of the solar atmosphere and material in the solar system, a guess at the helium abundance, taking some calculated values of the opacity, some measured and/or calculated nuclear reactions rates and evolving an initially homogeneous Sun to an age of 4.6×10^9 yr. This model in general will not have the observed luminosity and radius, but by adjusting the initial abundance of helium one can produce a model with the correct luminosity, and by adjusting the theory of convection – or to be more precise, by varying the entropy in the adabatic convective zone, the radius can be adjusted to agree with the present radius. These two free parameters – the helium abundance and the entropy of the convective zone are almost – but not entirely – independent. Not surprisingly given two free parameters one can produce a model that fits two observations!

This highlights the importance of the new observational constraints imposed by measurements of solar oscillations and the neutrino flux. A reliable measurement of the surface helium abundance would also be of great importance. Whilst the prescription given above is common to all 'standard models', different authors use different determinations of the abundances of elements other than helium and different determinations of opacity, they may differ in the number of reactions they include in their reaction network, use different values of nuclear cross sections and different methods of dealing with the change in composition produced by those reactions, and different numerical techniques for solving the equations. Nevertheless, there is a broad agreement that the standard model had an initial composition of

$$X_1 = 0.73 \pm 0.02, \quad X_4 = 0.25 \pm 0.02, \quad Z = 0.018 \pm 0.002,$$

where X_1, X_4, Z, are the initial abundances by mass of hydrogen, helium, and other elements. Details of such a standard model computed by Bahcall et $al.$ (1982) are given in Table I, where X_3 is the abundance of ^{3}He.

TABLE I

Standard solar model (Bahcall et $al.$, 1982)

M_r/M	$r/R_\odot$	T_ε	ρ	$L/L_\odot$	X_1	$10^4 X_3$
0	0	15.5	156	0	0.355	0.089
0.010	0.046	14.8	134	0.079	0.417	0.152
0.073	0.098	13.0	90.2	0.435	0.555	0.495
0.135	0.127	11.9	69.9	0.650	0.621	1.01
0.210	0.156	10.8	53.6	0.809	0.668	2.07
0.310	0.190	9.60	39.5	0.921	0.701	4.81
0.430	0.231	8.37	25.6	0.977	0.721	13.3
0.550	0.275	7.26	16.1	0.996	0.728	32.2
0.690	0.336	6.03	8.03	1.000	0.731	11.2
0.830	0.430	4.65	2.85	1.000	0.732	1.85
0.926	0.555	3.40	0.773	1	0.732	1.03
0.960	0.641	2.72	0.338	1	0.732	1.00
0.978	0.718	2.12	0.169	1	0.732	1.00
0.982	0.730	2.00	0.150	1	0.732	1.00
0.995	0.849	0.950	0.050	1	0.732	1.00
1.000	1.000	0.0058	–	1	0.732	1.00

Such models are, at best, only a first approximation, and at worst, totally misleading. The predictions can be compared with such additional empirical data as we possess, the solar neutrino flux – which is some 3 times the prediction of the standard models – and the oscillation data – which whilst in modest agreement with such models is now beginning to demonstrate certain possible errors; the convective zone probably extends to 0.7 $R_\odot$ which is slightly deeper than most predictions, and the inferred sound speed at $r = 0.5 R_\odot$ is about 2% larger than predictions of standard models (see Gough, 1985b). The properties of the very central regions are not yet probed with any accuracy so the observational constraints are weak, a situation that hopefully will be remedied in the next few years.

Details of the construction of standard models are given in Section 4, but for future reference I given here the simplest set of approximations that give the overall run of density, pressure, temperature and composition.

Equations of stellar structure:

$$\frac{\partial P}{\partial M_r} = -\frac{GM_r}{4\pi r^4}, \quad \frac{\partial r}{\partial M_r} = \frac{1}{4\pi r^2 \rho}, \quad \frac{\partial L_r}{\partial M_r} = \varepsilon,$$

$$\frac{\partial T}{\partial M_r} = -\frac{3\kappa L_r}{16\pi acT^3 r^4} \text{ (radiative)}, \quad P = KT^{5/2} \text{ (convective)},$$

$$\frac{\partial X_1}{\partial t} = -\frac{\varepsilon}{E}.$$

Equation of state: $P = \dfrac{\mathbf{R}}{\mu}\rho T, \quad \mu = \dfrac{4}{3 + 5X_1}.$

Opacity: $\kappa = 90(1 + X_1)^{0.5}(1 + 50Z)\rho^{0.5} T_6^{-2.8}.$

Energy generation: $\varepsilon = 2.8 \times 10^{-6} X_1^2 \rho T_6^{4.5}.$

Equilibrium ^{3}He: $X_3 = 8.9 \times 10^{-5} \left(\dfrac{13}{T_6}\right)^6 X_1.$

The variables have their usual meanings; M_r is the mass within a sphere of radius r; P, ρ, T are the pressure, density, and temperature; ε and κ are the energy generation rate and opacity per unit mass in cgs units; T_6 is the temperature in units of 10^6 K; L_r is the rate at which energy crosses a sphere of radius r; μ is the mean molecular weight; X_1, X_3, X_4, Z are respectively the fractions by mass of ^{1}H, ^{3}He, ^{4}He, and elements other than hydrogen and helium; G, $\mathbf{R}$, a, c the constant of gravity, the gas constant, the radiation constant and the velocity of light.

In arriving at this set of equations we have only considered the dominant ^{3}He + ^{3}He branch of the proton–proton chain with ^{2}H and ^{3}He in local equilibrium and have taken power law approximations to the reaction rates and opacity. The value of Z is taken from the determination by Ross and Aller (1981) which gives $Z/X_1 = 0.0228$. The value of X_4 is unknown and taken as a free parameter; likewise the entropic constant K is taken as a free parameter.

For an initial guess at X_4 and K an initially homogenous model is constructed, this model is integrated forward in time to determine the luminosity L and radius R of a model whose age is that of the present Sun taken to be 4.6×10^9 yr. In general this model will not have the observed values $L = L_\odot, R = R_\odot$; taking different initial values of X_4 and K this process is repeated until we find a model of the present Sun with the observed values $L_\odot, R_\odot$ (($X_4 = 0.25, K = 6.6 \times 10^{-3}$ cgs).

Of course, if we wish to study the solar neutrino problem, ^{3}He instability, a convective core or eigenmodes for oscillation then this system has to be refined to include the detailed reaction network, the change in entropy, refinements in the equation of state and as accurate opacities as are available.

3. Problems of Standard Models

I here consider a number of problems which in my view need further study – this section is a summary of problems some of which are considered in more detail in subsequent sections of this article.

3.1. MIXING IN THE SOLAR INTERIOR

In the standard solar model the only regions where matter is mixed are convective zones. Is this the correct assumption? The fact that the measured solar neutrino flux is only $\frac{1}{3}$ of the predictions of standard models (see Bahcall, 1985) has been taken by some authors as an indication that some mixing takes place in the regions of nuclear burning. The arguement is simple; the energy generation rate $\varepsilon \sim X^2 \rho T^{\eta_1}$ where $\eta_1 \approx 4$, since the total energy generated must equal the solar luminosity a higher central value of X_1 gives a model with a lower central temperature and hence a smaller number of reactions following the (Be, B) branch of the p–p chain. This leads to a reduction in the flux of neutrinos with sufficiently high energy to be detected by the ^{37}Cl experiment. Were the Sun to be kept homogeneous by some mixing process the central temperature would be 14.6×10^6 K as compared to 15.5×10^6 K in the standard unmixed model, the uniform hydrogen abundance X_1 would be 0.67 and the predicted flux of detectable neutrinos would be less than the measured value. With somewhat less efficient mixing a model can be obtained that fits the observations; e.g. a large mixed core (Shaviv and Salpeter, 1971), or some sort of diffusive mixing (Schatzman and Maeder, 1981).

Further evidence for some mixing comes from the observed surface abundance of lithium and the decrease in lithium abundance with age in solar type stars (cf. Cayrel et al., 1984). But this very observation indicates that mixing cannot be very effective in the outer regions of the Sun, otherwise all the lithium would have been burnt long ago, rather does it suggest a relatively weak 'diffusion' beneath the outer convective zone that carries the lithium down to regions where the temperature is $2.5–3 \times 10^6$ K on the time scale comparable to the solar age.

A fully mixed, chemically homogeneous model is anyway not viable; a model that has the observed luminosity and radius has a helium abundance of 0.31 and a shallow outer convective zone that extends to $0.8\,R_\odot$, it cannot produce the period spacing of the p-modes of oscillation and the variation of sound speed with depth is in contradiction with the results obtained by inversion of solar oscillations (Christensen-Dalsgaard et al., 1985; Roxburgh, 1986).

To maintain mixing inside the Sun against the chemical inhomogeneity that would be created be nuclear reactions requires a substantial energy input (Roxburgh, 1984b, c, 1985). In a stably stratified, chemically inhomogeneous region matter closer to the centre has a larger binding energy (that is a larger negative energy). To interchange matter close to the centre with matter further out requires an input of energy from some source, both to overcome the stable thermal stratification and the stable chemical stratification. Motion that is sufficiently slow and/or a sufficiently small length scale (eg. 'fingers') may be able to overcome the thermal stratification by the exchange of heat through radiation, but overcoming the chemical stratification is very much more difficult. Any steady mixing requires an energy input as great as the rate of increase of the 'energy gradient' due to nuclear reactions, and it is difficult to imagine any such energy source except the nuclear reactions themselves.

The gravitational binding energy of a standard model of the Sun has about doubled due to chemical evolution in the solar lifetime, to maintain the sun homogeneous against this effect of evolution requires an energy source comparable with the binding energy of the Sun (3×10^{48} ergs). The rotational energy for example is only 10^{-5} of the gravitational energy and it is therefore difficult to envisage a mixing mechanism driven by rotation. However the energy released by nuclear reactions during the solar lifetime is about 200 times the binding energy and is therefore able in principle to maintain mixing.

These arguments are related to the results of Mestel (1953, 1984) that rotationally driven circulation currents are only capable of mixing the central regions of a star provided the ratio of centrifugal force to gravity in the central regions is of the order of 0.03. The same is true of any other perturbing force that enables a fraction of the thermal energy to drive circulation currents (Roxburgh, 1984b). Were the Sun to be mixed by such circulation currents, the energy content of the perturbing fields (rotation, magnetic) would be sufficient to cause a very substantial oblateness of the solar surface of the order ot 10^{-3} to 10^{-4}.

One possible mechanisms to drive mixing is the ^{3}He instability discovered by Dilke and Gough (1972) and discussed below, were this to break down into some sort of turbulence or to drive circulation currents; there is more than enough energy available from the burning of ^{3}He through to ^{4}He in those regions where the proton–proton chain does not go to completion. Whether or not such turbulence is created is unknown and the instability may settle down to finite amplitude oscillations as discussed below.

A possibility that needs further study is the suggestion by Press (1981) that gravity waves generated by motions in the outer convective zone would break down into turbulence in the solar interior, such waves might just have enough energy to cause mixing (Spruit, 1985).

However there is not enough energy in differential rotation caused by spin down of the solar surface; and this shows itself in the suppression of the Goldreich-Schubert (1967)-Fricke (1968) instability by a very mild gradient of chemical composition. Whilst the Axisymetric Baroclynic Diffusive Instability (Shibahashi, 1980; Knobloch and Spruit, 1983; Roxburgh, 1984a) is not supressed by a chemical composition gradient, the resulting motions are almost horizontal and ineffective in producing mixing. Indeed this instability is suppressed by a very slight, almost horizontal redistribution of chemical composition (Roxburgh, 1984c, 1985).

Whilst chemical composition gradients suppress most mixing mechanisms they do not prevent almost horizontal mixing along surfaces of constant entropy and we may therefore expect shear type instabilities to redistribute angular momentum on such surfaces to achieve a rotation that is almost constant on spheres (see Zahn, 1983). The first results on the latitudinal variation of angular velocity in the solar interior (Brown, 1985) seem to support this conclusion, except for a possible differential rotation near $r = 0.35 R_\odot$. The radial variation in angular velocity shows an increase and then decrease in about the same region which perhaps suggest some fluids motions; perhaps there is a zone of mixing near the peak of the ^{3}He distribution.

As mentioned above the surface lithium abundance in the Sun and solar type stars strongly suggests some sort of mild diffuse mixing below the convective zone where the chemical composition gradient is very small. Whether this is driven by differential rotation, waves, penetrative convection, penetrative large scale circulation or even the very small fluctuations in gravitational potential caused by convective motions is unknown.

Fortunately we have a growing volume of increasingly accurate oscillation data to constrain the wilder speculations of the theoretician; the sound speed distribution could be used to place limits on the variation of hydrogen abundance if we accept the rest of the theory that goes into the standard model. This is turn could give us a better idea of the processes that determinated the hydrogen abundance distribution in the present Sun.

3.2. THE ^{3}He INSTABILITY

Several years ago, Dilke and Gough (1972) suggested that the young Sun would be unstable to excited g mode oscillations driven by the gradient of ^{3}He, and ^{1}H; the nuclear energy input during an oscillation exceeding the radiative damping. Detailed analysis of such global modes by Christensen-Dalsgaard *et al.* (1974) verifed this conjecture, and showed that after about 3×10^8 yr overstable oscillations set in. The consequences or this instability are not properly understood and the standard solar models just ignore it. This is not satisfactory. Whilst further study has shown that if the effects of the instability are ignored, the solar models become stable again at a time of about 3×10^9 yr, these models assume the instability has no effect for the previous 2.7×10^9 yr!

One suggestion advanced by Dilke and Gough was that this oscillation triggers a finite amplitude instability leading to an overturning of the core, thus maintaining approximate homogeneity, creating temporary thermal imbalance and producing phases of anomalously low neutrino emission. Others have conjectured that the oscillation breaks down into mild turbulence, diffusing ^{3}He into the centre and again producing a partially mixed Sun. Such mixed models lead to a somewhat lower neutrino flux but have difficulty in fitting oscillation data.

A further possibility analysed in this paper is that the oscillation settles down to a finite amplitude limit. In this limit the ^{3}He distribution is changed so that the energy input during a cycle is equal to the energy losses by radiation. The analysis given in

Section 5 gives an amplitude $\delta T/T$ such that

$$\left(\frac{\delta T}{T}\right)^2 = 0.066\left(\frac{t}{t_\odot} - 0.065\right)E(r),$$

where $E(r)$ is the eigenfunction normalised to unity as its first maximum.

The finite amplitude oscillation changes the mean rate of energy generation so that

$$\varepsilon_{\text{mean}} = \varepsilon_0\left[1 + 6\left(\frac{\delta T}{T}\right)^2\right];$$

this changes the evolutionary sequence leading to models of the present Sun. A preliminary investigation by the author suggests that this may produce the change in sound speed at $r = 0.5\,R_\odot$ found from inversion of the oscillation data.

Whether or not this turns out to be the case, the point I wish to emphasize is that the effect of these oscillations on the structure and evolution of the Sun cannot be ignored.

3.3. OVERSHOOTING FROM THE SOLAR CONVECTIVE ZONE

In standard solar models the outer convective zone is modelled by using the mixing length 'theory' of convection to determine the entropy of the deeper adiabatic layers. This theory of convection is certainly incorrect, it neglects the flux of kinetic energy and assumes that the properties of convection are determined locally rather than the whole convective layer settling down to some global equilibrium state (Roxburgh, 1978), but we have no valid theory to use instead. In practice the detailed theory does not matter and the entropy of the convective zone is taken as a free parameter to be adjusted so that the standard model of the present sun has the observed radius; this in turn is used to calibrate the mixing length.

In these models the base of the convective zone is taken to be that point where all the flux is carried by radiation and the temperature gradient equals the adiabatic gradient. This is unlikely to be correct; motion originating above the boundary will penetrate into the stable regions; this penetrative convection will carry energy inwards, producing a temperature gradient closer to the adiabatic gradient thus making penetration easier. This convective overshoot region may be the site of dynamo generation of the solar magnetic field, since the rise time of buoyant magnetic flux tubes is longer for field produced at this level.

The questions that needs further study is the nature and extent of this overshoot region and its effect on the deeper interior. No satisfactory analysis exists, nor in our present state of ignorance of the dynamics of convection are we able to make such an analysis. We have to fall back on mixing length studies and one such analysis is given in Section 6. This analysis assumes that the convective flux in the overshoot region is given by the properties of descending fluid originating at a height $H_p/2$ above the layer where $\Delta\nabla = 0$; $\Delta\nabla$ is the superdiabatic gradient $(\nabla_{\text{ad}} - \nabla)$, with $\nabla = \partial\ln T/\ln P$, and H_p is the pressure scale height. In this model convection penetrates a depth $H_p/2$ below the layer where

$F_{conv} = 0$; the overshoot region is almost adiabatic with a very thin transition region to the underlying radiative layers. (A similar result was obtained by the author on using an integral constraint on the structure of convective zones (Roxburgh, 1978).) When this 'theory' is incorporated into a solar model, normalised so that it has the observed luminosity and radius, the convective zone is 13% deeper than equivalent models without overshooting. Since the inversion of oscillation data suggests that the convective zone extends down for $0.3 R_\odot$ (Christensen-Dalsgaard *et al.*, 1985), whereas most standard models give a depth of nearer $0.27 R_\odot$; this 13% overshoot may improve the agreement between theory and observation.

Changes in the model of the outer layers have an effect on the internal structure; not a large effect by sufficient to produce a small change increase in temperature and hence sound speed at a depth of $0.5 R_\odot$.

The point I want to emphasize is not that the above results are correct – this author is sceptical about any results that depend on using the mixing length theory – but that the effects of overshooting must be included in our models if we wish to make accurate predictions to compare with current and future oscillation data.

3.4. A CONVECTIVE CORE

The standard solar models do not have a convective core, due in part to the relatively weak dependence on the energy generation rate on temperature and in part to the chemical composition gradients established during evolution. There is no simple criterion for the existence of such a core and one can only establish whether or not the centre is convective by constructing complete solar models. Nevertheless an approximate condition can be obtained by studying the variation of $\nabla = \ln T / \partial \ln P$ at the centre of a homogeneous radiative core where

$$\left(\frac{\partial \ln \nabla}{\partial \ln T} \right)_c = -\frac{1}{5} \left[(23 + 5\alpha + 5\beta - 3\eta) - \left(\frac{5\alpha + 8}{\nabla} \right) \right]$$

and the opacity and energy generation rate have been fitted to a local power approximation; $\kappa = \kappa_0 \rho^\alpha T^{-\beta}$, $\varepsilon = \varepsilon_0 \rho T^\eta$. Since $\nabla = 0.4$ in convective regions, and is of the order of 0.2 in radiative regions the condition that ∇ does not decrease too rapidly gives the condition for no convective core as

$$6\eta < 16 + 10\beta - 15\alpha.$$

This condition is related to, but different from the Naur–Osterbrock (1953) condition. On taking typical solar values we find models without a convective core provided $\eta < 4.9$. This condition is satisfied for the equilibrium p–p chain so standard models do not have a convective core.

However, the nuclear energy generation has not always been from the equilibrium p–p chain. In the initial Sun the abundances of ^{3}He and ^{12}C were much in excess of the equilibrium values and the young Sun went through a phase where a major part of its energy came from buring ^{3}He to ^{4}He and ^{12}C to ^{14}N. During these phases η was greater than 5 and a convective core was present. This core may be very important in influencing the subsequent dynamics of the solar interior, since there may well have been a dynamo generating a central magnetic field which subsequently diffused throughout the Sun.

The extent and lifetime of such a convective core needs very careful analysis and is dependent on assumptions made about convective overshooting. The value of the exponent η varies with the size of the core; if the p–p chain is in local equilibrium ^{3}He is produced locally at the rate at which it is destroyed and $\eta = 4.5$, the value from the ^{1}H + ^{1}H reaction. In a convective core the local balance is replaced by a global balance, ^{3}He (and ^{12}C) is convected into the central regions raising the ^{3}He : ^{1}H ratio and hence the value of η. This mixing sustains the convective core for about 3×10^8 yr with no overshooting and for a longer time with overshooting. Indeed with sufficient overshooting the present Sun could have an convective core (see Shaviv and Salpeter, 1971).

Indeed it is worth pointing out that the sequence of solar models constructed under the normal assumption that a region is stable unless is unstable may not always be valid; there could be a second sequence of models constructed under the assumption that the region is convectively unstable, this possibility being due to the different effective values of η in mixed and unmixed zones.

I am not arguing here that the Sun has a convective core or has had one for most of its lifetime – but I do wish to emphasize that until we are able to understand convective overshooting there remains some doubts about the validity of our standard models.

3.5. THE INTERNAL ROTATION AND MAGNETIC FIELD

At the time of writing the observational picture is confused. Duvall *et al.* (1984) deduce an internal profile for the angular velocity $\Omega(r)$, which is almost constant down to $0.4\,R_\odot$ then increases slightly between 0.4 and $0.3\,R_\odot$ than decreases to a minimum between 0.3 and $0.2\,R_\odot$ rising the centre to a value which may be twice the surface value. In contrast, Hill (1985) finds a rapid increase by a factor 3 just below the convective zone and almost constant Ω from there to the centre. Very recent results by Brown (1985) support the findings of Duvall *et al.*, and further suggest that the latitudinal differential rotation decreases with depth but that the region around $0.35\,R_\odot$ has enhanced differential rotation. The innocent theoretician (if there be such) may be forgiven if he/she takes the position that the issue is not yet settled.

It is difficult to reconcile the Duvall *et al.* (1984) rotational profile with even a tiny internal magnetic field unless this field is somehow separated into distinct zones, or if what we currently observe is a transient produced by a recent phase of mixing in the core, the internal magnetic field not having had enough time to equalise angular velocity. The decay time of the largest scale components of a global magnetic field is of order $t_\odot$ so in the age of the Sun all but the few lowest order modes will have decayed away under Ohmic diffusion, but unless there is some form of enhanced diffusion there should be some relict of the original field trapped in the Sun at the time of formation. Perhaps if the Sun had an initial field it floated to the surface during a fully convective pre-Main Sequence phase, or was dragged with the outer convective zone as a radiative core was forming during approach to the Main Sequence. However if the Sun had a small convective core during its initial Main Sequence phase a field would probably have been generated within this core, when convection ceased the field would diffuse outwards and decay, leaving the present Sun with the relict of the largest scale component of the dynamo field. If this dynamo field had no components from the lowest eigen decay modes then this field could by now have decayed away.

However even if a weak field remains trapped in the core, how this affects the rotation has yet to be fully understood. The condition that the angular velocity be constant on field lines is a steady state condition, a weak field being wound up by differential rotation until it is large enough to react back on the motion; a field of only $10 \, \mu G$ being large enough to produce uniform rotation in the life time of the Sun. Perhaps some kind of instability limits the increase in field before it can react back on the motion, a problem that remains to be resolved. However even if some dissipative mechanism intervenes, the energy dissipated originates in the kinetic energy of rotation so that whatever the dissipation mechanisms it reduces the kinetic energy of the rotation field (Mestel, private communication); for a given angular momentum the lowest rational energy state is that of uniform rotation so the mechanism must have the effect of equalizing angular velocity.

Since the present Sun is losing angular momentum in the solar wind it is reasonable to suppose that the surface layers and convective zone have been slowed down. Indeed the evidence from other solar type stars suggests that the Sun was originally rotating much more rapidly than at present and was slowed down during its early main sequence phase. How is the outer convective zone coupled to the deep interior? Since the surface field is oscillatory it can only penetrate a very small distance beneath the zone, so coupling to the interior is probably through some intermediate zone between the two regions. The low surface abundance of Lithium is strong evidence that there is some mild turbulent mixing beneath the convective zone down to the layers where lithium is burnt.

If there is no field in the interior then there is a range of instabilities that are driven by differential rotation produced by the spin down of the surface layers (Goldreich and Schubert, 1967; Fricke, 1968; Knobloch and Spruit, 1983; Roxburgh, 1984a). Spruit et al. (1983) used these instabilities to argue that the angular velocity distribution was close to marginal stability – which gave an almost constant $\Omega(r)$ down to $0.4 \, R_\odot$ and then an inward increase – not exactly that deduced by Duvall et al. (1984) but with the same overall behaviour except for the dip at $0.25 \, R_\odot$. If one assumes that angular momentum is transported by mild turbulence down this marginal gradient then the same turbulence must produce chemical mixing; this idea was developed by myself (Roxburgh, 1984) and by Law et al. (1984) in an attempt to solve the solar neutrino problem; such a model gave a reduced neutrino flux but has severe problems with the observed spacing of low order p modes of oscillation.

But such instabilities are not effective. Recently I showed (Roxburgh, 1984c, 1985) from consideration of the energetics that these instabilities would be unable to effectively mix the Sun, the motion being almost horizontal and leading to a latitude dependent distribution of molecular weight that could sustain a much larger angular velocity gradient. The same argument can be used to rule out any substantial mixing driven by rotation.

However, this may not be true in the layers beneath the convective zone and weak turbulence driven by a very small angular velocity gradient may be effective in carrying out angular momentum in this almost chemically homogenous region, and at the same time leading to a decay in the surface abundance of lithium (Dicke, 1972; Roxburgh, 1976). Since the energy to overcome the mild chemical composition gradient comes initially from the kinetic energy of differential rotation, such mixing will only be effective if the rotational energy is greater than the difference in gravitational energy due to the composition gradient. In standard solar models this is the case in the layers just beneath the convective zone, but recently Dziembowski and Paterno (private communication) have argued that gravitational diffusion leads to a larger composition gradient; if this is correct it is difficult to see how rotation can drive mixing except in regions of large differential rotation.

The drop in angular velocity at $0.3 R_\odot$ deduced by Duvall *et al.* (1984) is the most difficult observation to explain. One intriguing possibility (Gough, private communication) is that the central regions have recently turned over, low angular momentum material moving from near the centre to $0.3 R_\odot$ leading to a drop in angular velocity, and higher angular momentum material moving to the centre, thus leading to a rapidly spinning small core. The only energy source capable of driving such a turnover is the nuclear energy which revives the old conjecture of Dilke and Gough that the ^{3}He instability produces such a non linear instability. Were this to have taken place it would produce radically different solar models.

Were the Sun to have no internal magnetic field then there would be some spin up of the central core due to the increase in central density during evolution; this is large enough to explain the values given in Duvall *et al.* (1984) if the rotation were initially constant; but an initially constant rotation suggests a magnetic field which would prevent the spin up of the core!

Recent preliminary results by Brown (1985) suggest that whilst the rotation in most of the Sun is almost constant on spheres there may be a zone of latitudinal variation near the region where the angular velocity also varies with radius. Drawing on our understanding of theories of differential rotation of the outer convective zone this suggests that there is some fluid motion – either a turbulent shell or large scale laminar motion, which maintains a latitudinal variation in Ω. The location of such a shell is sufficiently close to the maximum in the ^{3}He distribution to put forward the conjecture that the Sun has a mixed shell at about $0.3–0.4 R_\odot$, driven by the burning of ^{3}He in that region where the distribution of ^{3}He has a very steep gradient. The resulting mild turbulence leads to a latitudinal variation in Ω and perhaps separates a more rapidly spinning core from a slower rotating outer radiative region (Roxburgh, 1986). This is however only a conjecture!

3.6. OTHER WORRIES ABOUT THE STANDARD MODEL

Were there to have been phases of significant mixing then the standard solar evolutionary sequence may be far from the truth. Ad hoc diffusive mixing has been proposed and investigated by Schatzman and Maeder (1981) but these models have serious difficulties in reproducing the observed properties of solar oscillations. Moreover, their driving mechanisms was thought to be small scale turbulence driven by rotation and the solar rotation field does not have enough energy to mix helium rich material in the centre with hydrogen rich matter further away (Roxburgh, 1984b). Were the ^{3}He instability to break down into turbulence then there is in principle an energy source although how this would drive the turbulence is unclear.

There is still considerable uncertainty over the opacity and the equation of state for a mixture of many ions, radiation and magnetic fields – do we understand the high velocity tail in the distribution functions which contain the particles that undergo nuclear reactions?

The nuclear reaction rates must also remain an area of uncertainty. Are the ^{3}He + ^{3}He and ^{3}He + ^{4}He rates sufficiently well determined to be sure that we have the right branching ratio and hence valid prediction of the flux of electron neutrinos? Or is there no astrophysical solar neutrino problem – the neutrinos mixing between states as predicted in most Grand Unified Theories, so that by the time they reach the Earth only $\frac{1}{3}$ of the neutrinos are electron neutrinos and thus detectable by their interaction with chlorine?

Then again the full set of reaction equations are non linear and coupled non linearly to the equations of Stellar Structure. We assume that the short time-scale reactions settle down to quasi equilibrium, but many high order non-linear dynamical systems settle down onto limit cycles or chaos. A preliminary, though very simple investigation by myself and R. Tavakol (1985) has not revealed any interesting dynamical behaviour, but the question is not yet settled.

There are other areas in which one could question the validity of the standard model, but I hope that the points that I have made convince the reader that we have not as yet achieved Eddington's (1926) hope. The difference is that through helioseismology we now have a new and exceedingly powerful experimental tool for probing the interior, which gives us hope that in the not too distant future we may be able to understand the interior of the Sun.

4. The Construction of Standard Solar Models

It may be of pedagogical value to the reader if I describe in a little detail the actual equations used in building such a model. If M_r is the mass within a sphere of radius r; P, ρ, T, the pressure, density, and temperature at M_r; ε, κ, the energy generation rate and opacity per unit mass, S the specific entropy and L_r the rate at which energy crosses a sphere of radius r, then these equations can be expressed as

$$\frac{\partial P}{\partial M_r} = -\frac{GM_r}{4\pi r^4}, \quad \frac{\partial r}{\partial M_r} = \frac{1}{4\pi r^2 \rho}, \quad \frac{\partial L_r}{\partial M_r} = \varepsilon - T\frac{\partial S}{\partial t},$$

$$\frac{\partial T}{\partial M_r} = \frac{3 \kappa L_r}{16 \pi a c T^3 r^4} \quad \text{(radiative)}$$

$$= \frac{\gamma}{\gamma - 1} \frac{T}{P} \frac{\partial P}{\partial M_r} \quad \text{(convective)} \quad \Rightarrow P = KT^{5/2}$$

with

$$\rho = \rho(P, T, X_i), \quad \kappa = \kappa(\rho, T, X_i), \quad \varepsilon = \varepsilon(P, T, X_i),$$

where $X_i(M_r)$ is the abundance by mass of element with atomic number i (e.g. X_1 hydrogen and X_4 helium) and γ the ratio of specific heats; these equations must be supplemented by the equations governing the rate of change of the X_i from nuclear reactions.

4.1. THE SURFACE CONVECTIVE ZONE

The outer layers of the Sun are convective and some theory is needed to model these layers. Attempts to produce such models indicate that except in a thin surface layer the convective zone is adiabatic and that over the bulk of the zone we can express the relationship between pressure and temperature as

$$P = KT^{5/2}, \quad K = K(R, L, M, X_i).$$

The theory of convection seeks to determine the entropic constant K in terms of the surface conditions and abundances; all such attempts are as inaccurate as the theory of convection used and cannot be believed.

For what is it worth I give below a simple approximation to detailed numerical calculations of Hobbs and myself (Hobbs and Roxburgh, 1986):

$$K = 6.6 \times 10^{-3} \left(\frac{R}{R_\odot}\right)^4 \left(\frac{L_\odot}{L}\right)^3 \left(\frac{X_1}{0.73}\right)^{1.91} \left(\frac{Z}{0.018}\right)^{-0.13}.$$

The constant has been chosen so that a standard solar model has the observed radius with the observed luminosity.

4.2. SIMPLIFIED NUCLEAR REACTION NETWORK

The energy generation in solar models comes promarily from the p–p chain with a small amount from the CN cycle. No author actually calculates the detailed time evolution of all reactions in these processes, they occur on such dramatically different time scales that it is not feasible to do so. For illustrative purposes I shall first consider just the dominant branch of the p–p chain with

$$^1\text{H} + {}^1\text{H} \rightarrow {}^2\text{H} + e^+ + \nu,$$

$$^2\text{H} + {}^1\text{H} \rightarrow {}^3\text{He},$$

$$^3\text{He} + {}^3\text{He} \rightarrow 2{}^4\text{He} + 2{}^1\text{H}.$$

If we define $R_{ij} X_i X_j$ as the number of reactions per unit mass per unit time between particles of species i and j, then the reaction rate equations are

$$\frac{N}{A_1} \frac{\partial X_1}{\partial t} = -2 R_{11} X_1^2 - R_{12} X_1 X_2 + 2 R_{33} X_3^2,$$

$$\frac{N}{A_2} \frac{\partial X_2}{\partial t} = R_{11} X_1^2 - R_{12} X_1 X_2,$$

$$\frac{N}{A_3} \frac{\partial X_3}{\partial t} = R_{12} X_1 X_2 - 2 R_{33} X_3^2,$$

$$\frac{N}{A_4} \frac{\partial X_4}{\partial t} = R_{33} X_3^2,$$

where A_i is the atomic mass of species i and N is Avogadro's number. The rate of release of energy per unit mass is

$$\varepsilon = E_{11} R_{11} X_1^2 + E_{12} R_{12} X_1 X_2 + E_{33} R_{33} X_3^2,$$

where E_{ij} is the energy released per reaction of i with j.

The reaction rates R_{ij}, and hence the reaction times per particle vary enormously; in typical solar conditions with $T = 14 \times 10^6$ K, $\rho \sim 100$ g cc^{-1}, the time scales for these three reactions are 10^{10} yr, 6 s and 10^6 yr. It is clearly not possible to follow by numerial integrations the detailed evolution of all the X_i. However since $R_{12} \gg R_{11}$, the abundance of X_2 rapidly adjusts so that the rate of creation of X_2 equals its rate of destruction, i.e.

$$X_{2E} = \frac{R_{11} X_1}{R_{12}}.$$

To be more accurate, treating X_1 as constant the equation for X_2 can be integrated to give

$$X_2(t) = \left(X_2(0) - \frac{R_{11} X_1}{R_{12}} \right) \exp\left(-\frac{R_{12} A_2 X_1 t}{R_{11}} \right) + \frac{R_{11} X_1}{R_{12}}$$

and X_2 approaches its equilibrium value on a time scale of order 10 s. The reaction network can then be reduced to

$$\frac{N}{A_1} \frac{\partial X_1}{\partial t} = -3 R_{11} X_1^2 + 2 R_{33} X_3^2,$$

$$\frac{N}{A_3} \frac{\partial X_3}{\partial t} = R_{11} X_1^2 - 2 R_{33} X_3^2,$$

$$\frac{N}{A_4} \frac{\partial X_4}{\partial t} = R_{33} X_3^2,$$

$$\varepsilon = (E_{11} + E_{12}) R_{11} X_1^2 + E_{33} X_3^2 R_{33}.$$

A similar argument can be applied to the abundance X_3 of ^{3}He; the assumption of equilibrium gives

$$X_{3E} = \left(\frac{R_{11}}{2 R_{33}}\right)^{1/2} X_1, \quad \frac{\partial X_1}{\partial t} = -2 \frac{A_1}{N} R_{11} X_1^2,$$

$$\varepsilon = \left(E_{11} + E_{12} + \frac{E_{33}}{2}\right) R_{11} X_1^2.$$

It is this approximation – together with a power law fit to the reaction rate R_{11} that gives the simple approximation given previously.

However, the assumption of equilibrium is not valid throughout the Sun; neglecting small correction factors the reaction rates are

$$R_{11} = \frac{\rho}{T_6^{2/3}} \exp\left(25.4 - \frac{33.8}{T_6^{1/3}}\right),$$

$$R_{33} = \frac{\rho}{T_6^{1/3}} \exp\left(81.1 - \frac{122.8}{T_6^{1/3}}\right),$$

and whilst X_3 reaches equilibrium in the central regions this is not true in the outer half of the Sun. Some authors, nevertheless take the equilibrium distribution throughout the Sun, whilst others seek to follow the evolution of X_3 in time – usually by separately integrating the equation for X_3 on a finer time grid than that used for the full equations (cf. Bahcall *et al.*, 1982). In my own evolution programme I integrate the equation for X_3 (and other minor species) analytically, by noting that where X_1 changes significantly over the age of the Sun X_3 evolves on a even shorter time scale, whereas when X_3 evolves on a long time scale, X_1 remains approximately constant, hence we can integrate the equation for X_3 taking X_1 constant which gives

$$X_3(t) = X_{3E} \tanh\left[(2 R_{11} R_{33})^{1/2} \frac{A_3}{N} t + \tanh^{-1} \frac{X_3(0)}{X_{3E}}\right].$$

These different approaches to dealing with ^{3}He abundance produce only slight differences in the 'standard models' obtained – although clearly if the distribution of ^{3}He is required the equilibrium approximation is invalid.

4.3. MORE DETAILED REACTION NETWORK

The full set of reactions included in the p–p chain and the CN cycle are as follows:

p–p chain	CNO cycle
${}^1H + {}^1H \rightarrow {}^2H + e^+ + \nu$	${}^{12}C + {}^1H \rightarrow {}^{13}N$
${}^1H + e^- + {}^1H \rightarrow {}^2H + \nu$	${}^{13}N \rightarrow {}^{13}C + e^+ + \nu$
${}^2H + {}^1H \rightarrow {}^3He$	${}^{13}C + {}^1H \rightarrow {}^{14}N$
${}^3He + {}^3He \rightarrow {}^4He + 2\,{}^1H$	${}^{14}N + {}^1H \rightarrow {}^{15}O$
${}^3He + {}^4He \rightarrow {}^7Be$	${}^{15}O \rightarrow {}^{15}N + e^+ + \nu$
${}^7Be + e^- \rightarrow {}^7Li + \nu$	${}^{15}N + {}^1H \rightarrow {}^{12}C + {}^4He$
${}^7Li + {}^1H \rightarrow 2\,{}^4He$	${}^{15}N + {}^1H \rightarrow {}^{16}O$
${}^7Be + {}^1H \rightarrow {}^8B$	${}^{16}O + {}^1H \rightarrow {}^{17}F$
${}^8B \rightarrow {}^8Be^* + e^+ + \nu$	${}^{17}F \rightarrow {}^{17}O + e^+ + \nu$
${}^8Be^* \rightarrow 2\,{}^4He$	${}^{17}O + {}^1H \rightarrow {}^{14}N + {}^4He$

Again the reaction rates vary enormously, in particular the Li, Be, B reaction times are very much shorter than that of ${}^3He + {}^4He$. Likewise, the time scales for ${}^{13}N$, ${}^{15}N$ and ${}^{15}O$ are much shorter than those of the ${}^{12}C$ and ${}^{14}N$ reactions. The ${}^{13}C + {}^1H$ reaction is somewhat faster than ${}^{12}C + {}^1H$, and the ${}^{16}O$ branch is hardly operative at the temperatures found in solar models (and is anyway only a small perturbation). Except in the very early phases where ${}^{13}C$ is not yet in equilibrium the abundance evolution equations can be condensed to

$$\frac{N}{A_1} \frac{\partial X_1}{\partial t} = -3R_{11}X_1^2 + 2R_{33}X_3^2 - R_{34}X_3X_4 - 2R_{112}X_1X_{12} - $$
$$- 2R_{114}X_1X_{14},$$

$$\frac{N}{A_3} \frac{\partial X_3}{\partial t} = R_{11}X_1^2 - 2R_{33}X_3^2 - R_{34}X_3X_4,$$

$$\frac{N}{A_4} \frac{\partial X_4}{\partial t} = R_{33}X_3^2 + R_{34}X_3X_4 + R_{114}X_1X_{14},$$

$$\frac{N}{A_{12}} \frac{\partial X_{12}}{\partial t} = -R_{112}X_1X_{12} + R_{114}X_1X_{14},$$

$$\frac{N}{A_{14}} \frac{\partial X_{14}}{\partial t} = R_{112}X_1X_{12} - R_{114}X_1X_{14}.$$

As in the simpler case considered above the evolution equations for X_3, X_{12}, X_{14} can be integrated to give:

$$X_3(t) = \frac{R_{34}X_4}{4R_{33}} + X^* \tanh\left(\frac{2A_3R_{33}X_t^*}{N} + \alpha\right),$$

$$X^* = \left(\frac{R_{11}X_1^2}{2R_{33}} + \frac{R_{34}{}^2X_4^2}{16R_{33}^2}\right)^{1/2}, \quad \alpha = \tanh^{-1}\left(\frac{X_3(0)}{X^*} + \frac{R_{34}X_4}{4R_{33}X^*}\right),$$

$$X_{12}(t) = R_{114}[A_{14}X_{12}(0) + A_{12}X_{14}(0)] \times$$

$$\times \left[1 - \exp\left\{-\left[\frac{A_{12}R_{112} + A_{14}R_{114}}{N}\right]X_1 t\right\}\right],$$

$$X_{14}(t) = X_{14}(0) + \frac{A_{14}}{A_{12}}[X_{12}(0) - X_{12}(t)].$$

The rate of generation of energy is

$$E = R_{11}E_{11}^* + R_{33}E_{33}^* + R_{34}E_{34}^* + R_{112}E_{12}^* + R_{114}E_{14}^*,$$

where E_{ij}^* is the energy released per reaction of i with j including subsequent reactions taken to be in equilibrium; e.g., $E_{11}^* = E_{11} + E_{12}$, $E_{112}^* = E_{112} + E_{113}$. The values of E_{ij}^* are given in Table II.

TABLE II
Energy release E_{ij}^* (10^{-6} ergs) for condensed reaction network

E_{11}^*	E_{33}^*	E_{34}^*	E_{112}^*	E_{114}^*
10.69	20.60	30.43	17.63	22.46

Finally some values of reaction rates must be taken. A detailed account of the current 'best values' is given by Bahcall *et al.* (1982); these can be written in the form:

$$R_{11} = C_{11}\frac{\rho}{T_6^{2/3}}\exp\left(25.44 - \frac{33.81}{T_6^{1/3}}\right),$$

$$R_{33} = C_{33}\frac{\rho}{T_6^{2/3}}\exp\left(81.10 - \frac{122.77}{T_6^{1/3}}\right),$$

$$R_{34} = C_{34}\frac{\rho}{T_6^{2/3}}\exp\left(72.36 - \frac{128.28}{T_6^{1/3}}\right),$$

$$R_{112} = C_{112}\frac{\rho}{T_6^{2/3}}\exp\left(73.725 - \frac{136.93}{T_6^{1/3}}\right),$$

$$R_{114} = C_{114} \frac{\rho}{T_6^{2/3}} \exp\left(74.457 - \frac{152.31}{T_6^{1/3}}\right),$$

where C_{ij} are correction factors of order unity which include the effects of electron screening and the variation of cross section with energy and are given in Table III.

TABLE III

Correction factors for reaction rates

$$C_{ij} = \left[1 + c_{ij_1} T_6^{1/3} + c_{ij_2} T_6^{2/3} + c_{ij_3} T_6\right]\left[1 + c_{ij_4} \frac{\rho}{T_6^{3/2}}\right]$$

	c_{ij_1}	c_{ij_2}	c_{ij_3}	c_{ij_4}
C_{11}	0.0123	0.011	0.00095	0.25
C_{33}	0.0034	-0.00067	-0.000016	1.0
C_{34}	0.0035	-0.002	-0.00046	1.0
C_{112}	0.0031	0.0066	0.00015	1.5
C_{114}	0.0027	-0.0078	-0.00015	1.75

4.4. THE OPACITY OF SOLAR MATERIAL

The calculation of stellar opacities is a complex problem and is undertaken by only a very small number of groups – they do not all agree with each other. The Los Alamos and Livermore codes, based on different approximation techniques, differ by about $\pm 10\%$ and calculations by Carson for temperatures around 10^6 K differ substantially from both these groups.

It would therefore be unwise to place great faith in any one set of calculations but rather to accept that at the present time there are uncertainties of the order of 20% and to ask what effect these uncertainties have on our models. In calculating a standard model different authors use different determinations of the opacity – introducing yet another variable between such models.

TABLE IV

Opacities for Ross–Aller mixture (c.g.s. units)

ρ/T_6^3	X	Z	T_6							
			15.7	12.8	11.3	10.0	7.0	4.5	3.0	1.8
0.035	0.75	0	0.71	0.77	0.81	0.86	0.99	1.20	1.57	2.23
0.041	0.75	0	0.77	0.83	0.88	0.93	1.08	1.33	1.72	2.50
0.035	0.75	0.0179	1.29	1.46	1.59	1.77	2.79	6.39	14.7	34.0
0.041	0.75	0.0179	1.38	1.58	1.72	1.90	2.98	6.78	15.5	37.5
0.035	0.35	0.0179	1.09	1.24	1.36	1.51	2.44	5.76	13.5	30.9
0.041	0.35	0.0179	1.17	1.34	1.46	1.63	2.62	6.12	14.4	34.0

Recent determinations by the Los Alamos group are given in Bahcall *et al.* (1982). For completeness I give below a condensed version of these opacities in Table IV. These values can be fitted by the approximation formula:

$$\kappa = 0.13(1 + X) + 20.5((1 + X)\rho)^{0.70}T_6^{-2.75} +$$

$$+ 15250Z(1 + X)^{1/2}\left(\frac{\rho}{T_6^3}\right)^{0.34}\left[\frac{\exp(0.1T_6)}{T_6^{1.78}} + 0.00022\,T_6\right].$$

5. ^{3}He Driven Instability

Several years ago Dilke and Gough (1972) pointed out that the Sun could be unstable to growing amptitude oscillations driven by the gradient of ^{3}He produced by the burning of hydrogen. Subsequent investigation by several authors (cf. Christensen-Dalsgaard *et al.*, 1974) have verified this result by detailed analysis of low order *g* modes, finding that a solar model is unstable after about 3×10^8 yr.

The exciting mechanism can be understood by studying the behaviour of a 'blob' of fluid of scale *l*, executing almost adiabatic oscillations with the natural bouyancy frequency. If radiative losses and energy generation are neglected such a blob executes simple harmonic motion about the equilibrium level where its density is the same as its surroundings; when it travels downwards from the equilibrium level it is lighter than its surroundings, it is therefore slowed down and its motion reversed. If radiative losses are include the motion is damped; when it is below the equilibrium level it is less dense and therefore hotter than its surroundings, on loosing heat to the surroundings the density difference is reduced which reduces the restoring force and the blob returns to its equilibrium level with a smaller velocity upwards than it had when travelling downwards.

Energy generation works in the opposite sense; in the phase when the temperature is hotter than the surroundings the energy generation is enhanced, this increases the temperature excess which leads to an enhanced restoring force so the blob returns to the equilibrium level with a larger upward velocity than it had on the downward phase. In the general case therefore the oscillation is excited if the energy gain from nuclear reactions exceeds the losses radiation. If the blob has dimensions *l*, this requires

$$\rho l^3 \delta\varepsilon > \frac{K\,\delta T}{l}l^2,$$

where the radiative energy flux is $F = -K\nabla T$, $K = 4acT^3/3\kappa\rho$.

If the energy generation comes from the proton–proton chain and if for the purpose of illustration we confine our attention to the ^{3}He + ^{3}He branch, the energy generation rate can be expressed as

$$\varepsilon = \varepsilon_1\rho X_1^2 T^{n_1} + \varepsilon_3\rho X_3^2 T^{n_3} = \varepsilon_{11} + \varepsilon_{33},$$

and then the perturbation in ε during an oscillation is $\delta\varepsilon$, where

$$\frac{\delta\varepsilon}{\varepsilon} = \left(2\frac{\varepsilon_{11}}{\varepsilon}\frac{\delta X_1}{X_1} + 2\frac{\varepsilon_{33}}{\varepsilon}\frac{\delta X_3}{X_3}\right) + \frac{\delta\rho}{\rho} + \left(\frac{\varepsilon_{11}}{\varepsilon}\eta_1 + \frac{\varepsilon_{33}}{\varepsilon}\eta_3\right)\frac{\delta T}{T}.$$

If we now use the relation $\mu = 4/(3 + 5X_1)$ and note that as the blob is in pressure equilibrium with its surroundings $\delta\rho/\rho = \delta\mu/\mu - \delta T/T$, the condition that energy generation exceed energy loss takes the form

$$\left(2\frac{\varepsilon_{11}}{\varepsilon} - \frac{5X_1}{3 + 5X_1}\right)\frac{\delta X_1}{X_1} + 2\frac{\varepsilon_{33}}{\varepsilon}\frac{\delta X_3}{X_3} > \left(\frac{KT}{\rho\varepsilon l^2} - \frac{\varepsilon_{11}}{\varepsilon}\eta_1 - \frac{\varepsilon_{33}}{\varepsilon}\eta_1 + 1\right)\frac{\delta T}{T}.$$

Since the oscillation is almost adiabatic when the blob is displaced at distance z, we have

$$\frac{\delta T}{T} = z\left(\frac{\gamma - 1}{\gamma}\frac{1}{P}\frac{\mathrm{d}P}{\mathrm{d}z} - \frac{\mathrm{d}T}{\mathrm{d}z}\right) = \frac{z}{P}\frac{\mathrm{d}P}{\mathrm{d}z}(\nabla_{\mathrm{ad}} - \nabla),$$

where $\nabla = (\partial\ln T/\partial\ln P)$, and noting that $\delta X_1 = z\,\mathrm{d}X_1/\mathrm{d}z$ the condition for growing amplitude oscillations is

$$\left(\frac{2\varepsilon_{11}}{\varepsilon} - \frac{5X_1}{3 + 5X_1}\right)\frac{1}{H_1} + \frac{\varepsilon_{33}}{\varepsilon}\frac{2}{H_3} > \left(\frac{KT}{\rho\varepsilon l^2} - \frac{\varepsilon_{11}}{\varepsilon}\eta_1 - \frac{\varepsilon_{33}}{\varepsilon}\eta_3 + 1\right)\frac{(\nabla_{\mathrm{ad}} - \nabla)}{H_p},$$

where H_p is the pressure scale height and H_1, H_3 are respectively the scale heights of X_1 and X_3.

As the Sun evolves both H_1 and H_3 decrease and this condition is easier to satisfy. This is found to be the case from detailed numerical calculations of global g mode oscillations and some results are given in Table V. It is interesting to note that just outside a convectively unstable region where $\nabla_{\mathrm{ad}} - \nabla \ll 1$ this relation is easier to satisfy which raises unanswered questions about the exitation of g modes in the neighbourhood of such a core.

TABLE V

Period and growth rates of g modes ($l = 1, 2$)

Age (yr)	Period (hr)	Growth rate (10^7 yr)	Period (hr)	Growth rate (10^7 yr)
0	1.74	− 1.02	2.42	− 2.42
10^8	1.72	− 0.15	2.40	− 0.51
2×10^8	1.71	0	2.38	− 0.35
3×10^8	1.70	0.11	2.36	− 0.22
6×10^8	1.66	0.47	2.28	0.25

The abundances of X_1 and X_3 are determined by the evolution equations

$$\frac{N}{A_1}\frac{\partial X_1}{\partial t} = -3R_{11}X_1^2 + 2R_{33}X_3^2,$$

$$\frac{N}{A_3} \frac{\partial X_3}{\partial t} = R_{11} X_1^2 - 2 R_{33} X_3^2,$$

and in the region where energy generation is significant X_3 is in local equilibrium so that

$$X_3 = X_1 \left(\frac{R_{11}}{2 R_{33}}\right)^{1/2}, \quad \frac{N}{A_1} \frac{\partial X_1}{\partial t} = -2 R_{11} X_1^2,$$

where to a good enough approximation we may take

$$R_{11} = a_1 \rho T^{\eta_1}, \quad R_{33} = a_3 \rho T^{\eta_3}, \quad \eta_1 \simeq 4, \quad \eta_3 \simeq 16;$$

it therefore follows that

$$\frac{1}{H_3} = \frac{1}{H_1} + \frac{6}{H_T}.$$

Using this relation to eliminate H_3, and using $H_T = H_p/\nabla$, the condition for instability reduces to

$$\left(2 - \frac{5X_1}{3 + 5X_1}\right) \frac{H_p}{H_1} > \left(\frac{KT}{\rho \varepsilon l^2} - 9\right)(\nabla_{ad} - \nabla) - 6\nabla.$$

In a homogenous solar model the right-hand side is positive, H_1 is infinite and the Sun is stable, as the Sun evolves H_1 decreases and instability sets in. The evolution of X_1 is governed by the equation

$$\frac{N}{A_1} \frac{\partial X_1}{\partial t} = -2 R_{11} X_1^2, \quad R_{11} \sim \rho T^{\eta}.$$

It is a satisfactory approximation to the variation of X_1 with time to take

$$\frac{1}{X_1(t)} - \frac{1}{X_1(0)} = 1.5 \left(\frac{T}{T_c}\right)^{\eta} \left(\frac{\rho}{\rho_c}\right) \left(\frac{t}{t_\odot}\right)$$

which gives a central value of 0.35 for $X_1 = 0.73$ at $t = 0$. It therefore follows that near the centre $\rho \simeq \rho_c$, $T \simeq T_c$ we have

$$\frac{1}{H_1} = 1.5 X_1 \left(\frac{\eta}{H_T} + \frac{1}{H_p}\right) \frac{t}{t_\odot}.$$

Eliminating H_ρ and H_T, and neglecting tha small contribution from the variation of the molecular weight gives

$$\left(\frac{t}{t_\odot}\right) > \frac{1}{1.5 X_1} \left[\left(\frac{KT}{\rho \varepsilon l^2} - 9\right)(\nabla_{ad} - \nabla) - 6\nabla\right] \left[\frac{1}{(\eta_1 - 1)\nabla + 1}\right] \left[\frac{6 + 5X_1}{3 + 5X_1}\right].$$

Taking $\nabla = 0.25$, $\nabla_{ad} = 0.4$, $X = 0.73$, $\eta = 4$ gives

$$\left(\frac{t}{t_\odot}\right) > 0.053 \left(\frac{KT}{\rho \varepsilon l^2} - 19\right).$$

Detailed numerical calculations give instability when $(t/t_\odot) = 0.065$.

5.1. FINITE AMPLITUDE LIMIT

The existence of the instability has been confirmed by several investigations but its consequences have not been fully explored – indeed it is usually ignored. Dilke and Gough originally suggested that this would trigger a finite amplitude instability in the central regions, the mixing of ^{3}He rich matter into the centre producing an excess energy production and destroying thermal equilibrium. Others have conjectured that this could lead to steady diffusive mixing in the central regions, or that the Sun settles down to finite amplitude g mode oscillations, and it is this finite amplitude limit I shall now consider.

Again confining our attention to just the ^{3}He + ^{3}He branch of the p–p chain, the evolution equations for ^{3}He is

$$\frac{N}{A_3} \frac{\partial X_3}{\partial t} = R_{11} X_1^2 - 2 R_{33} X_3^2,$$

where to a satisfactory approximation

$$R_{11} = \varepsilon_1 \rho T^{m_1}, \quad R_{33} = \varepsilon_3 \rho T^{m_3}.$$

If the Sun were static the equilibrium variation of ^{3}He is just

$$X_{3E}^2 = \left(\frac{R_{11}}{2 R_{33}}\right) X_1^2 \sim X_1^2 T^{(m_1 - m_3)}$$

and it is this profile that leads to the onset of instability.

However, the equilibrium value is changed by the oscillation and it is this feed-back mechanism that limits the growth. If the temperature variation during an oscillation is given by $\delta T = \delta T_0 \exp(i\omega t)$, then the equilibrium distribution of the ^{3}He is obtained by integrating Equation (5.1) over a period, hence

$$X_{3E}^* = \left(\frac{\overline{R}_{11}}{2\overline{R}_{33}}\right)^{1/2} X_1, \quad \overline{R}_{ij} = \frac{\omega}{2\pi} \int_0^{2\pi/\omega} R_{ij} \, dt.$$

If the oscillation is almost adidabatic then $\delta\rho/\rho = 1.5 \, \delta T/T$ and so

$$\overline{R}_{ij} = R_{ij}^0 \left[1 + \frac{(\eta + 1.5)(\eta + 0.5)}{4} \left(\frac{\delta T_0}{T}\right)^2\right].$$

The equilibrium distribution of the ^{3}He is then

$$X^*_{3E} = X_1 \left[\frac{R^0_{11}}{2R^0_{33}}\right]^{1/2} \left[1 + \frac{1}{32}[(2\eta_3 + 3)(2\eta_3 + 1) - \right.$$

$$\left. - (2\eta_1 + 3)(2\eta_1 + 1)]\left(\frac{\delta T_0}{T}\right)^2\right].$$

With $\eta_1 = 4$, $\eta_3 \simeq 16$ this gives

$$X^*_{3E} = X_1 \left[\frac{R^0_{11}}{2R^0_{33}}\right]^{1/2} \left[1 + 33\left(\frac{\delta T_0}{T}\right)^2\right]^{-1}$$

and the energy generation rate ε is increased to

$$\varepsilon^* = E_{11}\overline{R}_{11}X^2_1 + E_{33}\overline{R}_{33}X^2_3 = \varepsilon_0\left[1 + 6\left(\frac{\delta T_0}{T}\right)^2\right].$$

This finite amptitude oscillation has two important effects, it changes the rate of energy generation, and it produces a flatter ^{3}He profile in the central region. It is this change in ^{3}He profile that produces the feed back that limits the growth of the amplitude.

To determine the finite amplitude we write

$$X^*_3 = X_3 f(r), \quad \frac{1}{f(r)} = 1 + 33\left(\frac{\delta T_0}{T}\right)^2, \quad \frac{1}{H^*_3} = \frac{1}{H_3} + \frac{1}{H_f},$$

where H_f is the scale height of $f(r)$; then on replacing H_3 by H^*_3 in the previous analysis we obtain the condition

$$\left(\frac{2\varepsilon_{11}}{\varepsilon} - \frac{5X_1}{3 + 5X_1}\right)\frac{1}{H_1} + \frac{2\varepsilon_{33}}{\varepsilon}\frac{1}{H_3} + \frac{2\varepsilon_{33}}{\varepsilon}\frac{1}{H_f} >$$

$$> \left(\frac{KT}{\rho\varepsilon l^2} - \frac{\varepsilon_{11}\eta_1}{\varepsilon} - \frac{\varepsilon_{33}\eta_3}{\varepsilon} + 1\right)\left(\frac{\nabla_{ad} - \nabla}{H_p}\right).$$

Following the previous analysis the instability condition at time t becomes

$$\left(\frac{t}{t_\odot}\right) - 0.37\frac{H_p}{H_f} > 0.053\left(\frac{KT}{\rho\varepsilon l^2} - 19\right).$$

If we now approximate the amplitude δT_0 by

$$\frac{\delta T_0}{T} = \alpha \sin\left(\frac{2\pi r}{R}\right)$$

then in the central regions

$$\frac{Hp}{Hf} = 33\alpha^2\frac{6\pi P_c}{G\rho^2_c R^2} = 41\alpha^2,$$

where the numerical values are those of a polytrope of index 3. Since detailed calculations with $\alpha = 0$ give instability when $t/t_\odot = 0.065$, for larger t the amplitude α is given by requiring marginal instability, hence

$$\alpha^2 = 0.066 \left(\frac{t}{t_\odot} - 0.065 \right).$$

Of course, the above calculation is only indicative of the way the oscillation may evolve, a better approximation would be to use the actual eigenfunction for $\delta T/T$ rather than the simple approximation used here, to then compute the ^{3}He equilibrium profile and to test for marginal stability. However, even this is likely to prove inadequate since the expected amplitudes are large we need to consider the non-linear interaction with other modes. The observable surface effects also need careful investigation since the g modes are evanescent in the convective zone and whilst the first unstable mode has a relatively large surface amplitude this is not true of higher order modes which may be important in the non-linear limit. If overshooting beneath the convective zone is important the effect of a sharp change in the temperature gradient on the reflection of g modes may be important.

One prediction of this analysis is that the energy generation rate is enhanced, this will change the profile of chemical composition and thus affect the evolutionary sequence of models leading to the model of the present Sun; this in turn will change the predicted run of sound speed in the solar interior. It is interesting to note that the inversion of recent oscillation data gives a somewhat higher speed ($\simeq 2\%$) than the predicton of standard models in the region where $r/R_\odot \simeq 0.5$ (cf. Gough, 1985). The detailed theoretical prediction of a model that has undergone – or maybe still undergoing finite amplitude oscillations is not yet available.

6. Overshooting beneath the Outer Convective Zone

The properties of convection are normally determined using the mixing length theory in which 'parcels' of fluid are imagined to detach themselves from the 'mother layer', travel a distance l, the mixing length, and then mix with their surroundings. In the surface layers radiative losses from such 'parcels' are included, but deep in the convective zone such losses are unimportant and the parcels are taken to move adiabatically in pressure equilibrium with their surrounding and accelerated by buoyancy. This simple form of the theory gives

$$\frac{\delta T}{T} = \frac{1}{H_p} \int_0^l \Delta \nabla \, dz, \quad \frac{dv^2}{dz} = g \int_0^l \frac{\delta T}{T} \, dz,$$

and determines the convective flux from the turbulent average:

$$F_c = \langle c_p \rho v \delta T \rangle = c_p \rho T (gH_p)^{1/2} (\Delta \nabla)^{3/2}.$$

In these expressions $\nabla = \partial \ln T / \partial \ln P$, $\Delta \nabla = \nabla - \nabla_{ad}$ is the superadiabetic temperature gradient which is taken to be constant along the path of the parcel, and the mixing length l is taken as αH_p with α an adjustable constant of order unity.

This is not a real theory, at best it is a method of estimating the turbulent correlation F_c, and even so ignores the flux of kinetic energy (Roxburgh, 1976, 1978). It is by no means obvious to the writer that this 'theory' has any validity, and caution should be exercised in using results that depend on details of the formalism. Applied to the bulk of the convective zone the theory predicts that $\nabla - \nabla_{ad} \ll \nabla$, so that all but shallow surface layer is adiabatic stratified, a result that is almost theory independent.

The details of the theory are required however, to relate the adiabatic constant $K = P/T^{5/2}$ in the interior to the surface values of temperature and gravity. One such analysis by Hobbs and Roxburgh, using the 'fast' model of Belvedere et al. (1981) gives the value of K after ionisation of helium as

$$K = 6.59 \times 10^{-3} \left(\frac{R}{R_\odot} \right)^4 \left(\frac{L_\odot}{L} \right)^3 \left(\frac{X}{0.73} \right)^{1.9} \left(\frac{Z}{0.18} \right)^{-0.13}.$$

However, in constructing models of the Sun K is taken as a free parameter, adjusted so that the resulting solar model has the observed radius. The structure of the adiabatic layer is readily determined by neglecting the small variation in the mass M_r, using $P = KT^{5/2}$ and integrating the hydrostatic equation to yield

$$\frac{RT}{\mu} = \frac{2}{5} GM \left(\frac{1}{r} - \frac{1}{R_\odot} \right) \frac{2}{5} \frac{GMd}{R_\odot^2},$$

where d is the distance below the surface. Since $\gamma RT / \mu$ is the square of the sound speed all solar models will give more or less the same variation of sound speed with depth in the inner part of the convective zone.

The base of the convective zone is determined by the condition that the radiative flux in the adiabatic zone is equal to the total flux, the convective flux is therefore zero, this occurs at a base temperature T_b where

$$T_b^{3/2} = \frac{15 K L_\odot \kappa}{32 \pi ac GM_\odot},$$

where κ is the opacity. The standard model given in Table I has a base temperature $T_b = 2.0 \times 10^6$ K and extends to a depth of $0.27 R_\odot$, somewhat less than the $0.3 R_\odot$ inferred from the inversion of oscillation data (Duvall et al., 1985). Since the opacity in the base of the convective zone can be approximated by a power law in the form $\kappa_0 \rho T^{-3}$, the radiative flux inside the lower boundary varies as T^3, and hence the convective flux approaches zero at the base such that

$$F_{conv} = F - F_{rad} = F \left(1 - \frac{T^3}{T_b^3} \right) = \frac{3Fy}{d},$$

where y is the distance from the base of the zone.

The above results are unlikely to be correct. Motion originating in the unstable regions will 'overshoot' into the stable layers beneath; this will convect energy regions inwards and the radiative energy flux will exceed the total flux in a region that is maintained close to adiabaticity by the overshooting. Deeper inside the Sun the stable stratification will bring the fluid to rest and the temperature gradient adjusts to that required to carry all the energy by radiation (see Roxburgh, 1978). The question is not whether there is or is not an overshoot region, but how far does it extend into the stable region, what is its structure and how does it affect the rest of the solar interior? Attempts to answer these questions have to rely on some highly simplified model of convection such as a non local mixing length theory, and are little better than taking the adiabatic layer to extend an arbitrary distance into the stable region and then abruptly switching to an radiative zone. One such model by the present author simply determines the convective velocity in the overshoot region from

$$v^2 = v_0^2 + l^2 g H_p \Delta\nabla,$$

leaving the rest of the mixing length treatment unchanged and taking for v_0 the typical value one scale height into the unstable layer. This gives an almost adiabatic overshoot zone of 0.18 times the depth of the zone.

A marginally more sophisticated treatment is to expand $\Delta\nabla$ as a Taylor series about the layer where $\Delta\nabla = 0$, taking $\Delta\nabla = Az$; this variation is then used to evaluate the integrals given above, where previously $\Delta\nabla$ was taken as constant. The convective flux in the overshoot region is then estimated from the velocity and temperature deficiency of 'parcels' of fluid starting from a height $h = H_p/2$ inside the unstable layer; such 'parcels' first accelerate and then decelerate coming to rest in the stable layers. When at a distance z above the layer where $\Delta\nabla = 0$ such a 'parcel' has a temperature deficiency δT, and velocity v, where

$$\frac{\delta T}{T} = \frac{A}{2H_p}(h^2 - z^2), \quad v^2 = \frac{A}{3H_p} g(h - z)^2 (h + z/2).$$

Note that at $z = 0$ these downward moving parcels have a temperature deficiency so the outward convective flux is positive. The convective flux becomes

$$F_c = c_p \rho T \left(\frac{A}{2H_p}\right) \left(\frac{Ag}{3H_p}\right)^{1/2} (h - z)^2 (h + z)(h + z)^{1/2}$$

which is zero when $z = -h$, negative for $-h < z < -2h$, and again falls to zero at $z = -2h$. The convection penetrates to a depth $z = -2h$ below the layer where $\Delta\nabla = 0$, and a depth h below the layer where $F_c = 0$.

It is not quite valid to take the overshoot region as adiabatic down to the point where $z = -2h$, since the decreasing efficiency of convection requires an increase in A until the zone becomes strongly subadiabatic. However, this transition takes place in a very thin layer. To see this we note that in the neighbourhood of the region $z = -h$ where $F_c = 0$, the convective flux must equal $3Fy/d$ where $y = z - h$. But if we model this by allowing A to vary and take $A = A_0$ at $z = 0$, then the convective flux in the overshoot region can be written as

$$F_c = \frac{3y}{d} F \left(\frac{A}{A_0} \right)^{3/2} \left(1 - \frac{y}{2h} \right)^2 \left(1 + \frac{y}{h} \right)^{1/2}$$

so that for $F_c = 3yF/d$ we require A to be given by

$$A = A_0 \left(1 - \frac{y}{2h} \right)^{-4/3} \left(1 + \frac{y}{h} \right)^{-1/3} ;$$

hence as $y \to -h$, $A \to \infty$ and the layer is strongly subadiabatic. However, the left-hand side of this relation is a slowly varying function of y and A only becomes large in a very narrow transition zone close to $y = -h$, where the temperature gradient adjusts so that all the energy is carried by radiation.

We conclude from this analysis that overshooting penetrates a distance $h = H_p/2 = 0.2d$ below he bottom of the convective zone defined at the place where all the energy is carried by radiation. This is essentially the same as the value derived from the 'theory' I put forward some years ago to estimate uncertainty in mixing length formalisms (Roxburgh, 1976, 1978). Such overshooting affects the structure of the base of the convective zone since the bottom scale height is marginally stable rather than marginally unstable, and the transition to the underlying strongly stable layers is takes place in a very thin layer. This will have consequences for the properties of both pressure and gravity modes of oscillation that are not yet understood, and it could well be that this overshoot region is the site of the solar dynamo.

7. Condition for a Convective Core

Whether or not the young Sun had a convective core is of considerable importance not only for the chemical evolution which would depend on the extent of the mixed region, but also for the dynamical evolution of the internal angular velocity, since we might expect a dynamo to operate in such a core in much the same way as in the outer convective zone.

The condition for the onset of convection is the Schwarzschild criterion:

$$\nabla = \frac{d \ln T}{d \ln P} > \frac{\gamma - 1}{\gamma} = 0.4 ,$$

where γ is taken as $\frac{5}{3}$ the value for a fully ionised ideal gas. Some care needs to be exercised in using this condition. It is possible to imagine two equilibrium states, one without convection, the other with convection, the bifuration being caused by the change in central abundance of ^{3}He and/or ^{12}C in the two states; in the first the abundances are locally determined by equilibrium in the nuclear reaction network, in the second the abundances are determined by averaging over the whole core.

There is no simple criterion for the existence of a convective core since it depends on the complete structure of the solar model, but some insight can be obtained by examining the behaviour of ∇ in the central regions. If ∇ becomes to small the structure is almost isothermal and we know from studies of simple polytropes that there can be no solution of the structure equations that extends to the surface. If ∇ becomes to large, the Sun becomes convective.

The equations of stellar structure for a radiative core (neglecting the small term due to the change in entropy) are

$$\frac{\partial P}{\partial M_r} = -\frac{GM_r}{4\pi r^4}, \quad \frac{\partial r}{\partial M_r} = \frac{1}{4\pi r^2 \rho},$$

$$\frac{\partial T}{\partial M_r} = -\frac{3\kappa L_r}{16\pi acT^3 r^4}, \quad \frac{\partial L_r}{\partial M_r} = \varepsilon,$$

from which is follows that

$$\nabla = \frac{P}{T}\frac{dT}{dP} = \frac{3\kappa L_r P}{4acT^4}.$$

Differentiating this logarithmically with respect to T gives

$$\frac{\partial \ln \nabla}{\partial \ln T} = \frac{\partial \ln \kappa}{\partial \ln \rho}\frac{\partial \ln \rho}{\partial \ln T} + \frac{\partial \ln \kappa}{\partial \ln T} + \frac{1}{\nabla} - 4 - \frac{\partial \ln (L_r/M_r)}{\partial \ln T}.$$

If we now define the local power-law exponents

$$\alpha = \frac{\partial \ln \kappa}{\partial \ln \rho}, \quad \beta = -\frac{\partial \ln \kappa}{\partial \ln T}, \quad \eta = \frac{\partial \ln \varepsilon}{\partial \ln T}, \quad \frac{\partial \ln \varepsilon}{\partial \ln \rho} = 1,$$

then on using the gas law to give $(\partial \ln \rho)/(\partial \ln T) = (1/\nabla - 1)$

$$\frac{\partial \log \nabla}{\partial \log T} = \alpha\left(\frac{1}{\nabla} - 1\right) - \beta + \frac{1}{\nabla} - 4 + \frac{\partial \ln (L_r/M_r)}{\partial \ln T}.$$

To calculate the last term we first note that near the centre the energy equation can be written as

$$\frac{\partial L_r}{\partial M_r} = \varepsilon = \varepsilon_c\left[1 + \frac{\partial \ln \varepsilon}{\partial \ln \rho}\left(\frac{\rho - \rho_c}{\rho_c}\right) + \frac{\partial \ln \varepsilon}{\partial \ln T}\left(\frac{T - T_c}{T_c}\right) + \dots\right].$$

As $r \to 0$, $M_r \sim r^3$, $(T - T_c) \sim M^{2/3}$, $(\rho - \rho_c) \sim M^{2/3}$, and this can be integrated to give

$$\frac{L_r}{M_r} = \varepsilon_c\left[1 + \frac{3}{5}\left(\frac{\rho - \rho_c}{\rho_c}\right) + \frac{3\eta}{5}\left(\frac{T - T_c}{T_c}\right) + \dots\right];$$

hence

$$\frac{\partial \ln (L_r/M_r)}{\partial \ln T} = \frac{3}{5}\left(\frac{1}{\nabla} - 1\right) + \frac{3}{5}\eta\,.$$

Combining these results gives

$$\left(\frac{\partial \log \nabla}{\partial \log T}\right)_c = -\frac{1}{5}\left[(5\alpha + 5\beta + 23 - \eta) - \left(\frac{5\alpha + 8}{\nabla}\right)\right] \equiv s\,.$$

Were we to impose the condition that $\nabla_c = 0.4$ and ∇ decreases as T decreases we would obtain the Naur-Osterbrock (1953) condition for a small convective core:

$$6\eta > 6 + 10\beta - 15\alpha\,.$$

However, this is not the correct condition; even if $\nabla_c = 0.4$ it must not decrease to read too rapidly if the solution of the equations is to extend to the surface. Moreover it is not actually a sufficient condition for the existence of a convective core but rather a necessary condition if the Sun has a small convective core.

A more appropriate condition is to require that if $\nabla_c \leq 0.4$ then it must not decrease by more than a factor of 2 if the temperature decreases by a factor of 2; otherwise the solution approaches that of a polytrope of index greater than 5 which does not extend to the surface. This gives $s < 1$ or

$$6\eta < 16 + 10\beta - 15\alpha\,.$$

In the central parts of the Sun the local exponents in the opacity are $\alpha = 0.43$, $\beta = 2$ so that the core Sun be stable against convection if $\eta < 4.9$. For the equilibrium p–p chain at a temperature of 13×10^6 K, $\eta = 4.3$ and the solar model has no convective core.

But the proton–proton chain has not always been in equilibrium since the initial abundance of ^{3}He was in excess of the equilibrium value on the p–p chain, and the initial abundance of ^{12}C was much in excess of its equilibrium value on the CN cycle. Both ^{3}He and ^{12}C burning reactions have a high value of η (16 to 18) and can drive convection in the centre in the early phase of the Sun's life before the equilibrium abundances have been achieved. If the initial ^{3}He abundance was 10^{-4} then the p–p equilibrium value was achieved whilst the Sun was still contracting to the Main Sequence and the early convective core was due to the burning of ^{12}C to ^{14}N, although once started ^{3}He contributes to the maintainance of the convection.

The size of such a core, and its lifetime, depend on the assumptions made about convective overshooting, a problem which I discussed above. If overshooting is significant the core can be maintained up to the present time; the equilibrium abundance of ^{3}He is the value averaged over the core where mixing takes place and this is considerably larger than the local equilibrium value in the central regions. This raises the effective value of η and so maintains convection (see also Shaviv and Salpeter, 1971).

I am not here arguing that the Sun has a convective core, or even that it had one for much of its lifetime, but until we can reliably estimate the extent of convective overshooting there must remain doubts about the validity of the 'standard solar model' which ignores overshooting altogether.

References

Bahcall, J. N.: 1985, *Solar Phys.* **100**, 53 (this volume).

Bahcall, J. N., Huebner, W. F., Lubow, S. H., Parker, P. D., and Ulrich, R. K.: 1982, *Rev. Mod. Phys.* **54**, 767.

Belvedere, G., Paterno, L., and Roxburgh, I. W.: 1981, *Astron. Astrophys.* **91**, 356.

Brown, T.: 1985, paper presented at *Cambridge Conference on Oscillations*.

Cayrel, R., de Strobel, G., Campbell, B., and Dappen, W.: 1984, *Astrophys. J.* **283**, 205.

Christensen-Dalsgaard, J., Dilke, F. W. W. and Gough, D. O.: 1974, *Monthly Notices Roy. Astron. Soc.* **169**, 429.

Christensen-Dalsgaard, J., Duvall, T. L., Gough, D. O., Harvey, J. W., and Rhodes, E. J.: 1985, *Nature* **315**, 378.

Dicke, R. H.: 1972, *Astrophys. J.* **171**, 331.

Dilke, F. W. W. and Gough, D. O.: 1972, *Nature* **240**, 262.

Duvall, T. L., Dziembowski, W. A., Goode, P. R., Gough, D. O., Harvey, J. W., and Leibacher, J. W.: 1984, *Nature* **310**, 22.

Eddington, A. S.: 1926, *Internal Constitution of the Stars*, Cambridge University Press, p. 393.

Goldreich, P. and Schubert, G.: 1967, *Astrophys. J.* **150**, 571.

Gough, D. O.: 1985a, private communication.

Gough, D. O.: 1985b, *Solar Phys.* **100**, 65 (this volume).

Fricke, K.: 1968, *Z. Astrophys.* **68**, 317.

Hill, B. A., Yakowitz, D. S., Rosenwald, R. D., and Campbell, W.: 1985, in *Hydromagnetics of the Sun*, ESA SP–220.

Hobbs, N. and Roxburgh, I. W.: 1986, to be published in *Astron. Astrophys.*

Knobloch, E. and Spruit, H.: 1983, *Astron. Astrophys.* **125**, 59.

Law, W. Y., Knobloch, E., and Spruit, H. C.: 1984, in A. Meader and A. Renzini (eds.), *Observational Tests of the Stellar Evolution Theory*, D. Reidel Publ. Co., Dordrecht, Holland, p. 523.

Mestel, L.: 1953, *Monthly Notices Roy. Astronom. Soc.* **113**, 716.

Mestel, L.: 1964, in A. Meader and A. Renzini (eds.), *Observational Tests of the Stellar Evolution Theory*, D. Reidel Publ. Co., Dordrecht, Holland, p. 513.

Naur, P. and Osterbrok, D. E.: 1953, *Astrophys. J.* **117**, 306.

Press, W. H.: 1981, *Astrophys. J.* **245**, 286.

Ross, J. E. and Aller, L. H.: 1976, *Science* **191**, 1223.

Roxburgh, I.W.: 1976, in V. Bumba and J. Kleczeck (eds.), 'Basic Mechanisms of the Solar Activity', *IAU Symp.* **71**, 453.

Roxburgh, I.W.: 1978, *Astron. Astrophys.* **65**, 281.

Roxburgh, I.W.: 1984a, *Mem. Soc. Astron. Italy* **55**, 273.

Roxburgh, I.W.: 1984b, in A. Meader and A. Renzini (eds.), *Observational Tests of the Stellar Evolution*, D. Reidel Publ. Co., Dordrecht, Holland, p. 519.

Roxburgh, I.W.: 1984c, paper presented at *25th Liège Colloquium on Theoretical Problems in Stellar Stability and Oscillations*.

Roxburgh, I.W.: 1985, in M. L. Cherry, W. A. Fowler, and K. Lande (eds.), *Solar Neutrinos and Neutrino Astronomy*, Am. Inst. Physics, p. 88.

Roxburgh, I.W.: 1986, to be published in D. O. Gough (ed.), *Proceedings of the Cambridge Meeting on Seismology*.

Roxburgh, I.W. and Tavakol, R. K.: 1985, (to be published).

Schatzman, E. and Meader, A.: 1981, *Astron. Astrophys.* **96**, 1.

Shaviv, G. and Salpeter, E.: 1971, *Astrophys. J.* **165**, 171.

Shibahashi, H.: 1980, *Publ. Astron. Soc. Japan* **22**, 341.

Spruit, H. C.: 1985, in *The Hydromagnetics of the Sun*, ESA SP–220, p. 21.

Spruit, H. C., Knobloch, E. W., and Roxburgh, I.W.: 1983, *Nature* **304**, 320.

Zahn, J. P.: 1983, in A. N. Cox, S. Vauclair, and J. P. Zahn (eds.), *Astrophysical Processes in Upper Main Sequence Stars*, Geneva Observ., p. 253.

THE SOLAR NEUTRINO PROBLEM

JOHN N. BAHCALL

Institute for Advanced Study, Princeton, NJ 08540, U.S.A.

Abstract. The observed capture rate for solar neutrinos in the ^{37}Cl detector is lower than the predicted capture rate. This discrepancy between theory and observation is known as the 'solar neutrino problem'. I review the basic elements in this problem: the detector efficiency, the theory of stellar (solar) evolution, the nuclear physics of energy generation, and the uncertainties in the predictions. I also answer the questions of: So What? and What Next?

1. Introduction

There exists a long-standing and serious discrepancy between the standard theory of how the Sun shines and the most direct observational test of this theory using solar neutrinos. Is this discrepancy caused by an astronomical lack of understanding of how the Sun shines (neutrino production) or by something happening to the neutrinos on the way to the Earth from the Sun (neutrino propagation)? We don't know. In order to make progress in this subject, new experiments are required – and some are either underway or planned.

I will try to give you a feeling for solar neutrino astronomy without deluging you with the technical details of nuclear physics, large-scale chemistry, stellar evolution and particle physics that are used in carrying out and interpreting solar neutrino experiments. Since the subject has been reviewed many times in the technical literature, it may be useful to have a general account of the problem that is free of references and almost free of equations and tables.

Before getting into the more technical aspects of the subject, we might consider for a moment the relation of solar neutrino experiments to solar photon astronomy. These photons are emitted from the outer layers of the Sun and are most directly influenced by phenomena occurring in the outer few percent of the mass of the Sun. The neutrinos carry information about the nuclear reactions in the deepest parts of the solar interior, the inner 5%–20% by mass. Most phenomena that are of vital interest to photon solar astronomers do not influence models of neutrino production in the solar interior and, conversely, the effect of nuclear reactions in the interior is unimportant for models of solar activity on the surface.

Thus neutrino astronomy is a branch of solar physics, but a specialized one. Figure 1 illustrates the complimentarity of studies of p-mode oscillations of low degree and solar neutrino experiments. Nearly all of the neutrinos from ^{8}B decay (which are crucial for the ^{37}Cl experiment) originate in the inner 5% of the solar mass. Almost 70% of the p-mode splitting comes from the outer 10% of the solar mass. The generation of the solar luminosity, and the flux of neutrinos from the proton–proton reaction, are intermediate

Solar Physics **100** (1985) 53–63.   0038–0938/85.15.

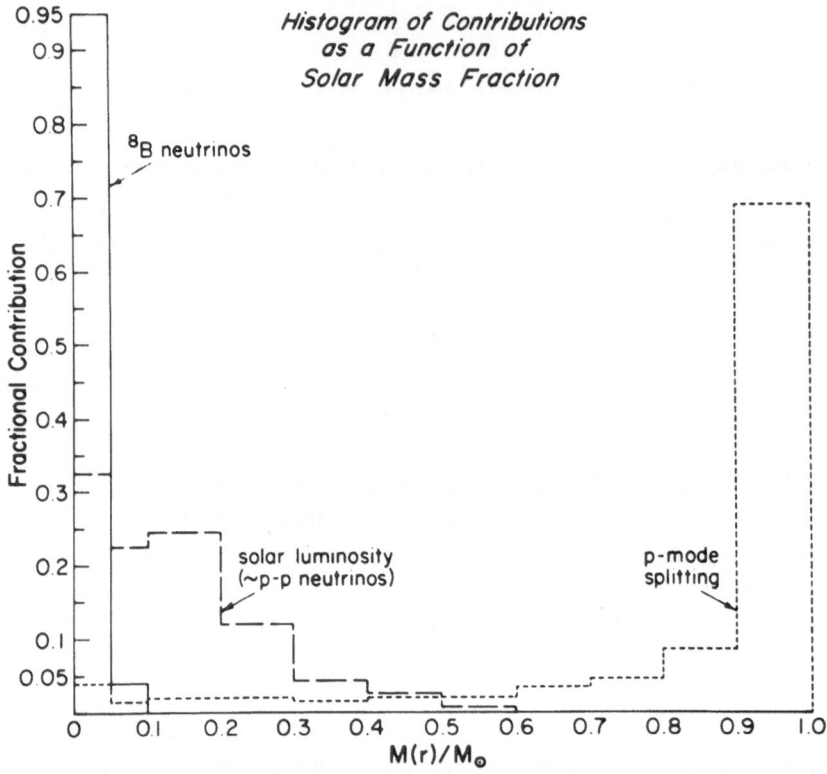

Fig. 1.   Histogram of the fractional contributions to the p-mode splitting, the flux of neutrinos from ⁸B decay, and the flux of neutrinos from the p–p reaction. Here $M(r)/M$ is the fraction of the solar mass interior to the point r. This figure is taken from the discussion of Bahcall (1981) *Neutrino* **81**, Vol. 2, p. 253, edited by R. J. Cence, E. Ma, and A. Roberts.

in distribution between the p-mode splitting and flux of ⁸B neutrinos. About 33% of the solar luminosity is produced in the inner $0.05M_\odot$ from which the neutrinos from ⁸B decay also originate, but a majority of the nuclear energy is produced in the intermediate region between $0.05M_\odot$ and $0.4M_\odot$. The fractional decomposition shown in Figure 1 is essentially independent of any plausible changes in nuclear cross sections. For all three of the quantities shown, the height of the histograms are the same to within ± 0.01 for the standard model and one calculated with extreme nuclear physics parameters.

The overlap between solar photon astronomy and solar neutrino astronomy will become larger if the characteristics of g-modes can be measured accurately. However, even assuming that g-modes are well-understood theoretically and observationally, photon and neutrino astronomy will remain distinct. Imagine trying to deduce the detailed isotopic abundances of radioactive elements that produce heat in the interior of the Earth by using the extensive terrestrial seismic data that are available. In the Sun, individual nuclear reactions affect directly the neutrino fluxes, but change g-modes only in a integral (and often insensitive) way.

An experiment designed to capture neutrinos produced by solar thermonuclear reactions is a crucial test of the theory of stellar evolution. As we all know, the theory of stellar evolution by thermonuclear burning is widely used in interpreting many kinds of astronomical information and is a necessary link in establishing such basic data as the ages of the stars and the abundances of the elements. The parameters of the Sun (its age, mass, luminosity, and chemical composition) are better known than those of any other star, and it is in the simplest and best understood stage of stellar evolution, the quiescent main sequence stage. Thus an experiment designed to capture neutrinos produced by solar thermonuclear reactions is a crucial one for the theory of stellar evolution. I hoped originally that the application of a new observing technique, using neutrinos as probes, would provide added insight and detailed information regarding stellar energy generation. Instead, it has created a surprising discrepancy between theory and observation, one which has spawned many speculative solutions.

In fact, it may be that the original motivation is no longer applicable and that what we are really investigating has to do more with the propagation of neutrinos than with their production by nuclear reactions in the solar interior. Because of this possibility, many physicists have devoted a great deal of effort to understanding the solar neutrino problem and have proposed a variety of experimental techniques and theoretical scenarios. The main reason that solar neutrino experiments are of such interest to physicists is that these experiments offer the opportunity of testing the stability of the neutrino over proper times that are inaccessible in the laboratory. In laboratory experiments, one can use distances that are at most a few km and which involve high energy neutrinos (tens or hundreds of GeV) or observe extremely high energy neutrinos that pass through the Earth's diameter. By contrast, solar neutrino experiments involve a source to detector distance of 10^{13} cm, with a characteristic energy of MeV. Thus elements of the neutrino mass matrix as small as 10^{-6} eV can be investigated with the aid of solar neutrino experiments, while it seems that the best one can do with terrestrially based experiments is about $10^{-1.5}$ eV. One of the major goals of modern theoretical physics is the unification of all of the known interactions in a Grand Unified Theory. A common characteristic of many existing models of unified theories is that neutrino mass matrix elements of order 10^{+1} to 10^{-6} eV are expected.

Many solutions to the solar neutrino problem, modifying either the physics or the astronomy (and in some cases both), have been proposed. Even if one grants that the source of the discrepancy is astronomical, there is no general agreement as to what aspect of the theory is most likely to be incorrect. As indicated above, many of the proposed solutions of the solar neutrino problem have broad implications for conventional astronomy and cosmology. Some of them would change the theoretical ages of old stars or the inferred primordial element abundances. On the other hand, some theories of the weak interactions have been proposed in which neutrinos may disappear by mixing or decay in transit from the Sun to the Earth, but for which there no terrestrially measurable consequences. It is conceivable that one of these theories of the weak interactions is correct and the standard solar model is not in conflict with observations.

I want to describe the main results and questions of solar neutrino astronomy. But, before doing that I will say just a little bit about a crucial aspect of the subject that I will not discuss in detail. Any neutrino experiment measures event rates, the product of flux times neutrino interaction cross sections. You might imagine that once you have calculated the cross sections for one nucleus it is routine to evaluate the cross section for any other nucleus. This, unfortunately, is not the case. Each nucleus has its own peculiarities and these can affect the capture rates dramatically. I will *not* discuss here the details of how one determines the cross sections using a combination of experimental nuclear physics information and atomic physics and weak interaction theories. The reason is: I am confident that this part of the subject is well understood; the calculations are correct to the stated accuracy. In fact, part of my personal definition of a 'good' solar neutrino experiment is that I can calculate the capture rate to an accuracy of 10% – for the target of interest (including the effects of nuclear excited states and all the relevant atomic physics effects).

2. Astronomy

Table I lists the most pertinent elements of the theory of stellar evolution. There are many crucial details that must be calculated with high accuracy in order to predict solar neutrino fluxes, but one doesn't have to have these details in mind to understand the main results of solar experiments. Table I summarizes the principles that are required for constructing solar models and the ideas that are tested by solar neutrino experiments.

The first principle is hydrostatic equilibrium, which in practice is used together with the special assumption of spherical symmetry (established to the accuracy required here by optical observations of the solar surface). The observationally-indicated solar rotation does not significantly influence the hydrostatic equilibrium in the interior. The second principle is that the energy source for sunlight is nuclear; the rates of the nuclear reactions depend on the density and the temperature and the composition. The practical part of this principle is that the rate at which the nuclear reactions produce energy when integrated over the whole Sun is equal to the observed solar luminosity *today*. The 'today' is an essential part of this principle.

The third principle is that the energy is transported from the deep interior to the surface via steady-state radiation and convection (and not by transient instabilities or waves). In practice, for most (but not quite all) of the models, the great bulk of the energy is transported by radiation. The key quantities are the gradient of the temperature and the opacity of the solar matter.

The characteristic densities in the solar interior are $150-100 \text{ g cm}^{-3}$ and the central temperature is about $15 \times 10^6 \text{ deg K}$. The primordial helium abundance by mass always turns to be $Y = 0.25 \pm 0.01$, for all the parameters that seem plausible for the standard solar model. I have done many numerical experiments in which I have varied the nuclear reaction rates to see how the inferred helium abundance is changed. In even the most extreme cases, the primordial helium abundance for a solar model with the present day luminosity of the Sun is always within a few percent of 0.25. I believe that examples

TABLE I

A menu of a theoretical model of the Sun

Hydrostatic equilibrium: approximately spherical Sun
Nuclear energy source
Energy transport by radiation and convection
Uniform primordial composition = surface composition
Evolution (age = 5×10^9 yr)

Bottom line: only ^{37}Cl experiment inconsistent with standard theory

---

in the literature in which larger variations are reported are the result of not iterating the stellar evolution models sufficiently accurately.

It is encouraging that the inferred helium abundance from the p-mode oscillation measurements is consistent with the most accurate calculations made using the theory of stellar evolution.

The assumption that the initial composition was uniform and is equal to the presently-observed surface composition is closely related to the question of which radiative opacity should be used. It is easy to show that the surface of the Sun has not changed much because of nuclear reactions since the Sun was formed. It is not quite so obvious that nothing much has been added to the solar surface since the Sun was born. However, the generalization of this assumption – primordial surface compositions for Main Sequence stars – is widely used in interpreting stellar and galactic observations and is the basis for making the standard calculations. One of the principal observational arguments supporting this assumption is the uniformity of surface compositions that are observed for stars throughout the galaxy.

The final principle is that the Sun evolves because it burns its own (limited) nuclear fuel. We believe that the Sun has been shining for something like 5 billion years so far. One mocks up this evolution by computing several quasi-static models which march along in time.

The bottom line of this brief course in stellar evolution is: within our store of observational information about stars, only the ^{37}Cl experiment of Ray Davis and his colleagues is inconsistent with the standard theory of stellar evolution. It is the only place where we don't see a way out of observational difficulties unless we modify something among the basic assumptions.

Table II lists some of the most important parameters of a standard solar model.

3. Nuclear Physics

I shall now outline the conventional wisdom regarding nuclear fusion as the energy source for Main Sequence stars like the Sun. We assume that the Sun shines because of fusion reactions similar to those envisioned for terrestrial fusion reactors. The basic solar process is the fusion of four protons to form an alpha particle, two positrons (e^+), and two neutrinos (v), that is, $4\,p \rightarrow \alpha + 2\,e^+ + 2\,v_e$. Almost all (98.5%) of the energy

TABLE II

Some important solar parameters

Parameter	Value
Moment of inertia	7.00×10^{53} g cm^2
Depth of convective zone	$0.27\,R$
Age	$\geq 4.55 \times 10^9$ yr
Central density	156 g cm^{-3}
Central temperature	15.5×10^6 K
Central hydrogen abundance by mass	0.355
Primordial helium abundance by mass	0.25 ± 0.01
Primordial ratio of heavy elements to hydrogen mass	0.0228
Neutrino flux from p–p reaction	6×10^{10} cm^{-2} s^{-1}
Neutrino flux from ^{8}B decay	4×10^6 cm^{-2} s^{-1}
Fraction of energy from p–p chain	0.985
Fraction of energy from CNO cycle	0.015

generation in the present day Sun comes from the p–p chain, with only about 1.5% from the CNO cycle. The principal reactions are shown in Table III with a column indicating in what percentage of the solar terminations of the proton–proton chain each reaction occurs. The rate for the initiating proton–proton (pp) reaction, number 1 in Table III, is largely determined by the total luminosity of the Sun. Unfortunately, these neutrinos are below the threshold, which is 0.81 MeV, for the ^{37}Cl experiment. Several of the proposed new experiments, especially the ^{71}Ga and ^{115}In experiments, will be primarily sensitive to neutrinos from the p–p reaction. The pep reaction (number 2), which is the same as the familiar pp reaction except for having the electron in the initial state, is detectable in the ^{37}Cl experiment. The ratio of pep to pp neutrinos is approximately independent of which model (see below) one uses for the solar properties. Two other reactions in Table III are of special interest. The capture of electrons by ^{7}Be (reaction 6) produces detectable neutrinos in the ^{37}Cl experiment. The ^{8}B beta decay, reaction 9, was expected to be the main source of neutrinos for the ^{37}Cl experiment because of their relatively high energy (14 MeV), although it is a rare reaction in the Sun (see Table III). There are also some less important reactions from the carbon–nitrogen–oxygen (CNO) cycle that give rise to the neutrino-producing isotopes, ^{13}N and ^{15}O, but I shall not discuss them in detail since the CNO cycle is believed to play a rather small role in the energy-production budget of the Sun.

A *minimum* event rate for solar neutrino experiments can be calculated that is consistent with the hypothesis that the Sun is currently producing nuclear energy by light element fusion at the same rate at which photons escape its surface and the additional hypothesis that nothing happens to the neutrinos on the way to the Earth from the Sun. This minimum rate would be obtained if the only nuclear reactions that occurred in the solar interior were numbers 1–4 of Table III (ending with the ^{3}He–^{3}He reaction). In this case, the p–p neutrino flux is 6.5×10^{10} cm^{-2} s^{-1} and the pep flux is 1.6×10^8 cm^{-2} s^{-1}.

TABLE III

The proton–proton chain in the Sun

Number	Reaction	Solar p–p terminations (%)	Maximum neutrino energy (MeV)
1	$p + p \rightarrow {}^2H + e^+ + \nu$ or	(99.75)	0.420
2	$p + e^- + p \rightarrow {}^2H + \nu$	(0.25)	1.44 (monoenergetic)
3	${}^2H + p \rightarrow {}^3He + \gamma$	(100)	
4	${}^3He + {}^3He \rightarrow {}^4He + 2\,p$ or	(88)	
5	${}^3He + {}^4He \rightarrow {}^7Be + \gamma$ with	(12)	
6	${}^7Be + e^- \rightarrow {}^7Li + \nu$		0.861 (90%); 0.383 (10%) (both monoenergetic)
7	${}^7Li + p \rightarrow 2\,{}^4He$ or		
8	${}^7Be + p \rightarrow {}^8B + \gamma$	(0.02)	
9	${}^8B \rightarrow {}^8Be^* + e^+ + \nu$		14.06
10	${}^8Be^* \rightarrow 2\,{}^4He$		

4. Chemistry

The chlorine solar neutrino detector is based on the neutrino capture reaction $\nu + {}^{37}Cl \rightarrow {}^{37}Ar + e^-$, which is the inverse of the electron capture decay of ${}^{37}Ar$. The radioactive decay occurs with a half-life of 35 days. This reaction was originally discussed by Pontecorvo because of its convenient experimental characteristics. Over the past 20 years, Ray Davis has devoted his scientific life to using chlorine as a detector of solar neutrinos. Chlorine was chosen for the first solar neutrino experiment because of its unique combination of physical and chemical characteristics, which were favorable for building a large-scale solar neutrino detector. Neutrino capture to form ${}^{37}Ar$ in the ground state has a relatively low energy threshold (0.81 MeV) and a favorable cross section, nuclear properties that are important for observing neutrinos from ${}^7Be$, ${}^{13}N$, and ${}^{15}O$ decay and the pep reaction.

The ${}^{37}Cl$ reaction is favorable from a chemical point of view. The material is cheap and abundant so that one can afford the hundreds of tons necessary to do a solar neutrino experiment. The chemical extraction is relatively simple, although the scale is astounding. Davis routinely searches for a few atoms of ${}^{37}Ar$ in a tank that is the size of an Olympic swimming pool. This is possible because the argon atoms that are produced form a noble gas. The most suitable chemical compound is perchloroethylene, C_2Cl_4, a pure liquid, which is manufactured on a large scale for cleaning clothes. The neutrino capture process produces an ${}^{37}Ar$ atom with sufficient recoil energy to break free of the parent perchloroethylene molecule and penetrate the surrounding liquid, where it reaches thermal equilibrium.

The ^{37}Cl detector was built by Ray Davis deep underground to avoid the production of ^{37}Ar in the detector by cosmic rays. This was done with the cooperation of the Homestake Gold Mining Company (Lead, South Dakota), who excavated a large cavity in their mine ($\sim$ 1500 m below the surface) to house the experiment. The final detector system consists of an $\sim$ 400 000 liter tank of perchloroethylene, a pair of pumps to circulate helium through the liquid, and a small building to house the extraction equipment.

A set of 59 experimental runs carried out by Bruce Cleveland, Ray Davis, and Keith Rowley in the ^{37}Cl experiment over the last 14 years show that the ^{37}Ar production rate in the tank is 0.47 $\pm$ 0.04 ^{37}Ar atoms per day in the 615 tons of C_2Cl_4. Even though the tank is nearly a mile underground, a small amount of ^{37}Ar is produced by cosmic rays. An evaluation of data obtained by exposing 7500 liters of C_2Cl_4 at various depths underground suggests that the cosmic-ray production rate in the detector may be 0.08 $\pm$ 0.03 ^{37}Ar atoms per day.

It is possible that all of the capture rate that has been observed is due to background effects. This possibility must be investigated by performing further background experiments with a more sensitive ^{37}K detector and by evaluating the ^{8}B neutrino flux with independent solar neutrino detectors.

If the small background rate cited above is used, then a positive signal corresponding to 2.1 $\pm$ 0.3 SNU (1-σ error) is inferred. Here, I use the convenient notation: 1 SNU = 10^{-36} captures per target particle per second (where SNU stands for Solar Neutrino Unit and is pronounced SNEW, the characteristic product of solar neutrino flux times cross section). You don't really need to remember the definition of a SNU, but just the fact that it is the characteristic value for the product of a nuclear weak interaction cross section and a solar neutrino flux.

The capture rate for the textbook assumptions of how the Sun shines has been computed and is about 6 SNU. Most of the capture rate for the chlorine experiment is expected to come from the rare ^{8}B neutrinos, which are produced only about once in every 10 000 terminations of the solar proton–proton chain.

5. Uncertainties in the Predictions

How serious is the discrepancy between theory and observation? Can the difference be explained by errors in some of the input parameters?

In order to answer these important questions regarding uncertainties, I have – together with my colleagues Walter Heubner, Steve Lubow, Peter Parker, and Roger Ulrich – calculated the partial derivative of the expected flux from each neutrino source with respect to every important input parameter. We used Roger's excellent solar evolution code, including all of the detailed physics that is necessary in order to get the interior results as correct as possible. These calculations required the construction of hundreds of 'standard' solar models that were iterated to obtain – with the changed parameters – final solar models with the observed luminosity and surface composition. Also did a few test cases of very large changes in input parameters to verify that the effect of conceivably big differences in input parameters could be estimated well with the aid of the partial derivatives.

The uncertainties in the experimental parameters were evaluated using the original data (which was reanalyzed in some cases). The uncertainties in the theoretical quantities (e.g., the opacity) were estimated by using the range in published evaluations by competent workers.

There are many different input parameters so the estimation of the final error is not easy. I defined an overall 'effective 3-σ' limit by evaluating the 3-σ range for each experimental quantity, taking the total published range of the theoretical estimates, and combining quadratically all of the independent sources of uncertainty.

The overall uncertainty in the predicted capture rate – computed with the aid of the rule of thumb stated above – is 2.2 SNU. Thus the discrepancy between theory and observation is serious.

Parameter changes do not seem likely to resolve the conflict that has existed for the past 17 years. The calculated results haven't changed much in the past decade and a half (after the initial large uncertainties in some of the nuclear parameters were reduced by laboratory experiments at low energies), despite the detailed investigation of many theoretical effects and the remeasurement of a number of the crucial nuclear cross sections. The inferred experimental capture rate has also remained unchanged for many years. The gap between theory and observation has remained approximately the same since it first became apparent in 1968.

6. So What?

The ^{37}Cl experiment tests theoretical ideas at different levels of meaning, depending on the counting rate being discussed. A value of 28 SNU's would have been expected if Bethe's original suggestion of the CNO cycle as the solar nuclear energy source were correct. We may even regard it as a triumph of the standard theory of nuclear energy generation in stars that the large rate predicted by the CNO cycle is not observed. Modern calculations had already suggested – before the chlorine solar neutrino experiment – that the CNO cycle was only important in stars more massive than the Sun.

More surprisingly, the best current models based on standard theory, which imply ~ 6 SNU's, are also inconsistent with the observations. This disagreement between standard theory and observations has led to many speculative suggestions of what might be wrong. A number of these *ad hoc* suggestions lead to expected counting rates in the range 1.5 to 2.5 SNU, which are consistent with the chlorine experiment. None of the suggestions are generally accepted. (In fact, nearly all of the explanations *must* be wrong, since they are mostly mutually exclusive.) In one 9-year period, 1969–1977, Davis and I counted 19 independent ideas of what might be wrong. New ideas have been suggested at the rate of 2 or 3 per year since then.

Present and future versions of the ^{37}Cl experiment are not likely to reach a sensitivity as low as 0.2 SNU, the *minimum* capture rate (from reaction 2 of Table III) that can be expected if the basic idea of nuclear fusion as the energy source for Main Sequence stars is correct (and electron neutrinos reach the Earth from the Sun with an undiminished flux).

7. What Next?

Another experiment is required to settle the issue of whether our astronomy or our physics is at fault. Fortunately, one can make a testable distinction. The flux of low energy neutrinos from the pp and pep neutrinos is almost entirely independent of astronomical uncertainties and can be calculated from the observed solar luminosity, provided only that the basic physical ideas of nuclear fusion as the energy source for the Sun and of stable neutrinos are correct. If these low energy solar neutrinos are not detected in a future experiment, we will know that the present discrepancy between theory and observation is due at least in part to a departure from classical weak interaction theory, not just poorly understood astrophysics.

Table IV lists the experiments that are currently being investigated by different groups around the world. Each of these experiments is difficult and will require many years of hard work. All of the experiments listed will provide valuable information about the solar interior or the propagation of neutrinos. We must have a variety of experiments in order to provide cross-checking and verification of observational results.

Over the past several years, a rare consensus has developed in which most astronomers, chemists, and physicists concerned with the question agree that the next step in resolving the solar neutrino problem should be a gallium experiment. This experiment has several advantages. The primary reason for the enthusiasm surrounding a gallium experiment is that a ^{71}Ga detector is mainly sensitive to the basic proton–proton neutrinos. Moreover, the chemistry for a GaCl solution is simple and has been demonstrated in a modular pilot experiment. The sensitivity to the background of the gallium detector is small and is self-monitoring (using ^{69}Ga which is more sensitive to the background than is the neutrino-detector, ^{71}Ga). A recent analysis of the amount of gallium that is required for a solar neutrino experiment shows that 30 tons is sufficient for a very good experiment and one may be able to answer the most important questions with as little as 15 tons.

The expected capture rate for a ^{71}Ga detector from different solar neutrino sources can also be calculated with the aid of a standard solar model. The answer is about 120 SNU with an uncertainty of about 10 SNU. The dominant contribution to this rate, almost 70%, comes from the basic p–p and pep reactions (numbers 1 and 2 in Table III). Most of the remaining neutrino capture rate comes from the ^{7}Be reaction (number 6 of Table III), which is intermediate in its sensitivity to solar interior details between the pp and the ^{8}B neutrinos. The pp flux is guaranteed to be present if nothing happens to the neutrinos on the way to the Earth from the Sun and if the Sun is currently supplying by nuclear fusion the energy it is radiating in photons. Since essentially all of

TABLE IV

Solar neutrino experiments being actively pursued by different groups[a]

Detector	Dominant neutrino source	Experimental group
^{37}Cl	^{8}B	U. Penn (U.S.S.R.)
^{71}Ga	p–p	LANL; MPI; U.S.S.R.
^{115}In	p–p	France; Oxford
^{7}Li	^{8}B, pep, ^{15}O	U.S.S.R. (ORNL)
^{81}Br	^{7}Be	ORNL
v_e-scattering	^{8}B	proton decay detectors Sydney
^{2}H	^{8}B	Canada-U.S.
^{97}Mo, ^{98}Mo	^{8}B	LANL (Mo)

[a] U. Penn = Ray Davis and collaborators; Canada-U.S. = currently forming Canadian-U.S. collaboration (H. Chen); France = several independent French groups; LASL = Los Alamos Scientific Laboratory under T. J. Bowles and R. G. H. Robertson; LASL (Mo) = Los Alamos Scientific Laboratory under G. A. Cowan; MPI = Max Planck Institute under T. Kirsten and W. Hampel; ORNL = Oak Ridge National Laboratory under G. S. Hurst; Oxford = Oxford University under N. Booth; Sydney = University of Sydney under L. Peak; U.S.S.R. = Soviet scientists under G. Zatsepin.

the solar energy generation is expected to be produced by reactions dependent upon the basic pp reaction, and since the luminosity of the Sun is observed accurately, most astrophysical uncertainties do not affect significantly the calculated pp flux. The theoretical rate of the pp reaction is tied closely to the observed solar constant. For the ^{7}Be neutrino flux, Table III indicates that approximately 12% of the solar luminosity is generated through this reaction branch. The flux of ^{7}Be neutrinos depends somewhat on the details of the solar interior, although not as sensitively as does the flux of rare ^{8}B neutrinos.

The *minimum* capture rate for the gallium experiment is 79 SNU if one assumes that the current solar energy production is equal to the surface (photon) luminosity and nothing happens to the neutrinos on the way to the Earth. This minimum rate is well separated from the *maximum* rate that can be expected for the gallium experiment if the discrepancy between theory and observation for the chlorine experiment is attributed to the propagation of neutrinos.

8. Conclusion

What this subject needs is a new experiment, hopefully several new experiments. We are in the uncomfortable position of not knowing whether the discrepancy that has existed for a decade and a half is due to our inadequate understanding of how the Sun shines or to an incomplete description of how neutrinos propagate. Neither pure theory nor laboratory experiments have resolved the problem. We must turn to nature for an answer.

INVERTING HELIOSEISMIC DATA

DOUGLAS GOUGH

*Institute of Astronomy and Department of Applied Mathematics and Theoretical Physics,
University of Cambridge, U.K.*

and

Observatoires du Pic-du-Midi et de Toulouse, Université Paul Sabatier, France

Abstract. Methods by which the observed frequencies of solar oscillations can be, and in some cases have been used to infer the internal structure of the Sun are discussed. Attention is confined to so-called inverse methods that identify and extract the information that is actually contained in the data. Because only a finite quantity of data can ever be acquired, the functions describing the interior stratification of the Sun can never be established completely without the acceptance of certain assumptions. Nevertheless, the assumptions that are required are simple to understand, and the results do not depend on the complicated and uncertain theory of stellar evolution which has traditionally been used to construct solar models. First results of the inversions have given us an estimate of the sound speed and the angular velocity throughout much of the solar interior. These estimates have already stimulated speculation which hopefully will encourage further theoretical and observational research that will improve our understanding of the Sun.

1. Introduction

The body of helioseismic data has become so extensive that inverse techniques can now be used to make measurements of certain aspects of the solar interior. Fortunately, the techniques for inversion have already been developed in other branches of science, and need merely to be adapted to suit the heliophysicists' needs. Therefore I make no attempt here to present a balanced account of the entire subject of inversion; for that the reader is referred to the extensive literature on what is now becoming known as remote sensing. The discussion here is confined to those techniques that have been or are being developed for use on solar data, and to the inferences about the structure and dynamics of the Sun that have already been made. Much of it is taken from a recent report of a conference organised by Ulrich *et al.* (1984). There is no discussion of the methods by which data are acquired.

The first step in any inversion procedure is to solve the so-called forward problem. In the present context, that is the computation of the frequencies of small oscillations about the equilibrium state of the Sun; it requires determining eigenvalues of a linearized elliptic system of differential equations with coefficients that depend on a prescribed equilibrium solar model. The second step is to infer from the measured eigenvalues what those coefficients are for the actual Sun. Evidently that cannot be carried out uniquely, because it is not possible to determine any function from what must necessarily be a finite amount of information. Indeed, even if an infinite amount of information were available, there are certainly some properties of the Sun that cannot be determined even

Solar Physics **100** (1985) 65–99. 0038–0938/85.15.

in principle from helioseismic data alone. It is an important aspect of the subject to ascertain what it is that can be learned.

Inversions can conveniently be divided into three categories. The simplest consists of the execution of the forward problem using solar models with a few adjustable parameters, and the calibration of those parameters by fitting theory to observation. The second is the use of analytical methods. These methods have the advantage of providing insight into both the model calibration and the techniques of the third category. In particular, they suggest what parameters can successfully be calibrated and how the data should be processed to do so effectively. They also indicate what aspects of the solar structure might be measurable by the formal techniques. Thirdly, there are the formal inversion techniques borrowed from geophysics that have been used on real and artifical solar data.

2. The Forward Problem

The stratification of the Sun appears to be very nearly spherically symmetrical and static. There are small deviations from this state on a large scale, produced mainly by rotation. On a much smaller scale there are quite substantial time-dependent deviations, caused by thermal convection and magnetic fields; these are probably important only immediately beneath the photosphere and in the atmosphere, where most modes of oscillation have very little energy. Therefore on the whole, for the purposes of computing oscillation frequencies, the Sun can be considered as being precisely static and spherically symmetrical in a first approximation. That permits oscillation eigenfunctions with well defined frequencies to be described as products of functions of the radial coordinate r and spherical harmonics. The eigenvalue problem then reduces to a system of ordinary differential equations, which are described in the standard texts (Ledoux and Walraven, 1958; Unno *et al.*, 1979; Cox, 1980).

It has been usual to calculate the oscillatory motion in the adiabatic approximation. The justification for this is partly that throughout most of the star nonadiabatic deviations are small. However, observations have now reached a precision (about 1 part in 10^4, according to Isaak (1985) and van der Raay *et al.* (1985)) that renders the influence on the eigenfrequencies of those small deviations significant: it has been estimated that the higher frequencies can be modified by as much as several parts in 10^3 (Christensen-Dalsgaard and Frandsen, 1983; Gough, 1984a). Consequently consistent nonadiabatic calculations should be performed. Unfortunately that is not yet possible, because a major contributor to heat flow in the region where nonadiabatic processes matter is convection, and there is no adequate theory to describe how the convective fluxes are modulated by the oscillations. This is an important limitation, since without an accurate solution to the forward problem one is severely hampered in solving the inverse problem. It does not necessarily prevent one for making deductions, however; as will be shown later, it is sometimes possible to find combinations of data from which certain uncertainties in the physics have been eliminated. It may also be possible subsequently to use other combinations that do depend on those uncertainties to learn more about the physics of the oscillations. But for the present I shall bypass this issue, and assume that for any theoretical model of the Sun the equations describing the oscillations are known. Their solution by numerical means, at least when convective fluctuations are ignored, is now a routine procedure.

When the equilibrium model of the Sun is spherically symmetrical, the displacement eigenfunction $\zeta(\mathbf{r}, t)$ may be written, with respect to spherical polar co-ordinates (r, θ, ϕ):

$$\boldsymbol{\xi} = (\Xi(r)P_l^m, r^{-1}H(r)\partial_\theta P_l^m, r^{-1} \operatorname{cosec} \theta\, H(r)P_l^m\partial_\phi) \cos(m\phi - \omega t), \qquad (2.1)$$

where $P_l^m (\cos \theta)$ is the associated Legendre function of the first kind and t is time; l and m are respectively the degree and the azimuthal order of the mode. For each value of l and m there is a discrete sequence of eigensolutions. These are labelled with an integer n, called the order of the mode, in such a way that ω increases with n. The modes fall into two classes called p modes and g modes, which when n or l is large are obviously identifiable as acoustic and internal gravity modes respectively. Separating them, when $l > 1$, is the f mode, which when l is large can be identified as a surface gravity wave. The f mode is assigned $n = 0$; p modes have $n > 0$ and g modes have $n < 0$, though it is common to label the modes with the letter p or g and replace n by $|n|$. I adopt this convention here. The frequencies ω are degenerate with respect to m; this must be so since m depends on the choice of the axis of co-ordinates, and in a spherically symmetrical system all choices are equivalent.

The dominant symmetry-breaking agent is the linear advection term coming from the angular velocity of the Sun. Its effect is to split the degeneracy of the eigenfrequencies of like order n and degree l. From the splitting one can hope to infer the internal angular velocity $\Omega(r, \theta, t)$. Quadratic terms, such as centrifugal forces acting on both the equilibrium state and the oscillations, are rather smaller. Therefore the frequency splitting is approximately a linear functional of Ω. The smaller quadratic effects, together with possible large-scale magnetic fields and meridional flow, induce changes in the eigenfrequencies that might be measurable. Also, large-scale convective currents (giant cells) should produce diagnostically useful perturbations.

3. Asymptotic Solutions of the Eigenvalue Problem

Before considering the calibration of solar models it is useful to discuss the asymptotic solution of the eigenvalue problem in the limit when the scale of variation of the eigenfunction is much less than the scale of variation of the equilibrium state. This condition is usually satisfied throughout much of Sun when either n or l is large. In particular, it is approximately the case for the five-minute oscillations, which constitute most of the oscillations that have been measured. It permits the application of the JWKB approximation or ray theory, either of which lead to quite simple analytical results. Those results help us to appreciate what questions we might hope to answer by analysing oscillation frequencies, even when n and l are only moderately large and the formulae are only roughly correct.

The first step is to transform the equations of motion into a form that is simple to handle. In the asymptotic limit of large n or l, perturbations to the gravitational potential can be neglected, and the equations can be reduced to a second-order system. It is then useful to remove the first derivative term, which can be achieved by deriving an equation for $\psi = \rho^{1/2} c^2 \operatorname{div} \boldsymbol{\xi}$, where ρ is the density and c the adiabatic sound speed in the equilibrium state. The manipulations are essentially the same as those carried out by Lamb (1932), who considered the oscillations of a plane-parallel atmosphere of gas with constant adiabatic exponent γ under constant gravitational acceleration g; they lead to

$$\psi'' + \kappa^2 \psi = 0,$$

where a prime denotes differentiation with respect to r, and

$$\kappa^2 = \frac{\omega^2 - \omega_c^2}{c^2} + \frac{L^2}{r^2}\left(\frac{N^2}{\omega^2} - 1\right) \tag{3.2}$$

is the square of the vertical component of the wave number. Also

$$\omega_c = \frac{c}{2H}(1 - 2H')^{1/2}, \tag{3.3}$$

H being the density scale height, is a generalization of Lamb's (1908) acoustical cutoff frequency,

$$N^2 = g\left(\frac{1}{H} - \frac{g}{c^2}\right) \tag{3.4}$$

is the square of the buoyancy frequency, and

$$L^2 = l(l + 1). \tag{3.5}$$

Equation (3.2) is essentially a local dispersion relation for acoustic-gravity waves. It may be used as the basis of a JWKB analysis, which leads to the resonance condition

$$\int_{r_1}^{r_2} \kappa \, dr = (n + \alpha)\pi, \tag{3.6}$$

where r_1 and r_2 are radii at which $\kappa = 0$ and between which $\kappa^2 > 0$, and α is a phase which depends on ω and the equilibrium state of the Sun in the vicinity of the turning points r_1 and r_2. The properties of the condition are discussed in some detail by Christensen-Dalgaard (1984), Deubner and Gough (1984), and Gough (1985a). Similar conditions had previously been derived in the limit $n/l \to \infty$ by Vandakurov (1967) and Zahn (1970), and more recently by Tassoul (1980) who paid much more attention to the properties of the solutions in the vicinity of the turning points. Essentially the same condition can also be obtained from ray theory when n and l are large; it was derived by Gough (1984b), who ignored ω_c and used a two-dimensional analysis (with L replaced by l), and it may be generalized to a sphere by the method of Keller and Rubinow (1960), leading to Equations (3.2)–(3.4) and (3.6) with L replaced by $l + \frac{1}{2}$.

It is interesting to approximate condition (3.6) for high and low ω. The former corresponds to the five-minute oscillations. In that case the upper turning point r_2 is very close to the surface $r = R$ (e.g. Tassoul, 1980; Christensen-Dalsgaard, 1984; Deubner and Gough, 1984); in that region the function ω_c can be replaced by its value ω_{cp} for a plane parallel polytrope and then subtracted out, its influence being absorbed into α (e.g. Christensen-Dalsgaard, 1984). Moreover, $\omega^2 \gg N^2$ and $\omega_c^2 \gg |\omega_c^2 - \omega_{cp}^2|$, so the condition (3.6) may be expanded for large ω, yielding

$$\frac{\pi(n + \alpha)}{\omega} + \frac{1}{\omega^2} \; \Psi(w) \simeq F(w) \equiv \int\limits_{\ln r_1}^{\ln R} a^{-1}(1 - a^2/w^2)^{1/2} \, \mathrm{d} \ln r \,, \qquad (3.7)$$

where $w = \omega/L$, $a = c/r$, and

$$\Psi(w, \omega) = \tfrac{1}{2} \int\limits_{\ln r_1}^{\ln R} a^{-1}(\omega_c^2 - \omega_{cp}^2 - N^2 a^2/w^2)(1 - \omega_{cp}^2/\omega^2 - a^2/w^2)^{-1/2} \, \mathrm{d} \ln r \,. \qquad (3.8)$$

The lower limit of integration is the point where $a = w$. It is evident that $\Psi \ll \omega$, so the left-hand side of Equation (3.7) is dominated by its first term. In the case of small ω, the condition approximates to

$$\frac{\pi(n + \alpha)}{L} \simeq \int\limits_{\ln r_1}^{\ln r_2} \left(\frac{N^2}{\omega^2} - 1 \right)^{1/2} \mathrm{d} \ln r \,. \qquad (3.9)$$

This form is valid for high-order g modes. Here α, r_1 and r_2 take quite different values from those in Equation (3.7); in particular, r_2 is close to the base of the convection zone. Moreover, r_1 tends to be closer to the centre of the Sun for g modes than it is for p modes of similar order and degree (cf. Figures 4 and 5).

It is evident from Equation (3.7) that the high-frequency p modes provide information principally about the sound speed c. Notice that if Ψ were ignored the quantity $(n + \alpha)/L$, which can be regarded as an observed quantity provided the modes can be identified, is a function of the single variable $w = \omega/L$, and does not depend on ω and L separately. This was first noticed to be the case by Duvall (1982) from an analysis of real solar data with α set to a constant. It is equivalent to saying that the frequencies depend only on a single (one-dimensional) function of the parameters n and l. That such a relation exists need not be surprising, once it is realised that in the limit of high frequency the dispersion relation (3.2) depends on the single function c of the single variable r. (Though it does not necessarily follow from this property that the relation must exist.) For low frequencies the combination $(n + \alpha)/L$ depends only on the single quantity ω.

An asymptotic approximation has also been developed for rotational splitting of degenerate p-mode eigenfrequencies. If angular velocity Ω is a function of r alone, the frequency of a mode is shifted by

$$\omega_1 \simeq m \int\limits_{\ln r_1}^{\ln R} K\Omega \, \mathrm{d} \ln r \bigg/ \int\limits_{\ln r_1}^{\ln R} K \, \mathrm{d} \ln r \,, \qquad (3.10)$$

where

$$K \simeq a^{-1}(1 - a^2/w^2)^{-1/2} \tag{3.11}$$

(Gough, 1984b). Note that the introduction of a new function of a single variable, namely $\Omega(r)$, has increased the parameter space upon which the frequencies depend; the observable quantity ω_1 depends on both m and w. Nevertheless, it depends on m in a particularly simple way, and does not increase the dimension of the parameter space; observation of ω_1 for only a single value of m would provide all the information about $\Omega(r)$ that could be extracted from the frequencies. When Ω is a function of both r and θ the splitting frequency genuinely depends on two independent functions of the parameters; according to ray theory these are most naturally $\omega/(l + \frac{1}{2})$ and $m/(l + \frac{1}{2})$.

I should point out at this stage that the introduction of a symmetry-breaking agent, such as rotation about a single axis with angular velocity $\Omega(r, \theta)$, distorts the eigenfunctions so that they are no longer precisely proportional to spherical harmonics. Nevertheless when Ω/ω is small, so is the distortion, and the mode can be associated with the spherical harmonic that it closely resembles. However, as Perdang (1985) has emphasized, true modes may cease to exist as Ω is imagined to increase away from zero, leading to quantum chaos and an associated smearing of the eigenfrequencies into finite continua. If this is the case it will evidently impose a fundamental limitation on the accuracy with which Ω can ever be determined. I shall not address this issue here; instead I assume that all symmetry-breaking agents are so small that for practical purposes there is a unique frequency associated with each combination of n, l, and m.

There is still scope for greater richness in the spectrum of eigenfrequencies. This can result directly from advection by fluid flow in the equilibrium state that is more complicated than pure rotation, involving meridional circulation or ϕ-dependent convective cells. The flow can lead to horizontal variations in sound speed, which also modify the frequencies. The contribution to ω from buoyancy, through the quantity Ψ in Equation (3.7), also adds to the complexity of the spectrum. Additional perturbations may be caused by magnetic fields. Nevertheless, the greatest observable richness that can occur is for ω to vary with the three parameters n, l, m independently. It seems highly unlikely that with only p, f, and g modes, complete three-dimensional information about c, N, flow velocity and magnetic field could ever be obtained from frequency data alone.

4. Testing Theoretical Solar Models

The most straightforward way to deduce information about the structure of the Sun from oscillation frequencies is to compute the eigenfrequencies of a sequence of solar models with certain adjustable parameters, and then to choose the values of the parameters that give the best fit to the observational data. I shall not discuss the details of the comparisons that have been made, for these have been reviewed elsewhere (e.g. Gough, 1983; Deubner and Gough, 1984; Provost, 1984; see also Christensen-Dalsgaard, 1985). I shall record simply that the principal parameters that have been estimated by this procedure are the initial helium abundance and the depth of the convection zone.

However, I should point out that there are uncertainties in the estimated values arising from the fact that no theoretical model has yet been constructed from stellar evolution theory whose eigenfrequencies agree with observation. Of course there will always be the additional uncertainty arising from the fact that once a model that does reproduce the data is found, it will then be possible to construct an infinite number of other data-reproducing models (see Section 7).

Although the model-fitting procedure requires eigenfrequencies to be calculated accurately by numerical techniques, it is nonetheless useful to use the asymptotic ideas in the previous section to guide the method by which frequencies are compared. Indeed, the first helioseismological inference that was made, concerning the depth of the convection zone, resulted from noticing how one would need to adjust the frequencies of a plane-parallel polytropic model to produce differences that were comparable with the discrepancies between numerically computed frequencies of high-degree p modes and observation.

The guidance given by the asymptotic analysis is particularly important for deciding what aspects of the theoretical frequency spectrum should be compared with observation to answer a particular question. For example, it was noticed, originally from Tassoul's (1980) asymptotic analysis, that the l dependence of low-degree five-minute modes is a signature of conditions in the energy-generating core. This is evident from Equation (3.7), where it is clear that the integral is most sensitive to w, and hence to l at fixed ω, only near the lower turning point $r = r_1$ where w is comparable with $a(r)$. Furthermore, amongst modes with nearly the same frequency it is those with the lowest values of l that penetrate the most deeply, where a is greatest. Thus for them it has been common practice to record the small frequency differences between modes (n, l) and $(n - 1, l + 2)$ (e.g. Fossat, 1985), which are a measure of how sound speed varies close to the centre of the Sun. In the core the sound speed depends most sensitively on chemical composition and thus the frequency differences put limits on the degree of material mixing that might have taken place (e.g. Gough, 1983; Provost, 1984; Christensen-Dalsgaard, 1985). The appreciation that the l dependence of ω stems chiefly from conditions near the lower turning point has also permitted Christensen-Dalsgaard and Gough (1984a) to assess from a direct comparison of the frequencies of p modes with observation where a particular solar model of Christensen-Dalsgaard (1982) was in error.

5. Asymptotic Inversions

So far the function F has been defined only on the set of discrete values of w corresponding to normal modes of oscillation. However, it may be regarded as being defined for all real positive w, though of course it is only on the discrete set that it can be measured. Let it be differentiable. Then the equality in (3.7) can be cast into Abel's integral equation, which can then be inverted to yield

$$r = R \exp\left\{ -\frac{2}{\pi} \int_{a_s}^{a} (w^{-2} - a^{-2})^{-1/2} \frac{\mathrm{d}F}{\mathrm{d}w} \, \mathrm{d}w \right\}, \tag{5.1}$$

where a_s is the value of a at the surface $r = R$ of the Sun. Thus the radius of the lower turning point at which a is equal to the observed quantity w is determined in terms of F, and can be evaluated from the discrete measurements of F by approximating the integral in Equation (5.1) with a suitable finite-difference formula.

A determination of F was first carried out by Duvall (1982) for the frequencies of high-degree modes. Duvall implicitly ignored Ψ and found the constant value of α that made the quantity $\pi(n + \alpha)/\omega$ approximately a function of w alone. That method was subsequently applied by Christensen-Dalsgaard *et al.* (1985) to a more extensive data set including modes of low and intermediate degree reported partially by Duvall (1982), Duvall and Harvey (1983), and Harvey and Duvall (1984a). The function F has been used by Gough (1985a) and Christensen-Dalsgaard *et al.* (1985) to determine $c(r)$ using Equation (5.1). The procedure is subject to both the inaccuracy of the asymptotic formula (3.7) and the errors introduced by ignoring Ψ and the variation of $\alpha(\omega)$. However, these errors are systematic, and it was found that they can be substantially reduced by using the difference between sound speeds computed with Equation (5.1) from observed and theoretical eigenfrequencies of identical sets of modes as an estimate of the difference between the actual sound speeds in the Sun and the theoretical model. The quality of that procedure can be assessed from Figure 1, where the inferred difference between two theoretical models is compared with the actual difference.

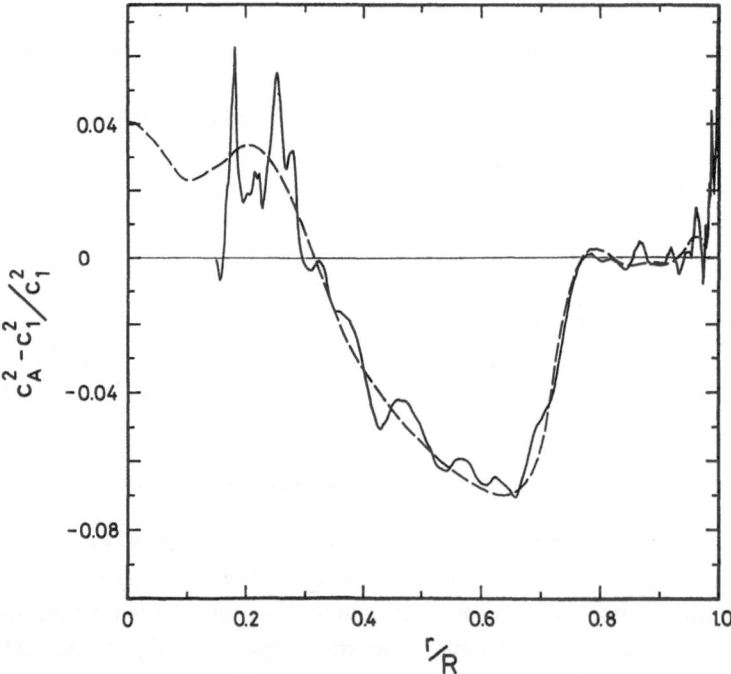

Fig. 1. The solid curve is the relative difference $(c_A^2 - c_1^2)/c_1^2$ between the squares of the sound speeds c_A^2 and c_1^2 inferred from Equation (5.1) using the frequencies of 363 corresponding modes of five-minute oscillations computed from two different theoretical models of the Sun. The dashed line is the actual difference (from Christensen-Dalsgaard *et al.*, 1985).

Improvements in the method are under way, partly by attempting to take into account Ψ and the frequency dependence of α. In practice it will no doubt be necessary to account for neglected higher terms in the asymptotic sequence, including contributions from the perturbation to the gravitational potential, before a significantly more accurate inversion is possible.

Notice that according to Figure 1 it has been possible to infer by this method the sound speed c_1, say, from a knowledge of c_A and the two sets of frequencies with an error of less than 1% in the region $0.3 \lesssim r/R \lesssim 0.9$, even though the sound speed is not accurately determined in the outer layers through which all the modes used in the inversion must travel. Thus it has been possible to eliminate much of the contribution to the frequencies from the uncertain upper layers where the physics of the oscillations is ill understood. A similar result was found by Christensen-Dalsgaard and Gough (1984b) using a formal inversion of numerically computed rotational splitting frequencies (see Figures 7 and 9). It is yet to be ascertained whether it will be possible to achieve a substantial improvement in the inversion at great depths without ever needing to know the structure of the superficial layers.

Low-degree g modes of high order are likely to improve our knowledge of the very central regions of the Sun and possibly the upper layers of the radiative interior immediately beneath the convection zone. The former will be manifest mainly in the l dependence, and the latter in the manner in which the asymptotic period-order relation is approached (e.g. Ellis, 1984, 1985; Berthomieu and Provost, 1985; Gabriel, 1985).

It is interesting to note that the inversion of Equation (3.9) is probably not very useful. Whereas $a(r)$ is a rapidly decreasing function of r, so that $w \gg a$ throughout most of the region of integration, N varies quite slowly throughout most of the radiative interior. Consequently N cannot be sampled in an easily understandable way by observable g modes alone. The reason is that only low-degree interior g modes can penetrate sufficiently well through the convection zone to have observable amplitudes in the photospheric layers (Dziembowski and Pamjataykh, 1978; Christensen-Dalsgaard et al., 1980), and of these there are too few with frequencies close to the typical values of $N(r)$ to sample the latter well. Moreover, any low-degree g modes with such high frequencies are also of low order, and for them the asymptotic analysis is not valid. Therefore, oddly enough, it seems more likely that $N(r)$ will first be measured by its small contribution to five-minute p modes.

One can see how this might be carried out within the framework of asymptotic theory. Since the variation of α depends principally on the approximation to the contribution of the surface layers near the upper turning point, it is essentially a function of frequency alone. Moreover, $\omega^2 \gg \omega_{cp}^2$ wherever $\omega_c^2 - \omega_{cp}^2$ and N^2 are comparable with their maximum values, so one might approximate the kernel $a^{-1}(1 - \omega_{cp}^2/\omega^2 - a^2/w^2)^{-1/2}$ by $a^{-1}(1 - a^2/w^2)^{-1/2}$. This renders Ψ a function of w alone. In view of its special functional form it can in principle be estimated from the data. One might then contemplate inverting Equation (3.8) with the help of Equations (3.3) and (3.4) and the equation for mass conservation to yield $N(r)$ and $\omega_c(r)$. Of course in practice this scheme may not work, because the neglected terms in the asymptotic expansion leading to Equation (3.6) may be comparable with $\omega^{-1}\,\Psi$. If that is the case, a direct inversion of the kind discussed in Sections 7–9 will be necessary.

The angular velocity Ω can be inferred similarly from the splitting frequencies. For example, if Ω were a function of r alone, Equations (3.10) and (3.11) would be valid. These can be cast directly into Abel's integral equation, and thence inverted. The accuracy of the inversion, for the small number of modes observed by Duvall and Harvey (1984), has been estimated by Gough (1984b).

6. Frequencies as Integral Constraints

Oscillations sample an extended region of the Sun, and so provide an integral measure of the structure over that region. This is most easily seen from the variational formulation of the adiabatic eigenvalue problem for a nonrotating star (e.g. Ledoux and Walraven, 1958):

$$\omega^2 \int \rho\boldsymbol{\xi}\cdot\boldsymbol{\xi}\,\mathrm{d}V = \int \left[\gamma p(\mathrm{div}\,\boldsymbol{\xi})^2 + 2\boldsymbol{\xi}\cdot\nabla p\,\mathrm{div}\,\boldsymbol{\xi} + \rho^{-1}\boldsymbol{\xi}\cdot\nabla p\,\boldsymbol{\xi}\cdot\nabla\rho\right]\mathrm{d}V -$$

$$- G \iint |\mathbf{r}-\mathbf{r}'|^{-1}\,\mathrm{div}(\rho\boldsymbol{\xi})\,\mathrm{div}'[\rho(\mathbf{r}')\boldsymbol{\xi}(\mathbf{r}')]\,\mathrm{d}V'\,\mathrm{d}V, \quad (6.1)$$

where p and ρ are the pressure and density of the equilibrium state, γ is the adiabatic exponent $(\partial \ln p/\partial \ln \rho)_{\mathrm{ad}}$, G is the gravitational constant and the integrals are over the volume of the star. Thus the frequencies are a combination of weighted averages of nonlinear functions of the equilibrium pressure and density and their derivatives, the weighting depending on the oscillation displacement eigenfunctions $\boldsymbol{\xi}$. Since different eigenfunctions weight the structure differently, the hope is that with a sufficient variety of data one can obtain an estimate of how p and ρ vary with r.

Most of the inversion procedures that have been developed apply only to linear integral equations. Therefore to make some progress it is expedient to develop an iterative procedure, like a generalized Newton–Raphson method, that improves (hopefully) upon a trial model of the Sun. One postulates an initial guess: p_0, ρ_0, γ_0, taken, for example, from a standard model of the Sun. Using that one carries out the forward problem, calculating the eigenfunctions $\boldsymbol{\xi}_0$ and eigenfrequencies ω_0 corresponding to the modes for which observational data are available. One presumes that the physics is correctly described by Equation (6.1), so that $\omega, p, \rho, \boldsymbol{\xi}$ refer to the actual Sun. Then one writes down the equation satisfied by $\delta\omega^2 \equiv \omega^2 - \omega_0^2$, by subtracting two equations of the type (6.1), and hopes that the trial model is sufficiently close to reality for linearization in the differences $\delta p \equiv p - p_0$, $\delta\rho \equiv \rho - \rho_0$, etc. to be valid. There results an equation in which $\delta\omega^2$ is expressed as a linear functional of the differences δp, $\delta\rho$, etc. It is at this point that one appreciates the variational formulation of the problem; since Equation (6.1) is stationary to variations in the functions $\boldsymbol{\xi}$ (that satisfy appropriate boundary conditions) about the true eigenfunctions, it follows that the terms that are linear in $\boldsymbol{\xi} - \boldsymbol{\xi}_0$ cancel, and $\delta\omega^2$ can be expressed in terms of $\boldsymbol{\xi}_0$ alone. Thus it is not necessary to perturb the forward problem.

One can proceed further by imposing the constraint of hydrostatic support

$$\frac{dp}{dr} = -\frac{G\tilde{m}\rho}{r^2},$$ (6.2)

where $\tilde{m}$ is the usual mass variable which satisfies

$$\frac{d\tilde{m}}{dr} = 4\pi r^2 \rho.$$ (6.3)

By substituting these equations into the equation for $\delta\omega^2$ and integrating by parts (and noticing that the surface integrals are negligible), it is possible to write the equation in the form

$$\frac{\delta\omega^2}{\omega_0^2} = \int_0^R S(\mathbf{X}_0, \xi_0, r) \frac{\delta\rho}{\rho_0} r^2 \rho_0 \, dr,$$ (6.4)

provided the equation of state is known. Here $\mathbf{X}_0(r)$ represents the equilibrium structure (p_0, ρ_0, etc.) of the trial model. It is a straightforward matter to calculate the formula for the differential kernel S. (The function S is sometimes called a Fréchet derivative or a sensitivity function.)

For each measured frequency one has an equation (6.4) to constrain the possible structure of the Sun. The set of constraints must be supplemented by

$$\int_0^R \frac{\delta\rho}{\rho_0} r^2 \rho_0 \, dr = 0,$$ (6.5)

to ensure that mass is preserved. This condition is of the same form as (6.4), having $S = 1$ and $\delta\omega^2 = 0$, and can therefore be treated as though it were a frequency constraint. If the system can be inverted to estimate $\delta\rho$, an improved estimate of ρ can be obtained, and the whole procedure could be repeated.

In deriving Equation (6.4) I presumed the equation of state was known. This is required for computing the variation in γ. What is needed is not only a knowledge of the microphysics of the solar material, but also the composition. Of course for the first iteration one has a trial composition taken from the standard solar model. But for subsequent iterations one has no such information, and the procedure must be generalized.

For the purposes of inferring the structure of the core (if it can be done without a detailed knowledge of the structure of the envelope) the uncertainty in the equation of state is not very important. There the material is highly ionised, and $\gamma \simeq \frac{5}{3}$. But in the convection zone where the abundant elements are only partially ionized, the problem is important. Formally it is quite straightforward to overcome the difficulty by generalizing Equation (6.4) to include additional integrals that are weighted averages of the abundances of the elements that influence γ, and appropriately generalizing the inversion procedure described in the following section. (The geophysical inverse problem originally formulated by Backus and Gilbert (1967) was posed for a vector function; a corresponding solar problem has been addressed by Denis and Denis (1984).) Alternatively, it may be simpler to separate the problem of determining the composition of the convection zone and solving that first. This should be possible because in the convection zone the stratification is close to being adiabatic and the chemical composition is presumably homogeneous.

The case of rotational splitting of nonaxisymmetrical modes can be treated similarly. The perturbations ω_1 to the eigenfrequencies produced by the angular velocity $\Omega(r, \theta, t)$ can be computed by linearizing with respect to Ω the variational principle of Lynden-Bell and Ostriker (1967). Alternatively it can be obtained in the form of a consistency condition in an expansion of the eigenvalue problem about the nonrotating state. The result is a functional of Ω which, in the special case when Ω is a function of r alone, can be written in the form (Gough, 1981; cf. Hansen et al., 1977)

$$\frac{\omega_1}{\omega_0} = m \int K(\mathbf{X}_0, \xi_0, r)\Omega(r)\, dr. \tag{6.6}$$

The asymptotic form of the kernel K for high-frequency p modes is given by Equations (3.10) and (3.11). Quadratic and higher-order terms in the expansion of ω_1 can be computed, if desired, but these are small compared with the linear term represented in Equation (6.6) (Dziembowski and Goode, 1984; Gough and Taylor, 1984). Once again, the data, namely the splitting frequencies ω_1, are weighted averages of the function to be determined (this time Ω), with weight functions that depend on the equilibrium model and its eigenfunctions. If the previous inverse problem has already been solved (using the observed frequencies of the axisymmetrical modes) K can be regarded as being known. This inverse problem is therefore linear, and requires no iteration.

Constraints of the type (6.6) have also been used in an attempt to measure horizontal flows associated with giant convective cells (Hill et al., 1984a, b). Provided the horizontal wavelength of the oscillation mode is very much less than the horizontal scale of the giant cell, the oscillations are advected locally by the horizontal flow whatever the large-scale variation of the latter. Consequently the perturbations to their frequencies are related to their apparent local wavenumber in the same way as if the flow did not vary horizontally.

7. Inverse Methods

In this section I consider explicitly the inversion of the idealized Equation (6.6) to obtain $\Omega(r)$ from M rotational splitting data. This typifies the linear inverse problem, which may be either complete in itself on an iteration of a nonlinear problem such as the determination of the hydrostatic density stratification. Now it is convenient to label the modes with a single index i to identify n, l and m, letting i take the integral values from

1 to M. Actually, in the special case considered here where Ω depends only on r, K is independent of m and therefore i need represent only n and l. Thus one writes

$$w_i = \int_0^R K_i(r)\Omega(r)\, dr, \tag{7.1}$$

where $w_i = \omega_{1i}/(m\omega_{0i})$ are observable quantities, K_i are known kernels and R is the radius of the Sun. In practice, observations contain errors, so if the w_i are regarded as the observations, Equation (7.1) holds only approximately.

As pointed out by Backus and Gilbert (1967), a solution to the problem represented by (7.1), if it exists, is not unique, even if strict equality is assumed. One reason is that since one can have only a finite number of observations, the kernels K_i cannot possibly span the space of functions on the interval $(0, R)$. Therefore there is an infinite number of functions, f_k, orthogonal to all the K_i, and any linear combination of them can be added to Ω without modifying any of the integrals. The inverse problem is said to be underdetermined. Let $\mathcal{O}$ denote the subspace spanned by the K_i. The functions f_k lie in its complement, $\mathcal{A}$, called the annihilator. Evidently the data contain no information about the projection of Ω in $\mathcal{A}$.

Permitting only approximate equality broadens the possibilities further. Therefore the problem is not simply to find an approximate solution to Equation (7.1), but to select which of the infinite number (if there are any at all) is the most likely. Here prejudice reigns, and opinions therefore differ. In almost all cases a condition related to smoothness is imposed, because variations on very short length scales cannot normally be detected. It is interesting to note, therefore, that this condition was automatically embodied in the asymptotic method described above, for although Equations (3.7), (3.8), and (3.10)–(3.11) have unique inverses, they are valid only when the scale of variation of the equilibrium state is much less than the wavelength of the oscillations.

7.1. Spectral Expansion

This is essentially an expansion of Ω in terms of the kernels K_i. The idea of using K_i as a basis is, at first sight, quite natural, since the K_i span the subspace $\mathcal{O}$ that is accessible to the observations. However, as I shall soon discuss, it is actually more useful to transform, at least conceptually, to a new basis, ψ_i, that takes into account the precision with which the data can measure the projection of Ω onto each basis function.

In geophysics, the expansion first arose out of a procedure formulated by Backus and Gilbert (1967). Suppose one has a preconceived idea, $W(r)$ say, of the function Ω. Let us assume that W is close to the truth, and seek that function Ω that minimizes the least-squares deviation

$$E \equiv \int_0^R (\Omega - W)^2\, dr. \tag{7.2}$$

Forgetting the errors for the moment, the minimization must be performed subject to the constraints (7.1), with exact equality. The result is

$$\Omega = W + \sum_i \alpha_i' K_i, \qquad (7.3)$$

with α_i' being the solutions of the Euler equations

$$\sum_j A_{ij}\alpha_j' = w_i - a_i, \qquad (7.4)$$

where

$$A_{ij} = \int_0^R K_i K_j \, dr, \qquad a_i = \int_0^R K_i W \, dr; \qquad (7.5)$$

the sums are over all values of the index from 1 to M. Thus the difference between Ω and W is expressed, in Equation (7.3), as a linear combination of the kernels K_i.

The foregoing analysis suggests that one might simply express Ω directly as an expansion in K_i:

$$\Omega = \sum_i \alpha_i K_i. \qquad (7.6)$$

If once again errors are ignored, one can determine the coefficients in the usual way by projecting the constraints (7.1) onto the basis, which leads to

$$\sum_i A_{ij}\alpha_j = w_i. \qquad (7.7)$$

The difference between the solutions (7.6)–(7.7) and (7.3)–(7.4) lies the annihilator $\mathscr{A}$, so the data w_i cannot distinguish between the two. Therefore, if w_i constitute the only information that is available about Ω, one has no sound basis for choosing between the two possibilities. An inherent advantage of the first formulation, however, is that one can write into Ω, via W, additional constraints (obtained from information other than w_i) that are not otherwise easily incorporated into a procedure for solving (7.1).

There are two related problems that one would encounter in trying to carry out the straightforward method outlined above. First, if there are redundant data, the matrix A_{ij} is singular, and some care must be taken in solving equations (7.4) or (7.7). If the data w_i were truly error-free, that would be possible, at least in principle: since by hypothesis the equations would be consistent, one would need only to reject the redundant equations and solve the reduced set that remains. In practice, however, the data are erroneous, and Equations (7.4) or (7.7) are formally inconsistent. One could still reject redundant data, but that, of course, would be unwise. Retention of redundancy is always important under these circumstances for reducing the influence of random errors. Therefore some kind of averaging procedure is required.

The second problem concerns error magnification. In addition to formal redundancy it is usually the case that there are different combinations of the data that give nearly but not strictly the same information. This leads to the matrix A_{ij} being ill-conditioned, or nearly singular. Once again, one wishes to average in some way the almost identical information that is contained in the two or more combinations, in much the same way as must be done for the genuinely redundant data. However, one must also reject the apparent information contained in the difference between the almost equivalent combinations, for that is dominated by the errors.

The second problem is analogous to trying to measure a vector $\mathbf{v}$ by making independent measurements of its components (α_1, α_2) in the directions of unit vectors $\mathbf{v}_1$ and $\mathbf{v}_2$ that are known to be nearly parallel. Thus if

$$\mathbf{v} = \alpha_1 \mathbf{v}_1 + \alpha_2 \mathbf{v}_2 = \tfrac{1}{2}(\alpha_1 + \alpha_2)(\mathbf{v}_1 + \mathbf{v}_2) + \tfrac{1}{2}(\alpha_1 - \alpha_2)(\mathbf{v}_1 - \mathbf{v}_2) \tag{7.8}$$

and the measurements of α_1 and α_2 have errors ε_1 and ε_2, one cannot measure the component of $\mathbf{v}$ in a direction roughly perpendicular to $\mathbf{v}_1$ and $\mathbf{v}_2$ (e.g. in the direction of $\mathbf{v}_1 - \mathbf{v}_2$) if either of the errors ε_1 or ε_2 has a magnitude comparable with the difference $\alpha_1 - \alpha_2$ between the two measurements. Even though formally the vectors $\mathbf{v}_1$ and $\mathbf{v}_2$ can be used as basis vectors for a plane, in practice the erroneous data provide information only along a line roughly parallel to $\mathbf{u}_1 = \mathbf{v}_1 + \mathbf{v}_2$. For practical purposes the component of $\mathbf{v}$ parallel to $\mathbf{u}_2 = \mathbf{v}_1 - \mathbf{v}_2$ is inaccessible. Thus the situation is essentially the same as if $\mathbf{v}_1$ and $\mathbf{v}_2$ were genuinely parallel, which is analogous to the first of the problems mentioned above.

The resolution of problems of this kind has been discussed by Lanczos (1961). The procedure is to find that part of the subspace $\mathcal{O}$ that is inaccessible to the data by virtue of the errors, and to relegate it to the annihilator. The solution Ω is the component of the actual function that is in the appropriately diminished subspace $\mathcal{O}'$ that remains. In terms of the simple vector analogy, one creates a new orthogonal basis $(\mathbf{u}_1, \mathbf{u}_2)$ of the plane, and recognizes that one can measure only the component $\tfrac{1}{2}(\alpha_1 + \alpha_2)$ in the direction of $\mathbf{u}_1$. Notice that in this analogy the result contains both measurements, so statistical errors are lower than had the second of the measurements, say, been rejected as being redundant and the resulting component estimated by α_1. The Lanczos inverse also has this property, providing a natural way of averaging data in the more complicated case when several different combinations provide similar information.

For the details and justification of the Lanczos method the reader is referred to the book by Lanczos (1961), and the discussions in the geophysical context by Jackson (1972), Wiggins (1972), and Parker (1977a). The following summary follows Parker (1977a).

Let σ_i be the standard errors of the data w_i. One weights the constraints (5.1) with σ_i^{-1}, yielding

$$w_i' = \int\limits_0^R K_i'(r)\Omega(r) \, dr, \tag{7.9}$$

where $w_i' = w_i/\sigma_i$ and $K_i' = K_i/\sigma_i$, so that w_i' has unit standard error. The matrix A_{ij}', defined in the same way as A_{ij} in Equations (7.5) but with K_k' replacing K_k, is positive-definite and symmetric, and can be diagonalized with an orthogonal matrix U_{ij} to give

$$\sum_{k,l} U_{ki} A_{kl}' U_{lj} = \lambda_i \delta_{ij}, \tag{7.10}$$

where λ_i are the (positive) eigenvalues of A'_{ij} and δ_{ij} is the Kronecker delta. Now consider the new basis $\psi_i(r)$ of $\mathcal{O}$ defined by

$$\psi_i = \lambda_i^{-1/2} \sum_j U_{ji} K'_j, \tag{7.11}$$

which has the property

$$\int_0^R \psi_i \psi_j \, dr = \delta_{ij}. \tag{7.12}$$

Regard the functions ψ_i to be ordered such that λ_i decreases with increasing i, and expand the solution Ω in terms of them:

$$\Omega = \sum_i \alpha_i \psi_i. \tag{7.13}$$

Of course the expansion excludes that part of Ω in the annihilator $\mathcal{A}$. In view of Equations (5.12) and (5.9), the expansion coefficients are determined by

$$\lambda_i^{1/2} \alpha_i = \lambda_i^{1/2} \int_0^R \psi_i \Omega \, dr = \sum_j U_{ji} w'_j. \tag{7.14}$$

Recalling that the w'_j have unit standard errors, it follows immediately from Equation (7.14) and the orthonormality of U_{ij} that if the errors in w'_j are statistically independent the errors in α_i are also statistically independent, with standard deviation $\lambda_i^{-1/2}$. Thus the uncertainty of the coefficients α_i increases with decreasing λ_i; and it is immediately evident from Equation (7.14) that the uncertainty is total when $\lambda_i = 0$. Notice that these statements incorporate the statistics of the errors in the data, as must be the case because each λ_i depends on the σ_j.

The final step in the procedure is to truncate the expansion (7.13), thereby effectively transferring to the annihilator the subspace spanned by those eigenfunctions ψ_i that correspond to small eigenvalues λ_i. It can easily be shown that these functions contribute little to the integrals in the contraints (7.9) compared to their contribution to the sum (7.13). It is necessary to decide where (and how) the expansion is to be truncated, and here several options are available (e.g. Wiggins, 1972; Jackson, 1973). Here I consider a pragmatic approach, which appears to be reliable when systematic errors in the data can be ignored. Moreover, it is particularly useful when the magnitudes of the random errors in the data are poorly estimated. It rests on the fact that as λ_i decreases, the functions ψ_i grow in amplitude and tend to develop more and more small-scale structure. Moreover, the coefficients increase (roughly as $\lambda_i^{-1/2}$) in a random way once they are dominated by errors. If one increases from unity the number of terms retained in the expansion (7.13), the result should first approach the correct solution. Then, once errors dominate, the successive approximations diverge, exhibiting structure generally on smaller and smaller scales and with larger and larger amplitude. The best estimate of Ω one can obtain by the method is the function to which the expansion appears to be converging before the divergence takes over.

Several expansions are illustrated in Figures 2 and 3, where artificial rotational splitting frequencies of 45 high-degree five-minute oscillations of a solar model were computed from an imposed angular velocity Ω. They are taken from Gough (1984c). The modes were distributed approximately uniformly along the nine lowest ridges in the $k - \omega$ diagram, with $100 \lesssim l \lesssim 1000$ and $1.5 \times 10^{-2} \, \text{s}^{-1} \lesssim \omega \lesssim 3.5 \times 10^{-2} \, \text{s}^{-1}$. In Figure 2 are displayed four expansions (7.13) obtained from data to which no random errors had been added. They include $I = 9$, 15, 20, and 33 basis functions, with eigenvalues satisfying $\lambda_i/\lambda_1 > 3 \times 10^{-2}$, 10^{-2}, 10^{-3}, and 10^{-8}, respectively. As

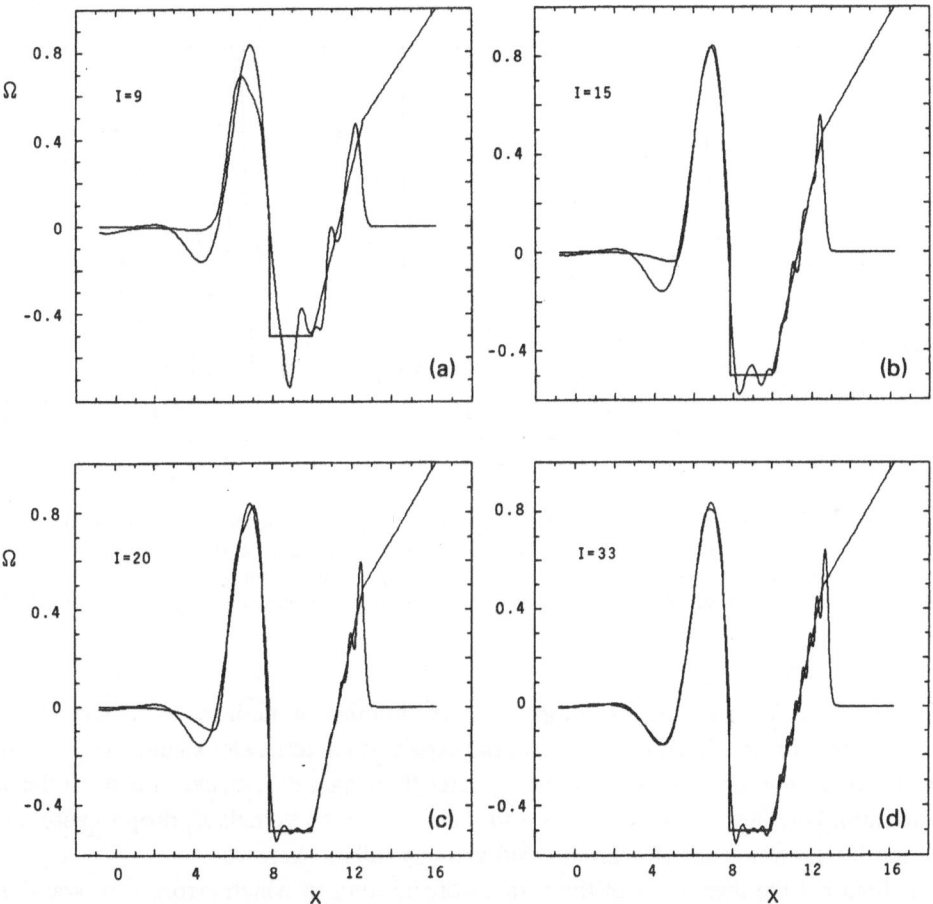

Fig. 2.   Inversions by spectral expansion of error-free frequency splittings ω_{1i} of 45 p modes of high degree. The results of the inversions and the imposed horizontal velocity Ω from which ω_{1i} were computed are shown; the latter is easily identifiable because it is common to all the panels. The eigenvalues λ_i retained in the expansions represented in panels (a)–(d) satisfy $\lambda_i > \lambda_m$, where $\lambda_m/\lambda_1 = 3 \times 10^{-2}$, 10^{-2}, 10^{-3}, and 10^{-8}, respectively; the corresponding numbers I of basis functions are 9, 15, 20, and 33. Since the problem is linear, the ordinate scale is arbitrary. The independent variable is $x = \log_{10} p$ (from Gough, 1984c).

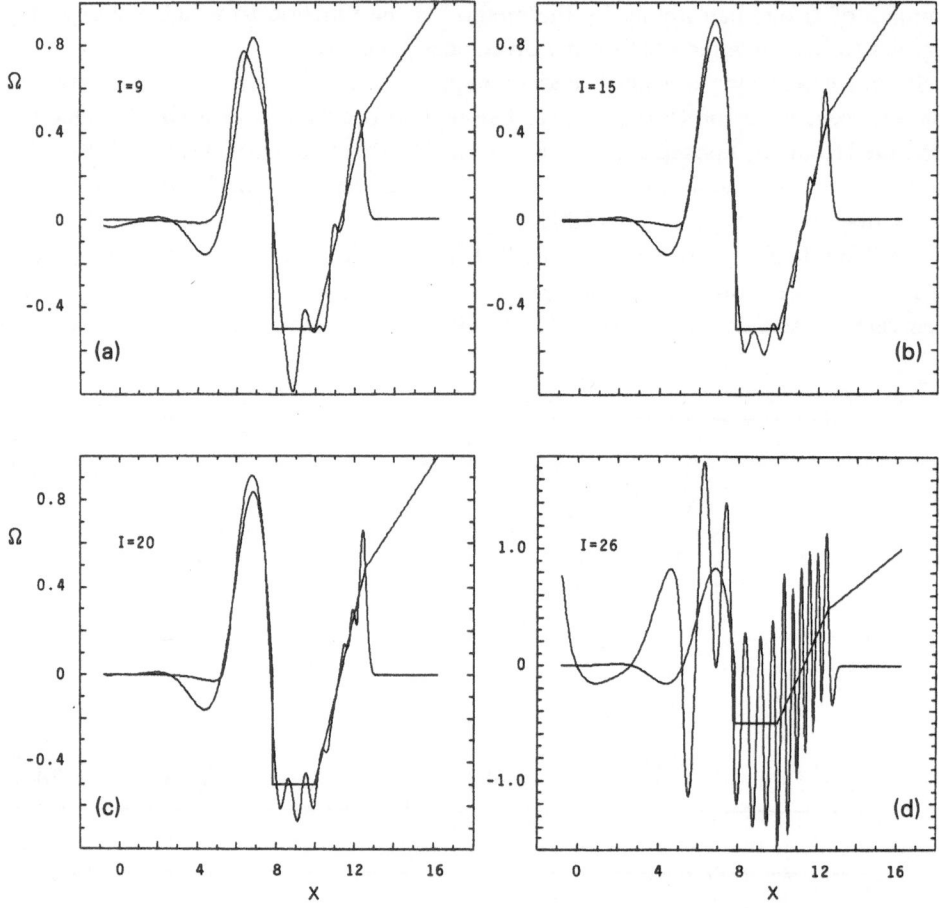

Fig. 3. Inversions by spectral expansion of frequency splittings to which independent gaussian-distributed errors with 5% standard deviation have been added. The modes are the same set that were used for Figure 2. The lower bounds λ_m to the eigenvalues retained in panels (a)–(d) are given by $\lambda_m/\lambda_1 = 3 \times 10^{-2}$, 10^{-2}, 10^{-3}, and 10^{-5}, respectively. Note the difference in the ordinate scale in panel (d) (from Gough, 1984c).

expected, convergence generally improves as the number of modes is increased. If even more modes are included the accuracy of the expansion deteriorates slightly, presumably as a result of rounding errors. At depths greater than $\log p \simeq 12.5$, even the most deeply penetrating oscillation eigenfunctions and their associated kernels K_i drop to zero, and the modes cannot sense Ω. The inferred velocity falls to zero too.

In Figure 3 are inversions of the same artificial data to which errors with standard deviations of 5% of the raw values have been added. The first two inversions are hardly distinguishable from the corresponding inversions with error-free data. The errors have been averaged to some degree by the inversion procedure, as a result of combining redundant information, leading to an inferred velocity with an error generally rather less than 5% of the mean magnitude of Ω. Evidently the error that is present is a result of imperfect resolution resulting from retaining no more than 15 basis functions. When 20 basis functions are included (Figure 3c), however, the errors are beginning to become important, and extraneous small-scale structure is introduced into the inferred velocity. Presumably this would not have happened at this level of truncation had substantially more than 45 data been included in the inversion. Figure 3d includes 26 modes, which have $\lambda_i/\lambda_1 > 10^{-5}$; now erroneous rapid oscillations dominate the expansion, though the general behaviour of Ω is still discernable. An example with 50% errors is illustrated by Gough (1984c). The errors in the inversion are greater, of course, and excessive spurious small-scale structure is present when $I \gtrsim 15$.

7.2. EXTRACTING LOCALIZED AVERAGES

The idea of finding localized averages was introduced by Backus and Gilbert (1968), partly as a means of assessing what they call the resolving power of the data; namely, the degree to which information about Ω that is contained in the data can be localized in space.

To begin, the constraints (7.1) are rescaled by dividing each by $\int_0^R K_i(r)\,dr$, yielding

$$w_i^* = \int_0^R K_i^*(r)\Omega(r)\,dr, \tag{7.15}$$

with K_i^* unimodular: i.e.

$$\int_0^R K_i^*(r)\,dr = 1. \tag{7.16}$$

Thus w_i^*, which is an observable quantity, is an average of Ω. The idea now is to seek, for a certain value r_0 of r, a set of coefficients $\beta_i(r_0)$ such that in the linear combination

$$\sum_i \beta_i(r_0)w_i^* = \int_0^R D(r_0, r)\Omega(r)\,dr \tag{7.17}$$

the averaging kernel

$$D(r_0, r) = \sum_i \beta_i(r_0)K_i^*(r) \tag{7.18}$$

resembles a Dirac delta function centred at $r = r_0$. One can assess the degree of success by inspecting D to see how localized it is. A quantitative assessment of the localization can be obtained from the spread, defined by Backus and Gilbert (1970) as

$$s(r_0; D) = 12 \int_0^R (r - r_0)^2 D^2(r_0, r)\,dr, \tag{7.19}$$

which, as Backus and Gilbert show, can be large either if D is not well localized or if

r_0 is far from centre $\bar{r}$ of a well-localized D, where

$$\bar{r} = \frac{\displaystyle\int_0^R r D^2(r_0, r)\, dr}{\displaystyle\int_0^R D^2(r_0, r)\, dr}.$$

(7.20)

The factor 12 is chosen in (7.19) because if $D = d^{-1}$ for $|r - r_0| < d/2$ and $D = 0$ otherwise, then $s(r_0; D) = d$. For any given D, $s(r_0; D)$ is minimized when $r_0 = \bar{r}$. That minimum, $\delta \equiv d[\bar{r}; D(r_0, r)]$, is the width of D, and measures the degree to which D resembles a delta function. Alternative measures have also been used.

The average (7.17) of Ω, which I denote by $\bar{\Omega}(\bar{r})$, is a useful diagnostic of the function Ω. Indeed, as Backus and Gilbert (1968) point out, it is only this that is in any real sense determined by the data. Thus one can regard $\bar{\Omega}$ as an estimate of Ω, bearing in mind that is is really a smoothed version obtained by averaging over a characteristic distance δ.

The determination of the coefficients $\beta_i(r_0)$ is discussed by Backus and Gilbert (1968, 1970). One chooses a function $J(r_0, r)$ that vanishes when $r = r_0$ and increases monotonically away from r_0, and one minimizes the functional

$$\Delta \equiv \int_0^R J D^2\, dr,$$

(7.21)

subject to the constraint

$$\int_0^R D(r_0, r)\, dr = 1.$$

(7.22)

When Δ is small the constraint (7.22) forces D to be small where J is large, and permits D to be large near $r = r_0$, where J vanishes.

In an application of this procedure to artificial high-degree solar p-mode data (Gough, 1978) it was found that the final result was not particularly sensitive to the form chosen for J, provided it rose steeply enough far from r_0. This is in accord with geophysicists' experience. I shall not discuss that issue further here. All the examples illustrated in the following section were computed with $J(r_0, r) = 12(r - r_0)^2$, which is the case that Backus and Gilbert (1970) discuss in detail when they consider erroneous data. Then

$$\Delta = s = \sum_{i,j} S_{ij}^* \beta_i \beta_j,$$

(7.23)

where

$$S_{ij}^* = 12 \int_0^R (r - r_0)^2 K_i^*(r) K_j^*(r)\, dr.$$

(7.24)

In practice, errors in the data cause the minimization of Δ subject to the constraint (7.22) not to provide a good measure $\overline{\Omega}$ of the average of Ω. The reason is that to obtain the most concentrated kernel D requires coefficients β_i with large magnitudes. The constraint (7.22) requires that $\sum_i \beta_i = 1$. Severe cancellations are therefore required in that sum, and also in the sum on the left-hand side of Equation (7.17). Correct cancellation in the latter does not actually take place when the data w_i^* contain errors. It is therefore necessary to modify the procedure, permitting slightly broader averaging kernels in return for a reduction in the influence of the errors.

Backus and Gilbert (1970) assume that an estimate E_{ij} of the covariance matrix of the errors in the data w_i^* is known, so that one can estimate the error ε in $\overline{\Omega}$:

$$\varepsilon^2 = \sum_{i,j} E_{ij} \beta_i \beta_j. \tag{7.25}$$

The idea then is to set a limit ε_0 on the error ε, and minimize s subject to the constraint $\varepsilon \lesssim \varepsilon_0$ and Equation (7.22). The result depends on the choice of ε_0. Alternatively one can set a limit s_0 on s, and minimize ε subject to $s \lesssim s_0$ and the contraint (7.22).

As Backus and Gilbert prove, the two formulations are essentially equivalent. They point out that S_{ij}^* and E_{ij} are both positive definite symmetric matrices, so that $s = s_0$ and $\varepsilon = \varepsilon_0$ each define hyperellipsoids in the parameter space spanned by β_i. The intersections of these hyperellipsoids with the hyperplane $\sum \beta_i = 1$ are also hyperellipsoids, which I denote by $s^\dagger = $ constant and $\varepsilon^\dagger = $ constant. (The functions $s^\dagger(\beta_i)$ and $\varepsilon^\dagger(\beta_i)$ are the right-hand sides of Equations (7.23) and (7.25) constrained by $\sum \beta_i = 1$.) In general, E_{ij} is not a scalar multiple of S_{ij}^*, so the centres of the ellipsoids are not coincident. Consequently, for any given value $\varepsilon_1 \leqslant \varepsilon_0$ of ε, s is minimized when the ellipsoid $s^\dagger = $ constant ($= s_2$, say) is tangent to the ellipsoid $\varepsilon^\dagger = \varepsilon_1$, and the solution β_i is the point of contact. The smallest value, s_0, of s is obtained when that point is as close to the centre of the ellipsoid $s^\dagger = s_2$ as is permissible, which implies that ε_1 is as large as is permissible: namely, $\varepsilon_1 = \varepsilon_0$. The reasoning is clearly symmetric; ε is minimized for $s = s_1 \leqslant s_0$ when the ellipsoid $\varepsilon^\dagger = \varepsilon_2$ is tangent to the ellipsoid $s^\dagger = s_1$, and the smallest permissible value ε_0 of ε is achieved when $s_1 = s_0$. The conditions for tangency are:

$$\sum_j (A_{ij}^* + \mu E_{ij}) \beta_j = v, \tag{7.26}$$

$$\sum_{i,j} S_{ij}^* \beta_i \beta_j = s_0, \tag{7.27}$$

$$\sum_i \beta_i = 1, \tag{7.28}$$

where A_{ij}^* is defined as in Equation (7.5) with K_k replaced by K_k^*. Equations (7.26)–(7.28) are to be solved simultaneously for the coefficients β_i and the unknown parameters μ and v. The solution is determined uniquely in terms of s_0 with $\mu > 0$, provided s_0 is in its allowable range. Evidently s_0 cannot be less than its minimum value obtained by ignoring errors; it is also bounded above, as can be appreciated from the definition (7.19) when it is recognized that the magnitude of D is bounded for finite values of ε.

In practice it is simpler to solve the problem implicitly, by choosing μ and solving for β_i, v and s_0. Backus and Gilbert replace μ by $\tilde{w} \tan \theta$ and v by $b \sec \theta$ with $0 < \theta < \pi/2$, and $\tilde{w} > 0$ chosen, for convenience, to make S_{ij}^* and $\tilde{w} E_{ij}$ of comparable numerical size. (Here θ is simply a parameter, and is unrelated to the spherical polar coordinate used earlier.) Then they introduce

$$W_{ij}(\theta) = S_{ij}^* \cos \theta + \tilde{w} E_{ij} \sin \theta. \tag{7.29}$$

Equation (7.26) now takes the form

$$\sum_j W_{ij}\beta_j = b. \tag{7.30}$$

The matrix W_{ij} is symmetric and positive definite, and therefore has a positive-definite inverse W_{ij}^{-1}. Therefore for any chosen value of θ, Equations (7.30) may be solved; whence s_0 and ε^2 may be evaluated from Equations (7.27) and (7.25).

The parameter θ determines the extent to which errors are to be restricted, at the expense of permitting D to be less well confined. When $\theta = 0$ the errors are ignored, and when $\theta = \pi/2$ no attempt is made to localize D. The aim is to find a tradeoff somewhere between.

The choice of θ is discussed by Backus and Gilbert (1970). As with the spectral method, the pragmatic approach is useful, especially when the estimate E_{ij} of the covariance matrix is uncertain. This is illustrated in the next section.

Examples of the application of the localized averaging procedure to real solar data are presented by Gough (1982), Duvall *et al.* (1984) and Hill *et al.* (1984a, b).

8. Localized Averages of an Artificial Angular Velocity

In an attempt to assess the value of optimal averages for the solar case, Christensen-Dalsgaard and Gough (1984b) performed several inversions to infer an artificially imposed angular velocity of the Sun. The procedure was superficially similar to that described above for the spectral method, except that modes that penetrated the entire volume of the Sun were used. Several different sets of modes were used, and it was hoped to be able to choose θ to obtain a good estimate Ω without prior knowledge of the errors that had been added to the data.

A selection of the splitting kernels for p modes and g modes are illustrated in Figures 4 and 5. None of them is sharply concentrated near any point. However, combinations can be successfully constructed that are. In Figure 6 is shown a set of optimal kernels using only the kernels of five-minute p modes, such as those in Figure 4. Even though the original kernels are all concentrated near the surface, it is indeed possible to construct localized averaging kernels D that are concentrated about points as close as $0.15R$ to the centre. However, it is not possible to get closer to the centre without the help of g modes.

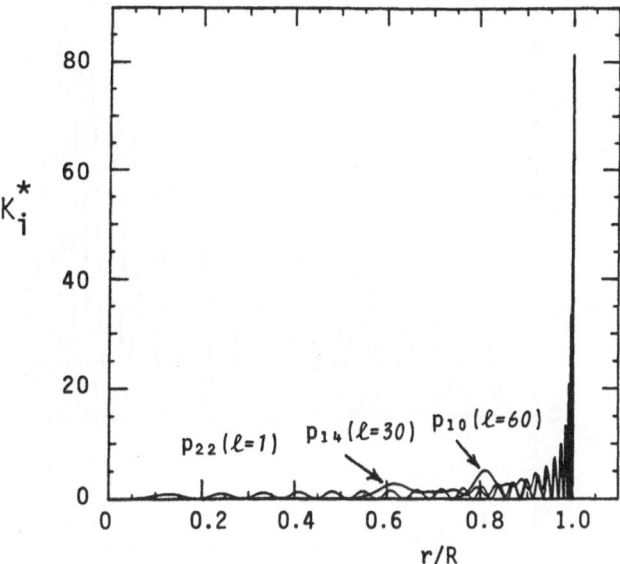

Fig. 4. Rotational splitting kernels K_i^* for three five-minute modes. The modes are p_{22} ($l = 1$), p_{14} ($l = 30$), and p_{10} ($l = 60$), having cyclic frequencies 3.24, 3.25, and 3.23 mHz, respectively (from Christensen-Dalsgaard and Gough, 1984b).

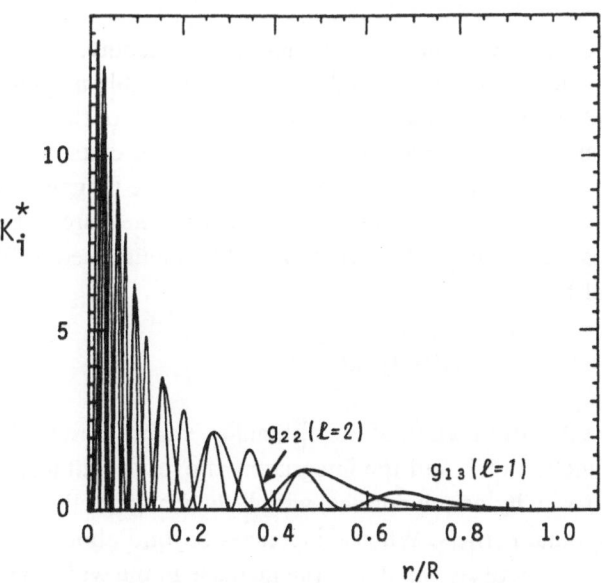

Fig. 5. Rotational splitting kernels for g_{13} ($l = 1$) and g_{22} ($l = 2$), which have cyclic frequencies 48.3 and 48.4 µHz (from Christensen-Dalsgaard and Gough, 1984b).

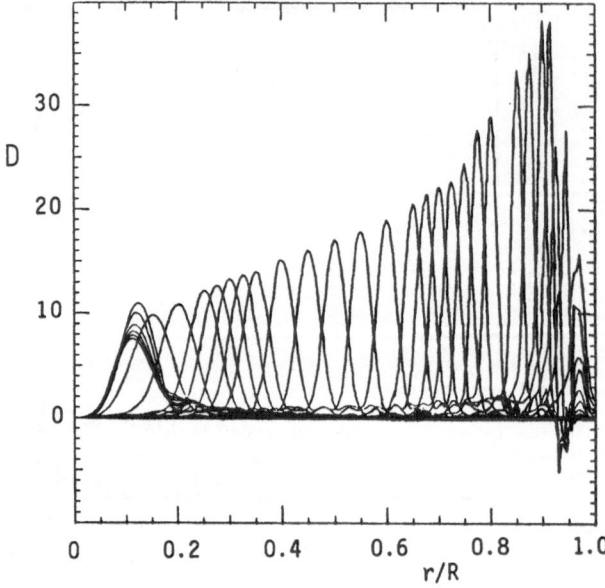

Fig. 6.   A selection of optimal averaging kernels $D(r, r_0)$, plotted against r/R, and computed from a set of 180 five-minute modes typical of those reported by Duvall and Harvey (1983) and Harvey and Duvall (1984). The modes have $1 \leqslant l \leqslant 1000$ with cyclic frequencies between 2.3 and 3.8 mHz (from Christensen-Dalsgaard and Gough, 1984b).

A tradeoff between resolution and accuracy is illustrated in Figures 7 and 8. The modes used this time all have $l \lesssim 10$, but include some low-frequency g modes. Figure 7 exhibits the results of three inversions with $\theta = 10^{-6}$, 1.5×10^{-3}, and 10^{-1}. Since no knowledge of the errors was incorporated into the procedure, $\tilde{w}$ was set to unity, with no attempt to render the terms S_{ij}^{*} and $\tilde{w}E_{ij}$ of comparable magnitude. Indeed, since it had been decided not to introduce any prior knowledge of the statistics of the errors into the inversion, E_{ij} was represented by the Kronecker delta. At the lowest value of θ resolution is potentially the greatest, but the errors are magnified to such an extent that spurious oscillations spoil the inversion. A coarse measure of the degree to which errors in the data are magnified by Equation (7.17) is illustrated in Figure 8, where the factor Λ, defined by

$$\Lambda^2 = R^{-1} \int \sum_i [\beta_i(r_0)]^2 \, d\bar{r}, \qquad (8.1)$$

is plotted against the mean width of the optimal averaging kernels. Here r_0 is regarded as an implicit function of $\bar{r}$, and the integration is over the attainable range of $\bar{r}$. The factor Λ declines rapidly as θ increases, with little change in the rms width (averaged over $\bar{r}$) of the optimal kernels. When θ increases beyond about 1.5×10^{-3}, however, the decrease in Λ is relatively small, but the increase in the width becomes substantial. Thus when $\theta = 10^{-1}$, the averages are unnecessarily broad, and the inversion lacks resolution. A finer indicator of the error magnification is provided by the integrand in

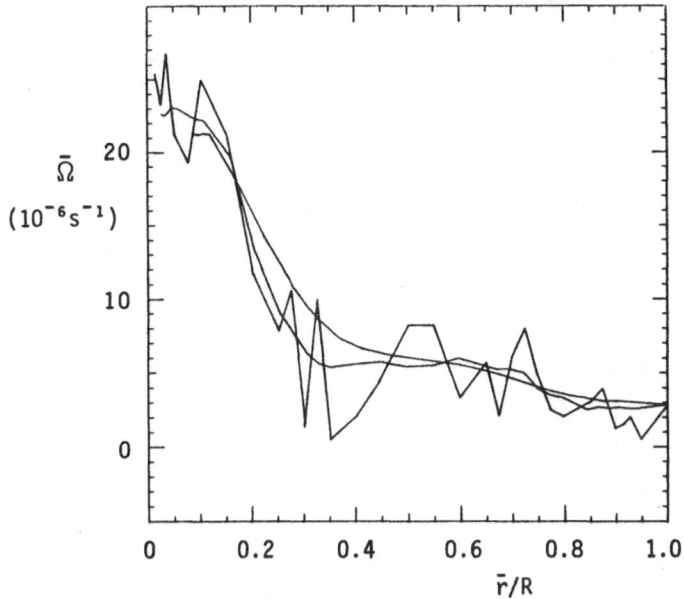

Fig. 7. Three sets of optimal averages of artificial splitting data computed from the angular velocity represented by the continuous curve in Figure 9. The averages at adjacent values of $\bar{r}$ are connected by straight lines; the sets were computed with $\theta = 10^{-6}$, 1.5×10^{-3}, and 10^{-1}. Smaller values of θ produce averaging kernels that are more localized, but which magnify the errors in the data to a greater degree and yield a more jagged inversion. Notice that it is possible to obtain averaging kernels closer to the centre of the Sun with smaller values of θ. The inversion with the largest value of θ is influenced the least by errors in the data, but it averages Ω with such broad kernels D as to conceal some of the structure. The inversion was carried out with all the modes with $1 \leqslant l \leqslant 3$ and 0.26 mHz $< \omega/2\pi < 3.8$ mHz, the modes with $4 \leqslant l \leqslant 10$ and 2.3 mHz $< \omega/2\pi < 3.8$ mHz, the modes with $l = 1$ and 2 and $0.05 < \omega/2\pi < 0.10$ mHz, and in addition the seven modes reported by Hill et al. (1982) (after Christensen-Dalsgaard and Gough, 1984b).

Equation (8.1), from which it can be seen how the errors influence the inversion at different radii. Of course when E_{ij} is known it is better to compute $\varepsilon(\bar{r}, \theta)$. Backus and Gilbert (1970) display contour plots relating ε, δ, and r_0 for a geophysical problem.

Plotted also in Figure 8 is the rms difference between the left and right sides of Equations (7.15). As expected it tends to rise with θ. The high value at $\theta = 10^{-6}$ is doubtless a result of rounding error.

The degree to which the inversion can faithfully represent the rotation curve from which the artificial data were generated is shown in Figure 9. The result of the inversion is shown as a sequence of crosses at selected radii, whose lengths are the widths δ of the optimal averaging kernels D and whose heights represent estimates of the rms errors resulting directly from errors in the data. The errors that were added to the splitting data were independent Gaussian-distributed random numbers with standard deviation $2\pi \times 10^{-8}$ s^{-1}, about 1% of the mean. They are substantially smaller than the errors in the current measurements of the splitting frequencies of the Sun, but somewhat larger than the random relative errors in the measurements of the absolute frequencies. The

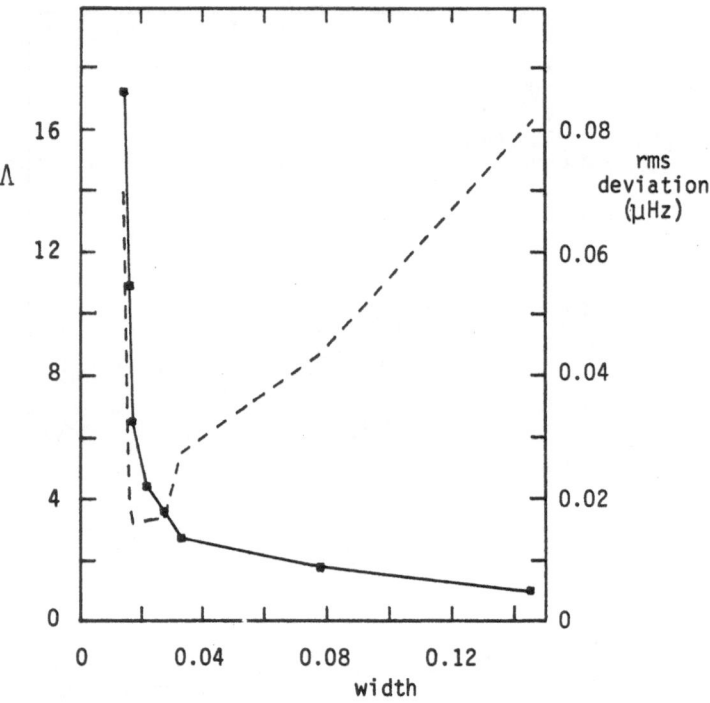

Fig. 8.   The filled squares connected by continuous straight lines define a tradeoff curve; they are error magnification factors Λ, defined by Equation (8.1), plotted against the mean width $\langle\delta\rangle$ of the localized optimal averaging kernels D for a selection of different inversions of the same artificial data. Three of the inversions are shown in Figure 7. The mean width $\langle\delta\rangle$ is the uniformly weighted average of δ over the range of $\bar{r}$ within which it is possible to centre the kernels D; it increases monotonically with θ. The values plotted are for $\theta = 10^{-6}, 5 \times 10^{-4}, 1.5 \times 10^{-3}, 5 \times 10^{-3}, 2 \times 10^{-2}$, and 10^{-1}. Notice that the tradeoff curve has a large negative slope at small θ; indeed it follows from the analysis of Backus and Gilbert (1970) that $d\Lambda/d\langle\delta\rangle \to -\infty$ as $\theta \to 0$. Thus there is always much to be gained in relaxing the demand for the smallest value of $\langle\delta\rangle$ by taking a nonzero value of θ. The dashed line represents the rms deviation from the splitting data ω_{1i} of the splittings calculated from Equation (6.6) using the approximation $\Omega(r) \simeq \bar{\Omega}(r)$, and thus represents the degree to which that approximation satisfies the data. The high values at small θ presumably result from numerical error in the inversion procedure, caused possibly by the limited spatial resolution in the computation of $\bar{\Omega}(\bar{r})$ (from Christensen-Dalsgaard and Gough, 1984b).

exercise clearly demonstrates that, like the spectral expansion, optimal averaging promises to be a useful tool for inverting helioseismic data.

9. Piecewise Constant Functions

One of the apparent disadvantages of both the spectral expansion and the optimal averaging procedure is that if, for example, one has only a few p modes of the type illustrated in Figure 4, it may not be possible to find combinations of kernels that do not oscillate with large amplitude in the surface regions. What that is telling us is that strictly speaking one does not have enough information to determine conditions at great depths. However, if one believes that Ω does not vary rapidly in the outer layers (in a

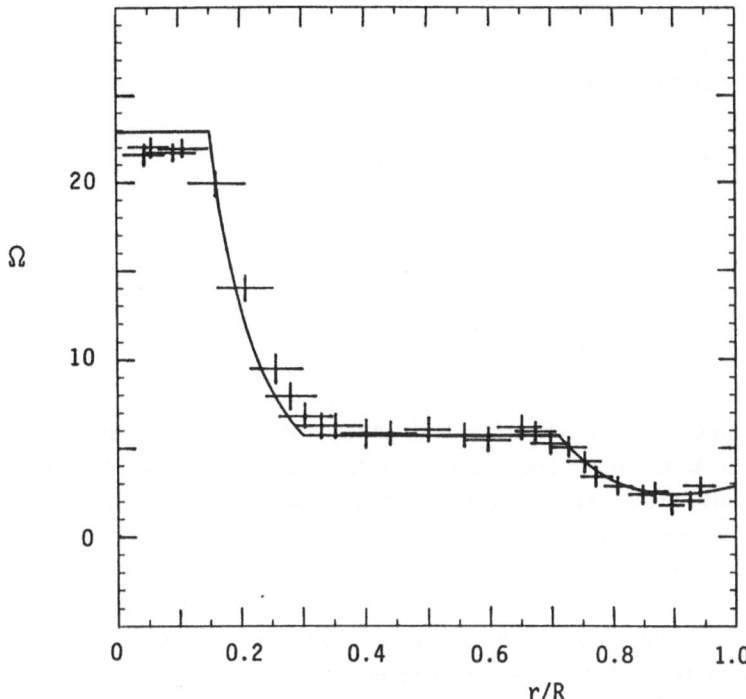

Fig. 9. The continuous line is the angular velocity $\Omega(r)$ from which the splitting frequencies ω_{1i} used in the inversions were computed. The crosses are centred at $(\bar{r}, \overline{\Omega}(\bar{r}))$ for a selection of optimal averages with $\theta = 1.5 \times 10^{-3}$; their horizontal components have length δ/R and their vertical components represent the uncertainty in the approximation $\Omega(r) = \overline{\Omega}(r)$, which was estimated by comparing different inversions. About 250 modes were used, comprising of those used to compute the averaging kernels in Figure 6 together with modes with $1 \leqslant l \leqslant 3$ and $0.36 \, \text{mHz} < \omega/2\pi < 2.3 \, \text{mHz}$, modes with $l = 1$ and 2 and $0.05 \, \text{mHz} < \omega/2\pi < 0.1 \, \text{mHz}$ and the seven modes reported by Hill *et al.* (1982) (from Christensen-Dalsgaard and Gough, 1984b).

way that corresponds to the variation of the kernels corresponding to those modes that happen to have been observed) one can extend the method by constructing a smooth estimate of Ω in the outer layers, down to $r = r_c$ say, and calculating its contribution to the integrals (7.15) for $r_c < r < R$. After those contributions are subtracted from the data w_i^* an inversion can be performed on the reduced data in the interval $0 < r < r_c$, where the variation of the kernels K_i^* is not so great. This procedure has been used with a modicum of success by Duvall *et al.* (1984) to invert rotational frequency splitting.

Another technique for avoiding the difficulty is to employ a basis other than ψ_i. The function Ω is thus represented as the sum of a finite number $I \ (<M)$ of functions $\tilde{\psi}_i$ which are chosen with some other criterion in mind. A common choice is the set of piecewise constant functions:

$$\tilde{\psi}_i(r) = \begin{cases} 1 & \text{if } r_i < r < r_{i+1} \\ 0 & \text{otherwise.} \end{cases} \qquad (9.1)$$

Then if Ω is represented by

$$\tilde{\Omega}(r) = \sum_{i=1}^{I} \Omega_i \tilde{\psi}_i(r) \tag{9.2}$$

the constraints (7.1) reduce to

$$w_i = \sum_{j=1}^{I} B_{ij} \Omega_j, \tag{9.3}$$

where

$$B_{ij} = \int_{r_j}^{r_{j+1}} K_i(r) \, \mathrm{d}r. \tag{9.4}$$

In this case the coefficients Ω_i estimate the mean value of Ω in the interval $r_i < r < r_{i+1}$. Of course in practice the spectral expansion and the optimal averaging outlined in Section 7 must be carried out numerically with Ω, K_i^* and ψ_i represented only at a finite number of discrete points. Therefore the inverse problem they solve is actually quite similar to that embodied in Equations (9.1)–(9.4), and is identical to it if the trapezoidal rule is used to evaluate the integrals. Indeed some authors (e.g. Wiggins, 1972) prefer to make this discretization at the outset, so that the inverse methods can be discussed in terms of finite-dimensional vector spaces rather than function spaces. In that case one adopts a mesh spacing that is everywhere smaller than the resolving length of the data, in order to be sure of attaining the maximum resolution; I is therefore greater than the number I' of effectively independent combinations of the data once errors are taken into account, so the problem is still essentially underdetermined. In contrast, the idea here is to choose the dissection $\{r_i\}$ to correspond to the resolution one hopes to achieve, so that $I \simeq I'$. Of course one cannot find a set of coefficients that satisfy all the constraints (9.3) precisely, for formally the constraints overdetermine the problem. Instead an appropriate best fit must be found, by minimizing, for example,

$$\chi^2 = \sum_{i=1}^{M} \sigma_i^{-2} \left(w_i - \sum_{j=1}^{I} B_{ij} \Omega_j \right)^2, \tag{9.5}$$

if the errors in the data are believed to be independent, σ_i being the standard deviations of those errors. This leads to the equation:

$$\sum_{i=1}^{M} \sigma_i^{-2} \left(\sum_{k=1}^{I} B_{ik} \Omega_k - w_i \right) B_{ij} = 0. \tag{9.6}$$

The method has been used by Duvall et al. (1984) and Leibacher (1984) to invert averaged rotational splitting data.

One of the immediate problems with the method is that one does not have an obvious simple way of choosing the dissection $\{r_i\}$. Duvall et al. (1984) took 10 intervals that were uniformly distributed except near the surface, where the Backus–Gilbert resolving power indicates that greater resolution is possible, and near the centre, where resolution is extremely poor. It was thus possible to obtain a smooth solution that was not highly sensitive to the representation of Ω in the surface layers. The result is shown in Figure 10.

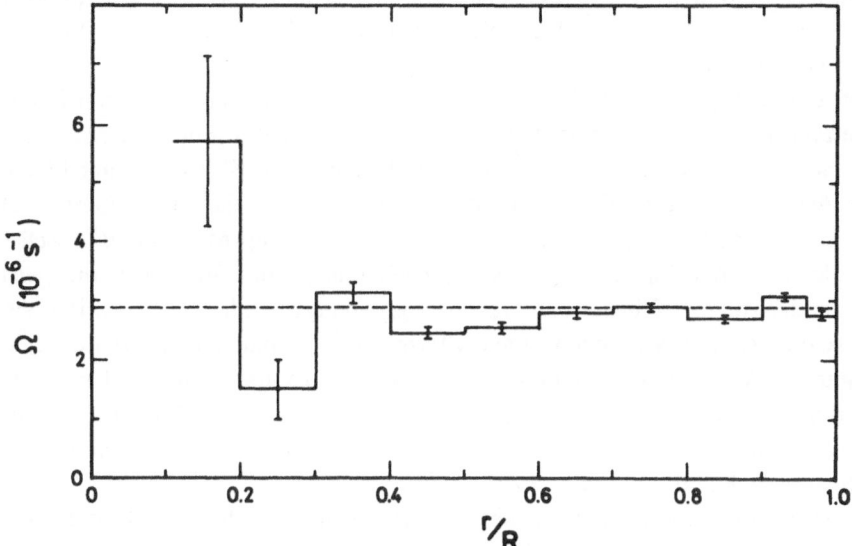

Fig. 10.   Piecewise constant estimate of the solar angular velocity in the equatorial plane. The dashed line is at the equatorial angular velocity of the photosphere. The error bars represent the standard errors of the averages Ω_i arising from the estimated errors in the data; the errors are a function of the dissection of r, and are not independent (after Duvall *et al.*, 1984).

It was essential to the success of the method that the dissection intervals in the surface layers were substantially greater than the characteristic scale over which the splitting kernels varied. This excluded unresolvable variation from the solution. The contribution from random errors in the data to the errors in Ω_α were estimated by a Monte Carlo calculation. In addition there are errors arising from the intrinsic deficiency of the representation, which may be rather greater than those indicated by the error bars in Figure 10.

10. Discussion

Two different principles underlie the inversion techniques described in this review. They can lead to rather different representations of the function Ω, say, that one is trying to measure. Which to choose is largely a matter of taste.

One of the principles is the basis of the spectral expansion. Here one isolates a function-subspace $\mathcal{O}'$ that can influence the data to a measurable extent, and determines the projection $\tilde{\Omega}$, say, of Ω in that subspace. In so doing, one takes due account of the inaccuracy of the data, excluding from $\mathcal{O}'$ any subspace into which the measured projection of Ω is dominated by the estimated errors. Evidently the procedure is not unique, because it depends on how one carries out the projection. Those discussed explicitly here are inner products on the radial coordinate r, but one can use other projections by stretching the independent variable, or equivalently by introducing a weighting function $\tilde{\mu}(r)$ into the definition of the inner product and appropriately modifying the kernels K' to leave the constraints (7.9) unchanged. This alters the subspace that is spanned by the basis functions ψ_i.

One of the properties of the spectral expansion is that the representation $\tilde{\Omega}$ of Ω tends to contain spurious small-scale structure. This is apparent in Figures 2 and 3. It results from the fact that the kernels K' from which the subspace $\mathcal{O}'$ is determined have much more detailed structure than can ever be resolved by the finite number of their projections. The Gibbs phenomenon in Fourier-series representations is a well known example of such behaviour. It presents problems particularly when one wishes to consider derivatives of the function Ω. There is usually a tendency for spurious structure to arise on a small scale when too many terms in the expansion (7.13) are retained, as in Figure 3d. These terms should have been assigned to the complement $\mathcal{A}'$ of $\mathcal{O}'$. One should not assume from this, however, that all the functions in $\mathcal{A}'$ are necessarily more rapidly varying than those in $\mathcal{O}'$. (A trivial counterexample can be found in attempting to infer the density distribution of a taut string with fixed ends from the odd overtones of transverse oscillation, when the measured frequencies are found to form an harmonic sequence.)

Evidently one needs to know the length scale below which one cannot rely on the details of the inversion. Thus Backus and Gilbert (1968) introduced the concept of resolving power, which is described in terms of the width δ of the optimal averaging kernel D. The average $\overline{\Omega}$ of Ω, weighted by the kernel D, is easy to interprete because it is a well defined function of the data. It is a smoothed representation of Ω which, therefore, in contrast to the spectral expansion, shows less fine structure than the function it represents (see Figure 9).

Omitting any feature that cannot be resolved by the data is the other principle which is commonly used to select a representation of Ω. In the context of this discussion the most natural choice is perhaps an optimal average $\overline{\Omega}(\bar{r})$. It has the advantage that its meaning can easily be appreciated, especially if it is viewed in conjunction with the averaging kernels D from which it was constructed. Notice that because the number of data is finite, the optimum width δ has a lower bound. Therefore even if the accuracy of the data were perfect, $\overline{\Omega}(r)$ would not in general be equal to $\Omega(r.)$ In particular, the approximation $\Omega(r) = \overline{\Omega}(r)$ would not satisfy the integral constraints (7.1).

Other recipes can be used to obtain smooth representations of Ω. For example, one can choose the flattest or the smoothest function Ω by minimizing for some specified weighting function $\tilde{\mu}(r)$ the quantities

$$S_1 = \int_0^R \left(\frac{d\Omega}{dr}\right)^2 \tilde{\mu}\, dr \quad \text{or} \quad S_2 = \int_0^R \left(\frac{d^2\Omega}{dr^2}\right)^2 \tilde{\mu}\, dr \tag{10.1}$$

subject to the constraints (7.1). As is the case with the other methods, in the face of errors one would prefer a tradeoff, permitting a relaxation in the precision to which the constraints are satisfied in return for a substantial reduction in S_1 or S_2; there is no good reason to demand that the constraints be satisfied more precisely than they can be measured. In common with the spectral expansion, the resulting representation is expressed as a series, but this time in terms of multiple integrals of the kernels K_i, which tend to be smoother than the raw kernels. Note that a degree of smoothness is written into piecewise constant functions at the outset. The asymptotic approximations discussed here automatically incorporate smoothness, for they approximate the correct solution well only when the scale of variation of the equilibrium state is much greater than the oscillation wavelengths. In this case errors are accommodated by taking smooth curves through appropriate plots of the data.

The inability of an optimal average $\overline{\Omega}$ to satisfy the constraints (7.1) arbitrarily closely when the data are presumed to be perfectly accurate has led to the idea that $\overline{\Omega}$ is inferior to other representations. This can hardly be true. Provided it is realised that $\overline{\Omega}$ is really an average of Ω, weighted with a known function D, and not a possible value of Ω, it stands as one of the most simply defined and easily understandable representations of what the data imply, especially when D is well localized. To find a solution that happens to satisfy the constraints precisely might even be misleading; after all, it is merely a single example of the infinitely many functions that share this property.

I am not advocating that one should compute only $\overline{\Omega}$. Optimal averages can be difficult and expensive to compute, especially when the constraints are numerous. Besides, other representations can offer additional useful information, provided the assumptions upon which they depend are consciously recognised. For example, with only 37 averaged rotational splitting data from five-minute modes with kernels similar to those illustrated in Figure 4, Duvall *et al.* (1984) were unable to obtain well localized averaging kernels D centred at radii less than about $0.7R$. That is no surprise, because one would hardly expect amongst so few kernels precise cancellation of the large contributions that come from the surface layers. To estimate the angular velocity at greater depths it was necessary to assume that in the convection zone Ω varied more smoothly than was warranted by the data alone. This illustrates what is undoubtedly a strength of the Backus–Gilbert approach: its failure to localize the information deep in the Sun forced an explicit recognition that an additional assumption was required to complete the inversion.

It is interesting to consider the computational effort required for the various kinds of inversion discussed in this review. Without any doubt the fastest and simplest is the asymptotic analysis, which requires no more than the fitting of a curve to the data and the evaluation of an integral. Even though it is probably the least accurate, the ease of execution makes it a particularly useful tool for a preliminary evaluation of the data. Next is the fitting of a piecewise constant function, which requires the computation and inversion of a matrix whose order is the number I of segments in the dissection of the independent variable. However, it must be recognised that it is also necessary to choose a suitable dissection, and to estimate the errors in the representation, and that can be a much more difficult and time-consuming task. Monte Carlo techniques can be used here, and when the number of data is very large may provide a much faster means of obtaining a good result than the resolving power calculation of Backus and Gilbert. Of course in addition one must compute the oscillation eigenfunctions, as is necessary also for the methods about to be mentioned; when there are many modes, that is quite a considerable undertaking. The spectral expansion and the evaluation of the series resulting from the minimization of S_1 and S_2 require the computation and inversion (or at least finding the largest eigenvalues and their associated eigenvectors) of matrices of order $M \times M$, where M is the number of constraints. This requires a substantial amount of computational effort when M is large. Computing optimal kernels also requires inverting $M \times M$ matrices, but this time at each value of the independent variable r_0; of the methods discussed in this review it is the most expensive, but on the other hand it is the one that gives the most information about the value of the inversion. It should be noted, however, that for linear problems such as determining the angular velocity of the sun when its hydrostatic stratification is assumed known, the matrices (and their inverses) are independent of the data (though they do depend on the estimated errors in the data). Therefore they need be computed only once, and can thence be used on all data sets with similar errors.

What are the tasks for the immediate future? The extensive data collected by Duvall and Harvey from which the sound speed has already been estimated by an asymptotic inversion must surely be analysed more fully. Inclusion of higher terms in the asymptotic expansion will improve the estimates. However, the data should also be subjected to one or more of the other inversions. This will involve a confrontation with the equation of state, which in the first instance should provide an estimate of the solar helium abundance; indeed steps in this direction have already been attempted using asymptotic inversions (Dappen and Gough, 1985). In the longer term, possibly with more precise data, it may permit the testing and calibration of theories of dense plasmas from which the equation of state is computed. The splitting data from a wide range of tesseral modes reported recently by Brown (1985a, b) promises for the first time the potential to infer the latitudinal dependence of the solar angular velocity, and possibly of other properties of the stratification. This is likely to be carried out first by expanding the θ dependence of $\Omega(r, \theta)$ in Legendre functions, or powers of $\cos \theta$, and inverting for the radial dependence of the coefficients by a direct generalization of the methods described above. Asymptotic expansions that can be inverted to yield $\Omega(r, \theta)$ as integrals in closed form are likely to be useful in this context. Large-scale magnetic fields, meridional flows and giant convective cells must also be studied in more detail. Further theoretical work on what can be inferred in principle from particular data sets, and conversely on what modes need to be measured to answer particular questions, must be undertaken, so that observers might assess what measurements would best be made to reap the greatest scientific rewards. Some emphasis should be placed on the importance of redundancy, particularly in connection with the exposure of misidentified modes.

This article has been concerned solely with the analysis of frequencies of oscillation. Analysis of the shapes and time variations of the lines in the power spectrum will no doubt in the future permit oscillation spectroscopy to shed light on the mechanisms that drive and damp the modes. That in turn will hopefully improve our understanding of the complicated nonlinear fluid dynamics of the Sun, particularly in the convection zone.

Of course the purpose of helioseismology is not merely to measure the stratification of the Sun; it is primarily to improve our theoretical understanding of the structure and evolution of the Sun, and of the other stars. Thus along with advances in the measurements must come new theoretical investigations that are posed by the results, which in turn will influence new measurements. The recent inferences of the sound speed and the angular velocity have already prompted theoretical speculation (e.g. Gough, 1985b; Rosner and Weiss, 1985) that relates to the rotational and thermal history of the Sun, the solar cycle and the neutrino problem. Hopefully this will stimulate the development of further sound theoretical investigation, with implications concerning the theory of stellar evolution in particular, and astronomy and cosmology in general. Of course observations of the oscillations of other stars will greatly assist this endeavour. The subject is yet in its infancy, but already it promises to expand into a fertile avenue of scientific enquiry.

Acknowledgements

Much of the content of this review has arisen, either directly or indirectly, from collaborations with J. Christensen-Dalsgaard, F. Hill, and J. Toomre, to whom I am greatly indebted. I thank S. Frandsen, M. E. McIntyre, J. Perdang, and K. Whaler for fruitful conversations.

References

Backus, G. and Gilbert, F.: 1967, *Geophys. J. Roy. Astron. Soc.* **13**, 247.
Backus, G. and Gilbert, F.: 1968, *Geophys. J. Roy. Astron. Soc.* **16**, 169.
Backus, G. and Gilbert, F.: 1970, *Phil. Trans.* **266A**, 123.
Berthomieu, G. and Provost, J.: *Astron. Astrophys.*, submitted.
Brown, T. M.: 1985a, *Nature*, in press.
Brown, T. M.: 1985b, in D. O. Gough (ed.), *Seismology of the Sun and the Distant Stars*, D. Reidel Publ. Co., Dordrecht, Holland, in press.
Christensen-Dalsgaard, J.: 1982, *Monthly Notices Roy. Astron. Soc.* **199**, 735.
Christensen-Dalsgaard, J.: 1984, in A. Noels and M. Gabriel (eds.), *Theoretical Problems in Stellar Stability and Oscillations*, Univ. Liège, p. 155.
Christensen-Dalsgaard, J.: 1985, in D. O. Gough (ed.), *Seismology of the Sun and the Distant Stars*, D. Reidel Publ. Co., Dordrecht, Holland, in press.
Christensen-Dalsgaard, J. and Frandsen, S.: 1983, *Solar Phys.* **82**, 165.
Christensen-Dalsgaard, J. and Gough, D. O.: 1984a, in R. K. Ulrich, J. W. Harvey, E. J. Rhodes, Jr., and J. Toomre (eds.), *Solar Seismology from Space*, Jet Propulsion Laboratory Publication 84–84, Pasadena, p. 199.
Christensen-Dalsgaard, J. and Gough, D. O.: 1984b, in R. K. Ulrich, J. W. Harvey, E. J. Rhodes, Jr., and J. Toomre (eds.), *Solar Seismology from Space*, Jet Propulsion Laboratory Publication 84–84, Pasadena, p. 79.
Christensen-Dalsgaard, J., Dziembowski, W. A., and Gough, D. O.: 1980, in H. A. Hill and W. A. Dziembowski (eds.), *Nonradial and Nonlinear Stellar Pulsation*, Springer, Heidelberg, p. 313.

Christensen-Dalsgaard, J., Duvall Jr, T. L., Gough, D. O., Harvey, J. W., and Rhodes Jr., E. J.: 1985, *Nature* **315**, 378.

Cox, J. P.: 1980, *Theory of Stellar Pulsation*, Princeton Univ. Press.

Däppen, W. and Gough, D. O.: 1985, in D. O. Gough (ed.), *Seismology of the Sun and the Distant Stars*, D. Reidel Publ. Co., Dordrecht, Holland, in press.

Denis, A. I. and Denis, C.: 1984, in A. Noels and M. Gabriel (eds.), *Theoretical Problems in Stellar Stability and Oscillations*, Univ. Liège, p. 270.

Deubner, F.-L. and Gough, D. O.: 1984, *Ann. Rev. Astron. Astrophys.* **22**, 593.

Duvall, Jr, T. L.: 1982, *Nature* **300**, 242.

Duvall, Jr, T. L. and Harvey, J. W.: 1983, *Nature* **302**, 24.

Duvall, Jr, T. L. and Harvey, J. W.: 1984, *Nature* **310**, 19.

Duvall, Jr, T. L., Dziembowski, W. A., Goode, P. R., Gough, D. O., Harvey, J. W., and Liebacher, J. W.: 1984, *Nature* **310**, 22.

Dziembowski, W. A. and Goode, P. R.: 1984, *Mem. Soc. Astr. Italiana* **55**, 185.

Dziembowski, W. A. and Pamjatnykh, A. A.: 1978, in Rösch (ed.), *Pleins Feux sur la Physique Solaire*, Centre National de la Recherche Scientifique, Paris, p. 135.

Ellis, A. N.: 1984, in A. Noels and M. Gabriel (eds.), *Theoretical Problems in Stellar Evolution and Oscillations*, Univ. Liège, p. 290.

Ellis, A. N.: 1985, *Monthly Notices Roy. Astron. Soc.*, submitted.

Fossat, E.: 1985, in E. Rolfe and B. Battrick (eds.), *Future Missions in Solar, Heliospheric and Space Plasma Physics*, ESA SP-235, European Space Agency, Noordwijk, p. 209.

Gabriel, M.: 1985, in D. O. Gough (ed.), *Seismology of the Sun and the Distant Stars*, D. Reidel Publ. Co., Dordrecht, Holland, in press.

Gough, D. O.: 1978, in G. Belvedere and L. Paternò (eds.), *Proc. Workshop on Solar Rotation*, Univ. Catania Press, p. 255.

Gough, D. O.: 1981, *Monthly Notices Roy. Astron. Soc.* **196**, 731.

Gough, D. O.: 1982, *Nature* **298**, 334.

Gough, D. O.: 1983, in P. A. Shaver, D. Kunth, and K. Kjär (eds.), *Proc. Primordial Helium Workshop*, European Southern Observatory, Garching, p. 117.

Gough, D. O.: 1984a, *Adv. Space Res.* **4**, 85.

Gough, D. O.: 1984b, *Phil. Trans.* **313A**, 27.

Gough, D. O.: 1984c, in R. K. Ulrich, J. W. Harvey, E. J. Rhodes, Jr., and J. Toomre (eds.), *Solar Seismology from Space*, Jet Propulsion Laboratory Publication 84–84, Pasadena, p. 49.

Gough, D. O.: 1985a, in B. Chen and C. de Jager (eds.), *Proc. International Workshop on Solar Physics and interplanetary Travelling Phenomena*, Kunming, November 1983, in press.

Gough, D. O.: 1985b, in E. Rolfe and B. Battrick (eds.), *Future Missions in Solar, Heliospheric and Space Plasma Physics*, ESA SP-235, European Space Agency, Noordwijk, p. 183.

Gough, D. O. and Taylor, P. P.: 1984, *Mem. Soc. Astron. Italiana* **55**, 215.

Hansen, D. J., Cox, J. P., and Van Horn, J. H.: 1977, *Astrophys. J.* **217**, 151.

Harvey, J. W. and Duvall, Jr., T. L.: 1984, in R. K. Ulrich, J. W. Harvey, E. J. Rhodes, Jr., and J. Toomre (eds.), *Solar Seismology from Space*, Jet Propulsion Laboratory Publication 84–84, Pasadena, p. 165.

Hill, F., Gough, D. O., and Toomre, J.: 1984a, *Mem. Soc. Astron. Italiana* **55**, 153.

Hill, F., Gough, D. O., and Toomre, J.: 1984b, in R. K. Ulrich, J. W. Harvey, E. J. Rhodes, Jr., and J. Toomre (eds.), *Solar Seismology from Space*, Jet Propulsion Laboratory Publication 84–84, Pasadena, p. 95.

Hill, H. A., Bos, R. J., and Goode, P. R.: 1982, *Phys. Rev. Letters* **49**, 1794.

Isaak, G. R.: 1985, in D. O. Gough (ed.), *Seismology of the Sun and the Distant Stars*, D. Reidel Publ. Co., Dordrecht, in press.

Jackson, D. D.: 1972, *Geophys. J. Roy. Astron. Soc.* **28**, 97.

Jackson, D. D.: 1973, *Geophys. J. Roy. Astron. Soc.* **35**, 121.

Keller, J. B. and Rubinow, S. I.: 1960, *Ann. Phys.* **9**, 24.

Lamb, H.: 1908, *Proc. London Math. Soc.* **7**, 122.

Lamb, H.: 1932, *Hydrodynamics*, Cambridge Univ. Press.

Lanczos, C.: 1961, *Linear Differential Equations*, Van Nostrand, London.

Ledoux, P. and Walraven, T.: 1958, *Handb. Physik* **51**, 353.

Leibacher, J. W.: 1984, in A. Noels and M. Gabriel (eds.), *Theoretical Problems in Stellar Evolution and Oscillations*, Univ. Liège, p. 298.

Lynden-Bell, D. and Ostriker, J.: 1967, *Monthly Notices Roy. Astron. Soc.* **136**, 293.

Parker, R. L.: 1977a, *Ann. Rev. Earth Planet Sci.* **5**, 35.

Parker, R. L.: 1977b, *Rev. Geophys. Space Phys.* **15**, 446.

Perdang, J.: 1985, in D. O. Gough (ed.), *Seismology of the Sun and the Distant Stars*, D. Reidel Publ. Co., Dordrecht, Holland, in press.

Provost, J.: 1984, in A. Maeder and A. Renzini (eds.), 'Observational Tests of Stellar Evolution Theory', *IAU Symp.* **105**, 47.

Rosner, R. and Weiss, N. O.: 1985, *Nature*, submitted.

Tassoul, M.: 1980, *Astrophys. J. Suppl.* **43**, 469.

Ulrich, R. K., Harvey, J. W., Rhodes, Jr., E. J., and Toomre, J. (eds.): 1984, *Solar Seismology from Space*, Jet Propulsion Laboratory Publication 84–84, Pasadena.

Unno, W., Osaki, Y., Ando, H., and Shibahashi, H.: 1979, *Nonradial Oscillations of Stars*, Univ. Tokyo Press.

Vandakurov, Yu. V.: 1967, *Astron. Zh.* **44**, 786.

van der Raay, H. B., Palle, P. L., and Roca Cortes, T.: 1985, in D. O. Gough (ed.), *Seismology of the Sun and the Distant Stars*, D. Reidel Publ. Co., Dordrecht, in press.

Wiggins, R. A.: 1972, *Rev. Geophys. Space Phys.* **10**, 251.

Zahn, J.-P.: 1970, *Astron. Astrophys.* **4**, 452.

THE 160 MINUTES OSCILLATIONS

V. A. KOTOV

Crimean Astrophysical Observatory, Nauchny, Crimea 334413, U.S.S.R.

Abstract. We describe basic observational data regarding the 160-min oscillations of the Sun as well as the implications for helioseismology. The most acceptable theoretical interpretation seems to be a resonant interaction of gravity *g*-mode oscillations of the solar model with a slight modification to the equilibrium structure (with low heavy element abundance).

It is noted also that there is significant 160-min commensurability over the solar system; e.g., spin rates of main and minor planets prefer, statistically, to be integer multiples of the 160-min period. The same period appears to be the most 'characteristic' period (in the range studied from about 110 to 830 min) for the distribution of orbital periods of close binary stars. Allowing for these facts various non-classical suggestions as to possible nature of the 160-min period are briefly reviewed.

1. Introduction

Observations of global oscillations of the Sun can provide direct information about the solar interior and therefore present a new type of diagnostics to test the properties of theoretical models; for reviews on helioseismology see, e.g., Leibacher and Stein (1981), Deubner and Gough (1984), Christensen-Dalsgaard (1984), Pecker (1984), Gough (1985). An attempt to detect low-frequency oscillations by measuring Doppler shifts in the solar spectrum was undertaken in 1974 by two groups of observers working in the Crimean Astrophysical Observatory and in Birmingham University. Using a solar magnetograph specially modified for these observations Severny *et al.* (1976) measured the difference in Doppler shift between the central portion of the solar disk and an outer annulus. Brookes *et al.* (1976) used an optical resonance technique to get a spectrum of oscillations with integral light from the entire solar disk.

In the same issue of *Nature* both groups reported the discovery of a particular oscillation with a dominant period near 160 min. The amplitude was claimed to be of the order of 1 m s^{-1}, i.e. near the limits of the observing techniques. The publications led to wide discussions regarding the true origin of the periodic signal, mainly due to the circumstances that (1) the period is too long to be the fundamental radial mode of solar oscillations and (2) it happens to be close to 1/9th of a day.

2. Observations from Different Sites

The first Crimean and Birmingham results agreed well with respect to the value of period P and amplitude A:

> Crimea: $P = 160.0 \pm 0.5$ min; $\quad A = 1.42 \pm 0.21$ m s^{-1},
> Birmingham: $P = 159 \pm 3$ min; $\quad A = 2.70 \pm 0.24$ m s^{-1},

with a satisfactory agreement also in phase (a discrepancy by factor of two in the amplitudes can easily be attributed to differences in calibration, averaging of the velocity

Solar Physics **100** (1985) 101–113. 0038–0938/85.15.

across the Sun's disk, etc., see Christensen-Dalsgaard and Gough (1982), Kotov *et al.* (1983)). On the basis of the observations the authors proposed that the damping is small ($Q \gtrsim 10^3$) as the oscillations seemingly kept an initial phase stable over many days and months.

The first author's suggestion – about the possible radial mode – was immediately given up since it requires a very drastic change in the solar model. Accordingly, a variety of terrestrial sources (e.g., distortions caused by the Earth's atmosphere and those associated with solar noise, etc.) of a potential 1/9-day period and errors were proposed by several authors: (a) low-frequency filtering of daily trends and day-night alternations of observations (Dittmer, 1977); (b) the influence of differential atmospheric extinction (Grec and Fossat, 1979); (c) supergranule velocity field rotating into and out of the instrumental field of view (Worden and Simon, 1976); (d) a by-product of 5-min solar oscillations through beating with 5-min averaging of the data (Delache, 1981); (e) an effect of a so-called quasi-persistency present in the data (Forbush *et al.*, 1983), etc. Consequently, thorough analyses of all spurious sources were performed, however, no sources for a false 160-min periodicity were found (Kotov *et al.*, 1978, 1983; Severny *et al.*, 1979; Koutchmy *et al.*, 1980; Rachkovsky, 1985).

The difficulties in the interpretation of the power spectra of low-frequency solar oscillations were stressed later by Connes (1984). However, his arguments can not explain the dominant character and long-time phase-coherency (see below) of the 160-min oscillation. The main and principal reasons in favour of a solar origin are well known: (a) a firmly established difference between the observed period P_0 = 160.010 min and 1/9 of a day; (b) the phase coherency of the 160-min oscillations observed from different sites on the Earth.

The differential velocity observations made in 1976–1980 at Stanford (Scherrer and Wilcox, 1983) showed the same period P_0 with a phase of maximum expansion velocity as predicted by the Crimean observations. The power spectrum (PS) of the combined, Crimea plus Stanford, data set (for the interval 1974–1980) calculated for the frequency range 60 to 235 µHz exhibited the highest peak at a period of exactly 160.010-min period (a portion of this PS is shown in Figure 1). The Crimean and Stanford results, thus, give the most credibility to solar origin of the 160-min period oscillation.

Further, to eliminate the difficulties caused by the day-night cycle and the daily trends inevitable at mid-latitudes, Grec *et al.* (1980) have taken advantages of observations from the geographic South pole where they succeeded in obtaining a 5-days interval of uninterrupted Doppler measurements. For the range 30 to 300 min the PS of these data showed an increase of power with decreasing frequency, with the spiked pattern consistent with noise fluctuations (we note however that one of four major peaks in the range $P < 170$ min corresponds to a period of about 161 min). Then Grec *et al.* applied a superposed epoch procedure and obtained the 160-min wave which fairly well matched the sinusoidal extrapolation of the average result of the Crimean and Stanford observations (Figure 2).

The earlier conclusion (Severny *et al.*, 1976; Brookes *et al.*, 1976) about the long-term

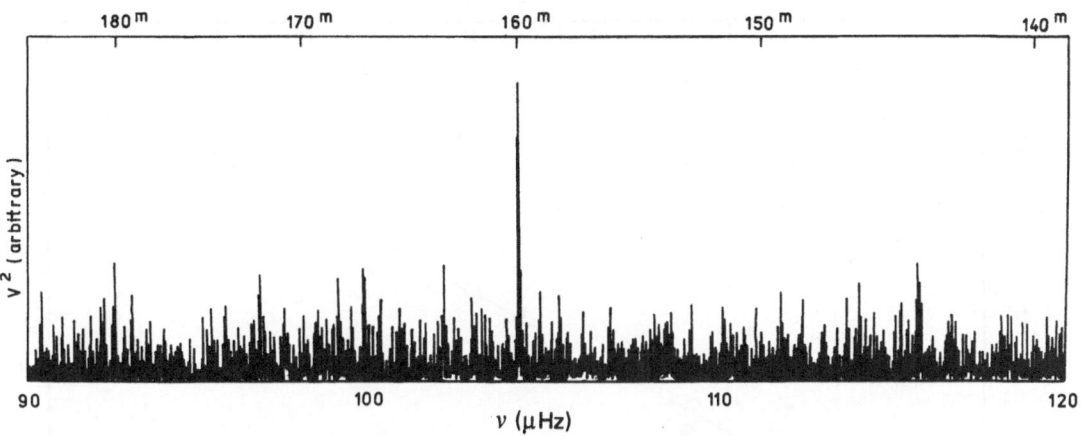

Fig. 1. The power spectrum of solar oscillations computed for combined, Crimea (1974–1979) and Stanford (1977–1980), data (courtesy of P. H. Scherrer).

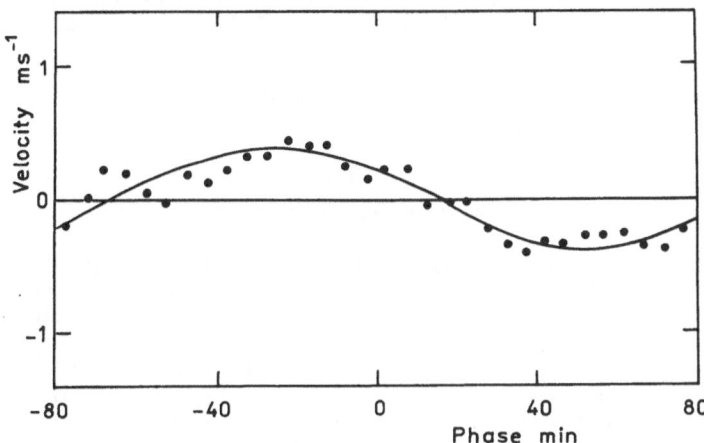

Fig. 2. Results of the superposed epoch analysis of the 160-min oscillation observed at the South pole (dots), Crimea and Stanford (average sinusoidal approximation for two latter observatories; solid line). (From Grec et al., 1980.)

phase stability is stressed by Figure 3 which demonstrates a continued agreement in phase between observations from different observatories. Using all the data (a summary of the observations is given in Table I) one can find a linear regression line yielding a period of

$$P_0 = 160.0102 \pm 0.0005 \text{ min.}$$

It was noted by Brookes et al. (1978) and Severny et al. (1979) that the observed amplitude varies significantly from day to day and also from year to year (Scherrer et al., 1980); the oscillations can almost disappear for some time intervals. It is thought that the variations are due to masking by velocity fields of supergranules, giant convective

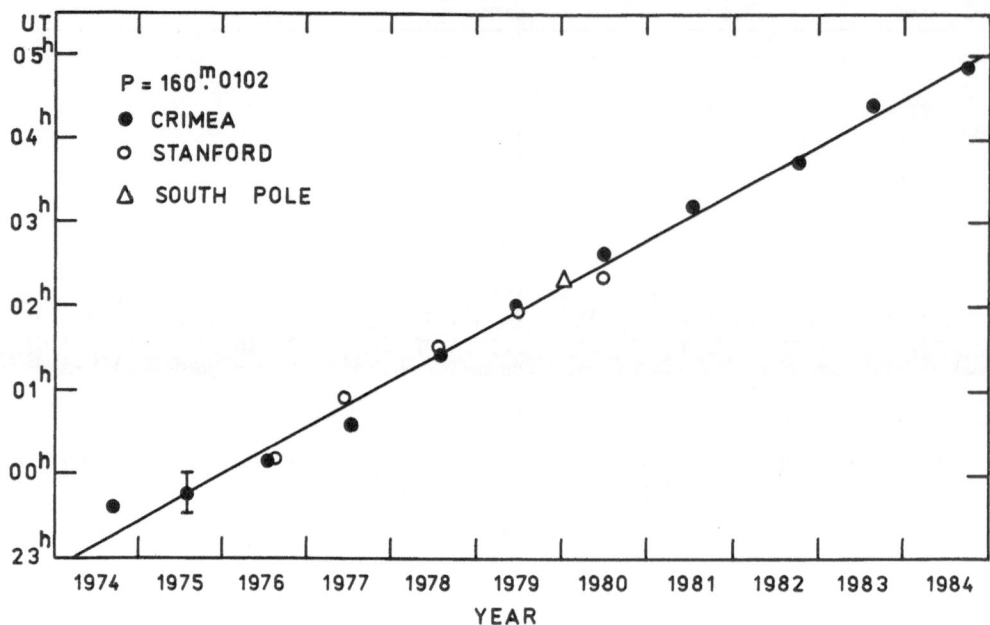

Fig. 3.   The phase of the 160-min oscillation observed at three different stations.

TABLE I

Summary of Doppler observations

Observatory	Time interval	Days	Hours	Harmonic amplitude[a] (m s^{-1})	UT of maximum velocity[a]
Crimea	1974–1984	510	2951	0.54 ± 0.06	01^{h}37^m (± 3 min)
Stanford	1976–1980	≈ 175	≈ 895	0.22 ± 0.02	01^{h}37^m (± 2 min)
South pole	1979–1980	5	121	0.33 ± 0.04	01^{h}41^m (± 3 min)

[a] The results of superposed epoch analysis with folding period of 160.010 min (Crimea: for the interval 1974–1982; Stanford: for the interval 1976–1979). The time of maximum velocity is determined with respect to a zero moment taken at 00^{h}00^m UT, on 1 January 1974.

cells, etc., and also due to modulation arising from the rotation of the Sun (Severny *et al.*, 1978). Several authors have announced the detection of the 160-min variations in the solar IR brightness (Koutchmy *et al.*, 1980), radioemission (Efanov *et al.*, 1983) and in some geomagnetic parameters (Toth, 1977; Vladimirsky *et al.*, 1983). (Notice that the IR measurements were recently recalibrated (Kotov *et al.*, 1984) and the new value of the amplitude, $\sim 9 \times 10^{-6}$ for the relative brightness variations does not severely contradict the upper limit of $\sim 5 \times 10^{-6}$ placed on 160-min amplitude by the SMM-spacecraft bolometric measurements, see Woodard and Hudson (1983).)

More recently, Hill and Tash (1984) reported the detection of the 160-min oscillation in the SCLERA solar diameter-type observations which implies that the spherical harmonic degree l and the azimuthal order m of the oscillation are both even. This finding may have important implications for the interpretation of Doppler shift observations.

Further details of the observations and discussions of the issues as regards the Sun's interior structure are given by Severny *et al.* (1979), Kotov *et al.* (1983) and references therein.

3. Interpretation

At present the solar nature of 160-min oscillations is generally accepted but there is a large controversy concerning theoretical explanation.

Under no circumstances can the 160-min oscillation be a p-mode for standard solar models (Christensen-Dalsgaard *et al.*, 1983). Discarding the interpretation of the oscillations as a fundamental radial mode of a nearly homogeneous star, Christensen-Dalsgaard and Gough (1976) indicated simultaneously that it can readily be ascribed to a high order ($n = 10$ or 11), quadrupole gravity g-mode oscillation of a nearly standard solar model with an abundance of heavy elements Z of about 0.02 to 0.04 (see also Vorontzov and Zharkov, 1978). Iben and Mahaffy (1976) suggested that the resonance of a beat of the fundamental and first overtone radial mode with some higher g-modes might also produce enhanced oscillations with a period near 160 min.

However, the interpretation in terms of normal g-modes suffers from a lack of knowledge about the proper mode (l and m) in the spherical harmonic expansion Y_l^m and the true mechanism responsible for the excitation. The major problem is connected with the necessity to explain the predominant character of this oscillation, while the theory usually predicts abundant modes within the frequency domain studied. This circumstance stimulated attempts to describe the 160-min period in other ways such as:

(a) a black hole oscillating around the center of the Sun;

(b) an isolated frequency resulted from coupling of the Sun to an external clock-mechanism (for the reference see Leibacher and Stein, 1981);

(c) a 'clock' of high precision inside the Sun which should exist in a fast rotating core, with period near 160 min (Kotov, 1979; unpublished);

(d) periodic surface disturbances induced by an object consisting of so-called y-matter and orbiting inside the Sun (Blinnikov and Khlopov, 1983);

(e) a g-mode driven by a 'seismonuclear' process in the solar core (Kopysov, 1983);

(f) a g-mode excited by close encounter of the Sun with a star (Kosovichev and Severny, 1983a);

(g) a 160-min oscillation related to periodic perturbations of the gravity field or gravitational radiation (GR; Kotov and Koutchmy, 1982, 1983) or induced by outer GR (Kotov and Koutchmy, 1983; Delache, 1983; Isaak, 1984; see also below).

Däppen and Perdang (1984) suggested that the oscillation might be a nonlinear phenomenon due to the interaction of higher-frequency p-modes. According to Dziembowski (1983) and Perdang (1983) the nonlinearity together with the effect of resonant coupling between gravity modes may lead to enhancement of low-degree g-mode oscillations and thus explain the 160-min period. We note, unfortunately, that the evidence for nonlinearity is still rather circumstantial.

Quite a different approach was developed by Kosovichev and Severny (1983b) who found that for a number of solar models the period of stationary oscillations determined by solutions of the gas-dynamical equations does not exceed 131 min and it seems improbable to get a 160-min period at all. Dolginov and Muslimov (1983) considered turbulent convection as a source for resonant excitation of large-scale nonradial oscillations and concluded that the mechanism in principle can explain the 160-min oscillations of the observed amplitude (see however Goldreich and Keeley, 1977). Childress and Spiegel (1981) discussed the possibility that the oscillation may be a solitary wave which depends on nonlinear interactions.

The most appropriate answer to the problem is now considered to be a resonant (possibly nonlinear) interaction between g-mode eigen frequencies (Christensen-Dalsgaard and Gough, 1976; Gough, 1977; Vandakurov, 1981; Dziembowski, 1983). This mechanism was recently studied in detail by Severny et al. (1984) on the basis of the Crimean 9-year observations. It was shown that the dominant 160-min peak in the PS might result from a close 'double' resonance of several g-modes of the solar model 'C' calculated by Christensen-Dalsgaard et al. (1979). The successful identification of 10 other g-modes of degree $l = 4$ of this model with the observed peaks in actual PS of oscillations strongly supports this model 'C' which evolved from low initial Z (from $Z = 0.001$ to 0.02) and now provides a reasonably small neutrino flux 1.7 SNU (but eventually it possesses too shallow a convective zone; current solar oscillation observations suggest that the convection zone extends to about 0.3 of the Sun's radius). The instability of the nonradial g-mode oscillations of the solar model with low Z interior was then investigated by Kosovichev and Severny (1984). Further evidence of the multiple resonant interaction of g-modes near 160-min period was presented also by Guenther and Demarque (1984).

4. A Tentative Application to the Solar System

Remarkable enough, the possibility of the existence of an 160-min period in the Sun has been advanced in literature long before its actual discovery in 1974. Namely, about 40 years ago Sevin (1946) stated that "la période propre de vibration du Soleil, c'est-à-dire la période de son infra-son (1/9 de jour), a joué un rôle essential dans la distribution des planètes supérieures". Presumably, his 'vibration period' was merely an issue of some speculations about resonances and distances inside the Solar system.

Later on, in 1981, it was noted that the spin periods of planets and asteroids (Gough, 1983; Kotov and Koutchmy, 1982, 1983) being expressed in units of 160 min seem to be closer to integer numbers than would be expected for a random sample.

To evaluate this apparent commensurability strictly, Kotov and Koutchmy (1982; see also Kotov, 1984) applied the so-called 'commensurability function' (CF) to various samples of rapidly rotating planets and asteroids:

$$F(v) = \frac{1}{\sigma}\left\{a - \left[\frac{1}{N}\sum_{i=1}^{N}[x_i - \text{INT}(x_i + 0.5)]^2\right]^{1/2}\right\},\tag{1}$$

where $x_i = v_i/v$ if $v \leqslant v_i$ and $x_i = v/v_i$ if $v > v_i$; v_i is the spin rate of the ith object, v is any tested frequency in the range studied, N the total number of objects in the sample; $a = 12^{-1/2}$ and $\sigma = (60N)^{-1/2}$. For a random sample of v_i the standard deviation of $F(v)$ equals one, and the $F(v)$-values are normally distributed around zero. The maximum of $F(v)$ corresponds to the best least-squares fit of the ratios of v_i and v by integers. By

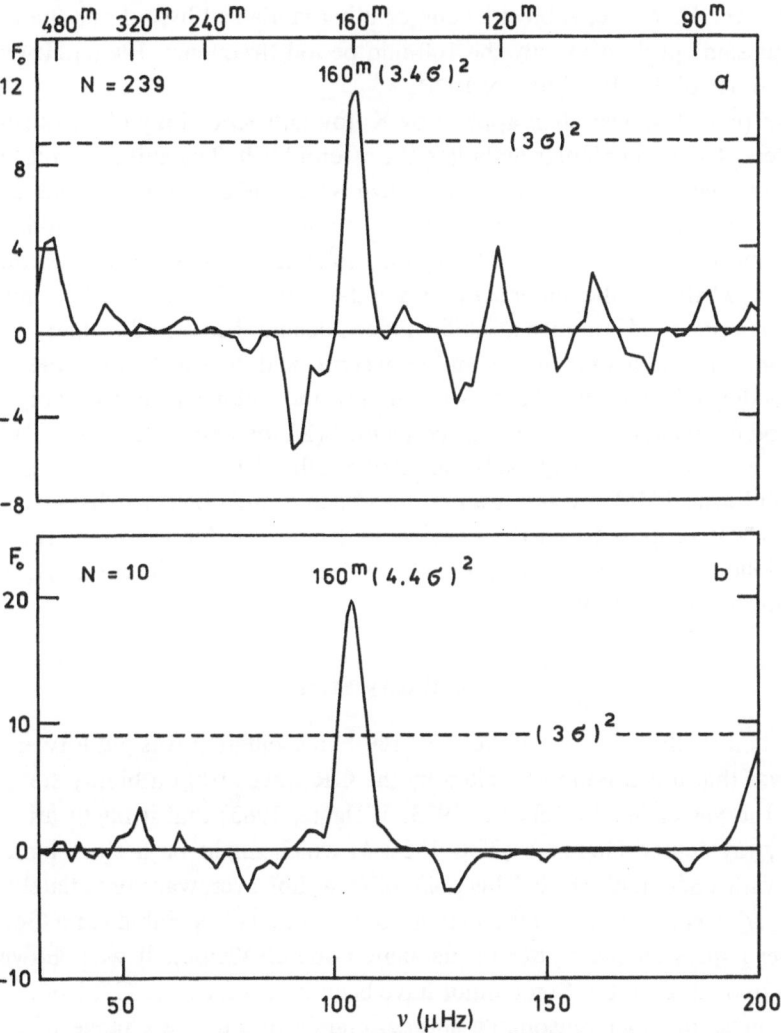

Fig. 4. (a) Commensurability spectrum computed for 239 bodies of the solar system. (b) Commensurability spectrum of 10 orbital parameters in the solar system (speed of light is set unity).

analogy with a usual PS analysis, we show $F(v)$ on the plots squared but with the sign accounted for:

$$F_0(v) = F^3(v)/|F(v)|. \tag{2}$$

For the illustration of a predominant 160-min commensurability of spin rates in the solar system we present in Figure 4a a 'commensurability spectrum' (CS) computed for a sample of 239 rapidly spinning bodies (6 major planets, 10 nuclei of short-periodic comets and 223 asteroids with diameters $D \geqslant 30$ km). The highest peak has an amplitude 3.4σ and reflects the preferable commensurability of the spin rates with the 160-min period. It is worthwhile to note that when we analyse several independent subsets of the same data and average the associated CS's, the peak reaches a 5.8σ-level of confidence. Therefore, subject to one or other statistical hypothesis the probability to get this same peak at exactly the 160-min period (frequency 104.2 µHz) by chance is of the order of 3×10^{-4} to 3×10^{-9}.

Similar procedure was then applied by Kotov and Koutchmy (1983) to the orbital parameters of 10 main planets including the asteroid belt. The authors have introduced an 'effective wavelength' $L = c \times P_0 = 19.24$ AU, where c is the speed of light (in computations it was set unity), and speculated that there might be a significant, in the average, quasi-commensurability between L and some planetary orbital parameters, a_i, $2a_i$ or $2\pi a_i$, where a_i is the major semi-axis of the orbit $i = 1, 2, \ldots 10$. They immediately found that a preferred commensurability indeed occurs: between $2\pi a_i$ and L for the five inner planets (asteroids included), and between L and $2a_i$ for the five outer ones. The corresponding CS (Figure 4b) shows again only one, unique, peak with amplitude 4.4σ at a period of 160.4 ± 1.5 min. It was concluded (Kotov and Koutchmy, 1983) that the L-commensurability may play a substantial role in the distribution of planetary distances (see again Sevin, 1946) and offers a new approach to the planetary distances problem. Further it was proposed that the physical mechanism enforcing these 160-min commensurabilities can be presumably related to the 160-min gravitational radiation from an unknown external source.

5. Binary Stars

A suggestion for the existence of the solar 160-min oscillation was put forward by Isaak (1984) was that it is a g-mode excited by the GR waves from a binary star, and soon after that it was claimed (Delache, 1983; Walgate, 1983) that it might arise from the intense γ-ray source Geminga (2CG 195 + 4) which might be a close binary system orbiting with a 320-min period. This point of view, however, was immediately subjected to severe criticism arising from theoretical considerations (see Fabian and Gough, 1984, and other papers on the matter in the same issue of *Nature*). It was shown that the 160-min oscillation of the Sun cannot have been induced by the GR from a binary of a stellar mass or other reasonable source, mainly due to the change of orbital rate predicted by the theory of general relativity.

Simultaneous was a conjecture (Kotov, 1984) that if there were a cosmical GR source with a frequency $v_0 \approx 10^4$ µHz, it might be reasonable to look for traces of v_0 in the observed distribution of periods of close binary stars in the galaxy. In other words, the idea was to consider the *total set* of galactic binaries as a 'detector' of the hypothetical 160-min GR: the action of monochromatic GR over sufficiently long time interval could produce a deficit of binaries with orbital periods $p_1 \approx 2zP_0$ and some excess of those with periods $P_2 \approx (2z + 1)P_0$, where z is an integer. In the frame of this hypothesis we focus attention on the pattern of the following 'generalized' commensurability function

$$F'(v) = F(v) - F(v/2) \tag{3}$$

and the associated CS, $F_1(v)$, similar to (2).

Calculations were performed for a sample of about 4000 binaries with orbital periods $P < 4$ days (Figure 5). One can see that the major peak, with an amplitude of about 4.1σ,

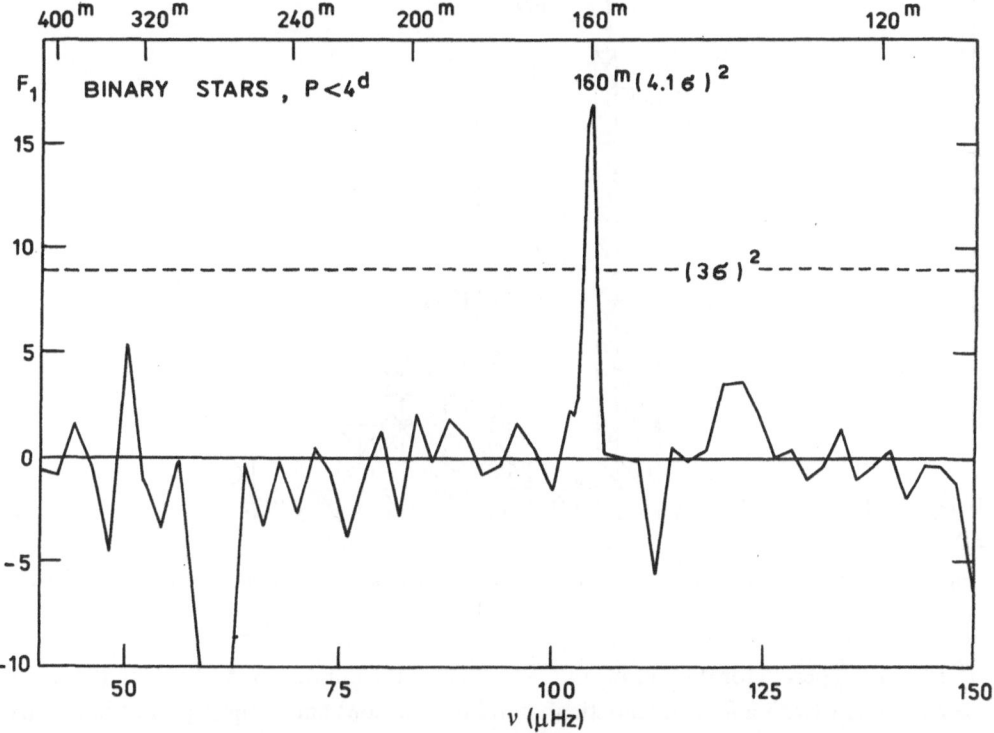

Fig. 5. Generalized commensurability spectrum for about 4000 binary stars with orbital periods less than 4 days (from Kotov, 1984).

corresponds to a period of 159.9 ± 0.5 min. Therefore, the conclusion may be drawn that in the range 20–150 µHz analysed (Kotov, 1984) the frequency 104 µHz is the most characteristic one for the distribution of orbital periods of galactic binary stars.

We notice then that the absolute majority of the binary stars analysed above have orbital periods in the range $0.3 \lesssim P < 4.0$ (days). Hence, one is tempted to see whether there are some indications of the presence of the P_0-period in the period-distribution of much more rapidly rotating binaries. We know that there is a class of compact binaries, called cataclysmic variables (CV's), see, e.g., Ritter (1983). It is generally believed that these systems consist of a white dwarf (or a neutron star, or a black hole) and a low-mass secondary which as a rule slightly overflows its Roche lobe. Typical orbital periods of CV's are a few hours. It is also widely accepted that GR, together with the accretion process, plays an essential role in the evolution of these binary systems.

The distribution of 97 CV's from Ritter's (1983) catalogue shown in Figure 6 demonstrates clearly that the 160-min period appears to be again a characteristic

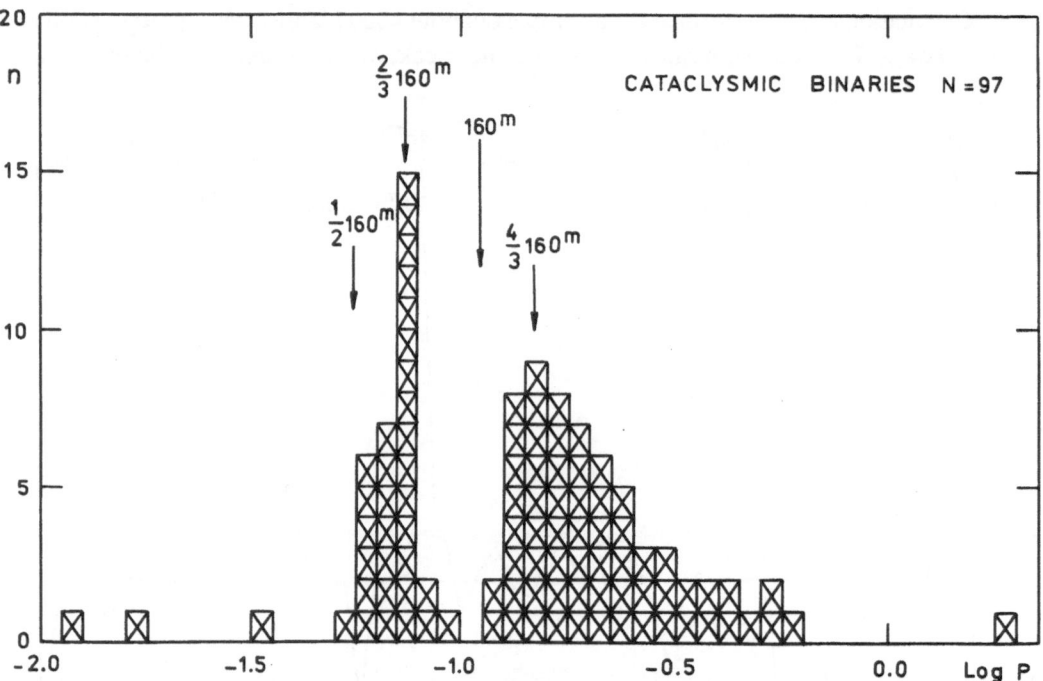

Fig. 6.   Distribution of 97 cataclysmic variables (period P is expressed in days). (See also, e.g., Ritter, 1983).

('resonant'?) period for the sample of CV's: there is the famous 'period gap' in the range ≈ 2 to ≈ 3 hr (with a formal statistical amplitude $\approx 3.6\sigma$) and a short-period boundary around 80 min (both pecularities of the distribution are well-known, see, e.g., Ritter, 1983), and two maxima corresponding approximately to periods $(2/3) \times 160$ min and $(4/3) \times 160$ min.

6. Conclusion

Observations made at different sites show good agreement in phase and amplitude of the 160-min oscillations of the Sun, which exhibit a remarkable long-term stability of an initial phase.

The most acceptable explanation of the oscillation seems to be a multiple resonant interaction of gravity g-mode oscillations with a combination of some modes due to possible nonlinearity effects. This interpretation favours the solar model with a low heavy elements abundance.

Simultaneously, it is noted that there is a statistically significant commensurability between the spin periods of different bodies in the solar system and the 160-min period. This intriguing fact is further reinforced by the apparently non-random distribution of planetary distances with respect to the geometric scale $L = cP_0 = 19.24$ AU.

At last, we find that the same period seems to be the most characteristic ('latent') period for the distribution of periods of close binary stars, and also for periods of RR Lyr variables in globular clusters (Kotov, 1985). From this standpoint the next thing to suggest is that a fundamental aspect of the physics of gravitation is not yet understood(?). However, it is so far difficult to accept any of the aforementioned interpretations without further supporting evidences and theoretical considerations.

It is interesting to note that the analysis of different subsets of data (on spin rates in solar system, orbital periods of binaries, etc.) did specify for the best value of the period:

solar system (spin rates):	159.9 ± 0.5 min,
solar system (geometric scaling):	160.4 ± 1.5 min,
binary stars:	160.0 ± 0.1 min,
RR Lyr variables in globular clusters:	161.4 ± 1.6 min,
highly evolved close binary systems:	158 ± 2 min,

which fairly well agrees with $P_0 = 160.0102 \pm 0.0005$ min for the Sun. Of course it seems impossible to imagine that these various periods are unrelated phenomena and that the agreement is just a coincidence. Still, a clear interpretation of this 'ubiquitous' behaviour of the 160-min periodicity does not exist at the present time.

Irrespective of other (different from the beforementioned) possible explanations, it is beyond doubt, on the basis of the solar observations and the facts discussed here, that the nature of the 160-min oscillation, firstly found in the Sun and then in the solar system as a whole and then amongst the stars, does present a new challenging problem for astrophysics.

At the present state of the interpretation of the 160-min oscillations, we feel some kind of comfort observing a number of successful suggestions; another set of proposals seemingly has been without any particular success; we witnessed sometimes hard criticism subsequently advanced to some speculations that were not sufficiently based on observations and solid theoretical considerations. However, all ideas, we believe, were (and are) fruitful for the attempts to fully understand the true nature of the 160-min oscillations.

Acknowledgements

I am very grateful to A. B. Severny for useful discussions and valuable comments on the manuscript, and to P. H. Scherrer for providing me with Figure 1.

References

Blinnikov, S. I. and Khlopov, M. Yu.: 1983, *Solar Phys.* **82**, 383.

Brookes, J. R., Isaak, G. R., McLeod, C. P., van der Raay, H. B., and Roca Cortes, T.: 1978, *Monthly Notices Roy. Astron. Soc.* **184**, 759.

Brookes, J. R., Isaak, G. R., and van der Raay, H. B.: 1976, *Nature* **259**, 92.

Childress, S. and Spiegel, E. A.: 1981, in S. Sofia (ed.), *The Solar Constant and Its Variation*, NASA Conf. Publ. 2191, Washington, D.C., p. 273.

Christensen-Dalsgaard, J.: 1984, in *The Hydromagnetics of the Sun*, Proc. 4th European Meeting on Solar Physics, ESA SP–220, p. 3.

Christensen-Dalsgaard, J., Cooper, A. J., and Gough, D. O.: 1983, *Monthly Notices Roy. Astron. Soc.* **203**, 165.

Christensen-Dalsgaard, J. and Gough, D. O.: 1976, *Nature* **259**, 89.

Christensen-Dalsgaard, J. and Gough, D. O.: 1982, *Monthly Notices Roy. Astron. Soc.* **198**, 141.

Christensen-Dalsgaard, J., Gough, D. O., and Morgan, J. G.: 1979, *Astron. Astrophys.* **73**, 121.

Connes, P.: 1984, personal communication.

Däppen, W. and Perdang, J.: 1984, *Mem. Soc. Astron. Ital.* **55**, 299.

Delache, P.: 1981, *Compt. Rend. Acad. Sci. Paris* **293**, 949.

Delache, P.: 1983, *J. Astron. Francaise* **19**, 13.

Deubner, F.-L. and Gough, D.: 1984, *Ann. Rev. Astron. Astrophys.* **22**, 593.

Dittmer, P. H.: 1977, Stanford Univ. IPR Rep., No. 686.

Dolginov, A. Z. and Muslimov, A. G.: 1983, Ioffe Inst. Phys. Technology, USSR Acad. Sci., preprint No. 840.

Dziembowski, W.: 1983, *Solar Phys.* **82**, 259.

Efanov, V. A., Moiseev, I. G., and Nesterov, N. S.: 1983, *Pis'ma Astron. Zh.* **9**, 741.

Fabian, A. C. and Gough, D. O.: 1984, *Nature* **308**, 160.

Forbush, S. E., Pomerantz, M. A., Duggal, S. P., and Tsao, C. H.: 1983, *Solar Phys.* **82**, 113.

Goldreich, P. and Keeley, D. A.: 1977, *Astrophys. J.* **212**, 243.

Gough, D. O.: 1977, in R. M. Bonnet, P. Delache, and G. de Bussac (eds.), *IAU Colloq.* **36**, 3.

Gough, D.: 1983, *Phys. Bull.* **34**, 502.

Gough, D. O.: 1985, *Solar Phys.* **100**, 65 (this volume).

Grec, G. and Fossat, E.: 1979, *Astron. Astrophys.* **77**, 351.

Grec, G., Fossat, E., and Pomerantz, M.: 1980, *Nature* **288**, 541.

Guenther, D. B. and Demarque, P.: 1984, *Astrophys. J.* **277**, L17.

Hill, H. A. and Tash, J.: 1984, *Bull. Am. Astron. Soc.* **16**, 1000.

Iben, I. and Mahaffy, J.: 1976, *Astrophys. J.* **209**, L39.

Isaak, G. R.: 1984, *Mem. Soc. Astron. Ital.* **55**, 45.

Kopysov, Yu. S.: 1983, Inst. Nucl. Res., USSR Acad. Sci., Preprint P-0317.

Kosovichev, A. G. and Severny, A. B.: 1983a, *Pis'ma Astron. Zh.* **9**, 424.

Kosovichev, A. G. and Severny, A. B.: 1983b, *Solar Phys.* **82**, 323.

Kosovichev, A. G. and Severny, A. B.: 1984, *Mem. Soc. Astron. Ital.* **55**, 129.

Kotov, V. A.: 1984, *Izv. Krymsk. Astrofiz. Obs.* **74**.

Kotov, V. A.: 1985, *Izv. Krymsk. Astrofiz. Obs.* **75** (in press).

Kotov, V. A. and Koutchmy, S.: 1982, *Izv. Krymsk. Astrofiz. Obs.* **70**.

Kotov, V. A. and Koutchmy, S.: 1983, *Izv. Krymsk. Astrofiz. Obs.* **72**.

Kotov, V. A., Koutchmy, S., Kononovich, E. V., Ryzhykova, N. N., and Tsap, T. T.: 1984, *Izv. Krymsk. Astrofiz. Obs.* **73** (in press).

Kotov, V. A., Severny, A. B., and Tsap, T. T.: 1978, *Monthly Notices Roy. Astron. Soc.* **183**, 61.

Kotov, V. A., Severny, A. B., and Tsap, T. T.: 1983, *Izv. Krymsk. Astrofiz. Obs.* **66**, 3.

Koutchmy, S., Koutchmy, O., and Kotov, V. A.: 1980, *Astron. Astrophys.* **90**, 372.

Leibacher, J. W. and Stein, R. F.: 1981, *The Sun as a Star*, NASA SP-450, Washington, D.C., p. 263.

Pecker, J.-C.: 1984, *La Vie de Sci., Comptes Rend., sér. gén.* **1**, 199.

Perdang, J.: 1983, *Solar Phys.* **82**, 297.

Rachkovsky, D. N.: 1985, *Izv. Krymsk. Astrofiz. Obs.* **71**, 25.

Ritter, H.: 1983, *Mitt. Astron. Gesellschaft*, No. 60, 159.

Scherrer, P. H. and Wilcox, J. M.: 1983, *Solar Phys.* **82**, 37.

Scherrer, P. H., Wilcox, J. M., Severny, A. B., Kotov, V. A., and Tsap, T. T.: 1980, *Astrophys. J.* **237**, L97.

Severny, A. B., Kotov, V. A., and Tsap, T. T.: 1976, *Nature* **259**, 87.

Severny, A. B., Kotov, V. A., and Tsap, T. T.: 1978, *Proc. 2nd European Meeting on Solar Phys.* CNRS, p. 123.

Severny, A. B., Kotov, V. A., and Tsap, T. T.: 1979, *Astron. Zh.* **56**, 1137.

Severny, A. B., Kotov, V. A., and Tsap, T. T.: 1984, *Nature* **307**, 247.

Sevin, E.: 1946, *Compt. Rend. Acad. Sci. Paris* **222**, 220.

Toth, P.: 1977, *Nature* **270**, 159.

Vandakurov, J. V.: 1981, *Astron. Circ. (USSR)*, No. 1173, 1.

Vladimirsky, B. M., Bobova, V. P., Bondarenko, N. M., and Veretennikova, V. K.: 1983, *Solar Phys.* **82**, 451.

Vorontzov, S. V. and Zharkov, V. N.: 1978, *Astron. Zh.* **55**, 84.

Walgate, R.: 1983, *Nature* **305**, 665.

Woodard, M. and Hudson, H.: 1983, *Solar Phys.* **82**, 67.

Worden, S. P. and Simon, G. W.: 1976, *Astrophys. J.* **210**, L163.

THE SUN AS A SYSTEM OF ELEMENTARY PARTICLES

JOSIP KLECZEK

Astronomical Institute, 25165 Ondřejov, Czechoslovakia

Abstract. No unusual observation of the Sun is described here. No new theory is proposed. The paper – based on known facts of solar physics – is a modest attempt to interpret the Sun as a selfgravitating system of about 10^{57} nucleons and electrons. These elementary particles are endowed with strong, electromagnetic, weak and gravitational interactions. Origin of the Sun, its evolution, structure and physiology are consequences of the four interactions.

1. Introduction

The progress of solar physics (and of sciences in general) is accompanied by increasing ramification and superspecialization. To penetrate deeper means to narrow one's subject and to concentrate on details. The present Volume is a good example of branching of our science. Were it not for the narrow specialization of solar physics, this volume's number would be substantially less than 100 and our knowledge of the Sun would be poor.

On the other hand, by stressing details the vision of the whole may be lost: for leaves and twigs we see neither trees nor the forest. By concentrating upon a detail we easily omit its relations to other parts of the whole, which give a 'raison d'être' to the detail. Hence the effort of various scientific disciplines to comprehend objects also in their integrity, i.e. to conceive them as systems. Medical science is an example: physician – be he a neurologist, cardiologist, ophtalmologist or a surgeon – knows the human body as a whole, i.e. as a system of interrelated organs, the organs as systems of tissues, tissues as systems of cells, etc.

The notion of system is of basic importance in any branch of science, in technology, economics, philosophy, etc. Astronomers use it in various contexts, e.g. Copernican system, Earth–Moon system, planetary system, solar system, ring system of Saturn, system of satellites, UBV-system, binary system, stellar system, Milky Way system, etc. Any student of astronomy therefore has an idea of what a system is, even if he does not know its exact definition.

General notions like a system, are difficult to define. In Klir (1969) twenty-four different definitions of the system notion are given. A simple definition will do for our purpose: it is a set of objects (parts, units, elements) with relationships between them. A mere agglomeration of parts (objects) is therefore not sufficient to be a system; interactions (relationships, orderliness) of the parts is the characteristic feature ('nota specifica') of each system. A system is more than an agglomeration of its parts, a whole is more than a sum of its parts, the Cosmos is more than the Universe. A system is a whole composed of its interacting parts.

Solar Physics **100** (1985) 115–123.   0038–0938/85.15.

2. Elementary Particles of the Sun

Elementary particles are the fundamental units of which all things in the Universe consist. Whatever the complexity of an object may be, its structure is based on elementary particles and their interactions. Protons, electrons and neutrons are the most important particles for the structure. Though they are called elementary, it is not yet clear whether they themselves do not consist of more simple entities. For us it would mean only introducing two quarks (u and d) instead of protons and neutrons. Protons and electrons are stable particles, hence their importance for the structure of all things (bodies). Proton is the lightest baryon and electron is the lightest charged particle at all; neither can decay due to conservation of baryon number and electric charge respectively. Neutrons are stable only in a nucleus.

The Sun is a gravitationally bound system of nucleons and electrons. It is a selfgravitating thermonuclear reactor, a star consisting of about

$$N = \alpha^{-3/2} \tag{1}$$

nucleons, where

$$\alpha^{-1} = \hbar c / G m_p^2 = 1.7 \times 10^{38} \tag{1a}$$

is the inverse of the 'gravitational fine structure constant' (Rees, 1984). In the above combination of constants, m_p is the rest mass of proton, c is the speed of light, $\hbar$ and G are Planck and gravitational constants, respectively. The vastness of this number reflects the weakness of gravitation on the elementary particle scale (see Figure 1). The above estimate of the number of nucleons in a star gives approximately $N = 2 \times 10^{57}$. Dividing the solar mass $M_\odot$ by the mass of one nucleon m_p we get $N = 1.2 \times 10^{57}$ nucleons. For the element abundances in the solar atmosphere (e.g. Withbroe, 1971), the number of protons in the Sun is 1.07×10^{57} and the number of neutrons 0.13×10^{57}. Due to electrical quasineutrality of plasma, the number of electrons is equal to the number of protons. If we take into account higher abundance of helium in the solar core, the number of protons should be 1.04×10^{57} and the number of neutrons 0.16×10^{57}.

3. Interactions of Particles in the Sun

Elementary particles interact with one another by means of their fields. Four fundamental forces of nature have been identified, viz. strong, electromagnetic, weak and gravitational interactions. This classification of interactions has been used since 1948. It is important for understanding the nature of the elementary particles themselves as well as the structure and evolution of all systems in the Universe. The weak and electromagnetic interactions can be grasped in the context of a single theory, but the interactions remain distinct.

The four interactions hold the elementary particles together and determine the structure and evolution of all the systems in the Universe from atomic nuclei to supergalaxies (Kleczek, 1976). The interactions differ in their strength and range as is

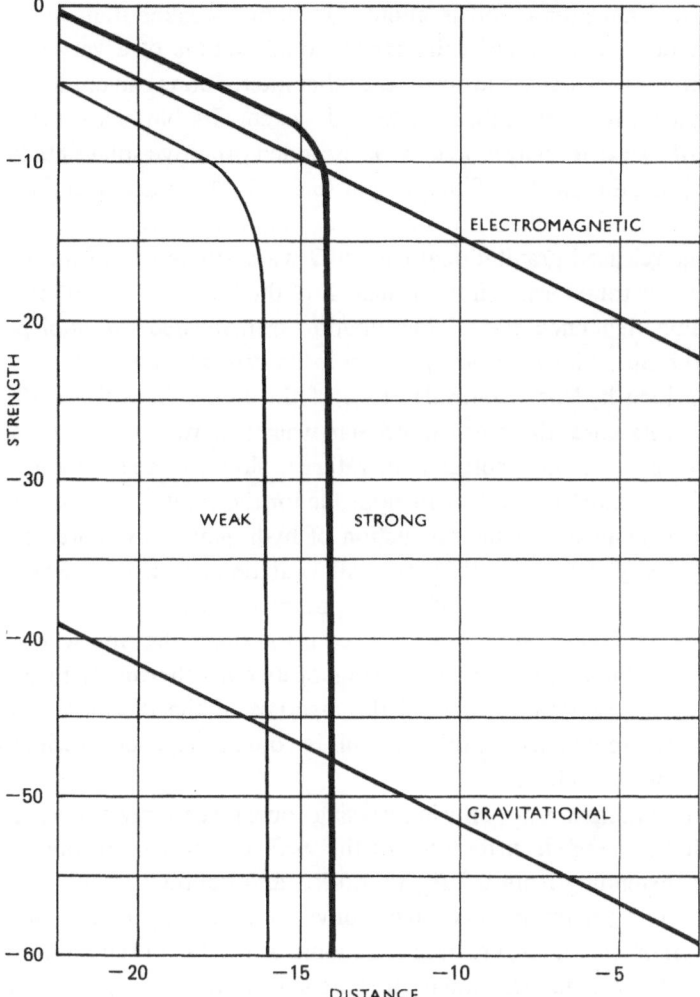

Fig. 1.   Interactions of elementary particles. Their strength in logarithmic scale is on vertical axis. Strength of strong interaction is put equal to one. Distances on horizontal axis are in centimetres; the scale is also logarithmic. Three interactions form systems: strong, electromagnetic and gravitation. Weak interaction is too short to form any system. There is a cutoff for the electric attraction (Debye length) in the solar plasma.
The weakest interaction, gravitation, attracts two particles at any distance without limitation.

shown in Figure 1 ('t Hooft, 1980). Their role depends on the size of the system: nuclei (size about 10^{-13} cm) are bound by strong interaction, while atoms (10^{-8} cm), molecules, hydromagnetic features (like fibres, prominences, coronal arches, etc.) are bound by electromagnetic interactions. Gravitation is the basic binding force in large bodies – roughly over 10^{47} nucleons), such as planets, the Sun and other stars, star clusters and galaxies. Weak interaction has the shortest range (10^{-15} cm) so that it is limited to the innermost part of a particle.

3.1. Gravitational interaction is about 10^{40} times weaker than nuclear force (i.e. strong interaction). Nevertheless the gravitational actions of a very large number of nucleons combine together while non of the other interactions can combine as efficiently. The Sun is therefore a gravitationally bound system. Its binding gravitational energy (about $GM_\odot^2/R_\odot$) was initially stored in the extended primitive solar nebula. Gravitation released the potential energy of the nebula (6.6×10^{48} erg according to Allen, 1973) during the contraction of the Protosun. According to the virial theorem ($2T + \Omega = 0$) one half of the released gravitational energy Ω was radiated from the Protosun while the other half was transformed into the heat T of the central core, so that the Protosun became a Main Sequence star. Selfgravitation transformed the cool primitive solar nebula into the Sun with hydrogen burning in its central core.

During the long hydrogen burning stage of the Sun (about 1.5×10^{10} yr), helium accumulates in its core, the mean molecular weight increases and the core is slowly compressed by gravitation. Central temperature is slowly increasing and so is the solar luminosity. Gravitation is therefore responsible for the evolution of the Sun away from the Zero Main Sequence. After exhaustion of hydrogen in the core, it will contract, gravitational energy being converted into both radiation and heat as was the case in the Protosun. At the temperature of 10^8 K 3-alpha reactions substitute gravitation as the solar luminosity source. During the short red giant stage, the Sun will draw ('squeeze out') energy from the rest energy of nucleons of its core alternately by gravitation and by thermonuclear reactions. When all the reserves of the thermonuclear and gravitational energies are exhausted, the Sun will become a degenerate white dwarf in the centre of a planetary nebula.

Gravitation participates not only in releasing some rest energy of nucleons, but it also participates in the energy transfer towards the surface. Through the outer portion of the solar interior, extending from $0.7 R_\odot$ to $1.0 R_\odot$ and containing only 3% of the solar mass, energy is overwhelmingly carried by convection. Plasma currents in the convective zone are driven by buoyancy, i.e. by gravitational attraction of the radiative zone which includes $0.97 M_\odot$. A small fraction of the kinetic (non-thermal) energy carried towards the photosphere is converted into heat of the chromosphere and corona by means of photospheric magnetic fields i.e. in the form of hydromagnetic waves. The high temperature corona then gives rise to a thermally driven solar wind. The chain of cause and effect starts often with gravitation. Another example would be solar g-mode of oscillations for which the gravitation provides the prevailing restoring force.

Each particle of the Sun is gravitationally attracted to all the other 10^{57} particles. By becoming part of the Sun each nucleon released about 6×10^{-9} erg (i.e. 3.7 keV) from its own rest energy $m_0 c^2$ (938.25 MeV or 939.55 MeV). Gravitation on the Sun is inefficient in releasing energy from nucleons. Their binding energy represents only $4 \times 10^{-6} m_0 c^2$. In order to escape from the Sun, a proton must regain its gravitational binding energy. The proton may be taken away by expanding magnetic field of a coronal event or of a spray prominence. It may also regain its energy defect from other particles in the corona by collisions, so as to be accelerated up to the escape velocity (618 km s^{-1}) or more.

3.2. Strong interactions act always when two nucleons (in general two hadrons) approach each other to a distance of 10^{-13} cm. Such a close approach of two protons (or of other two nuclei) is hampered by their mutual electric repulsion, so that high kinetic energy is necessary to overcome their repulsion. Coulomb energy e^2/r of two protons at the distance of 10^{-13} cm is 2×10^{-6} erg (about 1 MeV). Even in the solar core only a very small fraction of protons (in the Maxwellian tail) have kinetic energy comparable to the Coulomb energy. Fusion of protons is made easier by their very high collision rate in the core and by quantum mechanical tunneling.

In the core, strong interactions fuse four protons into an alpha particle. The final result of several fusion reactions is represented by the reaction

$$4p \longrightarrow \alpha + 2e^+ + 2v_e + \gamma \, \text{photons} \tag{2}$$

in which two protons p decay into neutrons n, positrons e^+ and electron neutrinos v_e, i.e.

$$p \longrightarrow n + e^+ + v_e. \tag{3}$$

While the proton fusion is a strong interaction, the decay of a proton in the nucleus (3) is a weak interaction. The energy released from one proton by the fusion (2) is about 10^{-5} erg which is $0.007m_p c^2$. Strong interactions in the Sun are therefore thousand times more efficient in releasing energy from nucleons than gravitation is; the latter releases only $4 \times 10^{-6} \, m_p c^2$ as indicated in Section 3.1. In some other systems the gravitation may be much more effective than in the Sun, e.g. in radio galaxies, supernovae and quasars.

During the whole life of the Sun (from primitive solar nebula to the white dwarf stage with radius $R \approx 10^9$ cm) the selfgravitation of its 10^{57} nucleons will release about $GM_\odot^2/R$ (i.e. 10^{50} erg) from their rest energy. Strong interactions (of about one tenth of nucleons) will transform protons and later alpha particles into carbon, oxygen etc. up to magnesium and release about $10^{57} \times 0.1 \times 10^{-5}$ erg i.e. 10^{51} erg. This order-of-magnitude estimates show that the overwhelming role in releasing energy from the selfgravitating Sun is played by strong interactions.

Strong interactions are sometimes also active in the solar atmosphere, but on a much smaller scale than in the core. Protons and alpha particles accelerated in flares interact with nuclei of abundant elements in the solar atmosphere, i.e. with H, He, C, N, and O. In the reactions positron-emitting nuclides and neutrons are produced, such as:

$$\begin{aligned} p + C^{12} &\longrightarrow C^{11} + p + n \\ C^{11} &\longrightarrow B^{11} + e^+ + v_e, \end{aligned} \tag{4}$$

$$\begin{aligned} p + N^{14} &\longrightarrow N^{13} + p + n \\ N^{13} &\longrightarrow C^{13} + e^+ + v_e, \end{aligned} \tag{5}$$

$$\begin{aligned} p + O^{16} &\longrightarrow O^{15} + p + n \\ O^{15} &\longrightarrow N^{15} + e^+ + v_e, \end{aligned} \tag{6}$$

and other reactions producing positrons and neutrons (Ramaty *et al.*, 1975; Hillier,

1984). Neutrons are then captured through strong interaction by protons to produce deuterons, i.e.

$$n + p \longrightarrow d + \gamma \text{ photon.} \tag{7}$$

Deexcitation of the deuterons (d) emits 2.2 MeV photons. Positrons from positron emitters in (4), (5), (6) etc., annihilate with electrons with emission of two 0.5 MeV photons, i.e.

$$e^+ + e^- \longrightarrow 2\gamma \text{ photons.} \tag{8}$$

3.3. Weak interaction and electromagnetic interaction are explained in the context of a single theory. Some high energy processes (e.g. $e^+ + e^- \rightarrow \mu^+ + \mu^-$) may be mediated by intermediate bosons (in our case by Z° boson) instead of by photons. On the Sun, however, there are no particles having sufficient energy: the solar cosmic rays have energies smaller than 50 GeV, while the intermediate bosons have energy over 80 GeV. The weak and electromagnetic interactions on the Sun are therefore considered separately.

The range of weak interactions is extremely short as may be seen in Figure 1. Their strength depends on distance r as

$$r^{-2}e^{-Kr}, \tag{9}$$

where K is considerably less than 10^{-15} cm. The weak interaction acts in the centre itself of an elementary particle. It cannot form any system of elementary particles, like the other three interactions do; it only changes their nature. The most important weak interaction in the Sun is the proton decay (3). It is a semileptonic weak interaction in which both leptons and hadrons participate. One process of an alpha particle synthesis of four protons produces two neutrinos and yields an energy of 4×10^{-5} erg (2). The solar luminosity therefore results from 3.8×10^{33} erg s^{-1} / 4×10^{-5} erg, i.e. about 10^{38} processes of the alpha-particle synthesis per second. Hence the neutrino luminosity of the Sun is 2×10^{38} v_e s^{-1}. The solar neutrinos have various energies dependent upon the proton-proton chain branch (Bahcall, 1979). Continuous neutrino spectrum is produced by processes with three products, such as

$$p + p \longrightarrow d + e^+ + v_e \tag{10}$$

and

$$B^8 \longrightarrow Be^8 + e^+ + v_e. \tag{11}$$

If there are only two products from a weak interaction, the neutrinos have discrete energy, i.e. they form a neutrino line in the solar neutrino spectrum, e.g.

$$p + e^- + p \longrightarrow d + v_e \tag{12}$$

with energy 1.442 MeV, and

$$Be^7 + e^- \longrightarrow Li^7 + v_e \tag{13}$$

with energies 0.862 MeV and 0.384 MeV. For detail discussion see Bahcall (1979). The wavelengths of neutrinos emitted from the reactions are only a few thousanths of an ångstrom and they have an extremely low probability of interacting with matter. About one ν_e of 10^{11} ν_e is absorbed when passing from the core to the solar surface.

Weak interactions in the solar atmosphere are extremely rare events, always associated with large flares. Collisions of accelerated high energy protons produce various particles and their antiparticles, which then decay through weak interactions. Let us mention two examples:

$$p + p \longrightarrow p + n + \pi^+, \tag{14}$$

then

$$\pi^+ \longrightarrow \mu^+ + \nu_\mu \tag{15}$$

and

$$\mu^+ \longrightarrow e^+ + \nu_e + \bar{\nu}_\mu. \tag{16}$$

The decay (15) is a semileptonic weak interaction (π^+ is a hadron) and the muon decay (16) is a purely leptonic weak decay (all the particles involved are leptons). Neutretto ν_μ and antineutretto $\bar{\nu}_\mu$ occur in the reactions as decay products.

Our second example of weak interactions in the solar atmosphere is a hyperon decay. High energy collisions during flares produce among other particles also hyperons and antihyperons – which is a strong interaction process. Hyperons then decay by weak interactions, e.g.

$$\Lambda^0 \longrightarrow p + \pi^- \tag{17}$$

which is a non-leptonic weak interaction (only hadrons are involved). The negative pion produced in (17) then decays similarly to π^+ in (15) by a semileptonic decay, while the following decay of the negative muon is a purely leptonic weak interaction, similar to (16).

We must stress again that the processes (14) to (17) are completely unimportant for the physiology of the Sun. They are mentioned only to indicate that weak interactions are not restricted to the proton decay (3), even though the latter is by far the most important weak interaction process in the Sun. Without the proton decay (3) the hydrogen burning (2) would not be possible.

3.4. Electromagnetic interactions regulate the rate of thermonuclear reactions in the solar core. They are of fundamental importance for energy transfer from the core to interstellar space. Protons and electrons – the basic constituents of the solar plasma – are endowed with electric charge. According to Coulomb's law (Figure 1) proton and electron continue to attract each other at infinite distances. However, Debye showed that there is a cutoff of this force where there are other particles in between. The Coulomb's field is screened at distances larger than

$$r_D \approx 6.9 \left(\frac{T_e}{n_e}\right)^{1/2} \text{cm}. \tag{18}$$

On the other hand, ordered collective motion of charged particles (i.e. electric current) may be felt at large distances on the Sun and in the interplanetary space (magnetic sectors). The Lorentz force binds electrons and ions to magnetic field lines; in other words, plasma is frozen into the magnetic field. This coupling of plasma and magnetic fields is the basic property of all the solar activity.

As already mentioned, the Coulomb's barrier (2×10^{-6} erg i.e. 1 MeV) drastically limits the rate of hydrogen burning. The average thermal energy of a proton in the core ($T = 10^7$ K) is only $kT = 1$ keV. Two circumstances make the fusion (10) of two protons less difficult: quantum mechanical tunnelling and fast protons in the Maxwellian distribution tail. However, the penetration probability decreases fast with decreasing particle energy, so that a proton in the solar core has to wait about 10^{10} yr for its fusion (10).

The energy released by thermonuclear reactions (i.e. through strong interactions assisted by weak and electromagnetic interactions) in the core, is transferred towards the surface by a very long chain of electromagnetic processes. Scattering by free electrons, free-free transitions of electrons in the field of ions, free-bound and bound-free as well as bound-bound transitions in heavier ions are involved in the radiative transfer. One initial gamma photon from a nuclear reaction has been crumbled by the many electromagnetic interactions into hundred thousands of photons leaving the photosphere through the last interaction

$$\text{H} + \text{e}^- \longrightarrow \text{H}^- + h\nu. \tag{19}$$

The emitted photons $h\nu$ with energies of a few electron-volts have a nearly Planckian distribution as a consequence of the Maxwellian distribution of the interacting electrons. The whole chain of electromagnetic interactions from the initial gamma photon (2) to the last emission of light photons (19) lasts about 10^6 yr.

The interactions of photons and electrons mentioned above and some other electromagnetic interactions (e.g. decay of neutral pions produced in flares, $\pi^0 \to \gamma + \gamma$) occur in a very small volume and last a very short time. Completely different types, i.e. large-scale electromagnetic interactions, occur in the convective zone and in various solar activity phenomena.

In the upper part of the solar interior (from about 0.7 $R_\odot$ to 1.0 $R_\odot$) the increased opacity of the solar plasma makes the radiative transfer ineffective. Convection transfers the hot plasma in bulk to the cooler surface. The convective zone is a large heat engine, in which a small fraction of the outflowing heat is converted into kinetic energy of the plasma flow. Magnetic field lines are frozen in the moving plasma of the convective zone, so that its kinetic energy is partly transformed into magnetic energy. This transformation takes place in the lowest layer of the convective zone, i.e. 1.5–2.0 $\times 10^5$ km under the photosphere (Parker, 1975; Spruit, 1974). From there the magnetic flux tubes rise upwards by buoyancy force to emerge through the photosphere. Most of the emerging flux tubes are small and even the general solar magnetic field of quiet regions is composed of flux tubes with about 400 km in diameter and an intensity of about 2000 G.

A part of the kinetic energy of the convective zone is transported by hydrodynamic and hydromagnetic waves through the photosphere, to heat the upper layers of the solar atmosphere. From there the energy is emitted as photons by various electromagnetic interactions (recombination, line emission, bremsstrahlung, cyclotron radiation, synchrotron radiation, and others).

4. Summary and Conclusions

The Sun consists of 1.2×10^{57} nucleons and 1.0×10^{57} electrons. Only gravitation was able to attract the distant parts of the primitive solar nebula, contract them into the Protosun and heat it up to hydrogen burning. Each elementary particle of the Sun is related by gravitation to all the others. Owing to the gravitational relationships of all the fundamental constituents, the Sun is a selfgravitation system.

Electromagnetic interaction is responsible for the existence of atoms, ions, properties of magnetized plasma, solar activity, emission and absorption of photons and hence also for transfer of energy and information. It moderates the rate of hydrogen burning in the solar core, although by itself it is very inefficient in releasing energy from matter (e.g. only 13.6 eV from recombination of an electron with a proton – the rest energy of which is 938.25 MeV + 0.51 MeV).

Owing to its very small range, the strong interaction is responsible for the existence of the smallest particle systems (10^{-13} cm), i.e. of atomic nuclei. It is very effective in releasing energy from hydrogen nuclei (about 7 MeV from 938 MeV of one proton). Through hydrogen burning the strong interaction releases all the solar luminosity. By the end of its evolution the Sun will have released about ten times more of its energy by strong interactions than gravitation will have.

Weak interaction acts at extremely short distances ($< 10^{-15}$ cm, smaller than the size of an elementary particle), so that it cannot form any system. It only transforms particles. Among others it transforms protons into neutrons (3) and thus it assists strong interaction in the hydrogen burning in the solar core.

Each structural property, every evolutionary process, any activity phenomenon or event on the Sun can be traced backwards to the four fundamental forces of nature, viz. to interactions of elementary particles.

References

Allen, C. W.: 1973, *Astrophysical Quantities*, The Athlone Press, University of London.
Bahcall, J. N.: 1979, *Space Sci. Rev.* **24**, 147.
Hillier, R.: 1984, *Gamma-Ray Astronomy*, Clarendon Press, Oxford.
't Hooft, G.: 1980, *American Scientist* **68**, 94.
Kleczek, J.: 1976, *The Universe*, D. Reidel Publ. Co., Dordrecht, Holland.
Klir, G. J.: 1969, *An Approach to General Systems Theory*, Van Nostrand, New York.
Parker, E. N.: 1975, *Astrophys. J.* **198**, 205.
Ramaty, R., Kozlovsky, B., and Lingenfelter, R. E.: 1975, *Space Sci. Rev.* **18**, 341.
Rees, M. J.: 1984, in X. Fustero and E. Verdaguer (eds.), *Relativistic Astrophysics and Cosmology*, World Scientific Publ. Co., Singapore, p. 3.
Spruit, H. C.: 1974, *Solar Phys.* **34**, 277.
Withbroe, G. L.: 1971, in the *Menzel Symposium on Solar Physics, Atomic Spectra, and Gaseous Nebulae*, Nat. Bur. of Standards, Wash. Publ. No. 353, p. 127.

THE SOLAR DYNAMO

A. A. RUZMAIKIN

*Keldysh Institute of Applied Mathematics, USSR Academy of Sciences,
Miusskaya pl. 4, Moscow 125047, U.S.S.R.*

Abstract. The basic features of the solar activity mechanism are explained in terms of the dynamo theory of mean magnetic fields. The field generation sources are the differential rotation and the mean helicity of turbulent motions in the convective zone. A nonlinear effect of the magnetic field upon the mean helicity results in stabilizing the amplitude of the 22-year oscillations and forming a basic limiting cycle. When two magnetic modes (with dipole and quadrupole symmetry) are excited nonlinear beats appear, which may be related to the secular cycle modulation.

The torsional waves observed may be explained as a result of the magnetic field effect upon rotation. The magnetic field evokes also meriodional flows.

Adctual variations of the solar activity are nonperiodic since there are recurrent random periods of low activity of the Maunder minimum type. A regime of such a magnetic hydrodynamic chaos may be revealed even in rather simple nonlinear solar dynamo models.

The solar dynamo gives rise also to three-dimensional, non-axisymmetric magnetic fields which may be related to a sector structure of the solar field.

1. Introduction

The equilibrium of the Sun as a star is determined by the balance between gravity and the pressure gradient. The solar activity is a realm of motions and magnetic fields. The magnetic field may be directly observed in active phenomena (spots, flares, prominences) or it is related to them (the coronal heating, coronal holes). It has a complex evolving structure specified as 'magnetic elements' (Zwaan, 1978). Thus, observations display the magnetic nature of the solar activity.

The theory of magnetism of a rest medium fails to explain this nature. Indeed, the conductivity of the plasma in the solar envelope is comparable with that of a poor metal. However, the effective time of magnetic field change, which is proportional to the product of the conductivity and the squared magnetic structure scale, greatly exceeds the times of observed variations in the magnetic phenomena since the characteristic scales of the magnetic structures are sufficiently large. The only and natural cause of the magnetic variations are hydrodynamic plasma motions first, rotation and convection. Thus, the solution to the solar (stellar) activity problem should be searched in the area of magnetohydrodynamics.

A fundamental feature of the magnetohydrodynamic processes is the ability of motions to amplify and support the magnetic field similar to the self-exciting dynamo-machine but without wires and coils (Larmor, 1919). The hydromagnetic dynamo is not only a possible mechanism of field amplification, but a general type process whose existence follows from basic principles. Specific examples (Moffatt, 1978; Parker, 1979) as well as theorems for a wide class of flows (Molchanov *et al.*, 1984) prove that there are undamped solutions for the magnetic field in a moving conducting fluid.

The solar plasma flow is a combination of large-scale motions (the differential

Solar Physics **100** (1985) 125–140. 0038–0938/85.15.

rotation, the meriodional circulation) with the stochastic one (turbulent convection). The magnetic field in such a medium is random subjected to complex stochastic equations. It is a more natural and simple way to look for smooth mean values: a mean field, a mean square and other statistic moments. The evolution of these quantities is described by simpler equations which are easier to deal with. The main achievements of the solar dynamo theory are due to the deduction and study of the equations for the mean magnetic field.

From the mathematical point of view the term 'mean' is understood as averaging over a statistical ensemble. In practice, one interprets the mean magnetic field as an average over times and scales exceeding the characteristic times and scales of basic energy-containing cells of turbulent motion (supergranules or giant cells if the latter exist). The mean magnetic field transport is governed by the mean flow values (Krause and Rädler, 1980). It is, first of all, the differential rotation which generates an azimuthal magnetic field from the poloidal one (Elsässer, 1946). The turbulent diffusion smoothes the gradients of the mean field. A remarkable feature of the solar stochastic motions is the mean helicity which generates the poloidal magnetic field from the azimuthal (and vice versa) by lifting and turning the omega-shape loops. The mean helicity defines a predominate left or right screw winding of convective motions in a non-homogeneous rotating fluid. An observational evidence of the mean helicity on the Sun would be of importance. It may be obtained by studying, for example, the sunspot motions with necessary involvement of *all three* velocity components. From the physical viewpoint the existence of the mean helicity is clear and natural: it arises due to the Coriolis force action upon the convective elements emerging and submerging in the non-homogeneous convective zone.

The above three mean characteristics of the solar motions (the differential rotation, the mean helicity and the turbulent diffusivity) mostly determine the mean field transport, which was demonstrated already in the pioneering models of the solar dynamo (Parker, 1955; Babcock, 1961; Leighton, 1969; Steenbeck and Krause, 1969). Later on, attention was given also to the turbulent permeability, the meridional circulation, the topological pumping and other features influencing the mean field transport.

By considering the mean field we obviously will know nothing about small-scale or quickly changing fields. In return we may explain large-scale and global variations of the magnetic field. The most prominent of these is the solar activity cycle. The term 'cycle' is usually understood as the period of activity recurrence, e.g., as the 11-years repetition of spot formation, or the 22-year variation of the magnetic field configuration. It may also be associated with the term 'the limiting Poincaré cycle' for the solar magnetohydrodynamic system (Ruzmaikin, 1981). The limiting cycle, as a specific feature of nonlinear dynamical systems, decorates the phase portrayal of the solar MHD-system whose nonlinear nature is easily seen from, say, the time variation of the spot number (the Wolf number). It may be obtained theoretically from the equation for the mean field by involving the nonlinear effect of the magnetic field upon the mean helicity (see, for example, Zeldovich *et al.*, 1983).

In reality, the phase space of the solar MHD-system is much richer. A regular quasi-periodic secular modulation observed (Gleissberg, 1958; Yoshimura, 1979) may be associated with the idea of a 2-frequency limiting cycle. Specialists and amateurs were amazed by the nonregular global minima of solar activity which took place in the past (Eddy, 1976). It is a direct indication of the global stochastic nature and the existence of a 'strange attractor' in the phase space of the solar MHD-system (Ruzmaikin, 1981; Zeldovich *et al.*, 1983; Weiss *et al.*, 1984). There are hopes that these phenomena can be described in terms of magnetic hydrodynamics of mean fields by adding to the mean field equation the one describing the evolution of the mean helicity (Kleeorin and Ruzmaikin, 1982) and/or the differential rotation (Jones *et al.*, 1984).

The purpose of this review is the discussion of the main results and problems of the solar dynamo theory for the mean magnetic field. Let us point out that the mean field is generated and transported together with small-scale fields which, unfortunately, will remain beyond the scope of this paper. The reason is not only the complexity of describing their behavior, but rather our poor knowledge of their motions. Even the theory of mean magnetic fields cannot be considered complete until a clear under-standing of the mechanism and a quantitative description are gained for the differential rotation, the mean helicity and other mean characteristics of the flow. Therefore, it is natural to begin this review with a brief discussion of the solar motion structure.

2. Triumph of Motions

The transport of the magnetic field in a moving medium is determined by the magnetic diffusion and the velocity field deformations. The magnetic diffusivity (inversely pro-portional to the conductivity) inside the Sun is so small that any magnetic fields with a scale exceeding the characteristic dimension of the granules may be considered on long time intervals to be frozen in the medium. Evolution of such fields is governed by motions.

The magnetic lines are stretched by the differential rotation. The latitude gradient of the solar angular velocity has been discovered long ago by direct observations of the spot motions (Carrington, 1863; Newton and Nunn, 1951) and the Doppler shifts of spectral lines in the surface layers (Howard and Harvey, 1970; Howard, 1984). Crudely, the drop in the angular velocity from equator to the pole is about 20%. Solar seismology is capable of evaluating the radial gradient of the angular velocity. While the data are not complete today and the quantitative results are contradictory, the observations give qualitative evidence that the radial gradient of the angular velocity is negative and its magnitude is a few times larger than that of the latitude gradient (Gough, 1982; Hill *et al.*, 1982; Dicke 1983).

Observations reveal also temporal variations of the solar rotation (Howard and LaBonte, 1980; Tuominen *et al.*, 1983; Howard, 1984). The torsional 11-year waves are a most remarkable discovery. It may be assumed that these variations of the angular velocity are due to the back action of the generated magnetic field upon the differential rotation (see p. 5).

According to standard models of the internal structure of the Sun its shell of about

0.3 $R_\odot$ in size is in a state of strong turbulent convection (the Rayleigh number is about 10^{12}; the scale is determined by the scale height; the Prandtl number is very small, about 10^{-9}, since the heat diffusion is radiative). The turbulent behavior of motions is usually determined by the value of the Reynolds number. In the convective zone Re $= 5.5 \times 10^{13}$ at a depth $d_1 = 10^{-2} R_\odot$ and Re $= 4 \times 10^{12}$ at a depth $d_2 = 0.3 R_\odot$. The cellular picture of granulation and supergranulation against the background of so strong turbulent motions appears as a surprising synergetic phenomenon. It is known that the characteristic scale of the granules is close to the density scale height. Simon and Weiss (1968) noted that in a density-stratified convective zone the appearance of two additional characteristic scales is natural. They identified one of the scales with the supergranular size, and the other with the size of giant cells so far unobserved. At first sight, the existence of convective cells of sizes exceeding the scale height looks improbable due to the fact that, because of flow continuity, a weak motion in the dense lower part of the cells results in enormous velocities in their upper parts. However, cells extending over several scale height are more effective heat transporters as compared with cells of one scale height in size in so far as they have a smaller superadiabatic gradient at the same temperature difference. There is a characteristic cell size at which the second effect is more important so that the appearance of such cells is prefered from the energy viewpoint. These interesting estimates have not yet been developed in a strict mathematical form. For the numerical study of the problem one should certainly refrain from the Boussinesq approximation. While the Mach number is small the fluid cannot be considered incompressible. One should take into account the effect of the density gradient, specifically, div $\mathbf{v} = 0$ must be replaced by div $\rho\mathbf{v} = 0$. The situation is unlike the case of laboratory turbulence which is usually characterized by two-dimensionless numbers, Re and Ma.

Turbulent convection acts on the mean magnetic field like magnetic diffusion or magnetic viscosity as was stressed by Leighton (1969) and other experts in the solar cycle. Crudely, the turbulent diffusivity is $v_T \simeq \frac{1}{3}\tau \langle v^2 \rangle$, where τ is a correlation time, say, the life time of supergranules, and $\langle v^2 \rangle$ is the mean square pulsation velocity. A more subtle effect which was first noted by Lebedinsky (1941) and Biermann (1951) is the anisotropy of the turbulent transport in the solar convective zone due to the dominant radial direction of motion. The different transport rates of the turbulent momentum in the radial and perpendicular directions is an essential factor in the mechanism supporting the solar differential rotation (Durney, 1976; Monin and Simuni, 1981; Rüdiger, 1982).

The mean magnetic field must diffuse anisotropically. For the sake of simplicity this effect is usually ignored in models of the solar dynamo.

Another feature of the solar turbulence is related to rotation. The Rossby number decreases from $R_0 = 16$ at d_1 to 4×10^{-2} at d_2, so that the direct influence of the rotation on the shape of the convective cells is important only near the bottom of the convective zone (Durney and Spruit, 1979). However, the rotation is of importance not only for the cell shapes. Owing to the rotation of stratified turbulent convection a qualitatively new feature – the mean helicity – arises; it plays an important role in the generation of the mean magnetic field (Moffatt, 1978; Parker, 1979; Krause and Rädler, 1980; Zeldovich *et al.*, 1983).

Let us give a demonstration of how the mean helicity arises. Consider a convective fluid element rising (or sinking) radially. It will expand (compress) due to the density gradient which results in additional velocity components $-v_\theta$, v_φ. The respective Coriolis torque gives the element the additional rotation. Crudely,

$$\frac{d}{dt} \operatorname{rot}_r \mathbf{v} \simeq -2\Omega \cos\theta \operatorname{div} \mathbf{v}. \tag{1}$$

The divergence is determined by the continuity equation $\operatorname{div} \mathbf{v} \approx -\nabla\rho\mathbf{v}/\rho \approx v/h$, where h is the density scale. In the upper part of the convective zone the correlation time $\tau = l/v$ is small in comparison with the period of revolution ($R_0 = v/l\Omega > 1$). Integrating (1) yields $-\operatorname{rot}_r \mathbf{v} = 2\Omega \cos\theta \tau(v/h)$. In the lower part of the zone $v/l\Omega < 1$, i.e. a convective element manages to make many revolutions in a time τ. Therefore one should take a mean value of $\operatorname{rot}_r \mathbf{v}$ over the period of revolution. Helicity is determined as the product of the speed of rise v_r with the additional vortex $\operatorname{rot}_r \mathbf{v}$. The equation of generation of the mean magnetic field contains the value $\alpha = -\frac{1}{3}\tau \langle \mathbf{v} \operatorname{rot} \mathbf{v} \rangle$, for which we estimate

$$\alpha \simeq \begin{cases} \dfrac{l^2\Omega}{h} \cos\theta, & R_0 > 1, \\[3mm] \dfrac{lv}{h} \cos\theta, & R_0 < 1. \end{cases} \tag{2}$$

The angular dependence is such that α vanishes on the equator and has different signs in the northern and southern hemispheres. According to models of the convective zone $l^2\Omega/h$ grows inward while lv/h increases outward. Thus the function $\alpha(r)$ has a maximum roughly at the point where $l\Omega \approx v$ ($d \approx 5 \times 10^9$ cm). The appearance of helicity means the violation of the mirror symmetry, the pseudoscalar $\alpha \sim \Omega\nabla\rho$ i.e. it is determined by the product of the pseudovector $\mathbf{\Omega}$ and the vector $\nabla\rho$. From this it is clear that just the rotation alone is not enough! Another vector, $\nabla\rho$ or $\nabla \langle v^2 \rangle$, is required. Note that the nonhomogeneous nature of the velocity fluctuations leads to still another interesting effect – the diamagnetic transport of the mean magnetic field (see, for example, Zeldovich *et al.*, 1983).

A hydromagnetic dynamo may arise only when the motions overcome the destructive action of magnetic diffusion. The simplest characteristic of this relation is the magnetic Reynolds number $R_m = lv/v_m$. In the convective zone it is very great, $R_m \simeq 2 \times 10^7$ at d_1 and 5×10^9 at the depth d_2. Therefore, the field of solar motions acts like a fast dynamo (Zeldovich *et al.*, 1983). The dynamo generates the mean (large-scale) magnetic field as well as the intermittent small scale structure (Molchanov *et al.*, 1984).

The main sources of the mean field generation are the differential rotation and the mean helicity. These sources operate against the background of the strong turbulent diffusion that smoothes gradients and accomplishes the diffusive transport and decay of the mean magnetic field. The back-action of the magnetic field upon the helicity and the differential rotation results in some interesting nonlinear effects: stabilization and modulation of oscillations, torsional waves, meridional circulation, and global minima.

3. The 22-Year Cycle

The magnetic hydrodynamics of mean field gives an explanation for the basic 22-year periodicity of the axisymmetric magnetic field. The respective equations are the simplest when the mean magnetic field is expressed through its azimuthal components $B_\varphi = B(t, r, \theta)$ and $A_\varphi = A(t, r, \theta)$ of the vector potential of the poloidal field (B_r, B_θ):

$$\frac{\partial A}{\partial t} = \alpha B + \eta \Delta A, \tag{3}$$

$$\frac{\partial B}{\partial t} = (\nabla \Omega \times \nabla A r \sin \theta)_\varphi + \eta \Delta B. \tag{4}$$

Here η is the magnetic diffusivity which includes the turbulent and molecular ones. In order to explain the nature of the solutions of the system (3)–(4) it is useful first to assume $\eta = 0$ and then to show the role of magnetic diffusion (Yoshimura, 1975). By introducing the variable $P = r \sin \theta A$ we obtain for it the equation

$$\frac{\partial^2 P}{\partial t^2} = r \sin \theta (\alpha \nabla \Omega \times \nabla P)_\varphi, \tag{5}$$

which is of the heat conductivity type if we change the positions of t and r. The solutions of (5) have the form of waves with increasing or decreasing amplitude that propagate along the surfaces $\Omega = $ const.

The direction of propagation of waves which Parker (1955) called dynamo-waves depends on sign of the product $\alpha \nabla \Omega$. Let Ω have only a radial component. Then the surface $\Omega = $ const are spherical and

$$\frac{\partial^2 P}{\partial t^2} = D \frac{\partial P}{\partial \theta},$$

where $D = r \sin \theta \alpha \, \partial \Omega / \partial r$ is the local dynamo number. We search for solutions of the form $P \sim \exp[qt]$ and obtain $q^2 + ikD = 0$, i.e. for $D < 0$

$$\omega = \operatorname{Im} q = (Dk/2)^{1/2},$$
$$\gamma = \operatorname{Re} q = (Dk/2)^{1/2}.$$

Returning to the field, we have a wave propagating from pole to equator:

$$A \sim e^{\gamma t} \cos(\omega t - k\theta),$$
$$B \sim e^{\gamma t} \cos(\omega t - k\theta + (\pi/4)). \tag{6}$$

For $D > 0$ the wave propagates into the opposite direction. The sign of $\partial\Omega/\partial r$ is the same in the northern and southern solar hemispheres, however, $\alpha \sim \cos\theta$, see (2), and so D changes sign. From this it follows that if $D < 0$ in the northern hemisphere the dynamo-waves propagate from the poles to the equator. The phase velocity of the waves is $v_{ph} = \omega/k = (D/k)^{1/2}$, the group velocity $v_g = d\omega/dk$ is a half of that. The phase shift between B and A is $\delta = \pi/4$, hence, B_r lags behind B_φ by $3\pi/4$.

In the other simple case of $\Omega(\theta)$ the dynamo-wave propagates radially. This situation is unlike the previous one. It does not resemble the behaviour of the field at the solar surface where the field moves from high latitudes to the equator as in the well-known Maunder butterfly diagram. However, the waves propagating radially may be connected if one takes turbulent diffusion into account (Yoshimura, 1975).

The involvement of turbulent magnetic diffusion allows one first of all to obtain a solution with constant amplitude, since $\gamma \to \tilde{\gamma} = \gamma - \eta\kappa^2$. The dynamo-number value corresponding to $\tilde{\gamma} = 0$ is usually called D_{cr} for the mean field generation. Besides, for $\eta \neq 0$, as is clear from (3), (4), the phase shift between the poloidal and azimuthal components is not exactly $\pi/4$.

The investigation of the dynamo equations with diffusion and other effects as well as the regular geometry and boundary conditions taken fully into account may be performed only by using computers. Numerical experiments about the solar dynamo theory were carried out by Steenbeck and Krause (1969), Stix (1976), Yoshimura (1979), Ivanova and Ruzmaikin (1977), and others.

In the linear theory an undamped or nongrowing solution may be obtained only for one distinguished value of the dynamo number $D = D_{cr}$. When back action of the magnetic field on turbulence is involved, nonlinear solutions of the limiting cycle type arise, at least for $D \geqslant D_{cr}$ (Figure 1 from the paper by Ivanova and Ruzmaikin, 1977).

The electromagnetic forces have a stronger effect on the mean helicity than the angular velocity. According to Kleeorin and Ruzmaikin (1982) $|\delta\Omega/\Omega| : |\delta\alpha/\alpha| \sim B_r/B_\varphi \, v/\Omega r \ll 1$, where v is the characteristic velocity amplitude of the turbulent pulsations. Therefore, one should first take into account the effect of the magnetic force upon helicity. For small nonlinearities it is natural to consider

$$\alpha_N = \alpha(r, \theta)(1 - \xi B^2), \tag{7}$$

where the parameter ξ is a measure of this effect.

An undamped solution of the nonlinear problem may be found as a series of eigen-vectors corresponding to the complex eigenvalues $p_n = \gamma_n + i\omega_n$:

$$\begin{pmatrix} A \\ B \end{pmatrix} = \sum_{n=1}^{\infty} F^n(t) \begin{pmatrix} a_n \\ b_n \end{pmatrix}.$$

After substituting the expression into Equations (3 and 4) with α_N instead of α we obtain a system of ordinary nonlinear differential equations for $F^n(t)$:

$$\frac{dF^n}{dt} = p_n F^n - \xi D \frac{dp_n}{dD} \sum_{l,m,s} K^n_{lms} F^l F^m F^s \tag{8}$$

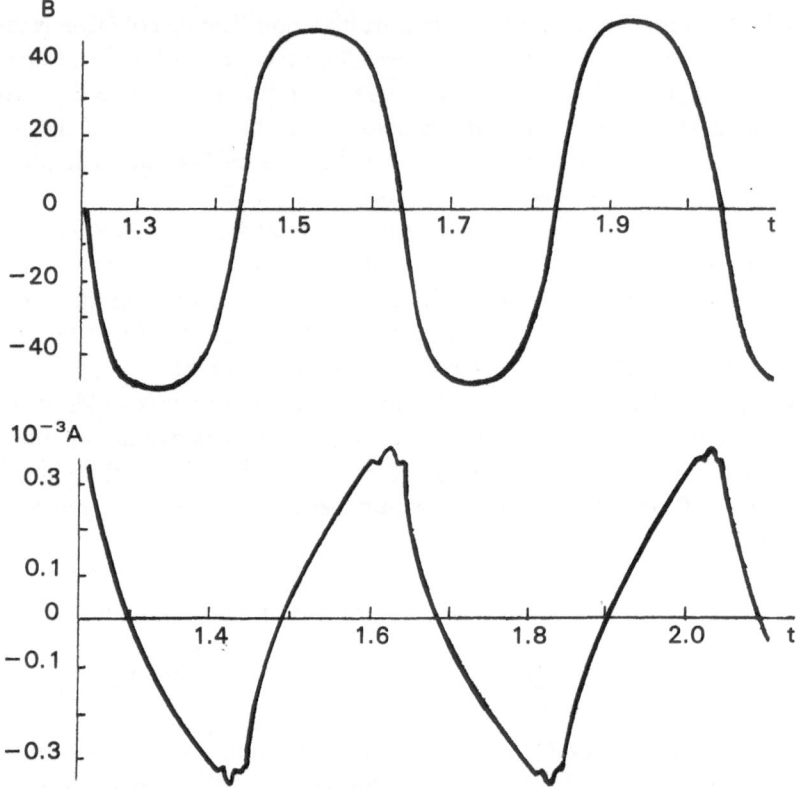

Fig. 1.   The time-dependence of the azimuthal magnetic field and the vector-potential of the poloidal field
at a given point (r, θ) in a strong non-linear case $D = 100\,D_{\mathrm{cr}}$ (Ivanova and Ruzmaikin, 1977).

(Kleeorin and Ruzmaikin, 1984a), where the coefficients K_{lms}^{n} are determined by the
eigen-vectors a^{n} and b_{m} of the linear problem.

In first approximation in $D - D_{\mathrm{cr}}$ it is sufficient to retain in Equation (8) two complex
conjugate functions $F^{1} = F$ and $F^{2} = F^{*}$ which correspond to a dipole type mode for
the mean magnetic field. The stationary solution for F can be interpreted as a limiting
cycle of circular form:

$$(\mathrm{Re}\,F)^{2} + (\mathrm{Im}\,F)^{2} = \mathrm{const}\,\frac{D - D_{\mathrm{cr}}}{D}.$$

The corresponding local limiting cycle in the $(B, \dot{B})$ plane has an elliptic form
(Ruzmaikin, 1981). It should be noted that by virtue of the non-conservativeness of the
dynamo the limiting cycles under consideration have no usual energetic sense inherent
in the dynamical systems similar to a pendulum clock.

The periodic solution has an interesting symmetry property $F(t + T/2) = -F(t)$, i.e.
it recurs with opposite sign after half a period (Bräuer, 1979; Kleeorin and Ruzmaikin,
1984a). The stability of the solution, as one may guess directly from (8), requires
$d\gamma/dD > 0$, i.e. the solutions with a decreasing dependence of γ on D are unstable.

Levy and Boyer (1982), Pudovkin and Benevolenskaya (1982), paid attention to the nonlinear feature of the Wolf number curve – alternation amplitudes of even and odd cycles (the Gnevyshev–Ohl's rule). This feature was explained by adding a (quasi) uniform component of the mean field to the periodic solution. In the even cycle it is added to the oscillating field, in the odd cycle it is subtracted. The origin of this component is still an enigma.

4. Regular Modulations of the Cycle

An increase in the dynamo number results in the excitation of several modes. The simplest case, two interacting modes with dipolar and quadrupolar symmetry, may be of interest in connection with the secular modulation of the solar cycle (Gleissberg, 1958; Yoshimura, 1979).

In first approximation it may be assumed that the dipole mode oscillates in the nonlinear regime as $f^d \exp(i\omega_d t)$. Two behaviours of the system are possible depending on the amplitude of the basic mode and the frequence difference $\Delta = \omega_d - \omega_q$. These are: (1) synchronization (the quadrupole frequency is captured by the dipolar one); and (2) nonlinear beats between the quadrupole and dipole modes. The quadrupole solution is searched for in the form $f(t) \exp i(\omega_q t + \psi(t))$; the amplitude and phase equations follow from (8):

$$\frac{df}{dt} = f[1 - f^2 - E(\cos\beta + d\sin 2\psi)],$$

$$\frac{d\psi}{dt} = \Delta - E(\sin\beta + d\cos 2\psi)$$

(Kleeorin and Ruzmaikin, 1984a), where

$$E = \frac{D - D_{cr}^d}{D - D_{cr}^q} \frac{K_{321}^3}{K_{211}^1} \sim |f^d|^2,$$

β is a sum of the dP/dD and K_{321}^2 phases, $d = \frac{1}{2}|K_{411}^3|/|K_{321}^3|$. Studying these equations allows one to mark the synchronization and beat regions in the (Δ, E) plane. The beats are more advantageous in the solar case, when $D > D_{cr}^q > D_{cr}^d$ and $E = 0(1)$. Their period can be interpreted as the period of cycle modulation. The real picture, however, is much more complicated (see Section 6).

5. Torsional Waves

The discovery of torsional waves imposed on the general differential rotation (Howard and LaBonte, 1980) has been one of the most remarkable events of recent years. The wave period is close to 11 years, the amplitude is about 3–6 m s^{-1}. In each hemisphere there are four zones of fast and slow rotation. A new zone arises near the pole and reaches the equator in 22 years. Along with the torsional waves periodic meridional flows are likely to exist as an analysis of the sunspot motions shows (Tuominen *et al.*, 1983).

The good correlation of the torsional waves with magnetic activity (Howard, 1984) indicates that the former may be explained as a backaction effect of the dynamo-waves on the differential rotation. This idea was proposed by Yoshimura (1981) and Schüssler (1981), and, after criticism by LaBonte and Howard (1982), developed further by Tuominen *et al.* (1984), and Kleeorin and Ruzmaikin (1984b).

The rotational velocity may be written as

$$V_\varphi = \Omega(r, \theta)r \sin \theta + u(r, \theta, t).$$

Near the solar surface the approximate solution for the additional contribution produced by the magnetic force is

$$u \simeq - \frac{h}{\rho v_T} \frac{B_r B_\varphi}{4\pi} \quad \text{at} \quad r = R_\odot - d,$$

where ρ is the density, h the heigh scale, v_T the turbulent viscosity and (B_r, B_φ) the mean magnetic field. By substituting the dynamo wave,

$$\begin{aligned}
B_\varphi &= b_\varphi(r, \theta) \cos(\omega_c t - g(r, \theta)), \\
B_r &= b_r(r, \theta) \cos(\omega_c t - g - (\pi/2) - \delta),
\end{aligned} \tag{9}$$

where ω_c and g are the cycle frequency and phase, we obtain two terms:

$$u = u_0(r, \theta) - u_1(r, \theta) \sin(2\omega_c t - 2g - \delta). \tag{10}$$

The first term (u_0) changes the stationary distribution of the differential rotation. In particular, if there are two modes (of the dipolar and quadrupolar type), a contribution of the type $\Delta\Omega \sim c_1 \cos \theta + c_2 \cos^3 \theta$ appear that is anti-symmetric with respect to the equator.

The time-dependent part (torsional waves) varies with a period equal to half the period of the cycle $2\pi/2\omega_c = T_c/2 = 11$ years. If we assume that near the surface $b \sim 100$ G, $b_r \sim 1$ G, we may roughly estimate the amplitude of the torsional wave:

$$u_1 \sim \frac{b_\varphi b_r}{8\pi\rho v_T} \sim 3 \text{ m s}^{-1}$$

(Kleeorin and Ruzmaikin, 1984b). Note that since $b_r(\theta)$ grows towards the poles and $b_\varphi(\theta)$ concentrates at low latitudes, the u_1 distribution is more uniform in θ. For example, for $b_r \sim \cos \theta$, $b_\varphi \sim \sin 2\theta$ we have $u_1 \sim \cos^2 \theta \sin \theta$.

According to observations by LaBonte and Howard (1982) the maximum of the torsional wave is shifted in phase as comparted with the maximum of the magnetic flow (in the zone of solar spots) by $\frac{1}{8}$ of the torsional wave length. It is easy to show from (8) and (9) that a phase shift of that value between u and $|B_\varphi|$ takes place for $\delta = \pi/4$ which agrees with the dynamo models (see Section 3).

6. Magnetohydrodynamic Chaos

In reality, the solar cycle is not simply a limiting cycle or the limiting cycle on which regular secular modulations are imposed. The discovery of the Maunder minimum and other global minima (Eddy, 1976) reveals the global stochastic nature of solar activity. From a modern point of view it is natural to connect it with a strange attractor (Ruzmaikin, 1981; Zeldovich and Ruzmaikin, 1983).

The strange attractor concept (Ruelle and Takens, 1971) arose from a study of the well known Lorenz system for the problem of convection. One of the first attempts to apply this concept to the solar dynamo theory was to reduce the leading equations, which involved back-action of the magnetic field on the helicity, to the Lorenz system:

$$A = -A + DB - CB,$$
$$B = -\sigma B + \sigma A, \tag{11}$$
$$C = -\nu C + AB.$$

Here A and B are the azimuthal components of the vector potential and the mean field, C is the helicity excess due to magnetic helicity (the direct effect upon α of type (2) is omitted for simplicity), σ and ν are positive constants, one of which, $\sigma = O(1)$, describes the difference between the diffusivities A and B, while the second, $\nu \ll 1$, determines the decay rate of the magnetic helicity due to molecular magnetic diffusion, hence, it is very small in the convective zone.

Rhetorically speaking, the activity terminates because the Sun may, in principle, end up without any magnetic field, since the system (11) has the singular solution $(A, B, C) = (0, 0, 0)$. For small dynamo numbers $D < 1$ the zero solution is stable. When $D > 1$ two additional singular points appear, $S_{\pm} = (\pm \nu(D-1)^{1/2}, \pm \nu(D-1)^{1/2}, D-1)$. In a certain region of the parameters, in the solar context for $\sigma \approx 1$, $\nu \ll 1$, $D_{cr} \approx 4(\sigma - 1)^{-1}$, the trajectory of the system (11) travels quasi-periodically in the space (A, B, C) from the vicinity of the point S_{+} to the vicinity of S_{-} and backwards. In this process the trajectory rarely and incidentally gets close to the singular point $(0, 0, 0)$. The representative point (solution) stays in this vicinity for about a time $\tau_M \sim \nu^{-1}(D - D_{cr})^{-1/2}$, i.e. very long. An approximate derivation of the equation for $\dot{C}$ with small ν is given by Kleeorin and Ruzmaikin (1982).

The simplified system of dynamo equations (11) has an essential defect: it does not contain the dynamo waves which are present in the original MHD equations of the mean field (see Section 3).

In using Equation (11) one may interpret the basic 11-year cycle as the transfer from S_{+} to S_{-} and backwards. The frequency of this process is of the order of $(\sigma D)^{1/2}$.

A more realistic simplification of Equations (3) and (4) for the mean magnetic field, involving dynamo waves, yields complex equations for A and B. Such an approach was developed by Weiss et al. (1984). However, they took into account the back-action of the magnetic field on the differential rotation instead of the mean helicity. A sixth order system of equations for the complex functions $A(t)$, $B(t)$, $\omega(t)$ was obtained as

$$\dot{A} = -A + DB,$$
$$\dot{B} = -B + iA - \tfrac{1}{2}\omega A^*, \tag{12}$$
$$\dot{\omega} = -\tilde{v}\omega - iAB,$$

which is a complex extension of the Lorenz system (11). The parameters $\tilde{v}$ and D are real and positive. For $\tilde{v} < 1$ and with D growing a sequence of bifurcations arises that eventually results in a chaotic behaviour of the system with periods of very low magnetic activity.

The trivial solution $(0, 0, 0)$ is stable for $D < 2$. The first bifurcation at $D_1 = 2$ gives rise to an oscillatory solution corresponding to the dynamo wave. Moreover, for $D \geqslant 2$ the system (12) has an exact periodic solution

$$B = |B|\,e^{ipt}, \qquad A = |A|\,e^{i(pt + \varphi)}, \qquad \omega = |\omega|\,e^{2ipt},$$

where for large D: $p \approx 4D/(2 + \tilde{v})$, $|A| \approx 8D/(2 + \tilde{v})$, $|B| \approx |\omega| \approx 16D/(2 + \tilde{v})^2$. This periodic solution is the limiting cycle in the six-dimensional phase space (A, B, ω). In the plane (Re A, Re B) it has the form of an ellipse. Numerical calculations for $\tilde{v} = 0.5$ show that the solution becomes unstable at $D_2 \approx 4.14$. After this bifurcation the solution becomes doubly-periodic with two different frequences (a 2-torus in the phase space). In turn the doubly-periodic solution becomes unstable at $D_3 \simeq 6.94$, when a three-frequency solution appears. After $D_4 \simeq 7.612$ a cascade of solutions with a double

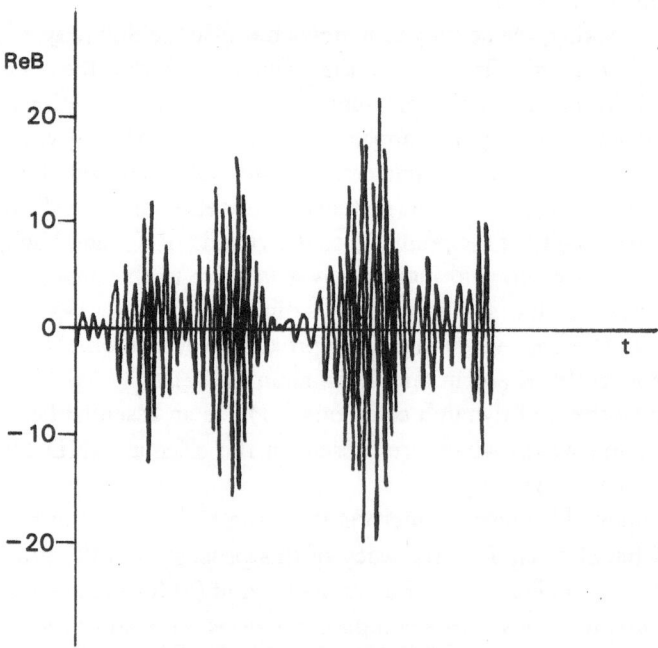

Fig. 2.   A fragment of the time behavior of the magnetic field with the episodes of reduced activity in the model by Jones *et al.* (1984), $D = 32$.

frequency originates (the Feigenbaum cascade), it converges at $D_{\text{chaos}} \simeq 7.68$. For large dynamo-numbers the solutions are chaotic in time with episodes of very low amplitude. An example of the solution for Re B and $D = 32$, taken from the paper by Weiss *et al.* (1984), is given in Figure 2.

It should be noted that, unlike the parameter v in the system (11), $\tilde{v}$ in the convective zone seems to be on the order of $O(h/l) > 1$, where h is the depth of the convective zone and l the density scale height, because $\tilde{v}$ is determined by the back time of the angular velocity decay due to turbulent viscosity. This emphasizes once more the model nature of the above example. Its actual merit is that it gives an indication that the global MHD chaos exists in the solar dynamo; besides the examples allow the evaluation of the order of magnitude. A detailed investigation of the problem, in which Equations (11) or (12) are replaced by partial differential equations, is possible only by using up-to-date computers.

7. Non-Axisymmetric Large-Scale Field

So far we have dealt with an axisymmetric φ-independent component of the large-scale field. It consists of a poloidal dipole-type field with a possible small addition of a quadrupolar field, and a subphotospheric azimuthal magnetic field.

Observations showed also the non-axisymmetric components of the mean magnetic field of the Sun (Altschuler *et al.*, 1974). There were first indications of complexes of solar activity concentrating on separate longitudes and existing for a few months. The sector structure of the interplanetary magnetic field connected with the Sun, and the coronal holes as the sources of the solar wind were then discovered. The spatial nature of the solar coronal structure and, in particular, the current sheet deformed with respect to equator (Svalgaard and Wilcox, 1978) confirmed the existence of the non-axisymmetric large-scale field. Unlike the axisymmetric field the three-dimensional component here does not noticeably change with the solar cycle period; the frequency of its variations is close to that of the solar rotation. The spatial distribution of the non-axisymmetric field is distinctly symmetrical with respect to the equatorial plane.

It is natural to associate the origin of this field with non-axisymmetric three-dimensional solutions of the solar dynamo. These solutions of the mean field equations exist despite the fact that the sources of the field generations (the differential rotation and the mean helicity) do not depend on φ (Stix, 1971; Krause, 1971). Stix (1974) was the first to discuss the sector structure of the solar magnetic field by using such an approach. A thorough numerical investigation of three-dimensional non-axisymmetric fields has been carried out quite recently (Ivanova and Ruzmaikin, 1985). The results are briefly given below.

Instead of the axisymmetric dynamo equations (3), (4), we consider a complete three-dimensional systems of the mean field equations

$$\frac{\partial \mathbf{B}}{\partial t} = \text{rot}\left(\mathbf{B} \times \mathbf{r} \times \boldsymbol{\Omega} + \alpha \mathbf{B} - \beta \, \text{rot} \, \frac{\mathbf{B}}{\mu} \right),$$

where Ω is the differential rotation, α is the mean helicity, β and μ are the turbulent diffusion and magnetic permeability, respectively. The non-axisymmetric case is characterized, instead of by one dynamo number, by two dimensionless numbers

$$D = \frac{\Omega_0 \alpha_0 R_\odot^3}{\beta_0^2} \quad \text{and} \quad Q = \frac{\alpha_0}{\Omega_0 R_\odot}.$$

Other combinations of these numbers may certainly be considered. In the solar convective zone $D \gtrsim 10^4$, $Q < 1$.

Instead of three components of the mean field it is convenient to use two scalar functions $\Phi(r, \theta, \varphi, t)$ and $\Psi(r, \theta, \varphi, t)$ by employing solenoidality (div $\mathbf{B} = 0$):

$$\mathbf{B} = \text{rot}(\Psi \mathbf{r} + R_\odot \text{ rot } \Phi \mathbf{r}).$$

The solution technique for the system of two equations involving $\Phi \sim \exp \text{ im } \varphi$ and $\Psi \sim \exp \text{ im } \varphi$ under various boundary conditions and with given functions $\Omega(r, \theta)$, $\alpha(r)$ cos θ, $\beta(r)$, $\mu(r)$ is described by Ivanova and Ruzmaikin (1984).

In the axisymmetric case $m = 0$ the solution coincides with the dynamo-wave which is obtained from (3), and (4). The curve separating the region of undamped solutions (top) from that of the damped ones for $m = 1$ is presented in Figure 3 for the mode antisymmetric (in B_r) with respect to equator. The maximum of differential rotation is concentrated near the bottom of the convective zone $r_\Omega = 0.7 R_\odot$, the maximum of mean helicity is at $r_\alpha = 0.9 R_\odot$. The critical dynamo numbers for all $Q \leqslant 1$ exceed D_{cr} for the axisymmetric solution. The solution oscillates with a frequency close to angular velocity near the bottom. A mode with B_r symmetric with respect to the equatorial plane has somewhat lower (however close) critical dynamo numbers.

The excitation of the second angular harmonics $m = 2$ prove to be harder to achieve.

A reduction in the critical dynamo number can be obtained by bringing together the maxima of the generation sources. At $r_\Omega = 0.85 R_\odot$, $r_\alpha = 0.9 R_\odot$, D_{cr} is even smaller for $m = 1$ than for the axisymmetric mode. However, models with close sources are hardly realistic since the period of oscillation of the mode $m = 0$ becomes very short ($\leqslant 22$ yr).

The non-axisymmetric solution has the form of a wave propagating azimuthally: $\cos(\omega t - m\varphi)$, and in its steady state it does not depend on the 22-year periodicity.

A comparison between the results and observations meets some difficulties. First of all, the observators stress the primary excitation of the four-sector field which corresponds to the equatorial quadrupole $m = 2$. However, according to the theory, the excitation of the harmonics $m = 1$ is preferable, i.e. the excitation of an equatorial dipole rotating with an internal angular velocity of the convective zone. The solution of this problem is obtained, possibly, by using the near-equality of the internal angular velocity and $2\Omega_0$ (see, for example, Dicke, 1983). Then the rotation of the equatorial dipole with a doubled angular velocity may be treated as the rotation of quadrupole with the surface angular velocity. They may be differentiated by simultaneous observations of the sector boundary rotations and independent tracers of the solar rotation, e.g., sunspots.

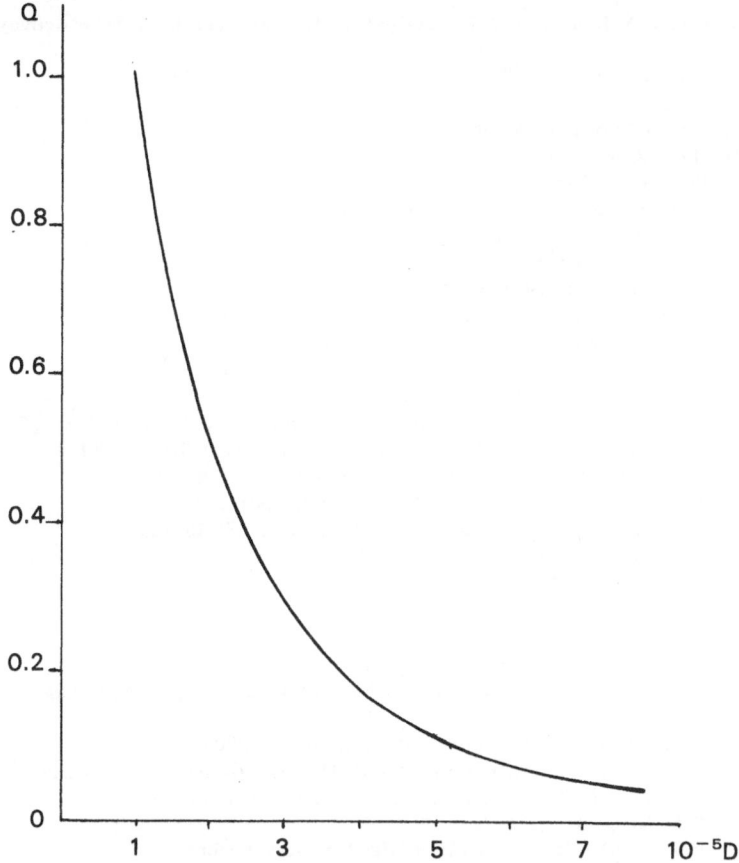

Fig. 3.   The curve separates the domain (top) of the undamped non-axisymmetric solutions with $m = 1$ from that of the damped ones (Ivanova and Ruzmaikin, 1984).

Another difficulty in the interpretation is due to the fact that the observations suggest a few non-axisymmetric harmonics (Altschuler *et al.*, 1974). In this connection one should bear in mind that the developed theory of dynamo is so far linear in the magnetic field. Excitation of a few harmonics and even of the harmonics $m = 1$ (because $m = 0$ is more preferable) means the establishment of nonlinear conditions. Under these conditions separation of the modes in m is impossible, although one of them, say $m = 1$, may be dominant. However, the solution of the nonlinear non-axisymmetric problem needs to overcome serious mathematical difficulties and is yet to be found.

References

Altschuler, M. D., Trotter, D. E., Newkirk, G. Jr., and Howard, R.: 1974, *Solar Phys.* **39**, 3.
Babcock, H. W.: 1961, *Astrophys. J.* **133**, 572.
Biermann, L.: 1951, *Z. Astrophys.* **28**, 304.
Bräuer, H.: 1979, *Astron. Nachr.* **300**, 43.
Carrington, R. C.: 1863, *Observations of the Spots on the Sun*, Williams and Norgate, London.
Dicke, R. H.: 1983, *Nature* **300**, 693.

Durney, B. R.: 1976, in V. Bumba and J. Kleczek (eds.), 'Basic Mechanisms of Solar Activity', *IAU Symp.* **71**, 243.

Durney, B. R. and Spruit, H. C.: 1979, *Astrophys. J.* **234**, 1067.

Eddy J. A.: 1976, *Science* **192**, 1189.

Elsässer, W. M.: 1946, *Phys. Rev.* **69**, 106.

Gleissberg, W.: 1958, *J. Brit. Astron. Assoc.* **68**, 148.

Gough, D. O.: 1982, *Nature* **298**, 334.

Hill, H., Bos, R., and Goode, P.: 1982, *Phys. Rev. Letters* **49**, 1794.

Howard, R.: 1984, *Ann. Rev. Astron. Astrophys.* **22**, 131.

Howard, R. and Harvey, J. W.: 1970, *Solar Phys.* **12**, 23.

Howard, R. and LaBonte, B. J.: 1980, *Astrophys. J.* **239**, L33.

Ivanova, T. S. and Ruzmaikin, A. A.: 1977, *Astron. J. USSR* **54**, 846.

Ivanova, T. S. and Ruzmaikin, A. A.: 1985, *Astron. Nachr.* **306**, 177.

Jones, C. A., Weiss, N. O., and Cattaneo, F.: 1985, *Physica* **D14**, 161.

Krause, F.: 1971, *Astron. Nachr.* **B293**, 187.

Krause, F. and Rädler, K.-H.: 1980, *Mean-field Magnetohydrodynamics and Dynamo Theory*, Pergamon Press.

Kleeorin, N. I. and Ruzmaikin, A. A.: 1981, *Geophys. Astrophys. Fluid Dyn.* **17**, 281.

Kleeorin, N. I. and Ruzmaikin, A. A.: 1982, *Magnetohydrodynamics* **18**, 116.

Kleeorin, N. I. and Ruzmaikin, A. A.: 1984a, *Astron. Nachr.* **B305**, 265.

Kleeorin, N. I. and Ruzmaikin, A. A.: 1984b, *Pisma Astron. J. USSR* **10**, 925.

LaBonte, B. J. and Howard, R.: 1982, *Solar Phys.* **75**, 161.

Levy, E. H. and Boyer, D.: 1982, *Astrophys. J.* **254**, L19.

Larmor, J.: 1919, Rep. Brit. Assoc. Adv. Sci., p. 159.

Leighton, R. B.: 1969, *Astrophys. J.* **156**, 1.

Lebedinsky, A. J.: 1941, *Astron. J. USSR* **18**, 10.

Moffatt, H. K.: 1978, *Magnetic Field Generation in Electrically Conducting Fluid*, Cambridge Univ. Press, Cambridge.

Monin, A. S. and Simuni, L. M.: 1982, *Proc. Nat. Acad. Sci.* **79**, 3903.

Molchanov, S. A., Ruzmaikin, A. A., and Sokoloff, D. D.: 1984, *Geophys. Astrophys. Fluid Dyn.* **30**, 242.

Newton, H. W. and Nunn, M. L.: 1951, *Monthly Notices Roy. Astron. Soc.* **111**, 413.

Parker, E. N.: 1955, *Astrophys. J.* **122**, 293.

Parker, E. N.: 1979, *Cosmical Magnetic Fields*, Clarendon Press, Oxford.

Pudovkin, M. I. and Benevolenskaya, E. E.: 1982, *Pisma Astron. J. USSR* **8**, 506.

Ruelle, D. and Takens, F.: 1971, *Comm. Math. Phys.* **20**, 167.

Rüdiger, G.: 1982, *Geophys. Astrophys. Fluid Dyn.* **21**, 1.

Ruzmaikin, A. A.: 1981, *Comm. Astrophys.* **9**, 85.

Simon, G. W. and Weiss, N. O.: 1968, *Z. Physik* **69**, 435.

Schüssler, M.: 1981, *Astron. Astrophys.* **94**, L17.

Stix, M.: 1971, *Astron. Astrophys.* **13**, 203.

Stix, M.: 1974, *Astron. Astrophys.* **59**, 73.

Stix, M.: 1981, *Solar Phys.* **74**, 79.

Steenbeck, M. and Krause, F.: 1969, *Astron. Nach.* **291**, 49.

Svalgaard, L. and Wilcox, J. M.: 1978, *Ann. Rev. Astron. Astrophys.* **16**, 429.

Tuominen, J., Tuominen, I., and Kyröläinen, J.: 1983, *Monthly Notices Roy. Astron. Soc.* **205**, 691.

Tuominen, I., Virtanen, A., Krause, F., and Rüdiger, G.: 1984, in *The Hydromagnetics of the Sun*, Fourth European Meeting on Solar Physics, The Netherlands, p. 225.

Weiss, N. O., Cattaneo, F., and Jones, C. A.: 1984, *Periodic and Aperiodic Dynamo Waves*, Preprint.

Yoshimura, H. A.: 1975, *Astrophys. J.* **201**, 740.

Yoshimura, H. A.: 1979, *Astrophys. J.* **227**, 1047.

Yoshimura, H. A.: 1981, *Astrophys. J.* **247**, 1102.

Zeldovich, Ya. B. and Ruzmaikin, A. A.: 1983, *Astrophys. Space Phys. Rev.*, Vol. 2, p. 333, Gordon and Breach, New York.

Zeldovich, Ya. B., Ruzmaikin, A. A., and Sokoloff, D. D.: 1983, *Magnetic Fields in Astrophysics*, Gordon and Breach.

Zwaan, C.: 1978, *Solar Phys.* **60**, 213.

THE SOLAR DIFFERENTIAL ROTATION:
PRESENT STATUS OF OBSERVATIONS*

E. H. SCHRÖTER

Kiepenheuer-Institut für Sonnenphysik, Schöneckstr. 6, 7800 Freiburg, F.R.G.

Abstract. The present status of observations regarding the solar differential rotation is reviewed from contributions published in the last two decades. The paper does not deal with the theory; it mentions theoretical aspects only where they are needed to guide and to understand observational efforts and results.

1. Overture

The determination of the Sun's rotation, its dependence on latitude and depth, its possible variations with the solar cycle activity as well as the search for meridional circulation patterns and/or the presence of Reynold's surface stresses has become an interesting branch of solar physics with important implications for stellar physics. In recent years solar-like magnetic activity has been detected in many stars. Our present understanding of the solar activity is that magnetic fields are built up in a periodically working dynamo process maintained within the convective zone by the interaction of differential rotation and turbulent motions. This leads us to speculate that each Main-Sequence star which has a convective zone should show a solar-like magnetic activity. The logical guiding line in this hypothesis is that rotation and turbulent motions result somehow in a differential rotation of the star, since this is essential for maintaining the turbulent dynamo process.

The Sun is the only star which gives us the key to understanding how differential rotation is produced and the way in which the strength and periodicity of activity depend upon the latitude and depth variation of the differential rotation.

There are several reviews on solar rotation in the literature (Solonsky, 1977; Howard, 1978; Paternó, 1978; Schröter and Wöhl, 1978; Gilman, 1980; Howard, 1984). Some of them contain historical retrospects which are omitted here for conciseness. In this report the discussion is restricted to the present status of observations and mentions theoretical aspects only when they are needed to guide or to understand observational results.

The modern era of solar rotation measurements started in the sixties. In retrospect, one identifies two more or less independent reasons: Photoelectric detectors became available at that time and the well-known Doppler-compensator method replaced the tiresome photographic method producing efficiently a bulk of velocity-data with rather high accuracy (see e.g. the famous paper of Howard and Harvey, 1970). Almost simultaneously with this instrumental development theorists came into a position to perform numerical computations of models of the solar differential rotation (e.g.

* Mitteilungen aus dem Kiepenheuer-Institut Nr. 250.

Solar Physics **100** (1985) 141–169.   0038–0938/85.15.

Kippenhahn, 1963; Köhler, 1970; Busse, 1970; Durney and Roxburgh, 1971). This stimulated observers greatly to prove or disprove predictions of such models.

The measurement of solar rotation rate as a function of latitude and height – $\Omega(\phi, h)$ – seems, at first glance, an easy observational task. Indeed, above a 5% accuracy level a unique result is obtained for $\Omega(\phi)$ from all types and methods of measurements.

ILLUSTRATION OF THE PROBLEM AREA BEHIND
THE WORDING „SOLAR DIFFERENTIAL ROTATION"

A) LATITUDE – DEPENDENCE OF Ω :   (EQUATOR-ON-VIEW)

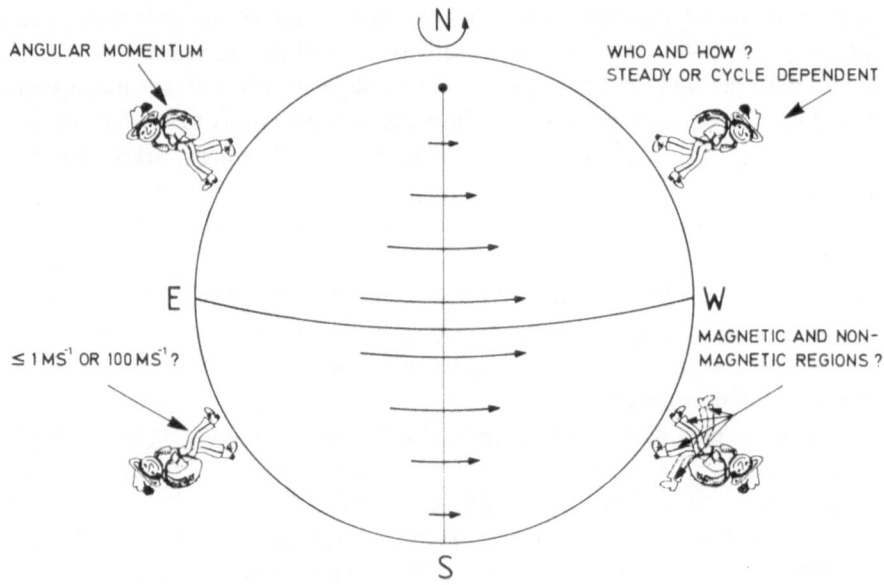

B) RADIAL - DEPENDENCE OF Ω : (POLE-ON-VIEW)

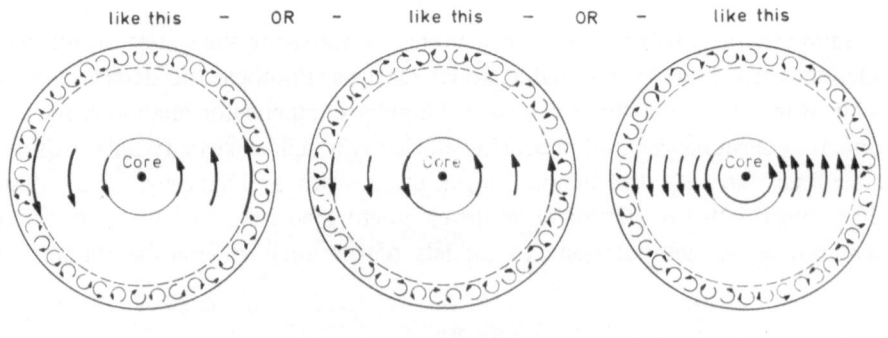

− DRAWINGS NOT ON SCALE −

Fig. 1.   Schematic illustration of the problem area behind the wording: solar differential rotation.

The solar sidereal angular rotation velocity of photospheric layers equals 14.1 to 14.4 deg day^{-1} (or 1.98–2.02 km s^{-1}) at the equator and decreases towards the poles where Ω (90°) is 10.07 deg day^{-1} (Howard *et al.*, 1983; Snodgrass *et al.*, 1984).

Setting the level to 1% changes the situation completely. Distinct variations of Ω occur now depending on whether spectroscopic or tracing methods are applied and what types of tracer structures are used and even the age of the tracers seems to play a significant role (see Sections 5 and 7). The basic problem area behind the wording 'solar differential rotation' is illustrated in Figure 1.

Neglecting inessential details, there are presently two particular classes of theories for solar differential rotation. In the first class of models, the transport of angular momentum towards the equator is ascribed to meridional circulations which are maintained either by an anisotropic viscosity or by a latitude dependent convective energy transport (Kippenhahn, 1963; Köhler, 1970; Durney and Roxburgh, 1971; Belvedere and Paternó, 1977; Belvedere *et al.*, 1980; Schmidt, 1982). Meridional motions of the order of a few m s^{-1} are sufficient to maintain the observed differential rotation. In the second class of models, the transport of angular momentum towards the equator is ascribed to Reynold stresses which originate when there is a correlation between two components of the turbulent velocities. Large-scale, long-lived cells ('giant-cells'), if existing, would show such a correlation caused by Coriolis forces (Busse, 1970; Busse and Carrigan, 1974; Gilman, 1978; Gilman and Glatzmaier, 1981; Rüdiger, 1982). Again motions of a few m s^{-1} are needed to produce at the surface Reynold stresses of the order of 2×10^3 m^2 s^{-2}, sufficient to maintain the observed solar differential rotation.

This short excursion into theory demonstrates clearly that the observer's task is presently not so much to establish a very precise form for Ω, but rather to search for systematic large-scale and long-lived motions on the Sun's surface (or if possible also in deeper layers) with typical velocities of less than 5 m s^{-1}, which is below 1% of the solar rotation rate. If the solar differential rotation is expected to show slight time-variations in connection with the solar cycle, the large-scale velocity system has to change its structure and in order to establish this we need an accuracy of 1–2 m s^{-1} or even better.

2. Reference System and Representation of the Solar Differential Rotation

In order to determine the Sun's rotation and to demonstrate the existence of meridional motions with the accuracy mentioned above, one needs a very accurate coordinate reference system. An obvious choice is to use heliographic coordinates. These are based on the position of the solar rotation axis within the ecliptic plane and specified by the two coordinates Ω'' and i (i is the inclination of the Sun's equator to the ecliptic plane and Ω'' is the longitude of the ascending node of the solar equator on the ecliptic). These two rotational elements determine the annual variations of the angle P (the angle between the north direction on the sky and the solar northpole) and B_0 (the annual tipping of the solar axis towards and away from the observer). The rotational elements

Ω'' and i have been derived and published by Carrington in 1863 by minimizing meridional motions of sunspots. They have been used uncritically by solar observers since this time. Carrington gave no error limits for Ω'' and i, but Stark and Wöhl (1981) showed that an error of $0°5-3°0$ in Ω'' and $0°3$ in i can not be excluded from his data and it is obvious that such an error can hide actually existing or produce artificial meridional motions. In 1963 Trellis recognized, analysing sunspot data from the Greenwich Photoheliographic Results, that the symmetry axis of the sunspot activity zone deviates from Carrington's rotation axis by $\approx 0°5$. Schröter and Wöhl (1975) attributed the observed remarkable difference in the rotation rate of Ca^+-mottles between the northern and southern hemisphere to an incorrectness of Ω'' and i.

Meanwhile several re-investigations of Ω'' and i have been undertaken, although the achieved accuracy is still not satisfactory. Wöhl's (1978) redetermination of Ω'' and i from his own plasma measurements are based on relatively few data. Clark et al. (1979) concluded from sunspot tracings Ω'' to be larger by $0°27 \pm 0°2$ and i to be smaller by $0°03 \pm 0°02$ as compared to Carrington. Stark and Wöhl (1981) arrived independently at $\Omega'' > \Omega''_{carr}$ and $i < i_{carr}$ within limits of one standard deviation for Ω and 3σ for i. LaBonte's (1981) determination of Ω'' and i from 1540 full disk velocity observations at Mount Wilson came out with relatively high errors of $1°2$ and $0°15$ respectively, as compared to determinations from sunspot data. This is indeed surprising and could be due to the very straightforward method he used. Reluctant to determine Ω'' and i independently from the Mt. Wilson data, he looked whether or not significant corrections to the Carrington values could be established within the range of the rather large error limits. Incorrectness of Ω'' and i should clearly result in an annual periodicity when looking at the power-spectra of residuals of measured Doppler data. Such an analysis from Mt. Wilson Doppler data and from sunspot proper motions (Greenwich material) has been undertaken by Kuhn and Warden (1979). They found from both data periodicities at 184^d and 364^d, attributing them to 'spurious seasonal effects'.

It has become increasingly common to represent the latitude dependence of the solar angular rotation by the formula

$$\Omega(\phi) = A + B \sin^2 \phi + C \sin^4 \phi, \tag{1}$$

where ϕ stands for the latitude, and the coefficients A, B, C are to be determined from measurements. The recognition that the two coefficients B and C are almost the same in magnitude and sign and that variations of B are strongly anticorrelated with variations of C, as well as the fact that the above formula does not represent an expansion in terms of orthogonal polynomials, were subjects of a lengthy discussion in the recent literature on the usefulness of this representation (Stenflo, 1977; Howard and Yoshimura, 1976; Duvall and Svalgaard, 1978). However, it has been pointed out by Howard et al. (1980) that the conversion of the line-of-sight velocity to angular velocity needs terms which introduce non-orthogonality (for sunspot position measurements it is generally assumed that $C = 0$).

Snodgrass (1984) very recently presented a very interesting attempt at orthonormal-ization with respect to the solar disc not only of the rotation function but also of the limb shift and meridional flow. He demonstrated clearly that by using Gegenbauer polynomials for the angular velocity function (normalized) the crosstalk between the coefficients A, B, C disappears almost completely for measurements of plasma rotation velocities.

In the following we use the conventional formula (1) to represent the solar differential rotation, whereby Ω, A, B, and C are given in units deg day^{-1} and Ω is considered as the sidereal angular velocity. Wherever comparison is needed with Doppler shift measurements, Ω is also discussed in terms of km s^{-1} or m s^{-1}.

3. The Spectroscopic Method; Sources of Errors, Limitations

We discuss here the most common photoelectrical Doppler-compensator method only. This is a very efficient and sensitive observational procedure in which the Doppler shift of a spectral line is measured via a pair of exit slits while scanning with the entrance slit of the spectrograph over the solar disc. At Mt. Wilson and Stanford full disc velocity measurements on the basis of this method are performed daily as routine work since more than a decade and meanwhile several millions of line-of-sight velocity data were obtained at these and other observatories like Kitt Peak, Crimea, Locarno, Capri, etc. For a typical medium strong Fraunhofer-line the intensity signals at the two exit slits have to be compensated below 5×10^{-4} in order to achieve an accuracy in the velocity of ≈ 5 m s^{-1}. This is in principle feasible, but there are many obstacles which jeopardize this goal. We shall distinguish between error sources and noise of solar and instrumental origin.

Among the known resolved solar effects, local motions cause mainly noise, while unresolved motions produce chiefly systematic errors. Noise of solar origin is produced by the following phenomena:

(a) granular motions;

(b) supergranulation flow pattern;

(c) oscillations.

The characteristics of these phenomena as well as the methods to minimize their influence on rotation data have been discussed in detail by Beckers and Canfield (1975) and by Beckers (1981). The result of this discussion is that even with sophisticated methods which lead to averages in space and time it is almost impossible to bring this noise down to < 10 m s^{-1} from one day observations only. The most severe obstacle is the supergranulation flow pattern because of its large size and long life-time. Rotation, meridional motions and large scale circulation patterns within this accuracy domain can only be extracted from full disc velocity data, when measurements of many consecutive days are compared.

Systematic errors are introduced into the rotation measurements by effects from unresolved motions, particularly by

(d) incorrectly eliminated limb shift (LS);

(e) the neglect of the variation of the line-asymmetry ('C-shape') across the solar disc;

(f) the neglect of the difference in the C-shape between quiet and active regions.

Regarding (d): the limb shift – the red shifting of the line centres of medium strong solar lines when approaching the solar limb – is a non-monotonic function of the heliographic angle ρ (between the measured point and the observer) as has been shown by various observations (Adam, 1959; Schröter, 1959; Appenzeller and Schröter, 1967; Adam *et al.*, 1976; Kubicela and Karabin, 1977; Duvall, 1979; Bruning, 1981; Brandt and Schröter, 1982; Koch, 1983, 1984; Andersen, 1984). The shape of the LS(ρ) function varies from line to line in a yet not fully understood manner. For the Fe I-line $\lambda 5250$ the following representations of the LS(ρ)-variation have been used by the Mt. Wilson observers in the last decade:

Howard and Harvey (1970):

$$\text{LS}(\rho) = e_1(1 - \cos \rho)^2 \quad \text{with} \quad e_1 = 332 \text{ m s}^{-1}, \quad \text{LS}(\cos \rho = 0.2) = 212 \text{ m s}^{-1};$$

Howard *et al.* (1980):

$$\text{LS}(\rho) = e_2(1 - \cos \rho)^N \quad \text{with} \quad e_2 = 740 \text{ m s}^{-1}, \quad N = 3, \quad \text{LS}(\cos \rho = 0.2) = 379 \text{ m s}^{-1}$$

LaBonte and Howard (1982):

$$\text{LS}(\rho) = -128(1 - \cos \rho) + 735(1 - \cos \rho)^3, \quad \text{LS}(\cos \rho = 0.2) = 274 \text{ m s}^{-1};$$

Snodgrass (1984):

$$\text{LS}(\rho) = -343(1 - \cos \rho) + 668(1 - \cos \rho)^2 - 190(1 - \cos \rho)^3 + 446(1 - \cos \rho)^4,$$
$$\text{LS}(\cos \rho = 0.2) = 239 \text{ m s}^{-1}.$$

Bruning (1981) found the best representation of the measured LS(ρ)-curve by a 5-term expansion which leads to $\text{LS}(\cos \rho = 0.2) = 355 \text{ m s}^{-1}$ and $\text{LS}(\cos \rho = 0.85) = -22 \text{ m s}^{-1}$, as compared to $+274$ and to -17 m s^{-1} respectively from LaBonte's and Howard's expression and to $+239$ and -38 m s^{-1} from Snodgrass' curve. Although the Mt. Wilson observers do not use data for $\cos \rho \leqslant 0.2$, there remains still a discrepancy of $\approx 100 \text{ m s}^{-1}$ at the edges of the data sets. The procedure of the Stanford observers (Scherrer *et al.*, 1980) to bypass the problem with the LS(ρ)-curve by omitting observations with $\cos \rho < 0.66$ is another compromise which still leads to systematic errors of approx. 15 m s^{-1} since $\text{LS}(\cos \rho = 0.66) \approx -15 \text{ m s}^{-1}$ (see LaBonte and Howard, Bruning, or Snodgrass).

It has, unfortunately, not been recognized until very recently how important the exact knowledge of the LS(ρ)-function is when searching for large scale cells or meridional flows. The problem became even more complicated in the recent two years. Brandt and Schröter (1982) reported to have found for the Fe I line $\lambda 5576$ a significant difference of the LS(ρ)-variation along the polar and equatorial diameters. This has been very recently confirmed by Andersen (1984) on the basis of much more accurate data sets. Although Andersen found a smaller difference than Brandt and Schröter reported, it seems now established that such a difference does exist. This difference may be caused by either a modulation of the granular properties along the latitudes (Beckers and Taylor, 1980; Brandt and Schröter, 1982) or by a meridional flow from the poles towards lower latitudes (Duvall, 1979; LaBonte and Howard, 1982; Andersen, 1984).

A choice between these two possible interpretations can only be made by studying the properties of granules (size, contrast, velocities) along solar latitudes (however see Snodgrass (1984), who claimed to have succeeded in decoupling these two effects, although he did not consider a pole-equator difference in the limb shift).

As to (e) and (f): the velocity-intensity correlation resulting from unresolved granular hydrodynamics causes not only a decreasing blue shift of the line-centres of medium strong lines when approaching the solar limb, but also a ρ-dependence of the line asymmetry, the so-called 'C-shape' (see Adam *et al.*, 1976; Brandt and Schröter, 1982; Balthasar, 1984). Such an asymmetry produces an artificial line-shift when the Doppler-compensator method is used, and this shift depends very critically on what position of the line profile the two exit slits are centred. The amount and the shape of this asymmetry varies sensitively with changes in the size, brightness and velocity of the granulation. One can estimate from the model of Beckers and Nelson (1978) that a change of one of these quantities by several percentages may yield artificial shifts of the order of 50 m s^{-1}. At first glance, the solar granulation appears to be rather uniform across the solar disc. But there are several studies, which show that the granulation pattern around and in active regions is different from that of the quiet Sun with regard to size, brightness and velocities (Macris, 1979; Schröter, 1962; Dunn and Zirker, 1973; Mattig and Nesis, 1974, 1976). Indeed, studies of the behaviour of the line asymmetry in quiet and below Ca$^+$-plages show considerable differences (Livingston, 1982; Kaisig and Schröter, 1983; Koch, 1984; Cavallini *et al.*, 1985). Livingston's and Cavallini's *et al.* measurements indicate a red shift of the line-bisectors in active regions of the order of 50–80 m s^{-1} as compared to quiet solar regions. Indeed, LaBonte and Howard (1982) in rediscussing the solar rotation measurements of the Mt. Wilson Observatory found red-shifted areas around and in active regions, which they preferred to interpret as radial inflows in magnetic regions.

The existence of a permanent meridional flow on the solar surface and the existence of systematic motions around and in active regions are closely connected with precise determinations of (a) the limb shift along polar and equatorial diameters: of (b) the variation of the C-shape across the solar disc; and (c) its difference between quiet and active regions of the Sun. Such observations are clearly very difficult and laborious, but they have to be made to reach a decision. Snodgrass (1984) has proclaimed future research on this problem with the Mt. Wilson data.

Noise and particularly spurious signals are caused by numerous instrumental effects. These effects have been discussed in the last decade extensively by many observers involved in solar rotation measurements. For conciseness we can therefore restrict our discussion to a listing of these effects and to the basic references, where a more detailed discussion can be found.

3.1. SPECTROGRAPH SEEING AND DRIFTS

Thermal inhomogeneities of the air inside a spectrograph, thermal stresses on its body or parts, long term changes of air pressure, temperature and water vapour content may well result in spurious Doppler-shifts of up to 40 m s^{-1} (Howard and Harvey, 1970; Livingston and Duvall, 1979; Howard *et al.*, 1980).

3.2. GRATING ILLUMINATION

An inexact re-imaging of the entrance pupil or a change of its intensity profile on the grating when scanning the Sun will change the grating illumination which via ruling errors leads to systematic wavelength shifts. The same effect arises when there is a brightness gradient across the width of the spectrograph due to limb darkening (Livingston, 1968; Brandt *et al.*, 1978; Howard *et al.*, 1980).

3.3. NONLINEARITY OF DETECTORS

When two detectors are used for the Doppler-compensator servo system, differential nonlinearities cause spurious signals. Brandt *et al.* (1978) found an error of 25 m s^{-1} resulting from a 0.1% difference in the response-curve of the two photomultipliers. Howard *et al.* (1980) measured a spurious shift of 60 m s^{-1} when lowering the intensity signals by a factor of 100.

3.4. INTERFERENCE FRINGES

If one of the components of the telescope-spectrograph-analysing optics-system acts as a Fabry–Perot Etalon, interference fringes are produced in the spectrum. This fringe pattern varying with room temperature and other quantities results in an incorrect centering of the line on the exit slits and hence leads to systematic errors (LaBonte and Howard, 1981).

3.5. SCATTERED LIGHT

Scattered light – whether of atmospheric or instrumental origin – with a characteristic width of the spread function of the order of one solar diameter, has generally a spectrum which is less affected by solar rotation as the actual point under measurements is. Hence, the presence of scattered light in the spectrum reduces the measured solar rotational rate accordingly. Svalgaard *et al.* (1978) were the first who redetected the importance of scattered light corrections for solar rotation measurements, a wellknown and widely discussed subject in the Twenties (see IAU Commission 15 before 1928).

The procedure of correction for scattered light is well known (see e.g. Mattig, 1983) and it is also wellknown that more than one measurement of scattered light outside the limb is needed for a precise knowledge of the spread function. The correction depends sensitively on the assumption on the core and the wings of the spread function. Generally one uses measurements of the aureola from the solar limb up to 2 solar radii to determine the shape of the spread function which is assumed to be either a sum of two Gaussians or a sum of a Gaussian (core) and a dispersion function (wings). Such a detailed treatment of scattered light correction is quite laborious and seems almost inapplicable for full disc velocity measurements. A linear dependence of the rotational velocity on scattered light is assumed by both the Mt. Wilson and Stanford observers whereby the slope is derived from only *one* aureola point outside the limb. Scherrer *et al.* (1980) arrived from their data at the relation:

$$A_{\text{meas}} + 0.228 \times I_{S_c} \ (\% \text{ of } I_c^\odot) = A_{\text{corr}}.$$

The corresponding coefficient amounts to 0.168 deg day^{-1} from Mt. Wilson data (Howard *et al.*, 1983). Both plots show an enormous scatter ($\Delta A_{rms} = \pm 0.8$ deg day^{-1} for Mt. Wilson and ± 0.5 deg day^{-1} for Stanford). Besides, the distance of 12.5 arc sec outside the limb, where the aureola intensity at Mt. Wilson is measured, is certainly too close to the limb to be free of telescopic aberrations and atmospheric blurring. It is also not far enough from the limb in order to describe correctly the important long-distance action of scattered light. The larger distance of 2 arc min from the limb as used by the Stanford observers is a better solution. It meets, according to experience with tower telescope optics and common atmospheric conditions, the spread function somewhere around its half-width value.

We treated the scattered light problem somewhat more extensively since we shall see very soon how important the judgment of the correctness of the scattered light correction is when comparing the photospheric rotation rate from plasma measurements with that from sunspot tracings.

4. Proper Motions (Tracing Solar Features); Sources of Errors, Limitations

Only rather stable and long-lived solar features, which do not evolve greatly can be used as tracers of the solar angular velocity. One also has to permanently keep in mind that all long-lived solar features are of magnetic origin and that their longitudinal proper motion components do not necessarily represent the angular velocity 'in situ' but more likely one at a depth where the magnetic fields of these solar features are 'anchored'. They often are also subject to internal motions due to magnetohydrodynamical forces within the magnetic structure. With these precautions in mind, tracing solar features can yield a powerful method to determine the solar rotation rate. Tracing solar features has the advantage over the spectroscopioc method, that one can principally measure two components of the velocity field. This allows, for example, a measurement of surface Reynold's stresses, as has been shown in several publications (see e.g. Ward, 1965; Belvedere *et al.*, 1976; Schröter and Wöhl, 1976; and others).

We consider first the accuracy which can be achieved by this method. From tracings of solar features like sunspots using heliographic charts or Sun's photographs, an internal accuracy of better than 20 m s^{-1} (or 0.14 deg day^{-1}) can be achieved. Let us assume a single positioning to be accurate within 0.5 mm on a 150 mm solar image (6 arc sec around disc centre). Repeating this positioning on 7 consecutive days during the feature's disc passage ($-50° \leqslant L \leqslant +50°$) yields: $4.3 \times 10^6/8.6 \times 10^4 \sqrt{7}$ or 19 m s^{-1}. An accuracy of better than 4 m s^{-1} or 0.03 deg day^{-1} for the rotation rate is obtained by tracing of hundreds of sunspots (Newton and Nunn, 1951; Ward, 1965, 1966).

Computer-controlled tracings of solar features, where maximum or minimum intensity is used to determine the position, arrive at an accuracy for a single positioning of $\leqslant 1$ arc sec. Such measurements can well achieve an internal accuracy of $\leqslant 2$ m s^{-1} or $\leqslant 0.1\%$ from measurements of a sunspot disc passage on 4 consecutive days (Koch *et al.*, 1981) or $\leqslant 4.5$ m s^{-1} for features like Ca$^+$ fine-mottles, as has been demonstrated by Schröter and Wöhl (1976), and Koch *et al.* (1981). Tracing features like prominences during their visibility at the east and west limb, even when each single positioning accuracy is assumed to be 10 arc sec only yields an accuracy of $\lesssim 10$ m s^{-1}.

Similar to the spectroscopic methods there are numerous obstacles which prevent an achievement of this hypothetical accuracy. These are again of solar and instrumental origin and we shall list them for completeness as follows:

(a) intrinsic proper motions due to changes within the magnetic field patterns;

(b) evolution of the structure, e.g. changes in the brightness distribution, very likely due to the same or similar physical processes;

(c) Wilson-depression for sunspots and faculae (see e.g. Balthasar and Wöhl, 1983) and the inverse effect for chromospheric and coronal features (see e.g. Bruzek, 1961);

(d) atmospheric differential refraction (correction of it may well raise v_{rot} by ≈ 4 m s^{-1} as has been stated by Koch (1984);

(e) locally variable distortion of the image scale due to atmospheric seeing or telescopic aberrations;

(f) inaccuracy of positioning the solar axis (P and B_0) on heliographic charts or solar photographs (an error of $0°1$ in P yields an error of 4 m s^{-1} for v_{rot} and introduces an artificial meridional flow);

(g) personal errors, particularly when observers change.

All these error sources depend very much on the type of tracers, the tracing procedure, the instrumentation, the observational method. Therefore we cannot treat them to the same extent as we did for the spectroscopic method.

5. Differential Rotation in Photospheric Layers

In this section we shall discuss and compare results from Doppler shift measurements (basically Doppler-compensator method) and from tracings of sunspots.

5.1. PLASMA ROTATION MEASUREMENTS

Livingston (1969) was the first within the new era who published comprehensive measurements of solar differential rotation taken at the Kitt Peak Observatory, covering the period 1966–1968. Soon afterwards Howard and Harvey (1970) followed with a presentation of Mt. Wilson data for the years 1966 through the end of 1968 based on 350 magnetogram (or dopplergram) observations. Howard *et al.* (1980) performed an extensive analysis of possible instrumental errors and published new values for the coefficients A, B, C from measurements in the period 1973 to 1977. Almost simultaneously Scherrer *et al.* (1980) presented results from Doppler measurements performed at the Stanford Observatory in the period of 1976 to 1979. They were the first who corrected for scattered light. In the following 4 years further results from Mt. Wilson as well as from other observatories have been published. The results are summarized in Table I.

Considering only the very recent measurements of $\Omega(\phi)$, one arrives at an rms error for the coefficient A of ± 0.008 deg day^{-1} or 1.2 m s^{-1} from measurements for periods

TABLE I

Differential rotation of the photospheric plasma

Reference	A	B	C	Period	Remarks
Livingston (1969)	13.74	–	–	1966–1968	
Howard and Harvey (1970)	13.76	– 1.74	– 2.19	1966–1968	(1)
Snider et al. (1979)	13.5	–	–	1977	(2)
Howard et al. (1980)	13.95	– 1.61	– 2.63	1973–1977	(3)
Scherrer et al. (1980)	14.44	– 1.98	– 1.98	1976–1979	(4)
Perez Garde et al. (1981)	14.32	–	–	1978	(5)
Duvall (1982)	14.14	–	–	1978–1980	(6)
LaBonte and Howard (1982)	14.23	– 1.54	– 2.80	1967–1980	(7)
Howard et al. (1983)	14.192	– 1.70	– 2.36	1967–1982	(8)
Snider (1983)	13.8	–	–	1979–1982	(9)
Küveler and Wöhl (1983)	$\left\{\begin{matrix}14.15\\13.90\end{matrix}\right\}$	–	–	$\left\{\begin{matrix}1981\\1982\end{matrix}\right\}$	(10)
Snodgrass et al. (1984)	14.112	– 1.69	– 2.35	1967–1982	(11)
Snodgrass (1984)	14.049	– 1.492	– 2.605	1967–1984	(12)
Koch (1984)	14.20	–	–	1980–1981	(13)
Pierce and Lopresto (1984)	14.07	– 1.78	– 2.68	1979–1983	(14)

Remarks:

(1) Very large daily scatter due to several instrumental error sources.

(2) Measurements with an atomic-beam resonance scattering apparatus; observations of 3 days only.

(3) After removal of several instrumental error sources; still left with unexplained 'ears'.

(4) First correction for scattered light within the modern literature; they assume $B \hat{=} C$.

(5) Observations on 30 consecutive days with a linear photodiode array; rather large rms errors: ± 0.14 deg day^{-1}.

(6) Observations with the McMath Telescope on 14 days only; finds scattered light negligible.

(7) Apply for the first time a non-monotonic limb shift; find meridional flow of 6 m s^{-1} (poleward) and are left with radial inflow in active regions.

(8) Apply scattered light correction; low rms error, now ΔA down to 0.006 deg day^{-1} only.

(9) Method as mentioned for (2); measurements at Oberlin College and at Mt. Wilson, $\Delta A \approx 0.04$ deg day^{-1}.

(10) Observations on 5 different days only with a two-dimensional photodiode array; low rms errors (± 0.02 or ± 0.03 deg day^{-1} respectively).

(11) The same data as for (8); but correction for erroneous dispersion used before at Mt. Wilson.

(12) New fit of Mt. Wilson data with (normalized) Gegenbauer polynomials for the angular velocity function; demonstrates the disappearance of the crosstalk between B and C.

(13) Few measurements only; finds the rotation rate in active region to be higher by 60 m s^{-1}.

(14) Average from 6 different lines; find no differences of Ω for NaD-lines and medium strong Fe lines (no depth dependence of v_{rot} within photospheric layers).

covering at least several years. The rms errors for the coefficients B and C are larger by approximately one order of magnitude (0.08 deg day^{-1}).

The angular rotation velocity close to the solar poles has been investigated by Beckers (1978) and Cram et al. (1983). Beckers found around the poles $\Omega = 9.85 \pm 0.34$ deg day^{-1}. This average value refers to approximately 7°.5 distance from the pole for which Ω_{MW} equals to 10.2 deg day^{-1} and $\Omega_{SF} = 10.60$ deg day^{-1}. Cram et al. found for a latitude of 75° $\Omega = 10.28$ whereas the values Ω_{MW} and Ω_{SF} are 10.34 and 10.83, respectively. This is in fair agreement (within the error limits) demonstrating that the formulae of Mt. Wilson and Stanford are usable even close to the poles and that the so-called polar vortex of the Sun, as predicted by Gilman (1977) does not exist.

Several attempts to measure spectroscopically a height dependence of Ω in photospheric and chromospheric layers have been reported in the literature, but they show either disagreements or insignificant results. In 1911, Adams measured in Hα a faster rotation by 0.5 deg day^{-1} as compared to photospheric lines (scattered light??). Livingston (1969a, b) and Gasanalizade (1980) found no systematic variation of Ω within the photosphere, but Livingston (1970) reported a faster rotation from chromospheric lines. Aslanov (1963) and Solonsky (1972) extracted from their measurements on photospheric lines a decrease of Ω with optical depth. Balthasar's (1983) result from measurements on 63 Fraunhofer lines indicated also a decrease of Ω with optical depth, but the variation of 50 m s^{-1} per 1000 km was not significant within the error limits. A remeasurement of 143 lines (Balthasar, 1984) yielded no deviation from solid rotation. Pierce and Lopresto (1984) investigated 13 lines of different strength covering photospheric and lower chromospheric layers and found no significant variation of Ω with height.

Hence, one has to conclude that a height variation of Ω within photospheric and chromospheric layers is not yet an established fact.

5.2. Rotation rates from sunspot measurements

Much effort has been spent in the last decade (see Table II) to derive from measurements of sunspot positions the solar differential rotation, since hundreds of sunspots have to be measured in order to achieve an accuracy comparable to that from plasma velocity measurements. Only very few data sources for these treatments are available. Until very recently the 'Greenwich Photoheliographic Results' (1874–1976) was the only basic source used extensively for the determination of $\Omega(\phi)$. In 1983 Lustig applied sunspot drawings performed daily since 1947 at the Kanzelhöhe Observatory to determine $\Omega(\phi)$. In 1984 Howard et al. presented new measures of the solar rotation rate determined from sunspot positions using the white light plate collection of the Mt. Wilson Observatory (since 1921). Very recently, Ribes and Mein (1984) started to digitize Ca$^+$ (Iv)-spectroheliograms, recorded at Meudon Observatory since years on a daily routine basis, with the objective to determine long-term variations of $\Omega(\phi)$ from sunspots and Ca$^+$-faculae. Solar patrol data of the Catania Observatory published since 1967 have been used for several investigations of the solar rotation, e.g. by Ternullo et al. (1981) who found the slope of $\Omega(\phi)$ to depend on the age of the sunspot groups used as tracers.

All recent measures confirm basically the result of Ward (1966), that the rotation rate derived from sunspot positions depends on certain characteristics of the sunspots used. The following characteristics have been considered so far in the literature: structure (single, bipolar, follower, leader, complex) area, age (short-lived, long-lived, recurrent), Zürich classification (which basically is a mixture of all these characteristics). It seems now established that recurrent sunspots (essentially H- and J-type spots) show the slowest rotation rate and that sunspots in their early stage of development (essentially

B-, C-, D-type spots) rotate considerably faster (see e.g. Balthasar and Wöhl, 1980; Ternullo *et al.*, 1981; Baltharsar *et al.*, 1985).

In Table II we summarized the results obtained so far from sunspot position measurements. We have distinguished between two classes: 'single, long-lived, and recurrent sunspots' and 'all sunspots'. The difference of the equatorial rotation rate between these two classes is certainly significant. There is an indication of a somewhat

TABLE II

Differential rotation from tracings of sunspots

Reference	A	B	Period	Remarks
Single, long-lived and recurrent sunspots:				
Newton and Nunn (1951)	14.368	− 2.69	1878–1944	(1)
	± 0.004	± 0.04		
Ward (1966)	14.378	− 2.69	1878–1944	
	± 0.003	± 0.08		
Balthasar *et al.* (1982)	14.34	–	1940–1969	(2)
	± 0.08			
Lustig (1983)	14.38	− 2.57	1947–1981	(3)
	± 0.01	± 0.07		
Howard *et al.* (1984)	14.393	− 2.95	1921–1982	(4)
	± 0.010	± 0.09		
Lustig and Dvorak (1984)	14.23	− 2.36	1948–1976	
	± 0.02	± 0.24		
Balthasar *et al.* (1985)	14.37	− 2.86	1874–1939	(5)
	± 0.01	± 0.12		
All sunspots:				
Ward (1966)	14.523	− 2.69	1905–1954	(6)
	± 0.006	± 0.06		
Godoli and Mazzucconi (1979)	14.58	− 2.84	1944–1954	(7)
Balthasar and Wöhl (1980)	14.525	− 2.83	1940–1968	
	± 0.009	± 0.08		
Arévalo *et al.* (1982)	14.626	− 2.70	1872–1902	
	± 0.014	± 0.16		
Howard *et al.* (1984)	14.552	− 2.84	1921–1982	
	± 0.004	± 0.04		
Balthasar *et al.* (1985)	14.551	− 2.87	1874–1976	
	± 0.006	± 0.06		

Remarks:
(1) Averages from his own data reduction and from previously published data.
(2) Average of the two values of the two classes of recurrent sunspots, considered by the author.
(3) Data from sunspot drawings; for corrections of the image scale see Balthasar *et al.* (1984).
(4) 'Spot-group' rotation; the positions are not centers of gravity but those of individual sunspots.
(5) Average values for H- and J-type and La Laguna-type 3 sunspots.
(6) From a least-square fit of the numerical data reported by the author.
(7) From a least-square fit of the data presented in a figure by the authors; no error bars available.

steeper latitude gradient for 'all sunspots' but this difference is within error limits. Balthaser *et al.* (1982) found that stable, recurrent sunspots show during their lifetime a decreasing braking of the rotation velocities from 0.8 to 0.3 m s^{-1} per day which supports the finding of a slower rotation of long-lived sunspots. Such a decrease of the angular velocity with age can also be inferred from the recent work of Howard *et al.* (1984) when one assumes small sunspots to be statistically short-lived and the largest to be long-lived. They measured the equatorial rotation to 14.55, 14.44, and 14.28 deg day^{-1} for smallest, medium size and largest sunspots, respectively. Neidig (1980) measured during the years 1967–1974 a rotation velocity of 14.38 $\pm$ 0.02 deg day^{-1} for single sunspots and 14.71 $\pm$ 0.05 deg day^{-1} for bipolar groups, confirming an earlier result of Kearns (1979). Gilman and Howard (1984) found from Mt. Wilson plates leader spots to rotate faster than follower spots by about 0.1 deg day^{-1} (or 14 m s^{-1}). The fact that the leader and follower sunspots recede from one another, particularly in the first stage of evolution, is well known since years (see e.g. also Godoli and Mazzucconi, 1982; Mein and Ribes, 1984).

5.3. Results from Other Tracers in Photospheric Layers

Duvall (1980) cross-correlated supergranulation velocity patterns observed at different days and derived a rotation rate of 14.72 $\pm$ 0.07 deg day^{-1} which is remarkably higher than that for the photosphere and even higher than that for young sunspots. Using daily magnetograms of Mt. Wilson (1967–1982) and cross-correlating them over successive days, Snodgrass (1983), on the other hand, found the angular velocity of these magnetic patterns to be very close to the rotation rate of recurrent sunspots. The equatorial velocity derived by Snodgrass is in fair agreement with that from earlier treatments of this type (Wilcox and Howard, 1970; Wilcox *et al.*, 1970; Stenflo, 1977), but the slope of $\Omega(\phi)$ is steeper than in the previous investigations in which a more rigid rotation has been measured.

5.4. Comparison between Plasma and Sunspot Rotation Measurements

As can bee seen from Table I and Table II the equatorial rotation rate of recurrent (and long-lived single) sunspots equals (within the limits of accuracy) that of the photospheric plasma. One could argue that a more detailed and precise treatment of scattered light as compared to the procedure applied at the Mt. Wilson and Stanford Observatories could well raise the plasma rotation rate to a value in accordance with that measured from all sunspots. Beckers (1977), Adam (1979), and Koch (1984) reported on simultaneous measurements of plasma rotation in umbrae of large single sunspots and those from tracings of their positions. They found the two rotation rates to coincide within the limits of errors ($\pm$ 20 m s^{-1} or $\pm$ 0.14 deg day^{-1}). On the other hand, Foukal (1979) found from similar observations a significant difference between these two rotation rates.

Depending on whether or not one believes that the scattered light problem has been sufficiently correctly treated by the Mt. Wilson and Stanford observers, one is confronted with two alternative findings: either (a) sunspots in their early stage of evolution exhibit a rotation being considerably faster (0.2–0.3 deg day^{-1}) than that of the surrounding photospheric plasma and approach – when aging – the plasma rotation velocity; or (b) sunspot groups co-rotate in their early stage of development with the surrounding plasma and slow down their rotation velocity permanently, exhibiting – when sufficiently old and recurrent – an rotation rate being considerably slower than that of the photospheric plasma.

A choice between these two alternatives and in addition important conclusions can only be obtained from more extensive observations of the type performed by Adam, Beckers, Foukal, and Koch and from a more sophisticated treatment of scattered light for plasma rotation velocity measurements.

6. Meridional Flow and/or Large-Scale Velocity Pattern

As already stated in Section 1, one important task for solar observers is to test the two presently existing theories of the solar differential rotation; i.e. to search for the existence of a meridional flow or for Reynold's surface stresses. Basically three different methods have been applied to this problem: (i) residuals from full-disc dopplergrams were analyzed searching for long persisting systematic poleward or equatorward directed Doppler shifts; (ii) longitudinal and latitudinal components of proper motions of tracers were correlated to study whether or not systematic surface Reynolds stresses can be established; (iii) from daily magnetograms the long-term behaviour of the background magnetic fields was studied looking for the existence of persistent north-south drifts or for regular cell-patterns.

Beckers (1978) found from scans across the solar disc an apparent meridional flow towards the poles of 42 ± 9 m s^{-1}. But he already warned that this could also be interpreted as a difference in the limb shift along the polar and equatorial diameter. Duvall (1979) analyzed Stanford's full disc Doppler measurements and found a signal consistent with a poleward directed flow of 20 ± 5 m s^{-1}. Since he subtracted from the data a limb shift curve which was measured along the equatorial diameter, one has to consider this result cautiously as was done by Beckers.

Howard (1979) reported the presence of giant velocity features associated with solar activity and a meridional flow towards the poles of the order of 20 m s^{-1}. Somewhat later Howard and LaBonte (1980) rediscussed these Mt. Wilson data with the result that no such patterns (with amplitudes $\geqslant 10$ m s^{-1}) could be discerned. Finally, LaBonte and Howard (1982) explained the 'ears' by the none monotonic limb shift, the velocity patterns associated with activity as downflows in active regions. Subsequently, residual systematic Doppler shifts were left which they attributed to a poleward meridional flow of 16 m s^{-1}. Since again a possible latitude dependence of the limb shift has not been considered, the same statements as for Beckers and Duvall must be made here.

Perez Garde *et al.* (1981) reported on Doppler shift measurements of the Fe I line

λ 6301.5 Å with respect to the next telluric O_2-line at many positions of the solar disc. They found residuals leading to an equatorward meridional flow of 20 ± 10 m s^{-1}.

In a very recent rediscussion of Mt. Wilson Doppler shift measurements Snodgrass (1984) presented several interesting conclusions which demonstrate clearly the strong nexus between the assumptions about the shape across the disc of LS(ρ) and the residual Doppler shifts which then are generally left for an interpretation as 'meridional flow'. Recognizing that the two coefficients L_1 and L_2, which represent in his formalism of data reductions the limb shift, showed variations with the solar cycle, he concluded that these variations result from the fact that the granulation induced limb shift may well change with solar activity. Hence, he accepted that the limb shift may well be affected by background solar magnetic fields. But he did not go so far as to consider the limb effect to be different along the equatorial and polar diameter, as found by Brandt and Schröter (1982) and confirmed principally by Anderson (1984) using more precise measurements. The latter author came to the conclusion that the difference of the limb shift, for the line Fe λ 5576 may either be explained by the presence of a meridional flow < 50 m s^{-1} from both the equator and from poles towards $\phi = \pm 45°$ or alternatively by a difference in the structure and dynamics of granulation parameters along polar and equatorial diameters.

In summarizing the existing attempts to detect in photospheric layers the existence of a meridional flow or of a large-scale circulation pattern it appears that there is presently no established evidence for the existence or non-existence of such a large scale flow pattern, which could be attributed to the maintenance of the solar differential rotation.

Ward (1965) was the first who correlated observed longitudinal and latitudinal components of sunspot proper motions (u and v respectively) to search for global eddies or waves in the solar photosphere. From the integral

$$I = \int_0^t \oint \rho u v \, \mathrm{d}x \, \mathrm{d}t = \rho \, \langle [u, v] \rangle \, Lt,$$

where the square brackets denote a space average over a length L and the angular brackets a time average over the interval t, it is evident that the covariance is directly proportional to the amount of angular momentum transported. He obtained for $\langle [u, v] \rangle$ a value of 2×10^7 cm^2 s^{-2} which seemed large enough and with the correct sign to maintain the observed differential rotation.

In 1976, Belvedere *et al.* published their results of an attempt to measure Reynold's stresses from proper motions of faculae, using Catania data. They also found a correlation $\langle [u, v] \rangle$ of the correct sign and even somewhat higher in magnitude $(4 \times 10^7$ cm^2 s$^{-2})$.

Almost at the same time, Schröter and Wöhl (1976) analyzed the proper motion components of Ca$^+$-network fine mottles measured at Locarno in 1974 and 1975 on the basis of a computer controlled method and found $\langle [u, v] \rangle = 4 \times 10^7$ cm^2 s^{-2}, twice the value of Ward and in good agreement with Belvedere *et al.* and again with the correct sign.

Very recently Gilman and Howard (1984) reported on a new attempt to measure again the covariance of sunspot proper motions from the Mt. Wilson white light plate collection. They found for sunspot groups a value being similar to that of Ward, but for individual sunspots the covariance was less by a factor of 3. They concluded that there is a real correlation between u and v components of individual spots, but not sufficient to account for the maintenance of the solar differential rotation.

At first glance, these results look very promising. But, questions and doubts emerge immediately. Why were further attempts of the Locarno group (Schröter *et al.*, 1978) unsuccessful to re-measure this covariance? Why has such a correlation not been found from other measurements?

Do such global, long-lived eddies or waves really exist and if the correlation found from sunspots and Ca^+-features really mirrors their velocity structure why have they not yet been detected from plasma measurements? A covariance of the order of 4×10^7 cm^2 s^{-2} results from $[u] \approx [v] \approx 60$ m s^{-1} and this leads to Doppler shifts of the order of 20–30 m s^{-1} at $\sin \rho = 0.5$ which is well above the noise-level of such measurements.

On the other hand, as stated at the beginning of Section 4, all these magnetically related tracers may well be subject to intrinsic motions due to internal magnetohydrodynamic forces. Indeed, the well known investigations of sunspot proper motions by Tuominen and coworkers (for references see e.g. Tuominen *et al.*, 1983) as well as the results obtained by Balthasar and Wöhl (1980) and by Arévalo *et al.* (1982) suggest that these proper motions are possibly related to global magnetohydrodynamic forces in connection with the dynamo excitation mechanism (Tuominen *et al.*, 1984).

Hence, we are left with very few tracers which are believed to be less affected by intrinsic magnetic forces. Schröter and Wöhl (1975, 1976), Schröter *et al.* (1978), have looked at the movements of Ca^+-network mottles in order to find evidence of global circulations. Observations in summer 1975 reveal strong evidence of the existence of four giant circulation cells crossing the equator. Schwan and Wöhl (1978) applied a more sophisticated mathematical method to the data and could confirm this result. However a further attempt of the same group to test this result from more observations in 1976 failed completely.

Nevertheless, the very existence of a large-scale pattern is suggested from observations gained from coronal holes, particularly from those which survive several rotations. The arrangement of the pattern of large filaments on the Sun is also suggestive of an underlying global disturbance structure (Wagner and Gilliam, 1976); the same holds for the evolution of the Hα neutral lines (McIntosh, 1979).

Recently Topka *et al.* (1982) reported on an observational study of the proper motions of polar zone Hα filaments ($\phi > 30°$) and compared these motions with the behaviour of the polar magnetic field pattern. They concluded that the observed poleward drift of the filaments of ≈ 10 m s^{-1} along with the magnetic regions of both polarities is due to a meridional poleward flow and not only due to the well known tendency of the 'following'-polarity flux to move polewards.

Again the results from tracings of magnetically related features regarding the existence of a persistent meridional flow or a large-scale velocity pattern are rather meagre.

Howard and LaBonte (1980) analyzed velocity data from 12 years of full-disc Mt. Wilson observations with the aim to study horizontal east–west motions. They found – after subtracting from the data a smooth curve for the differential rotation– zones with faster than and slower than average rotation velocity. These zones being symmetric about the equator drift from the poles towards the equator with a speed of ≈ 3 m s^{-1} and they need apparently 22 years (two solar cycles) to arrive at the equator, where they disappear. The authors found further that solar active regions are formed at the boundaries between the fast and slow velocity zones. They named this evidently deep-seated circulation pattern, the first one which has been established to be really a persistent phenomenon, a travelling torsional wave pattern. But it has been already perceived by several theorists that this travelling torsional wave pattern cannot be the one that is needed to maintain the solar differential rotation.

7. Solar Rotation above Photospheric Layers

Excellent and rather complete reviews of chromospheric and coronal rotation measurements have been published by Antonucci (1978) and Noci (1978). For conciseness I can therefore restrict my report to a few basic investigations characterizing the present status.

The rotation rate of solar faculae has been determined by Newton (1924). He found $A = 14.54$ deg day^{-1} and $B = -2.81$ deg day^{-1}, which is in fair agreement with previous measurements by Stratanoff and Chevalier and close to the rotation rate for *all* sunspots. Polar faculae were investigated by Müller (1954) and Waldmeier (1955). They found for the latitude range $65°–70°$ an angular velocity of 10 deg day^{-1}, which is somewhat less than that measured by Beckers (1978) and Cram *et al.* (1980) for the plasma, close to the poles. Due to the large scatter of the former data this difference may not be significant.

It became evident from measurements within the last decade that different chromospheric and coronal tracers show a different shape of $\Omega(\phi)$. It seems established that particularly short-lived (small scale) and long-lived (large scale) features exhibit the largest difference regarding the latitude dependence of Ω. Figure 2 which is a rescaled version from a paper by Antonucci *et al.* (1977) demonstrates this for chromospheric layers. We therefore distinguish in the following Table III between results obtained from short-lived and long-lived chromospheric and coronal structures. Of course in a few cases a certain ambiguity can not be avoided.

It seems clear from these results that the gradient of the differential rotation depends very much on the type of tracers used, regarding size and life-time. The long-lived structures show less steep gradients and the very long-lived coronal holes rotate almost rigidly. Presently it is not clear whether the large variation of the coefficient A represents a height gradient, since it is generally not known with sufficient accuracy at what height the tracers originate. The result of Golub and Vaiana (1978) that X-ray emission

CHROMOSPHERIC ROTATION CURVES

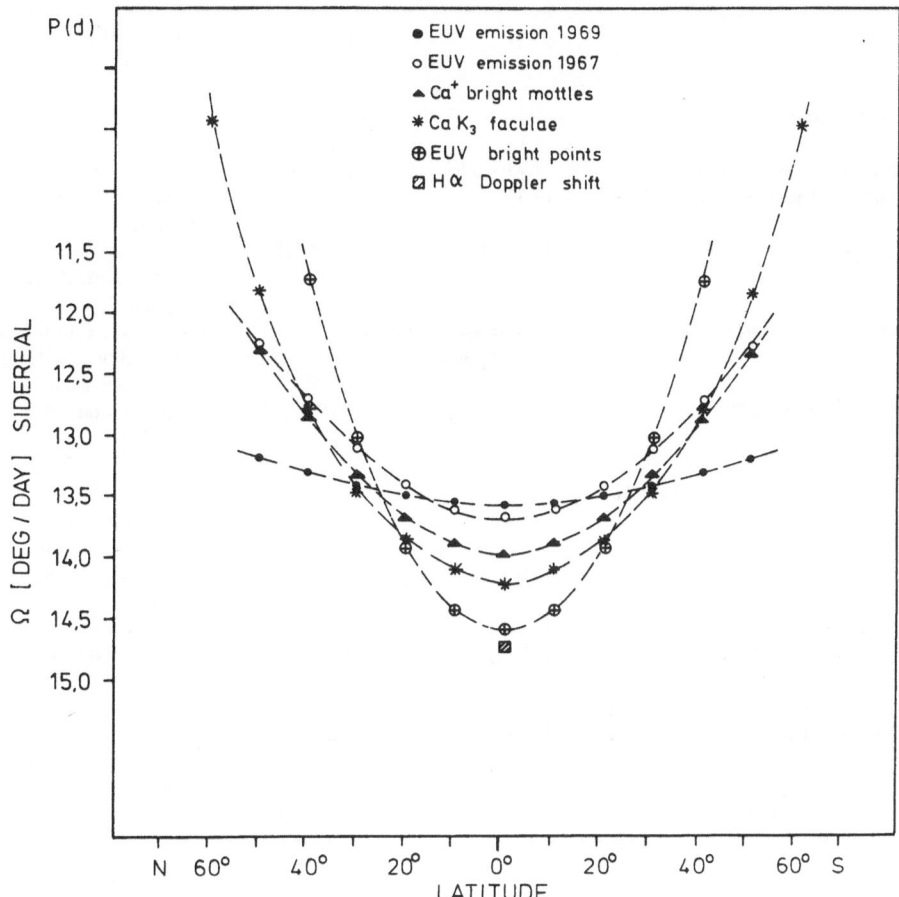

Fig. 2. Comparison of various determinations of the solar differential rotation obtained from small-scale short-lived and large-scale long-lived features of the chromosphere (rescaled figure from Antonucci *et al.*, 1977).

features, doubtless of coronal origin, rotate very much like the photosphere speaks against a significant height-gradient. Parker *et al.* (1982) studied the recurrence of white-light coronal features (K-coronameter observations at Hawaii) and concluded the corona at distances of $1.12\,R_\odot$ and $1.5\,R_\odot$ to rotate faster than the photospheric plasma. However, they compared their measurements with the $\Omega(\phi)$-curve from Howard and Harvey (1970). Taking for a comparison the very recent plasma rotation values from Mt. Wilson and Stanford, the difference is not significant anymore, particularly if one considers that then $\Omega(0)$ for $R = 1.12\,R_\odot$ is larger than $\Omega(0)$ for $R = 1.5\,R_\odot$ with $\Omega(\phi)_{\mathrm{phot}}$ in between. Hence, we have no reason to anticipate a significant gradient of

TABLE III

Differential rotation from chromospheric and coronal structures

Reference	A	B	Type of structure, remarks
Short-lived features:			
Milosevic (1955)	14.14	− 3.18	Ca^+-K_3 facylae
Schröter and Wöhl (1975, 1976)	13.93 ± 0.08	− 2.9 ± 0.73	Ca^+ bright mottles
Dupree and Henze (1972, 1973)	13.54	− 1.5	Lyman continuum emission
Simon and Noyes (1972)	14.7 ± 0.2	− 7.1 ± 1.1	Lyman continuum, bright points in active regions
Golub and Vaiana (1978)	as photospheric plasma as sunspots		X-ray emission features: small, short-lived, larger and longer-lived
Liu and Kundu (1976)	14.5 ± 0.27	− 4.19 ± 3.0	radio mm-emissive regions
Long-lived features and Doppler shifts:			
Livingston (1969)	14.90	−	Hα Doppler shifts
Antonucci and Dodero (1977)	14.33	− 0.34	green corona line
Antonucci et al. (1977)	14.09	− 0.37	long-lived Ca^+K_3 regions
d'Azambuja[2] (1948)	14.48	− 2.16	Hα filaments
Liu and Kundu (1976)	14.73 ± 0.28	− 1.05 ± 1.6	radio mm-absorption regions
Wagner (1975)	14.33	− 0.39	EUV coronal holes
Adams (1976)	14.48	− 0.29	magnetic fields surrounding coronal holes
Timothy et al. (1975)	14.23 ± 0.03	− 0.4 ± 0.1	coronal holes

$\Omega(0)$ within the corona and are left with the observational result that $\Omega(\phi)$ depends sensitively on size and lifetime of the tracers under study.

In a series of papers Antonucci and co-workers (Antonucci and Svalgaard, 1974; Antonucci and Dodero, 1977; Antonucci et al., 1977; Antonucci et al., 1979; Antonucci and Dodero, 1979) studied the dependence of $\Omega(\phi)$ on lifetime of chromospheric and coronal structures. Three basic findings can be extracted from these investigations:

(i) very short-lived ($< 1^d$) features of the $Ca^+ - K_3$ chromosphere rotate at the same rate as the chromospheric plasma, i.e. somewhat faster than the photospheric layers;

(ii) long-lived Ca^+-K_3 chromospheric patterns rotate very much like the green corona;

(iii) short-lived and long-lived features of the green corona may well coexist in a wide range of latitudes showing rather different shapes of $\Omega(\phi)$.

Doubtless, coronal holes observed in Fe xv λ 284 Å and in soft X-rays exhibit the most rigid rotation (Wagner, 1975; Timothy *et al.*, 1975). However, very recently Shelke and Pande (1985) found from KNPO helium λ 10830 Å synoptic charts (Carrington rotations 1716 through 1739) coronal holes to show substantial differential rotation. From the numbers published by the authors one arrives at the coefficient $B \approx -1.5$ deg day^{-1} well above B from Wagner and Timothy *et al.*

Some other controversial results can be found in the literature. From earlier investigations of Wilcox and Howard (1970), Wilcox *et al.* (1970) and Stenflo (1977) it was believed that the photospheric background magnetic field patterns rotate almost as rigidly as the coronal holes do. But recently Snodgrass (1983) arrived, using a larger set of Mt. Wilson data, at the conclusion that the rotational profile of the magnetic background field pattern is very similar to that of recurrent sunspots for lower latitudes and similar to that of the Mt. Wilson plasma measurements for higher latitudes. Furthermore d'Azambuja's observations of Hα-filament rotation rates yielded a latitude gradient which is almost identical to that of the photosphere although these are long-lived structures of the upper chromosphere. According to Adams and Tank (1977) short-lived filaments exhibit a flatter $\Omega(\phi)$-curve as compared to long-lived ones.

It is generally believed that the sector-structure of the interplanetary field and that of the solar wind are strongly correlated with the large-scale, long-lived coronal magnetically related features. The latter are known to show a fairly rigid rotation. But Blums and Lotova (1979) reported from a statistical analysis of the recurrence of solar wind properties derived from multisite observations of interplanetary scintillations, that streams of solar wind take part in the differential rotation of the Sun.

This discussion shows that we are still far away from having a clear picture of the differential rotation of chromospheric and coronal layers. The particularly confusing fact is that we do not have at present the key to understanding why short-lived and long-lived chromospheric and coronal features show a considerably different shape of $\Omega(\phi)$.

8. Cycle Dependence of the Solar Differential Rotation

It is evident from the reported observational facts that if such a cycle dependence of $\Omega(\phi)$ exists, then the amplitude of its variation is expected to be less than ≈ 20 m s^{-1} (or 1% of $\Omega(0)$). The corresponding accuracy can only be achieved by a complete elimination of effects from other solar dynamical processes, i.e. by long-term measurements on a routine basis. There are at present, in principle, three types of sources of material available which meet this condition: (i) the *Greenwich Heliographic Results*, the Mt. Wilson plate collection (and to a certain extent data from Kanzelhöhe and Catania); (ii) collections of Ca$^+$-spectroheliograms or filtergrams (Meudon, Sacramento Peak, Wendelstein, etc.) or continuous corona observations; (iii) plasma rotation data from Mt. Wilson and Stanford.

Plasma rotation data from Mt. Wilson and Stanford suffer from the fact that several systematic error sources of solar and instrumental origin had to be removed to reach the present accuracy. Therefore, a continuation of these observations at both observatories is very important. Presently these data can only be used for verification tests of results achieved from other sources.

In the last section we saw that our picture regarding the differential rotation of chromospheric and coronal structures is rather complicated and confusing. The main obstacle is the size and lifetime dependence of $\Omega(\phi)$. As long as we don't understand this connexion it seems to be futile and premature to try to extract from such data a cycle dependence of $\Omega(\phi)$. During a solar activity cycle the relative ratio of short-lived to long-lived magnetic features may well change (in a manner not yet known by us) and produce an artificial cycle dependence of $\Omega(\phi)$.

Hence, we are left with sunspot data as the only reliable source. Attempts to derive $\Omega(\phi) = f(t)$ prior to 1978 have been reviewed by Schröter and Wöhl (1978) and shall not be rediscussed here. They did not yield a homogeneous and clear picture of $\Omega(\phi) = f(t)$.

Results regarding $\Omega(\phi) = f(t)$ from sunspot tracings have to be considered with great precaution since they may easily be misinterpreted. The sunspot occurrence during a cycle is latitude dependent (butterfly-diagram). As a result, the rotation of the various latitudes (and the extrapolation towards the equator) is covered with different numbers of the samples at different phases of the cycle. Hence, there is an unpleasant correlation of the phase of the cycle and the achieved accuracy for the various latitudes. Moreover, rotation rates derived from sunspots show also a dependence on lifetime. Although to a much lesser extent, the situation is similar to that in the chromosphere and corona. Furthermore, it is known that a great fraction of emerging solar magnetic flux occurs in preferred or so-called 'active longitudes' which are more evident during maximum activity and less evident during the minimum phase (Schröter, 1984; Mein and Ribes, 1984). Hence, tracing magnetic related features does not necessarily mean the tracing of a surface fluid flow but perhaps rather tracing the rotation of a subphotospheric magnetic instability.

Bearing these precautions in mind we shall now review the most recent attempts to search for a cycle dependence of $\Omega(\phi)$.

Clark *et al.* (1979) selected from Greenwich data sunspot records centered near the maxima of the solar cycles 12 and 19. The first one showed very small activity ($R_{max} = 60$), the latter a rather high activity ($R_{max} = 180$). They used for their analysis only long-lived single sunspots and found the equatorial rotation rate to remain independent of R_{max} when standard errors were accounted for. But the coefficient B – the gradient of Ω – changed drastically within these two periods, in the sense that for the low activity cycle 12 the rotation was more rigid than for the high activity cycle 19. They concluded that any future theory must account for an invariant equatorial rotation rate and significant variation of the latitude-gradient with the sunspot number.

The author prefers to conclude from this investigation that when comparing the solar rotation for cycles of different activity a more rigid rotation is indicated for lower activity cycles. But, the result of Clark *et al.* may also well be an artifact since Tuominen and Kröläinen (1982) and Balthasar *et al.* (1985) found in the Greenwich data a drop of both the coefficients A and B between cycle 13 and 14 which seems unlikely to be of solar origin but rather caused by some changes in the observation and reduction procedure at Greenwich.

Balthasar and Wöhl (1980) analysed Greenwich data from 1940 to 1968 and found somewhat faster rotation for the equator-zone around sunspot minimum, but not significant if a 3σ-level error limit is considered. Arévalo *et al.* (1982) using the same data for the period 1874–1902 found a similar variation, but again below the 3σ-level. Lustig (1983) concluded, from an analysis of sunspot records of the Kanzelhöhe Observatory covering 4 solar cycles (1947–1981), besides other findings that equatorial rotation rates are significantly higher and the rotation is slightly more rigid during sunspot minima. Higher angular velocities around ± 15 degree in latitude were also found by Tuominen and Kyröläinen (1982).

Very recently Gilman and Howard (1984) have undertaken an analysis of the Mt. Wilson white-light plate collection obtained within the period 1921 through 1981 (6 solar cycles) with the aim of searching for a variability of the solar differential rotation. They used individual sunspots only within the belt of $30°$ N to $30°$ S. They found a strong peak of $\Omega(-30° \leq \phi \leq 30°)$, in average 0.1 deg day^{-1} near sunspot minimum, and a somewhat weaker one near sunspot maximum. For the period 1967–1982 they claimed to have found a similar peak distribution from Mt. Wilson Doppler data, although it is known that these data suffer from much larger systematic errors, which were not recognized before 1981. They also found a third peak of Ω in high latitudes 3 years before the end of the cycle. At least the peak around sunspot minimum seems to be significant, taking a 3σ-level, and is in agreement with previous findings by Balthasar and Wöhl (1980), Arévalo *et al.* (1982), Tuominen and Kyröläinen (1982) and Lustig (1983). Very recently Balthasar *et al.* (1985) used the complete sample of the Greenwich data covering the period 1874 to 1976 and also found the coefficient A to be significantly higher ($0.1 \leq \Delta A \leq 0.2$ deg day^{-1}) around sunspot minimum.

A faster rotation of the equatorial belt around sunspot minimum seems therefore to be the first established fact regarding a cycle dependence of the solar differential rotation. One could object against this interpretation with the following argument: let us assume, that it is established that young sunspots of the Zürich type A, B, C rotate faster than the larger and older ones of the Zürich-type F, G, H, and J. If around minimum the young short-lived sunspots were to occur more frequently on average, this would yield an artificial increase of Ω around activity minimum. This argument of the author has been disproved by Balthasar *et al.* (1985) who investigated the cycle dependence of the rotation of the Zürich-groups A, B, C on the one hand and G, H, J on the other hand and demonstrated convincingly that both classes show almost the same time behaviour of Ω. Only a small fraction of the increase of Ω at the sunspot minimum was left for this 'selection effect'.

Less significant appear (to the author's opinion) the secondary peaks reported by Gilman and Howard (near sunspot maximum and 3 years before the end of the cycle) and by Balthasar *et al.* (in the fourth year after minimum). The amplitude of those peaks may well be produced by the selection effect mentioned above, namely by a temporary variation in the relative ratio between the occurrence of short-lived and long-lived sunspots within the solar cycle.

9. Rotation of the Solar Interior

The knowledge of the rotation of the solar interior is of fundamental interest not only with a view to studying the internal structure of the Sun but also with a view to testing various basic theories, e.g. the solar gravitational quadrupole moment, the $\alpha - \omega$-dynamo process and the history of the solar angular momentum.

The key which enables us to gain an insight into the rotation of the solar interior is the rotational splitting of global solar oscillations. Hitherto three different types of observations have been analyzed:

(i) oblateness measurements or low-order, low-degree global oscillations in the limb-darkening function;

(ii) 5-min p-mode oscillations in the power spectrum of the solar total irradiance;

(iii) frequency shifts of the solar acustic (p-mode) oscillations of high and low degree.

The determination of the angular velocity in subphotospheric layers is based on the following properties of the p-mode oscillations:

Oscillation frequencies of a mode $v_{n,l,m}$ (n = radial order, l = spherical harmonic degree, and m = angular order) are shifted by rotation. Since modes of different spherical harmonic degree and radial order are trapped in cavities of different depth ranges, the study of their frequencies enables us to estimate the depth dependence of Ω provided the depth dependence of the internal structure of the Sun is known from a solar model. Modes of different m span over different latitude zones allowing for estimates of $\Omega(\phi)$. For a more detailed description of the method see e.g. Gough (1985), Hill (1985), or Deubner and Gough (1984).

The first investigation of this kind has been published by Deubner et al. (1979) from observations performed with a diode-array and consisting of scanning the solar surface diametrically parallel to the equator every 100 s. The corresponding contribution functions have been computed by Ulrich et al. (1979). Since these observations represent p-modes of high l-values they give information about the rotation of the subphotospheric layers (down to approx. 20 Mm). A faster rotation by some 80 m s^{-1} was found below 10 Mm. However, the error bars were rather large and the result quite ambigous. A few years later Rhodes et al. (1983) reported on results based on high degree ($l > 150$) p-mode measurements at the KPNO. These observations failed to show a significant deviation from a solid body rotation.

Within the domain of very low-degree modes ($l = 0, 1, 2$) Claverie et al. (1981) announced, from their full disc Doppler measurements performed by means of optical scattering in the K-resonance line a splitting of roughly $0.75\,\mu$Hz (or $\Omega = 23.3$ deg day^{-1}, or $T = 15.4$ days). This would yield a core rotation being 2 to 9 times faster than that of the surface, depending on the assumption about the extension of the 'core'. The correctness of the identification of the modes observed by Claverie et al. has been doubted by several authors (see e.g Hill, 1984). To meet this criticism – the Birmingham group observed the $m = 0$ components which ought to be absent in a global view of the Sun – Isaak (1982) postulated the existence of a strong magnetic field in the solar core. This is very similar to a proposal of Dicke (1982) to explain the

12.2 day (0.95 μHz) variations in the Princeton oblateness measurements (Dicke and Goldenberg, 1967). Isaak's explanation has not been accepted universally (see e.g. Gough, 1982).

Furthermore, Woodard and Hudson (1983) could not confirm, from their low-degree oscillation measurements with the ACRIM on board of the SMM spacecraft (solar total irradiance measurements), the rotational splitting deduced by the Birmingham group, since their frequency peaks were observed to be much shallower than one would expect from a rotation rate of 0.75 μHz.

In a very recent paper, Woodard (1984) concluded, from a m-state splitting analysis based on the width of the ($n = 16$–26, $l = 0$–2) multiplets in the spectrum of ACRIM, that the mean interior rotation rate cannot exceed the surface rotation by more than a factor of 2.2.

Very recently Duvall and Harvey (1984) and Duvall et al. (1984) reported on new results from observations at the KPNO performed in summer 1983. These authors focussed their observational program on the rotational splitting of oscillation modes with $l = 1$–200 and with $m = \pm l$. For any given l, such 'sectorial' modes show the largest possible splitting, $2 \times l \times \Omega$, but on the other hand prevent a determination of $\Omega(\phi)$. The authors thus aimed to investigate the radial variation of $\Omega(0)$ over most of the solar radius. The results are very interesting and impressive. For the highest-degree modes measured ($l = 80$–100), the rotation splitting shows a slightly faster rotation than the surface equatorial rotation; this is in a certain sense in agreement with Deubner et al. (1979), since the latter authors used for their analysis only high degree modes. For modes between $l = 30$ to $l = 4$ a rotation slower than that of the surface is indicated, although one should bear in mind the observational limitations. For $l < 3$ the rotation increases drastically leading to 660 nHz or $\Omega = 20.5$ deg day^{-1} or $T = 17.6$ days per revolution (sidereal) close to the solar centre although the error bars increase drastically too. This increase of rotation starts well below the convective zone roughly at a depth where the energy-generating core is encountered. The degree $l = 1$ splitting agrees well with the result of the Birmingham group (Claverie et al., 1981) and within the limits of errors also with the findings of Woodard and Hudson (1983) from ACRIM data. There is a disagreement with the result of Hill et al. (1982). However, one has to bear in mind that Duvall and Harvey have not observed the same modes as Hill et al. and a comparison of these two results is not straightforward. Hence the most recent observations seem to confirm a rapidly rotating solar core with a rotation rate between 12 and 17 days per revolution, consistent with previous speculations by Dicke (1976) from solar oblateness measurements.

The splitting of the degree $l = 2$ mode disagrees with the Birmingham result, but is consistent with the findings of Woodard and Hudson. Furthermore, Grec et al. (1982) found from their full-disc oscillation measurements at the south-pole that the measured width of the $l = 2$ modes is inconsistent with the large rotational splitting reported by the Birmingham group. In the range between $l = 4$ and $l = 30$ the rotational splitting yields a significant slower rotation of the Sun for those parts of the solar radius mirrored by these modes. It is important to note (see Duvall et al., 1984) that between $0.9 \, R_{\odot}$ and

0.6 $R_\odot$ the observations imply no significant change in the rotation rate at the base of the convective zone. However, the relative peak of rotation splitting around $l = 11$ remains unexplained. Gough (1984) arrived at a very similar $\Omega(r)$ by a different reduction procedure.

Hill (1984) and Hill *et al.* (1984) inferred from 100 nonradial multiplets of acoustic and internal gravity modes (SCLERA diameter type measurements) an increase of Ω below the surface by approximately 30% and at $R = 0.75\,R_\odot$ a sharp transition to 3.0 μHz (or 93 deg day^{-1}, or $T = 3.9$ days) for almost the whole interior. This is more than six times the surface rotation rate and in disagreement with Woodard's and Duvall's and Harvey's results. Hill (1984) gave also an estimate of the gravitational quadrupole moment $J_2 = (5.5 \pm 1.3) \times 10^{-6}$ and the upper limit of the internal magnetic field ($B \approx 1$–3 MG). Gough (1982) arrived at $J_2 = 3.6 \times 10^{-6}$ from a similar analysis. Whereas Hill's value is in conflict with the Einsteinian general relativistic interpretation of Mercury's perihelion motion, Gough's value of J_2 is not large enough to seriously conflict with it.

The present results with regard to the rotation of the solar core are still quite discrepant. But in view of the newness of this branch of solar physics we are justified in expecting very soon new exciting and more coherent results.

Acknowledgement

I acknowledge Dr S. S. Hasan's assistance in rephrasing certain portions of the manuscript.

Refrences

Adam, M. G.: 1959, *Monthly Notices Roy. Astron. Soc.* **119**, 460.
Adam, M. G.: 1979, *Monthly Notices Roy. Astron. Soc.* **188**, 819.
Adam, M. G., Ibbetson, P. A., and Petford, A. D.: 1976, *Monthly Notices Roy. Astron. Soc.* **177**, 687.
Adams, W. M.: 1976, *Solar Phys.* **47**, 601.
Adams, W. M. and Tang, F.: 1977, *Solar Phys.* **55**, 499.
Anderson, B. N.: 1984, *Solar Phys.* **94**, 49.
Antonucci, E.: 1978, *Proc. Workshop on Solar Rotation*, Publ. Catania No. 162, p. 204.
Antonucci, E. and Dodero, M. A.: 1977, *Solar Phys.* **53**, 179.
Antonucci, E. and Dodero, M. A.: 1979, *Solar Phys.* **62**, 107.
Antonucci, E. and Svalgaard, L.: 1974, *Solar Phys.* **34**, 3.
Antonucci, E., Azzarelli, L., Casalini, O., and Cerri, S.: 1977, *Solar Phys.* **53**, 519.
Antonucci, E., Azzarelli, L., Casalini, O., Cerri, S., and Denoth, F.: 1979, *Solar Phys.* **61**, 9.
Appenzeller, I. and Schröter, E. H.: 1967, *Astrophys. J.* **147**, 1100.
Arévalo, M. J., Gomez, R., Vazquez, M., Balthasar, H., and Wöhl, H.: 1982, *Astron. Astrophys.* **111**, 266.
Aslanov, I. A.: 1963, *Astron. Zh.* **40**, 1036.
Balthasar, H.: 1983, *Solar Phys.* **84**, 371.
Balthasar, H.: 1984, *Solar Phys.* **93**, 219.
Balthasar, H. and Wöhl, H.: 1980, *Astron. Astrophys.* **92**, 111.
Balthasar, H. and Wöhl, H.: 1983, *Solar Phys.* **88**, 71.
Balthasar, H., Schüssler, M., and Wöhl, H.: 1982, *Solar Phys.* **76**, 21.
Balthasar, H., Lustig, G., and Wöhl, H.: 1984, *Solar Phys.* **91**, 55.
Balthasar, H., Vazquez, M., and Wöhl, H.: 1985, submitted to *Astron. Astrophys.*

Beckers, J. M.: 1977, *Astrophys. J.* **213**, 900.

Beckers, J. M.: 1978a, *Astrophys. J.* **224**, L143.

Beckers, J. M.: 1978b, *Proc. Workshop on Solar Rotation*, Publ. Catania No. 162, p. 166.

Beckers, J. M.: 1981, *The Sun as a Star*, NASA–SP **450**, 11.

Beckers, J. M. and Canfield, R. C.: 1975, *CNRS Colloq.* **25**, 200.

Beckers, J. M. and Nelson, G. D.: 1978, *Solar Phys.* **58**, 243.

Beckers, J. M. and Taylor, W. R.: 1980, *Solar Phys.* **68**, 41.

Belvedere, G., Godoli, G., Motta, S., Paternò, L., and Zappalà, R. A.: 1976, *Solar Phys.* **46**, 23.

Belvedere, G. and Paternò, L.: 1977, *Solar Phys.* **54**, 289.

Belvedere, G., Paternò, L., and Stix, M.: 1980, *Geophys. Astrophys. Fluid Dyn.* **14**, 209.

Blums, D. F. and Lotova, N. A.: 1979, *Astron. Z.* **56**, 872.

Brandt, P. N. and Schröter, E. H.: 1982, *Solar Phys.* **79**, 3.

Brandt, P. N., Deubner, F.-L., Schröter, E. H., and Soltau, D.: 1978, *Proc. Workshop on Solar Rotation*, Publ. Catania No. 162, p. 126.

Brunning, D. H.: 1981, *Solar Phys.* **71**, 233.

Bruzek, A.: 1961, *Z. Astrophys.* **51**, 75.

Busse, F. H.: 1970, *J. Fluid Mech.* **44**, 441.

Busse, F. H. and Carrigan, C. R.: 1974, *J. Fluid Mech.* **62**, 579.

Cavallini, F., Ceppatelli, G., and Righini, A.: 1985, *Astron. Astrophys.*, in press.

Clark, D. H., Yallop, B. D., Richard, S., Emerson, B., and Rudd, P. J.: 1979, *Nature* **280**, 299.

Claverie, A., Isaak, G. R., McLeod, C. P., van der Raay, H. B., and Roca Cortes, T.: 1981, *Nature* **293**, 443.

Cram, L. E., Durney, B. R., and Guenther, D. B.: 1983, *Astrophys. J.* **267**, 442.

d'Azambuja, M. and d'Azambuja, L.: 1948, *Ann. Obser. Paris* **6**, 1.

Deubner, F.-L. and Gough, D.: 1984, *Ann. Rev. Astron. Astrophys.* **22**, 593.

Deubner, F.-L., Ulrich, R. K., and Rhodes, E. J.: 1979, *Astron. Astrophys.* **72**, 177.

Dicke, R. H.: 1976, *Solar Phys.* **47**, 475.

Dicke, R. H.: 1982, *Solar Phys.* **78**, 3.

Dicke, R. H. and Goldenberg, H. M.: 1967, *Phys. Rev. Letters* **18**, 313.

Dunn, R. B. and Zirker, J. B.: 1973, *Solar Phys.* **33**, 281.

Dupree, A. K. and Henze, W.: 1972, *Solar Phys.* **27**, 271.

Durney, B. R. and Roxburgh, I. W.: 1971, *Solar Phys.* **16**, 3.

Duvall, T. L.: 1979, *Solar Phys.* **63**, 3.

Duvall, T. L.: 1980, *Solar Phys.* **66**, 213.

Duvall, T. L.: 1982, *Solar Phys.* **76**, 137.

Duvall, T. L. and Harvey, J. W.: 1984, *Nature* **310**, 19.

Duvall, T. L. and Svalgaard, L.: 1978, *Solar Phys.* **56**, 463.

Duvall, T. L., Dziemboroski, W. A., Goode, P. R., Gough, D. O., Harvey, J. W., and Leibacher, J. W.: 1984, *Nature* **310**, 22.

Foukal, P.: 1979, *Astrophys. J.* **234**, 716.

Foukal, P. and Jokipii, J. R.: 1975, *Astrophys. J.* **199**, L71.

Gasanalizade, A. G.: 1980, *Izvest. Pulkovo Obs.* **197**, 145.

Gilman, P. A.: 1977, *Geophys. Astrophys. Fluid Dyn.* **8**, 93.

Gilman, P. A.: 1978, *Geophys. Astrophys. Fluid Dyn.* **11**, 157, 181.

Gilman, P. A.: 1980, *Highlights of Astronomy* **5**, 91.

Gilman, P. A. and Glatzmaier, G. A.: 1981, *Astrophys. J. Suppl.* **45**, 335.

Gilman, P. A. and Howard, R.: 1984, *Solar Phys.* **93**, 171.

Godoli, G. and Mazzucconi, F.: 1979, *Solar Phys.* **64**, 247.

Golub, L. and Vaiana, G. S.: 1978, *Astrophys. J.* **219**, L55.

Gough, D. O.: 1982, *Nature* **298**, 350.

Gough, D. O.: 1985, *Phil. Trans. Roy. Soc. London* **A113**, 27.

Gough, D. O.: 1985, in H. S. Hudson (ed.), *Proc. 25th COSPAR-Plenary Meeting*, Graz, in press.

Grec, G., Fossat, E., and Pomerantz, M. A.: 1983, *Solar Phys.* **82**, 55.

Henze, W. and Dupree, A. K.: 1973, *Solar Phys.* **33**, 425.

Hill, H. A.: 1984, *Inter. J. Theoret. Phys.* **23**, 683.

Hill, H. A.: 1985, submitted to *Astrophys. J.*

Hill, H. A., Bos, R. J., and Goode, P. R.: 1982, *Phys. Rev. Letters* **49**, 1794.

Hill, A., Yakowitz, D. S., Rosenwald, R. D., and Campbell, W.: 1984, Proc. 4th EPS-meeting *Hydrodynamics of the Sun*, p. 187.

Howard, R., Tanenbaum, A. S., and Wilcox, J. M.: 1968, *Solar Phys.* **4**, 286.

Howard, R.: 1978, *Rev. Geophys. Space Phys.* **16**, 721.

Howard, R.: 1979, *Astrophys. J.* **228**, L45.

Howard, R.: 1984, *Ann. Rev. Astron. Astrophys.* **22**, 131.

Howard, R., Boyden, J. E., and LaBonte, B. J.: 1980, *Solar Phys.* **66**, 167.

Howard, R. and Harvey, J. W.: 1970, *Solar Phys.* **12**, 23.

Howard, R. and LaBonte, B. J.: 1980, *Astrophys. J.* **239**, 738.

Howard, R. and Yoshimura, H.: 1976, in V. Bumba and J. Kleczek (eds.), 'Basic Mechanisms of Solar Activity', *IAU Symp.* **71**, 19.

Howard, R., Adkins, J. E., Boyden, J. E., Gragg, T. A., Gregory, T. Y., LaBonte, B. J., Padilla, S. P., and Webster, L.: 1983, *Solar Phys.* **83**, 321.

Howard, R., Gilman, P. A., and Gilman, P. I.: 1984, *Astrophys. J.* **283**, 373.

Isaak, G. R.: 1982, *Nature* **296**, 130.

Kaisig, M. and Schröter, E. H.: 1983, *Astron. Astrophys.* **117**, 305.

Kearns, M.: 1979, *Solar Phys.* **62**, 393.

Kippenhahn, R.: 1963, *Astrophys. J.* **137**, 66.

Koch, A.: 1983, Thesis University Göttingen.

Koch, A.: 1984, *Solar Phys.* **93**, 53.

Koch, A., Wöhl, H. and Schröter, E. H.: 1981, *Solar Phys.* **71**, 395.

Köhler, H.: 1970, *Solar Phys.* **13**, 3.

Kubicela, A. and Karabin, M.: 1977, *Solar Phys.* **52**, 199.

Küveler, G. and Wöhl, H.: 1983, *Astron. Astrophys.* **123**, 29.

Kuhn, J. R. and Worden, S. P.: 1979, *Astrophys. J.* **228**, L119.

LaBonte, B. J.: 1981, *Solar Phys.* **69**, 177.

LaBonte, B. J. and Howard, R.: 1981, *Solar Phys.* **73**, 3.

LaBonte, B. J. and Howard, R.: 1982a, *Solar Phys.* **75**, 161.

LaBonte, B. J. and Howard, R.: 1982b, *Solar Phys.* **80**, 361.

Liu, S. J. and Kundu, M. R.: 1976, *Solar Phys.* **46**, 15.

Livingston, W. C.: 1968, *Astrophys. J.* **153**, 929.

Livingston, W. C.: 1969a, *Solar Phys.* **7**, 144.

Livingston, W. C.: 1969b, *Solar Phys.* **9**, 448.

Livingston, W. C.: 1970, Contr. KPNO No. 531.

Livingston, W. C.: 1982, *Nature* **297**, 208.

Livingston, W. C. and Duvall, T. L.: 1979, *Solar Phys.* **61**, 219.

Lustig, G.: 1983, *Astron. Astrophys.* **125**, 355.

Lustig, G. and Dvorak, R.: 1984, *Astron. Astrophys.* **141**, 105.

Macris, C. J.: 1979, *Astron. Astrophys.* **78**, 186.

Mattig, W.: 1983, *Solar Phys.* **87**, 187.

Mattig, W. and Nesis, A.: 1974, *Solar Phys.* **38**, 337.

Mattig, W. and Nesis, A.: 1976, *Solar Phys.* **50**, 255.

Mazzucconi, F. and Godoli, G.: 1984, preprint.

McIntosh, P.: 1979, World Data Center A. Rep. UAG 70, NOAA, Boulder, Colo.

Milosevic, K. M.: 1950, *Czech. Acad. Sci.* **201**, 666.

Müller, R.: 1954, *Z. Astrophys.* **35**, 61.

Neidig, D. F.: 1980, *Solar Phys.* **66**, 205.

Newton, H. W.: 1924, *Monthly Notices Roy. Astron. Soc.* **84**, 431.

Newton, H. W. and Nunn, M. L.: 1951, *Monthly Notices Roy. Astron. Soc.* **111**, 413.

Noci, G.: 1978, *Proc. Workshop on Solar Rotation*, Publ. Catania No. 162, p. 55.

Parker, G. D., Hansen, R. T., and Hansen, S. F.: 1982, *Solar Phys.* **80**, 185.

Paternó, L.: 1978, *Proc. Workshop on Solar Rotation*, Publ. Catania No. 162, p. 11.

Pérez Garde, M., Vàzquez, M., Schwan, H., and Wöhl, H.: 1981, *Astron. Astrophys.* **93**, 67.

Pierce, A. K. and Lopresto, J. C.: 1984, *Solar Phys.* **93**, 155.

Rhodes, E. J., Harvey, J. W., and Duvall, T. L.: 1983, *Solar Phys.* **82**, 111.

Ribes, E. and Mein, P.: 1984, preprint, private communication.

Rüdiger, G.: 1982, *Geophys. Astrophys. Fluid Dyn.* **21**, 1.

Scherrer, P. H., Wilcox, J. M., and Svalgaard, L.: 1980, *Astrophys. J.* **241**, 811.

Schmidt, W.: 1982, *Geophys. Astrophys. Fluid Dyn.* **21**, 27.

Schröter, E. H.: 1959, *Monatsver. Akad. Berlin* **1**, 738.

Schröter, E. H.: 1962, *Z. Astrophys.* **56**, 183.

Schröter, E. H.: 1984, *Astron. Astrophys.* **139**, 538.

Schröter, E. H. and Wöhl, H.: 1975, *Solar Phys.* **42**, 3.

Schröter, E. H. and Wöhl, H.: 1976, *Solar Phys.* **49**, 19.

Schröter, E. H. and Wöhl, H.: 1978, *Proc. Workshop on Solar Rotation*, Publ. Catania No. 162, p. 35.

Schröter, E. H., Wöhl, H., Soltau, D., and Vázquez, M.: 1978, *Solar Phys.* **60**, 181.

Schwan, H. and Wöhl, H.: 1978, *Astron. Astrophys.* **70**, 297.

Shelke, R. N. and Pande, M. C.: 1985, *Solar Phys.* **95**, 193.

Simon, G. W. and Noyes, R. W.: 1972, *Solar Phys.* **22**, 450.

Snider, J. L., Howald, A. M., Kearns, M. D., Thomas, S. W., and Tinker, P. A.: 1979, *Solar Phys.* **61**, 3.

Snodgrass, H. B.: 1983, *Astrophys. J.* **270**, 288.

Snodgrass, H. B., Howard, R., and Webster, L.: 1984, *Solar Phys.* **90**, 199.

Snodgrass, H. M.: 1984, *Solar Phys.* **94**, 13.

Solonsky, Y. A.: 1972, *Solar Phys.* **23**, 3.

Solonsky, Y. A.: 1977, *Trudy Astron. Obs. Leningrad* **33**, 112.

Stark, D. and Wöhl, H.: 1981, *Astron. Astrophys.* **93**, 241.

Stenflo, J. O.: 1977, *Astron. Astrophys.* **61**, 797.

Svalgaard, L., Scherrer, P. H., and Wilcox, J. M.: 1978, *Proc. Workshop on Solar Rotation*, Publ. Catania No. 162, p. 151.

Ternullo, M., Zappala, R. A., and Zuccarello, F.: 1981, *Solar Phys.* **74**, 111.

Timothy, A. F., Krieger, A. S., and Vaiana, G. S.: 1975, *Solar Phys.* **42**, 135.

Topka, K., Moore, R., LaBonte, B. J., and Howard, R.: 1982, *Solar Phys.* **79**, 231.

Trellis, M.: 1963, *Compt. Rend. Acad. Sci. Paris* **256**, 2300.

Tuominen, J. and Kyröläinen, J.: 1982, *Solar Phys.* **79**, 161.

Tuominen, J., Tuominen, I. V., and Kyröläinen, J.: 1983, *Monthly Notices Roy. Astron. Soc.* **205**, 691.

Tuominen, J., Krause, F., Rüdiger, G., and Virtanen, H.: 1984, Proc. 4th EPS-Meeting *Hydrodynamics of the Sun*, p. 225.

Ulrich, R. K., Rhodes, E. J., and Deubner, F.-L.: 1979, *Astrophys. J.* **227**, 638.

Wagner, W. J.: 1975, *Astrophys. J.* **198**, L141.

Wagner, W. J. and Gilliam, L. B.: 1967, *Solar Phys.* **50**, 265.

Waldmeier, M.: 1955, *Z. Astrophys.* **38**, 37.

Ward, F.: 1965, *Astrophys. J.* **141**, 534.

Ward, F.: 1966, *Astrophys. J.* **145**, 416.

Wilcox, J. M. and Howard, R.: 1970, *Solar Phys.* **13**, 251.

Wilcox, J. M., Schatten, K. H., Tanenbaum, A. S., and Howard, R.: 1970, *Solar Phys.* **14**, 255.

Wöhl, H.: 1978, *Astron. Astrophys.* **62**, 165.

Woodard, M. and Hudson, H. S.: 1983, *Nature* **305**, 589.

Woodard, M.: 1984, *Nature* **309**, 530.

EIGHT DECADES OF SOLAR RESEARCH AT MOUNT WILSON

ROBERT HOWARD

*National Optical Astronomy Observatories**

Abstract. The Mount Wilson solar program has figured prominently in the field of solar physics throughout this century. This review describes the development of the instrumentation and the progress of the research at Mount Wilson from 1904 to 1984.

1. The Early Years

George Ellery Hale was the driving force behind the founding of the Mount Wilson Observatory. He was without a doubt one of the most dynamic scientists working in the early years of this century. His accomplishments in the field of astronomy and in general scientific and cultural promotion are staggering. His influence in scientific matters at Mount Wilson in the first decades of this century was enormous, and even more recently his influence has been felt. Hale stands as one the giants of this century in astronomical research.

The Mount Wilson Solar Observatory was founded in 1904 with an initial grant from the Carnegie Institution of Washington. The Carnegie Institution has continued to provide the major support for the Observatory until this year. The name was changed to the Mount Wilson Observatory in 1917.

The first instrument erected at Mount Wilson was the Snow telescope which Hale had brought from Yerkes Observatory with the hope of getting better solar seeing in the mountains of Southern California. This horizontal reflecting telescope, completed in 1905, was built with a 60-cm objective and a focal length of 18 m. Figure 1 shows the Snow telescope shortly after it was completed. The original optics are still in use. The building was built originally with piers to accommodate a 46 m focal length objective, but the plans for such a long focal length telescope were dropped because of the difficulties encountered with the solar seeing at the Snow telescope. The horizontal beam was sensitive to convective flows along its path, and in spite of the many precautions taken to minimize thermal air currents, the solar image was not considered to be satisfactory much of the time. The early years saw many experiments intended to improve the seeing, but apparently little success was achieved with these efforts and it was decided that a whole new type of telescope should be experimented with – a vertical telescope.

In spite of the problems with the Snow telescope, a great deal of work was done with this instrument. In particular the '5-foot spectroheliograph' (1.5 m) was much used in the early years, along with a 5.5 m spectrograph described in an early paper (Hale and Ellerman, 1906) and a wooden focal-plane shutter built by Ferdinand Ellerman for white-light photographs (and still in daily use at the 60-foot tower telescope at Mount

* The National Optical Astronomy Observatories are operated by the Association of Universities for Research in Astronomy, Inc., under contract to the National Science Foundation.

Fig. 1.   The Snow telescope shortly after its completion in 1905. This instrument was named after the original donor of the funds for the optical and mechanical components of the telescope at the Yerkes Observatory. Note that the sides of the building are louvered for protection against the heat and covered with white canvas. Vents at the top of the roof allow the warm air in the walls to escape. The coelostat mirrors are protected when not in use by a small structure that can be moved to the south to cover them. Guy wires support the flimsy outer structure, which is separate from the piers that support the optics. The trees and shrubs are much more numerous in this photograph than they are now.

Wilson). In the plate vault at the Mount Wilson offices in Pasadena there are thousands of plates from the early years in white-light, Hα, the various components of the H- and K-lines of Ca ii, and some other lines. There are also many spectroheliograms that Hale brought from the Kenwood Observatory in Chicago – some dating from as early as 1892. The results of this early work were published by Hale (1903).

It is difficult for us today to appreciate the hardships faced by the astronomers of that era. Hale and his coworkers were struggling to establish an observatory with sophisticated instrumentation in what was then a remote and hostile environment. Perhaps we can appreciate the magnitude of their problem when we read the first Annual Report of the Mount Wilson Solar Observatory in which Hale thought it necessary to record that the Snow telescope and the other buildings built on the mountain in the first year or so were 'supplied with electric lights'.

Much of the work in the early years centered on attempts to determine the relative heights in the atmosphere that contributed to the various components of the H- and

K-lines and Hα. In part this was carried out using an early version of the blink comparator to examine Hα- and Ca-plates. Also a stereo viewing system was built to determine heights of features, using Ca-plates taken from 1 to 10 hr apart. In the early days, positions of solar features were at times determined by projecting the image onto a white sphere having a printed coordinate grid. In addition, a number of sunspot spectra were exposed, and from a study of the lines that were strengthened and those that were weakened, Hale concluded that a sunspot was cooler than the surrounding photosphere. This and other spectroscopic work was accompanied by an active laboratory spectroscopy program, both on the mountain and in Pasadena. Some preliminary solar Doppler rotation work was carried out by Walter S. Adams, who also compared photospheric and sunspot spectra with those of a few of the brighter stars. (An early spectrum in the first order blue of Arcturus required 14 hr of exposure at the Snow telescope.) Also in these early years, Mr C. G. Abbott of the Smithsonian Institution made some of his first bolographic observations of the Sun at Mount Wilson.

2. The 60-foot (18 m) Tower Telescope

All along, Hale had in mind to construct a very long focal length solar telescope for high dispersion spectroscopy on a large solar image. The Snow telescope building was made large enough to accommodate a 46 m unfolded optical system, and the optical shop in Pasadena was made large enough to test optics with such long focal lengths. In fact, a reflecting mirror with that focal length and a diameter of 60 cm was actually fabricated and never used. When a horizontal system proved a disappointment, Hale designed an unfolded tower lens system having a focal length of 18 m. At first, they erected only the simple tower (adapted from a standard agricultural water-tank tower) to hold the optics. A few years later, an outer tower and dome were added as protection against the wind while observing and against rain and snow during storms.

This 60-foot (18 m) tower telescope was completed in 1907. Care was taken in the design of the telescope and optical system so as to give the best possible seeing. For example, the mirrors were exceptionally thick (~ 30 cm) in order to slow down the thermal effects on the glass. Also, the backs of the mirrors were silvered in addition to the front surfaces, and sunlight was reflected to these back surfaces so as to minimize the deflection of the mirror faces from uneven heating. Later, heating coils were used on the rear surfaces. For a time (a bit later) they experimented with cooling the mirrors with circulating water. Also experiments were carried out with speculum metal mirrors, but they suffered from low reflectivity. It is difficult nowadays to appreciate the technical obstacles encountered by these hardy pioneers of solar research. The problem was not just that the effective focal lengths of these large instruments changed a great deal (by as much as 30 cm for the 18 m system) and that astigmatism was introduced, but that the changes came during the early part of the day when the seeing was best, and that these variations were significant during the long exposure times necessitated by the slow photographic emulsions in use at that time. But the solution was clear even then. In the first Annual Report (1905) Hale mentioned that experiments were under way at the

Throop Institute (later to become the California Institute of Technology) to construct
solar mirrors of quartz. This work was done in collaboration with G. W. Ritchey, who
was on the Mount Wilson staff in the early years and was active in optical design.
Unfortunately, many years would pass before low-expansion materials would be feasible
for solar mirrors. In late 1926 the old glass (18 m) tower flat mirrors were replaced by
pyrex mirrors, and in 1960 the pyrex mirrors were replaced by fused quartz mirrors.

Another innovation of the first tower telescope was a vertical spectrograph in a deep
pit. At first, a 9 m Littrow spectrograph was installed in this pit at the 60-foot tower.
This spectrograph probably achieved a spectral resolution at least as good as any
spectrograph up to that time. This instrument is shown in Figure 2. Large prisms were

Fig. 2.   The 30-foot (9 m) spectrograph at the 60-foot (18 m) tower telescope. A Nicol prism and rhomb
are in position for measuring magnetic fields. This instrument was used by Hale to detect magnetic fields
in the Sun for the first time in 1908. This spectrograph was later converted to a spectrohelioscope, but has
not been used for many years. The building is a temporary sheetmetal structure, which was replaced in 1914
by a permanent concrete house.

used as the dispersing element. Early experiments with large glass prisms had proved
to be less successful than had been hoped, so the first work was done with liquid prisms.
The pit offered two advantages over the horizontal spectrographs that had been used
until that time. The vertical light beam resulted in improved spectrograph seeing because

the bottom of the pit was generally cooler than the top, and the pit stayed remarkably constant in temperature, which reduced the tendency for deflection of the instrument during the long exposures. These characteristics make both Mount Wilson pit spectrographs excellent, stable instruments to this day.

Fig. 3. Andrew Carnegie (on the left) and George Ellery Hale at the dedication of the 60-inch (1.5 m) telescope in late 1908 or early 1909. Mr. Carnegie, one of the great philanthropists of that era, came on a foggy day to view one of the results of his generosity – what was then the largest telescope in the world. Mr. Hale, a very successful scientific entrepreneur, stands slightly downhill from the diminutive Mr Carnegie.

In the early years there were several distinguished international visitors to Mount Wilson – evidence that the reputation of the new observatory was fast growing. J. C. Kapteyn from Groningen spent several summers at Mount Wilson gathering parallax and proper motion data for studies of the structure of the galaxy. Mr John Evershed visited briefly on his way to take up his duties at the Kodaikanal Observatory. Professor

W. H. Julius from Utrecht also visited in the summer of 1907. He was a physicist who was active in the study of anomalous dispersion, which had recently been discovered and was in fashion then. Julius hypothesized that the flash spectrum, the chromosphere at the limb, and even prominences did not really exist as they were observed but were due entirely to anomalous dispersion. Many physicists of that day were impressed with this notion, but the solar observers were skeptical.

One other phenomenon that Julius proposed to explain by means of anomalous dispersion was the broadening of absorption lines in the sunspot spectrum. This puzzling effect had been known for some years, and it remained a problem for the hypothesis favored by Hale and others that sunspots were cooler than the photosphere. Hale and his coworkers had already noticed that the swirls in the chromospheric patterns seen in spectroheliograms resembled magnetic field lines around poles of a magnet. They pointed out that often the direction of the swirls was opposite around adjacent spots and they suspected that this represented opposite polarity magnetic fields (Hale, 1908a). It turned out later that the polarity of a spot has nothing to do with the direction of the swirls in the chromospheric fibrils surrounding it, but this was not the first time a false clue had led to an important discovery, and in any case the fibrils themselves *are* linked to the magnetic fields. Moreover, the solution to the problem was associated with yet another Dutch scientist, Prof. P. Zeeman, who in 1896 had discovered the effect named after him. On June 25, 1908, Hale, using the newly completed 60-foot tower telescope with its 9 m Littrow spectrograph, detected the Zeeman effect in a photograph of a sunspot spectrum. This was the first time that a magnetic field had been detected outside the Earth, and it remains one of the most important discoveries in astronomy in this century (Hale, 1908b).

Hale was greatly elated by this spectacular development. Not only was a vitally important scientific discovery made – a discovery that was of significance to both physics and the fledgling and rather insecure field of 'astro-physics' – but also what was for that time an enormous expenditure of money on a scientific program had been to a large extent justified. In Hale's reports and correspondence following the announcement of the result, one can detect a sense of elation and confidence in the future of solar research. He and others at that time felt that the final key to understanding solar activity had just been discovered, and that the answer to the riddle should be evident in a short time. Their optimism, as it turned out, was premature.

3. The 150-Foot (46 m) Tower Telescope

In the meantime, encouraged by the superb images of the Sun obtained regularly at the 60-foot tower telescope, Hale set about to plan an instrument with a longer focal length. The design of a 46 m focal length unfolded vertical telescope was completed in the summer of 1909 by a Chicago architectural firm. Hale obtained the money for its fabrication (about $ 50 000) from the Carnegie Institution of Washington, and construction was begun late in 1909.

The 150-foot tower telescope, which was built by the Mount Wilson construction crew, introduced some innovative and ingenious design concepts. The optics at the top of the tower are supported by a thin metal framework that rests at ground level on four concrete piers. Surrounding each beam of this inner tower are square, hollow beams that are mounted on separate massive piers and support the structure and dome at the top. Figure 4 shows details of the construction. The clearance between these two separate

Fig. 4.   Early stages of the construction of the 150-foot tower telescope. The lowest section shows the completed outer beams in place. The upper section shows the inner tower structure before being sheathed with the outer beams. The large concrete piers that will support the outer tower are not yet poured. The 23 m pit, which has already been dug (by hand), is covered in this photograph. The small building in the background is a visitor's gallery, later replaced with a larger structure.

towers at the top is only a few centimeters. Standing on the optical framework at the top of the tower, one can force the inner tower to oscillate, and with enough effort, the amplitude of this oscillation at the top can be made great enough that the inner and outer towers will touch. The purpose of having two towers, of course, is to prevent the optics from shaking in the wind. This works very well, and only in a very severe wind storm will the outer tower actually strike the inner tower. Normally the solar image at the base of the tower is very steady.

The spectrograph of the 150-foot tower telescope is in a vertical pit. It is currently a Littrow system with a 15-cm diameter triplet lens with an f ratio of 150. At the start, both prisms and gratings were used interchangeably for various tasks. The gratings available in those days were crude and inefficient by modern standards.

The tower structure was completed in 1910, in time for a meeting of the International Union for Cooperation in Solar Research (precursor of the I.A.U.) at the end of August of that year in Pasadena. Figure 5 shows the group photograph from this meeting. The 80 or so attendees of the meeting included astronomical luminaries from around the world. Among them were Deslandres, Schwarzschild, Kapteyn, Larmor, Pickering, Russell, and Rydberg. Although the tower had been completed in time for the meeting, it had not been possible to complete by that time the $f/150$ lens (diameter 30 cm). In fact, great difficulties were encountered in fabricating such a lens. At first a triplet apochromat was made, but this was unsatisfactory. It suffered from astigmatism and chromatic aberration. During the interval when the tower was without a long focal length lens, the 18 m lens from the 60-foot tower telescope was installed (part way up the larger tower) so that some observations could begin there. To accommodate this optical system, the spectrograph optics were lifted up to a level 9 m below the entrance slit.

In the first months at the 150-foot tower telescope there was no means to get to the top of the tower other than the ladder, which is attached to one leg of the tower. Daily trips to the top of the tower were (and are still) necessary to set the coelostat mirrors. This was a single ladder, 46 m high, with no landings or safety equipment. Such a ladder would not be permitted today, and it is now equipped with safety mechanisms. The elevator (or 'manlift') that was installed about 1911 worked well until the gear train was changed for more modern equipment in 1983.

Originally a spectroheliograph was fabricated for the 150-foot tower telescope, but apparently little was ever done with this instrument, and today very few of the parts remain.

The Snow telescope and both tower telescopes, along with the spectrographs, are described in detail in a publication by Hale and Nicholson (1938). This monograph also describes the method of measurement of magnetic fields at the 150 foot tower telescope and presents a number of illustrations of early observations and instrumentation, including the spectroheliograph mentioned above.

Starting in 1911 a doublet achromat lens was used in the tower. Even this at first was unsatisfactory and had to be returned to the manufacturer for refiguring. This 30 cm lens served well until it was replaced by a cemented triplet apochromat in December 1971. The original lens had a small chip, about 1 cm square, missing from the bottom surface, just in the center of the lens. The story about that chip that was often repeated but possibly apocryphal was that when the lens was first installed, Walter S. Adams, who was doing the initial alignment, placed a postage stamp at the center of the bottom surface of the lens as an aid in the alignment. Later, when he removed the stamp, a thin

FOURTH CONFERENCE

INTERNATIONAL UNION FOR COOPERATION IN SOLAR RESEARCH

August 30 - September 3, 1910

Mount Wilson

1. ELLERMAN	23. CORTIE	45. KAPTEYN
2. H. C. WILSON	24. TURNER	46. MRS. FLEMING
3. ST. JOHN	25. RUSSELL	47. WATSON
4. LARKIN	26. KAYSER	48. SCHLESINGER
5. TOWNLEY	27. ADAMS	49. HUMPHREYS
6. V. M. SLIPHER	28. MILLER	50. MADRILL
7. FOWLE	29. AMES	51. J. F. SANFORD
8. COBLENTZ	30. BACKLUND	52. CHRETIEN
9. FROST	31. KONEN	53. DE LA BAUME PLUVINEL
10. IDRAC	32. PICKERING	54. FABRY
11. PUISEUX	33. FOWLER	55. MC ADIE
12. HARTMANN	34. LAMPLAND	56. HILLS
13. KÜSTNER	35. HALE	57. LARMOR
14. SLOCUM	36. BELOPOLSKY	58. COTTON
15. HAMY	37. DESLANDRES	59. DYSON
16. KNIGHT	38. SCHUSTER	60. BARNARD
17. WOLFER	39. CAMPBELL	61. KING
18. FATH	40. RICCO	62. NEWALL
19. RYDBERG	41. MRS. KAPTEYN	63. PRINGSHEIM
20. HERPFERGER	42. BOSLER	64. LEUSCHNER
21. FOX	43. K. SCHWARZSCHILD	65. J. S. PLASKETT
22. HAUSSMANN	44. MC ADIE	66. GALE
		67. CHANT
		68. EVERSHEIM
		69. ROTCH
		70. W. MITCHELL
		71. STRATTON
		72. H. D. BABCOCK
		73. RITCHEY
		74. BRACKETT

Fig. 5. Fourth Conference of the International Union for Cooperation in Solar Research – later to become the International Astronomical Union. This picture was taken on August 30, 1910, while the group was touring Mount Wilson.

Fig. 6.   The Snow telescope, 60-foot Tower Telescope, and 150-foot Tower Telescope. The small structure covers the Snow coelostat mirrors and the 150-foot dome is closed. This photograph was taken after 1911, when the canvas covering of the Snow was replaced with sheet metal and before 1914 when a dome was added to the 60-foot tower telescope.

layer of glass came with it. In spite of this slight flaw the lens proved to be an excellent one.

Soon a grating was obtained for the spectrograph from Michelson, and a good Littrow lens from Brashear. With these additions the tower was finally completed in May 1912. For half a century, until the completion of the McMath telescope at Kitt Peak, it was the solar telescope with the longest focal length in the world. It still remains an excellent instrument with an extremely stable spectrograph: a credit to Hale and his coworkers.

One of the principal reasons for building such a telescope was to find the 'general magnetic field' of the Sun. Hale and others in those years believed that the coronal streamers seen above the polar regions of the Sun during a solar eclipse signaled the presence of a significant dipole magnetic field that could be found if one could muster sufficient spectral resolution; hence the 23 m spectrograph. Although in these later years Hale had little time to devote to solar research himself, his colleagues at Mount Wilson set about to find the general magnetic field with much zeal and confidence, inspired by the fine example set by George Ellery Hale. Unfortunately, zeal and confidence can at times be a disadvantage in scientific pursuits, and when people are convinced that an effect exists in a noisy signal they can often find it whether it is there or not. (There have been much more recent examples of this effect in solar research.) This is apparently what happened to the Mount Wilson observers because the rather large dipole magnetic field they claimed to find (Hale, 1913; Hale *et al.*, 1918) is now known not to exist – the magnetic field situation is much more complicated than that. An objective modern measurement of many of the original plates (Stenflo, 1970) also showed no measurable dipole field. It is interesting to note that the Mount Wilson astronomers set about at the same time to measure the general magnetic fields of galaxies.

An early rotation study by Adams (1911) gave very accurate spectroscopic rotation results and even hinted at the velocity field now known as supergranulation. This was the first of a long line of Mount Wilson solar rotation measures. During the 20's and 30's C. E. St. John continued the work of Adams, and later work with the magnetograph also was concentrated in part on this topic.

In 1914, near the minimum of the activity cycle, Hale discovered (Hale and Nicholson, 1938), that polarities of sunspots in the north and south hemispheres reversed from one cycle to the next. The full 22-year cycle is therefore referred to as the 'Hale cycle'.

As a means of keeping track of the polarities of the sunspots, a series of sunspot drawings was begun early in 1917 at the 150-foot tower telescope. Each spot was recorded and its magnetic field was measured by eye, using the 23 m Littrow spectrograph and a polarizing analyzer. Using the ~ 10 mm Å^{-1} dispersion of that spectrograph, it is possible today to achieve an accuracy of about 100 gauss with such a procedure. In the early configuration the dispersion was about $\frac{1}{3}$ the modern value. In recent years the observers have taken pride in the neatness and accuracy of their drawings, but in the earliest years the drawings were only intended as a guide to which spot each magnetic determination referred. The first 20 years of the drawings and spot magnetic field measurements were published (Hale and Nicholson, 1938). The data since that publication are available on microfilm from the World Data Center in Boulder, Colorado, U.S.A.

4. The Middle Years

Hale's influence at Mount Wilson waned even before his retirement on July 1, 1923. He suffered from poor health, which kept him inactive for months at a time, and when he was well his time was often spent on other matters. Adams, who succeeded Hale as Director of the Mount Wilson Observatory in 1918, developed interests in stellar astronomy and without Hale's driving force the solar group lost some of its momentum. There were competent people in the solar group – including Seth Nicholson, Edison Petit, and Charles E. St. John, but the exciting results of the early years were not followed up. To a large extent, of course, this was due to the fact that the theoretical tools for understanding the behavior of magnetic fields in an ionized gas would not be ready for many years. Also to some extent further progress would have to wait for

technological advances that would not be available until after the Second World War
– notably the photomultiplier tube.

But the solar staff did carry on solar work in these middle years. Petit contributed
a great deal to the study of prominences. He also initiated one of the first automated
film flare patrols in about 1936, in the K-line of Ca II, using the spectroheliograph at
the 60-foot tower telescope. The 35 mm films from the early years of the patrol (on a
highly inflammable cellulose base) are still preserved. St. John put a great deal of effort
into determining the gravitational red shift of the Sun, using the spectrograph of the
150-foot tower telescope. The difficulties of the limb red shift, caused by local,
granulation-related velocities limited the effectiveness of this program.

In September 1923 an ambitious eclipse expedition was mounted to view a total solar
eclipse from Point Loma, near San Diego. Photographs of the star field around the Sun
were planned along with coronal spectra and other types of observations. Unfortunately,
the day was overcast.

During this interval, the various synoptic observing programs – the sunspot drawings
and magnetic field measurements and the Ca–K and Hα spectroheliograms and
white-light photographs at the 60-foot tower telescope – continued. During many
intervals these were the only solar observations at Mount Wilson. Although there may
have been some feeling that long synoptic data sets would eventually be of value in
sorting out many of the puzzling phenomena of solar activity, it was probably inertia
as much as anything else that kept these programs going over the years. It seems
inconceivable in today's world of short-term research grants and tight budgets that such
programs were permitted to continue for decades, unreviewed. It is fortunate for us now
that they were.

Among the observers who kept the solar program going over many years was
Ferdinand Ellerman, who worked for Hale at the Kenwood Observatory in Chicago and
accompanied him to California when he moved there in 1903. Later Joseph O. Hickox,
Thomas A. Cragg and Larry Webster were among the skillful observers who amassed
eight decades of careful observations of the Sun.

5. The Modern Era

With the start of the Second World War, the pace of solar research at Mount Wilson
slowed even more as most of the staff scientists became engaged in war-related work.
The synoptic solar observations continued, however, during the approach to a very active
solar maximum.

After the end of the war, much of the new technology that was developed as a part
of the war effort was adapted to the field of astronomy, and a number of young men
who came back from military service into the field of astronomy brought with them a
thorough knowledge of the wartime advances in the field of electronics. In those years
it was not unusual to see a photomultiplier amplifier circuit published in the Astrophysical Journal. This was a time of rapid instrumental development in astronomy –
rather similar to the present period, with its strong emphasis on area detectors.

Starting in 1949 and continuing for about 10 years the Snow telescope was used by a group from the McMath Hulbert Observatory of the University of Michigan. Leo Goldberg, Orren Mohler and Keith Pierce were involved in the project. Figure 7 shows

Fig. 7. The coelostat mirrors of the Snow telescope. This photograph was taken in about 1951. The student observer is Mr. Dale Vrabec. At the lower right may be seen a cable going down below the observing floor. This is a weight-driven clock drive controlled by a governor. The clock drives the mirror through a sector gear – a portion of a very large gear – which can also be seen above the cable.

the Snow coelostat mirros about this time. An infrared spectrograph was installed at the Snow telescope. At first a lead sulfide tube was used as a detector, and later a lead telluride cell was installed. Infrared spectra were obtained for an atlas, and a great deal of work on the energy distribution in the infrared and limb darkening was carried out. Also some abundance determinations were made. By this time this was the only work being done at the Snow telescope. Among the students hired by the University of Michigan to carry out the observations in the early years were Dale Vrabec, William

Livingston and Walter E. Mitchell, Jr. Mitchell, who joined the faculty of Ohio State
University, carried on the Snow telescope observations through the 1960's and on a few
occasions later than that. In 1961 the spectrograph was converted to a double pass
mode. The Snow telescope has seen very limited use since the last of these observations
was completed.

In 1950, Dr K. O. Kiepenheuer from the Fraunhofer Institute in Freiburg, West
Germany, spent six weeks at Mount Wilson. One of the projects he worked on while
he was there was a device to measure magnetic fields on the solar surface photoelectri-

Fig. 8.   The original magnetograph built by H. W. Babcock at the Hale Solar Laboratory near Pasadena.
The solar image may be seen on the white table. Below the table is the ADP analyzing crystal. Forward
and to the left of this are the photomultiplier tubes. The spectrograph, a replica of the unfolded 23 m Littrow
spectrograph at Mount Wilson, is in a pit below this level. The meter stick in front driven by a variable-radius
cam tilts a block of glass to give correction for the line shift due to solar rotation. The amplifier is to the
left of the phototubes. To the right of the ADP crystal is the CRT display, and to the right of that is the
circuitry for the synchro control of the image scanning.

cally. (Actually, Hale and A. E. Whitford had attempted to measure solar magnetic fields photoelectrically about 20 years earlier, but the instrumentation of that day was just too crude.) Kiepenheuer's efforts showed no general solar field with a limit of about 0.6 gauss, but he did find weak fields surrounding sunspots. Within a short time after that, H. W. Babcock, of the Mount Wilson staff, developed the principle of the modern magnetograph. This work was done at the Hale Solar Laboratory near Pasadena – the observatory built by Hale to serve as his private laboratory during his retirement and left to the Carnegie Institution of Washington at his death in 1938. In the first years this instrument was operated daily by H. D. Babcock (the father of H. W. Babcock and one of the early Mount Wilson staff members – by that time retired) and Thomas Cragg, one of the Mount Wilson solar observers. This instrument fell into disuse after about 1960. It is shown in its original configuration in Figure 8.

H. W. Babcock built an improved copy of the Hale Laboratory magnetograph and installed it at the 150-foot tower telescope on Mount Wilson early in 1957 – in time for the IGY. The instrument at the Hale Laboratory had already shown (Babcock and Babcock 1954) that the Sun possessed weak magnetic fields in the polar regions and that at lower latitudes active regions showed bipolar magnetic configurations. In addition, large areas of unipolar magnetic fields could be identified. After a few years H. D. Babcock (1959) demonstrated from his observations at the Hale Laboratory that the polar magnetic fields reversed sign at about the time of solar activity minimum.

The development of the magnetograph was the beginning of the modern era at Mount Wilson, but soon advances were to be made in another direction. Robert B. Leighton was active in cosmic ray physics in the mid 1950's at the California Institute of Technology. Much of his work was done at a cosmic ray site that had been established on Mount Wilson near the 60-foot tower telescope. He became interested in this telescope and the possibilities for advances in solar research, and when the cosmic ray research ended, he shifted to solar physics. He rebuilt the slit jaws of the spectrohelio-graph and experimented with difference pictures between two spectroheliograms taken in the wing of a Zeeman-sensitive line in opposite polarizations. Actually, this was a technique suggested by Hale many years earlier. Leighton met with spectacular success in this work and was soon producing photographic magnetograms with angular resolution that had never been achieved before – and has not often been achieved since. About that time H. W. Babcock and I modified the magnetograph at the 150-foot tower telescope so that it could be used with relatively high resolution to scan a limited region of the solar surface. Both of these instruments were used in early studies of the relationship of small-scale magnetic fields to chromospheric features.

Leighton extended the photographic subtraction technique to the measurement of velocity fields, and soon discovered that essentially the whole solar surface was undergoing periodic oscillations in what seemed then to be small random patterns (Leighton et al., 1962). These '5-minute oscillations', as they were known then, have more recently been shown to be global modes of acoustic oscillations with great potential for analysis of the structure and dynamics of the solar interior.

The concept of supergranulation also arose at this time from Leighton and his students (Leighton *et al.*, 1962). This velocity pattern showed up in early cancelled velocity maps from the 60-foot tower telescope.

These advances, made at the Hale Solar Laboratory and Mount Wilson from the early 1950's to the early 1960's form the basis for much of the field of solar physics today. It was an exciting decade, when fundamental discoveries followed each other rapidly. For a time the atmosphere was like that of the first decade at Mount Wilson, with instruments being developed and exciting results following one after the other.

The magnetograph at the 150-foot tower telescope was gradually improved over the years. Probably the first digital data to be obtained at a solar telescope (recorded on punched paper tape) were taken in the summers of 1962 and 1963 (Howard, 1967). These were detailed observations of the 5-min oscillations at a single point. The data were reduced by calculating autocorrelations and power spectra at an early digital computer at Caltech.

Daily synoptic full-disk magnetograms had begun at the 150-foot tower telescope at Mount Wilson in 1957 at the start of the IGY. They were continued as a part of the Mount Wilson synoptic solar observations after the IGY ended. Improvements made to the instrument included a guider in about 1964 and soon thereafter a digitized signal recorded on magnetic tape. The software for this analysis required many months to develop, and it was not until late 1966 that the daily scans could be preserved on tape. Doppler velocity data were a part of the daily scans from the start, and an early paper presented the first results (Howard and Harvey, 1970).

Perhaps the most unique aspect of the modern Mount Wilson observatories has been the full-disk velocity data. With the very stable spectrograph, these data have been of good quality for many years, but in 1983 the quality improved dramatically because of two improvements. A new exit slit assembly was installed that gave more accurate positioning of the exit slit than had been possible previously and also provided fiber-optic coupling to the photomultiplier tubes (Howard *et al.*, 1983). Secondly, a new grating was installed in the 23 m spectrograph. The old grating had been submerged in dirty water on one occasion during a heavy rain storm when a sump pump at the bottom of the pit failed, and in any case it was 20 years old and generated a great deal of scattered light. The new grating is in a hermetically sealed box with a window. The box is filled with dry nitrogen. The grating has remained in pristine condition since it was installed several years ago, and there is good reason to believe that it will remain clean almost indefinitely.

A further improvement in 1984 was the installation of a stable wavelength reference source – a temperature controlled iodine absorption cell. This cell, installed just ahead of the spectrograph focus so as to avoid contaminating the solar line with I_2 lines, gives an excellent reference to correct spectrograph seeing and long-term spectrograph drifts. A separate exit slit assembly detects the position of the reference line. This enables one to obtain long-term relative velocity measurements with an accuracy of a few m s^{-1}.

The long synoptic series of velocity data led to the discovery of torsional oscillations (Howard and LaBonte, 1980). This very low-amplitude systematic velocity feature could only have been detected with a long series of observations taken with a very stable, low-noise system. Similarly the search for giant cells has required exceptional stability and low noise. The current upper limit (~ 1 m s^{-1}/wave number) (Snodgrass and Howard, 1984) is testimony to the precision of the instrument and the skill and dedication of the observers.

6. The Future

At this time (March 1985) the future of Mount Wilson is uncertain. It appears likely that very shortly solar observations will cease there. The days of George Ellery Hale's influence at the observatory administration and in the Carnegie Institution of Washington are long past. This is a sad time for Mount Wilson and for the field of solar physics.

Programs should not be allowed to continue out of force of habit, but part of the value of the solar program at Mount Wilson has been the long, uninterrupted series of synoptic observations of fundamental solar parameters. If these cease, the goals set forth by Hale only eight decades ago will remain unfulfilled.

Acknowledgement

All of the figures in this review are supplied through the courtesy of the Mount Wilson and Las Campanas Observatories.

References

Adams, W. S.: 1911, CIW Publication No. 138, 'An Investigation of the Rotation Period of the Sun by Spectroscopic Methods'.

Babcock, H. D.: 1959, *Astrophys. J.* **30**, 364.

Babcock, H. W. and Babcock, H. D.: 1954, *Astrophys. J.* **121**, 349.

Hale, G. E.: 1903, *Publ. Yerkes Obs.*, Vol. III, Part 1, 3.

Hale, G. E.: 1905, *Annual Reports of the Director*, The Mount Wilson Observatory, Carnegie Institution of Washington.

Hale, G. E.: 1908a, *Astrophys. J.* **28**, 114.

Hale, G. E.: 1908b, *Astrophys. J.* **28**, 315.

Hale, G. E.: 1913, *Astrophys. J.* **38**, 37.

Hale, G. E. and Ellerman, F.: 1906, *Astrophys. J.* **23**, 54.

Hale, G. E. and Nicholson, S.: 1938, CIW Publication No. 498, 'Magnetic Observations of Sunspots'.

Hale, G. E., Seares, F. H., Van Maanen, A., and Ellerman. F.: 1918, *Astrophys. J.* **47**, 206.

Howard, R.: 1967, *Solar Phys.* **2**, 3.

Howard, R. and Harvey, J.: 1970, *Solar Phys.* **12**, 23.

Howard, R. and LaBonte, B. J.: 1980, *Astrophys. J. Letters* **239**, 233.

Howard, R., Boyden, J. E., Bruning, D. H., Clark, M. K., Crist, H. W., and LaBonte, B. J.: 1983, *Solar Phys.* **87**, 195.

Leighton, R. B., Noyes, R. W., and Simon, G. W.: 1962, *Astrophys. J.* **135**, 474.

Snodgrass, H. B. and Howard R.: 1984, *Astrophys. J.* **284**, 848.

Stenflo, J. O.: 1970, *Solar Phys.* **14**, 263.

MEASUREMENTS OF MAGNETIC FIELDS AND THE ANALYSIS OF STOKES PROFILES

J. O. STENFLO

Institute of Astronomy, ETH-Zentrum, CH-8092 Zürich, Switzerland

Abstract. Recent advances in polarimetry allowing the recording of polarized line profiles with high spectral resolution and signal-to-noise ratio over large portions of the solar spectrum offer rich new diagnostic possibilities. Thus we can now in a systematic way build models of the height variation of the magnetic field, temperature, density, and mass motions in the spatially unresolved subarcsecond magnetic structures. The analysis of the Stokes spectra also allows us to build a foundation for proper diagnostics of vector magnetic fields, a goal that cannot be achieved before the intrinsic properties of the spatially unresolved magnetic fields have been determined. Another new diagnostic tool is the Hanle effect. A recent exploratory survey of coherence effects through the recording of the linear polarization with high spectral resolution throughout the whole visible solar spectrum aims at establishing a foundation for the exploitation of the Hanle effect on the solar disk.

This review describes these developments, most of which have taken place in the 1980s, and summarizes the results obtained so far.

1. Introduction

The structuring of the solar atmosphere, which we may see as X-ray loops, coronal streamers, Hα fibrils, etc., is. evidence of magnetic fields at work, and provides information on the magnetic-field geometry and topology. This information is however quite indirect, since it is the temperature-density structure that is diagnosed, from which we via our limited knowledge of solar MHD try to draw conclusions about the magnetic field.

Fortunately the effect of the magnetic field on the solar spectrum is not only this indirect. The magnetic field influences the atomic emission process directly via the Zeeman and Hanle effects, which results in polarization effects in spectral lines. The observation and interpretation of this polarization allows us to diagnose the magnetic field directly.

The Zeeman effect was first used on the Sun by Hale (1908), who discovered magnetic fields in sunspots. 45 years later the photoelectric magnetograph was developed by Thiessen (1952), Babcock (1953), and Kiepenheuer (1953), and was applied to recordings of the longitudinal (line-of-sight) magnetic field across the solar disk. The magnetograph concept was extended by Stepanov and Severny (1962) for the systematic mapping of vector magnetic fields through the simultaneous recording of the transverse and longitudinal Zeeman effect in various spectral lines. More recently such mapping has also been done by Hagyard *et al.* (1982), using filter techniques.

These maps of the vector magnetic field, which were produced on a routine basis since about 1959 with relatively high spatial resolution (a few sec of arc), seemed very successful. The conversion of polarization into magnetic field vectors was based on the

Solar Physics **100** (1985) 189–208.   0038–0938/85.15.

theory developed by Unno (1956), extended to include magneto-optical effects by Rachkovsky (1962a, b), and this is the theory that is still used to-day.

The application of the Unno–Rachkovsky theory was however valid only if the magnetic field were spatially resolved. We now know that this is far from being the case (except inside sunspots), even with the most powerful telescopes available to-day. The unresolved fine structure 'distorts' the results differently in different spectral lines. This may seem to cause great difficulties, but it also brings remarkable opportunities in diagnosing the field.

Discrepancies between the apparent line-of-sight magnetic fluxes recorded in different spectral lines were noted by Harvey and Livingston (1969), who interpreted this in terms of line weakenings (temperature effects) alone. By a proper combination of spectral lines it is however possible to untangle the various effects of temperature, magnetic field, height of formation, etc., to get around the limitations of spatial resolution and derive the intrinsic properties of the magnetic elements. By developing and applying such an interpretative scheme, it was discovered that the photospheric magnetic flux occurs in the form of kG fluxtubes, even in the quiet network far from active regions, where the average field is only a few G (Stenflo, 1973).

In all this previous work the polarization recordings were limited to a few fixed and narrow wavelength bands in some selected spectral lines. A major advance in the observational possibilities occurred during the 70 s, when techniques to record the polarization throughout the line profiles in the four Stokes parameters (I, Q, U, and V) were developed (Harvey et al., 1972; Baur et al., 1980, 1981). The main breakthrough came however with the conversion of the Fourier transform spectrometer (FTS) of the McMath telescope at Kitt Peak into a polarimeter. As compared with other instruments, the FTS is superior in spectral resolution and wavelength coverage. The FTS records all the spectral lines within the passband of the prefilter strictly simultaneously. Typical passbands used are 1000 Å wide. Generally the spectrum is completely resolved, i.e., one does not have to consider any instrumental broadening, and there is no straylight to worry about.

The new possibilities for fluxtube diagnostics opened up by the FTS were explored by Stenflo et al. (1984), using FTS recordings of Stokes I and V made in 1979. This material was subsequently exploited by Solanki and Stenflo (1984, 1985) through statistical analysis of simultaneously recorded Stokes I and V profiles of 400 Fe I and 50 Fe II lines, to derive the intrinsic temperature and magnetic-field structure of the subarcsec magnetic fluxtubes. Such FTS data allow us to build detailed empirical models of the height variations of the physical parameters in fluxtubes (the modern counterpart of facular models). Although the observational data are now available, such a construction of models which are self-consistent in a magnetohydrodynamic sense is a major undertaking, which has only been started. For further details on the fine-scale structure of the field, we refer to a recent review (Stenflo, 1984).

Through the analysis of Stokes profiles in many different spectral lines, it has now also become possible to sort out the diagnostic problems that have prevented the vector magnetic field to be determined, due to the spatially unresolved structures (Stenflo,

1985). Simultaneous recordings of Stokes I, Q, and V in the visible and infrared, made in May 1984 with the Kitt Peak FTS, are now being analyzed with the aim of developing a foundation for future determinations of the vector magnetic field (Stenflo *et al.*, 1985).

The main application of the Hanle effect to the Sun was during most of the 1970s limited to determinations of the magnetic field in solar prominences, using the HeI D_3 line (Sahal-Brechot *et al.*, 1977; Bommier, 1980; Bommier *et al.*, 1981; Landi Degl' Innocenti, 1982; Leroy *et al.*, 1983; Athay *et al.*, 1983). Through Stokesmeter recordings in 1978 it was for the first time possible to see the Hanle effect in a line (CaI $\lambda 4227$ Å) formed on the solar disk (Stenflo, 1982), and to use the Hanle effect to obtain information on the strength of a possible 'turbulent' magnetic field in the solar atmosphere.

The Hanle effect represents the modification by magnetic fields of the linear polarization produced by coherent scattering in the line formation process. Before we can start to systematically exploit the Hanle effect on the solar disk, we need to properly understand the basic physics of scattering polarization, in the absence of magnetic fields. For this purpose a survey of the linear polarization throughout the solar spectrum, 3165–9950 Å, was made with the Kitt Peak FTS (Stenflo *et al.*, 1983a, b), which revealed a number of unexpected coherence effects. The Hanle effect has important diagnostic potential, but because of the complicated physics involved, and the present lack of large 'polarization-free' telescopes, there will be some time before we can make full use of it.

In the present review we will try to summarize the various recent developments in diagnosing the magnetic field from Stokes line profiles.

2. Analysis of Circular Polarization

To illustrate the type of information that we can read out of Stokes spectra, we show in Figure 1 a small portion of a Stokes I and V recording made in a plage near disk center with the FTS in 1979. Some of the main diagnostic features are:

– The Stokes V amplitudes provide information on the magnetic flux.

– The Stokes V amplitude *ratios* between different lines provide information on the intrinsic field strengths, filling factors, and temperatures (with respect to the surroundings).

– The Stokes V line width contains information on the intrinsic field strengths and the non-thermal Doppler broadening.

– The asymmetries between the blue and red Stokes V peaks of the various lines provide information on the internal mass motions.

– The wavelength position of the Stokes V zero-crossing point provides information on the systematic mass flows inside the fluxtubes.

It should be noted that regardless of how small the filling factor α inside the spatial resolution element is, the Stokes V spectrum has its contributions exclusively from this small fraction α of the surface covered by magnetic fields. The Stokes I spectrum on the other hand has contributions from the whole spatial resolution element, but as the

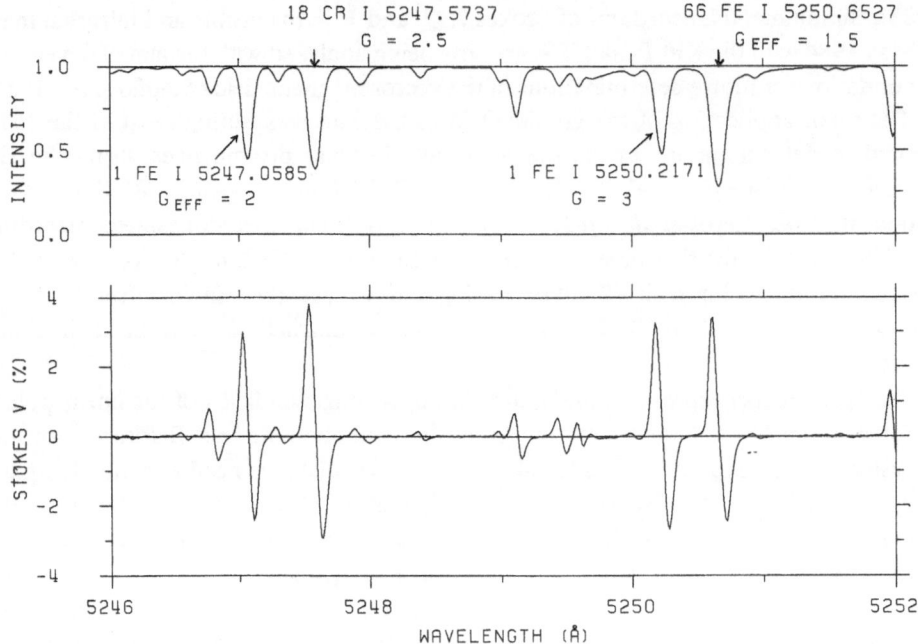

Fig. 1.   Stokes I and V spectra over the range 5246–5252 Å, recorded in a strong plage near disk center with the NSO McMath FTS polarimeter (Stenflo *et al.*, 1984). V is given in units of the adjacent continuous spectrum.

magnetic filling factor is usually $\ll 1$, the Stokes I spectrum is more representative of the non-magnetic regions.

For a normal Zeeman triplet formed in a magnetic field oriented parallel to the line of sight, the intensities of the σ components can be written

$$I_{\sigma_{1,2}} = \tfrac{1}{2}(I \pm V)\,. \tag{2.1}$$

If the field is homogeneous, and as the Zeeman splitting is

$$\Delta\lambda_H = 4.67 \times 10^{-13}\, g\lambda^2 B\,, \tag{2.2}$$

where g is the Landé factor, B is in G, and the wavelengths are in Å, the following relation is valid:

$$I_{\sigma_{1,2}}(\lambda) = \tfrac{1}{2}I(\lambda \pm \Delta\lambda_H)\,. \tag{2.3}$$

(2.1) and (2.3) give

$$V = \tfrac{1}{2}[I(\lambda + \Delta\lambda_H) - I(\lambda - \Delta\lambda_H)]\,, \tag{2.4}$$

which we may expand as

$$V = \Delta\lambda_H \left[\frac{\partial I}{\partial \lambda} + \frac{1}{6}\,(\Delta\lambda_H)^2\, \frac{\partial^3 I}{\partial \lambda^3} + \dots \right]. \tag{2.5}$$

For weak fields ($\Delta\lambda_H \ll$ line width), only the first term in the Taylor expansion needs to be retained.

Let us for pedagogical purposes introduce a two-component model, in which the fraction α (filling factor) of the resolution element is covered by line-of-sight magnetic fields of strength B and line intensity I_α (that the spectral line would have if the Zeeman splitting, i.e., the Landé factor, were zero), whereas the remaining fraction $1 - \alpha$ is field free. Assume further that we observe in two spectral lines of equal strength (I_α equal) and almost equal wavelength λ, and which are formed in the same way at the same height in the atmosphere (B equal). If we represent the two lines by indices 1 and 2, and define

$$k = 4.67 \times 10^{-13} \, \lambda^2 \tag{2.6}$$

in the units used for (2.2), we obtain the line ratio

$$V_1/V_2 = \frac{g_1}{g_2} \left(\frac{\partial I_\alpha}{\partial \lambda} + \frac{k^2}{6} \, g_1^2 B^2 \, \frac{\partial^3 I_\alpha}{\partial \lambda^3} + \dots \right) \Bigg/ \left(\frac{\partial I_\alpha}{\partial \lambda} + \frac{k^2}{6} \, g_2^2 B^2 \, \frac{\partial^3 I_\alpha}{\partial \lambda^3} + \dots \right). \tag{2.7}$$

Note that the filling factor α does not appear in the ratio. If the field were intrinsically weak,

$$V_1/V_2 \approx g_1/g_2 \,, \tag{2.8}$$

the ratio between the Landé factors.

The two lines Fe I $\lambda\lambda\,5247.06$ and 5250.22 Å with Landé factors of 2.0 and 3.0, respectively, which were shown in Figure 1, are used in Figure 2 to illustrate how the line ratio works. In the diagram to the left the Stokes V profiles in the blue line wings are plotted for a strong and a weak plage near disk center, differing in amount of flux (filling factor) by a factor of 6. The diagram to the right shows the line ratio $(g_2/g_1) \, (V_1/V_2)$, in the notation of (2.8). The deviation of the curves from unity demonstrates the importance of the higher order terms in the expansions of (2.7), and can be used to determine the field strength B (*independent* of what the filling factor is). The determination of B however requires the use of the intensity profile I_α.

Due to temperature effects inside the fluxtubes, most spectral lines are strongly weakened in magnetic regions. I_α is however not a directly observed quantity (since $\alpha < 1$). The observed intensity is I, which for small α more represents the non-magnetic atmosphere. I_α in (2.7) is the intensity profile inside the fluxtube in the case that the Landé factor of the line were zero. If we assume that the only fluxtube effect on I_α is line weakening, without any change in broadening or relative line shape, we can write

$$1 - I_\alpha/I_{c,\,\alpha} = w(1 - I/I_c) \,, \tag{2.9}$$

where I_c is the continuum intensity, and w is the line weakening factor. In this case we can replace I_α by I in (2.7). Since I is observed, there are no unknowns in the problem of determining B from V_1/V_2.

The only influence of the fluxtube thermodynamic structure on the determination of B occurs if I_α cannot be represented by (2.9), in particular if the combined thermal and

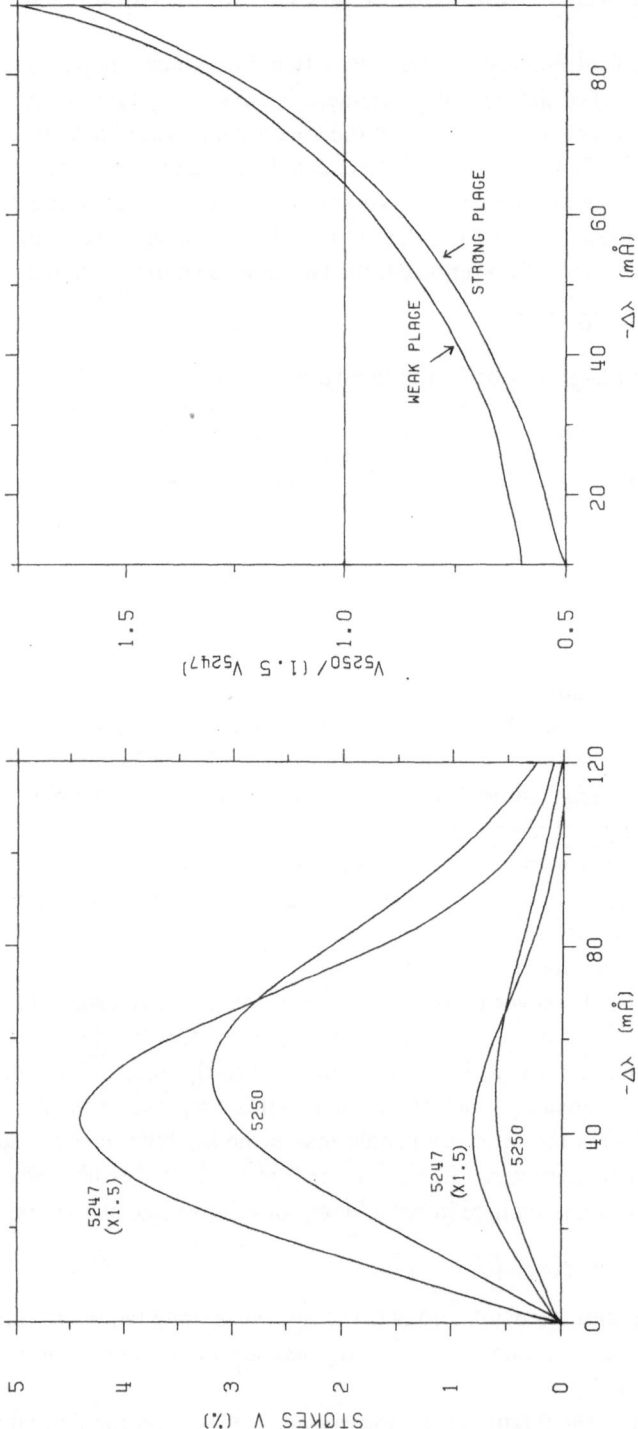

Fig. 2. Illustration of the line-ratio technique for the Stokes *V* parameter (Stenflo and Harvey, 1985). Left diagram: Stokes *V* profiles for the blue wings of the Fe I λλ 5247.06 and 5250.22 Å lines. The two upper curves refer to recordings in a strong plage (the same as in Figure 1), the two lower curves to a weak plage. The factor of 1.5 is the ratio between the Landé factors of the two lines. Right diagram: ratio between the curves of the left diagram. If the intrinsic field strength were small (less than a few hundred G), the curves would coincide with unity.

non-thermal line broadening in the fluxtube differs from the exterior conditions. These effects are however small, so in a first approximation the determination of B from the line ratio is model independent. The previous use of this line-ratio technique with a Babcock-type magnetograph to determine the kG field strengths has been reviewed before (Stenflo, 1976, 1978a), but the more powerful FTS data illustrate the principle with greater clarity. As the line ratio is now obtained as a function of wavelength, and as the line core is formed higher in the atmosphere than the outer line wings, the line ratio in a single line even contains information on the height variation of B (cf. Stenflo, 1984). Such a height variation can also be established by using observations at various center-to-limb distances, or by using lines of various line strengths.

If the two lines used for the line ratio have different temperature sensitivities (e.g. different excitation potentials) but the same relative I line shape, (2.7) is changed into

$$V_1/V_2 = \frac{w_1}{w_2} \frac{g_1}{g_2} \left(\frac{\partial I_1}{\partial \lambda} + \frac{k_1^2}{6} g_1^2 B^2 \frac{\partial^3 I_1}{\partial \lambda^3} + \cdots \right) \Big/ \left(\frac{\partial I_2}{\partial \lambda} + \frac{k_2^2}{6} g_2^2 B^2 \frac{\partial^3 I_2}{\partial \lambda^3} + \cdots \right).$$

$$(2.10)$$

where I_1 and I_2 are the observed non-magnetic profiles, and we neglect the change of B due to the difference in height of formation. If we choose a line pair for which the second and higher order terms are small (g_1 and g_2 small), the B dependence disappears, and the line ratio gives the ratio between the temperature weakenings of the two lines. Note that this is the temperature weakening occurring *inside* the fluxtube, completely independent of how small the filling factor α happens to be.

If one compares the Stokes V amplitude of the Fe I $\lambda 5250.65$ Å line with that of the Fe I $\lambda\lambda 5247.06$ and 5250.22 Å lines, one notices its larger polarization signal in spite of its smaller Landé factor. Only part of this effect can be explained in terms of Zeeman saturation (higher order terms in the Taylor expansion), but much of it is due to the smaller temperature sensitivity (higher excitation potential) of the 5250.65 Å line.

When the line ratio V_1/V_2 is determined for a given line pair like the 5250.22 and 5247.06 Å lines, one finds (near disk center) practically the same values wherever one samples, in the quiet region network or in active-region plages with an order of magnitude or more difference in the flux per resolution element. This small spread in the observed values of V_1/V_2 suggests that the spread in the fluxtube intrinsic properties is also small. This has led to the concept of 'unique' fluxtube properties (Stenflo, 1976). Recent Stokesmeter measurements show that this is correct in a first approximation, but that in a second approximation there is a slow dependence of field strength on filling factor, such that the kG field strength increases by about 20% when the filling factor is increased by a factor of 6 (Stenflo and Harvey, 1985). The temperature structure appears to vary more with filling factor, which has become evident when comparing network and plages (Solanki and Stenflo, 1984, 1985). Nevertheless the statistical spread in these properties for a given filling factor appears to be small, but it is not yet known how small it really is.

Since the FTS allows us to record many hundreds of spectral lines strictly simultaneously, one has the opportunity of using a sample of lines formed at different heights, to determine the internal height structure of the subarcsec fluxtubes. Lines like Mg I b$_2$ in Figure 3 indicate that it may be possible to pursue this structure beyond the level of the temperature minimum. Most of the strongest lines like Ca II H and K and the Balmer lines are very broad and therefore have a weak Stokes V signal, not so suitable for detailed analysis. The Doppler core of the Mg I b$_2$ line is however quite narrow, which should make an analysis of the shape of its Stokes V profile profitable (cf. Stenflo *et al.*, 1984). For such a strong line a non-LTE treatment is obviously required in the interpretation to refer the measured field to the correct height in the atmosphere.

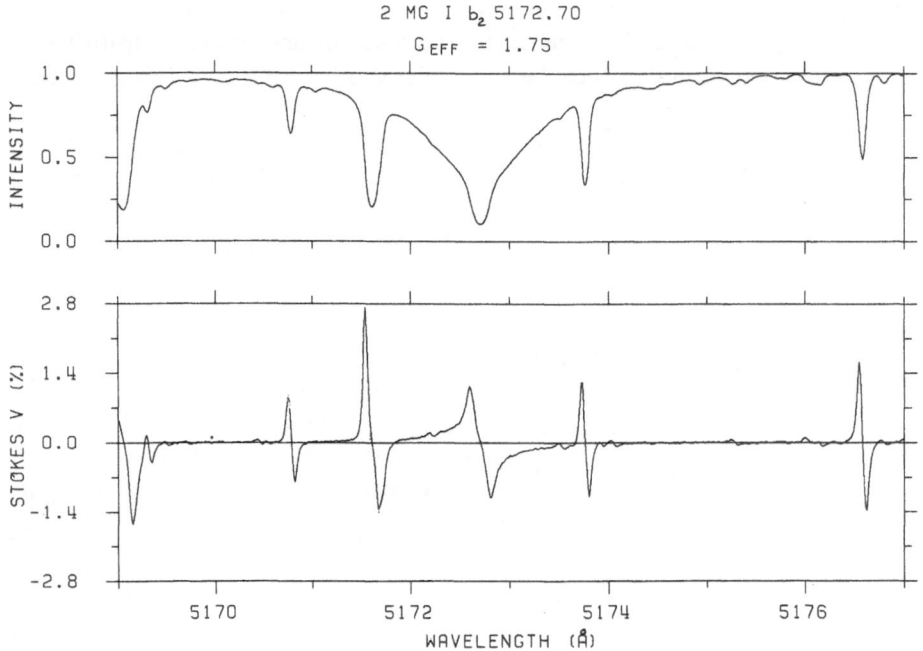

Fig. 3.   Stokes I and V around the Mg I b$_2$ line (Stenflo *et al.*, 1984). This is a portion of the same FTS recording, of which Figure 1 is a part.

As the Zeeman splitting $\Delta\lambda_H$ according to (2.2) is proportional to λ^2, whereas the line width varies approximately as λ, the Zeeman effect becomes more prominent in the infrared. Thus Harvey and Hall (1975; cf. Harvey, 1977) used the infrared line Fe I λ 1.56486 μm to measure the separation between the σ components directly, finding field strengths of 1.5–2.0 kG outside active regions. With the FTS this spectral region is now easily accessible for detailed Stokes analysis. Figure 4 shows two Stokes I and V recordings of the 1.56486 μm line, one in a sunspot at μ (cosine of the heliocentric angle) = 0.47, one in the network at disk center (μ = 1.00) (Stenflo *et al.*, 1985). The triplet pattern of Stokes I in the sunspot is very conspicuous, but there appears to be a blend in the red σ component, which is also seen in Stokes V. In the network, the σ components in the far Stokes I line wings are hard to discern because they are so weak.

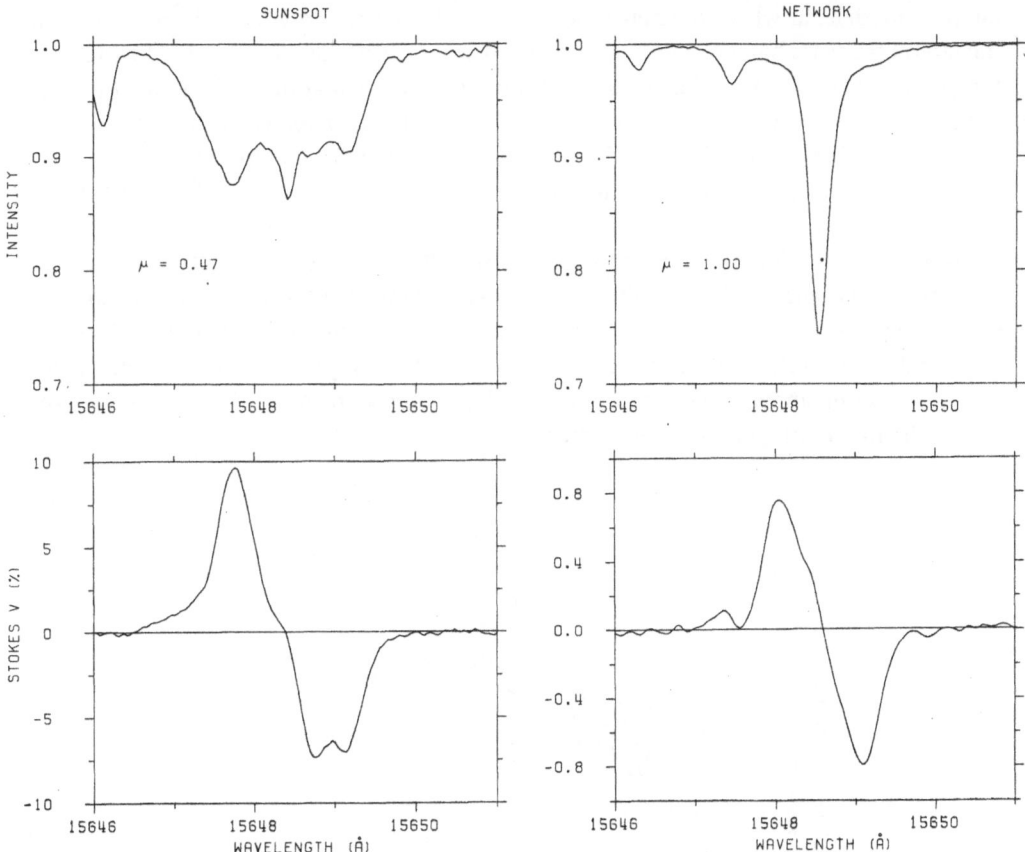

Fig. 4. Stokes I and V profiles of the infrared Fe I $\lambda 1.56486$ µm line, recorded with the FTS in May 1984. The left profiles were obtained in a sunspot at μ (cosine of the heliocentric angle) = 0.47, the right profiles in a network element at disk center ($\mu = 1.00$).

Also they mix with weak blend lines, but their presence shows up strongly in Stokes V. The determination of weak σ components in the unpolarized Stokes I stellar spectra has become an important tool to determine the magnetic fine structure on the surfaces of other stars (Robinson *et al.*, 1980; Marcy, 1983).

With its broad spectral coverage, the FTS allows a statistical approach to Stokes analysis for fluxtube diagnostics. Thus Solanki and Stenflo (1984, 1985) have analysed about 400 Fe I and 50 Fe II lines to determine how the Stokes I and V line parameters (line strengths, widths, asymmetries, etc.) depend on Stokes I line strength, excitation potential, and Landé factor. The aim is to systematically build empirical models of the intrinsic magnetic and thermodynamic structure of the fluxtubes. One of the results that have so far come out of this analysis concerns the temperature structure. In particular it was found that network fluxtubes are substantially hotter in their lower layers as compared with plage fluxtubes. A detailed comparison with previous facular models has been presented by Solanki (1984). The next step is to develop improved empirical fluxtube models, in which the magnetic pressure and tension are taken into account in the lateral pressure balance in a self-consistent way. Eventually one would like to relate the empirical temperature-density structure to an energy equation. This difficult undertaking has drastically gained in importance due to the availability of the FTS data.

One intriguing result of the FTS data has concerned the fluxtube dynamics. The Stokes V profiles are asymmetric, in the sense that the amplitude a_b of the blue-wing polarization peak in always larger than the amplitude a_r of the red wing. Also the areas A_b and A_r of the blue and red-wing peaks are different. Figure 5, from Solanki and Stenflo (1984), shows the amplitude and area asymmetries $(a_b - a_r)/(a_b + a_r)$ and $(A_b - A_r)/(A_b + A_r)$ as functions of line strength S_I (Fraunhofer), defined as the area of the intensity profile below the half-level chord. The amplitude asymmetries of the network data lie above the dashed line, the plage data lie below it. Such a distinction cannot be made for the area asymmetry.

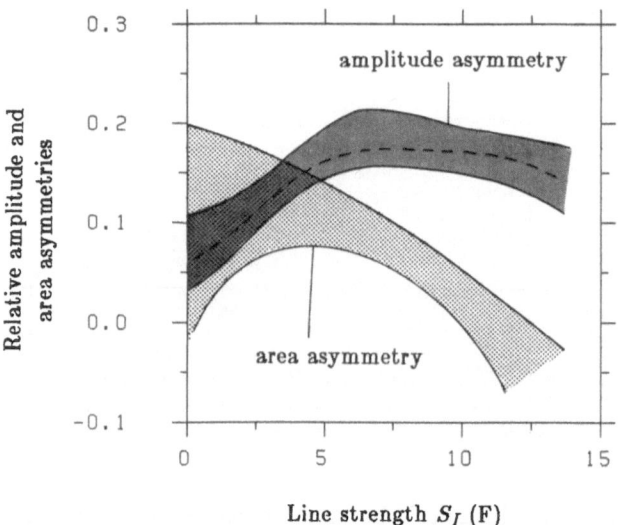

Fig. 5.   Relative amplitude and area asymmetries in the Stokes V line profiles, plotted vs line strength S_I
(Solanki and Stenflo, 1984). For details, see text.

Although it is quite clear that such asymmetries can only arise if there are mass motions within the fluxtubes, correlated with the height gradient of the magnetic field, it has not been possible to develop an adequate model for the fluxtube dynamics. The reason is that any model involving quasi-stationary flows leads to large wavelength shifts of the Stokes V zero crossing position relative to the position of the Stokes I profile (which mainly represents the surrounding non-magnetic atmosphere). The observations (Stenflo and Harvey, 1985) show that there is no such shift to within about 0.2 km s^{-1}. This suggests that the fluxtube dynamics is basically non-stationary in nature (e.g. in the form of fluxtube oscillations).

3. Determination of Vector Magnetic Fields

To clarify the problems in determining the vector magnetic field we will expand the transverse Zeeman effect as we did for the longitudinal Zeeman effect in (2.1)–(2.5). Thus for a normal Zeeman triplet formed in a magnetic field oriented perpendicular to the line of sight, the intensities of the π and σ components become

$$I_{\pi,\sigma} = \tfrac{1}{2}(I \pm Q),\tag{3.1}$$

if we define the positive Q direction to be parallel to the magnetic-field direction. As

$$I_\pi = \tfrac{1}{2}I(\lambda),$$
$$I_\sigma = \tfrac{1}{4}[I(\lambda + \Delta\lambda_H) + I(\lambda - \Delta\lambda_H)],\tag{3.2}$$

we obtain

$$Q = \tfrac{1}{2}\{I(\lambda) - \tfrac{1}{2}[I(\lambda + \Delta\lambda_H) + I(\lambda - \Delta\lambda_H)]\},\tag{3.3}$$

which can be Taylor-expanded as

$$Q = -\frac{1}{4}(\Delta\lambda_H)^2\left[\frac{\partial^2 I}{\partial\lambda^2} + \frac{1}{12}(\Delta\lambda_H)^2\frac{\partial^4 I}{\partial\lambda^4} + \dots\right].\tag{3.4}$$

When we compare this expression with (2.5), keeping (2.2) in mind, we see the fundamental difference between the longitudinal and transverse Zeeman effects. Whereas for weak fields the circular polarization is proportional to the average field B, the linear polarization is proportional to the average B^2 or magnetic energy. A filling factor α different from unity will thus enter into the interpretation of the transverse and longitudinal Zeeman effects in quite different ways, which results in the derivation of an incorrect field inclination angle if α is not known and taken into account. The effect is always in the sense that the magnetic field orientation appears to be much more transverse (perpendicular to the line of sight) than it really is. None of the vector magnetic field maps published in the past have been corrected for the filling factor. These interpretative problems of the transverse Zeeman effect were first described by Stenflo (1971), and have recently been clarified in great detail by calculating the resulting error for various field configurations (Stenflo, 1985).

Figure 6 serves to illustrate this behaviour of the linear polarization. We have used the analytic solution found by Rachkovsky (1962b) of the transfer equations for a Milne–Eddington atmosphere, including magneto-optical effects. The field inclination $\gamma = 45°$, the field azimuth $\chi = 0$. The Milne–Eddington parameters have been chosen to simulate the behaviour of the Fe I $\lambda 5250.22$ Å line.

Since the field azimuth is zero, the entire signal in Stokes U is due exclusively to magneto-optical effects. Let us however focus our attention on the relative magnitudes of Q and V. For strong fields Q and V are of similar magnitude. The polarization scales for Q and V are identical in the two upper diagrams, for $B = 2000$ and 1000 G. The two lower diagrams, for $B = 100$ and 10 G look identical, but the scales are now quite

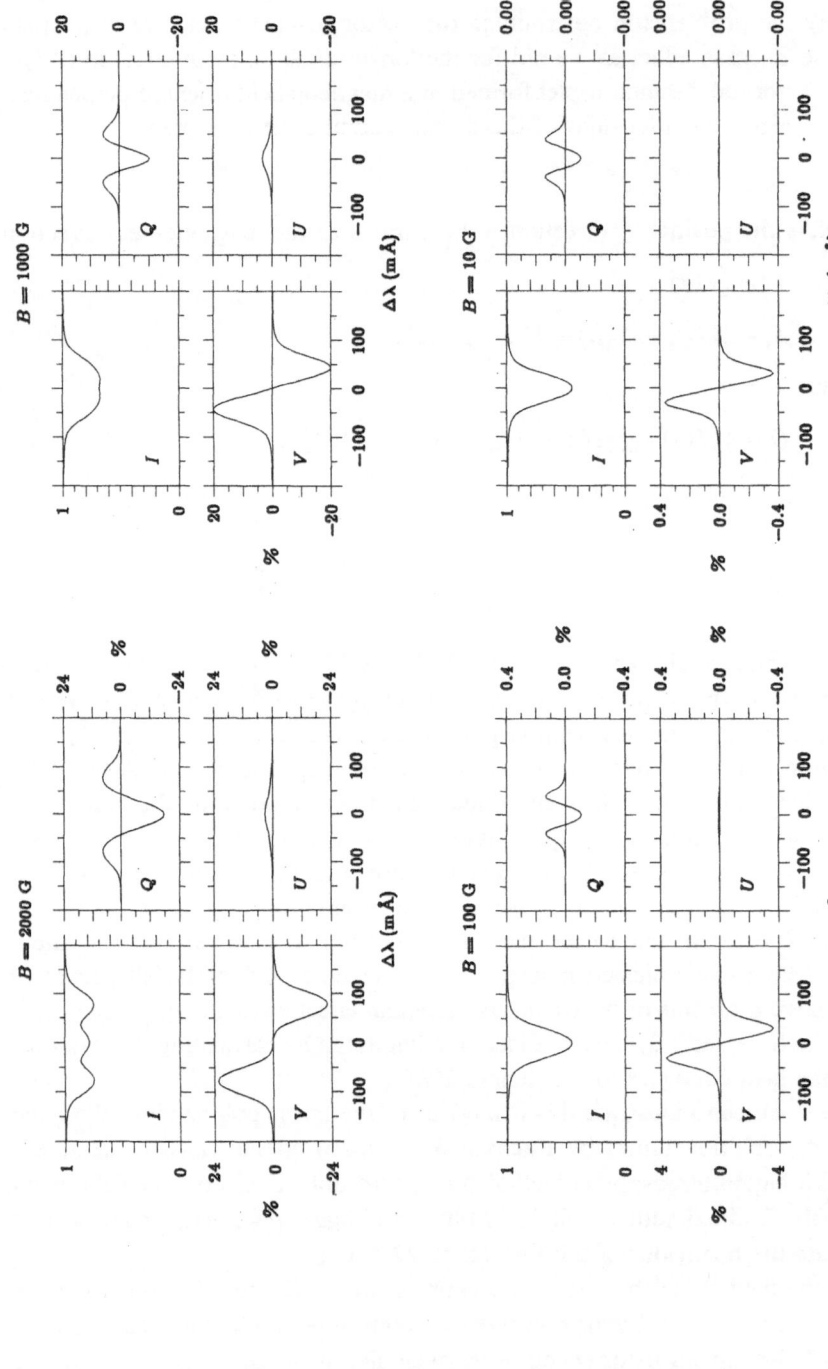

Fig. 6. Stokes line profiles, calculated with a Milne–Eddington model atmosphere including magneto-optical effects (Stenflo, 1985). The field inclination $\gamma = 45°$, the field azimuth $\chi = 0$, the Landé factor $= 3.0$. Q, U, and V are given in units of the adjacent continuous spectrum.

different. In the 100 G diagram, the V and Q scales are in proportion $10:1$, in the 10 G diagram the proportion is $100:1$. This is due to the B^2 scaling of Q, while V scales as B.

When the magnetic field is spatially resolved (filling factor $\alpha = 1$), the field inclination and magnitude can easily be determined from Q and V. When however $\alpha < 1$, the filling factor couples in a complicated way with the derived vector field, as shown explicitly in Stenflo (1985). A relatively large ratio Q/V could for instance mean that either (i) the angle γ between the field vector and the line of sight is large if the filling factor is also large, or that (ii) γ is much smaller if the filling factor is also small.

In Figure 7 we illustrate what an actual recording of I, V, and Q looks like. It was made in a plage at $\mu = 0.28$, in May 1984 with the 1-m FTS of the McMath telescope (Stenflo *et al.*, 1985). The wavelength range illustrated is identical with that of Figure 1 for the disk center, to allow us to make a direct comparison.

If we come back to the line ratio discussed in connection with Figure 1, we note that the corresponding V line ratio in Figure 7 ($V_{5250.22}/V_{5247.06}$) is considerably larger as compared with that of Figure 1. This shows that the intrinsic field strength inside the plage subarcsec fluxtubes decreases substantially with decreasing μ, i.e., with increasing height in the atmosphere. The μ variation of the V profiles allows us to derive the height variation of the fluxtube parameters. This is also possible at disk center by using lines of different strengths, but the center-to-limb observations provide an independent data set.

The expected Q line ratio of the 5250.22 and 5247.06 Å lines is for the case of weak fields $1.5^2 = 2.25$, but Figure 7 shows that the actual ratio is smaller, as expected when the fields are not weak. The deviation of the Q line ratio from 2.25 also provides information on the intrinsic field strength, independent of the V information.

We also notice in Figure 7 that the negative Q π-component is more suppressed for the lines with smaller Landé factors, because of more overlap between the π and σ components, as also demonstrated by the theoretical diagrams of Figure 6 when the Zeeman splitting is decreased. The ratio between the Q amplitudes of the π and σ components of a given spectral line thus also provides information on the intrinsic field strength.

These examples indicate that the extention of the FTS Stokesmeter recordings to include not only I and V but also Q greatly enriches our possibilities for fluxtube diagnostics.

The interpretation of the Q recordings however has to be done with great care, since a number of unexpected anomalies in the Q profiles have been revealed. We would expect that the situation should be the cleanest in the infrared due to the larger Zeeman splitting there. A glance at the infrared spectrum in Figure 8, which was obtained with the FTS in May 1984 in a sunspot at $\mu = 0.47$ (portion of this spectrum was shown in the left part of Figure 4 above) shows that the situation is not so simple. In the Q spectrum the π component of the FeI $\lambda 1.56486$ µm line is anomalously strong, much stronger than expected from theory. As a contrast, the π components of some of the other lines in Figure 8 are missing. We have no explanation for this mysterious behaviour yet.

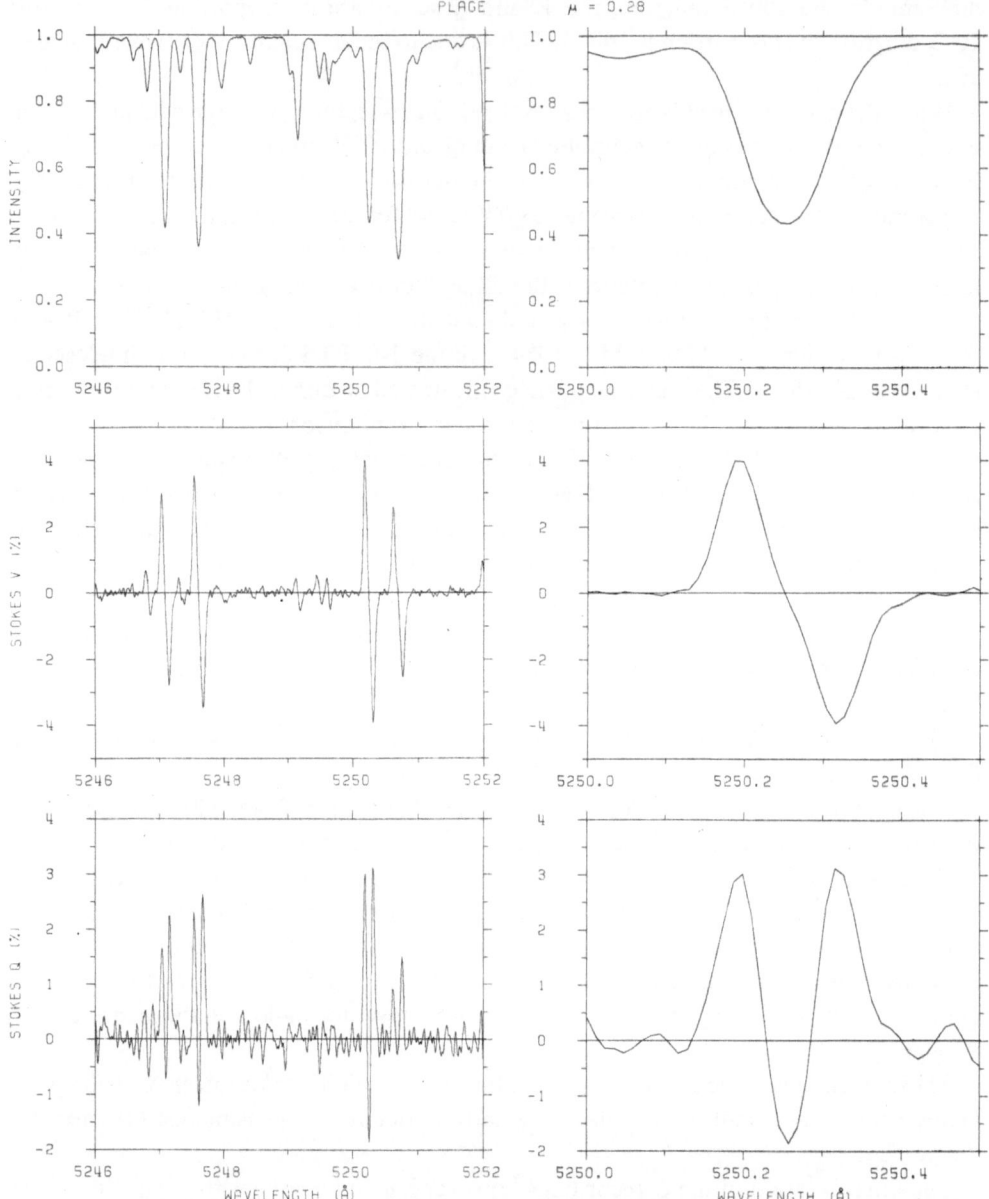

Fig. 7. Left set of diagrams: Stokes I, V, and Q spectra of the range 5246–5252 Å (same as Figure 1), recorded with the FTS in May 1984 in a plage at $\mu = 0.28$. Right set of diagrams: blown-up portion of the left diagrams, to show the 5250.22 Å line profiles in greater detail.

Notice also in the V spectrum of Figure 8 that although the 1.56486 μm line has a remarkably large Zeeman splitting, many of the surrounding lines have larger polarization amplitudes. Much information on fluxtube physics is encoded in these relations.

Through an exploratory analysis of the FTS I, Q, and V recordings we hope to

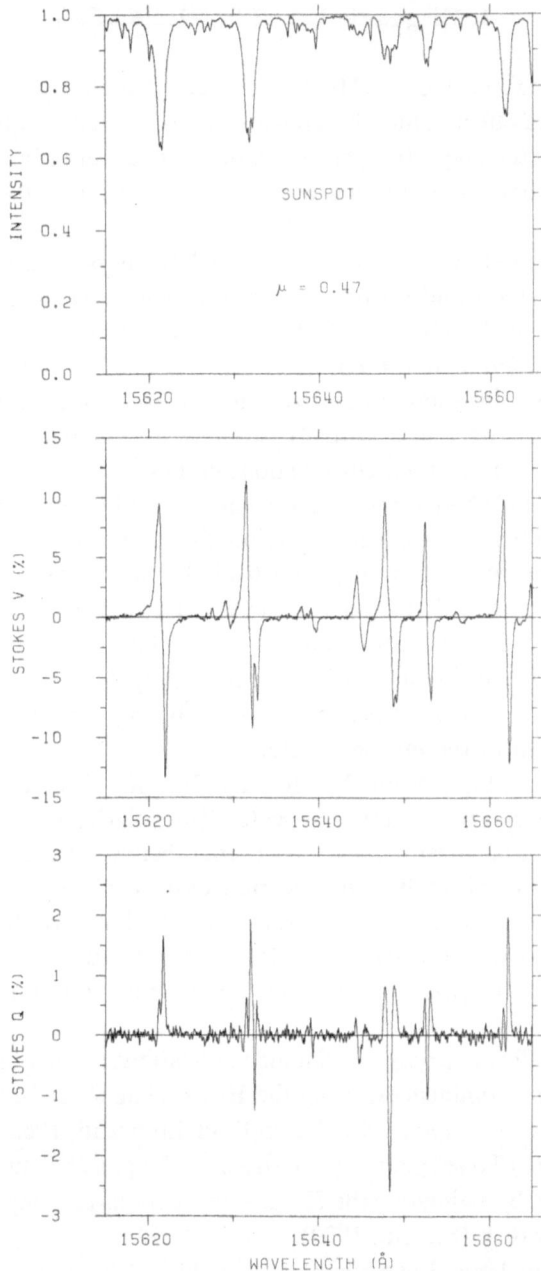

Fig. 8. Stokes I, V, and Q spectra of the range 15615–15665 Å, recorded with the FTS in a sunspot at $\mu = 0.47$. A detail of this recording was illustrated in the left part of Figure 4.

develop a deeper understanding of the Zeeman effect in the solar atmosphere, with the aim of building a foundation on which vector magnetic fields can be diagnosed properly. A next step would be to include Stokes U in the FTS recordings.

4. Diagnostic Use of the Hanle Effect

The polarization phenomena of the Hanle effect are qualitatively entirely different from those of the Zeeman effect. Thus the Hanle effect is caused by quantum-mechanical interference between the magnetic sublevels of the excited state of a given transition, and can only occur if coherent scattering contributes significantly to the line formation process.

The coherently scattered radiation is generally linearly polarized, depending on the type of atomic transition and on the scattering geometry. The maximum amount of polarization occurs in the absence of magnetic fields, in contrast with the ordinary Zeeman effect, for which there is no polarization in the absence of magnetic fields. The Hanle effect is the modification of the scattering polarization by a magnetic field. This modification manifests itself in essentially two ways: (1) Depolarization. (2) Rotation of the plane of linear polarization. Observations of these two manifestations give us two observable parameters, whereas the vector magnetic field has three spatial components. A basic limitation of the Hanle effect is therefore that it does not allow an unambiguous determination of the full vector magnetic field. In many cases, however, we have supplementary information for the system of interest, either through model conside-rations or through additional observational constraints, such that the Hanle-effect parameters provide the information needed to determine the vector magnetic field. This is for instance the case for prominence observations and for the determination of the 'turbulent' component of the magnetic field.

The polarization resulting from the ordinary Zeeman effect depends on the ratio between the Zeeman splitting and the line width. The polarization phenomena resulting from the Hanle effect however depend on the ratio between the Zeeman splitting and the damping width (inverse undisturbed life time of the excited level) of the line. As the damping width is generally much smaller than the actual line width, the Hanle effect is generally sensitive to weaker magnetic fields than the ordinary Zeeman effect. The Zeeman and Hanle effects thus complement each other (cf. Landi Degl'Innocenti, 1983b).

So far the Hanle effect has only had limited applications to the Sun: (a) To determine the magnetic field in prominences, using the He I D_3 line (Sahal-Brechot et al., 1977; Bommier, 1980; Bommier et al., 1981; Landi Degl' Innocenti, 1982; Leroy et al., 1983; Athay et al., 1983). (b) To set limits on the strength of a possible 'turbulent' component of the magnetic field, by estimating the Hanle-effect depolarization in a number of lines formed on the solar disk (Stenflo, 1982).

The reason for this limited use has been twofold:

– It is hard to measure weak linear polarization ($<0.1\%$) properly, since all the largest solar telescopes in the world have serious instrumental polarization. The planned next-generation solar telescope LEST is therefore designed to be 'polarization-free'.

– The theory of radiative transfer that includes the Hanle effect has only been partially developed (Stenflo, 1978b; Landi Degl' Innocenti, 1983a), and it is complicated to apply.

The main reason why previous applications of the Hanle effect have focused on prominences is that the radiative transfer problem can then be avoided entirely. One does not have to solve for the incident radiation field, since this is given as the radiation from the underlying solar disk, which is assumed to be known. The problem then reduces to the calculation of the polarization for single scattering processes, and averaging over different scattering angles.

As the main physics involved is the process of coherent scattering, and as the Hanle effect can be regarded as a modification of this process by magnetic fields, it is imperative for later use of the Hanle effect to first understand the basic physics of the scattering polarization in the absence of magnetic fields. For this purpose a survey of the non-magnetic scattering polarization in the solar spectrum, over the range 3165–9950 Å, was undertaken with the vertical grating spectrometer and the FTS of the Kitt Peak McMath telescope (Stenflo, 1983a, b). This survey revealed a number of unexpected effects: (i) Quantum-mechanical interference between widely separated fine-structure components. One example is represented by the Ca II H and K lines (Stenflo, 1980), but there are also many other cases corresponding to different combinations of quantum numbers. (ii) Polarization due to fluorescent scattering, both within and between multiplets. (iii) Many cases of anomalous and as yet unexplained behaviour of the polarization.

The line that shows the largest scattering polarization in the entire visible solar spectrum is the Ca I λ4226.74 Å line. A recording of Stokes I and Q/I with the FTS in a non-magnetic region 10 arc sec inside the solar polar limb is shown in Figure 9. Notice

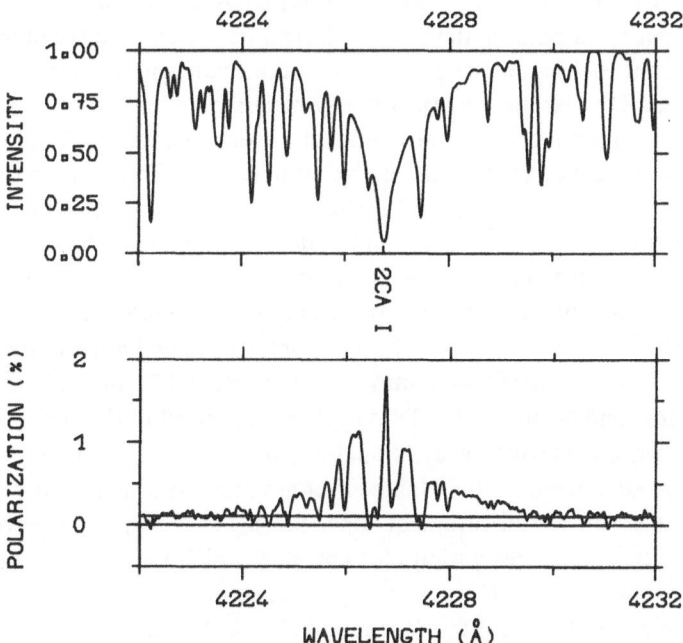

Fig. 9.   Stokes I and Q/I spectra around the Ca I λ4226.74 Å line, recorded with the FTS in a non-magnetic region 10 arc sec inside the solar limb (Stenflo *et al.*, 1983b).

Hanle and Zeeman effects
Ca I λ4227 Å

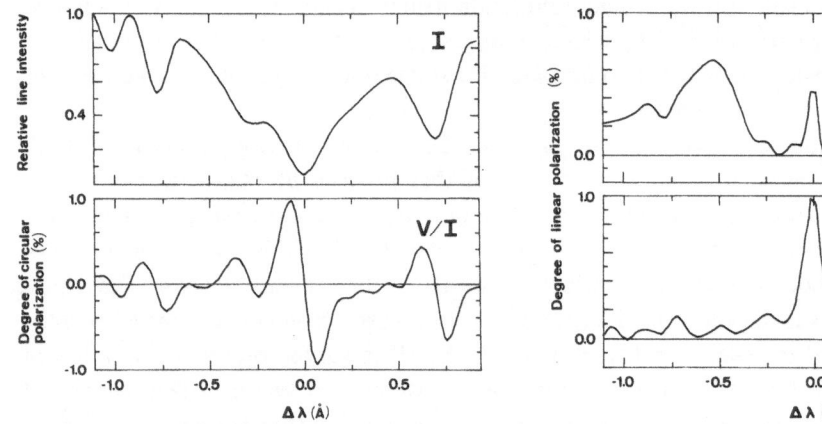

Fig. 10. Stokes profiles of the Ca I λ4226.74 Å line, recorded in an active region 70 arc sec inside the solar limb with the HAO Stokesmeter (Stenflo, 1982). Stokes *V* exhibits the Zeeman effect, *Q* and *U* the Hanle effect.

the large polarization peak in the Doppler core of the line. Figure 10 shows a recording of the same line in all Stokes parameters, this time in an active-region plage 70 arc sec inside the solar limb obtained with the HAO Stokesmeter at the Sacramento Peak Observatory in February 1978 (Stenflo, 1982). Whereas the circular polarization shows the usual Zeeman-effect pattern, the linear polarization shows no trace of the transverse Zeeman effect, but is completely determined by the scattering process. In the absence of magnetic fields *U* would be zero (positive *Q* means polarization parallel to the solar limb). According to the theory of the Hanle effect, the rotation of the plane of polarization due to the magnetic field occurs in the Doppler core alone, not in the dispersion wings (Omont *et al.*, 1973; Stenflo, 1978b). This is exactly what has happened in Figure 10. We notice how the relative amplitude of the *Q/I* peak has been reduced as compared with Figure 9, while a core peak has appeared in *U/I*, but nothing in the wings.

To be able to properly interpret such polarization recordings, we need to extend the radiative-transfer theory to include the case intermediate between the Hanle and Zeeman effects. Partial redistribution may also be important for lines like Ca II λ4227 Å and Ca II K (Rees and Saliba, 1982, 1985). We further need to develop procedures for handling the always occurring cases of spatially unresolved structures.

The two main advantages of the Hanle effect are that it is sensitive to weak magnetic fields, and that it can reveal turbulent magnetic fields which have zero net flux as integrated over the spatial resolution element, and which therefore do not contribute to any Zeeman-effect signal. The sensitivity to weak magnetic fields should make the Hanle effect particularly useful for magnetic-field diagnostics in the EUV (magnetic fields in the chromosphere and corona), since the λ^2 dependence of the Zeeman splitting makes the polarization effects of the ordinary Zeeman effect extremely small. In the case of the turbulent magnetic field, the first crude attempt to apply the Hanle effect indicated that the rms turbulent field that presumably fills the photospheric volume has a strength in excess of 10 G and a scale of less then one arc sec (Stenflo, 1982). Such a magnetic field may carry more flux and average magnetic energy than the field revealed in high-resolution magnetograms (based on the longitudinal Zeeman effect).

5. Concluding Remarks

During the past few years rich new possibilities of diagnosing the Sun's magnetic field have opened up. The main observational advances are: (a) Recordings of polarized line profiles with high S/N ratio. (b) Recordings of many spectral lines simultaneously. (c) Simultaneous recordings of all the four Stokes parameters (I, Q, U, and V).

This sudden wealth of qualitatively new data permits us to untangle the many interrelated parameters of the spatially unresolved magnetic fields, like: (1) field strengths, fluxes, and filling factors, (2) temperature-density structure, (3) dynamic structure, (4) height variations, and (5) properties of the chaotic or 'turbulent' component of the magnetic field. It is an enormous undertaking to make full use of this great diagnostic potential, and we are only at the beginning of trying to do this.

The use of these new diagnostic tools is of particular importance for the determination of vector magnetic fields, since the deduced field direction will be severely in error (and practically useless) unless the effect of the spatially unresolved structures are properly accounted for. It has long been well known that outside sunspots the magnetic structures are always spatially unresolved by all existing telescopes. We now however have the means of handling these problems in a systematic way.

The next major observational advance would come when high spectral resolution of the full Stokes vector in many spectral lines is combined with high spatial resolution using large two-dimensional diode arrays. For proper polarimetric accuracy for investigations of the vector magnetic field the instrumental polarization must be eliminated. These different requirements would be met by LEST, an international next-generation solar telescope, since its design is 'polarization free', and the foreseen use of adaptive optics would bring the spatial resolution to 0.1 arc sec (Andersen *et al.*, 1984). Its large ($\varnothing$2.4 m) aperture would ensure high polarimetric and photometric accuracy combined with high time resolution. The use of large CCD detectors with slow read-out rate in combination with fast (50–100 kHz) piezoelastic modulation of the full Stokes vector is made possible by an optical demodulation system (piezoelastic modulators instead of lock-in amplifiers) in front of the detectors (Stenflo and Povel, 1985).

References

Andersen, T. E., Dunn, R. B., and Engvold, O.: 1984, LEST Foundation Technical Report No. 7, Institute of Theoretical Astrophysics, University of Oslo.

Athay, R. G., Querfeld, C. W., Smartt, R. N., Landi Degl' Innocenti, E., and Bommier, V.: 1983, *Solar Phys.* **89**, 3.

Babcock, H. W.: 1953, *Astrophys. J.* **118**, 387.

Baur, T. G., House, L. L., and Hull, H. K.: 1980, *Solar Phys.* **65**, 111.

Baur, T. G., Elmore, D. E., Lee, R. H., Querfeld, C. W., and Rogers, S. R.: 1981, *Solar Phys.* **70**, 395.

Bommier, V.: 1980, *Astron. Astrophys.* **87**, 109.

Bommier, V., Leroy, J. L., and Sahal-Brechot, S.: 1981, *Astron. Astrophys.* **100**, 231.

Hagyard, M. J., Cumings, N. P., West, E. A., and Smith, J. E.: 1982, *Solar Phys.* **80**, 33.

Hale, G. E.: 1908, *Astrophys. J.* **28**, 100.

Harvey, J. W.: 1977, in E. A. Müller (ed.), *Highlights of Astronomy* **4**, 223.

Harvey, J. and Hall, D.: 1975, *Bull. Am. Astron. Soc.* **7**, 459.

Harvey, J. W. and Livingston, W.: 1969, *Solar Phys.* **10**, 283.

Harvey, J., Livingston, W., and Slaughter, C.: 1972, in *Proc. Conf. Line Formation in a Magnetic Field*, NCAR, Boulder, Colorado, p. 227.

Kiepenheuer, K. O.: 1953, *Astrophys. J.* **117**, 447.

Landi Degl' Innocenti, E.: 1982, *Solar Phys.* **79**, 291.

Landi Degl' Innocenti, E.: 1983a, *Solar Phys.* **85**, 3.

Landi Degl' Innocenti, E.: 1983b, *Solar Phys.* **85**, 33.

Leroy, J. L., Bommier, V., and Sahal-Brechot, S.: 1983, *Solar Phys.* **83**, 135.

Marcy, G. W.: 1983, in J. O. Stenflo (ed.), 'Solar and Stellar Magnetic Fields: Origins and Coronal Effects', *IAU Symp.* **102**, 3.

Omont, A., Smith, E. W., and Cooper, J.: 1973, *Astrophys. J.* **182**, 283.

Rachkovsky, D. N.: 1962a, *Izv. Krymsk. Astrofiz. Obs.* **27**, 148.

Rachkovsky, D. N.: 1962b, *Izv. Krymsk. Astrofiz. Obs.* **28**, 259.

Rees, D. E. and Saliba, G. J.: 1982, *Astron. Astrophys.* **115**, 1.

Rees, D. E. and Saliba, G. J.: 1985, preprint.

Robinson, R. D., Worden, S. P., and Harvey, J. W.: 1980, *Astrophys. J.* **236**, L155.

Sahal-Brechot, S., Bommier, V., and Leroy, J. L.: 1977, *Astron. Astrophys.* **59**, 223.

Solanki, S. K.: 1984, in T. D. Guyenne and J. J. Hunt (eds.), *The Hydromagnetics of the Sun*, ESA SP−220, p. 63.

Solanki, S. K. and Stenflo, J. O.: 1984, *Astron. Astrophys.* **140**, 185.

Solanki, S. K. and Stenflo, J. O.: 1985, *Astron. Astrophys.* **148**, 123.

Stenflo, J. O.: 1971, in R. Howard (ed.), 'Solar Magnetic Fields', *IAU Symp.* **43**, 101.

Stenflo, J. O.: 1973, *Solar Phys.* **32**, 41.

Stenflo, J. O.: 1976, in V. Bumba and J. Kleczek (eds.), 'Basic Mechanisms of Solar Activity', *IAU Symp.* **71**, 69.

Stenflo, J. O.: 1978a, *Rep. Prog. Phys.* **41**, 865.

Stenflo, J. O.: 1978b, *Astron. Astrophys.* **66**, 241.

Stenflo, J. O.: 1980, *Astron. Astrophys.* **84**, 68.

Stenflo, J. O.: 1982, *Solar Phys.* **80**, 209.

Stenflo, J. O.: 1984, *Adv. Space Res.* **4**, 5.

Stenflo, J. O.: 1985, in M. J. Hagyard (ed.), *Measurements of Solar Vector Magnetic Fields*, NASA Conf. Publ. 2374, p. 263.

Stenflo, J. O. and Harvey, J. W.: 1985, *Solar Phys.* **95**, 99.

Stenflo, J. O. and Povel, H.: 1985, *Applied Optics*, in press.

Stenflo, J. O., Twerenbold, D., and Harvey, J. W.: 1983a, *Astron. Astrophys. Suppl. Ser.* **52**, 161.

Stenflo, J. O., Twerenbold, D., Harvey, J. W., and Brault, J.: 1983b, *Astron. Astrophys. Suppl. Ser.* **54**, 505.

Stenflo, J. O., Harvey, J. W., Brault, J. W., and Solanki, S.: 1984, *Astron. Astrophys.* **131**, 333.

Stenflo, J. O., Solanki, S. K., and Harvey, J. W.: 1985, in preparation.

Stepanov, V. E. and Severny, A. B.: 1962, *Izv. Krymsk. Astrofiz. Obs.* **28**, 166.

Thiessen, G.: 1952, *Z. Astrophys.* **30**, 185.

Unno, W.: 1956, *Publ. Astron. Soc. Japan* **8**, 108.

SOLAR CONVECTION

ÅKE NORDLUND

Copenhagen University Observatory, Øster Voldgade 3, Dk-1350 Copenhagen K, Denmark

Abstract. The hydrodynamics of solar convection is reviewed. In particular, a discussion is given of convection on the scale of granulation; i.e., the energy carrying convection patterns in the solar surface layers, and its penetration into the stable layers of the solar photosphere. Convection on global and intermediate scales, and interaction with rotation and magnetic fields is discussed briefly.

1. Introduction

The dynamics of convection in the solar plasma is governed by the equations of (magneto-)hydrodynamics, and also, in the solar surface layer, by the equations of radiative transfer. Because of the extremely weak molecular dissipative processes (viscosity, heat conductivity, electrical resistivity), the resulting behavior of the plasma is quite complicated. The non-linear transport terms (representing advection of plasma properties) dominate the behavior 'almost everywhere', forcing dissipative processes to become significant only in the fractionally small volume where extreme gradients develop. This kind of behavior is next to impossible to describe by analytical methods, but is marginally tractable on present-day computers. Thus, recent advances in the understanding of solar convection, and its interaction with rotation and magnetic fields, have come partly through the development of computer codes capable of numerically simulating the behavior of these systems.

Traditionally, one refers to convection on three size scales on the Sun; *global* convection, on the scale of hundreds of Mm, extending throughout the entire convection zone; *supergranular* convection, on the scale of tens of Mm, and *granular* convection, on the scale of Mm. Recently (November *et al.*, 1981, 1982), *mesogranulation* has been introduced as a notation for observed velocity and temperature fluctuations on a scale intermediate between those of granulation and supergranulation.

There is a corresponding set of size scales for magnetic activity on the Sun; with *activity complexes* (Bumba, 1975) on the scale of global convection, *active regions* on the scale of supergranulation, and the solar *filigree* (Dunn and Zirker, 1973; Mehltretter, 1974; Muller, 1983) on the scale of granulation. Most likely, these magnetic phenomena are dynamically related to the corresponding scales of convection, but details of these relations are not yet well understood. This review concentrates mostly on convection on a granular scale, but larger scale convection, and interaction with magnetic fields are also touched upon.

The equations of hydrodynamics and their basic physical interpretation are discussed in Section 2. The application to convection on a granular scale, and its influence on the solar photosphere is reviewed in Section 3. Section 4 briefly reviews some recent

Solar Physics **100** (1985) 209–235.   0038–0938/85.15.

numerical work on convection on a global scale and also discusses the less well studied intermediate scales ('super-' and 'meso-granulation'). The interaction of convection on various scales with magnetic fields is discussed in Section 5.

2. Hydrodynamics of Solar Convection

2.1. BASIC EQUATIONS

For the purpose of the following discussion, we write the equations of (non-relativistic) fluid dynamics with radiative energy transport in a form suitable to describe convection on the scale of granulation, mesogranulation, and supergranulation on the Sun (rotation, radiation pressure, and v/c terms unimportant):

$$\frac{\partial \rho}{\partial t} + \nabla \cdot (\rho \mathbf{u}) = 0 \,, \tag{1}$$

$$\rho \frac{\partial \mathbf{u}}{\partial t} + \rho \mathbf{u} \cdot \nabla \mathbf{u} = -\nabla P + \rho \mathbf{g} + \rho \mathbf{f}_{\text{visc}} \,, \tag{2}$$

$$\rho \frac{\partial H}{\partial t} + \rho \mathbf{u} \cdot \nabla H = Q + \frac{\partial P}{\partial t} + \mathbf{u} \cdot \nabla P \,, \tag{3}$$

$$\rho = \rho(P, T) \,, \tag{4}$$

$$H(P, T) = E(P, T) + P/\rho \,, \tag{5}$$

$$Q = Q_{\text{rad}} + Q_{\text{visc}} \,, \tag{6}$$

$$Q_{\text{rad}} \equiv -\nabla \cdot F_{\text{rad}} = \int_{0}^{\infty} \int_{4\pi} \rho \kappa_\nu (I_{\nu\Omega} - S_\nu) \, d\Omega \, d\nu \,, \tag{7}$$

$$\frac{d I_{\nu\Omega}}{d s} = \rho \kappa_\nu (I_{\nu\Omega} - S_\nu) \,, \tag{8}$$

with standard notation (cf. Nordlund, 1982, 1984b).

Although this set of coupled differential equations may appear complicated (and may require substantial effort to solve numerically), the individual equations are *conceptually* simple, and the *qualitative* properties of solar convection may be fruitfully discussed with reference to the individual equations.

2.2. THE CONTINUITY EQUATION

It turns out to be convenient, both for the purpose of numerical work (Nordlund, 1982) and for a qualitative discussion, to apply a horizontal Fourier transform to the continuity equation, and to the equation of motion. For a discussion of the advantages of such *spectral* methods in (non-shocking) hydrodynamics calculations, see Orzag (1971) and

Herring *et al.* (1974). Symbolically, we may write, for example,

$$\rho\mathbf{u}(x, y, z) = \sum_{lm} \psi_{lm}(z)\, e^{i(lx + my)} \tag{9}$$

(although, in practice, explicit sine and cosine transforms are numerically more efficient). With this representation, it is easy to show (Nordlund, 1978) that, for slowly changing flows, the ratio of the horizontal to vertical velocities of a certain Fourier component is inversely proportional to the horizontal wavenumber (which we may denote k):

$$\frac{u_{\text{hor}}}{u_{\text{vert}}} \approx \frac{1}{kH_\psi}, \tag{10}$$

where H_ψ is the scale height of the vertical mass flux, $(\psi_z)_k = (\rho u_z)_k$. In other words, the larger the convective cell, the larger the horizontal velocities must be, for given vertical velocities. The reason is that the rapidly decreasing vertical mass flux in the ascending parts of the cell must be balanced by a horizontal outflow, and for larger cells a given scale height H_ψ becomes a smaller fraction of the size of the cell. In the case of granulation, the ratio of horizontal to vertical velocity in the photosphere is just slightly larger than unity, but for supergranular flows the ratio must be an order of magnitude larger. This is the reason why supergranulation is seen primarily as a horizontal velocity field.

2.3. THE ANELASTIC APPROXIMATION

The qualitative difference between pressure waves, gravity waves, and slowly changing flows is apparent through a consideration of the continuity equation, Equation (1). In pressure waves, the converging flow ($\nabla \cdot (\rho\mathbf{u}) < 0$, $\partial\rho/\partial t > 0$) causes an overpressure that halts and reverses the compression phase of the wave. In slowly changing flows, and in slow gravity waves, the continuity equation acts only as a *constraint* on the flow, allowing only flows with (approximately) divergence-free mass fluxes:

$$\nabla \cdot (\rho\mathbf{u}) \approx 0. \tag{11}$$

Adopting this *anelastic approximation* (Batchelor, 1953; Charney and Ogura, 1960) one finds that the pressure is (apart from boundary conditions) uniquely determined by the divergence of the (volume) forces (including the 'intertial' force-term $-\rho\mathbf{u} \cdot \nabla\mathbf{u}$):

$$\nabla^2 P = \nabla \cdot \{\text{volume forces}\} = \nabla \cdot \{\rho(\mathbf{g} + \mathbf{f}_{\text{visc}} - \mathbf{u} \cdot \nabla\mathbf{u})\} \tag{12}$$

(cf. Nordlund, 1982). This particular pressure has a gradient that conspires with the force field to keep

$$\frac{\partial}{\partial t}(\nabla \cdot (\rho\mathbf{u})) = 0, \tag{13}$$

i.e., the mass flux divergence remains zero. By Equation (12), the pressure responds instantaneously to changes in the force field; i.e., the system behaves as if the sound speed was infinite. From the foregoing discussion is clear that the anelastic approximation will serve to exclude the possibility of pressure waves from the system, while still allowing (divergence-free) flows.

The anelastic approximation has been used in many forms (Batchelor, 1953; Charney and Ogura, 1960; Ogura and Phillips, 1962; Gough, 1969; Latour *et al.*, 1976; Nordlund, 1978; Gilman and Glatzmaier, 1981), often with an expansion of the equations in terms of some (small) measure of the fluctuations.

Note, however, that according to the discussion above, the validity of the anelastic approximation has to do with how slowly or rapidly the density changes in an *Eulerian* frame, which is not necessarily directly related to the magnitude of the velocity. Indeed, a *stationary* hydrodynamical system with arbitrary velocites satisfies Equation (11) exactly. Criteria having to do with velocity or perturbation amplitudes come (directly or indirectly) from a translation of velocity amplitude to time scale, using some characteristic scale length. In the case of the solar granulation, the quasi-stationary radiative cooling of the ascending gas in the granules make the patterns evolve on a much longer time scale (tens of minutes) than, e.g., the travel time over a pressure scale height (of the order of a minute).

2.4. THE EQUATION OF MOTION

The horizontal flows required by the continuity equation (from ascending to descending parts of the convection cells) must be supported by a corresponding pressure pattern since, by the equation of motion, only pressure gradients are available to accelerate the horizontal flow away from updrafts and to decelerate horizontal flows coming towards downdrafts. Thus, we may expect pressure excess over both updrafts and downdrafts, with pressure deficiency in between. (This discussion applies to layers near the upper boundary of an unstable layer, where the strongest horizontal velocities occur, because of the rapid decrease of the vertical mass flux.) The pressure excess over ascending and descending material decreases the buoyancy of ascending material ('buoyancy braking', Massaguer and Zahn, 1980; Latour *et al.*, 1983) but also increases the weight of the descending material, an effect that is not brought out by single-mode calculations such as those of Massaguer and Zahn and Latour *et al.* This asymmetric buoyancy modification is one effect that contributes to the asymmetry between updrafts and downdrafts in numerical simulations of convection in stratified media (Hurlburt, 1983; Hurlburt *et al.*, 1984, 1985; Toomre *et al.*, 1984; Hurlburt and Toomre, 1985). Another effect (Nordlund, 1978) is producd by the advection of fluid properties by the non-linear term in the equation of motion; with horizontal flows *from* updrafts *to* downdrafts in a stratified medium, ascending material will tend to spread with time, while descending material will tend to narrow. (This is again a consequence of the advection of properties by a low-viscosity fluid flow; in this case the flow topology itself is the advected quantity.)

The non-linear terms in the equation of motion and the energy equation produce a steepening of the gradients of advected quantities in regions where the fluid velocity decreases downstream. Because the viscosity of the solar plasma is very small, viscous effects that halt the steepening set in on scales that cannot (by many orders of magnitude) be resolved in numerical models. To prevent the buildup of steeper gradients than can be numerically resolved, the numerical scheme must contain suitable dissipative (diffusive) terms. Sufficiently large ordinary diffusion terms ('turbulent viscosity') will do the job, but will also influence the resolved motions appreciably. Other types of diffusive terms, which influence resolved motions less, while still affecting the smallest resolved scales sufficiently to prevent excessive buildup of gradients, are to be preferred. Such terms are often referred to as 'sub-grid-scale eddy diffusivities' (e.g., Smagorinsky, 1963). A particular form of such diffusive terms suitable for the semi-spectral representation Equation (9) was derived by Nordlund (1982, Section 3.3).

2.5. WAVES AND WAVE GENERATION

The interaction of waves (pressure waves, gravity waves) and flows in the solar envelope is, for several reasons, a difficult problem:

(1) Wave generation by the turbulent convective flow is a complicated process, where so far only exploratory work based on order of magnitude estimates has been made (e.g., Goldreich and Keeley, 1977b).

(2) The large fluctuations in temperature, pressure, and thermodynamic quantities (e.g., heat capacity and sound speed) associated with the convection. Sufficiently large fluctuations must influence the propagation and dispersion properties of waves; e.g., because the propagation speed becomes a strong function of horizontal position. Some remaining systematic discrepancies between the observed and calculated frequencies of solar p-modes in the 5-min period range (Christensen-Dalsgaard, 1984, 1985) may possibly be related to the question of the wave propagation speed in a very inhomogeneous medium. The upper approximately 500 km of the solar convection zone is very inhomogeneous, and presumably large scale waves propagate with a different speed than in a plane-parallel medium with the same average temperature stratification.

(3) The convective velocity field, which may cause scattering and aberration of waves. Smaller scale waves, especially of a size comparable to the size of granulation, may be expected to suffer refraction, reflection, and advection by the velocity field associated with the convection. To my knowledge, this problem has not been studied in the solar context.

(4) The modulation of the convective energy flux by large scale waves, which changes the stability properties of standing wave modes. Attempts to estimate modulation of convection by pressure waves have been made by, e.g., Unno (1976), Gough (1977), and Goldreich and Keeley (1977a). However, these attempts were based on a mixing length description of convection, and thus the results are of uncertain validity.

(5) Radiative transfer effects in the surface layers, which are crucial both to the convection and to wave propagation and reflection. Non-adiabatic effects on pressure wave propagation in the solar photosphere have been studied by Christensen-Dalsgaard and Frandsen (1983), who concluded that effects of departures from radiative

equilibrium in the mean state may change the stability properties of standing waves drastically. Non-adiabatic effects on internal gravity waves have been studied (in the Newtonian cooling approximation) by Mihalas and Toomre (1982).

It would be perfectly possible to study these problems by solving the linear wave equations separately in a background medium obtained, for example, from anelastic simulations of the convection. This would not be a very useful approach, though, because the solution of the wave equations under these conditions (large amplitude fluctuations of the thermodynamic quantities and flows with significant non-linear terms) would be about as time consuming as a solution of the full set of equations. A separate solution of the linear wave equations would not, however, provide any information on the back reaction of the waves on the convection. Thus there is really no point in aiming for less than a solution of the full problem, as the next logical step. The main complication in doing the full problem probably lies with implementing boundary conditions that behave reasonably. The time steps would not have to be reduced very much compared to the anelastic simulations of granular convection, since the convective velocities are a significant fraction of the sound speed in some places.

2.6. CONVECTIVE PENETRATION INTO STABLE LAYERS

The penetration of motions from a convectively unstable zone into neighboring stable layers may have significant consequences for the temperature structure of these stable layers. A qualitative understanding of the extent to which such flow penetration occurs may be gained from a consideration of the Poisson equation for the pressure, Equation (2) (cf. Nordlund, 1982, Section 3.6).

In terms of the horizontal Fourier components P_{lm} of the pressure, the *relative* pressure fluctuations,

$$p_{lm}(z) = P_{lm}(z)/P_{00}(z) ,$$ (14)

obey the ordinary differential equation

$$\frac{d^2 p_{lm}}{dz^2} + \alpha \frac{dp_{lm}}{dz} - k^2 p_{lm} = O ,$$ (15)

where α is the inverse of the pressure scale height,

$$\alpha = \frac{d \ln P_{00}}{dz} ,$$ (16)

where α is the inverse of the pressure scale height,

$$k^2 = (l^2 + m^2) .$$ (17)

For the purpose of the discussion, it is convenient to disregard induced temperature perturbations in the stable layers, and consider only the penetration of pressure perturbations induced from the unstable layers. This is not unrealistic for the stable layer above the convection zone, where the radiative transport of energy efficiently reduces the temperature perturbations, and where relatively large pressure fluctuations in the top layers of the unstable zone have a strong influence in the stable layers. Below the convection zone, conditions are different, and here the induced temperature perturbations may be important in reducing the penetration of motions into the stable layer (undershoot). However, as we shall see, even disregarding temperature perturbations, the penetration is much weaker below a strongly stratified convection zone than above. The general solutions to the differential equation (15) are of the type

$$p_{lm} \approx e^{\alpha_{\pm} z} , \tag{18}$$

where

$$\alpha_{\pm} = -\alpha/2 \pm (\alpha^2/4 + k^2)^{1/2} . \tag{19}$$

In the limit of large k (disturbances small compared with the scale height; $k \gg \alpha$), we have

$$\alpha_{\pm} \to \pm k , \tag{20}$$

i.e., small scale pressure disturbances at the convection zone boundary decay away over a distance comparable with their size.

In the limit of large scale disturbances, there is a qualitative difference in the situation below and above an unstable layer: In a stable layer above an unstable layer, large scale fluctuations have a long range influence (since $\alpha_{+} \to k^2/\alpha$), whereas in a stable layer below an unstable layer large scale disturbances lose importance over about one pressure scale height ($\alpha_{-} \to -\alpha$). Qualitatively, this is because in a stratified atmosphere the mean state pressure increases rapidly below the unstable layer, which rapidly diminishes the *relative* important of a certain (absolute) pressure disturbance, while the opposite is true above an unstable layer.

This result has important consequences for the character of the motions generated in stable layers above and below convection zones. A forced motion in a stable layer, if sufficiently slow, sets up propagating internal gravity waves. Internal gravity waves are characterized by a back-and-forth motion, where the restoring force is due to the excess weight of the displaced fluid. However, for sufficiently strong perturbations (as for example in the case of the solar photosphere), the induced motion will have the character of a *flow*, where the fluid goes from A (an upflow) to B (a downflow), never to return.

The essence of the results cited above is that it is easier to drive (mixing) penetration into stable layers above than into stable layer below a convection zone. The character of the penetration, if sufficiently strong, is one of large scale flows (Nordlund, 1984b) rather than the excitation of internal gravity waves (Mihalas and Toomre 1981, 1982). The qualitative difference between penetration above and below an unstable layer is nicely illustrated in the work by Hurlburt et al. (1985).

Superficially, the results of Massaguer et al. (1984) would seem to be in conflict with the conclusions on penetration stated above. Massaguer et al. conclude, on the basis of numerical solutions of anelastic one- or two-mode equations for a strongly stratified medium, that the penetration below an unstable layer is more pronounced than that above. For three reasons, this conclusion cannot be directly confronted with the results quoted above: (1) because they compare the penetration in terms of physical depth, rather than in terms of the local scale height; (2) because their strong downward penetration occurs only for a planform with downward directed flow at cell center, a topology that is not found in three-dimensional multi-mode calculations; (3) because they do not regard a flow with strong horizontal velocities but weak vertical velocities as penetrating. However, *in terms of local density or pressure scale heights* the penetration is actually much stronger in their stable layer above than in their stable layer below, especially for the planform with upward directed flow at cell center, and for the case with two coexisting oppositely oriented planforms (cf. their Figures 1, 3, and 5).

2.7. THE ENERGY EQUATION AND RADIATIVE ENERGY TRANSFER

The energy equation, Equation (3), describes the evolution of temperature (as measured by the enthalpy) as a result of pressure changes ($\rho\, \partial H/\partial t + \rho\mathbf{u}\cdot\nabla H - \partial P/\partial t - \mathbf{u}\cdot\nabla P = 0$ describing adiabatic advection; $DS/Dt \equiv 0$), and non-adiabatic heating/cooling, due to contributions from, e.g., radiation and viscous dissipation, Equation (6).

The most important non-adiabatic effect in the solar photosphere is due to the radiative term Q_{rad}. Indeed, the granulation pattern results from the release (through the Q_{rad} term) of excess heat carried to the surface by ascending material in the convection flow. The Q_{rad} term is also responsible for a substantial heating of the optically thin parts of the photosphere (cf. below, Section 2.8).

From Equtions (3) and the leftmost part of (7), one may estimate the ascent velocity necessary for the advection of enthalpy to balance the solar luminosity:

$$\rho u_z \Delta H/\Delta z = F_\odot/\Delta z\,, \tag{21}$$

$$u_z \approx \frac{F_\odot}{\rho\, \Delta H}\,, \tag{22}$$

which comes out to be of the order of 2 km s^{-1} for solar conditions (Nordlund, 1982). Thus, material ascending slower than about 2 km s^{-1} will suffer net cooling (in an Eulerian frame) at the surface. This is one of the limiting factors in the growth of granules with time; the growth of individual granules (caused by the advection by the horizontal flows from updrafts to downdrafts) is accompanied by a steady increase in pressure excess in the granule with time. The consequent loss of buoyancy leads to a decreasein the ascent velocity. When the ascent velocity falls below the critical value, Equation (22), the temperature excess can no longer be upheld in the center of the granule. This process is most obvious in exceptionally undisturbed granules that grow symmetrically, where the retardation/cooling at the center results in a bright granule with a dark center ('exploding granules', Namba and van Rijsbergen, 1977; Bray *et al.*, 1984, Section 2.3.7).

In optically thin layers the radiative flux changes relatively little, and it is more enlightening to discuss the radiative energy exchange in terms of the rightmost part of Equation (7); i.e., the radiative heating/cooling is expressed as the difference between absorption $\rho\kappa_v I_{v\Omega}$ and emission $\rho\kappa_v S_v$, integrated over frequencey and angle.

2.8. THE ENERGY BALANCE OF THE UPPER PHOTOSPHERE

In one-dimensional hydrostatic models of the solar photosphere, the temperature structure of most of the photosphere is determined by *radiative equilibrium*, which may be written as an equilibrium between absorption ($\kappa_v I_{v\Omega}$) and emission ($\kappa_v S_v$) of radiation:

$$Q_{\text{rad}} = \int_v \int_\Omega \rho\kappa_v (I_{v\Omega} - S_v)\, d\,v\, d\,\Omega = 0, \tag{23}$$

where κ_v is the monochromatic absorption coefficient (m^2 kg^{-1}), and $I_{v\Omega}$ and S_v are the monochromatic radiation intensity and source functions, respectively (W Hz^{-1} sterad^{-1}).

Qualitatively, the positive (heating) contributions to this integral come from frequencies and angles where the radiation temperature is higher than the local kinetic temperature, and *vice versa*. In an atmosphere with frequency independent (grey) absorption coefficient, this is basically a balance between heating in out-going rays, and cooling in in-coming rays. The exact solution for the temperature structure is well known (e.g., Mihalas, 1978, p. 72), and would, in the solar case, correspond to an essentially constant temperature (approximately equal to 4600 K) in most of the photosphere.

The presence of hundreds of thousands of spectral lines in the solar spectrum changes the energy balance qualitatively, and complicates the calculation of the temperature structure considerably (cf. the review by Carbon, 1979).

Qualitatively, for models in radiative equilibrium, the energy balance in the layers that are optically thin in the continua is one between heating in the continua and cooling in spectral lines: at a certain heigth in the atmosphere, the integrand in Equation (23) is positive for all frequencies where the monochromatic optical depth is much smaller than unity (i.e. in the continua, in weak lines, and in the wings of strong lines). However, in frequencies where the optical depth is of the order of unity (at the same certain height in the atmosphere), the out-going intensity is close to the local source function, while the in-coming intensity is significantly lower. Thus, for these frequencies, the integrand in Equation (23) is negative (cooling). As a result, the overall temperature of the upper layers of an atmosphere in radiative equilibrium is reduced ('surface cooling by spectral lines'), such that the cooling in the spectral lines is compensated for by a corresponding heating in the continuum.

In frequencies where the optical depth is large, the radiation intensity approaches the source function (thermalizes) exponentially with optical depth, and consequently such frequencies contribute little to the integral in Equation (23). On the other hand, since the continuum radiation is 'blocked' in these frequencies, the temperature of the continuum forming layers has to be increased (relative to a grey model), to conserve the total emitted radiation flux (effective temperature). This is usually referred to as 'backwarming'.

The surface cooling by spectral lines (and the backwarming) determines the temperature structure of the upper photosphere in one-dimensional, static models of stellar atmospheres in radiative equilibrium (cf. Athay, 1970c; Gustafsson *et al.*, 1975; Kurucz, 1970). In particular, such ('line blanketed') models of the solar photosphere have a temperature structure much closer to empirical models than models with only continuum absorption included.

The energy balance of the upper photosphere is qualitatively different in 3-D hydrodynamical models of stellar atmosphere (Nordlund, 1984b; Nordlund and Dravins, 1986). Because of the presence of a velocity field in the photosphere, the temperature structure is no longer determined by radiative effects alone. Rather, the energy balance is one between *convective cooling* and *radiative heating*. Ideal gas ascending adiabatically from the continuum forming layers at the base of the photosphere (where the temperature is of the order of 6000 K and the pressure is of the order of 10 kPa) up to the temperature minimum (where, empirically, the temperature is of the order of 4100–4300 K and the pressure is of the order of 0.1 kPa) would have its temperature reduced by a factor of about $10^{2(2/5)} = 10^{0.8} = 6$! Thus, clearly, significant radiative heating is required to heat gas ascending from the continuum layers up to the temperature minimum.

In the calculation of one-dimensional, plane-parallel model stellar atmospheres, it is possible to take into account line blanketing from hundreds of thousands of spectral lines, either by statistical methods ('opacity sampling'), or by a reordering of the absorption into monotonic functions of frequency ('opacity distribution functions'), cf. the review by Carbon (1979). In the 3-D hydrodynamical models, the radiative transfer has to be treated along a large number of rays, for a large number of time steps and both of the above mentioned methods become prohibitively expensive. A further simplification is possible if one assumes that the depth-dependence of the monochromatic opacity is approximately the same for all frequencies. Then, one may classify the opacities into bins (as in the ODF method), but then *average over the source function in each bin before solving the transfer equation*, Equation (8). Thus, the radiative transfer equation needs to be solved only for a small number of bins (e.g., four) along each ray at any one time in the simulation, which makes the computational cost of the radiative transfer problem affordable (Nordlund, 1982, 1984b).

Though the assumption of frequency independent depth-dependence of the monochromatic opacity is a reasonable first approximation, there are important opacity contributions for which this is not a good approximation. Weak lines of neutral iron are one such case, since neutral iron is abundant in the (cool) upper photosphere, whereas iron is essentially ionized in the lower photosphere. This makes the dept-dependence of the weak iron line absorption radically different from that of, e.g., the H^- continuum absorption. The absorption of continuum radiation by optically thin iron lines may, in fact, contribute significantly to the heating of the upper photosphere (cf. Nordlund, 1984b, 1985b).

3. Numerical Simulations of Granular Convection

Numerical simulations of granular convection have been performed by myself (Nordlund, 1982, 1984b), by Cloutman (1979), and by Gigas and Steffen (1984). Cloutman used a two-dimensional model, which unfortunately had a horizontal extension too small for a realistic representation of solar granulation. Steffen and Grigas used a model with cylindrical symmetry, with radiative transfer included, and attemted to find stationary solutions. They used a method of characteristics, but did not include any viscous effects (viscosity or shocks). Probably due to the lack of viscous effects, and/or because of the symmetry constraints, they were able to find stationary solutions only for cases where the size of the cylinder was sufficiently small.

My own simulations were performed using a three-dimensional model with $32 \times 32 \times 32$ degrees of freedom per variable, using a semi-spectral representation with a horizontal period of 3 Mm, and a vertical extent of 1.6 Mm. The anelastic approximation was used to filter out pressure waves, and the radiative transport of energy was approximated by the binning of frequencies according to opacity as described above (Section 2.8).

3.1. THE GRANULATION PATTERN

The granulation pattern, visible on the Sun on scales of the order of a Mm (cf. Bray *et al.*, 1984), consists of bright granules formed by ascending hot gas, and dark intergranular lanes of descending cool gas, with relatively strong horizontal flows from the granules to the intergranular lanes. The intensity contrast between the hot granules and the cool intergranular lanes is closely linked to the ΔH necessary to carry the solar flux $F_{\odot}$ (cf. Equation (21)). This ΔH corresponds to a ΔT (peak to peak) of some 5000 K immediately below the visible surface! Most of this temperature contrast is masked by the strong temperature sensitivity of the continuum opacity, which raises (lowers) the surface of continuum optical depth unity in granules (inter-granular lanes).

The amplitude of the visible intensity fluctuation depends on details of, e.g., the temperature dependence of the opacity. Also, in the numerical simulations, the resulting intensity fluctuations may depend on the numerical resolution, and on uncertainties in the absorption coefficients. The tests carried out thus far, however, have not indicated that such uncertainties could influence ΔI_{rms} significantly.

The actual intensity fluctuations obtained in the numerical simulations are of the order of 25–30% (ΔI_{rms}) at 500 nm, and 20–25% at 600 nm (Nordlund, 1984a). This would seem to be in conflict with empirical determinations of ΔI_{rms}, which generally fall in the range 5–15% at 500 nm (based on actually observed values of typically 2–8%) (cf. the review by Wittman, 1979). These measurements are typically corrected for the influence of 'seeing' (atmospheric + instrumental image degradation) by assuming gaussian point spread functions, with a subjectively estimated width. Empirically determined spread functions (Levy, 1971; Deubner and Mattig, 1975; Ricort and Aime 1979; Ricort *et al.*, 1981) have shapes that are far from gaussian. Instead, empirical spread functions – usually one-dimensional ('line') spread functions are displayed – have extended wings which contain a substantial fraction of the light. The smallest discernable detail ('resolution') is determined by the width of the 'core' part of the spread function, while the degradation of the contrast is determined by the amount of light in the wings of the spread function ('scattering'). It follows that it is *not* possible to estimate and correct for the degradation of ΔI_{rms} from an estimate of the 'resolution' alone. The determination of a spread function must be done independently, for example from the intensity transition across the lunar limb during a partial eclipse (Levy, 1971; Deubner and Mattig, 1975), or else we must await high resolution observations from space (cf. Spruit and van Ballegoiijen, 1985).

Deubner and Mattig (1975), who estimated $\Delta I_{rms} \approx 12.8\%$ at $\lambda = 607$ nm, fitted their observations of the intensity transition across the lunar limb with the sum of two gaussians, a procedure that may severely underestimate the 'wing' component of the spread function. A re-evaluation of the same data, using the sum of two Lorenzians to fit the spread function resulted in a revised ΔI_{rms} of at least 20% at $\lambda = 607$ nm (Nordlund, 1984a), largely in agreement with the results of the numerical simulations.

The blueshift and widths of photospheric spectral lines (cf. discussion below) together constitute another, independent measure of the intensity fluctuations in the granulation: since the velocity amplitude determines the spectral line widths (of heavy elements like iron), and the 'product' of velocity and intensity fluctuation determines the spectral line shift, the shift and the width together constitute an indirect measure of the intensity fluctuation. The accuracies of laboratory wavelengths of weak Fe lines are (marginally) sufficient for a comparison of observed and synthetic spectral line shifts (Nordlund, 1980; Dravins *et al.*, 1981; Nordlund, 1984b). The comparison shows good agreement of the blueshifts and widths (cf. Figure 3 and 4 below, and Figure 7(a)–(d) of Dravins *et al.*, 1981). Detailed analysis of *limb-effect* curves (cf. Nordlund, 1984b, Figure 6.11) could provide another test, although the behavior at the extreme limb is uncertain (both observationally and theoretically).

The geometrical properties of synthetic granulation images obtained from the numerical simulations (Nordlund, 1984b) and granulation images obtained from high resolution observations were compared by Wöhl and Nordlund (1985) and found to be in basic agreement.

3.2. TEMPERATURE FLUCTUATIONS IN THE PHOTOSPHERE

The typical characteristics of the velocity field and the temperature fluctuations in the photosphere are illustrated in Figure 1, which shows contour plots of the temperature and vector plots of the velocity in selected horizontal and vertical planes in the numerical model. The horizontal temperature fluctuation pattern at the $z = 0$ level is a good proxy of the observable intensity fluctuation.

At larger heights, the temperature fluctuation pattern looks very different. This is the result of two competing effects; adiabatic expansion cooling (or compression heating), and radiative energy exchange, which is typically a heating contribution. The combined result is to produce relative temperature minima above small granules (where the ascent velocities are high), and just outside the edges of larger granules (where the horizontal

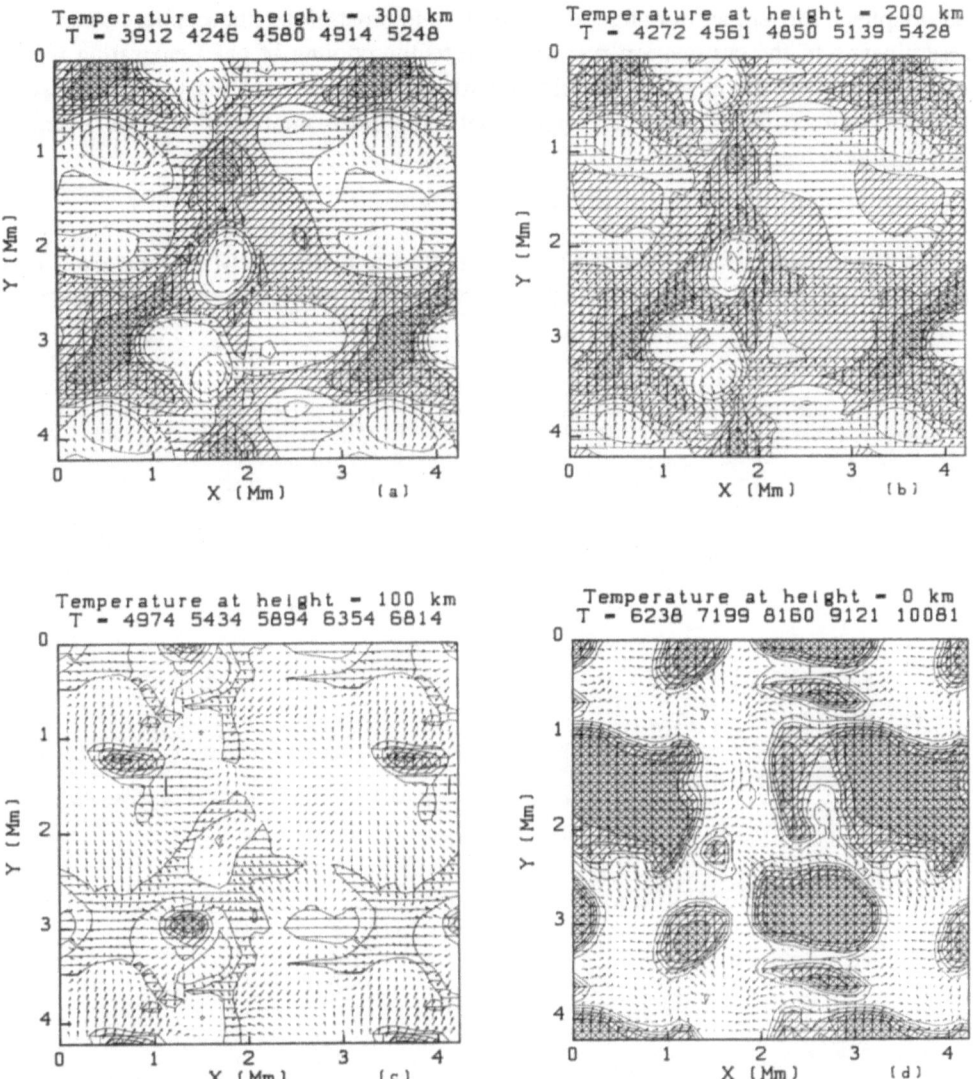

Fig. 1.   Shaded contour plots overlaid with vector plots, showing a snapshot of the temperatures and horizontal velocities in four horizontal planes, from numerical simulations of the solar granulation (after Nordlund, 1984b). The panels show the temperature at heights 300 (a), 200 (b), 100 (c), and 0 km (d). The zero-point of the depth scale corresponds approximately to $\tau_{500\,nm} = 1$ in a plane parallel model. The contour levels are chosen to span the interval between maximum and minimum temperature with equidistant levels.

velocities are high and, therefore, the pressure is low, and where also the ascent velocities are large). The vertically inclined temperature inhomogeneities resulting from these competing effects leave characteristic signatures in slit-spectra of photospheric spectral lines. Such signatures were observed and correctly interpreted by Evans (1964).

Thus, while the velocity field associated with the granulation penetrates the entire photosphere, the temperature fluctuation pattern in most of the photosphere bears little resemblance to the granulation pattern. Due to the cooling of gas penetrating into a stable layer, there is a negative correlation with the continuum intensity, but because of the temperature fluctuations associated with the horizontal motion, this negative correlation is weak.

3.3. THE SUB-SURFACE LAYERS

Below the visible surface, the character of the flow is qualitatively different from that at and above the surface. This is illustrated in Figure 2, which shows the temperatures

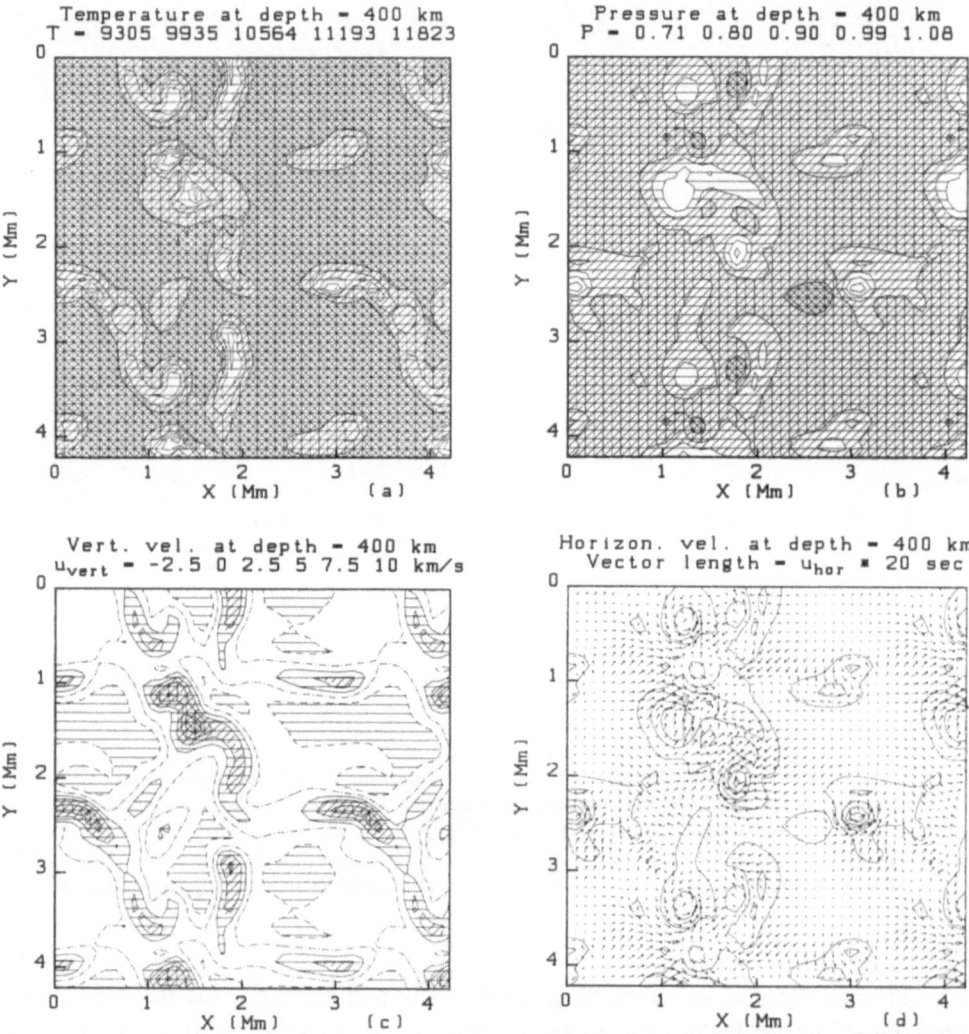

Fig. 2. Shaded contour plots showing snapshots of the temperature (a), pressure (b), and vertical velocity (c) in a horizontal plane at a depth of 400 km. In panel c, the contours corresponding to $u_z > 0$, $u_z = 0$ and $u_z < 0$ are shown full drawn, dashed-dotted, and dashed, respectively. The last panel (d) shows a vector plot of the horizontal velocities, overlaid with the contour lines from panel b.

and horizontal velocities at depth, at one instant in time in the numerical simulations of granular convection (Nordlund, 1984b). The downflow below the surface is concentrated into narrow 'fingers', with unstructured upflow in between. This is similar to the narrow structures that develop in experiments and simulations of Rayleigh–Taylor instabilities. In addition, the strong density stratification aids in concentrating the downflows and dispersing the upflows.

The fluid that goes down into these finger-shaped downdrafts typically has a non-zero angular momentum with respect to the center of the downdraft (as a result of the random positions and shapes of the surrounding granules which supply the inflow). This sets up a circular motion around the center of the downdraft, and the circular velocity is amplified as the downdraft narrows ('bath-tub' or 'inverted tornado'). The centripetal force that balances the acceleration of the fluid in its path around the center of the downdraft is supplied by a negative pressure gradient directed towards the center, corresponding to a pressure deficiency at the center of the downdraft. These pressure minima can become large enough to cause difficulties in the numerical simulations. These 'inverted tornados' are interesting from a purely hydrodynamics point of view, but the fact that any vertical magnetic field lines in the surrounding photosphere must be carried towards, and 'sucked into', these downdrafts also makes the phenomenon potentially very important as a source of hydromagnetic disturbances.

3.5. SYNTHETIC SPECTRAL LINE PROFILES

Time and area averaged synthetic spectral line profiles provide an important diagnostic for granular convection, particularly since such profiles do not require high spatial resolution. Indeed, using full disc profiles, comparisons of synthetic full disc spectral line profiles with observed stellar spectral line profiles are possible (Dravins, 1982; Dravins and Lind, 1984; Nordlund and Dravins, 1986).

Figure 3 (from Nordlund, 1984b) shows a comparison of synthetic and observed solar center disc spectral line profiles for four Fe I lines. Accurate oscillator strengths from the Oxford group (Blackwell *et al.*, 1976, 1979a, b, 1982) were used, and the iron abundance was calibrated on the weakest line. The iron ionization equilibrium was calculated from the radiation intensities at the most important ionization edges of Fe, using data from Lites (1972; see also Athay and Lites, 1972), and scaled H^- absorption coefficients, calibrated on the observed solar disc center intensities in the blue and the near UV.

The comparison shows basic agreement, but with a noticeable discrepancy in the cores of the stronger lines. This discrepancy is probably due to too low temperatures in the uppermost layers of the model caused by approximations in the treatment of spectral line blanketing effects (cf. Nordlund, 1984b, Section 7, and Nordlund, 1985b). On the other hand, a quite cool component of the upper photosphere is necessary to account for the strength of the infrared bands of carbon monoxide (Ayres and Testerman, 1981, Ayres *et al.*, 1985), and Fe I line cores may be substantially brightened by velocity induced non-LTE effects (Nordlund, 1985a).

Figure 4 shows the spectral line bisectors corresponding to the line profiles in

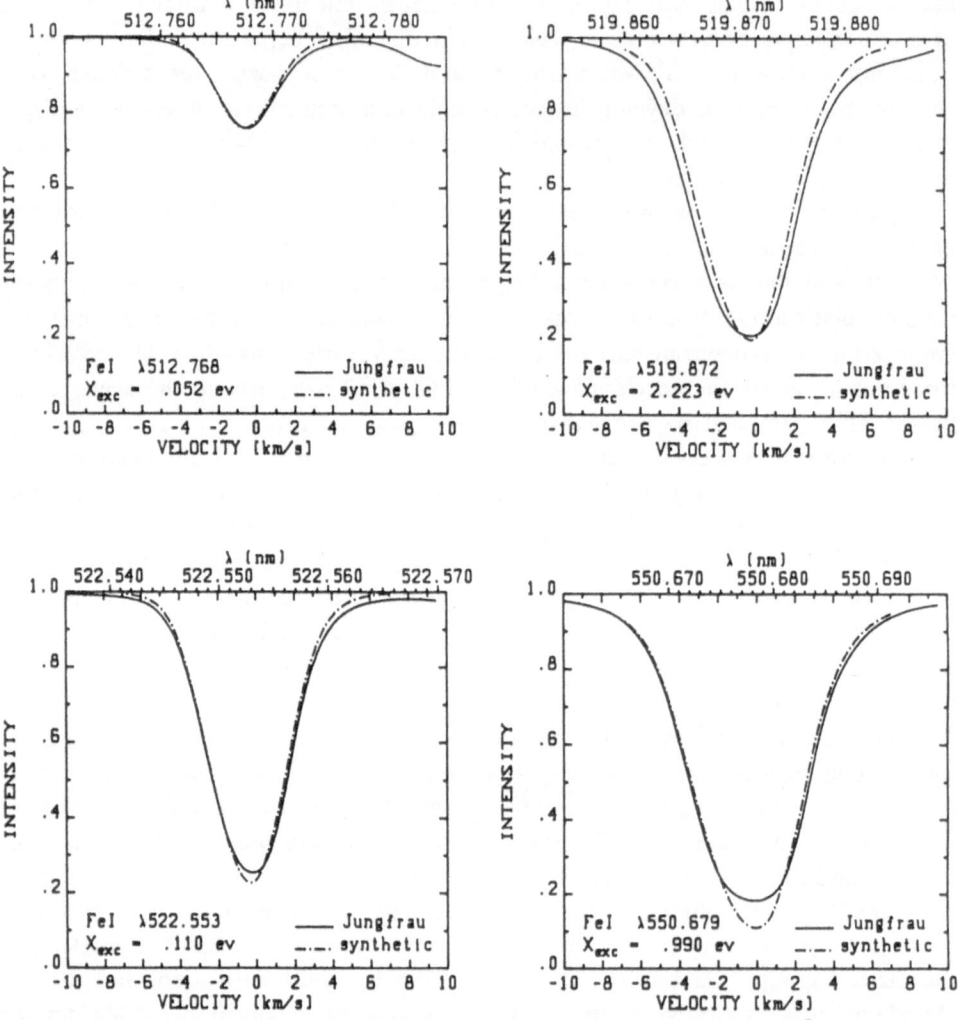

Fig. 3. Synthetic (dash-dotted) and observed (full drawn) Fe I spectral lines (from Nordlund, 1984b). The observed line profiles are from the Jungfraujoch atlas (Delbouille *et al.*, 1973), with line center positions derived from the Kitt Peak Tables (Pierce and Breckinridge, 1973). The synthetic spectral lines are averages over a 110 solar min simulation sequence. Accurate *gf*-values from Blackwell *et al.* (1976, 1979a, b, 1982) were used, and the iron abundance was chosen to fit the central depth of the weakest line.

Figure 3. The bisectors of the synthetic spectral lines fit the bisectors of the observed lines remarkably well, especially after the slight broadening due to the 5-min oscillations has been taken into account (this changes the asymmetry of the synthetic spectral lines even if one assumes that the broadening itself is symmetric, because the unbroadened spectral line is asymmetric).

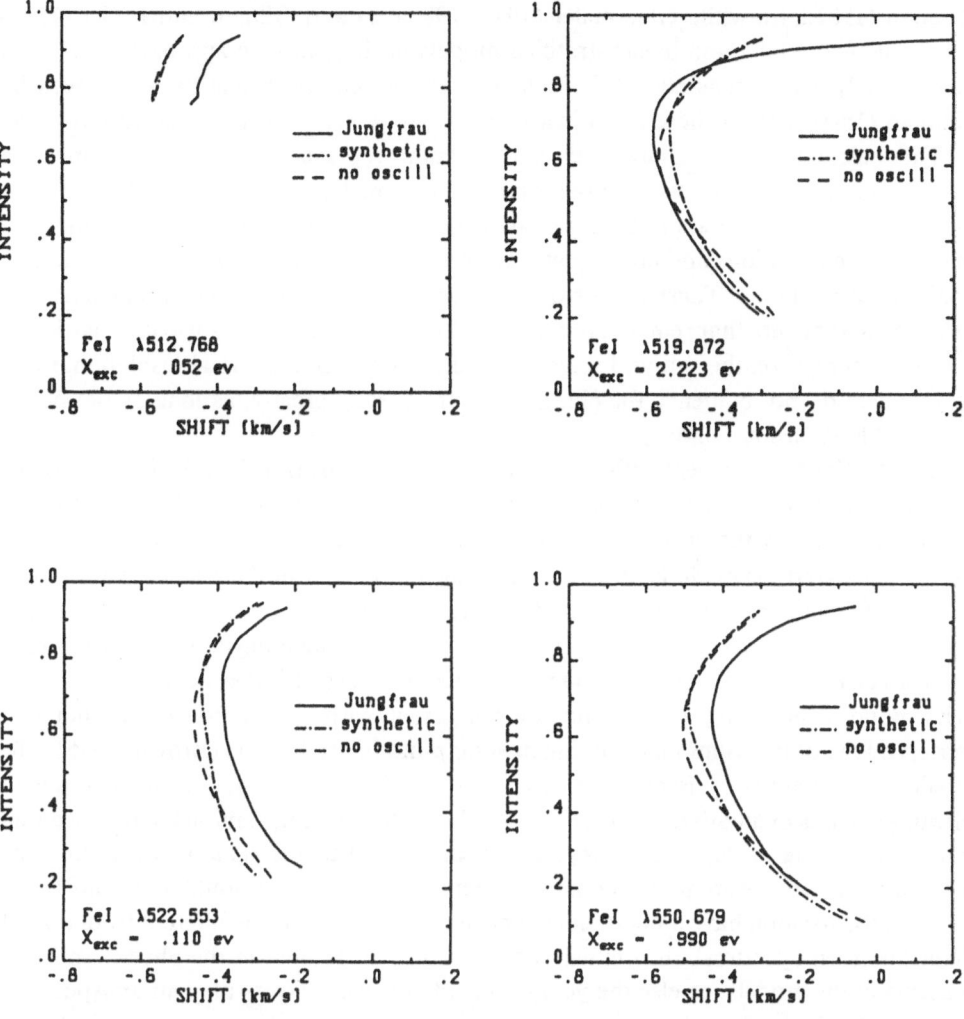

Fig. 4. Spectral line bisectors (cf. Dravin *et al.*, 1981) of the synthetic (dashed-dotted) and observed (full drawn) Fe I lines from Figure 3. The dashed curves show the bisectors of the synthetic spectral lines before folding with an oscillatory (and instrumental) symmetric broadening with $v_{rms} = 0.33$ km s^{-1}.

4. Convection on Global and Intermediate Scales

4.1. GLOBAL CONVECTION

Numerical simulations of global convection have been performed quite extensively. Non-linear convection in a spherical shell has been simulated in the Boussinesq approximation by Durney (1970), Young (1974), Gilman (1977, 1978a, b), and Marcus (1979, 1980a, b). Similar models with magnetic fields included were calculated by Gilman and Miller (1981) and Gilman (1983b). Linear and second-order non-linear stratified models in the anelastic approximation have been calculated by Glatzmaier and

Gilman (1981a, b), while Glatzmaier (1984, 1985a, b) and Gilman and Miller (1985) have calculated fully non-linear stratified models in the anelastic approximation, with dynamically active magnetic fields included. The earlier work has been reviewed by Gilman (1980, 1983a); here I will just make a few remarks on the most recent work.

The calculations by Glatzmaier and Gilman represent a logical evolution of increasingly realistic models of global solar convection. The most recent of these models have a differential rotation in basic agreement with the observed one, and have meridional circulations and pole-equator temperature differences consistent with obser-vational upper limits. These results do *not* involve any tuning of free parameters. The few free parameters that remain are related to the numerical methos used, in particular to what happens on the sub-grid scale. For sufficiently high numerical resolution. these parameters do not influence the character of the results greatly, as long as physically reasonable values are chosen.

Substantial disagreement with observations remains only (for the velocity field) in the amplitudes of the horizontal velocities near the upper boundary (which however is several Mm below the visible surface) and more indirectly, with the time dependence of the generated magnetic fields, which propagate *away* from the equator with time.

The differential rotation in these models tends to be constant on cylinder surfaces (cf. Glatzmaier, 1984, Figure 5; 1985a, Figure 2(a)). This is something that may be expected from an extension of the Taylor-Proudman theorem to a stratified system (Busse, 1977). The model results may or may not be consistent with recent Fourier Tachometer observations of the rotational splitting of solar *p*-mode oscillations (Brown, 1985). The preliminary analysis by Brown indicates that the differential rotation decreases with depth (which is qualitatively consistent with the hydrodynamical models), but also that the rotation rate at depth becomes more nearly equal to the *equatorial* rotation rate (which does not seem to be consistent with the models). Should the empirically determined rotation have a depth dependence that is qualitatively different from that of 'constancy on cylinders', then either some essential ingredient in the physics must be missing in the models, or else the yet not included surface layers play an unexpectedly important role.

Since the most prominent disagreement between the behavior of the numerical models and the Sun is the failure of the numerical models to predict magnetic cycles of the right period, and with the right sense of magnetic fields propagation, further discussion of the numerical models of global convection is deferred to Section 5.3.

4.2. SUPERGRANULATION AND MESOGRANULATION

The intermediate scales (3–10 Mm), with supergranulation (Leighton *et al.*, 1962; Simon and Leighton, 1964; Worden and Simon, 1976) and mesogranulation (November *et al.*, 1981, 1982) have received little attention in numerical work, compared with granular convection and global convection. This may be partially due to the difficulties of describing the interaction of convection on a supergranular scale with granular convection and radiation at the solar surface. In view of the very small limits on any horizontal temperature variation associated with supergranulation, and the results of granulation simulations showing that (on the time-scale of supergranulation) the energy is removed at an approximately constant rate in a layer at, say, 1 Mm below the surface, it would seem possible to use a 'constant heat flux' upper boundary condition for numerical simulations of convection on a horizontal scale of 3–100 Mm, over a depth range covering perhaps 25–50% of the solar convection zone.

The term 'mesogranulation' was introduced by November *et al.* (1981, 1982) for a velocity and temperature fluctuation field observed on the scale of 5–10 Mn, with a characteristic time-scale of an hour or so. The apparent velocity magnitude is of the order of 100 m s^{-1}. The true character of these fluctuations is somewhat unclear, since apparent velocity shifts and temperature fluctuations of this order of magnitude, and with this characteristic time scale are produced also by a modulation of the 'vigor' of granulation, even though there is no net mass flux involved (cf. Dravins *et al.*, 1981, Figures 11(a)–(b)). The separation of a real velocity/temperature fluctuation field from this apparent one is a difficult observational problem. However, with superb spatial resolution, it might be possible to observationally determine, and correct for, the apparent fluctuations caused by the slow modulation of granular convection.

5. Interaction with Magnetic Fields

The observation of magnetic fields on the Sun and interpretation of their structure is reviewed elsewhere in this volume (Stenflo, Zwaan, Muller). For reviews on magneto-convection, see Proctor and Weiss (1982) and Nordlund (1984c). Dynamo theory has recently been reviewed by (e.g.) Cowling (1981), Gilman (1983a), and Schüssler (1983). Spruit (1983, 1984a, b) has reviewed the magnetic flux tube picture of solar magnetic fields. However, since the interaction of magnetic fields with convection is crucial in determining the structure of the solar magnetic fields, some mention of magnetic fields is appropriate at this point.

5.1. INTERACTION WITH GRANULAR CONVECTION

The 'convective collapse instability' as a mechanism for flux concentration has been discussed extensively (Nordlund, 1976; Parker, 1978; Webb and Roberts, 1978; Spruit and Zweibel, 1979; Spruit, 1979; Unno and Ando, 1979; Hasan, 1983, 1984, 1985; see also Parker, 1979, Section 10 and references therein).

These investigations qualitatively illustrate how a superadiabatic (convectively unstable) mean stratification leads to a 'collapse' of an initially weak field into a strong flux concentration. Proctor (private communication) has pointed out that, rather than being the effect of an instability, this is the quenching of a (somewhat modified) convective instability: when the magnetic field becomes sufficiently concentrated, it *stops* something that looks very much like a normal convective instability.

The 'convective collapse' carries a weak initial magnetic field into a configuration where the interior of the magnetic flux structure has a temperature that is lower than the mean surroundings over several pressure scale heights, and where, consequently, the gas pressure is much reduced inside the flux concentration, relative to outside

(Parker, 1955; Spruit, 1976). The gas pressure difference is then only slightly smaller than the external pressure and can support a magnetic field pressure almost corresponding to the surrounding gas pressure. In the Sun, the gas pressure is about 13 kPa at the level where the continuum optical depth is unity, and this gas pressure corresponds to a magnetic flux density of some 0.18 mT (1.8 kG). Thus, if the evacuation is significant down to this level, *but not much further*, one gets flux densities which are similar to the observed ones.

Most of the investigations quoted above have assumed adiabatic perturbations (no radiative transfer effects). There has also been some confusion about the choice of boundary conditions, and one must conclude that the balance of effects that determines the actual degree of evacuation (and, hence, the observable field strength) in the small scale solar flux concentrations is still not well understood (cf. Nordlund, 1984c, Section 4).

Numerical simulations of the interaction of convection and magnetic fields in strongly stratified models have been performed by myself (Nordlund, 1983, 1986), and by Hurlburt and Toomre (1985).

The calculations of Hurlburt and Toomre are two-dimensional, with a polytropic model covering 2.4 density scale heights, fully compressible, but are not directly relevant to the solar case, since no description of radiative transfer effects were included. Concentration of magnetic flux towards the downdrafts occurs initially because of kinematic effects, but is further enhanced because of the superadiabatic stratification. Since the temperature gradient excess relative to an adiabatic stratification is not very large, the temperature deficiency in the interior of the flux concentrations is not big enough to produce a strong evacuation of the flux structures.

My own calculations are extensions of the numerical simulations of solar granulation, complemented with appropriate equations for the magnetic field. Thus, these calculations are three-dimensional, anelastic, cover about 5.5 density scale heigths, and include a realistic description of radiative transfer effects. The behavior of the magnetic flux concentrations obtained in these simulations is in qualitative agreement with the observed properties of small scale flux concentrations of the Sun (Muller, 1983). The magnetic flux, which is unipolar with a conserved average vertical component, is concentrated in the inter-granular lanes, where the presence of the magnetic field helps to reduce the temperature (by screening off the flow, and hence allowing the radiative cooling at the surface to cool the interior further). The resulting gas pressure difference is large enough to concentrate the magnetic field to flux densities of 0.1–0.15 mT (1–1.5 kG) at the observable level. The volume in-between the magnetic flux concentrations are swept free of magnetic field, and here convection proceeds to produce new granules and new intergranular lanes. The magnetic flux concentrations 'creep' into the newly formed downdrafts, since the gas pressure in the downdrafts is insufficient to balance the magnetic pressure in the flux concentrations.

Thus, these simulations are capable of illustrating the qualitative properties of the interaction of granular convection and unipolar magnetic fields. However, the numerical resolution (50 km vertically, 190 km horizontally) is insufficient to realistically model the steep walls of the flux concentrations, and the numerical diffusion of plasma through the magnetic field is a significant factor in the determination of the internal temperature structure of the flux concentrations in these simulations.

5.2. MAGNETIC FIELD TOPOLOGY

Numerical simulations, although still with insufficient numerical resolution, provide an important basis for a qualitative understanding of the interaction of convection, magnetic fields, and radiation. For one thing, they serve as a reminder of the complicated topology of the magnetic field lines in the solar photosphere. This may be a crucial factor, both in the determination of (quasi-) equilibrium field strengths in the flux concentrations, and for the role of the magnetic field as a 'channel' for mechanical energy to the solar chromosphere and corona.

Torsional disturbances due to the 'bath-tub' effect mentioned above, or 'buffeting' by the granulation, are possible sources for 'sudden disturbances' in the photosphere which, according to Hollweg *et al.*, (1982) and Hollweg (1982), may produce spicule-like ejections into the chromosphere. In this connection it should be noted that the horizontal velocities and, especially, the relative pressure fluctuations obtained in the numerical simulations of solar granulation are significantly larger than the values assumed by Spruit (1984a) in his discussion of tube mode generation ($u_{hor}^{rms} \approx 3$ km s^{-1} and $\delta P/P \approx 0.5$, rather than 2 km s^{-1} and 0.05, respectively).

The finite lifetime of the individual flux concentrations, and the strong horizontal velocity field present, cause the field lines in the photosphere to wander around in a random-walk fashion. Such field line motions have been discussed by Parker (1983) as a possible heating mechanism ('topological heating') for the quiet solar corona. The actual energy dissipation due to the foot point motion depends on details of the dissipation mechanism and is hard to estimate quantitatively, as discussed by Spruit (1984b).

5.3. SOLAR DYNAMO MODELS

The failure of even the best self-consistent models so far to produce the observed equatorward drift of magnetic field with cycle phase (Gilman, 1983; Glatzmaier, 1984, 1985a, b; Gilman and Miller, 1985; cf. also the discussion by Gilman, 1984), is telling us that something essential is still missing in the way these models describe the interaction of convection and magnetic fields. (The apparent success of kinematic models (e.g., Yoshimura 1978a, b, 1983) results from the abundance of free parameters controlling the behavior in these models; e.g., the freedom to choose the depth dependence of the rotation profile.) Since the same physical situation (magneto-convection in a rotating spherical shell) has been modeled with several independent numerical methods, with largely the same results, we can be quite confident that – within the given assumptions of these numerical models – the basic magnetohydrodynamics has been correctly modeled, and that the remaining disagreement with the behavior of the solar dynamo is not due to any pecularity of the numerical methods used.

Glatzmaier's models were calculated with a semi-spectral method, where the (θ, ϕ) dependence is represented with spherical harmonics $Y_l^m(\theta, \phi)$ and the radial dependence

is represented by Chebyshev polynomials $T_n(r)$. These calculations (with
17×1024 degrees of freedom per variable) are close to the limits of what is practical
with present day computers. Nevertheless, it would be highly desirable to either increase
the number of degrees of freedom per variable horizontally, or else use the available
degrees of freedom in such a way as to increase the horizontal resolution. This may
perhaps be achieved by using a representation where the ϕ dependence has a period
equal to $2\pi/M$, where M is larger than 1. The need for increased horizontal resolution
is clear if we consider the 'aspect ratio'

$$\frac{2\pi R/L}{0.43R/N} \approx 15N/L \approx 7.5 . \tag{24}$$

Of course, this is not a true aspect ratio (the Δr spacing varies with r, from a maximum
of the order of $(\pi/2)0.43\ R/N$, to a minimum which is much smaller), but the ratio (24)
serves to illustrate the relatively poor horizontal resolution compared to the vertical one.
The broad maximum in the kinetic energy spectrum around $l = 15$ has only a factor 2
margin to the maximum horizontal wavenumber (31); i.e., the dominant scales modeled
are only twice as big as the smallest ones that could be represented. Thus, for example,
up/down asymmetries of the energy carrying cells can not be well represented in the
model, since the harmonics of the dominating modes lie outside the wavenumber limits
of the model. As discussed in Section 2.4, up/down asymmetries are a prominent feature
of resolved compressible convection, and the failure to resolve such asymmetries (and
corresponding asymmetries in the magnetic field) may be one reason for the short-
comings of present models of global convection and dynamo action. The separation of
convective motions and magnetic fields ('flux expulsion'), e.g., Galloway and Wiess,
1981) found in well resolved models of magnetoconvection may be an important factor
in determining the period of the solar cycle, since such a separation would be expected
to decrease th coupling between the convective motions and the magnetic field. Such
a separation can only be modeled if convective cells are well resolved.

The Chebyshev representation automatically produces an increase in the numerical
resolution close to the boundaries, which is very desirable, since the scale height
decreases rapidly at the upper boundary, and boundary effects near the transition from
unstable to stable stratification at the bottom need to be resolved. This, however,
requires a correspondingly increased horizontal resolution. Thus, ideally, the code
should be able to (at least marginally) resolve supergranulation at wavenumbers of 100
to 200.

Several suggestions as to what else might be missing in the modeling have been put
forward: the suggestion by Galloway and Weiss (1981) and by Golub *et al.* (1981) that
the solar dynamo might be operating at the base of the convection zone, in the transition
zone between the stable interior and the turbulent convection zone, was tested by
Glatzmaier (198b) by including a stable region below the convection zone, and by
decoupling the velocity field and the magnetic field in the bulk of the convection zone,
to simulate magnetic flux structures that are too small and too concentrated to be

affected by the helicity in the convection zone. Only part of a cycle has been simulated, and the results are not conclusive.

Glatzmaier also suggested that the solar dynamo could be operating in the outer 5% or so of the solar radius, where angular velocity is probably decreasing with depth, and where the helicity has the correct sign for producing equatorward drift of the magnetic field. Considerations of magnetic buoyancy (Parker, 1975) suggested that a magnetic cycle with a period as long as 22 years may be hard to achieve if the dynamo is not deep seated, but these considerations neglected the possibly important role of radiative cooling of flux structure interiors at the solar surface.

6. Conclusions and Recommendations for Future Work

Solar convection, and the interaction of convection and magnetic fields on the Sun serves as an important test for our understanding of the dynamics of the (magneto-) hydrodynamics of astrophysical plasmas in general. From the interaction of large-scale convection and magnetic fields in the solar interior, through the interaction of convection, radiation, and magnetic fields in the photosphere, to the plasma physics of the chromosphere and corona, there is a wealth of physical phenomena, likely to have counterparts in non-solar astrophysical systems we cannot resolve observationally. For this reason, improvements in our understanding of the dynamics and energy transport in the solar plasma is of great importance.

I would like to conclude this review with some hopes and recommendations as to where we should push for progress in research on solar convection in the near future:
– The numerical simulations of global convection and dynamo action in the solar convection zone need further development, to resolve the discrepancy in the temporal behavior of the magnetic fields generated by the convective dynamo. Part of such a development should go in the direction of improved numerical resolution, to allow a better representation of processes on small scales (close to the upper and lower boundaries). This may perhaps be feasible with further improvement of the numerical methods, or else must await the next generation of gigaflop super-computers.

– 'Cartoon' models of the solar dynamo/solar activity phenomenon (Schüssler, 1980, 1983; Van Ballegoiijen, 1982a, b) should be developed further. 'Cartoons' may provide important inspiration for improved numerical models, and/or may suggest critical observational tests.

– There is a need for numerical simulations of convection on the scale of supergranulation, both to model supergranulation itself, and to model its interaction with magnetic fields on the scale of active regions. This is probably feasible with present day computers and numerical methods.

– Our understanding of granular convection, and its direct and indirect influence on the solar photosphere, chromosphere, and corona needs improvement both on the theoretical and observational sides. As regards the chromosphere and the corona, we need a better understanding of how the energy is transferred from motions in the convection zone (on all scales), through stresses in the magntetic field that extends out into the chromosphere and corona, where dissipative processes and magnetic field instabilities convert the energy into heat and non-thermal particle energy. As regards the photosphere, we need a better understanding of the influence of spectral lines on the energy balance in the optically layers. Also, one would like to see fully compressible codes developed, ideally covering a sufficient depth range to study the interaction of the 5-minute oscillations with convection.

 – Observationally, we may look forward to the Solar Optical Telescope, which will be able to resolve solar granular convection and its interaction with small scale magnetic fields in great detail (cf. Spruit and van Ballegoiijen, 1985). Apart from the direct advantage of improved spatial resolution, a great advantage of observations from space as compared to ground based observations is the temporal stability of the picture degradation, and the consequent possibility to obtain long time series of observations with a well determined spread function.

Acknowledgements

I would like to thank Tom Ayres, Tim Brown, Jörgen Christensen-Dalsgaard, Peter Gilman, Gary Glatzmaier, Neal Hurlburt, Mike Proctor, and Juri Toomre for stimulating discussions, and for providing previews of work in progress.

This work was supported by the Danish Natural Science Research Council and the Danish Space Board.

References

Athay, R. G..: 1970, *Astrophys. J.* **161**, 71.
Athay, R. G. and Lites, B. W.: 1972, *Astrophys. J.* **1976**, 809.
Ayres, T. R. and Testerman, L.: 1981, *Astrophys. J.* **245**, 1124.
Ayres, T. R., Testerman, L. and Brault, J.: 1985, preprint.
Batchelor, G. K.: 1953, *Quart. J. Roy. Meteor. Soc.* **79**, 224.
Blackwell. D. E., Ibbetsen, P. A., Petford, A. D., and Willis, R. B.: 1976, *Monthly Notices Roy. Astron. Soc.* **177**, 219.
Blackwell, D. E., Ibbetsen, P. A., Petford, A. D., and Shallis, M. J.: 1979a, *Monthly Notices Roy. Astron. Soc.* **186**, 633.
Blackwell, D. E., Petford, A. D., and Shallis, M. J.: 1979b, *Monthly Notices Roy. Astron. Soc.* **186**, 657.
Blackwell, D. E., Petford, A. D., Shallis, M. J., and Simmons, G. J.: 1982, *Monthly Notices Roy. Astron. Soc.* **199**, 43.
Bray, R. J., Loughhead, R. E., and Durrant, C. J.: 1984, *The Solar Granulation,* Cambridge University Press, Cambridge.
Brown, T. M.: 1985, preprint.
Bumba, V.: 1975, in V. Bumba and J. Kleczek (eds.), 'Basic Mechanisms of Solar Activity', *IAU Symp.* **71**, 47.
Busse, F. H.: 1977, in E. A. Spiegel and J.-P. Zahn (eds.), 'Problems of Stellar Convection', *Proc. IAU Colloq.* **38**, 156.
Carbon, D. F.: 1979, *Ann. Rev. Astron. Astrophys.* **17**, 513.
Charney, J. G. and Ogura, Y.: 1960, in *Proc. Int. Symp. on Num. Weather Pred.*, Japan Meteorological Society, Tokyo.
Christensen-Dalsgaard, J.: 1984, in 'The Hydromagnetics of the Sun', *Proc. 4th European Meeting in Solar Physics*, ESA Special Publ. 220, p. 3.
Christensen-Dalsgaard, J.: 1985, in 'Theoretical Problems in Stellar Stability and Oscillations', *Proc. 25th Liége Intern. Astrophys. Colloq.*, Institut d'Astrophysique, Liège.

Christensen-Dalsgaard, J., and Frandsen, S.: 1983, *Solar Phys.* **82**, 165.

Cloutman, L. D.: 1979, *Astrophys. J.* **227**, 614.

Cowling, T. G.: 1981, *Ann. Rev. Astron. Astrophys.* **19**, 115.

Delbouille, L., Neven., N. and Roland, G.: 1973, *Photometric Atlas of the Solar Spectrum from λ3000 to λ10000*, Institut d'Astrophysique de l'Université de Liège.

Deubner, F. and Mattig, W.: 1975, *Astron. Astrophys.* **45**, 167.

Dravins, D.: 1982, *Ann. Rev. Astron. Astrophys.* **20**, 61.

Dravins, D. and Lind, J.: 1984, in S. L. Keil (ed.), *Small Scale Dynamical Processes in Quiet Stellar Atmospheres'*, Sacramento Peak Observatory, Sunspot, N. M. 88349, p. 414.

Dravins, D., Lindegren, L., and Nordlund, Å.: 1981, *Astron. Astrophys.* **96**, 345.

Dunn, R. B., and Zirker, J. B.: 1973, *Solar Phys.* **33**, 281.

Durney, B.: 1979, *Astrophys. J.* **161**, 1115.

Evans, J. W.: 1984, *Astroph. Norvegica* **4**, 33.

Galloway, D. J., and Weiss, N. O.: 1981, *Astrophys. J.* **243**, 945.

Gigas, D. and Steffen, M.: 1984, *Publ. Sternwarte Kiel*, No. 368.

Gilman, P. A.: 1977, *Geophys. Astrophys. Fluid Dyn.* **9**, 93.

Gilman, P. A.: 1978a, *Geophys. Astrophys. Fluid Dyn.* **11**, 157.

Gilman, P. A.: 1978b, *Geophys. Astrophys. Fluid Dyn.* **11**, 181.

Gilman, P. A.: 1980, *Highlights Astron.* **5**, 91.

Gilman, P. A.: 1983a, in J.-O. Stenflo (ed.) 'Solar and Stellar Magnetic Fields: Origin and Coronal Effects', *IAU Symp.* **102**, 247.

Gilman, P. A.: 1983b, *Astrophys. J. Suppl.* **53**, 243.

Gilman, P. A.: 1984, *Solar Seismology from Space*, JPL Publ. 84–84, p. 41.

Gilman, P. A. and Glatzmaier, G.A.: 1981, *Astrophys. J. Suppl.* **45**, 335.

Gilman, P. A. and Miller, J.: 1981 *Astrophys. J. Suppl.* **46**, 211.

Gilman, P. A. and Miller, J.: 1985, in preparation.

Glatzmaier, G. A.: 1984, *J. Comp. Phys.* **55**, 461.

Glatzmaier, G. A.: 1985a, *Astrophys. J.* **291**, 300.

Glatzmaier, G.A.: 1985b, *Geophys. Astrophys. Fluid Dyn.* **31**, 137.

Glatzmaier, G. A. and Gilman, P. A.: 1981a, *Astrophys. J. Suppl.* **45**, 351.

Glatzmaier, G. A. and Gilman, P. A.: 1981b, *Astrophys. J. Suppl.* **47**, 103.

Goldreich, P. and Keeley, D. A.: 1977a, *Astrophys. J.* **211**, 934.

Goldreich, P. and Keeley, D. A.: 1977b, *Astrophys. J.* **212**, 243.

Golub, L., Rosner, R., Vaiana, G. S., and Weiss, N. O.: 1981, *Astrophys. J.* **243**, 309.

Gough, D. O.: 1969, *J. Atmos. Sci.* **26**, 448.

Gough, D. O.: 1977, *Astrophys. J.* **214**, 196.

Gustafsson, B., Bell, R. A., Eriksson, K., and Nordlund, Å.: 1975, *Astron. Astrophys.* **42**, 407.

Hasan, S. S.: 1983, in J.-O. Stenflo (ed.), 'Solar and Stellar Magnetic Fields: Origin and Coronal Effects', *IAU Symp.* **102**, 73.

Hasan, S. S.: 1984, *Astrophys. J.* **285**, 851.

Hasan, S. S.: 1985, *Astron. Astrophys.* **143**, 39.

Herring, J. R., Orzag, S. A., Kraichnan, R. H., and Fox, D. G.: 1974, *J. Fluid Mech.* **66**, 417.

Hollweg, J. V.: 1982, *Astrophys. J.* **257**, 345.

Hollweg, J. V., Jackson, S., and Galloway, D. J.: 1982, *Solar. Phys.* **75**, 35.

Hurlburt, N. E.: 1983, Thesis, University of Colorado.

Hurlburt, N. E. and Toomre, J.: 1985, *Astrophys. J.* (submitted).

Hurlburt, N. E., Toomre, J., and Massaguer, J. M.: 1984, *Astrophys. J.* **282**, 557.

Hurlburt, N. E., Toomre, J., and Massaguer, J. M.,: 1985, *Astrophys. J.* (submitted).

Kurucz, R. L.: 1979, *Astrophys. J. Suppl.* **40**, 1.

Latour, J., Spiegel, E. A., Toomre, J., and Zahn, J. P.: 1976 *Astrophys. J.* **207**, 233.

Latour J., Toomre, J., and Zahn, J. P.: 1983, *Solar Phys.* **82**, 387.

Levy, M.: 1971, *Astron. Astrophys.* **14**, 15.

Leighton, R. B., Noyes, R. W., and Simon, G. W.: 1962, *Astrophys. J.* **135**, 474.

Lites, B. W.: 1972, Thesis, University of Colorado.

Marcus, S.: 1979, *Astrophys. J.* **231**, 176.

Marcus, S.: 1980a, *Astrophys. J.* **239**, 622.

Marcus, S.: 1980b, *Astrophys. J.* **240**, 203.

Massaguer, J. M. and Zahn, J.-P.: 1980, *Astron. Astrophys.* **87**, 315.

Massaguer, J. M., Latour, J., Toomre, J., and Zahn, J.-P.: 1984, *Astron. Astrophys.* **140**, 1.

Mehltretter, J. P.: 1974, *Solar Phys.* **38**, 43.

Mihalas, B. W. and Toomre, J.: 1981, *Astrophys. J.* **249**, 349.

Mihalas, B. W. and Toomre, J.: 1982, *Astrophys. J.* **263**, 386.

Mihalas, D.: 1978, *Stellar Atmospheres,* 2nd ed., W. H. Freeman, San Francisco.

Muller, R.: 1983, *Solar Phys.* **85**, 113.

Namba, O. and van Rijsbergen, R.: 1977, in 'Problems of Stellar Convection', *Proc. IAU Colloq.* **38**, 119.

Nordlund, Å.: 1976, in E. Muller (ed.), *Highlights of Astronomy,* Vol. 4, part 2, remark on p. 272.

Nordlund, Å.: 1977, in 'Problems of Stellar Convection', *Proc. IAU Colloq.* **38**, 237.

Nordlund, Å.: 1978, in A. Reiz and T. Anderson (eds.), *Astronomical Papers Dedicated to Bengt Strömgren,* Copenhagen Univ. Obs., Copenhagen, p. 95.

Nordlund, Å.: 1980, in D. F. Gray (ed.), 'Stellar Turbulence', *Proc. IAU Colloq.* **51**, 213.

Nordlund, Å.: 1982, *Astron. Astrophys.* **107**, 1.

Nordlund, Å: 1983, in J.-O. Stenflo (ed.), 'Solar and Stellar Magnetic Fields: Origin and Coronal Effects', *IAU Symp.* **102**, 79.

Nordlund, Å.: 1984a, in S. L. Keil (ed.), *Small Scale Dynamical Processes in Quiet Stellar Atmospheres,* Sacramento Peak Observatory, Sunspot, N.M. 88349, p. 174.

Nordlund, Å.: 1984b, in S. L. Keil (ed.), *Small Scale Dynamical Processes in Quiet Stellar Atmospheres,* Sacramento Peak Observatory, Sunspot, N.M. 88349, p. 181.

Nordlund, Å.: 1984c, in 'The Hydromagnetics of the Sun', *Proc. 4th European Meeting in Solar Physics,* ESA Spec. Publ. 220, p. 37.

Nordlund, Å.: 1985a, in J. O. Beckman and L. Crivellari (eds.), *Problems in Stellar Spectral Line Formation Theory,* D. Reidel, Dordrecht, Holland, p. 215.

Nordlund, Å.: 1985b, in *Proceedings from the MPA/LPARL Workshop on 'Theoretical Problems in Solar Physics',* Munich 16–18 September, 1985.

Nordlund, Å.: 1986, in preparation.

Nordlund, Å., Dravins, D.: 1986, in preparation.

November, L. J., Toomre, J., Gebbie, K. B., and Simon, W.: 1981, *Astrophys. J.* **245**, L123.

November, L. J., Toomre, J., Gebbie, K. B., and Simon, W.: 1982, *Astrophys. J.* **258**, 846.

Ogura, Y., Phillips, N. A.: 1982, *J. Atmos. Sci.* **19**, 173.

Orzag, S. A.: 1971, *J. Fluid Mech* **49**, 75.

Parker, E. N.: 1955, *Astrophys. J.* **121**, 491.

Parker, E. N.: 1975, *Astrophys. J.* **198**, 205.

Parker, E. N.: 1978, *Astrophys. J.* **221**, 368.

Parker, E. N.: 1979, *Cosmical Magnetic Fields: Their Origin and Their Activity,* Clarendon Press, Oxford.

Parker, E. N.: 1983, *Astrophys. J.* **264**, 642.

Proctor, M. R. E., and Weiss, N. O.: 1982, *Rep. Prog. Phys.* **45**, 1317.

Ricort, G. and Aime, C.: 1979, *Astron. Astrohys.* **76**, 324.

Ricort, G., Aime, C., Deubner, F. and Mattig, W.: 1981, *Astron. Astrophys.* **97**, 114.

Schüssler, M.: 1980, *Nature* **288**, 150.

Schüssler, M.: 1983, in J.-O. Stenflo (ed.), 'Solar and Stellar Magnetic Fields: Origin and Coronal Effects', *IAU Symp.* **102**, 213.

Simon, G. W. and Leighton, R. B.: 1964, *Astrophys. J.* **140**, 1120.

Smagorinsky, J.: 1963, *Monthly Weather Rev.* **91**, 99.

Spruit, H. C.: 1976, *Solar Phys.* **50**, 269.

Spruit, H. C.: 1979, *Solar Phys.* **61**, 363

Spruit, H. C.: 1981, *Astron. Astrophys.* **98**, 155.

Spruit, H. C.: 1983, in J.-O. Stenflo (ed.), 'Solar and Stellar Magnetic Fields: Origin and Coronal Effects', *IAU Symp.* **102**, 41.

Spruit. H. C.: 1984a, in S. L. Keil (ed.), *Small Scale Dynamical Processes in Quiet Stellar Atmospheres,* Sacramento Peak Observatory, Sunspot, N.M. 88349, p. 249.

Spruit, H. C.: 1984b, *Mitt. Astron. Ges.* **60**, 83.

Spruit, H. C. and Zweibel, E. G.: 1979, *Solar Phys.* **62**, 15.

Spruit, H. C. and van Ballegoiijen, A. A. (eds.): 1985, *Proceedings from the MPA/LPARL Workshop on 'Theoretical Problems in Solar Physics',* Munich, 16–18 September, 1985.

Toomre, J. Hurlburt, N. E., and Massaguer, J. M.: 1984, in S. L. Keil (ed.), *Small Scale Dynamical Processes in Quiet Stellar Atmospheres*, Sacramento Peak Observatory, Sunspot, N.M. 88349, p. 222.

Unno, W.: 1967, *Publ. Astron. Soc. Japan* **27**, 81.

Unno, W., Ando, H.: 1979, *Geophys. Astrophys. Fluid Dyn.* **12**, 107.

van Ballegoiijen, A. A.: 1982a, *Astron. Astrophys.* **106**, 43.

van Ballegoiijen, A. A.: 1982b, *Astron. Astrophys.* **113**, 99.

Webb, A. R. and Roberts, B.: 1978, *Solar Phys.* **59**, 249.

Wittman, A.: 1979, *Mitt. Kiepenheuer Inst.*, No. 179, 29.

Worden, S. P. and Simon, G. W.: 1976, *Solar Phys.* **46**, 43.

Wöhl, H. and Nordlund, Å.: 1985, *Solar Phys.* **97**, 213.

Yoshimura, H.: 1978a, *Astrophys. J.* **220**, 692.

Yoshimura, H.: 1978b, *Astrophys. J.* **220**, 706.

Yoshimura, H.: 1984, *Astrophys. J. Suppl.* **52**, 363.

Young, R. E.: 1974, *J. Fluid Mech.*, **63**, 695.

THE FINE STRUCTURE OF THE QUIET SUN

R. MULLER

Observatoires du Pic-du-Midi et de Toulouse, 65200 – Bagnères de Bigorre, France

Abstract. The observed properties of the small-scale features visible in the quiet photosphere – the granulation, of convective origin, and the network bright points, associated with kG magnetic fields – are described. The known properties of the magnetic flux tubes associated with network bright points are also presented. Empirical models derived from the observations are discussed, as well as a few theoretical models of particular importance for the understanding of the origin of the small-scale features of the quiet photosphere. Finally, the observational evidences showing that the structure of the granulation and of the photospheric network are varying over the solar cycle are reported.

1. Introduction

If one observes the quiet surface of the Sun under high resolution conditions, through a wide bandpass filter (or in continuum windows), the most striking feature one will watch is the granulation pattern. It consists of convective cells, $1''.5$ of typical size, originating at the top of the convection zone (where the temperature gradient is superadiabatic) and overshooting into the photosphere. If then one observes through a narrow filter centered on a photospheric line one will see many tiny bright points embedded in the intergranular lanes, their apparent size being close to the theoretical resolution of the telescope. They are arranged so that they form a network, called the photospheric network, which is cospatial with the supergranule boundaries. It is well known that the network bright points are associated with kG magnetic fields. The magnetic flux of the network apparently comes from the flux which emerges in both the large and small active regions (the so-called 'ephemeral active regions') and is dispersed into the network by convective motions.

In this paper we will review the morphological, dynamical and brightness properties of granules and network bright points; as the later are associated with strong magnetic fields, their magnetic structure will also be presented. Empirical models derived from the observations will be discussed, as well as a few theoretical models of particular importance for the understanding of the origin of the small-scale features of the quiet photosphere. The granulation is reviewed in Section 2, network bright points and associated flux tubes in Section 3. The variation of the structure of the granulation and photospheric network over the solar cycle, recently discovered, is presented in Section 4. The questions of the stability of flux tubes and interaction of magnetic field and convection will not be tackled.

2. The Solar Granulation

2.1. THE WHITE-LIGHT STRUCTURE

The solar granulation is visible at the surface of the Sun as a cellular pattern of bright elements (the granules) on a darker background (the intergranular spaces). Their shape

Solar Physics **100** (1985) 237–255.   0038–0938/85.15.

may be very irregular and *apparently*, most of the granules have a diameter in the range 1″–2″ (Figure 1).

Until very recently all the analysis of high-resolution observations (down to 0″.3) were converging to find that the solar granulation has a well defined or a characteristic horizontal length scale, as shown by the numerous centre-to-centre distances of granules histograms and by the unique size histogram published (Namba and Diemel, 1969). The mean granule size and distance between granules, averaged over the published values, are found to be of 1″.35 and 1″.9 respectively (Bray *et al.*, 1984).

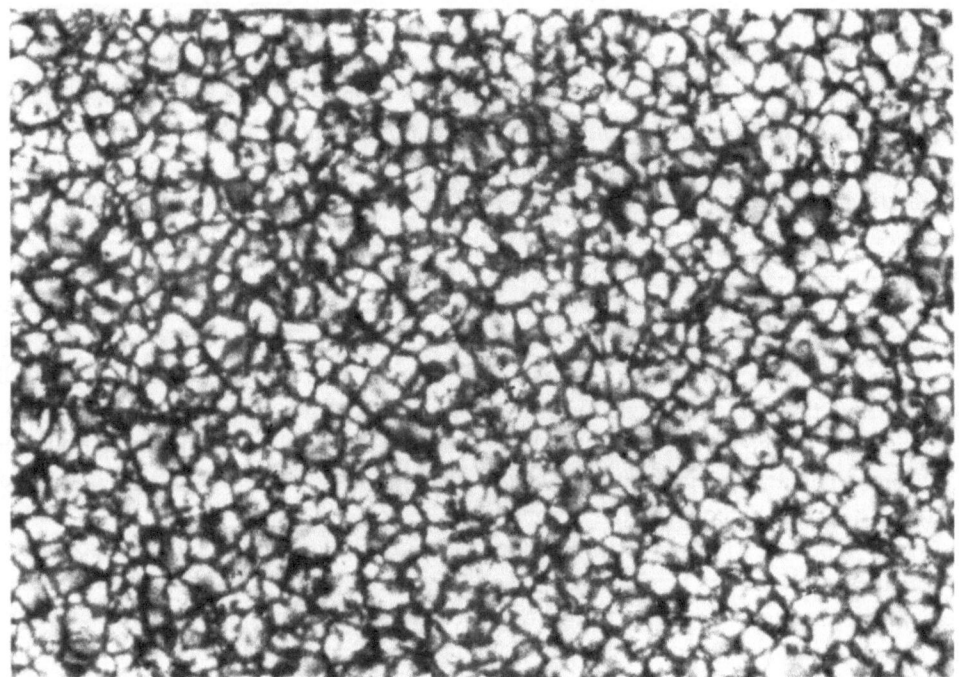

Fig. 1.   The solar granulation observed with a resolution of 0″.25 (Pic-du-Midi Observatory).

Analysing a picture of the solar granulation obtained with a resolution of 0″.25, Roudier and Muller (1985) found a quite different result, namely that the number of granules actually increases continuously toward the smaller scales; that means that the solar granulation has no characteristic scale. On the other hand the granular pattern has a characteristic centre-to-centre distance between granules of 1″.8. In addition they found that the solar granulation appears to have a *critical* size of 1″.37, at which drastic changes in the granule properties occur. The fractal dimension D of granules, defined by $P \sim A^D$, where P is the granule perimeter and A the granule area, changes abruptly from a value of about 2 for the granules larger than 1″.37, to a value of 1.25 for the smaller ones (this means that the granules larger than 1″.37 are much more contorted than the smaller ones); the value $D = 1.25$ is an indication that the granules smaller than 1″.37

could be of turbulent origin (Mandelbrot, 1977; Lovejoy, 1982); the critical size of $1\overset{''}{.}37$ is close to the characteristic size of fragmentation of granules which is found to be of about $1\overset{''}{.}5$ by Kawaguchi (1980); the granules of size $1\overset{''}{.}37$ are the main contributors to the granule radiation. The increasing number of small granules and their fractal dimension are arguments in favor of a turbulent origin of the smaller granules. However Karpinsky (1985) comes to an opposite conclusion: analysing a two-dimensional spatial spectrum of granulation brightness fluctuation, he does not find evidence of the inertial domain where the turbulent cascade should occur in the classical Kolmogorov's description of a highly turbulent medium as the convection zone is.

2.2. EVOLUTION

The first correct description of a typical granule was given by Rösch (1962). A detailed description was later published by Mehltretter (1978) and Kawaguchi (1980), which can be summarized as follows. Granules are born, almost without exception, from previous granule fragments or by merging of neighbouring granules. They are destroyed by essentially three processes: fragmentation (into 2 to 5 fragments); fading away; merging with one or more neighbouring granules. About 50% of granules split; small granules tend to fade away, while all large granules fragment or merge. Once a granule is formed, usually it expands; its fragmentation is announced by the appearance of a dark dot in its centre, then by the formation of a dark notch which connects the dark dot to the intergranular lanes (Kitai and Kawaguchi, 1979). Exploding granules (Carlier *et al.*, 1969) are spectacular and violent extremes of the process of fragmentation. The group of granules produced from a single granule by fragmentation or merging forms a family of granules, with a typical size of 3–5" and lifetimes ranging from 5 to more than 45 min. (Kawaguchi, 1980). Oda (1984) defines as 'active granules', those experiencing repeated fragmentations and mergings. They tend to form a network (brighter than the average), of spatial scale (11") comparable to that of the meso-granulation. The clumpy structure of the active granules, which tend to expand almost simultaneously inside a group, may be the result of convective motions originating in the first ionization of the atoms in the outer envelope of the Sun; their impacts in higher layers may then become new sources of rising blobs of gas visible as granules at the surface of the Sun.

2.3. LIFETIME

The characteristic lifetime of granules is not yet accurately known, varying from less than 6 min to more than 16 min, according to the authors. It depends of the quality of the observation and of the definition of the lifetime. Those authors who define the lifetime as twice the decay time of the surviving granules from a 'master frame', found a lifetime of 16 min (Simon, 1967; Mehltretter, 1978); according to Mehltretter this value is typical for granules which fragment. Kawaguchi (1980), defining the lifetime as the time interval from the moment of formation of a granule by fragmentation or merging, until the moment of the next fragmentation or disappearance, finds a much shorter lifetime of 5.7 min. There is a correlation between the maximum size and the lifetime for those granules which do not fragment or merge: the smaller the size, the smaller the lifetime; the lifetime for granules of maximum size $0\overset{\prime\prime}{.}5$ is 2.5 min, while it is 10 min for the granules of maximum size $1\overset{\prime\prime}{.}5$ (Kawaguchi, 1980).

2.4. CONTRAST AND BRIGHTNESS FLUCTUATIONS

The contrast is defined by the quantity

$$C = 2\frac{I_{max} - I_{min}}{I_{max} + I_{min}},$$

where I_{max} and I_{min} represent the brightness of the granules and the intergranular material respectively, while δI_{rms} is the root mean square of the brightness fluctuations. At the centre of the disk, the values corrected for seeing and instrumental scattered light, obtained by various authors, fall within the range 15–35% for the contrast and 8–18% for δI_{rms}: these brightness parameters are still poorly known. The most currently adopted values are 25% for C and 14% for δI_{rms} at 5500 Å. However, Nordlund (1984a), re-evaluating the δI_{rms} of 12.8% obtained by Deubner and Mattig (1975), by fitting the observed spread function by the sum of two Lorentzians rather than two Gaussians, finds a value of 20%, which is in good agreement with its tri-dimensional numerical model (Nordlund, 1984b). δI_{rms} decreases from 4000 Å to 20 000 Å, as if granules and intergranules were radiating like black bodies (Karpinsky and Pravdjuk, 1972; Albregtsen and Hansen, 1977; Wittmann, 1979). On the other hand, the corrected contrast determined at different wavelengths is still too uncertain, to provide a reliable variation with λ (Bray, 1982).

The centre-to-limb variation of δI_{rms} is rather flat from the centre of the disk to $\mu = 0.5$, and then slowly decreasing towards the limb (Pravdjuk *et al.*, 1974; Keil, 1977; Schmidt *et al.*, 1979; Wiesmeier and Durrant, 1981; Durrant *et al.*, 1983). This decrease towards the limb is accompagnied by an apparent increase in the horizontal scale of the pattern (Schmidt *et al.*, 1979; Durrant *et al.*, 1983).

2.5. LINE INTENSITY FLUCTUATIONS AND TEMPERATURE STRUCTURE OF THE GRANULATION

The root mean square of intensity fluctuations, δI_{rms}, measured in the core of photospheric lines decreases with increasing height of formation of the lines (Keil and Canfield, 1978; Durrant *et al.*, 1981). In the strong line Mg b_2, δI_{rms} decreases towards the centre of the line (Kneer *et al.*, 1980). These line intensity fluctuations as well as continuum δI_{rms} measured at various heliocentric angles, are used to derive temperature models. They all agree to find a steep decrease of temperature fluctuations, δT_{rms}, with increasing height in the lowest layers of the photosphere, and a constant δT of about 100 K in the upper layers (Altrock, 1976; Keil and Canfield, 1978; Durrant *et al.*, 1981; Figure 3). Such a kind of model is confirmed by a coherence analysis of brightness fluctuations in the continuum and at various positions in the line Mg b_2, which sharply drops at $\Delta\lambda = 0.35$ Å from line center (Kneer *et al.*, 1980; Durrant and Nesis, 1981). Such a steep decrease reflects the penetration of temperature fluctuations of convective origin (associated to the granulation) into the lower photosphere not higher than about 80 km, presumably because they are dissipated by radiative relaxation (Kneer *et al.*, 1980; Durrant and Nesis, 1981).

2.6. LINE DOPPLER SHIFTS AND VELOCITY STRUCTURE OF THE GRANULATION

Granular velocities are measured from the Doppler effect in Fraunhofer lines: broadening, asymmetry, line shifts; on spatialy resolved spectra, the lines show the characteristic wiggly shape. The vertical structure of the velocity is usually constructed from line shifts measurements on photospheric lines of different strengths, or at different distances from the centre of strong lines. The problem is that the measured Doppler shifts are the results of several velocity fields of different origin: the granulation, the oscillations and the supergranulation. The granulation field has to be separated from the other components and this is not a trivial problem. The results are usually given in a statistical form, namely the root mean square of the velocity fluctuations versus altitude in the photosphere, $\delta v(h)$.

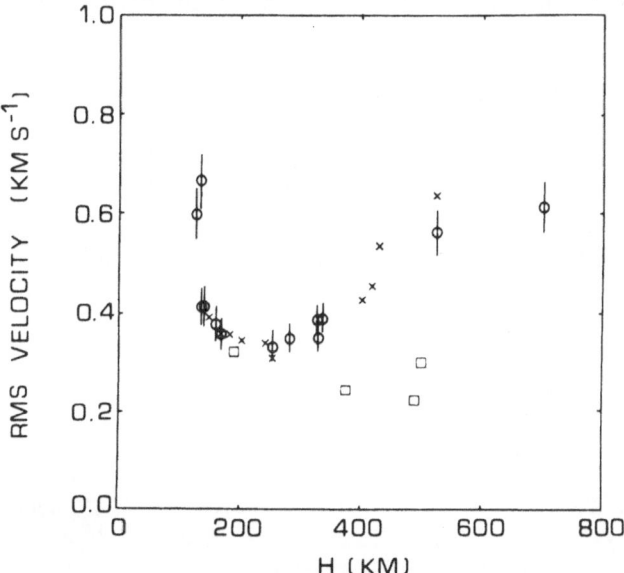

Fig. 2.   Variation with height of observed rms line shifts. (ϕ): averaged over time values derived by Keil (1980a); ($\times$): values derived from a single high spatial resolution plate by Keil and Canfield (1978); ($\square$): eclipse observations by Durrant *et al.* (1979). The plotted heights are in km above $\tau_{5000} = 1$.

Due to the limited angular resolution, even of the best spectrograms (0".75), only the behavior of granules larger than 1" can be analysed. While the temperature fluctuations vanishes at about 80 km above the continuum level, presumably by radiative relaxation, the velocity fluctuations penetrate much higher into the photosphere, at least up to 300 km, driven by the residual momentum of the ascending gas. But the height of penetration or the steepnes of the gradient of the vertical component of the velocity field fluctuations is a controversial question. Empirical models, based upon velocity fluctuations measured in lines of different heights of formation, published by the Sac Peak group (Canfield, 1976; Keil and Canfield, 1978; Keil, 1980a, b), markedly differ from those published by the Kiepenheuer Institut group (Durrant *et al.*, 1979; Bässgen and Deubner, 1982). In the first kind of models, $\delta v(h)$ decreases steeply with height (steep gradient models); no power is left above 300 km. In the second kind of models, $\delta v(h)$ decreases smoothly (flat gradient models) having still a large value at 500 km (Figure 3).

The discrepancy between these models has several causes: different observed Doppler shifts versus height (Figure 2); different methods of correction for instrumental and atmospheric blurring and, probably the main one, different procedures used to separate the convective and oscillatory components. Empirical models deduced from line asymmetries of spatially unresolved (Kaisig and Schröter, 1983) or resolved (Keil and Yacovitch, 1981; Keil, 1984) granules and intergranules favor a steep gradient model. On the other hand the coherence of vertical convective velocities (of the larger granules) is maintained at least up to 300 km in the photosphere (Durrant and Nesis, 1982; Pravdjuk, 1982; Nesis *et al.*, 1984), supporting a flat gradient model. Durrant and Nesis (1982) state that the velocity coherence and the correlation with the intensity

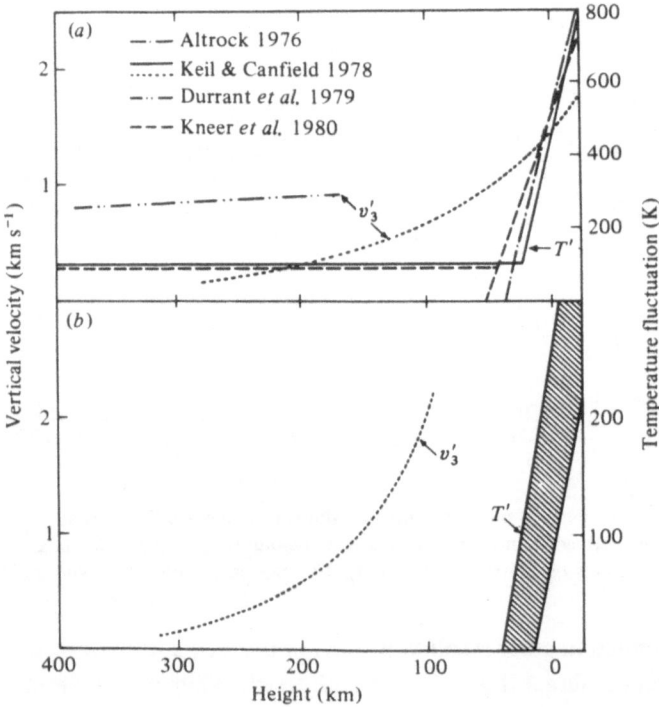

Fig. 3.   Empirical height dependence of the granular temperature fluctuations and vertical velocities in the solar atmosphere (from Bray *et al.*, 1984). (a) Rms values. (b) Amplitude (half peak-to-peak). The band of temperature fluctuations is taken from Altrock and Musman (1976), the run of vertical velocity from Keil (1980b).

immediately above the continuum of the structures of sizes in the range 2″–6″, corresponding to the largest granules and to families of granules, is maintained high up into the photosphere while it is lost much more rapidly for the smaller ones. It thus appears that the question of the height of penetration of the vertical component of the convective velocities fluctuations associated with the granulation is still open. Anyhow, its amplitude at 100 km above the continuum level (where it cannot be measured) falls, depending of the authors, in the range 1.0–2.0 km s^{-1}. Velocities up to 2 km s^{-1} (and up to 3 km s^{-1} after correction for seeing) have been measured directly in upflows and downflows on partially resolved granules and intergranules (Beckers and Morrison, 1970; Bray *et al.*, 1977; Namba *et al.*, 1983; Altrock, 1984). Downflows seem to be larger than upflows.

The horizontal component of the velocity fluctuations are deduced from centre-to-limb measurements of line Doppler shifts (Beckers and Morrison, 1970; Keil and Canfield, 1978; Mattig *et al.*, 1981). In the lower photosphere its amplitude appears to be 2 to 3 times larger than the vertical component. The horizontal velocity patterns are coherent only in the low atmosphere. They are not coherent with the vertical velocity and brightness patterns, indicating a lack of horizontal heat transport. It is not possible to conclude how the horizontal velocity and its variation with the height in the atmosphere are related to the structure of granules (Nesis *et al.*, 1984). It has to be stressed that, by lack of spatial resolution, the available empirical models refer only to the granules of size larger than 1″.5; the smaller ones, which are of particular importance for the understanding of the turbulent convection, still escape observations. Spectrographic observations of improved angular resolution, better than 0″.5, are urgently needed. In order to improve our understanding of the granulation dynamics (and consequently of the solar surface convection), the velocity structure of the low photosphere, as well as the associated temperature structure, should be described for various granular scales ranging from the smallest observable ones to about 5″, corresponding to families of granules.

2.7. SHIFT AND ASYMMETRY OF SPATIALLY UNRESOLVED PHOTOSPHERIC LINE PROFILES

Spatially unresolved line profiles are asymmetric (*C*-shape of the line bisector) and blueshifted relatively to their laboratory reference wavelength; these properties are signatures of the presence of convection, as first shown by Voigt (1956) and Schröter (1957). They are used as a mean to derive informations about the vertical and horizontal temperature and velocity structure of the photosphere (Beckers and Nelson, 1978; Dravins *et al.*, 1981; Dravins, 1982; Kaisig and Durrant 1982; Nordlung, 1982; Kaisig and Schröter, 1983; Kostik, 1983). But this is not a trivial problem, as demonstrated by Kaisig and Durrant (1982), because the depth dependences of temperature, opacity and velocity fluctuations altogether contribute to the shaping of line profiles.

Lines with cores formed at higher layers of the solar atmosphere show wavelength shifts of about the gravitational redshift: in other words they are not blue-shifted. The blue-shift progressively decreases as one moves away from the disk centre and may even become a gravitational redshift at the extreme limb: this is known as the limb effect. The limb effect is a combination of shifts of the line center and of the centre-to-limb variation of the shape (Beckers and Nelson, 1978; Brandt and Schröter, 1982).

Line profiles in partially resolved granules and intergranules also show asymmetries and shifts of convection origine (Keil and Yacovitch, 1981; Namba *et al.*, 1983; Kavetsky and O'Mara, 1984; Keil, 1984).

2.8. THEORETICAL CONSIDERATIONS

It is not the aim of this section to present an exhaustive review of the theoretical papers dealing with the granular convection, but, more restrictively, to mention a few papers which are felt important by an observer, as they try to reproduce and to explain some of the major observed features of the granulation. Several suggestions have been made to explain the apparently well defined granular scale. But is appears now that the granulation has no characteristic scale (Roudier and Muller, 1985). Nevertheless the convective granules (those driven by buoyancy) might have a characteristic scale, the increasing number of granules towards the smaller scales being a result of the turbulent cascade. Simon and Leighton (1964) have suggested that the scale of the solar granulation may be a consequence of the specific depth (2000 km) at which hydrogen is ionized. On the other hand attempts have been made to identify the observed length scale with the linear convective mode of maximum growth rate, within the frame of the mixing-length theory. These rates were found by Böhm (1963, 1967) and Spiegel (1964, 1965) to increase monotonically with increasing length scale well past the observed cut-off. Subsequently, including the pressure turbulence as well as the mechanical and thermal effects of turbulence, and taking into account the radiative exchange in the governing hydrodynamical equations, a maximum growth rate has been obtained (Narasimha *et al.*, 1980; Antia *et al.*, 1981; Narasimha and Antia, 1982; Antia *et al.*, 1983). Nevertheless it should be noted that the resulting computed length scale could be inferred from the vertical scale height of the convection zone introduced into the models. The resultant vertical velocities are found in reasonable agreement with observed granular velocities. Nordlung (1982) developed a full three-dimensional numerical simulation of the solar granulation. He solved the hydrodynamic equations, including radiative transfer equations and viscous terms, in the anelastic approximation. The results of these simulations show most of the features observed in the granulation (Dravins *et al.*, 1981; Nordlung, 1984b): granules surrounded by intergranular lanes; downward motions in the lane of relatively larger amplitude than the upward motion in the granules, granules increasing their horizontal dimension in time; breaking apart into smaller ones, which in turn grow larger and merge. Synthetic spectral lines are in good agreement with the widths, strengths and shapes of observed spectral lines. In this model, granules are driven by the buoyancy force; horizontal motions are driven by pressure gradients and the centre of granules is cooled while they expand and brake. It is to be noted that it is inherent to the model not to be able to predict a turbulent cascade towards the smaller granules. We should also mention that Nelson (1978) refined a previous two-dimensional, steady state semi-empirical model of the solar granulation (Nelson and Musman, 1977) solving the hydrodynamic equations and the transfer equation. The input of the models is a mean temperature and pressure stratification and two parameters: the mean cell size and a length which scales the turbulent drag. The two-dimensional model adequately reproduces the mean limb darkening, the magnitude and centre-to-limb variation of the intensity fluctuations, and the penetration of temperature and velocity fluctuations (of steep gradient type) into the photosphere.

3. Network Bright Points and Associated Magnetic Flux Tubes

High resolution filtergrams taken in the core of photospheric lines or in the wings of strong lines like Ca II K 3933 or Ca II H 3968 reveal a network of tiny bright elements, sub-arcsecond in size, concentrated at supergranules boundaries. The network bright elements are responsible for the so-called 'gaps' in photospheric lines (Sheeley, 1967), also present in spectra of plages (Beckers and Schröter, 1968, Stellmacher and Wiehr, 1971, 1973); they are also visible in white light near the limb. Following Mehltretter (1974) the network bright elements were called facular points (Muller, 1983, and subsequent papers) by analogy with the bright elements in faculae; but I find more appropriate the terminology recently introduced by Stenflo and Harvey (1985), 'network bright point', which will be used in this review paper, because it makes the difference between the bright points in quiet regions of the Sun and in active ones, which, although quite similar in nature, exhibit some differences (see Section 3.6).

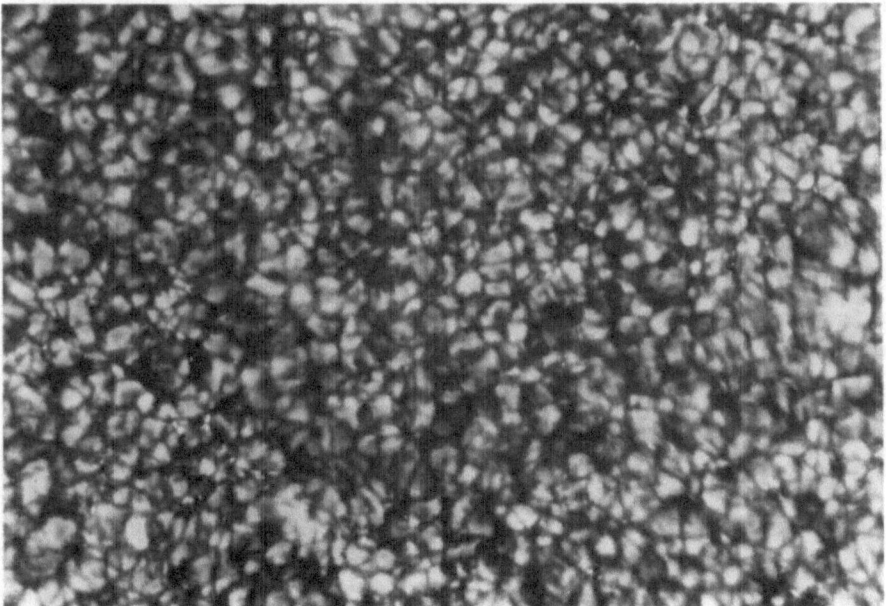

Fig. 4.   Network bright points embedded in intergranular lanes. Numerous bright points are particularly well visible near the lower left corner of the figure (Pic-du-Midi Observatory).

In the places where a large number of network bright points are gathered, they can be alined in the intergranular lanes, forming chains of several adjacent points. Such a filamentary looking feature has been called 'filigree' by its discoverers, Dunn and Zirker (1973). The filigree is more apparent in active regions and in remnants of active regions, particularly when it is observed in the wings of Hα. On white light pictures, the granulation in these regions seems to be washed out (abnormal granulation). However, on the best Pic-du-Midi pictures (resolution of 0".25), the 'abnormal granulation' still appears to have a granular structure, resulting from the intergranular lanes being filled by tiny bright points. A slight decrease of image quality is then enough to give to such a region a washed out appearance.

3.1. MORPHOLOGICAL AND EVOLUTION PROPERTIES

Network bright points (as well as bright points in active regions) are embedded in the convective pattern of the granulation, but always located in the intergranular lanes (Mehltretter, 1974; Muller, 1983; Muller and Keil, 1983); this is clearly shown on Figure 4. Their characteristic size, corrected for blurring, is found to be of 0".22 (150 km), and their characteristic white light (5750 Å) brightness of 1.3–1.5, relative to the mean photosphere (Muller and Keil, 1983). These values are consistent with those reported by Mehltretter (1974) for a few well-defined bright points observed with a 60 Å bandpass filter centered on the Ca II k 3933 Å: 0".2 and 1.43, respectively, as well as with the width of 0".25 of the 'crinkles' measured by Dunn and Zirker (1973) in the filigree. The average brightness of a few crinkles measured in the wings of Hα (+2 Å from line center) is found to be 1.10 (Dunn and Zirker, 1973), while the corrected brightness of one peculiar isolated point (probably too much peculiar to be representative of a typical point) is found as high as 2.00 by Koutchmy (1978).

A detailed description of the behaviour of the network bright points has been given by Muller (1983): relatively to the supergranular pattern, they appear at the supergranular boundaries, seldom inside the cells; relatively to the pattern of the granulation, they appear in spaces at the junction of several granules, never inside a granule nor in a space between two granules only; their mean lifetime is 18 min; their size never significantly exceeds 0".5; they remain in intergranular lanes during their whole life; they have a strong tendency to form very close to an already existing point; about 15% of them seem to split into two facular points; they disappear simply by fading away in an intergranular space; no merging with another network point was observed.

3.2. ASSOCIATION OF NETWORK BRIGHT POINTS AND MAGNETIC FLUX TUBES

Earlier observations by Chapman and Sheeley (1968) have shown a high degree of association between longitudinal magnetic fields and the observed photospheric network, with a resolution of 1"–2"; the association was later confirmed indirectly by Mehltretter (1974) and more directly by numerous authors measuring magnetic field strengths on network features with a higher resolution. The photospheric network is also closely associated with the coarser Calcium network (Chapman and Sheeley, 1968), indicating that the network elements are diverging throughout the photosphere; the associated flux tubes are also diverging due to the decrease of the gaz pressure, and nearly vertical. In view of the close association network bright points – magnetic field, it is puzzling to observe almost no bright points inside the supergranules, whereas Harvey (1977) reported a large number of inner network bipolar magnetic fields.

3.3. MAGNETIC FIELD STRENGTHS

Due to the small size of the flux tubes compared with the spatial resolution of spectrographic or magnetographic observations, which in the best cases are hardly better than 1″, it is very difficult to measure directly the true strengths of the magnetic field in the network. However, field strenghts as high as 2 kG were derived indirectly from weak fields, low resolution measurements, by Stenflo (1973), using two lines of different Landé factor belonging to the same multiplet (the so-called line-ratio technique). High field strengths in the photospheric network were confirmed subsequently by various indirect methods (Mehltretter, 1974; Chapman, 1974; Tarbell and Title, 1977; Wiehr, 1978, 1979; Solanki and Stenflo, 1984; Stenflo and Harvey, 1985). On the other hand, high field strengths have to be introduced in flux-tube models in order to reproduce line profiles (Stenflo, 1975; Koutchmy and Stellmacher, 1978; Chapman, 1977, 1979). In addition, the interpretation of those indirect observations requires the uniqueness of field strength in the network. In a few cases, magnetic field strengths higher than 1 kG have been directly measured on high resolution spectrograms (Simon and Zirker, 1974; Koutchmy and Stellmacher, 1978), or using an infrared line (Harvey and Hall, 1975). Of course, kilogauss magnetic fields are derived also directly or indirectly, in active regions; it is thus believed that the quiet network and active-region plages are made up of practically the same type of flux tubes (Frazier and Stenflo, 1972; Wiehr, 1978), although some differences in their properties may exist (see Section 3.6). The existence of a unique magnetic field strength, although widely accepted, is questionned by Semel (1985), who rather believes in a distribution of strengths, ranging from 600 to 1600 G.

3.4. FLUX-TUBE SIZE

While there is no doubt about the small size of network bright point (< 300 km), the true diameter of the associated flux tubes, at the photospheric level, is not yet clearly established. The flux-tube size inferred by indirect methods is usually comparable to the network bright-point size (Stenflo, 1973; Mehltretter, 1974; Chapman, 1974; Wiehr, 1978, 1979). On the contrary, the size measured directly on high resolution spectrograms is found to be significantly larger (1″–3″), according to Simon and Zirker (1974) and Daras–Papamargaritis and Koutchmy (1983). It is to be noted that the theoretical models predict the same (small) size for flux tubes and for associated brightenings (see Section 3.8), while Koutchmy and Stellmacher (1978) were forced to introduce in their modeling a magnetic area much larger than the flux-tube area, in order to reproduce the line weakening they observe in the network.

3.5. Mass motions associated with magnetic flux tubes

Earlier observations have shown that the photospheric lines are Doppler-shifted in the photospheric network and plages, indicating that a downdraft up to 1.5 km s^{-1} is associated with magnetic (or network) elements outside of sunspots (Beckers and Schröter, 1968; Frazier, 1970, 1974; Stenflo, 1973; Simon and Zirker, 1974; Harvey and Hall, 1975; Tarbell and Title, 1977; Frazier and Stenflo, 1978). But due to the insufficient resolution of the observations, it is not clear whether this downdraft is closely associated to the magnetic flux tubes or is simply the intergranular convective downdraft since the flux tubes are embedded into the intergranular lanes.

Subsequently, the method of the wavelength position of the zero crossing of the Stokes V profile has been used as a measure of the Doppler shift of the line profile inside the flux tubes. In that way, Giovanelli and Slaughter (1978) reported downward motions of 0.6 km s^{-1} in the deep photosphere, vanishing in the upper photosphere. This result has not been confirmed by Stenflo *et al.* (1984) and Stenflo and Harvey (1985) who, using high spectral resolution observations, did not find a downflow within the magnetic structure. They have ascribed the result of Giovanelli and Slaughter to the lack of spectral resolution of their observation which, combined with the asymmetry of the V profiles, may cause a fictitious redshift of the zero-crossing point. Koutchmy and Stellmacher (1978) reported a strong magnetic field observation in a rosette with no associated Doppler shift. The asymmetry of the V profiles indicates that there must be mass motions with a vertical gradient within the flux tubes (Stenflo *et al.*, 1984; Stenflo and Harvey, 1985); the observed downflow associated with magnetic fields could be located into the field-free plasma surrounding the tubes (Stenflo and Harvey, 1985).

The existence or not of an associated mass motion is of fundamental importance for the theory of flux tubes, but this question does not appear to be solved, at the moment, by the observations.

3.6. Centre-to-limb variation of the contrast of facular points

Centre-to-limb variations (CLV) of the continuum contrast of bright points are important in deriving empirical models and checking theoretical models of flux tubes. The contrast is defined as the brightness excess of faculae ΔI, relative to the surrounding quiet photospheric brightness I. The problem is that there is not yet a CLV of network bright points available, all the published variations referring to facular elements in active regions (which include facular points and facular granules, of size respectively smaller and larger than $0''5$). As it is usually assumed that flux-tube properties are the same in quiet and active regions (it seems that it is really the case for the magnetic field strength), facular variations are used in practice. However, some differences exist, making this use somewhat hazardous: the size distribution of bright points is wider in active than in quiet regions, extending up to $1''5$ and $0''5$ respectively (Spruit and Zwaan, 1981; Muller and Keil, 1983); the temperature excess is higher in the network than in faculae (Solanki and Stenflo, 1984). In addition there are selection effects near the limb, as only 5% of the points visible in the quiet network at the centre of the disk are still visible near the limb

(Muller and Roudier, 1984a); on the other hand the observations there may often refer not to a single point, but to a cluster of adjacent individual points. Nevertheless, because they are the only ones available and widely used to construct and to check flux tube models, we shall review hereafter centre-to-limb variations and empirical models of facular elements.

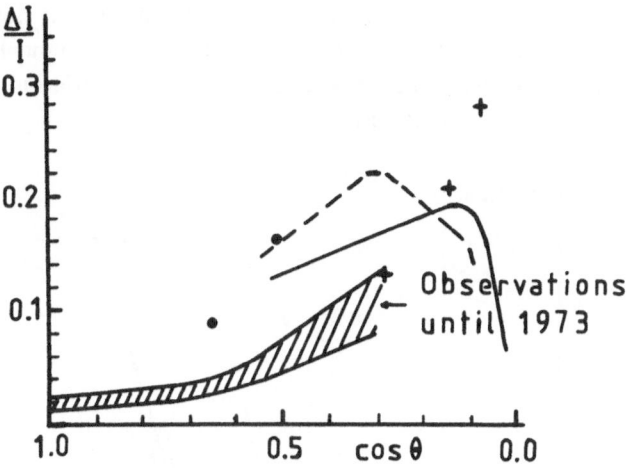

Fig. 5. Observed centre-to-limb variations of the facular contrast. (----): Muller (1975) + Hirayama (1978); (——); Libbrecht and Kuhn (1984); ●: Hirayama and Moriyama (1979); +: Chapman and Kabunde (1982), normalized to the older observations.

All measurements agree to find that the facular contrast increases with increasing $\mu = \cos\theta$ out to $\mu \approx 0.2$, from $\Delta I/I \approx 0$ at the center of the disk to $\Delta I/I \approx 0.2$ at $\mu = 0.2$. Closer to the limb, it is not yet known whether the contrast is increasing or decreasing with μ (Figure 5). While measurements made with a resolution better than $1''$ exhibit a decrease of the contrast toward the limb (Muller, 1975; Hirayama, 1978), low resolution measurements exhibit either a decrease (Badalyan and Prudkovskii, 1973; Libbrecht and Kuhn, 1984) or an increase (Muller, 1975; Chapman and Klabunde, 1982). The observed differences may result from the various observing procedures used: low and high resolution observations; measure of average excess (Chapman and Klabunde, 1982; Libbrecht and Kühn, 1984) or maximum brightness; selection of the brightest facular elements.

3.7. EMPIRICAL MODELS OF FACULAR AND NETWORK BRIGHT POINTS

Empirical models are derived either from the centre-to-limb continuum contrast variation (Chapman, 1970; Badalyan and Prudkovskii, 1973; Muller, 1975; Ingersoll and Chapman, 1975; Hirayama, 1978; Hirayama and Moriyama, 1979) or from (weakened) photospheric line profiles (Stenflo, 1975; Chapman, 1977, 1979; Koutchmy and Stellmacher, 1978; Stellmacher and Wiehr, 1979; Solanki and Stenflo, 1984); the weakening is prominent in the core but vanishing in the wings. The later class of models often make use of additional informations about continuum brightness at the center of the disk and include magnetic field and finite resolution effects on the observed line profiles. A few of them introduce a Wilson depression, which is a consequence of the presence of magnetic field; the tube is then depressed compared to the neighbouring photosphere (Chapman, 1977, 1979; Solanki and Stenflo, 1984). The later authors, instead of fitting the line profile, fit the V Stokes profile which, formed in the magnetic elements alone, is independent of the spatial resolution. In the two classes of models, flux tubes are assumed to be in hydrostatic and local thermodynamical equilibrium. It is to be noted that models derived from continuum contrast variation refer to facular elements, that is to say to the active Sun.

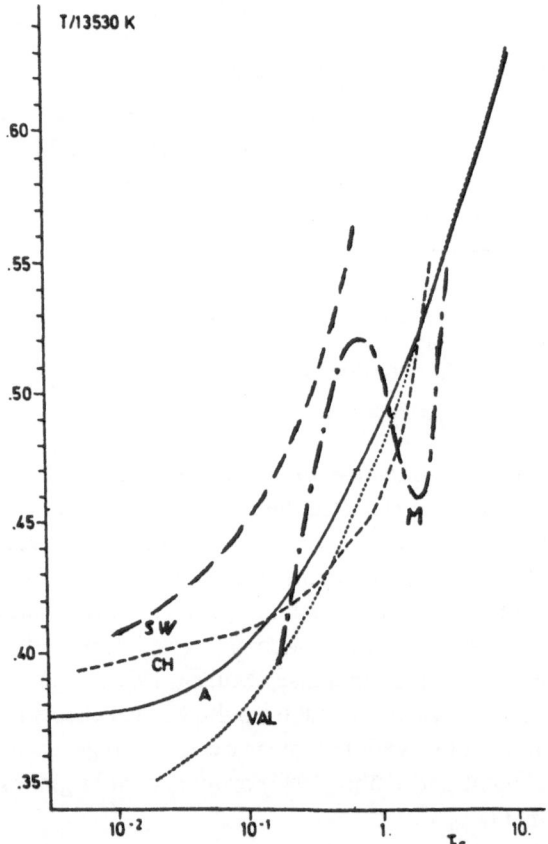

Fig. 6.   Temperature stratifications of network bright points. Empirical models of Stellmacher and Wiehr (1979) (SW) and of Chapman (1979) (CH); the axis of the theoretical model A of Deinzer *et al.* (1984) (A); VAL/Spruit model of the undisturbed atmosphere (VAL, see Deinzer *et al.*). Also shown the model of facular granules of Muller (1975) (M).

In all the empirical models, facular brightening toward the limb and line weakening are interpreted as a temperature excess ΔT in the photospheric layers, relative to the surrounding unperturbed atmosphere (Figure 6), although NLTE effects may significantly contribute to the observed brightening and weakening (Stenholm and Stenflo, 1977). In the second class of models, ΔT increases with the height above the level $\tau = 1$; ΔT, which does not exceed at few hundred K at $\tau = 1$, increases to 1000 K or more at $\tau = 10^{-4}$. Such a kind of models involve CLV increasing toward the limb. However we have seen above that NLTE effects are important and that the actual CLV is not necessary increasing beyond $\mu = 0.2$. If the actual CLV of the continuum contrast of bright points peaks and then decreases beyond $\mu = 0.2$, such a behavior may be explained by: (a) a temperature excess peaking around $\tau = 0.3$, then decreasing (Badalyan and Prudkovskii, 1973; Muller, 1975; Hirayama, 1978; Hirayama and Moriyama, 1979); (b) geometrical effects due to the foreshortening (Stellmacher and Wiehr, 1973, 1978); (c) to the enhanced visibility of the deeper (hence hotter) layers of the photosphere seen through the optically thin flux tubes (Spruit, 1976; Chapman, 1979; Deinzer et al., 1984).

None of those empirical models, in fact, is realistic and satisfying, because they are computed for LTE atmospheres, when it is well known that NLTE effects of the horizontal radiative transfer in such evacuated thin features are important, as shown by the theoretical models discussed in the next section. On the other hand, hydrostatic equilibrium is assumed, although not yet firmly established by the observations.

3.8. THEORETICAL MODELS OF FLUX TUBES

A number of detailed flux-tube models have been computed and derived quantities compared to observed properties of flux tubes, network bright points and facular elements. Magnetohydrostatic (Spruit, 1976; Osherovich et al., 1983) as well as magnetohydrodynamic models (Unno and Ribes, 1979; Deinzer et al., 1984a, b) have been considered. The tubes are partially evacuated, due to the magnetic pressure, involving a Wilson depression of 100–200 km. Radial influx of heat into such tubes of diameter less than 300 km increases the temperature inside the tube. Calculated emergent intensities show a bright ring of diameter close to the flux-tube diameter (Spruit, 1976; Caccin and Severino, 1979; Osherovich et al., 1983; Deinzer et al., 1984b); in the Deinzer et al. model, the bright ring is surrounded by an extended dark ring, from which energy is taken to heat the tube interior. When the tubes are away from the centre of the disk, their hot walls radiate through their evacuated, optically thin, interior, resulting in a brightness excess (the 'hot wall effect': Spruit, 1976; Caccin and Severino, 1979; Chapman and Gingel, 1984; Deinzer et al., 1984b). The hot wall effect accounts, at least partly, for the CLV of the facular contrast exhibiting a maximum away from the disk centre. In the model of Deinzer et al. the dark ring, by reducing the hot wall effect, points to the necessity of an additional heating process for the upper layers of the flux-tube atmosphere. The hot wall effect shows that the interpretation of the observed positive facular contrast by a simple heating is probably not valid.

In the thermodynamical model of Unno and Ribes (1979), a strong downdraft provides a consistent mechanism for the facular heating by transporting the entropy excess from the chromospheric layers. But this model does not account for the kG accepted fields, while the presence of a strong downdraft inside the tube is still a much debated question. In fact, the presence or not of a systematic downdraft is of crucial importance in understanding the physics of flux tubes, network and faculae. Finally, the Deinzer *et al.* model predicts a baroclinic downflow around the tubes, which might be related to the downflow reported by Stenflo and Harvey (1985).

4. Variation of the Properties of Solar Granulation and Photospheric Network over the Solar Cycle

The mean size of granules, measured in quiet regions of the Sun, at the disk centre, decreases with increasing activity (Macris and Rösch, 1983; Macris *et al.*, 1984; Muller and Roudier, 1984b). On the other hand, the granule-intergranule lane intensity ratio increases with increasing activity (Alissandrakis *et al.*, 1982). A decrease of the equivalent width of photospheric lines, ranging from 0 to 2.3% is observed from 1976 to 1980, that is during the ascending phase of the last cycle (Livingston and Holweger, 1982); on the other hand, bisectors of strong lines in the full-disk Fraunhofer spectrum are observed to diminish in curvature as the activity cycle proceeds from minimum (1976) to maximum (1979–1982) (Livingston, 1983, 1984).

The number of network bright points per surface unity (network density), counted at disk centre, is found to vary in antiphase to the sunspot number: it increases by a factor 3 or more from maximum to minimum of activity (Muller and Roudier, 1984a). In addition, the network density is not uniform over the surface of the Sun. As a quantum of magnetic flux is associated to each network element, density variations of the photospheric network express in fact variations of the quiet Sun magnetic flux.

The observed variation of the properties of the granulation could be a consequence of a periodic change of the structure of the convection zone in response to the 11-year cyclic variation of the solar magnetic flux. But we cannot yet decide whether this is under the influence of the global flux, the active flux or the quiet flux.

5. Conclusion

Our knowledge of the fine structure of the solar photosphere and our understanding of the origin of the small-scale features, have considerably progressed with increasing resolution of the observations. However a still improved resolution is needed, especially of spectrographic observations, in order to solve a number of yet open crucial questions. Concerning the granulation, the (turbulent?) origin of the smaller granules and the height of penetration of granules of various sizes into the photophere are probably among the most important ones. Concerning the network bright points and associated flux tubes, we would like to know particularly: whether systematic flows, or oscillating flows, are associated or not with magnetic flux tubes; the time evolution of the magnetic field strength and of the flows and how they interact with the convective patterns; the actual magnetic and temperature structure.

References

Albregsten, F. and Hansen, T. L.: 1977, *Solar Phys.* **54**, 31.

Alissandrakis, C. E., Macris, C. J., and Zachariadis, G.: 1982, *Solar Phys.* **76**, 129.

Altrock, R. C.: 1976, *Solar Phys.* **47**, 517.

Altrock, R. C.: 1984, in S. L. Keil (ed.), *Small-Scale Dynamical Processes in the Solar and Stellar Atmosphere*, Sacramento Peak Observatory, Sunspot, New Mexico.

Antia, H. M., Chitre, S. M., and Pandey, S. K.: 1981, *Solar Phys.* **70**, 67.

Antia, H. M., Chitre, S. M., and Narasimha, O.: 1983, *Monthly Notices Roy. Astron. Soc.* **204**, 865.

Badalyan, O. G. and Prudkovskii, A. G.: 1973, *Soviet Astron.* **17**, 356.

Bässgen, M. and Deubner, F. L.: 1982, *Astron. Astrophys.* **111**, L1.

Beckers, J. M. and Morrison, R. A.: 1970, *Solar Phys.* **14**, 280.

Beckers, J. M. and Nelson, G. D.: 1978, *Solar Phys.* **58**, 243.

Beckers, J. M. and Schröter, E. H.: 1968, *Solar Phys.* **4**, 142.

Böhm, K. H.: 1963, *Astrophys. J.* **137**, 881.

Böhm, K. H.: 1967, in R. N. Thomas (ed.), 'Aerodynamic Phenomena in Stellar Atmospheres', *IAU Symp.* **28**, 366.

Brandt, P. N. and Schröter, E. H.: 1982, *Solar Phys.* **79**, 3.

Bray, R. J.: 1982, *Solar Phys.* **77**, 299.

Bray, R. J., Loughead, R. E., and Tappere, E. J.: 1976, *Solar Phys.* **54**, 319.

Bray, R. J., Loughead, R. E., and Durrant, C. J.: 1984, *The Solar Granulation*, 2nd edition, Cambridge University Press.

Caccin, B. and Severino, G.: 1979, *Astrophys. J.* **232**, 297.

Canfield, R. C.: 1976, *Solar Phys.* **50**, 239.

Carlier, A., Chauveau, F., Hugon, M., and Rösch, J.: 1968, *Compt. Rend. Acad. Sci. Paris* **226**, 199.

Chapman, G. A.: 1970, *Solar Phys.* **14**, 315.

Chapman, G. A.: 1970, *Astrophys. J.* **191**, 255.

Chapman, G. A.: 1970, *Astrophys. J. Suppl.* **33**, 35.

Chapman, G. A.: 1970, *Astrophys. J.* **232**, 923.

Chapman, G. A. and Gingell, T. W.: 1984, *Solar Phys.* **91**, 243.

Chapman, G. A. and Klabunde, D. P.: 1982, *Astrophys. J.* **261**, 387.

Chapman, G. A. and Sheeley, N. R.: 1968, *Solar Phys.* **5**, 442.

Daras–Papamargaritis, H. and Koutchmy, S.: 1983, *Astron. Astrophys.* **125**, 280.

Deinzer, W., Hensler, G., Schüssler, M., and Weisshaar, E.: 1984a, *Astron. Astrophys.* **139**, 426.

Deinzer, W., Hensler, G., Schüssler, M., and Weisshaar, E.: 1984b, *Astron. Astrophys.* **139**, 435.

Deubner, F. L. and Mattig, W.: 1975, *Astron. Astrophys.* **45**, 167.

Dravins, D.: 1982, *Ann. Rev. Astron. Astrophys.* **20**, 61.

Dravins, D., Lindegren, L., and Nordlung, A.: 1981, *Astron. Astrophys.* **96**, 345.

Dunn, R. B. and Zirker, J. B.: 1973, *Solar Phys.* **33**, 281.

Durrant, C. J. and Nesis, A.: 1981, *Astron. Astrophys.* **95**, 221.

Durrant, C. J. and Nesis, A.: 1982, *Astron. Astrophys.* **111**, 272.

Durrant, C. J., Mattig, W., Nesis, A., Reiss, G., and Schmidt, W.: 1979, *Solar Phys.* **61**, 251.

Durrant, C. J., Kneer, F., and Maluck, G.: 1981, *Astron. Astrophys.* **104**, 211.

Durrant, C. J., Mattig, W., Nesis, A., and Schmidt, W.: 1983, *Astron. Astrophys.* **123**, 319.

Frazier, E. N.: 1970, *Solar Phys.* **14**, 89.

Frazier, E. N.: 1970, *Solar Phys.* **38**, 69.

Frazier, E. N. and Stenflo, J. O.: 1972, *Solar Phys.* **27**, 330.

Frazier, E. N. and Stenflo, J. O.: 1978, *Astron. Astrophys.* **70**, 789.

Giovanelli, R. G. and Slaughter, C.: 1978, *Solar Phys.* **57**, 255.

Harvey, J. W.: 1977, in E. A. Müller (ed.), *Highlights of Astronomy*, D. Reidel Publ. Co., Dordrecht, Holland, p. 223.

Harvey, J. W. and Hall, D.: 1975, *Bull. Ann. Astron. Soc.* **7**, 459.

Hirayama, T.: 1978, *Publ. Astron. Soc. Japan* **30**, 337.

Hirayama, T. and Moriyama, F.: 1979, *Solar Phys.* **63**, 251.

Ingersoll, A. P. and Chapman, G. A.: 1975, *Solar Phys.* **42**, 279.

Kaisig, M. and Durrant, C. J.: 1982, *Astron. Astrophys.* **116**, 332.

Kaisig, M. and Schröter, E. H.: 1983, *Astron. Astrophys.* **117**, 305.

Karpinsky, V. N.: 1985, in R. Muller (ed.), *High Resolution in Solar Physics*, 8th Regional Meeting of the IAU, Toulouse, France.

Karpinsky, V. N. and Pravdjuk, L. M.: 1972, *Soln. Dann.*, No. 10, 79.

Kavetsky, A. and O'Mara, B.: 1984, *Solar Phys.* **92**, 47.

Kawaguchi, I.: 1980, *Solar Phys.* **65**, 207.

Keil, S. L.: 1977, *Solar Phys.* **53**, 359.

Keil, S. L.: 1980a, *Astrophys. J.* **237**, 1024.

Keil, S. L.: 1980b, *Astrophys. J.* **237**, 1035.

Keil, S. L.: 1984, in S. L. Keil (ed.), *Small-Scale Dynamical Processes in the Solar and Stellar Atmospheres*, Sacramento Peak Observatory, Sunspot, New Mexico, p. 149.

Keil, S. L. and Canfield, R. C.: 1978, *Astron. Astrophys.* **70**, 169.

Keil, S. L. and Yacovitch, F. H.: 1981, *Solar Phys.* **69**, 213.

Kitai, R. and Kawaguchi, I.: 1979, *Solar Phys.* **64**, 3.

Kneer, F. J., Mattig, W., Nesis, A., and Werner, W.: 1980, *Solar Phys.* **68**, 31.

Kostik, R.: 1983, *Publications of the Academy of Science of Ukraine* **83**, 63.

Koutchmy, S.: 1978, *Astron. Astrophys.* **61**, 397.

Koutchmy, S. and Stellmacher, G.: 1978, *Astron. Astrophys.* **67**, 93.

Libbrecht, K. G. and Kuhn, J. R.: 1984, *Astrophys. J.* **277**, 889.

Livingston, W.: 1983, in J. O. Stenflo (ed.), 'Solar and Stellar Magnetic Field', *IAU Symp.* **71**, 149.

Livingston, W.: 1984, in S. L. Keil (ed.), *Small-Scale Dynamical Processes in the Solar and Stellar Atmosphere*, Sacramento Peak Observatory, Sunspot, New Mexico, p. 330.

Livingston, W. and Holweger, H.: 1982, *Astrophys. J.* **252**, 375.

Lovejoy, S.: 1982, *Science* **216**, 185.

Macris, C. J. and Rösch, J.: 1983, *Compt. Rend. Acad. Sci. Paris* **296**, 265.

Macris, C. J., Muller, R. Rösch, J., and Roudier, Th.: 1984, in S. Keil (ed.), *Small-Scale Dynamical Processes in the Solar and Stellar Atmospheres*, Sacramento Peak Observatory, Sunspot, New Mexico, p. 265.

Mandelbrot, B.: 1977, *Fractals*, Freeman, San Francisco.

Mattig, W., Mehltretter, J. P., and Nesis, A.: 1981, *Astron. Astrophys.* **96**, 96.

Mehltretter, J. P.: 1974, *Solar Phys.* **38**, 43.

Mehltretter, J. P.: 1978, *Astron. Astrophys.* **62**, 311.

Muller, R.: 1975, *Solar Phys.* **45**, 105.

Muller, R.: 1983, *Solar Phys.* **85**, 113.

Muller, R. and Keil, S. L.: 1983, *Solar Phys.* **87**, 243.

Muller, R. and Roudier, Th.: 1984a, *Solar Phys.* **94**, 33.

Muller, R. and Roudier, Th.: 1984b, *4th European Meeting on Solar Physics*, Noordwijk, The Netherlands, p. 239.

Namba, O. and Diemel, W. E.: 1969, *Solar Phys.* **7**, 167.

Namba, O., Hafkenscheid, G. A., and Koyama, S.: 1984, *Astron. Astrophys.* **117**, 277.

Narasimha, D. and Antia, H. M.: 1982, *Astrophys. J.* **262**, 358.

Narasimha, D., Pandey, S. K., and Chitre, S. M.: 1980, *J. Astrophys. Astron.* **1**, 165.

Nelson, G. D.: 1978, *Solar Phys.* **60**, 5.

Nelson, G. D. and Musman, S.: 1977, *Astrophys. J.* **214**, 912.

Nesis, A., Durrant, C. J., and Mattig, W.: 1984, in S. L. Keil (ed.), *Small-Scale Dynamical Processes in the Solar and Stellar Atmospheres*, Sacramento Peak Observatory, Sunspot, New Mexico, p. 243.

Nordlund, Å.: 1982, *Astron. Astrophys.* **107**, 1.

Nordlund, Å.: 1984a, in S. L. Keil (ed.), *Small-Scale Dynamical Processes in the Solar and Stellar Atmospheres*, Sacramento Peak Observatory, Sunspot, New Mexico, p. 174.

Nordlund, Å.: 1984b, in S. L. Keil (ed.), *Small-Scale Dynamical Processes in the Solar and Stellar Atmospheres*, Sacramento Peak Observatory, Sunspot, New Mexico, p. 181.

Oda, N.: 1984, *Solar Phys.* **93**, 243.

Osherovich, V. A., Fla, T., and Chapman, G. A.: 1983, *Astrophys. J.* **268**, 412.

Pravdjuk, L. M.: 1982, *Soln. Dann.*, No. 2, 103.

Pravdjuk, L. M., Karpinsky, V. N., and Andreiko, A. V.: 1974, *Soln. Dann.*, No. 2, 70.

Rösch, J.: 1962, in D. H. Sadler (ed.), *Transactions of International Astronomical Union*, Vol. 11B, London Academic Press.

Roudier, T. and Muller, R.: 1985, submitted to *Solar Phys.*

Schmidt, W., Deubner, F. L., Mattig, W., and Mehltretter, J. P.: 1979, *Astron. Astrophys.* **75**, 223.

Schröter, E. H.: 1957, *Z. Astrophys.* **41**, 141.

Semel, M.: 1985, in R. Muller (ed.), *High Resolution in Solar Physics*, 8th Regional Meeting of the IAU, Toulouse, France, p. 180.

Sheeley, N. R.: 1967, *Solar Phys.* **1**, 171.

Simon, G. W.: 1967, *Z. Astrophys.* **65**, 345.

Simon, G. W. and Leighton, R. B.: 1964, *Astrophys. J.* **140**, 1120.

Simon, G. W. and Zirker, J. B.: 1974, *Solar Phys.* **35**, 331.

Solanki, S. K. and Stenflo, J. O.: 1984, *Astron. Astrophys.* **140**, 185.

Spiegel, E. A.: 1964, *Astrophys. J.* **139**, 959.

Spiegel, E. A.: 1965, *Astrophys. J.* **141**, 1068.

Spruit, H. C.: 1976, *Solar Phys.* **50**, 269.

Spruit, H. C. and Zwaan, C.: 1981, *Solar Phys.* **70**, 207.

Stellmacher, G. and Wiehr, E.: 1971, *Solar Phys.* **18**, 220.

Stellmacher, G. and Wiehr, E.: 1973, *Astron. Astrophys.* **29**, 13.

Stellmacher, G. and Wiehr, E.: 1979, *Astron. Astrophys.* **58**, 273.

Stenflo, J. O.: 1973, *Solar Phys.* **32**, 41.

Stenflo, J. O.: 1975, *Solar Phys.* **42**, 79.

Stenflo, J. O. and Harvey, J. W.: 1985, *Solar Phys.* **95**, 99.

Stenflo, J. O., Harvey, J. W., Brault, J. W., and Solanki, S.: 1984, *Astron. Astrophys.* **131**, 333.

Stenholm, L. G. and Stenflo, J. O.: 1977, *Astron. Astrophys.* **58**, 273.

Tarbell, T. D. and Title, A. M.: 1977, *Solar Phys.* **52**, 13.

Unno, W. and Ribes, E.: 1979, *Astron. Astrophys.* **73**, 314.

Voigt, H. H.: 1956, *Z. Astrophys.* **40**, 157.

Wiehr, E.: 1978, *Astron. Astrophys.* **69**, 279.

Wiehr, E.: 1979, *Astron. Astrophys.* **73**, L19.

Wiesmeier, A. and Durrant, C. J.: 1981, *Astron. Astrophys.* **104**, 207.

Wittmann, A.: 1979, in Proceedings of the Freiburg Colloquium *Small Scale Motions on the Sun*, p. 29, Freiburg.

THE CHROMOSPHERE AND TRANSITION REGION –
CURRENT STATUS AND FUTURE DIRECTIONS OF MODELS

R. GRANT ATHAY

High Altitude Observatory, National Center for Atmospheric Research*, Boulder, CO 80307, U.S.A.

Abstract. This essay on the chromosphere and transition region begins with a general discussion of mechanically heated atmospheres and goes on to discuss both descriptive and physical models. The descriptive models include the thermodynamic and fluid dynamic properties of the atmosphere as deduced from observations. Particular features of the models are identified with the properties of the radiation loss associated with the ionization of hydrogen and with the properties of thermal conduction. The role of spicules in chromosphere and transition region properties is emphasized. Physical models that attempt to predict the basic features of the descriptive models are reviewed.

1. Perspective

1.1. MECHANICALLY HEATED ATMOSPHERES

The Sun's outer atmosphere is heated by nonradiative processes commonly referred to as 'mechanical'. It is widely believed that the mechanical heating derives its origin either directly from the convection zone or through coupling between the convection zone and differential rotation. In this paper, we discuss those portions of the outer solar atmosphere in which the dominant energy loss is by radiation in spectral lines, viz., the chromosphere and the transition region to the corona. Our concern is with the physical system represented by these layers and not with a detailed review of the many contributions found in individual papers. The intent, therefore, is to focus the discussion on what the chromosphere and corona tell us about mechanically heated, radiatively cooled atmospheres with emphasis on where the problem stands and what the future directions should be.

Regardless of whether heat is transferred to the outer atmosphere by waves or other mechanisms, the very presence of mechanical heating carries certain implications concerning the nature of the atmosphere. Among the more obvious characteristics expected for mechanically heated atmospheres are the following:

(a) the heating rate will fluctuate both spatially and temporally;

(b) the fluctuations in heating rate will produce correlated changes in radiation intensity, which will give rise to a variety of fluctuating structural features; and

(c) the differential heating will give rise to gradients in gas pressure resulting in widespread motions, both vertical and horizontal.

These anticipated characteristics, taken in a broad sense, describe much of what we observe in the chromosphere, transition region and corona. Horizontal and vertical motions are widespread, and intricate and fluctuating structural patterns cover the entire solar disk.

* The National Center for Atmospheric Research is operated by the University Corporation for Atmospheric Research under sponsorship of the National Science Foundation.

The challenge of studying a remote, constantly changing atmosphere is one that must be guided by a carefully charted course. In the coming decade we face the prospect of observing the Sun with unprecedented angular resolution. It is clear from current data that we have not yet resolved the fine structure of the solar atmosphere. Hence, it is highly probable that observations made with much improved resolution will both reshape our concepts of features that are currently identified and reveal new classes of features previously unrecognized.

It is incumbent on us as solar physicsts to make proper and efficient use of high-resolution data when it becomes available. The temptation to pursue 'solar dermatology' by which I mean the study of fine structure without regard for its importance to a major problem in solar physics, must be deliberately avoided. Otherwise, we run the risk of having little of significance to show for our efforts and little to justify our continued work.

Perhaps no one will disagree with the preceding plea to concentrate research on problems of recognized importance rather than proceeding on an unstructured approach. Also, we will perhaps mostly agree on at least a few of the major problems of solar physics. Where we will disagree most, sometimes sharply, is in identifying those aspects of observable phenomena that have greatest bearing on the major problems. However, the difficulty of making such identifications only emphasizes the need to make them. It does not provide an excuse for following a haphazard course.

1.2. MAJOR PROBLEMS

Having stated the case for a planned approach to high-resolution solar physics, I will suggest what I consider to be some of the key elements of such a plan. Certain major objectives are readily identified as follows:

(i) What are the primary mechanisms for heat deposition and energy transport?

(ii) Once energy is dissipated from the primary heating mechanism, what is its disposition? What fraction is radiated locally and what fraction is transported elsewhere before being radiated?

(iii) What gives rise to momentum? How much of it is inherent in the heating mechanisms and how much arises from pressure gradients due to differential heating?

(iv) How is energy stored in magnetic fields before its catastrophic release in solar flares?

(v) What gives rise to the cyclic variation in the solar output?

(vi) What is the internal rotation and structure of the Sun?

These are fundamental questions. The purpose in defining them is to establish the guide posts and the goals towards which we are working. Thus, our objective is not to construct a model atmosphere or to accurately describe such phenomena as sunspots, plages and spicules, or even flares. Rather, our objective is to study these phenomena for the purpose of reaching the more fundamental goals.

The major goals of solar physics, by definition, are too broad to be the immediate objective of individual research efforts. It is both instructive and practical, however, to attempt to relate more narrowly defined and more readily reachable research objectives leading to one of the major goals. In the following, I will illustrate some elements of such an approach beginning with studies of global properties.

1.3. GLOBAL STRUCTURE

In the past, we have expended considerable effort in establishing standard reference models for the solar atmosphere. Such models average out all spatial fluctuations and, equivalently, all temporal fluctuations, except those of long term. Thus, they average over different heating rates and possibly over different physical mechanisms of heating. Based on current empirical knowledge of chromospheric and transition region structure, it appears that any given location in the atmosphere may be near the average condition for that particular height only a small fraction of the time. If the fine structure is associated with large amplitude fluctuations in thermodynamic variables, as it clearly is in the case of spicules, for example, major excursions from the mean may be commonplace. It is necessary, therefore, that we understand the excursions from the mean as well as the mean itself.

None of the foregoing remarks implies that average models are not valuable. We must have such models to provide the baseline from which we judge the importance of individual fluctuations. On the other hand, the average model is not extremely useful as a means of reaching any of the major goals stated in Section 1.2 unless we also understand the individual components of which the average is composed. For example, the radiation loss at a given location in the chromosphere, which provides the measure of the heat input, depends critically upon both the local electron density and temperature and the overlying atmosphere as well. Thus, the average model may provide a poor guide to the magnitude of local heating. Similar reservations can be voiced concerning the utility of average models concerning heat transport and momentum. In this respect, it is noteworthy that the extensive modeling program of the Center for Astrophysics has expanded to include a number of different classes of solar features (Vernazza *et al.*, 1981).

In a somewhat different vein, the standard chromospheric models exhibit a thin temperature plateau near 2×10^4 K in which most of the Lα flux is generated. In contrast with this thin plateau picture, the Lα spectroheliograms obtained by Bonnet *et al.* (1980) show a pattern of fibrils and loops. Similarly, Feldman (1983) has drawn attention to the clear evidence of an unresolved fine structure with considerable vertical extent at temperatures below about 10^5 K. It seems inescapable, therefore, that the Lα plateau in the standard models is partly an artifact of the plane-parallel assumption employed in constructing the models. What, then, is the real structure?

In still a different context, we must ask the question as to whether any part of the outer solar atmosphere is in quasi-equilibrium. Or is it that the mechanical heating leads to disequilibrium either because heating fluctuations are too large and too short lived or because the heating mechanism itself precludes the possibility of establishing a balance of forces. These questions need to be answered, but it is unlikely that they will be answered using average models. Thus, we re-iterate the conclusion that the global models must be accompanied by models of the primary elements of which the average is constituted.

1.4. MESOSCALE STRUCTURE

In the category of mesoscale structure, we include such features as large sunspots, plages, supergranule cells, and network. The structure is representative of the large-scale features of the magnetic field and the supergranule circulation cells. Sunspots are unique in that their thermodynamic structure in the lower atmosphere departs sharply from the average conditions. In each of the remaining cases, the thermodynamic properties are much nearer the global averages.

Certainly in the case of sunspots, energy balance is achieved differently than elsewhere in the atmosphere, and possibly different energy mechanisms are present. For plages, supergranule cells and the network, however, it is unclear whether their different radiation properties are indicative of different heating mechanisms or whether a single mechanism is present but to a different degree in each case. The tendency of the plages and network to lie selectively in regions of magnetic field concentration whereas the supergranule cells are more typified by weak magnetic fields suggests that different heating mechanisms may be present in the strong field regions. This is by no means a necessary conclusion, however.

It is probable that the mesoscale features themselves do not represent the primary departures from the mean properties of the global atmosphere. In other words, the network, supergranule cells, and plages appear to have different radiative signatures more because of the different make-up of their fine structure than because of differences in their large scale properties. The evidence for this lies in the observed amplitudes of intensity fluctuations. The relatively large amplitude of the fine structure fluctuations in such spectral lines as $H\alpha$ and K tend to obscure the larger scale intensity fluctuations associated with the network and supergranule cells to such an extent that the latter structure generally was unrecognized prior to the observation of definitive velocity patterns in the supergranule cells by Leighton *et al.* (1962). Sunspots are an exception in the sense that the entire sunspot may deviate more from the global average than even the large amplitude fluctuations in quiet Sun regions.

A corollary to the argument that the mesoscale structure does not represent the largest deviations from the average radiative brightness is that the mesoscale structure per se is not the primary fluctuation in the heat input. To the contrary, we have strong reasons to suppose that it is the fine structure of the atmosphere that represents the largest fluctuation in heat input. It is plausible, therefore, that the fine structure contains essential clues related to the mechanical heating.

On the other hand, it is a mistake to study fine structure out of context from the mesoscale structure. The very existence of mesoscale structure is evidence that the fine structure is collectively organized. In certain respects, the organizational patterns override even the mesoscale structure. Unipolar magnetic regions, for example, cover many supergranule cells. The same is true to even greater degree in coronal holes. One cannot assume, therefore, that all mesoscale features of a given class are the same just as one cannot assume that the fine structure is independent of the mesoscale feature in which it is located.

1.5. Fine Structure

The intricate and all pervasive fine structure of the chromosphere and transition region probably contains the essential clues to the mechanical heating of the solar atmosphere. Many of the important physical interactions of the plasma and gases of the solar atmosphere with the magnetic field, including such phenomena as the pressure fluctuations in compressive waves and magnetic reconnection, occur on spatial scales comparable to or shorter than those of the observed fine structure. Observational studies of the energy input mechanisms, therefore, must be carried out at high spatial resolution.

As noted in the preceding discussion, the study of a constantly fluctuating atmosphere presents a challenge as to how to proceed. Certain aspects of the problem are clear, however. Since both motions and magnetic fields are prevalent, the basic structures are magnetohydrodynamic in character. This requires that measurements of magnetic fields, fluid velocities and the pressure forces driving the flow be closely integrated. Furthermore, the dominant energy loss is by radiation, which requires that measurements of temperature and density be integrated as closely as possible with the magnetic and velocity observations.

It is not necessary, of course, that we study every feature of the fine structure. What is necessary is to study representative samples of each class of fine structure in each of the mesoscale features. By specifying the thermodynamic and MHD structure as a function of time, we can hopefully identify the important physical processes at work. This hope is dependent, however, on our ability to determine physical parameters such as magnetic field and velocity vectors and temperature and pressure with sufficient precision. In order to do so, we may have to develop more accurate diagnostic methods than we are accustomed to, particularly in the case of the magnetic and velocity fields.

Thus, we face a difficult challenge, but a rewarding one if we do it properly. High resolution studies must be carried out if we are to make progress in understanding the primary physics of the solar atmosphere. The challenge to be kept before us is to discover and understand the physics and to avoid the temptation of doing whatever happens to be easiest to do.

2. Descriptive Models

2.1. Physical Variables

Often in solar physics we refer to descriptions of the physical variables (e.g., temperature, pressure, etc.) in a system as a 'model'. We continue this usage in this manuscript, but we broaden the meaning to include physical models that include a specified set of physical processes together with the set of physical variables they produce. The most direct route to a physical model usually starts with a set of observables, which are then converted to a set of physical variables. From the physical variables, one attempts to identify the mechanisms producing them either by trial and error or by deductive reasoning. In those cases where theory successfully predicts the physical variables, we are fortunate. It was not too difficult, for example, to deduce that the photosphere is governed mainly by the conditions of radiative, hydrostatic and local thermodynamic equilibrium. All that remained to specify the model was to properly identify and quantify all important sources of opacity (a task that is not yet completed). The same process has not been sufficient for sunspots, however. Nor do we have such models for the chromosphere, transition region and corona. In each case, there are important physical processes that are still unspecified.

In a mechanically heated, fluctuating atmosphere, there are couplings between physical variables that are not present in static atmosphere. The state of motion, which is now a physical variable, influences the local radiation loss thereby coupling with the local temperature. In a different vein, the presence of a magnetic field can alter the nonradiative energy transport and gas pressure, which, in turn, couple with the radiation loss and, hence, the temperature. These are only examples of a variety of possible couplings, but they serve to illustrate the need to obtain descriptions of the physical variables that are as complete as possible rather than attempt a piecemeal approach. The only reason we have been able to follow the piecemeal approach in the past is that we have been forced by lack of spatial resolution to work with average properties that hide the detailed correlation between the various physical parameters. We cannot ignore such correlations in the high-resolution studies, however.

2.2. CHROMOSPHERE

Much progress has been made in recent years in developing thermodynamic models of the global and mesoscale chromospheric structures. Such models owe their existence to two major developments in solar physics: (i) the availability of infrared data and far ultraviolet spectral data from space experiments, and (ii) the availability of large computers for solving the non-LTE radiative transfer equations. Most extant models are still based on the plane parallel atmosphere approximation and the assumptions of hydrostatic equilibrium and steady state.

For purposes of this review, we are interested in what we can learn from the models concerning the physical processes that influence the chromosphere. In particular, we will restrict our attention to features of the models that are not likely to be seriously modified by the assumptions imposed on the models. Models by different authors, such as Basri et al. (1979) and Vernazza et al. (1981), differ in detail but are similar in overall characteristics. Again, our concern will be with the salient features rather than the details, and we will make no attempt, therefore, to reconcile differences between various models.

Figure 1 shows the temperature versus height ($h = 0$ corresponds to $\tau_{5000} = 1$) for the global, network and cell models of Vernazaa et al. (1981). Figure 2 shows a similar plot for the sunspot model of Lites and Skumanich (1982) and the plage model of Basri et al. (1979). The global model from Figure 1 is replotted in Figure 2 for comparison. Those familiar with solar observations will recognize that the mesoscale features differ substantially within a given species. Different sunspots, for example, may have very different intensities in chromospheric spectral lines, which indicates that their temper-

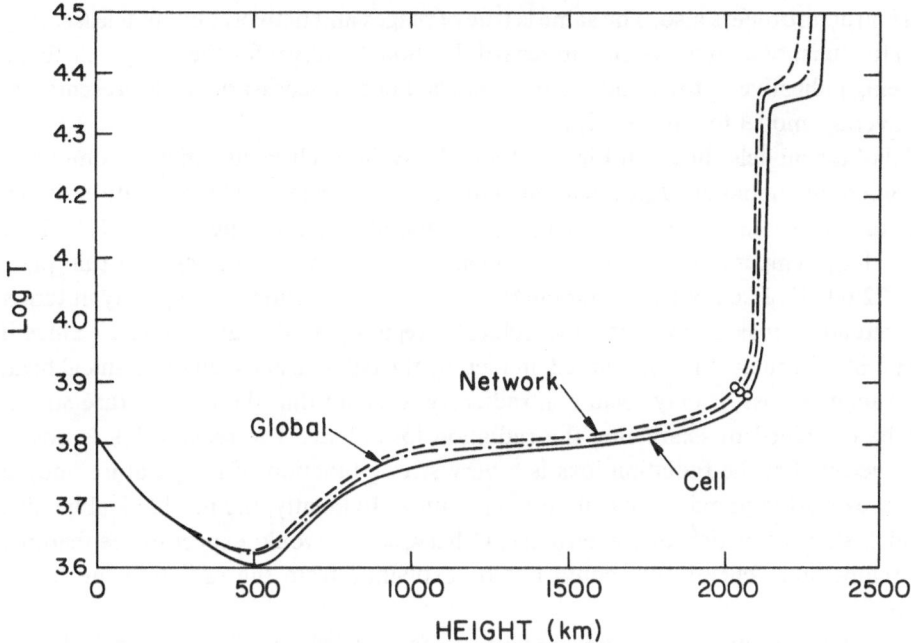

Fig. 1.   Chromospheric temperature models for the average (global) Sun, supergranule cells and network from Vernazza *et al.* (1981). The circled points on the curves indicates depths at which hydrogen is 50% ionized.

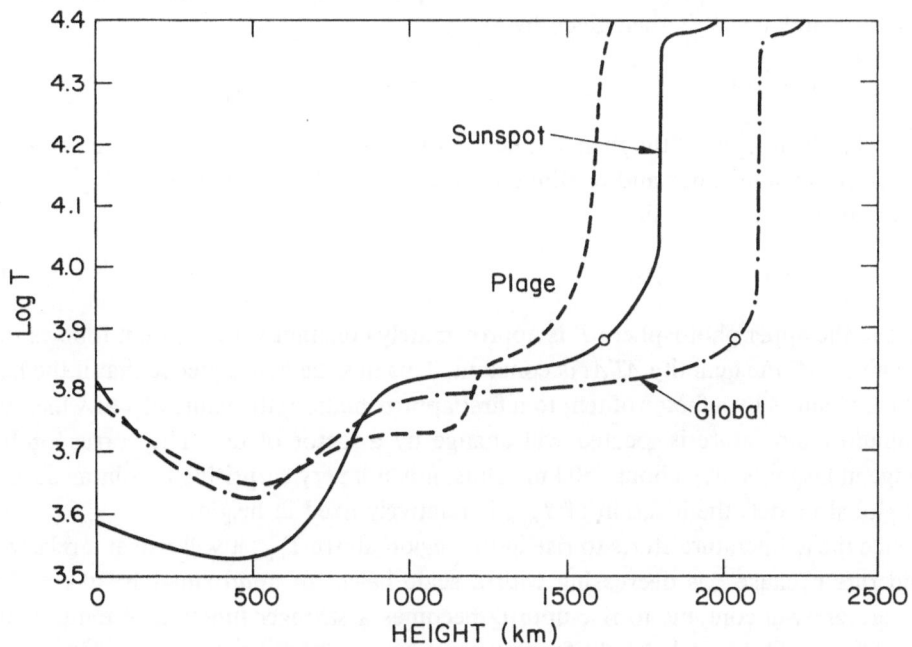

Fig. 2.   Representative chromospheric temperature curves for a plage (Basri *et al.*, 1979) and a sunspot (Lites and Skumanich, 1982). The global model from Figure 1 is shown for comparison. The circled points indicate the depths at which hydrogen is 50% ionized.

ature structure differs also. The same is true of plages and network, but to a lesser degree than for sunspots. Thus, we should regard the models shown for the mesoscale features as being indicative of the trends in the models, but not necessarily as representative of the average model for the species.

All of the models shown in Figures 1 and 2 have four characteristics in common: (1) a temperature minimum T_{min}, near 500 km, (2) a moderately fast temperature rise from T_{min} to approximately 5500–6000 K, (3) a temperature plateau near 6000–7000 K, and (4) a sharp temperature rise beginning near 8000 K and terminating in a thin plateau near 22 000 K. Each of these four characteristics can be understood purely in terms of the radiation loss rates from the relevant regions of the atmosphere. Since the temperature structure is determined, in fact, by the balance between mechanical heating and radiation loss, it may seem contradictory to claim that the temperature structure can be explained by examining the radiation loss alone. The reason this is possible, however, is that the radiation loss is a very strong function of temperature and, as a result, exerts the primary control on temperature. Evidently, the mechanical heating is a much slower function of temperature. Otherwise, we would expect to see features in the temperature curves that could not be explained from the radiation mechanisms alone.

The temperature minimum region occurs near continuum optical depth $\tau_{5000} = 3 \times 10^{-4}$. Radiation losses from this region include a mixture of H^- continuum and spectral lines of atoms, ions and molecules. For simplicity, we consider the continuum and the lines to be in LTE and optically thin. The radiative loss rate Q between τ and $\tau + \Delta\tau$ is then given by

$$Q = B\Delta\tau, \tag{1}$$

where B is the integral of the Planck function over frequency and $\Delta\tau$ is the mean change in opacity including lines and continuum. Replacing B by σT^4 and assuming $\Delta\tau$ to be a slow function of T, we find

$$\Delta Q = 4\sigma T^3 \Delta T \Delta\tau. \tag{2}$$

Since in the upper photosphere T is approximately constant with depth, it follows that for a given ΔQ the quantity $\Delta T \Delta\tau$ is constant. This has the consequence that if the heat input L changes by a factor of ten, to a first approximation, the value of τ at which the minimum temperature is located will change by a factor of ten. The corresponding change in height is only about 150 km. Thus, it is not very surprising that in mesoscale and global models the location of T_{min} is relatively fixed in height.

Once the temperature starts to rise in the region above T_{min} it will rise at a relatively rapid rate because τ is decreasing with a scale height of approximately 70 km. The temperature will continue to rise until Q becomes a stronger function of temperature than T^4 or until the scale height for the mean opacity undergoes a major increase. It is primarily the former condition that limits the initial temperature rise.

In the region above T_{min} the spectral lines are predominantly formed in non-LTE, and the radiation loss rate due to lines is given by

$$Q = \sum_i h\nu_i C^i_{LU} n^i_L,$$ (3)

where C^i_{LU} is the collisional excitation rate from the lower to the upper level of the spectral transition and n_L is the population density of the lower state. To a first approximation C^i_{LU} may be written in the form

$$C^i_{LU} \sim n_e \exp[-h\nu_i/kT],$$ (4)

where n_e is the electron density.

Two effects tend to make C^i_{LU} a very strong function of temperature. Firstly, most of the spectral lines lie in the violet end of the spectrum, which makes the exponential term a strong function of temperature. Secondly, chromospheric electrons come mainly from the ionization of hydrogen once the temperature rises above about 5500 K. As soon as this occurs, n_e becomes a rapidly increasing function of temperature. The result is that even a small rise in T will produce a large increase in Q. This provides a strong thermostatic effect and severely limits the rate of temperature rise (Athay, 1981b).

The thermostatic effect of hydrogen ionization begins at a temperature near 5500 K and continues until hydrogen is mainly ionized. Once hydrogen is mainly ionized, n_e becomes a slow function of temperature and the thermostatic effect is strongly diminished. This is illustrated by the open circles placed on the curves in Figures 1 and 2 at the locations where hydrogen is half ionized. In all cases, the temperature gradient increases markedly just above the circled points.

At the depths where hydrogen is half ionized, the atmosphere is optically thin at all wavelengths except at radio wavelengths and in the resonance series of hydrogen and helium. Since these are the most abundant elements, their resonance lines are potentially important sources of additional radiation loss. The chromosphere becomes effectively thin in Lα at a temperature near 20 000 K, and the escaping Lα photons create a second thermostatic effect in the upper layers of the chromosphere.

The second thermostatic plateau near 22 000 K differs fundamentally from the lower plateau. Most of the temperature dependence of Q in the lower plateau arises from the n_e term. However, in the upper plateau n_e varies inversely with temperature (gas pressure is constant) and the temperature control arises solely from the factor $n_1 \exp[-h\nu_i/kT]$. To illustrate the approximate temperature dependence of this factor, we write the equilibrium equations for hydrogen in the approximate forms

$$n_1 A_{12} = n_2 A_{21},$$ (5)

and

$$n_2(A_{21} + A_{2c}) = n_c A_{c2} + n_1 A_{12},$$ (6)

which reduce to

$$n_2 A_{2c} = n_c A_{c2}$$ (7)

and

$$n_1 = \frac{n_c A_{c2} A_{21}}{A_{2c} A_{12}}.$$ (8)

The A_{ij}'s represent the radiative excitation and de-excitation rates and level c denotes the continuum state.

Since A_{21} and A_{2c} are constants and both n_c and A_{c2} are proportional to n_e (we neglect the slow dependence of A_{c2} on T), we can replace Equation (8) with

$$n_1 \sim \frac{n_e^2}{A_{12}}. \tag{9}$$

In the effectively thin regime ($\tau \ll (A_{21}/C_{12}) e^{-h\nu/kT}$), we may represent A_{12} as

$$A_{12} = C_{12}N_s, \tag{10}$$

where N_s is the mean number of scatterings before escape. For plane parallel geometries (cf. Hummer, 1964), N_s is proportional to n_1. Hence, from (9) and (10) we find

$$n_1 C_{12} \sim n_e C_{12}^{1/2}. \tag{11}$$

It follows from Equations (3) and (4) that

$$Q \sim n_e^{3/2} \exp[-\tfrac{1}{2}(h\nu/kT)], \tag{12}$$

and if we set $n_e T \approx$ const., we find

$$Q \sim \frac{1}{T^{3/2}} \exp[-\tfrac{1}{2}(h\nu/kT)]. \tag{13}$$

This still leaves Q as a moderately strong function of T. For example, an increase in T from 20 000 to 40 000 K is accompanied by an increase in Q by a factor of 6.4. On the other hand, further increases in T produce much smaller relative gains in Q.

It is clear from the preceding arguments that Lα has a large enough thermostatic effect to produce at least a minor temperature plateau. Thus, it appears that the plateaus near 22 000 K in Figures 1 and 2 can be at least partially explained by the onset of Lα photon escape. There is, however, a difficulty with this interpretation.

The problem with the upper temperature plateau is in the energy balance. Figure 3 shows a plot of Q versus height for the global model in Figures 1, as computed by Vernazza *et al.* (1981). It is clear from this plot that in the temperature plateau near 6000 K there is a logical expectation for a balance between Q and L suggesting that L reaches a maximum in the low chromosphere then decreases more-or-less exponentially with height. Are we to believe, however, that L increases suddenly and fortuitously by an order of magnitude in order to accommodate the rapid escape of Lα photons? Before attempting to discuss this issue further, it is useful to consider the temperature region beyond the Lα plateau, viz., the transition region.

2.3. TRANSITION REGION

The interpretation of spectral data from the transition region is simplified by the optical thinness of the transition region but complicated by the complexity of its structure. Following the well-known method of analysis beginning with Equation (3), we convert

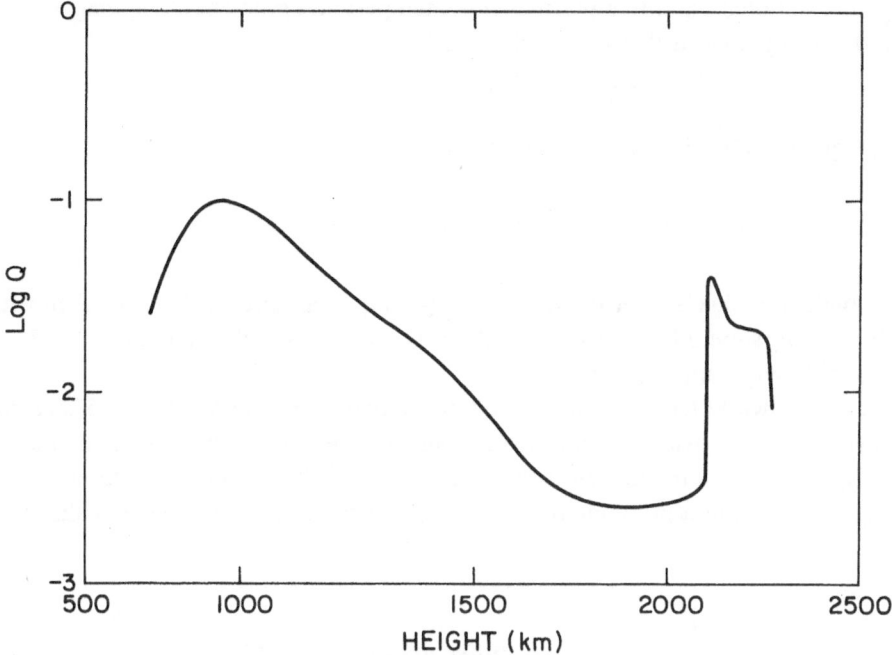

Fig. 3.   The net energy lost by radiation for the global model shown in Figure 1. Below 2000 km most of energy loss is in spectral lines of Mg II and Ca II. Above 2000 km Lα is the dominant radiation. The plot could be misleading below 1000 km because of the combined effects of many lines that have been omitted from the computations.

observed spectral line intensities into a differential emission measure E defined by

$$E = n_e^2 (d \ln T/dh)^{-1}. \tag{14}$$

Each line is then assigned a temperature that characterizes the maximum relative abundance of the particular ion producing the line. In this way it is possible to construct a curve giving $E(T)$. It was soon recognized by this process that the transition region is geometrically thin with only a small decrease in gas pressure across it. This allows the approximation

$$P_0 = n_e T = \text{const.}, \tag{15}$$

from which Equation (14) becomes

$$E(T) = P_0^2 T^{-2} (d \ln T/dh)^{-1}. \tag{16}$$

There are two ways of interpreting the term $d \ln T/dh$. If the atmosphere is plane-parallel, $d \ln T/dh$ is a true temperature gradient. On the other hand, if the atmosphere is highly structured dh must be interpreted as a measure of the volume of space in which $\ln T$ lies between $\ln T$ and $\ln T + d \ln T$, and $d \ln T/dh$ can no longer be interpreted as a temperature gradient.

The question of whether $d \ln T/dh$ represents a temperature gradient can be tested using observational data. In a plasma with a temperature gradient along magnetic field lines, the energy flux in thermal conduction is

$$F_c = 1.1 \times 10^{-6}\, T^{7/2}\, d \ln T/dh. \tag{17}$$

Thus, Equation (16) can be rewritten as

$$E(T) = 1.1 \times 10^{-6} \frac{P_0^2 T^{3/2}}{F_c}. \tag{18}$$

If the conduction flux is large compared to $\int Q\, dh$ over an interval h_1 to h_2, then within that interval F_c should be relatively constant. Hence, we should expect $E(T) \sim T^{3/2}$ to a reasonably good approximation.

Figure 4 shows $E(T)$ as a function of T for network and supergranule cells as given by Raymond and Doyle (1981). Between temperatures of 10^5 and 10^6 K the network data give $Q \sim T^{3/2}$ and the cell center data give $Q \sim T^{2.2}$. These results are clearly suggestive that the interpretation of $d \ln T/dh$ as a temperature gradient is realistic. The

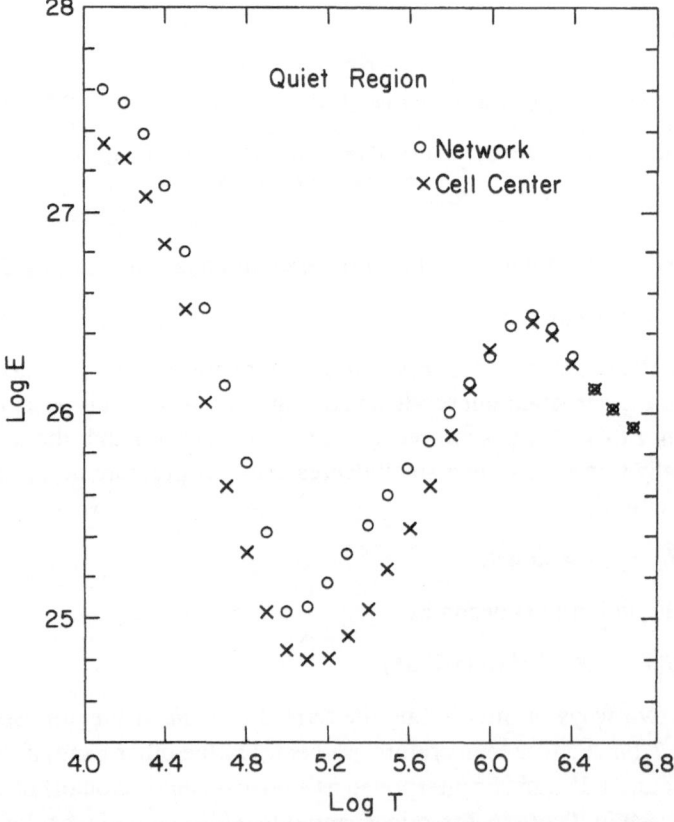

Fig. 4.   Plots of the quiet Sun emission measure as a function of T for network and supergranule cells after Raymond and Doyle (1981). Between $\log T = 5.2$ and 6.2 the network curve is broadly consistent with a constant thermal conduction flux of approximately 10^6 ergs cm^{-2} s^{-1}.

conductive flux indicated by the network branch of the curve is near 10^6 ergs cm^{-2} s^{-1}, which, indeed, is much larger than the observed radiation flux in this temperature range.

By the same logic, the branches of the $E(T)$ curves in Figure 4 for $T > 10^6$ K and $T < 10^5$ K are inconsistent with the assumption of a plane-parallel atmosphere carrying a large energy flux as thermal conduction. The coronal portion of $E(T)$ $(T > 10^6$ K) is readily understood in terms of a structured corona in which the material at temperatures above the average coronal temperature T_c occupies a decreasing fraction of the total volume as the temperature increases. This includes the case for which T_c is the maximum temperature reached. Unfortunately, no such simple explanation accounts for the rapid rise in $E(T)$ with decreasing temperature below 10^5 K.

It is instructive to use the observed $E(T)$ together with Equation (16) to determine values of $\Delta h(T)$. The results are shown in Figure 5 as solid lines for both network and

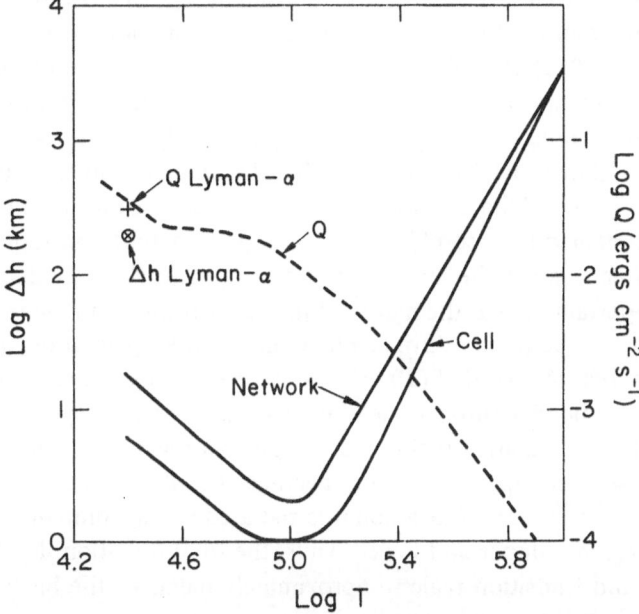

Fig. 5.   Plots of Δh derived from $E(T)$ curves shown in Figure 4 and the theoretical radiation loss rate Q from Summers and McWhirter (1979). The separate points plotted for Lα are taken from the global models shown in Figures 1 and 3.

cells. Also shown is the value of Δh for the Lα plateau in the global VAL model. Note that Lα appears to be unique in that the Δh required to match observation is an order of magnitude larger than the average Δh derived for other lines said to be formed at the same temperature. It is within the freedom of models of the VAL-type to change the value of Δh to some extent. By concentrating the Lα emitting volume in regions of local density enhancement and cooler temperatures, the mean Δh would be reduced. However, the Lα plateau influence other spectral lines, notably the C II lines near $\lambda 1335$, and this limits the types of models that can be considered.

We also show in Figure 5 a segment of the theoretical radiation loss function Q from Summers and McWhirter (1979). The computed Lα loss rate for the VAL global model is indicated by the + sign, and is in good agreement with the theoretical curve. However, the segment of the theoretical curve in Figure 5 between log $T = 4.3$ and 4.5 has been recomputed using the Lα collisional excitation rate given by Athay *et al.* (1975). The Summers and McWhriter curve rises much higher at log $T = 4.3$.

One cannot help but note the striking dissimilarity between the Q curve in Figure 5 and the $E(T)$ curves in Figure 4. The arguments advanced in discussing the chromospheric model lead to the expectation that the largest emission measures should correlate with the largest values of Q. Such a correlation exists up to $T = 10^5$ K, but reverses to a negative correlation at greater T. This dichotomy suggests a change in the basic physical mechanisms producing the observed structure.

We have already noted that the upper transition region ($T > 10^5$ K) is consistent with a thermal conduction model. A second mechanism advanced by Pneuman and Kopp (1978) is the flow of enthalpy carried by subsiding high temperature matter. Either model, or a combination of the two, can produce an emission measure similar to that observed (Athay, 1981a). It is of interest to note, that the thermal conduction mechanism relies on energy transport to establish the temperature structure rather than energy deposition. Also, the enthalpy flow model depends only on the radiative cooling of pre-heated coronal material. This suggests that the difference in structure between the upper and lower halves of the transition region arises because the upper transition region structure is determined by a combination of energy transport and radiative cooling of pre-heated matter whereas the lower half is determined by energy deposition.

The most important clue to the nature of the lower transition region, and very likely to the Lα region of the upper chromosphere, lies in the upper transition region. The inference from the observed $E(T)$ curves of conduction energy fluxes of order 10^6 ergs cm^{-2} s^{-1} flowing through the upper transition region suggests that the energy source for the lower transition region and Lα plateau may be by conduction from the corona aided by enthalpy flow. The average Lα flux from the solar disk is 3×10^5 ergs cm^{-2} s^{-1}. A similar amount is radiated in the combined lines formed in the transition region – upper and lower. Thus, the total radiation flux from the upper chromosphere and transition region approximately balances the back flow of energy from the corona.

Although there is no problem with balancing the total conduction flux against radiation losses, there is a major problem in distribution. The challenge is how to get the bulk of the conduction energy into the Lα region. We will return to this problem in the following section on physical models. For the moment, we simply draw attention to the particular issues raised by the observations, viz., the evidence for two distinct layers of primary heating – one in the lower chromosphere and one in the corona – with the intervening region heated by conduction back from the corona. There is no established need for primary heating in the upper chromosphere or transition region. Indeed, the addition of primary heating to these intervening layers, as has been suggested by some authors, may only compound the problem of total energy balance.

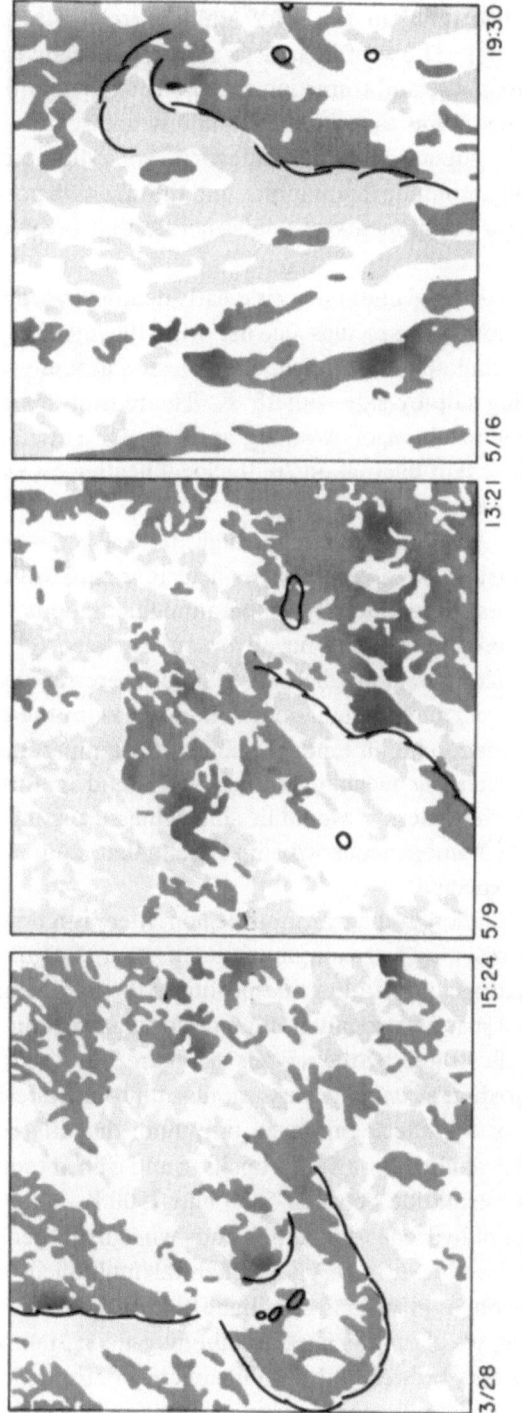

Fig. 6. Carbon IV Dopplergrams from the UVSP experiment on SMM. Solar north is at the top: east is to the left. The grey shading indicates velocities in excess of 4 km s^{-1}; dark grey is blue shift and light grey is red shift. Sunspots umbrae are indicated by the smaller encircled areas and Hα dark filaments are shown as tufted lines. The radius vectors and azimuth angles (measured counterclockwise from solar north) are 0.8, 51° (left raster); 0.69, 294° (center raster); and 0.88, 101° (left raster). Dates (1980) and times are indicated under the figures. (After Athay *et al.*, 1986.)

2.4. MULTICOMPONENT MODELS

Thermal multiplicity (bifurcation) in the solar atmosphere has been suggested in a number of different contexts. Those of interest in this paper relate to the temperature minimum region, chromosphere and transition region. The type of multiplicity referred to is a large amplitude fluctuation associated with fine structure rather than mesoscale structure of the network, supergranule cell category. Solar fine structure obviously includes a variety of small amplitude fluctuations, and the question posed by multiplicity is whether there are major components of fine structure that involve large amplitude changes in T.

The only case for which thermal multiplicity is clearly demonstrated is in the transition region where relatively cool $H\alpha$ spicules coexist with the much hotter interspicule plasma. It is highly likely that additional multiplicity occurs at these same heights with, say, 10^5 K plasma existing side-by-side with 10^6 K. The reason this is highly probable is that there are no strong mechanisms working against such structure, whereas there are good reasons to believe that fluctuations in the local heating rate will tend to create temperature inhomogeneities.

Magnetic fields in the solar atmosphere strongly reduce cross-field thermal conduction and plasma diffusion. As a result, cross-field thermal gradients can be long-lived. Also, radiation effects tend to amplify temperature fluctuations. The dashed curve in Figure 5 giving Q as a function of T has negative slope between $2 \times 10^4 \leq T \leq 10^6$ K. Because of the negative slope, regions of lower T cool more rapidly than regions of higher T. We note, however, that thermal conduction works in the opposite sense to dampen temperature fluctuations. This tends to stabilize the temperature structure along the field lines. Also, the heating mechanism itself may tend to further stabilize the temperature structure. Nevertheless, it would be somewhat surprising if the mechanical heating rate were spatially homogeneous, which suggests that multicomponent temperature structure is to be expected.

By contrast with the situation in the transition region, there is a powerful mechanism in the chromosphere for resisting large amplitude temperature fluctuations. The strong thermostatic effect of hydrogen ionization that produces the thick temperature plateau near 6000–7000 K works just as strongly in the horizontal direction as in the vertical direction. This is clearlt illustrated by the models in Figure 2 in which plages, sunspots and the global chromosphere exhibit very similar chromospheric structure. To appreciate the strength of the effect, one need only note that the collision rate C_{LU} occuring in Equation (3) evaluated for a 4 eV transition and a fixed total density changes by about four orders of magnitude between 5500 and 7500 K.

The thermostatic effect of hydrogen-ionization ends when hydrogen is predominantly ionized. The effect this produces in terms of thermal multiplicity is to reduce the thickness of the chromosphere in regions of enhanced heating and extend it in regions of reduced heating. Thus, we should expect an inhomogeneous temperature structure in the sense that the boundary between the chromosphere and transition region is highly convoluted. This can, and probably does, occur even on the scale of the fine structure.

Nevertheless, the middle layers of the chromosphere are very likely to be relatively homogeneous in temperature.

The only observational evidence, to the author's knowledge, that has been advanced in favor of temperature multiplicity in the chromosphere is the presence of helium lines in the middle chromospheric spectrum. Eclipse observations show strong emission in the lines of both He I and He II down to heights well below 1500 km on the scale shown in Figures 1 and 2. Hirayama and Irie (1984) have recently demonstrated that the widths of the He I lines are consistent with the widths of hydrogen and metal lines formed at the same heights suggesting quite unambiguously that they are formed in a similar temperature regime. This is consistent with the assumption that the excitation mechanism in He I is primarily the diffusion of EUV photons from the transition region.

It cannot be concluded, however, that local hot spots do not occur in the middle chromosphere. EUV and X-radiation is highly localized on the solar disk, and this localization will be reflected in the middle chromosphere in the density of energetic photons diffusing downwards. In this respect, the analysis of the width of the He II line at $\lambda 4686$ by Hirayama and Irie (1984) suggests that the line is formed at temperature near 20 000 K in the middle chromosphere. Again, this temperature is too low to produce the He II emission by thermal excitation and suggests that the local hot spots are most likely produced by energetic photons penetrating from above. The scale of this structure in the chromosphere is expected to be larger than the scale of the EUV and X-ray fluctuations in the transition region and corona because of photon diffusion. Thus, it may be considerably coarser than the normal fine structure of the chromosphere.

Thermal multiplicity in the temperature minimum region remains an open question. At these depths, free electrons come mainly from metal ionization and n_e is not strongly dependent on temperature. Thus, the thermostatic effect present in the chromosphere is not operative in the temperature minimum region.

Observations of the CO spectrum suggest a considerably different temperature structure than shown in Figure 1. In the model proposed by Ayres and Testerman (1981), temperature decreases monotonically outwards reaching values below 4000 K at heights above 1000 km in the VAL model. Thus, there is no chromosphere in the usual sense.

The interpretation of the CO spectrum by Ayres and Testerman (1981) is based on the assumption of LTE. Admittedly there are good arguments to support this assumption in the case of the ground electronic states observed by Ayres and Testerman. Nevertheless, we should take note of the fact that nearly all strong lines lead to models with monotonically decreasing temperature in the chromospheric layers when LTE is assumed. It is only when LTE is abandoned that the chromospheric structure emerges. This is not a definitive argument against LTE in molecular bands, but it raises a warning that should not be ignored.

Further discussion of possible temperature multiplicity in the temperature minimum region will be given in Section 3.

2.5. Dynamics

Limitations on the length of this article do not allow an in-depth discussion of chromospheric and transition region dynamics. We have already noted that these regions are expected to be dynamic and, indeed, observations bear this out. Here, we wish to draw attention to two particular aspects of the motions; viz., mass balance and large scale organization.

The questions raised by the inferred upwards mass flux associated with spicules are long standing and among the most basic facing those who study the chromosphere and transition region. Questions regarding the return downward mass flux were partially resolved with the discovery of an average downflow in the transition region over the network (cf. Doschek *et al.*, 1976). Both the spicule upflow and the network downflow indicate average mass flows of about 1.5×10^{15} protons cm^{-2} s^{-1}. Since this exceeds the net mass flux in the solar wind by some two orders of magnitude, the network downflow very nearly cancels the spicule upflow.

An important aspect of the vertical flow is that the upflow appears to be at a temperature near 16 000 K (Beckers, 1972) whereas the downflow is observed at temperatures an order of magnitude higher. A second important aspect is that the upflow is concentrated in high velocity jets whereas the downflow is more diffuse and at lower velocity. Since the network where the flows are observed are regions of high field strength, the flow undoubtedly follows magnetic lines of force. Thus, continuity requires that the downflow be longer-lived than the upflow. The composite picture has a short duration spicule outburst followed by a gradual downflow. During the course of the ejection the material is apparently heated so that the return downflow is observed at higher temperatures.

The above picture, of course, is only descriptive and gives no insight to the mechanism accelerating the spicules outwards. A number of mechanisms have been suggested but will not be reviewed here. It is worthy of note, however, that the domain in which the spicules are observed is the upper chromosphere and transition region where a number of other problems are unresolved. The estimated energy flux required to lift the spicular material against gravity is approximately 6×10^6 ergs cm^{-2} s^{-1} averaged over the solar surface (Athay and Holzer, 1982). This is higher than either the downward conduction flux or the total radiation loss from the chromosphere and transition region. Thus, in terms of the overall energy balance the spicules play a significant role.

A surprising feature of the velocity field in the transition region is the existence of large scale patterns of predominantly horizontal flow near active regions (cf. Athay *et al.*, 1982). Figure 6 shows a C IV Dopplergram made with the UVSP experiment on the SMM satellite. Blue shifts in excess of 4 km s^{-1} are shaded dark grey and red shifts in excess of 4 km s^{-1} are shaded light grey. The positions of sunspots, and prominent Hα dark filaments are indicated. The remarkable association of the dark filaments with velocity reversal lines is characteristic of several such examples (Athay *et al.*, 1986). In cases where the dark filaments parallel the line of sight, the oppositely directed velocities either side of the filaments persists indicating that the motion has a significant component of flow paralleling the filament. Thus, the filament apparently resides in a zone of velocity shear and, presumedly, magnetic shear.

3. Physical Models

3.1. TEMPERATURE MINIMUM

Ayres (1981) has drawn attention to an opacity effect associated with carbon monoxide concentration in the upper photosphere that may have major consequences on the thermal structure in these layers. The basis of Ayres argument is that the increased association of CO molecules with decreasing temperature leads to strong surface cooling in the outer layers of the atmosphere that can be overcome only by heating substantially in excess of that needed for the chromosphere. By implication, therefore regions of insufficient heating will be cooled much below the level found in radiative equilibrium models that omit the CO opacity.

The details of the arguments for excess cooling given by Ayres (1981) are based on an incorrect interpretation of how the cooling occurs. In non-LTE the cooling rate is given by Equation (3), and there is no reason to suppose that either C_{LU} or n_L is decreased from its LTE value. Indeed, it has been shown by detailed computations (Athay, 1970) that the cooling effect due to collisions is just as strong in non-LTE as in LTE.

Although, Ayres uses the assumption of LTE to justify his arguments, it is not necessary to do so. As long as there is an abundance of CO, there will be strong cooling. To fully understand the influence of this cooling on the atmosphere, however, will require a careful treatment of all the major sources of opacity. It is not clear, at this time, whether any extant solar model meets the necessary requirements so the question of temperature structure remains open.

3.2. CHROMOSPHERE

Kneer (1983) has argued recently that CO formation in the temperature minimum region produces a radiatively driven instability that precludes the possibility of radiative equilibrium. The essence of the argument is that d κ_{CO}/dT is negative, where κ_{CO} is the opacity due to carbon monoxide. Thus, a decrease in T results in an increase in κ_{CO}, which further decreases T through surface cooling effects.

Kneer postulates that the thermal instability resulting from d $\kappa_{CO}/dT < 0$ results in a transfer of radiation flux into kinetic energy flux, which subsequently heats the chromosphere. No specific mechanisms by which the depleted radiation energy is first transferred to kinetic energy then dissipated as heat in the chromosphere are suggested. In spite of the uncertainties in the reality of the effects and in the existence of appropriate energy transfer mechanisms, Kneer's suggestion poses interesting and challenging possibilities.

The more classical mechanism for chromospheric heating is by compressive waves generated in the convection zone. In a recent review, Ulmschneider and Stein (1982) conclude that heating by compressive waves in consistent with the mesoscale structure of the chromosphere. Sound waves are generated in supergranule cells where magnetic fields are weak through quadrapole interactions in the convection zone. Since quadrapole wave generation is relatively inefficient, chromospheric heating over the supergranule cells is minimal. In the strong magnetic fields of the network, however, slow-mode MHD waves, which are compressive, are generated by monopole interactions. This more efficient conversion of convective motion into wave energy produces enhanced heating in the network regions, as required by the VAL models.

The attractiveness of the wave heating hypothesis remains largely qualitative, however. Most of the heating is believed to be done by waves whose frequencies are too high to allow their ready detection. In order for the waves to dissipate, they must travel at least a wavelength from their place of origin, which has the consequence for high frequency waves that the range of depths over which a spectral line forms includes a substantial fraction of a wavelength. This makes it difficult to measure wave amplitudes and energy fluxes. Additionally, the theory of wave generation by turbulent convection leaves considerable uncertainty in the energy fluxes.

Whatever its origin, there are certain aspects of chromospheric heating that are worthy of note. On the one hand, the heating is enhanced by the presence of strong magnetic fields as indicated by the network, plage and, possibly, sunspot models. On the other hand, the heating has a major component that is independent of either solar latitude or of phase in the sunspot cycle. The polar chromosphere has much the same appearance as the equatorial chromosphere, and, aside from active regions, the sunspot maximum chromosphere is little different from the sunspot minimum chromosphere. This suggests that the heating mechanism is enhanced somewhat by strong magnetic fields but does not require magnetic fields generally. The wave heating mechanism predicts just these effects and, therefore, remains very attractive.

3.3. TRANSITION REGION

Much of the work on transition region modelling has been done in the context of coronal loop geometries. Since the basic physical processes are to a large extent independent of the particular magnetic geometry and since there is a separate review of coronal structure in this issue, the problems peculiar to coronal loops will not be reviewed here.

Considerable work was done in the 1970's attempting to construct transition region models including energy dissipation from propagating sound waves and shock waves. Although such models still hold some promise at chromospheric heights, they encounter major difficulties in the transition region (cf. McWhirter et al., 1975; Vanbeveren and De Loore, 1976; and Leibacher et al., 1982). For details, the reader should consult the indicated references.

It seems clear from observational data that the upper transition region is heated by thermal conduction from the corona in combination with a downward flow of enthalpy (cf. Dupree and Goldberg, 1967; Pneuman and Kopp, 1978). Also, the conduction-enthalpy models work as well as might be expected when the energy flow is constrained to follow magnetic lines of force even in such extreme cases as isolated magnetic loops (cf. Withbroe, 1981) or the rapidly diverging fields of the chromospheric network

(Gabriel, 1978; Athay, 1981a). There is little incentive at the present time, therefore, to seek alternative mechanisms.

This is not to argue that all problems associated with the upper transition region are solved, however. Why, for example, is there so little difference in the transition region from coronal holes to closed field regions. Both the coronal pressure and the magnetic field differ markedly in the two cases but somehow conspire to leave the transition region relatively unmolested. Until we understand such problems, we cannot be content with the models.

In contrast with the situation in the upper transition region, there is little, if any, concensus on the physical mechanisms operative in the lower transition region. One thing seems evident, however. If we accept the conduction model for the upper transition region, we must accept the possibility of a downward conduction flux that is ample for the lower transition region, including $L\alpha$. Thus, it is not at all clear that we should seek mechanisms that produce additional heating.

Very recently, three independent suggestions have been made for explaining the behaviour of the emission measure $E(T)$ in the lower transition region. Rabin and Moore (1984) proposed a model consisting of finely filamented electric currents flowing in magnetic loops. Through a suitable choice of parameters they are able to reproduce $E(T)$ satisfactorily. The two main objections to the model are that the origin of the currents is not specified and the surplus conduction energy flux entering the lower transition region is ignored.

Antiochos (1984) proposes a geometrical model in which most of the lower transition region emission occurs in magnetic loops whose maximum temperature is less than 10^5 K. Apparently, such loops do not suffer from the thermal instability problems of hotter coronal loops, and, in addition, have the capability for reproducing $E(T)$ for $T < 10^5$ K. Again, the mechanism for heating the loops is unspecified.

Athay (1984) has shown that the shape of $E(T)$ in the lower transition region is consistent with the radiation from a system of initially cool filaments (possibly loops) heated at a constant rate. The value of $E(T)$ is then proportional to the time required for heating between T and $T + \Delta T$. Because of the increased value of Q at lower temperature (Figure 5) a greater time is required for heating, which results in a negative slope for $E(T)$ as observed. Also, the magnitude of $E(T)$ is consistent with the mass flux ejected in spicules, if it is assumed that it is the spicular material that is being heated. A possible scenario is one in which the thermal conduction flux from the upper transition region passes through a thin lower transition region and is then consumed in heating the upper chromospheric gases to transition region temperatures. Such a situation will undoubtedly give rise to dynamic effects as suggested in the first volume of *Solar Physics* by Kuperus and Athay (1967).

In this connection, Athay and Holzer (1982) have suggested that the lifting of spicular material represents work done against the solar atmosphere in the form of gravitational potential energy, and that the work done will eventually appear as heat energy. As noted in Section 2 of this paper, the average rate of doing work in lifting spicule material exceeds either the conduction energy from the corona or the net radiation loss from the chromosphere and transition region. It is very likely, therefore, that the questions of spicule acceleration and energy balance in the transition region are intimately connected. Once again, however, the specific mechanisms are unclear.

Finally, we note that Shoub (1983) has proposed, as a result of detailed computations, that the shape of $E(T)$ in the lower transition region may be explained by non-LTE in the free electrons distribution function. An attempt by Owocki and Canfield (1985) to model the transition region using an approximation to the non-LTE heat conduction gives encouraging results in a qualitative sense. Quantitative applications are still uncertain, however.

In the steep temperature gradients of the lower transition region, energetic electrons in the tail of the Maxwellian distribution at temperature T_1 penetrate to depths of much lower temperature T_2. Thus, at T_2 the electron distribution function has an excess of energetic electrons. These energetic electrons increase the thermal conductivity and alter its dependence on local temperature. In addition, they spread the temperature range over which a given ion is populous. Both effects act in the sense to extend the agreement between $E(T)$ and the conduction models to lower temperature. Whether they can fully resolve the problems of the lower transition region is still problematical.

In summary, it seems likely that no one mechanism will satisfy the needs of the lower transition region. The region is observed to be highly dynamic and highly structured thermally. Certainly there are magnetic loops, and certainly there are non-LTE effects in electron distribution functions. Just as certainly, each will play a role in the structure and in the energy balance. The challenge ahead is to discover the relative magnitudes of these and other important effects.

Acknowledgements

The author is indebted to T. Holzer, B. Lites, and T. Woods for valuable comments on the manuscript.

References

Antiochus, S. K.: 1984, *Bull. Am. Astron. Soc.* **16**, 928.
Athay, R. G.: 1970, *Astrophys. J.* **161**, 713.
Athay, R. G.: 1981a, *Astrophys. J.* **249**, 349.
Athay, R. G.: 1981b, *Astrophys. J.* **250**, 709.
Athay, R. G.: 1984, *Astrophys. J.* **287**, 412.
Athay, R. G. and Holzer, T.: 1982, *Astrophys. J.* **255**, 743.
Athay, R. G., Gurman, J. B., Henze, W., and Shine, R. A.: 1982, *Astrophys. J.* **261**, 684.
Athay, R. G., Mihalas, D., and Shine, R. A.: 1975, Solar Phys. **45**, 15.
Athay, R. G., Klimchuk, J. A., Jones, H., and Zirin, H.: 1986, in preparation.
Ayres, T. R.: 1981, *Astrophys. J.* **244**, 1064.
Ayres, T. R. and Testerman, L.: 1981, *Astrophys. J.* **245**, 1124.
Basri, G. S., Linsky, J. L., Bartoe, J.-D. F., Brueckner, G., and van Hoosier, M. E.: 1979, *Astrophys. J.* **230**, 924.
Beckers, J. M.: 1972, *Ann. Rev. Astron. Astrophys.* **10**, 73.
Bonnet, R. M., Bruner, E. C. Jr., Acton, L. W., Brown, W. A., and Decaudin, M.: 1980, *Astrophys. J.* **237**, L47.

Doschek, G. A., Feldman, U., and Bohlin, J. D.: 1976, *Astrophys. J.* **295**, L177.
Dupree, A. K. and Goldberg, L.: 1967, *Solar Phys.* **1**, 229.
Fedlman, U.: 1983, *Astrophys. J.* **275**, 367.
Gabriel, A. H.: 1976, *Phil. Trans. Roy. Soc. London* **A281**, 339.
Hirayama, T. and Irie, M.: 1984, *Solar Phys.* **90**, 291.
Hummer, D. G.: 1964, *Astrophys. J.* **140**, 276.
Kneer, F.: 1983, *Astron. Astrophys.* **128**, 311.
Kuperus, M. and Athay, R. G.: 1967, *Solar Phys.* **1**, 361.
Leibacher, J., Gouttebroze, P., and Stein, R. F.: 1982, *Astrophys. J.* **258**, 393.
Leighton, R. B., Noyes, R. W., and Simon, G. W.: 1962, *Astrophys. J.* **135**, 474.
Lites, B. W. and Skumanich, A.: 1982, *Astrophys. J.* **49**, 293.
McWhirtyer, R. W. P., Thoneman, P. C., and Wilson, R.: 1975, *Astron. Astrophys.* **450**, 63.
Owocki, S. P. and Canfield, R. C.: 1985, *Astrophys. J.* (submitted).
Pneuman, G. W., and Kopp, R. A.: 1978, *Solar Phys.* **57**, 49.
Rabin, O. and Moore, R.: 1984, *Astrophys. J.* **285**, 359.
Raymond, J. C. and Doyle, J. G.: 1981, *Astrophys. J.* **247**, 686.
Shoub, E. C.: 1983, *Astrophys. J.* **266**, 339.
Summers, H. P. and McWhirter, R. W. P.: 1979, *J. Phys.* **B14**, 2387.
Ulmschneider, P., and Stein, R. F.: 1982, *Astron. Astrophys.* **106**, 9.
Vanbeveren, D. and De Loore, C.: 1976, *Solar Phys.* **50**, 99.
Vernazza, J. E., Avrett, E. H., and Loeser, R.: 1981, *Astrophys. J. Suppl.* **45**, 635.
Withbroe, G. W.: 1981, in F. Q. Orrall (ed.), *Solar Active Regions*, Univ. of Colorado Press, Boulder, p. 199.

PROGRESS IN CORONAL PHYSICS

JACK B. ZIRKER

National Solar Observatory/Sacramento Peak, Sunspot, NM 88349, U.S.A.

Abstract. This paper reviews research highlights of the past five years. Considerable progress has been made in observing and interpreting coronal mass ejections. The stability of coronal loops is much better understood and new observations of the onset of wind streams in coronal holes have been made. Observations from the Solar Maximum Mission should helpt to clarify the physics of the active corona.

The mechanisms that heat the corona and accelerate the high-speed wind streams remain to be identified, however.

1. Introduction

This review focusses on the past five years. It has been a period of transition between Skylab and the Solar Maximum Mission. New ideas in coronal physics that were generated during the Skylab era (say 1973 to 1979) have been elaborated and consolidated. Intensive theoretical efforts have been made to understand coronal stability and to follow the dynamical consequences of an instability. Radio astronomers have begun to make significant contributions to the study of coronal structure, especially the magnetic field, by using the high-resolution VLA and Westerbork Telescope.

New data from the Solar Maximum Mission will certainly change our ideas when they are fully assimilated, but it is still a bit early to feel SMM's impact. One big exception concerns the subject of coronal mass ejections, which has matured quite rapidly.

Two big problems remain to be solved: the identification of the mechanisms that heat the coronal gas (see J. A. Ionson's chapter in this issue) and that accelerate the solar wind to supersonic speed (see L. Hartmann). Many problems also remain in prominence physics (see T. Hirayama).

I have selected three topics in coronal physics: coronal mass ejections, coronal loops and active regions and coronal holes. Only the non-flaring corona has been considered. The reader may wish to consult the recent review articles and monographs that are listed among the references.

2. Coronal Mass Ejections (CME)

Since their discovery during the Skylab era, CME's have been the focus of intense observational and theoretical research. The coronagraphs aboard the SMM and Solwind satellites, as well as the Mark III polarimeter at Mauna Loa and the Culgoora radio heliograph, have revealed a complex phenomenon in intricate detail. Theoretical models that attempt to interpret these events are gradually improving, but a complete understanding of the dynamics, much less the origins of CME's, remains a distant goal.

* Operated by the Association of Universities for Research in Astronomy, Inc., under contract with the National Science Foundation.

Solar Physics **100** (1985) 281–287.   0038–0938/85.15.
© 1985 *by D. Reidel Publishing Company*

The subject has been recently reviewed by Wagner (1984) whose excellent paper contains extensive references to the literature. MacQueen (1980) has summarized CME results obtained during the flight of Skylab (i.e., near solar minimum).

2.1. EMPIRICAL RESULTS

Let us consider some recent empirical results that seem to constrain theoretical models most directly.

2.1.1. *Association with Activity*

Munro *et al.* (1979) found that 40% of 110 CME's observed with Skylab were associated with flares, 50% with eruptive prominences (without flares) and 70% with eruptive priminences (with or without flares). Since the flare rate increases by at least an order of magnitude between solar minimum and maximum, Hildner *et al.* (1976) expected the frequency of CME's also to increase dramatically.

That expectation has not been borne out. Hundhausen *et al.* (1984) examined 65 CME's observed from SMM. The frequency of CME's during March–September, 1980, a period of high solar activity, was 0.9 per day, compared with 0.75 per day during Skylab, a period of declining activity. Thus although the flare rate increased sixfold, the CME rate increased only by 20%.

Two factors help to explain this result. First, a significant fraction of all CME's (48% during 1973–1974 and 30% during 1980) are unassociated with *any* activity (Hundhausen *et al.*, 1984) and the fraction associated with eruptive prominences remains roughly constant at about 30%. Thus although the flare rate increased, it produced a relatively minor overall increase in the CME rate.

Moreover, evidence has accumulated for a finite 'recovery time' (about ten hours), in which the coronal magnetic field closes after a CME event (Rust, 1983; Wagner, 1984). Although several flares may occur during this interval, they produce weak or no CME's. A flare may be neither necessary nor sufficient to produce a CME.

The previously unrecognized category of spontaneous CME's and the existence of a recovery time should emphasize to theorists the importance of magnetic rather than explosive forces in a mjaority of CME's.

2.1. SHAPE

The question whether a typical CME is a planar loop or an ellipsoidal bubble has not been definitely answered. However, recent empirical density models (e.g. the 'ice-cream cone' of Fisher and Munro, 1984) tend to favor the latter. More studies are needed of the evolution of the density distribution, based on the assumption of one or the other geometry.

2.1.3. *Motion*

The three categories of CME are characterized by different maximum speeds. Thus flare-associated CME's range up to 1200 km s^{-1}, those associated with prominences up to 600 km s^{-1} and spontaneous CME's only up to about 500 km s^{-1}. Moreover, the prominence CME's accelerate up to a height of 2–3 $R_\odot$, while the flare CME's achieve their high, constant speed well below this height range (MacQueen and Fisher, 1983). These results may suggest that magnetic forces predominate in driving prominence CME's but may be supplemented by high gas pressures in flare CME's.

2.1.4. *Source of Mass*

Only a few estimates have been made of the mass present in the corona before a CME occurs. To within a factor of two, the pre-existing mass seems sufficient to account for the mass in the CME. This statement can mean different things to different people, however. It can be interpreted to say that half the mass must be injected during the event (e.g., Fisher and Munro, 1984). Better estimates are needed to guide theoretical models, especially those that include a realistic energy equation.

2.1.5. *Energy Partition and Magnetic Field*

A particularly revealing study was carried out by Wagner *et al.* (1981), who combined optical and radio observations of a huge CME (April 7, 1980) to derive estimates of the density of different types of energy. They found that the convected magnetic energy density (0.5 erg cm^{-3}) exceeded the kinetic energy density (0.1 erg cm^{-3}) and greatly exceeded the thermal energy ($\beta \simeq 0.06$). From an observed moving type-IV burst, located at the leading edge of the CME, they deduced a field strength of 4 G. No type-II burst (e.g., shock) preceded this event, presumably because of a high Alfvén speed in the background corona.

2.2. MODELS OF CME's

Low (1984) classifies current models of CME's into two groups. One group represents the CME as a wavelike, nonlinear, MHD response to impulsive energy input at the base of the corona (e.g., Dryer *et al.*, 1979; Steinholfson, 1982). The other group represents the CME as a magnetically propelled flux loop (e.g., Mouschovias and Poland, 1978; Pneuman, 1980; Anzer, 1978; Yeh and Dryer, 1981a, b). Low introduces a third view: the CME is a fully-developed MHD outflow. He has found self-similar solutions to the nonlinear, time-dependent MHD equations by exact analytic methods, assuming axisymmetry, a polytropic gas and r^{-2} gravity. His technique can be generalized to non-axisymmetric flows. This is an achievement of high order.

Low suggests that a CME results from the instability of a coronal system, in which magnetic tension and gravity (which tend to stabilize the system) are overcome by the tendency of a magnetized plasma to expand into the surrounding medium. Although waves are involved, they are distinct from the transient itself. (They might be identified with the 'forerunners' that are occasionally observed: Jackson and Hildner, 1978). The evolving MHD flow generates features that resemble such observed phenomena as voids, loops and blobs. Low's ideas deserve a careful comparison with time-dependent empirical modes of density or brightness.

Such a comparison has been carried out (Sime *et al.*, 1984) for a class of models (Dryer *et al.*, 1979) that postulate impulsive energy release, (e.g., a flare) as the origin of the CME. These models fail in two respects: (a) they predict that the maximum brightness enhancement appears at the top of the 'loop', rather than at the flanks and legs, and (b) they predict that the legs move laterally with speeds comparable to the vertical speed, in contradiction to the observations.

3. Coronal Loops

The Skylab Workshop on Solar Active Regions (Orall, 1981) greatly stimulated the interest of solar physicists in the physics of coronal loops. That monograph contains the best current reviews on the observations and theory. Of special interest for this section are the chapters on *Active Region Structure*, D. F. Webb; *Physics of Static Loops*, G. L. Withbroe; *Theory of Loop Flows and Instability*, E. R. Priest.

During the past five years a large number of theoretical studies have been published on the formation, stability, dynamics, and thermodynamic structure of coronal loops in active regions. The critical issues of the distribution of heating and the dynamic response of a loop to a change in energy supply have been explored without, however, yielding direct clues on the nature of the heating mechanism itself.

A scaling law, that relates the length, maximum temperature and pressure of a static cylindrical loop was derived by Rosner *et al.* (1978), who used plausible assumptions and simple stability arguments. They showed that stable hydrostatic equilibrium obtains only in a loop with the temperature maximum at the top, and with an energy deposition scale length larger than the loop length. Their scaling law, $T_{max} = 1.4 \times 10^3 (pL)^{1/3}$, fits the observations reasonably well, but, like more elaborate quasi-static models (see Withbroe, 1981), is too insensitive to the longitudinal temperature distribution to select a unique heating mechanism (Chiuderi *et al.*, 1981). Generalized scaling laws were developed by Galeev *et al.* (1981), Serio *et al.* (1981), and Hood and Priest (1979). Hearn and Kuin (1981) point out that Rosner's original scaling laws are fully consistent with their concept of a minimum energy flux corona.

EUV and X-ray observations of loops show they are remarkably stable over most of their lifetimes. Many authors have studied the conditions necessary for thermal and dynamic stability. Static, thermally isolated loops are unstable to radiative cooling (Antiochos, 1979) but become stable if the conductive flux into the chromosphere is sufficiently large (Hood and Priest, 1980). Craig and McClymont (1981a) suggested that the radiatively stable chromosphere acts as an energy sink that buffers fluctuations in the loop's energy supply. This idea is similar to the 'evaporation' of chromospheric plasma (Krall and Antiochos, 1978) that follows rapid heating, with a consequent increase in loop density and radiating power at coronal temperatures. Since such evaporation implies transient flows and since such flows are occasionally observed, dynamic models of loops have become popular (Craig and McClymont, 1981b; Craig *et al.*, 1982; Oran *et al.*, 1982; Kuin and Martens, 1982; McClymont and Craig, 1984). These studies have shown that the exchange of energy and mass at the footpoints provides a powerful thermostat for stabilizing coronal loops.

Another aspect of stability concerns the growth of perturbations in a loop's magnetic field. The 'tying' of a loop's field lines in the dense photosphere provides a strong stabilizing influence (Hood and Priest, 1979). Other stabilizing factors are the finite length of a loop (Van Hoven *et al.*, 1981), the surrounding plasma (Kattenberg and Sillen, 1982), gravity, and field line compression (An, 1984). Sufficiently large or rapid distortions of the magnetic field can lead to a catastrophic instability, however. Hood and Priest (1980) found an elegant criterion for the critical uniform twist (2.49π) in a force-free loop.

Several groups have combined observations of active regions at radio, optical, EUV and X-ray wavelengths in order to explore the temperature, density and magnetic field distributions. The availability of the VLA, the Westerbork telescope and the instruments aboard the Solar Maximum Mission has provided fresh data of high spatial resolution. Older investigations have used Skylab XUV data effectively. Cheng *et al.* (1980) find the brightest regions at X-ray wavelengths correspond to low compact and stable loops. Hot loops ($T > 2$–3×10^6) are thick, while cool loops are slender (aspect ratios of 6×10^{-3}). The dominant emission mechanism at 6 cm over spots is due to gyro-resonance, although free-free emission can occur in large loops that connect spots (Pallavicini *et al.*, 1981a, b; Shibasaki *et al.*, 1983). Field strengths in the corona above spots range up to 900 G.

Away from spots, the identification of emission mechanisms at 6 cm becomes more difficult. Webb *et al.* (1983) and Schmahl *et al.* (1982) agree that the peaks of brightness at 6 cm and in the XUV do not coincide, but at best, overlap. This result rules out free-free emission but gyroresonance emission is not completely satisfactory either. Schmahl *et al.* (1982) suggest that electrical currents are present in some loops.

In a recent paper, McConnell and Kundu (1983) interpret VLA maps at 20 cm of an active region loop system. Bremsstrahlung predominates in the feet and gyroresonance in the tops. The tops have the following properties: $B = 130$–170 G, $T = 1.7 \times 10^6$ K, $N = 5 \times 10^8$ cm^{-3}, $\beta = 3 \times 10^{-4}$.

Hurford (1983) has developed a clever technique for determining coronal magnetic fields in active regions. Microwave spectra (1–18 GHz) change slope at locations within an active region where the gyroresonance opacity decreases abruptly, because of a rapid change with height of the magnetic field and temperature. The frequency at which the slope changes is related directly to the magnetic field strength at the base of the corona.

4. Coronal Holes

Progress in understanding the physics of coronal holes and their associated high-speed wind streams has slowed somewhat since the publication of the Skylab Workshop monograph (Zirker, 1977). The basic question of the mechanism that accelerates the streams remains open, although momentum deposition by some form of MHD wave continues as the prime candidate.

Synoptic observations of holes have continued during the rise to solar maximum. Peak speeds in the streams associated with the holes have declined to 500–600 km s^{-1} from their high values (600–700 km s^{-1}) during the decline of spot cycle 20 (Sheeley and Harvey, 1981). Harvey *et al.* (1982) find that low-latitude holes contained three times more magnetic flux near spot maximum than near minimum, although their sizes were comparable. Average field strengths range from 3–36 G near maximum, to 1–7 G near minimum.

The contrast of coronal holes in He I λ10830 and in soft X-rays also declined between 1974 and 1981, which suggests to Kahler *et al.* (1983) that the mix of open and closed magnetic field lines in holes may depend upon the phase of the solar cycle.

F. Orrall and co-workers have attempted to detect a systematic plasma outflow in coronal holes that could be attributed to the onset of the wind streams (Rottman *et al.*, 1981, 1982; Orrall *et al.*, 1983). EUV transition zone and coronal lines in low latitude holes are blue-shifted relative to the rest of the disk. The shift increases with formation temperature (e.g., height): O v λ629 is shifted by 7 km s^{-1} and Mg x λ625 by 12 km s^{-1}. In a polar hole, Mg x λ625 shows a shift of 8 km s^{-1}. These values should help to guide models of wind acceleration.

Fast-mode MHD waves have been relatively neglected (as compared with Alfvén waves) in the search for an agent that accelerates wind streams to high speed. Fla *et al.* (1984) consider the possibility that fast-mode waves propagate outward, depositing momentum and energy in the supersonic wind. If this scheme is to work, however, the fast mode must be generated in the mid-corona, i.e., a region where the Alfvén speed decreases with height. Otherwise the positive height gradient of Alfvén speed in the low corona will tend to drive the waves out the sides or base of a coronal hole, leaving little momentum available for acceleration.

5. Summary

As this very brief survey indicates, the last five years have produced significant advances in our understanding of certain aspects of coronal physics. We can look forward to the results that new data from the Solar Maximum Mission will stimulate.

References

An, C.: 1984, *Astrophys. J.* **276**, 352.
Antiochos, S.: 1979, *Astrophys. J.* **232**, L125.
Anzer, U.: 1978, *Solar Phys.* **57**, 111.
Cheng, C., Smith, J., and Tandberg–Hanssen, E.: 1980, *Solar Phys.* **67**, 259.
Chiuderi, C., Einaudi, G., and Torricelli-Ciampone, G.: 1981, *Astron. Astrophys.* **97**, 27.
Craig, I. and McClymont, A.: 1981a, *Nature* **295**.
Craig, I. and McClymont, A.: 1981b, *Solar Phys.* **70**, 97.
Craig, I., Robb, T., and Rollo; 1982, *Solar Phys.* **331**.
Dryer, M., Wu, S., Steinolfson, R., and Wilson, R.: 1979, *Astrophys. J.* **227**, 1059.
Fisher, R. R. and Munro, R. H.: 1984, *Astrophys. J.* **280**, 428.
Fla, T., Habbal, S. R., Holzer, T. E., and Leer, E.: 1984, *Astrophys. J.* **280**, 382.
Galeev, A., Rosner, R., Serio, S., and Vaiana, G.: 1981, *Astrophys. J.* **243**, 301.
Harvey, K. L., Sheeley, N. R., and Harvey, J. W.: 1982, *Solar Phys.* **79**, 149.
Hearn, A. G. and Kuin, N. P. M.: 1981, *Astron. Astrophys.* **98**, 248.

Hildner, E., Gosling, J. T., MacQueen, R. M., Munro, R. H., Poland, A. I., and Ross, C. L.: 1976, *Solar Phys.* **48**, 127.

Hood, A. and Priest, E.: 1979, *Solar Phys.* **64**, 303.

Hood, A. and Priest, E.: 1980, *Astron. Astrophys.* **87**, 126.

Hundhausen, A. J., Sawyer, C., House, L., Illing, R. M. E., and Wagner, W. J.: 1984, *J. Geophys. Res.* **89**, 2639.

Hurford, G.: 1983, *Bull. Am. Astron. Soc.* **15**. 703.

Jackson, B. and Hildner: 1978, *Solar Phys.* **60**, 155.

Kahler, S., Davis, J., and Harvey, J.: 1983, *Solar Phys.* **87**, 47.

Kattenberg, A. and Sillen, R.: 1982, *Solar Phys.* **79**, 343.

Krall, K. and Antiochos, S.: 1978, *Astrophys. J.* **108**, L1.

Kuin, N. and Martens, P.: 1982, *Astron. Astrophys.* **108**, L1.

Low, B.: 1984, *Astrophys. J.* **281**, 392.

MacQueen, R. M.: 1980, *Phil. Trans. Roy. Soc. London*, **A297**, 605.

MacQueen, R. M. and Fisher, R. R.: 1983, *Solar Phys.* **89**, 89.

McClymont, A. and Craig, I.: 1984, *Bull. Am. Astron. Soc.* **16**, 928.

McConnell, D. and Kundu, M. R.: 1983, *Astrophys. J.* **269**, 698.

Mouschovias, T. and Poland, A.: 1978, *Astrophys. J.* **220**, 675.

Munro, R., Gosling, J., Hildner, F., MacQueen, R., Poland, A., and Ross, C.: 1979, *Solar Phys.* **61**, 201.

Oran, E., Mariska, J., and Boris, J.: 1982, *Astrophys. J.* **254**, 349.

Orrall, F., Rottman, G., and Klimchuk, J.: 1983, *Astrophys. J.* **266**, L65.

Pallavicini, R., Sakurai, T., and Vaiana, G.: 1981a, *Astron. Astrophys.* **98**, 316.

Pallavicini, R., Peres, G., Serio, S., Vaiana, G., Golub, L., and Rosner, R.: 1981b, *Astrophys. J.* **247**, 692.

Pneuman, G.: 1980, *Solar Phys.* **65**, 369.

Rosner, R. Tucker, W., and Vaiana, G.: 1978, *Astrophys. J.* **220**, 643.

Rottman, G., Orrall, F., and Klimchuk, J.: 1981, *Astrophys. J.* **247**, L135.

Rottman, G., Orrall, F., and Klimchuk, J.: 1982, *Astrophys. J.* **260**, 326.

Rust, D. M.: 1983, *Space Sci. Rev.* **34**, 21.

Schmahl, E., Kundu, M., Strong, K., Bentley, R., Smith, J., and Krall, K.: 1982, *Solar Phys.* **80**, 233.

Serio, S., Peres, G., Vaiana, G., Golub, L., and Rosner, R.: 1981, *Astrophys. J.* **243**, 288.

Sheeley, N. and Harvey, J.: 1981, *Astrophys. J.* **237**, 1981.

Shibasaki, K., Chiuderi-Drago, F., Melozzi, M., Slottje, C., and Antonucci, E.: 1983, *Solar Phys.* **89**, 307.

Sime, D., MacQueen, R., and Hundhausen, A.: 1984, *J. Geophys. Res.* **89**, 2113.

Steinholfson, R.: 1982, *Astron. Astrophys.* **115**, 39.

VanHoven, G., Ma, S. and Einaudi, G.: 1981, *Astron. Astrophys.* **97**, 232.

Wagner, W. J.: 1984, *Ann. Rev. Astron. Astrophys.* **22**, 267.

Wagner, W., Hildner, E., House, L., Sawyer, C., Shelidan, K., and Dulk, G.: 1981, *Astrophys. J.* **244**. L123.

Webb, D., Davis, J., Kundu, M. and Velusamy, T.: 1983, *Solar Phys.* **85**, 267.

Withbroe, G.: 1981, in F. Orrall (ed.), *Solar Active Regions*, Colorado University Assoc. Press, p. 199.

Yeh, T. and Dryer, M.: 1981a, *Astrophys. J.* **245**, 704.

Yeh, T. and Dryer, M.: 1981b, *Solar Phys.* **71**, 141.

Reviews

J. Zirker (ed.): 1977, *Coronal Holes and High Speed Wind Streams, Skylab Workshop I*, Colorado Associated University Press.

G. S. Vaiana and R. Rosner: 1978, Recent Advances in Coronal Physics, *Ann. Rev. Astron. Astrophys.* **16**, 393.

F. O. Orrall (ed.): 1981, *Solar Active Regions, Skylab Workshop* III, Colorado Associated University Press.

S. Jordan (ed.): 1981, *The Sun as a Star*, NASA SP-450.

THE HEATING OF CORONAE

JAMES A. IONSON

Department of Physics and Astronomy, University of Maryland, College Park, MD 20742, U.S.A.

Abstract. Probably the most significant breakthroughs in understanding the heating of coronae have emerged from a global electrodynamics approach. Although this approach ties together A.C. (resonant) and D.C. (non-resonant) heating mechanisms, the detailed physics of the dissipation process is still not clearly understood.

1. Introduction

One of the most exciting observational discoveries of this decade has been that tenuous, X-ray emitting plasmas ($T > 10^6$ K) are surprising common, found in association with a wide variety of gravitationally (or inertially) confined, mechanically active astrophysical systems. They range from the well-known solar-coronal complex, early and late-type stars (Vaiana *et al.*, 1981), accretion disks (Galeev *et al.*, 1979; Ionson and Kuperus, 1984; Kuperus and Ionson, 1985), interstellar clouds (Rosner and Hartquist, 1979), and galaxies (Sturrock and Stern, 1980) to possibly even inertially confined galactic jets and double radio sources (Scott, 1982). These systems are mechanically active in the sense that thay all possess zones of differentially convective velocity fields. The logical supposition that will adopted here is that the site of mechanical activity couples to and energetically maintains a spatially distinct yet contiguous site of X-ray activity.

One of the most outstanding problems faced by solar and stellar physicists is an understanding of the cause and effect relationship between inner (e.g., photospheric) and outer (e.g., coronal) atmospheric dynamics. Central to this problem of global coupling within a stellar atmosphere is the role played by magnetic fields. Magnetic fields typically thread the inner and outer atmosphere of a star, and thus one cannot disregard the potentially important role that associated electrodynamic processes could play in effecting an 'electrodynamic coupling' of these two regions.

A useful parameter that identifies the dominance of 'electrodynamic coupling' over, for example, mechanical coupling of the inner and outer atmosphere by acoustic waves, is the plasma beta, $\beta \equiv v_s^2/v_A^2$ ($v_s \equiv$ sound speed; $v_A \equiv$ Alfvén speed). Specifically, electrodynamic coupling within a stellar atmosphere dominates mechanical coupling whenever the plasma beta is less than unity within the outher atmosphere, i.e., $\beta_{outer} < 1$, and of the order of unity within the inner atmosphere, i.e., $\beta_{inner} \gtrsim 1$. Thus, mechanical dynamics such as convective and/or differential rotation velocity fields within the $\beta_{inner} \gtrsim 1$ atmosphere can couple to and drive causally related phenomena, such as heating, within the $\beta_{outer} < 1$ outer atmosphere through various electrodynamic processes associated with the interconnecting magnetic field. This is in contrast to a mechanically coupled stellar atmosphere which would require a $\beta_{outer} \gtrsim 1$ outer atmosphere.

Solar Physics **100** (1985) 289–308.   0038–0938/85.15.
© 1985 *by D. Reidel Publishing Company*

It is now widely believed that the $\beta \gtrsim 1$ inner atmosphere (photosphere) and $\beta < 1$ outer atmosphere (corona) of the Sun are electrodynamicallly coupled since extrapolations of the observed photospheric magnetic energy into the corona exceeds the, *in situ*, thermodynamic energy (i.e., $\beta < 1$).

The solar-coronal complex is the only astrophysical system whose X-ray structure can to some degree be spatially resolved. The most important lesson that we have learned from these relatively high-resolution observations ($\sim 10^{8.5}$ cm) is that the solar corona is highly structured, comprising a variety of closed, loop-like regions of enhanced radiative output. Furthermore, coronal radiation loops are now known to be spatially coincident with coronal magnetic loops which confine the radiating plasma (Withbroe and Noyes, 1977; Vaiana and Rosner, 1978). There is no reason to suspect that the basic X-ray morphology of the solar-coronal complex is unique. This morphology could, in fact, be fundamental in developing physical models of other as yet spatially unresolved X-ray emitting astrophysical systems – an important possibility that clearly rationalizes the goals of this section. In general, the goals are to identify and discuss the primary submodels that are involved in the global coupling between a mechanical energy reservoir with $\beta \gtrsim 1$ and a contiguous site of $\beta < 1$ X-ray acitivity (see Figure 1).

Ideally, a 'dynamo model' could establish a quantitative connection between properties of the mechanical driver (e.g., differential and convective velocity fields) and the dimensions (e.g., l_B and l), field strength (i.e., B), and number density distribution of loops (i.e., fill factor). As pointed out by Gilman (1980) and Golub *et al.* (1980), the mechanical energy reservoir comprises two basic regions that are differentiated in terms

GLOBAL COUPLING SCENARIO

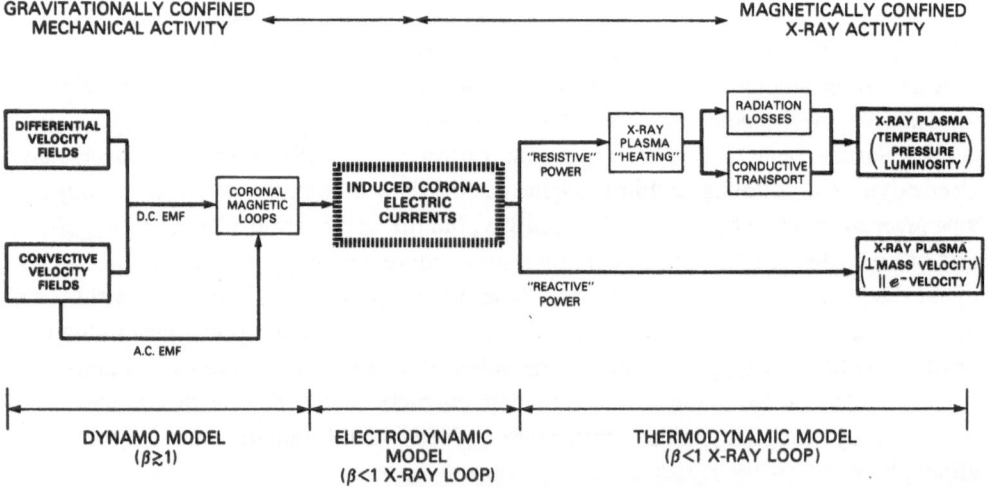

Fig. 1.   Flow diagram illustrating the global coupling between $\beta \gtrsim 1$ site of mechanical activity and $\beta < 1$ site of magnetically confined, quasi-steady X-ray activity.

of the velocity field's convective time- and size-scales. Specifically, the 'surface layer' velocity field is characterized by relatively small convective time- and size-scales compared to the underlying 'subsurface layer'. As pointed out by Ionson (1982), the overall potential magnetic filed structure is established by large-scale subsurface velocity fields whose characteristic time scale exceeds the magnetic loop's Alfvén transit time-scale (i.e., DC emfs). In contrast, the relatively smaller scale surface convection, whose time-scale is of the order of the magnetic loop's Alfvén transit time, results in AC emfs which drive an elemental loop into a nonequilibrium electrodynamic state characterized by the flow of electrical currents along (i.e., force-free) and across (i.e., non-force-freee) the coronal B field. The electrodynamical model describes how these currents are generated as well as how they couple to the magnetically confined plasma. This coupling results in heating and a confined, cross-field coronal velocity field. The thermodynamic model demonstrates how radiation and conduction losses determine the X-ray loop's temperature, pressure and, hence, luminosity.

This article does not attempt to review the current status of dynamo models. This extremely complicated problem has been the subject of articles too numerous to discuss here (see review by Cowling, 1981), many casting doubt upon various versions of the canonically accepted $\alpha - \omega$ dynamo accomplished by large-scale differential motion (i.e., the ω-effect), with field reversal occurring as a result of the diffusive effects of helical convection (i.e., the α-effect). The result of this interaction is a temporal variation in magnetic flux generation, along with a corresponding spatial propagation of this flux across the $\beta \gtrsim 1$ mechanical driving zone. A major observational problem with this model is that, at least for the Sun, the total magnetic flux appears to be constant throughout the activity cycle (Golub $et\ al.$, 1979). This is contrary to the predictions of various $\alpha - \omega$ dynamo models. In addition, the hydrodynamically inferred solar differential rotation is directed opposite to that required by kinematic $\alpha - \omega$ dynamos (Gilman, 1981). Today's conventional dynamo models are thus the weakest link in the global coupling chain presented here. Therefore, this article will focus upon the electrodynamic model – the essential link connecting X-ray observations to the basic mechanical properties of, for example, a star.

2. Coronal Loop Electrodynamics

Two classes of coronal radiation loops have been identified, specifically, the so-called 'cool loops' with temperatures of the order of 10^4–10^5 K and hot 'X-ray loops' with temperatures of $\sim 2.5 \times 10^6$ K. Since their discovery, an enormous amount of research has been devoted to determine how or if these two classes of radiating loops are correlated. One need only refer to the monograph from Skylab Solar Workshop III, Solar Active Regions (1981) in order to appreciate the efforts that have gone into the investigation of this problem. Since it is impossible to review the evolution of this topic here, I will simply point out the current state of affairs based upon reviews by Van Hoven (1981) and more recently by Raymond and Foukal (1982).

Figures 2(a)–(b) illustrates the topological similarity of the thermodynamically

SOLAR CORONAL RADIATION LOOPS

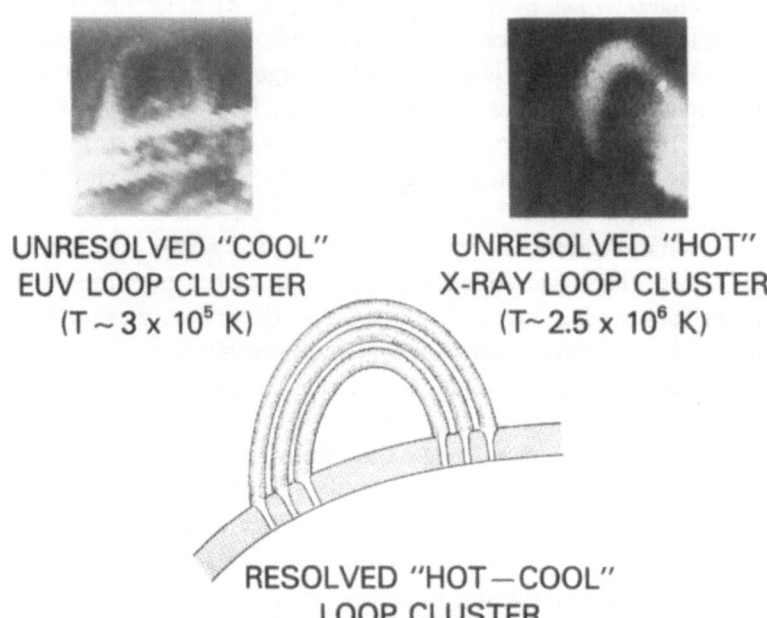

UNRESOLVED "COOL"   UNRESOLVED "HOT"
EUV LOOP CLUSTER   X-RAY LOOP CLUSTER
(T ~ 3 x 10⁵ K)   (T ~ 2.5 x 10⁶ K)

RESOLVED "HOT — COOL"
LOOP CLUSTER

Fig. 2.   Solar coronal radiation loops in the extreme ultraviolet (i.e., Foukal's 1976 'cool' loops) and in broad-band soft X-rays (i.e., 'hot' loops, courtesy L. Golub). These two different classes of loops could represent an unresolved cluster of elemental radiation loops with each elemental loop containing both 'hot' and 'cool' plasma (Raymond and Foukal, 1982)

dissimilar 'cool' and 'hot' coronal loops. These two classes of loops typically have the same basic dimensions (i.e., lengths $\lesssim 10^{10}$ cm and diameters $\sim 10^{8.5}$ km). As pointed out by Van Hoven (1981), the 'point spread' function of X-ray telescopes tends to widen artificially the image of the hot X-ray loops. In support of Foukal's (1975, 1976, 1978) original suggestion, Raymond and Foukal (1982) have recently presented evidence that 'hot' and 'cool' loops are cospatial within the resolution limits of the observing instruments (viz., $\sim 10^{8.5}$ cm). A distinct possibility is that the currently observed 'cool' and 'hot' loops actually represent an unresolved cluster of topologically distinct radiation loops whose diameter is significantly less than $10^{8.5}$ cm (see Figure 2(c)). This possibility is logically consistent with the observed existence of small diameter ($\lesssim 10^{7.5}$ cm) magnetic flux elements within the photosphere which extend into the corona, confining the radiating plasma. Raymond and Foukal's (1982) observations suggest that each of these unresolved radiation loops contains both 'cool' and 'hot' plasma. This is also consistent with many contemporary coronal heating theories, the primary point of contention beigng the relative volumes of cool and hot plasma (see Wentzel, 1978; Kuperus *et al.*, 1981; Chiuderi, 1981, 1982). For example, Ionson (1978, 1982) has pointed out that electrodynamics coupling between a radiation loop with $\beta < 1$ and the underlying $\beta \gtrsim 1$ velocity field occurs within a spatial resonance shell where effective electrodynamic energy absorption and dissipation occurs (i.e., cool cores and hot envelopes).

Based upon previous observational and theoretical studies of solar coronal loops, the radiation and magnetic morphologies of an elemental loop are illustrated in Figure 3.

ELEMENTAL CORONAL MAGNETIC LOOP

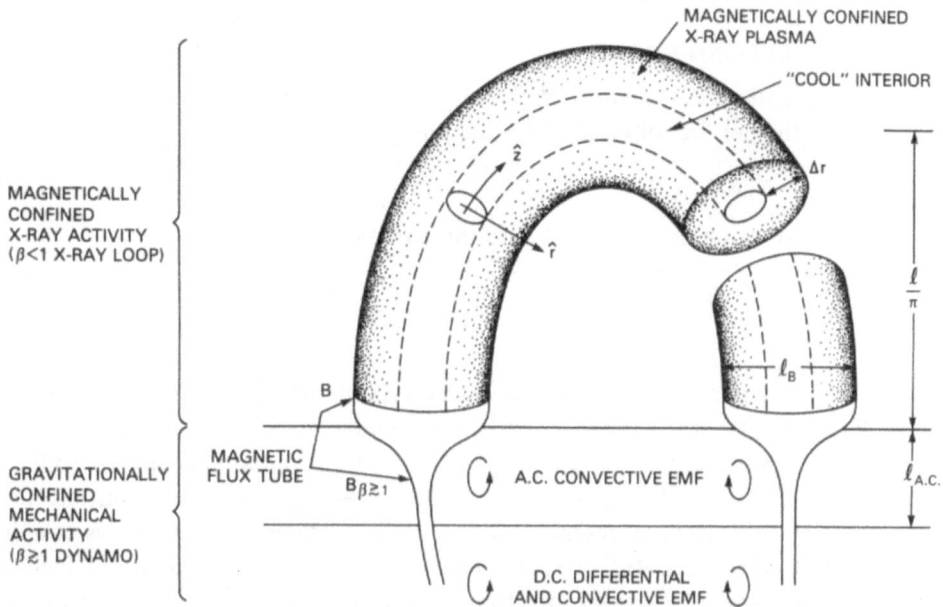

Fig. 3.   Illustration of a hypothesized elemental coronal loop of the cluster sketched in Figure 2.

This illustration corresponds to a single radiation loop from the cluster illustrated in Figure 2(c). The hot X-ray emitting plasma is magnetically confined within a shell of volume $V_{X-ray} = \pi l_B l \Delta r$, where l_B is the magnetic scale diameter, l is the magnetic scale length of the loop, and Δr is the X-ray shell thickness. The central core of this strtucture contains relatively cooler plasma. Two heating possibilities have been considered. The first possibility is that an as yet unknown transport mechanism allows the electrodynamic energy dissipated within the spatial resonance shell to efficiently diffuse throughout the loop volume (i.e., $\Delta r \approx l_B/4$). There is no reason to expect that such a transport mechanism exists, and, in fact, many mechanisms have been investigated with little succes (Hollweg, 1981). The second and perhaps more realistic possibility is that the dissipation shell (e.g., a spatial resonance shell) is energetically isolated. In this case, the X-ray plasma volume would strongly depend upon the dissipation mechanism.

Electrodynamic coupling between a $\beta < 1$ coronal magnetic loop and its $\beta \gtrsim 1$ mechanical driver (i.e., the convection zone) involves three basic aspects. Namely, generation of an electrodynamical Poynting flux, propagation of this flux into the coronal loop and its subsequent dissipation. We recognize that identification of the dissipation processes is an important aspect of any electrodynamic model, but feel that it is just as important and perhaps even a pre-requisite that we understand the physics of electrodynamic coupling in toto. Specifically, how does the power input scale with intrinsic loop properties such as the Alfvén transit time t_A, dissipation time t_{diss}, as well as intrinsic properties of the $\beta \gtrsim 1$ velocity driver? In this regard, the development of an electrodynamic model should include studies of global coupling mechanisms as well as investigations of specific dissipation mechanisms which can be built into the global model.

2.1. GLOBAL COUPLING MODEL

In general, a quantitative statement of global electrodynamic coupling is the following expression for the net flux of electrodynamic energy F_H that heats a coronal loop:

$$F_H = F_0 \, \varepsilon, \tag{1}$$

where F_0 is the Fresnel flux of energy into the corona, i.e.,

$$F_0 = 4 \left(\frac{B}{B_{\beta \gtrsim 1}} \right) \left(\frac{4 v_A^{\beta \gtrsim 1}}{v_A} \right) v_A^{\beta \gtrsim 1} (\tfrac{1}{2} \varrho v_{tot}^2)_{\beta \gtrsim 1} \tag{2}$$

and ε an electrodynamic coupling coefficient describing modifications resulting from heating. Equation (2) for the Fresnel flux F_0 can be interpreted by noting that the first factor of four accounts for two footpoints per loop and two excitation polarizations (i.e., 'twist' and 'shake'): the $B/B_{\beta \gtrsim 1}$ term results from magnetic field expansion into the corona; and the third term represents a transmission coefficient which in this case is given by the Fresnel relation, $4 v_A^{\beta \gtrsim 1}/v_A$. Note that in general the transmission coefficient is frequency dependent whenever the transition zone thickness is larger than the wavelength of the excited magnetic fluctuation (cf. Hollweg, 1983). Since the $\beta \gtrsim 1$ convection is typically peaked at low frequencies (i.e., long wavelengths), the frequency-independent Fresnel relation accurately represents the transmission coefficient in many applications of interest. The fourth term $v_A^{\beta \gtrsim 1}(\tfrac{1}{2}\rho v_{tot}^2)_{\beta \gtrsim 1}$, represents the maximum available flux of electrodynamic energy whitin the $\beta \gtrsim 1$ region.

The electrodynamic coupling coefficient is actually the ratio of the frequency integrated Poynting flux entering the loop to the maximum possible energy flux, i.e.,

$$\varepsilon \equiv \int\limits_{-\infty}^{\infty} \frac{(c/4\pi) \langle (\mathbf{E}' \times \mathbf{B}')_{\parallel} \rangle_v^{\beta \gtrsim 1}}{[(1/2)\rho v_{tot}^2]_{\beta \gtrsim 1} \, v_A^{\beta \gtrsim 1}} \, dv, \tag{3}$$

where the primes denote fluctuations and where the parallel sign refers to the direction parallel to the ambient magnetic field. Noting that $\mathbf{E}'_{\perp} \approx \mathbf{v}_{\perp} \times \mathbf{B}_{\beta \gtrsim 1}/c$, Equation (3) reduces to

$$\varepsilon = \int\limits_{-\infty}^{\infty} \left(\frac{\langle v_{\perp}^2 \rangle_v}{v_{tot}^2} \right)_{\beta \gtrsim 1} \left(\frac{2 \langle v_{\perp} B'_{\perp} \rangle_v^{\beta \gtrsim 1} v_A^{\beta \gtrsim 1}}{B_{\beta \gtrsim 1} \langle v_{\perp}^2 \rangle_v^{\beta \gtrsim 1}} \right) dv \tag{4}$$

which illustrates that the interaction between the $\beta \gtrsim 1$ energy flux spectrum, $\langle \frac{1}{2}\rho v_\perp^2 \rangle_\nu^{\beta \gtrsim 1} v_A^{\beta \gtrsim 1}$, and the corresponding magnetic stressing spectrum, $\langle v_\perp B'_\perp \rangle_\nu^{\beta \gtrsim 1}/4\pi$, is described by

$$\frac{\langle v_\perp B'_\perp \rangle_\nu^{\beta \gtrsim 1} B_{\beta \gtrsim 1}}{4\pi} = \langle \tfrac{1}{2}\rho v_\perp{}^2 \rangle_\nu^{\beta \gtrsim 1} v_A^{\beta \gtrsim 1} e(\nu; t_A, t_{loss}, t_{diss}), \tag{5}$$

where $e(\nu; t_A, t_{loss}, t_{diss})$ is a dimensionless velocity magnetic field interaction parameter which is a function of frequency ν and parameterized in terms of a loop's typical Alfvén transit time $t_A = 2l/v_A$ (i.e., v_A is the typical Alfvén speed) and electrodynamic energy 'loss' time-scale t_{loss}, which is a combination of dissipation, magnetic stress leakage and phase mixing time-scales, and the dissipation time-scales t_{diss}. The electrodynamic coupling coefficient is, therefore, a convolution of the velocity-magnetic field interaction parameter, $e(\nu; t_A, t_{loss}, t_{diss})$ with the normalized $\beta \gtrsim 1$ convection power spectrum, $(\langle v_\perp^2 \rangle_\nu / v_{tot}^2)_{\beta \gtrsim 1}$. That is, from Equation (4) and (5) one finds that ε can be written as

$$\varepsilon = \int_{-\infty}^{\infty} \left(\frac{\langle v_\perp^2 \rangle_\nu}{v_{tot}^2} \right)_{\beta \gtrsim 1} e(\nu; t_A, t_{loss}, t_{diss}) \, d\nu. \tag{6}$$

The primary objective of any global coupling model is to identify, from first principles, a general (parameterized) velocity-magnetic field interaction parameter, $e(\nu; t_A, t_{loss}, t_{diss})$.

Ionson's (1982, 1983, 1984, 1985) previous work centered around the problem of identifying a general form for $e(\nu; t_A, t_{loss}, t_{diss})$. The major result of these work was that the linear electrodynamic coupling between high and low β-loop systems can be modeled by a simple, yet equivalent, LRC circuit analog. The derived analog points to the existence of global structure oscillations, such as Alfvenic surace waves (Ionson, 1977, 1978), which resonantly excite field line oscillations within a spatial resonance shell of thickness Δr. These localized B-field oscillations result from induced currents that are driven by $\beta \gtrsim 1$ AC-stressing emfs. Although the width of this spatial resonance, as well as the magnitude of the induced currents, explicitly depends upon irreversible processes (e.g., viscosity and resistivity), the resonant form for the heating rate (i.e., the 'resistive power') is virtually independent of irreversibilities (see Heyvaerts and Priest, 1983). This is a classic feature of high 'quality' resonant systems that are driven by a broad-band source of spectral power. In this case the source of spectral power is the power spectrum of the AC-stressing velocity, $\langle v_{\beta \gtrsim 1}^2 \rangle_\nu$. The essential feature of resonant heating mechanism is that magnetic loops with different lengths and field strengths and, hence, different global resonance frequencies, are heated at a rate that critically depends upon the amount of $\beta \gtrsim 1$ spectral power at the resonance frequency ν_{res} as well ast the value of B.

Whereas Ionson's work focused upon the concept of 'resonant' (i.e., A.C.) heating processes, Heyvaerts and Priest (1984) investigated 'non-resonant' (i.e., D.C.) heating processes. Ionson (1985) has pointed out that the distinction between A.C. and D.C.

heating mechanisms can be made by noting whether the generated coronal current closes within the corona or only within the photosphere. For example, A.C. currents have a cross-field component ($j_\perp$) within the corona allowing closure to occur outside of the photosphere whereas D.C. currents close only within the photosphere. This is an extremely important difference since current closure within the corona (i.e., A.C. currents) requires that the polarization properties of the coronal plasma play an important role in determining the incoming electrodynamic Poynting flux. The cross-field component of an A.C. current is essentially the 'polarization current', $j_\perp = j_{pol} = c^2 \dot{E}_\perp /4\pi v_A^2$, and plays a key role in the periodic exchange of mechanical and magnetic energy within the corona. This energy exchange occurs simply because there is a Lorentz force associated with the polarization current, and this force drives the coronal plasma into motion at the expense of previously stored magnetic energy. Such an exchange process causes magnetic energy to propagate as a wave which immediately brings to mind the concept of 'resonance'. In general, resonant systems can be 'under-damped' (i.e., high Q) or 'overdamped' (i.e., low Q). Also, if a resonant system is excited by a relatively broad band (i.e., in frequency) driver, then the induced currents include contributions from 'underdriving' (i.e., $v < v_{res}$ is the resonance frequency) and 'over-driving' (i.e., $v > v_{res}$) portions of the spectrum. Furthermore, resonant systems can in general possess more than one resonance frequency. For example, the resonance frequencies for coronal loops correspond to harmonics of the inverse Alfvén transit time, $t_A = 2l/v_A$, along the loop (i.e., $v_{res} = m/t_A$ where m is the harmonic number). Although Ionson (1982, 1983, 1984) only considered the fundamental (i.e., $m = 1$) it is a simple matter to extend the general theory to higher harmonics as described later. The important point made by Heyvaerts and Priest (1984), however, is that the 'zeroth' harmonic (i.e., $m = 0$) corresponds to a D.C. Poynting flux. This can be understood by noting that the zeroth harmonic (i.e., $m = 0$) implies a non-wavelike, or more accurately, a 'non-resonant' coupling between the convective driver and the responding coronal loop simply because the Lorentz restoring force goes to zero within the corona. In the language of Ionson's equivalent circuit theory the $m = 0$ harmonic decouples the field-aligned LR circuit from the cross-field LRC circuit.

It is important to note that the field-aligned and cross-field circuits only couple together into an equivalent LRC circuit when $m = 1, 2, 3, \ldots$, etc.

From the MHD equations and using an averaging procedure Ionson (1982, 1983, 1984, 1985) reduced the loop electrodynamics to the following equivalent LRC circuit equation:

$$L \frac{d^2 I(t)}{dt^2} + R \frac{d I(t)}{dt} + \frac{I(t)}{C} = \frac{d \mathscr{E}(t)}{dt} . \tag{7}$$

If one includes the harmonic number, m, which appears by allowing $\partial/\partial z \to im\pi/l$ where l is the coronal loop length, then the equivalent capacitance, $C = c^2 l/4\pi v_A^2 m^2$, goes to infinity for the zeroth harmonic (i.e., $m = 0$) indicating that the field aligned currents are not influenced by the dielectric properties of the coronal plasma. This implies that for the $m = 0$ harmonic, j_z and $j_\perp$ completely decouple within the corona forcing the

induced currents to close within the photosphere. In Equation (7), the equivalent induc-
tance, $L = 4l/\pi c^2$, $I(t)$ represents the total net current (i.e., as opposed to the current
density, $\mathscr{E}(t) = (2v_\perp Bl_B)_{\beta \gtrsim 1}/c$ in the driving emf and $R = L/t_{\text{loss}}$ is related to electrodynamic energy 'loss' resulting from dissipation and magnetic stress leakage out of the
loop. Since the time-averaged coronal heating flux F_H, is given by

$$F_H = \frac{\langle I^2 R_{\text{diss}} \rangle}{(\pi l_B^2/4)}.$$

$$(8)$$

Equation (8) immediately results in Equations(1) and (2) for F_H and F_0, with the full
electrodynamic coupling coefficient being given by

$$\varepsilon = \sum_m \int_{-8}^{\infty} \left(\frac{\langle v_\perp^2 \rangle_v}{v_{\text{tot}}^2} \right)_{\beta \gtrsim 1} \frac{Q_{\text{loss}}^2}{Q_{\text{diss}} \left(1 + \left(vt_A - \frac{m^2}{vt_A} \right)^2 Q_{\text{loss}}^2 \right)},$$

$$(9)$$

where

$$Q_{\text{loss}} \equiv \frac{2\pi t_{\text{loss}}}{t_A} \quad \text{and} \quad Q_{\text{diss}} \equiv \frac{2\pi t_{\text{diss}}}{t_A}.$$

Comparing Equation (9) with Equation (6) one readily finds that the parameterized
velocity-magnetic field interaction parameter $e(v; t_A, t_{\text{loss}}, t_{\text{diss}})$ is given by

$$e(v; t_A, t_{\text{loss}}, t_{\text{diss}}) = \frac{Q_{\text{loss}}^2}{Q_{\text{diss}}(1 + v^2 t_A^2 Q_{\text{loss}}^2)} +$$

$$+ \sum_{m > 0} \frac{Q_{\text{loss}}^2}{Q_{\text{diss}} \left[1 + \left(vt_A - \frac{m^2}{vt_A} \right)^2 Q_{\text{loss}}^2 \right]},$$

$$(10)$$

where the first term represents non-resonant (D.C.) coupling and the second term
represents resonant (A.C.) coupling. Note that the resonance terms peak at the
resonance frequencey, $v_{\text{res}} = m/t_A$, with a maximum value of $Q_{\text{loss}}^2/Q_{\text{diss}}$ (per harmonic)
and can be thought of a measure of the magnitude of the magnetic stressing rate
spectrum, $\langle v_\perp B_\perp' \rangle_{v_{\text{res}}}^{\beta \gtrsim 1} B_{\beta \gtrsim 1}/4\pi$, with respect to the $\beta \gtrsim 1$ energy flux spectrum,
$\langle \frac{1}{2} \rho v_\perp^2 \rangle_{v_{\text{res}}}^{\beta \gtrsim 1} v_A^{\beta \gtrsim 1}$. The characteristic width of $e(v; t_A, t_{\text{loss}}, t_{\text{diss}})$ about this maximum
can be thought of as an interaction bandwidth $\Delta v_{\text{interaction}} = m\pi/t_A Q_{\text{loss}}$. Since the
electrodynamic coupling coefficient ε is a convolution of the normalized $\beta \gtrsim 1$ convection spectrum, $(\langle v_\perp^2 \rangle_v/v_{\text{tot}}^2)_{\beta \gtrsim 1}$ with the velocity-magnetic field interaction parameter, a general form for the normalized convection spectrum that parameterizes the
details will be assumed, viz.,

$$\left(\frac{\langle v_\perp^2 \rangle_v}{v_{\text{tot}}^2} \right)_{\beta \gtrsim 1} = \sum_{t_c, \bar{v}^2} \frac{\bar{v}^2 t_c/\pi}{\left[1 + \left(vt_c - \frac{1}{vt_c} \right)^2 \right]},$$

$$(11)$$

where I have assumed that the period of a local peak is equal to the inverse frequency half-width t_c of the local peak. In addition $\bar{v}^2$ is the fraction of the total spectrum associated with a local peak. The sum in Equation (11) allows the possibility of multiple peaks in the normalized velocity spectrum. Using this parameterized form for the velocity spectrum, and evaluating the integral in Equation (9) by contour integration, results in the following general expression for the $m = 0$ and $m > 0$ electrodynamic coupling coefficients (i.e., $\varepsilon \equiv \varepsilon_0 + \sum_{m>0} \varepsilon_m$):

$$\varepsilon_0 = \sum_{t_c,\, \bar{v}^2} \left(\frac{t_{\text{loss}}}{t_{\text{diss}}}\right)\left(\frac{2\pi t_{\text{loss}}}{t_A}\right)\left[\frac{x^2 - x + 1}{x^4 + x^2 + 1}\right]; \tag{12}$$

for the $m = 0$ case t_{loss} is probably equal to t_{diss};

$$\varepsilon_m = \sum_{t_c,\, \bar{v}^2} \left(\frac{t_{\text{loss}}}{t_{\text{diss}}}\right)\left(\frac{2\pi t_{\text{loss}}}{t_A}\right) \times$$

$$\times \begin{cases} \dfrac{1}{x}\left[1 + \left(\dfrac{t_c m}{t_A} - \dfrac{t_A}{t_c m}\right)^2\right]^{-1} & \text{if } \dfrac{2\pi m t_{\text{loss}}}{t_A} > 1, \\[2em] \dfrac{x^2 + 1}{x^4(1 + m^8 t_c^8/t_A^8) + x^2(1 + m^4 t_c^4/t_A^4) + 1} + \\[2em] + \dfrac{x^7 m^{10} t_c^{10}/t_A^{10}}{x^4 t_c^8 m^8/t_A^8 + x^2 t_c^4 m^4/t_A^4 + 1} - \\[2em] - \dfrac{x}{x^4 + x^2 + 1} & \text{if } \dfrac{2\pi m t_{\text{loss}}}{t_A} < 1, \end{cases} \tag{13}$$

where

$$x \equiv 2\pi t_{\text{loss}}/t_c . \tag{14}$$

Note that the solution when $2\pi t_{\text{loss}}/t_A \sim 1$ for A.C. mechanisms (ie., $m > 0$) does not have a convenient analytic representation. For this critically damped category it is best to evaluate the integral in Equation (9) numerically.

Equations (12) and (13) clearly embody two basic global coupling categories, viz., non-resonant (D.C.) and resonant (A.C.) coupling. It is important to note that all D.C. heating theories such as Parker (1983) and Heyvaerts and Priest (1984) emerge from Equation (1), (2), and (12) when $t_{\text{loss}} \sim t_{\text{diss}} \ll t_c$ where t_c can be thought of as the correlation time of the driving convection. In addition all A.C. heating theories are contained by Equation (13) (cf. Ionson (1984, 1985) for a comparison with other A.C. coupling schemes).

A point of immediate interest is that the electrodynamic coupling coefficient for the zeroth harmonic exhibits a maximum when the dissipation time is of the order of the convective correlation time and smaller than the magnetic flux leakage timescale.

Furthermore, although higher harmonics of the fundamental (i.e., $m = 1$) A.C. processes play a role in heating the coronal loop plasma, the $m = 1$ harmonic appears to dominate for solar coronal loops. Furthermore, in many cases of interest the heating rate is not sensitive to the details of the dissipation physics.

In order to illustrate these points as well as apply this extended theory to solar coronal loops, one must first adopt a set of appropriate solar conditions. Specifically, let us assume that $v_A = 5 \times 10^7$ cm s^{-1}, $t_c = 250$ s (750 s) and $\bar{v}^2 = 0.67$ (0.33). The numbers in parentheses refer to $\beta \gtrsim 1$ granulation while those outside refer to p-mode oscillation (cf. Figure 4). I have also incorporated Hollweg's (1984) latest estimate for

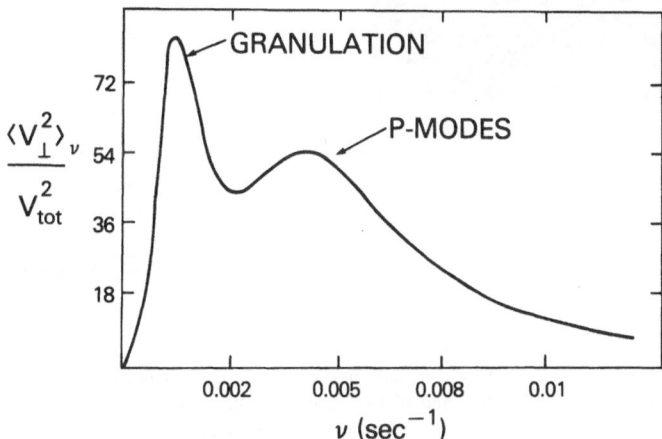

SOLAR CONVECTION SPECTRUM

Fig. 4. Model solar convection spectrum illustrating *P*-mode oscillation and granulation.

the leakage time-scale, i.e., $t_{\text{leak}}/t_A \approx 25$. Since a dissipation model does not exist at this time, it is difficult to estimate a viable dissipation time-scale, t_{diss}. I have, therefore, allowed for an eleven order of magnitude variation in the dissipation factor, $2\pi t_{\text{diss}}/t_A$, when numerically evaluating the electrodynamic coupling coefficient (i.e., Equations (12) and (13)). The results are illustrated in Figures 5–7 for young active region loops ($l = 1.2 \times 10^9$ cm), active region loops ($l = 6 \times 10^9$ cm) and large scale loops ($l = 1.6 \times 10^{10}$ cm). Note that the top half of these figures represents the total electrodynamic coupling coefficient (i.e., summed over the first twenty harmonics) while the bottom half highlights the relative contribution to the total from the first five harmonics. The point of major interest illustrated by this set of figures is that young active region loops (cf. Figure 5) are most likely heated by D.C. mechanisms while active region (cf. Figure 6) and large scale loops (cf. Figure 7) are heated by A.C. mechanisms. Furthermore, it is important to note that this conclusion is viable for a variation in the dissipation time-scale of over eleven orders of magnitude. However, one should also

YOUNG ACTIVE REGION LOOPS
($\ell = 1.2 \times 10^9$cm)

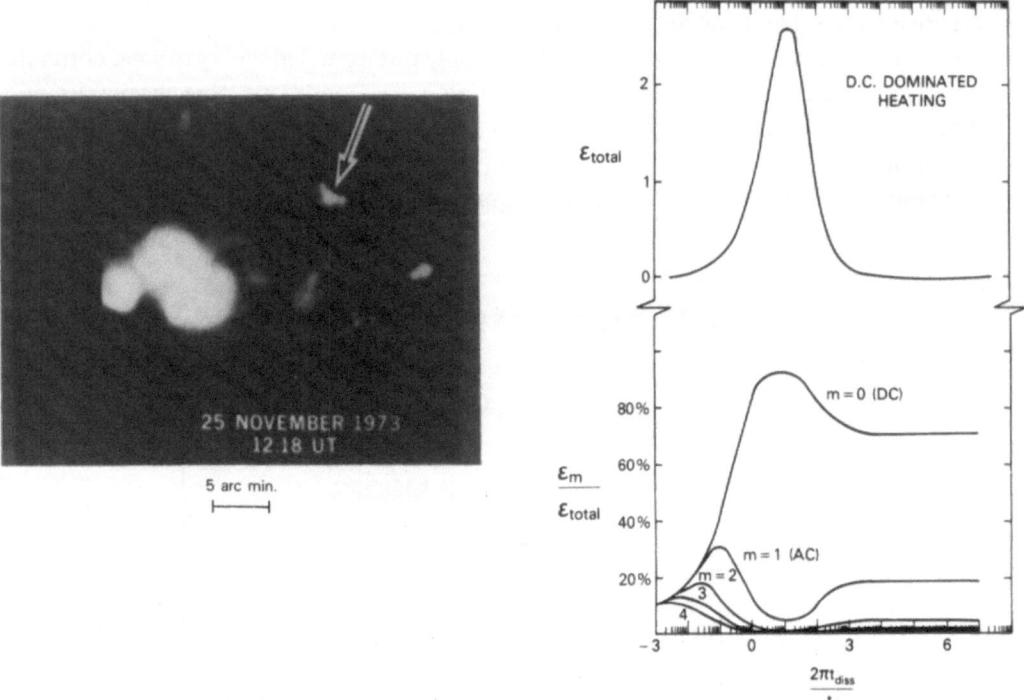

Fig. 5. Functional dependence of the electrodynamic coupling coefficient with the dissipation time-scale for young active region loops. The top half of the right-hand figure represents the total electrodynamic coupling coefficient (i.e., summed over the first twenty harmonics) while the bottom half highlights the relative contribution to the total from the first five harmonics.

note that this conclusion is relatively sensitive to the assumed form for the convection spectrum (cf. Figure 4) since the efficiency of D.C. mechanisms is limited by the constraint that the dissipation time be comparable to the convection correlation time. Assuming that this convection spectrum is reasonable for the Sun, it appears that the total electrodynamic coupling coefficient for young active region loops can be about a factor of two larger than that for active region and large-scale loops. Although this is not a substantial difference in the coupling efficiency, it could result in a considerable difference in the heating flux. Specifically, if since F_0 scales with B^2 and since the magnetic field for young active region loops can be a factor of ten larger than that for active region and large-scale loops, one comes to the interesting conclusion that D.C. heating of small loops can exceed that of larger loops by at least two orders of magnitude! This important result confirms Heyvaerts and Priest's (1982) suggestion that there could be a relationship between coronal heating and flaring activity.

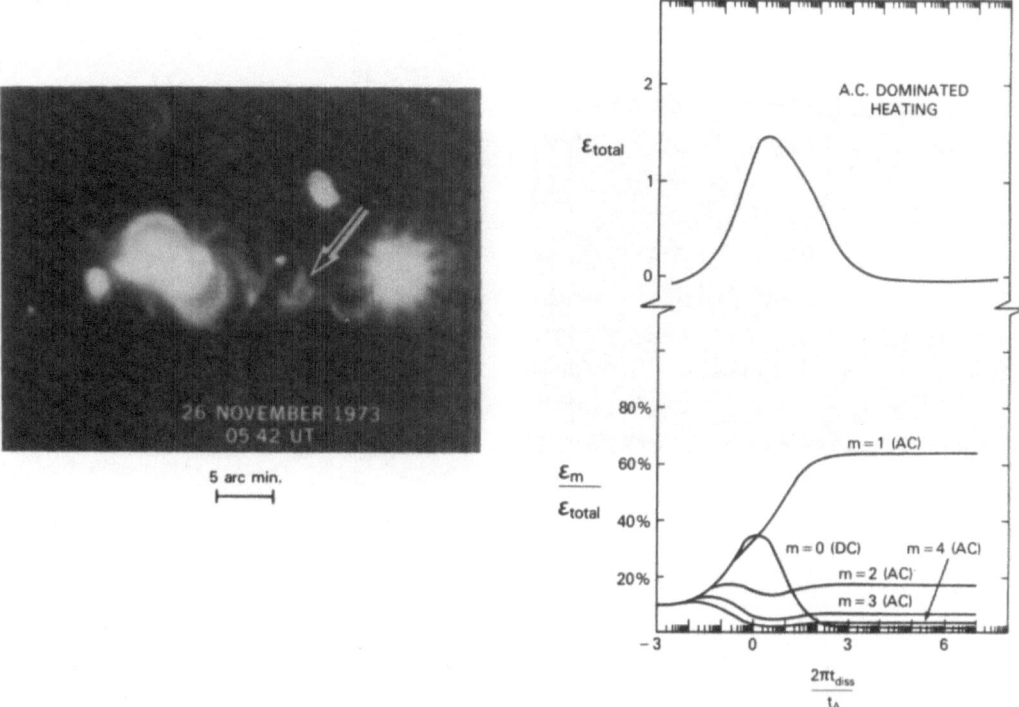

Fig. 6.   Same as Figure 5 but for active region loops.

2.2. THE DISSIPATION MODEL

One should recognize that a complete model of electrodynamic coupling involves more than the development of global coupling scenario. To the extent that the electrodynamic energy flux F_H depends explicitly upon t_{loss} and t_{diss} it is important that we also investigate these various processes. Even if F_H is explicitly independent of the dissipation time, it's important to note that irreversibilities still play a vital role. In fact, the rms amplitude of the induced current, which is associated with the microscopic (i.e., electronic) and macroscopic (i.e., ionic) reactance of the plasma contained by the magnetic loop, does indeed depend upon irreversible processes such as electron-ion collisions. The microscopic reactance is related to the field-aligned current density $j_{\parallel}$ which is carried by the electrons drifting at velocity v_d along the ambient field. The macroscopic reactance is related to the cross-field current density $j_{\perp}$ which through Lorentz forces drives an azimuthal flow of coronal plasma with velocity $v_{\perp}$. The appearance of irreversible phenomena is also explicit in expressions for the electron drift speed v_d, the plasma's azimuthal flow velocity $v_{\perp}$, and the width Δr of the coronal field lines that are subject to electrodynamic activity.

302   JAMES A. IONSON

LARGE SCALE LOOPS
($\ell = 1.6 \times 10^{10}$ cm)

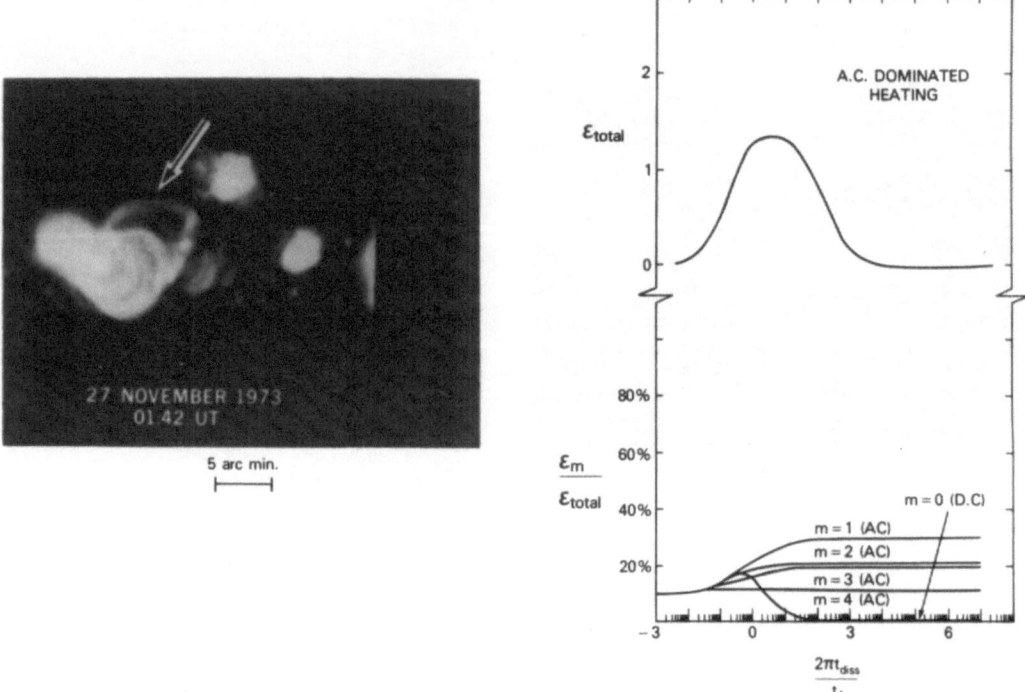

Fig. 7.   Same as Figure 5 but for large-scale loops.

Knowledge of the plasma reactance (i.e., the electron drift velocity v_d, and plasma flow velocity $v_\perp$) derived by a linear formalism is very useful in determining the relevance of various nonlinear processes. For example, microscopic nonlinearities associated with the destabilization of the field-aligned electron current could result in an anomalous increase in the electron collision frequency. This effect generally occurs when $(v_d/v_{sound})_{corona} > 1$ and would result in both anomalous current dissipation and anomalous viscous dissipation. Thus the ratio of ion to electron heating could be significantly affected by the onset of microscopic plasma turbulence. In addition, the thermal versus nonthermal energy partitioning of the 'heated' electrons is affected, with the nonthermal component becoming increasingly enhanced as the ratio $(v_d/v_{sound})_{corona}$ increases beyond unity (cf. Ionson, 1982).

Other nonlinearities that could develop would be driven by the macroscopic flow field $(v_\perp)_{corona}$. Specifically, when $(v_\perp/v_A)_{corona}$ becomes of order unity, the induced magnetic field becomes comparable to the ambient field B_0. Thus, in addition to the possible development of shocks, wave-wave interactions resulting in the nonlinear generation of sound waves (cf. Wentzel, 1976) become increasingly efficient. It should, therefore, be evident that a linear analysis is very useful in determining the relevance of various nonlinear processes.

Since the observed coronal line broadening implies a velocity field of at most 100 km s^{-1} (Acton et $al.$, 1981), the condition $B'/B < 1$ is generally satisfied for non-flaring loops (i.e., B' is a magnetic perturbation). Therefore, linearization is a reasonable starting point. The linearized equation for the magnetic field perturbation $\mathbf{B}'$ and associated coronal velocity perturbation $\mathbf{v}'$ are

$$\frac{\partial \mathbf{B}'_\perp}{\partial t} = (\mathbf{B} \cdot \nabla)\mathbf{v}'_\perp - \frac{\mathbf{B}'_\perp}{\tau_{\text{Joule}}} , \tag{15}$$

$$\frac{\partial \mathbf{v}'_\perp}{\partial t} = \frac{1}{4\pi\rho}[-4\pi\nabla_\perp P' + (\nabla \times \mathbf{B}') \times \mathbf{B} + (\nabla \times \mathbf{B}) \times \mathbf{B}']_\perp - \frac{\mathbf{v}'_\perp}{\tau_{\text{viscous}}} , \tag{16}$$

where the subscript $\perp$ refers to directions that are perpendicular to the ambient field $\mathbf{B}$ and where P' is the thermal pressure fluctuation. Joule and viscous dissipation are represented by diffusion time-scales t_{Joule} and t_{viscous}, i.e.,

$$t_{\text{Joule}} = \left(\frac{t_A}{2\pi^2}\right) \text{Re}_{\text{mag}} \left(\frac{\Delta r}{l_B}\right)^2_{\text{Joule}} , \tag{17}$$

$$t_{\text{viscous}} = (t_{\text{shear}}^{-1} + t_{\text{comp}}^{-1})^{-1} , \tag{18}$$

with

$$t_{\text{shear}} = \frac{t_{\text{Joule}}}{m \, \text{Pr}_{\text{mag}}^\perp} \left(\frac{\Delta r_{\text{shear}}}{\Delta r_{\text{Joule}}}\right)^2 , \tag{19}$$

$$t_{\text{comp}} = \frac{3\pi}{2} \left(\frac{l}{l_B}\right) \left(\frac{t_{\text{Joule}}}{m \, \text{Pr}_{\text{mag}}^\parallel}\right) \left(\frac{\Delta r}{l_B}\right)^{-2}_{\text{Joule}} . \tag{20}$$

Note that the total dissipation time depends upon the resistivity and viscosity coefficients as well as the degree of current localization within coronal filaments of thickness Δr. As discussed in Ionson (1977, 1978, 1982, 1983, 1984), the thickness of these filaments scales with the resistivity and viscosity. The fundamental form of this scaling, however, depends upon the mechanisms responsible for the filamentation. For example, Ionson (1982) has discussed filamentation of the currents by a resonance absorption process which linearly couples large-scale fluctuations into smaller scales localized about a resonance absorption layer whose thickness is of the order of the dissipation scale length. As discussed by Heyvaerts and Priest (1983) this process is very closely related to phase mixing. Other possible filamentation processes include Joule mode instabilities (Ferrari et $al.$, 1982), magnetic tearing (Spicer, 1981), dynamical reconnection (Parker, 1981, 1983) and nonlinear cascades (Van Ballegooijen, 1985). For now, these important details will be absorbed within t_{Joule}, t_{shear}, and t_{comp} which combine to form the total dissipation time, $t_{\text{diss}} = t_{\text{Joule}}^{-1} + t_{\text{shear}}^{-1} + t_{\text{comp}}^{-1}$, where Δr_{Joule},

Δr_{shear} in Equations (17)–(19) refer to filamentation based upon Joule and shear viscous dissipation driven processes, respectively. Note also that the magnetic Reynolds numbers, Re_{mag}, and magnetic Prandtl numbers, $\text{Pr}_{\text{mag}}^{\perp}$, are given by

$$\text{Re}_{\text{mag}} \equiv \frac{4\pi v_A l_B^2}{\eta c^2 l}, \tag{21}$$

$$\text{Pr}_{\text{mag}}^{\perp} \equiv \frac{4\pi \mu_{\perp}}{\rho c^2 \eta}, \tag{22}$$

$$\text{Pr}_{\text{mag}}^{\parallel} \equiv \frac{4\pi \mu_{\parallel}}{\rho c^2 \eta}. \tag{23}$$

In Equations (21)–(23) the resistivity and viscosity transport coefficients are given by

$$\eta = \frac{4\pi v_e}{\omega_{pe}^2}, \tag{24}$$

$$\mu_{\perp} = \rho \rho_{ci}^2 v_i / (1 + v_i^2/\Omega_i^2), \tag{25}$$

$$\mu_{\parallel} = \mu_{\perp}(1 + \Omega_i^2/v_i^2), \tag{26}$$

with v_e representing an effective electron collision frequency and where the effective ion collison frequency, $v_i \approx (m_e/m_i)^{1/2} v_e$. It is important to also bear in mind that kinetic effects could become important (cf. Ionson, 1978; regarding kinetic Alfvén waves) if the filamentation width Δr becomes comparable to the ion gyro radius.

Using the identity

$$[\nabla \times \mathbf{B}') \times \mathbf{B} + (\nabla \times \mathbf{B}) \times \mathbf{B}']_{\perp} = -\nabla_{\perp}(\mathbf{B} \cdot \mathbf{B}') + (\mathbf{B} \cdot \nabla)\mathbf{B}'_{\perp}, \tag{27}$$

Equation (16) becomes

$$\frac{\partial \mathbf{v}'_{\perp}}{\partial t} = \frac{1}{4\pi\rho}(\mathbf{B} \cdot \nabla)\mathbf{B}'_{\perp} - \frac{\mathbf{v}'_{\perp}}{\tau_{\text{viscous}}} - \frac{1}{\rho}\nabla_{\perp}\left(P' + \frac{\mathbf{B}' \cdot \mathbf{B}}{4\pi}\right). \tag{28}$$

The third term on the right-hand side of Equation (28) represents a nondissipative coupling between magnetic field lines. This term vanishes for purely incompressible fluctuations. In general, however, the field line oscillations are slightly compressible, as pointed out by Wentzel (1979) and Ionson (1982). Therefore, the coupling term, $\rho^{-1}\nabla_{\perp}(P' + \mathbf{B}' \cdot \mathbf{B}/4\pi)$, must be retained in the analysis, resulting in a component of the Poynting flux across field lines $B'_{\parallel} \neq 0$. This is an important effect because it allows 'phase' information to be communicated between field lines, thereby coupling global modes such as Alfvénic surface waves to local field line resonance (Ionson, 1977, 1978, 1982, 1983). Heyvaerts and Priest (1984) did not include this coupling term because they assumed purely imcompressible fluctuations (see their Equation (5)). Thus, their conclusion that an inhomogeneous loop should be thought of as a collection of oscillators weakly coupled by friction is correct in the limit of zero compressibility. If the field lines are also coupled by magnetic pressure, then a loop does indeed behave as a single oscillator as emphasized by Ionson. Although Ionson's (1982) mathematical analysis did not clarify this point, the coupling between global 'modes' and local field line resonances was implicit within the physical analysis.

Since there is an assumed gradient in the Alfvén speed across the ambient field, it follows that each field line has it own natural oscillation frequency. The bandwidth of this continuum of frequencies is given by $(\Delta\omega)_{\text{continuum}} \approx \Delta v_{\text{A}}/v_{\text{A}} t_{\text{A}}$ where $\Delta v_{\text{A}}/v_{\text{A}}$ represents the change in the coronal Alfvén speed across a loop and where t_{A} is the average Alfvén transit time along the loop. An interesting consequence of this inhomogeneity in the Alfvén speed is that the relative phase difference between field lines 'mixes' in a time that is proportional to $1/(\Delta\omega)_{\text{continuum}}$. The actual phase mixing time t_{phase} must also depend upon the degree of initial phase locking at the coronal base measured by $(l_c/l_B)_{\beta \gtrsim 1}$ where $(l_c)_{\beta \gtrsim 1}$ is the cross-field correlation scale length of the $\beta \gtrsim 1$ convection and where $(l_B)_{\beta \gtrsim 1}$ is the Alfvén speed's cross-field gradient scale length at the coronal base. This phase mixing time t_{phase} is, therefore, expected to have the following form:

$$t_{\text{phase}} \approx \left(\frac{l_c}{l_B}\right)_{\beta \gtrsim 1} \left(\frac{v_{\text{A}}}{\Delta v_{\text{A}}}\right) t_{\text{A}}, \tag{29}$$

which also follows indirectly from the coupling term, $\rho^{-1}\nabla_\perp(P' + \mathbf{B}' \cdot \mathbf{B}/4\pi)$, in Equation (28).

To emphasize the physics of phase mixing it is instructive to replace the dissipationless field line coupling term by $\mathbf{v}'_\perp/t_{\text{phase}}$, reducing Equation (28) to the following model form:

$$\frac{\partial \mathbf{v}'_\perp}{\partial t} = \frac{1}{4\pi\rho}(\mathbf{B} \cdot \nabla)\mathbf{B}' - \left(\frac{1}{\tau_{\text{viscous}}} + \frac{1}{\tau_{\text{phase}}}\right)\mathbf{v}'_\perp. \tag{30}$$

It is important to note that t_{phase} represents the time it takes for individual field line oscillations associated with an initially coherent global fluctuation to become incoherent. Therefore, t_{phase} is best thought of as a coherency 'loss' time with the dissolution of coherency resulting in a corresponding decrease in the amplitude. Thus, t_{phase} is a dissipationless (i.e., reversible) damping time which, not surprisingly, equals the damping time of an Alfvénic surface wave in an ideal plasma (i.e., t_{phase} as given by Equation (29) is equal to the inverse of the surface wave damping rate, γ_S^{-1} given by Ionson (1978, Equation (64)).

Ionson's (1977, 1978) theory of coronal heating by Alfvénic surface waves was the first theory that depended explicity upon an Alfvén speed variation across the coronal loop. In these articles, Ionson assumes that the cross-field correlation scale length $(l_c)_{\beta \gtrsim 1}$ of the $\beta \gtrsim 1$ convection is comparable to the Alfvén speed's gradient scale length, $(l_B)_{\beta \gtrsim 1}$, at the base of a coronal loop. This condition results in phase locking between different field lines at the coronal base. Thus, individual field lines are not driven independently, which allows excitation of what Ionson calls standing 'Alfvénic surface waves'. This is somewhat of a misnomer because these waves do not require a discontinuous surface. Therefore, in a subsequent paper Ionson (1982) refers to global structure oscillations which according to his analysis can also be thought of as *LC* oscillations. The reader should be aware that standing Alfvénic surface waves, global structure oscillations, and *LC* oscillations are different names for the same physical manifestation – namely, fluctuations in the magnetic field and velocity that are coherent across a coronal loop for some period of time. Since the Alfvén speed varies across the loop, field line oscillations associated with an Alfvénic surface wave communicate phase information across field lines at different speeds. Therefore, as already discussed, the phases mix over a time-scale given by $t_{\text{phase}} \approx \gamma_S^{-1}$, implying a corresponding reduction of the global amplitude, even in the absence of dissipation. Ionson's main point, however, was that in a real plasma phase mixing also represents an intermediate mechanism that linearly transfers large-scale fluctuations into smaller cross-field scales suitable for dissipation. Specifically, linear superposition of phase-mixed field line oscillations results in a degradation of the fluctuation's cross-field correlation length until finally the correlation length is of the order of the dissipation scale. Ionson postulated that in steady state, the rate at which energy is transferred from large to small scales by phase mixing (i.e., resonance absorption) equals that dissipated within a layer centered upon the resonant field. As stated by Heyvaerts and Priest (1983), 'phase mixing can always proceed to a sufficiently advanced stage that the gradients match any dissipative efficiency no matter how small it is' (cf. the second sentence following their Equation (65)).

Linear filamentation processes such as those described above have been studied extensively. Although they can, in theory, provide the required heating rates, the associated coronal velocities are quite larger than those observed. Specifically, for the Sun about 30 km s^{-1} is observed while about 100 km s^{-1} is predicated based upon linear filamentation mechanisms (Ionson, 1983). Therefore, we have still not unravelled the physics of the dissipation mechanism.

3. Conclusion

Probably the most significant breakthroughs in understanding the heating of coronae has emerged from a global electrodynamic approach. In this regard, it is appropriate that I quote some of L. Cram's summary remarks of the *IAU Symposium* **102** on 'Solar and Stellar Magnetic Fields: Origins and Coronal Effects' (1982): "As reviewed by Chiuderi in this volume, there has been a major conceptual breakthrough in the related problem of coronal heating, provided by the 'electrodynamic coupling' model of Ionson (1982). This is not an entirely new approach to problems in cosmic electrodynamics, since 'circuit models' have been used widely in magnetospheric studies for several decades (e.g., Alfvén, 1979). The application of this approach to coronal heating is new, however, and the explicit derivation of the equivalent circuit is more firmly based on physical principles than the picturesque models that are sometime used in these studies.

Since several heated debates regarding this model arose, both inside and outside the

Symposium, it is appropriate to make a few comments at this point. First, it may be noted that the existing theories of coronal heating were unfruitful. These theories focussed on specific processes that might transport and dissipate energy in the corona, but they could not be tested because either (i) the theories were too abstract to be applicable to the solar atmosphere (which is highly inhomogeneous on all scales, or (ii) the crucial tests could not be made because of observational constraints (e.g., the coronal magnetic field cannot be measured accurately). In addition to these objections, the existing theories have generally assumed that the problem could be divided into one part dealing with the propagation and dissipation of energy, and another dealing with the generation of the necessary Poynting flux. This approach could never explain the coupling between coronal heating and the primary energy source in the convection zone.

It is important to judge Ionson's model in the spirit of its objectives, and not in terms of the older, failed approaches. For, example, by focussing on the global electrodynamic coupling, Ionson cannot answer questions regarding the details of specific generation, propagation, or dissipation modes. This is seen as a shortcomming by many commentators, but future work within the spirit of the global model will be able to make specific predictions that may prove to be more readily tested than those made by previous workers. Finally, Ionson's model provides a close analogy between the electrodynamic behaviour of the solar atmosphere and the physics of electrical circuits, and thus provides a new way of looking at several old problems. It is quite enlightening to reverse the analogy, and to seek parallels in the solar atmosphere for several processes that are familiar in electrical circuit theory (the reader may speculate on the parallels between open transmission lines and coronal holes, or electrical discharges and solar flares, or analogies to feedback processes)."

My own view is that since we now have a relatively firm grasp of the global coupling process, special emphasis should be placed upon the detailed dissipation physics. Although linear filamentation theories have not been successful, they do point the way towards the need to investigate nonlinear filamentation (i.e., shredding) mechanisms. For example, it may be that linearly driven resonance absorption layers nonlinearly break up into many narrower filaments. This would enhance the dissipation efficiency, but at the same time keep the reactive coronal velocity field at an acceptable level.

So far, I do not feel that anyone has presented a viable dissipation mechanism. I am, however, quite certain that the answer will involve localized nonlinearties associated with current filamention, and that the best way to confirm the dissipation model will be from a global coupling model that predicts associated heating and velocity fields (e.g., resulting in line broadening).

Acknowledgement

I would like to take this opportunity to thank Alice Gehl for her help in preparing this manuscript.

References

Acton, L. W., Culhane, J. L., Gabriel, A. H., Wolfson, C. J., Rapley, C. G., Phillips, K. J. H., and 12 co-authors: 1981, *Astrophys. J.* **244**, L137.

Chiuderi, C.: 1981, in R. M. Bonnet and A. K. Dupree (eds.), *Solar Phenomena in Stars and Stellar Systems*, D. Reidel Publ. Co., Dordrecht, Holland, p. 267.

Chiuderi, C.: 1982, in J. O. Stenflo (ed.), *Solar and Stellar Magnetic Fields: Origins and Coronal Effects*, D. Reidel Publ. Co., Dordrecht, Holland, p. 375.

Cowling, T. G.: 1981, *Ann. Rev. Astrophys.* **19**, 115.

Ferrari, A., Rosner, R., and Vaiana, G. S.: 1982, *Astrophys.* **263**, 944.

Foukal, P.: 1975, *Solar Phys.* **43**, 327.

Foukal, P.: 1976, *Astrophys. J.* **210**, 575.

Foukal, P.: 1978, *Astrophys. J.* **223** 1046.

Galeev, A. A., Rosner, R., and Vaiana, G. S.: 1979, *Astrophys. J.* **229**, 318.

Gilman, P. A.: 1980, *Highlights Astrophys.* **5**, 19.

Gilman, P. A.: 1981, *The Sun as a Star*, NASA SP-450, p. 1.

Gilman, P. A. and Miller, J,: 1981, *Astrophys. J. Suppl.* **46**, 211.

Golub, L., Davis, J. M., and Krieger, A. S.: 1979, *Astrophys. J.* **229**, L145.

Golub, L., Maxson, C., Rosner, R., Serio, S., and Vaiana, G. S.: 1980, *Astrophys. J.,* **238**, 343.

Heyvaerts, J. and Priest, E. R.: 1982, *Astron. Astrophys.* **117**, 220.

Heyvaerts, J. and Priest, E. R.: 1984, *Astron. Astrophys.* **137**, 63.

Hollweg, J.: 1981, in F. Orrall (ed.), *Proc. Skylab Workshop on Active Regions*, Chap. 8.

Hollweg, J.: 1984, *Solar Phys.* **91**, 269.

Ionson, J. A.: 1977, Ph. D. Thesis, University of Maryland.

Ionson, J. A.: 1978, *Astrophys. J.* **226**, 650.

Ionson, J. A.: 1982, *Astrophys. J.* **254**, 318.

Ionson, J. A.: 1983, *Astrophys. J.* **271**, 778.

Ionson, J. A.: 1984, *Astrophys. J.* **276**, 357.

Ionson, J. A., and Kuperus, M.: 1984, *Astrophys. J.* **284**, 389.

Ionson, J. A.: 1985, *Astron. Astrophys.* **146**, 199.

Kuperus, M. and Ionson, J. A.: 1985, *Astron. Astrophys.* **148**, 309.

Kuperus, M., Ionson, J. A., and Spicer, D. S.: 1981, *Ann. Rev. Astron. Astrophys.* **19**, 7.

Parker, E. N.: 1981, *Astrophys. J.* **224**, 644.

Parker, E. N.: 1983, *Astrophys. J.* **264**, 642.

Raymond, J. C. and Foukal, P.: 1982, *Astrophys. J.,* **253**, 323.

Rosner, R. and Hartquist, T. W.: 1979, *Astrophys. J.* **231**, L83.

Scott, J. S.: 1982, private communication.

Spicer, D. S.: 1981, in F. Praderie, D. S. Spicer, and G. L. Withbroe (eds.), Activity and Outer Atmospheres of the Sun and Stars', *11th Advanced Course Swiss Society of Astronomy and Astrophysics*, p. 89.

Sturrock, P. A. and Stern, R.: 1980, *Astrophys. J.* **238**, 98.

Vaiana, G. S. and 15 co-authors: 1981, *Astrophys. J.* **245**, 163.

Van Hoven, G.: 1981, in E. R. Priest (ed.), *Solar Flare Magnetohydrodynamics,* Gordon and Breach, New York, p. 217.

van Ballegooijen, A. A.: 1985, *Astrophys. J.* (in press).

Wentzel, D. G.: 1976, *Solar Phys.* **50**, 343.

Wentzel, D. G.: 1978, *Rev. Geophys. Space Phys.* **16**, 757.

Wentzel, D. G.: 1979a, *Astrophys. J.,* **227**, 319.

Wentzel, D. G.: 1979b, *Astron. J.* **76**, 20.

SOME RECENT DEVELOPMENTS IN THE THEORETICAL DYNAMICS OF MAGNETIC FIELDS

B. C. LOW

High Altitude Observatory, National Center for Atmospheric Research*, Boulder, CO 80307, U.S.A.

Abstract. This article describes recent developments in the theoretical investigation of magnetostatic equilibrium in the presence of gravity, nonequilibrium in hydromagnetics, and classical problems in hydromagnetic stability. The construction of magnetostatic equilibria has progressed beyond geometrically idealized systems, such as the axisymmetric system, to fully three-dimensional systems capable of modelling realistic solar structures. Nonequilibrium in a magnetic field with an arbitrary interweaving of lines of force due to random footpoint motion is a novel and subtle property with important implications for the solar atmosphere. Work begun by Parker and subsequent developments are described. To the extent quasi-static solar structures are approximated by stable equilibrium, ideal hydromagnetic stability theory provides a first insight into how stability is achieved in the solar environment. A qualitative physical picture based on recent stability analyses is given. The article places emphasis on understanding basic principles and issues rather than detailed results which can be found in the published literature.

1. Introduction

Most physical phenomena in the solar atmosphere involve magnetic fields. In attempting to understand the role of magnetic fields in these phenomena, we have learned much about the dynamics of magnetic fields, particularly in the hydromagnetic approximation. This approximation is physically the simplest to adopt for the solar atmosphere. It treats the magnetized plasma as a fluid and neglects a variety of plasma processes as higher order effects (e.g., Friedberg, 1982). Even within such an approximation, the dynamics of magnetic fields is highly nonlinear, with properties not readily anticipated by intuition. Therefore, a great deal of effort has concentrated on working out the basic principles. In measured small steps, we broaden our physical intuition and bring ourselves closer to more sophisticated modelling capable of relating directly to observed phenomena. In the following sections, I will describe recent developments in modelling quasi-static magnetic structures, phenomena of hydromagnetic nonequilibrium and problems in linear hydromagnetic stability. These topics form a certain wholeness but do not exhaust all the hydromagnetic concerns in solar physics. Other important topics such as the dynamo theory, the mechanics of concentrating magnetic flux tubes in the photosphere, the dynamics of coronal mass ejections, etc., can easily occupy articles of similar lengths. With the three topics chosen for review in this article, I hope a coherent view will emerge as to what we can accomplish with hydromagnetic theory in solar physics.

2. Static Magnetic Structures

The solar atmosphere is an excellent electric conductor. Its never-ending evolution is accompanied with substantial induction of electric currents. The persisting electric

* The National Center for Atmospheric Research is sponsored by the National Science Foundation.

currents interact with the plasma medium, through the Lorentz force, to create the rich diversity of large-scale, quasi-static structures that populate the solar atmosphere. Structures like sunspots and prominences have long been studied and they continue to fascinate us. With the recent availability of observations of UV and X-ray emission, plasma loops of widely varying thermodynamic states, as well as coronal holes, have also attracted attention. As a first approximation in building models, we may neglect electric current dissipation and take a long-lived structure to be in static equilibrium. The governing equation for force balance is

$$\frac{1}{4\pi}(\nabla \times \mathbf{B}) \times \mathbf{B} - \nabla p - \rho g \hat{z} = 0, \tag{1}$$

where $\mathbf{B}$, p, ρ, and g are the magnetic field, pressure, density and the gravitational acceleration, respectively. For purpose of discussion, we have set the gravitational acceleration to be uniform and vertical. To close this equation, we need an equation for steady energy balance. The structure of an equilibrium state depends on how force equilibrium is achieved by its embedded magnetic field and electric current, and how various energy processes in the plasma operate to maintain thermodynamic equilibrium. These two aspects of the equilibrium state are usually strongly coupled and a complete treatment is not tractable at the present (but see Milne *et al.*, 1979; Low and Wu, 1981). Previous work has largely ignored the energy equation and concentrated on using Equation (1) to relate plasma structures to specific magnetic fields. Various artificial assumptions have been introduced for closure. For example, one may take the atmosphere to be isothermal. In this manner, models of quiescent prominences and sunspots have been constructed (e.g. Kippenhahn and Schlüter, 1957; Schlüter and Temesvary, 1958; Lerche and Low, 1980; Low, 1980a, 1981; Osherovich, 1982). Models have also been constructed for the large-scale global structures in the low corona, taking into account the $1/r^2$ gravity (Hindhausen *et al.*, 1981). The issues in this kind of theoretical work are illustrated by the modelling of sunspots as described below.

2.1. STATIC SUNSPOT MODELS

A popular model for an isolated sunspot is an axisymmetric system satisfying Equation (1) in cylindrical coordinates (R, z, φ). A magnetic field with no azimuthal component can be expressed in terms of a scalar function H:

$$\mathbf{B} = \frac{1}{R}\frac{\partial H}{\partial z}\hat{R} - \frac{1}{R}\frac{\partial H}{\partial R}\hat{z} = \hat{\varphi} \times \nabla H. \tag{2}$$

The condition $\nabla \cdot \mathbf{B} = 0$ is automatically satisfied. Equation (1) then decomposes into the two components

$$R\frac{\partial}{\partial R}\left(\frac{1}{R}\frac{\partial H}{\partial R}\right) + \frac{\partial^2 H}{\partial z^2} + 4\pi R^2 \frac{\partial p(z, H)}{\partial H} = 0, \tag{3}$$

$$\frac{\partial p(z, H)}{\partial z} + \rho g = 0, \tag{4}$$

where it is evident that another equation is needed for closure (Low, 1975). Equation (4) is the hydrostatic relation along a magnetic field line given by a contour of constant H. Equation (3) deals with balancing the Lorentz force in a perpendicular direction. If we include thermal conduction, steady energy balance is described by

$$\nabla \cdot (\varkappa \cdot \nabla T) = - S(p, \rho, H) + L(p, \rho, H), \tag{5}$$

where $\varkappa$ is the thermal conduction tensor, S and L the heat source and energy loss in the medium. The energetics that goes into defining $\varkappa$, S, and L is exceedingly difficult to treat. It touches on the central question of what cools the sunspot, a puzzle that is being debated to this day (e.g., Biermann, 1941; Cowling, 1953; Parker, 1974; Roberts, 1976; Spruit, 1977). Even if $\varkappa$, S, and L were prescribed by various idealized assumptions, the coupled system (3)–(5) is formidable. The problem is much more difficult to treat than the one-dimensional problem based on the thin flux-tube approximation (e.g., Roberts and Webb, 1978). In Equations (3)–(5), a flux tube of a finite thickness is considered, as is appropriate for modelling sunspots.

Instead of the complete problem, we may scrutinize all the solutions admitted by Equations (3) and (4), without Equation (5), seeking solutions with properties resembling those of observed sunspots. A convenient way of generating these solutions is to prescribe the function H, which fixes the magnetic field, and use Equations (3) and (4) to compute the equilibrium pressure p and density ρ (Low, 1980a). In a further development, we may insert a solution, which compares favorably with observation, into Equation (5) to derive implications on the transport of energy that maintains thermodynamic equilibrium. The Schlüter–Temesvary model and the return-flux models are based on specific prescriptions of magnetic fields and they have been compared with observation (Landman and Finn, 1979; Flå et al., 1982). In these works, observational input determines the background atmosphere and the plasma conditions along the axis of the idealized sunspot, defining the temperature in terms of p and ρ by the ideal gas law. A test with observation then rests on the continuum photometric appearance of the model sunspot. The Schlüter–Temesvary sunspot of Landmann and Finn was found to have a bright ring around it and it was suggested that the assumption of static equilibrium was inadequate. The return-flux model of Flå et al. fared better, with optimism about how the magnetic-field topology could account for important sunspot features, including the penumbra. More recently, a different approach to the problem was taken by Pizzo (1985). Instead of prescribing a specialized magnetic-field geometry, the function $p(z, H)$ is prescribed and Equation (3) is solved numerically as a Dirichlet boundary-value problem for H. Once H is determined, p and ρ can be computed as functions of space. An interesting result obtained is that slight changes in the functional forms of $p(z, H)$ lead to, as to be expected, slight changes in the spatial structure of the magnetic field given by H. However, the changes in p and ρ as functions of space can alter significantly the photometric signature of the model sunspot. A fine tuning of the magnetic field can adjust the $\tau = 1$ layer so as to remove or create a bright ring around the model sunspot. It follows that Landman and Finn (1979) could have removed their unwanted bright ring without changing their equilibrium in a major way. A relaxation of their self-similar magnetic-field geometry, as Pizzo pointed out, could achieve that result. By the same token, the optimism of Flå *et al.* (1982) is on weaker ground than previously realized. The important point of Pizzo's result is that the photometric signature is a poor guide to the validity of a given magnetic field structure in a magnetostatic model sunspot. For a more discriminating comparison with observation, determining all four Stokes parameters in the sunspot emission seems an obvious next step. It should be emphasized that static equilibrium is only a working assumption for a complex physical object. The above development shows the usefulness of this working assumption in providing a starting point for putting physical concepts and issues on a quantitative basis. More sophisticated models, in particular, models with flows (Tsinganos, 1982a; Morozov and Solov'v, 1980), can come later as we progress with the understanding of simpler models.

2.2. THREE-DIMENSIONAL EQUILIBRIA

Much of what we know about magnetostatic equilibrium comes from studying systems with an ignorable coordinate, such as the axisymmetric system. Real solar structures are, of course, never symmetric to such a degree. Until recently, building fully three-dimensional equilibria has been passed off as intractable except for the potential and linear force-free magnetic fields. For these two classes of magnetic fields, three-dimensional boundary-value problems with prescribed boundary distributions of normal magnetic flux are tractable by standard numerical techniques (Altschuler and Newkirk, 1969; Nakagawa and Raadu, 1972). These models are popular because magnetograph measurements of the photospheric magnetic fields can be used as inputs. However, they neglect the Lorentz force completely. Introducing a non-vanishing Lorentz force in three-dimensional geometry complicates the problem immensely. From a separate concern about nonequilibrium, Parker (1972, 1979) pointed out that the introduction of an ignorable coordinate does more than mathematically simplifying the equation for force balance. The symmetry associated with the ignorable coordinate has an essential physical role in ensuring the existence of equilibrium. The Lorentz force, being perpendicular to the local magnetic field everywhere, has an anisotropic structure. To balance this force with pressure forces and plasma weight, derivable from isotropic scalar potentials, a high degree of symmetry in the magnetic field is required. This requirement can be expressed as a local relation involving Euler potentials (Low, 1980b). For purpose of discussion, consider a less general form of that relation:

$$\frac{\partial}{\partial x}[(\mathbf{B} \cdot \nabla)B_y] = \frac{\partial}{\partial y}[(\mathbf{B} \cdot \nabla)B_x], \tag{6}$$

obtained by eliminating the pressure from the x and y components of Equation (1). A general magnetic field must satisfy this relation in order that equilibrium with suitable distributions of pressure and density is admissible. If Equation (6) is not satisfied, $\partial^2 p/\partial x\, \partial y \neq \partial^2 p/\partial y\, \partial x$, and equilibrium requires a pressure distribution not single-valued in space, which is unphysical. It should be emphasized that Equation (6) is a necessary but not sufficient, condition. For systems with an ignorable coordinate, Equation (6) is trivial as can be easily verified. Thus, this necessary condition never came into question in the study of systems with ignorable coordinates.

Equation (6) can be used to uncover an infinite variety of magnetic fields which are compatible with equilibrium and recent work has shown that fields with an ignorable coordinate are only a sub-set of all admissible fields. Fully three-dimensional equilibrium states are possible. Yet, the original postulate of Parker on the need to have magnetic-field symmetry in order to have equilibrium is valid, as we will see later in this section and in Section 3. The first three-dimensional equilibria were built by treating the departure from axisymmetry as linear static perturbations (Hu *et al.*, 1983). The analytic solutions were employed to model the penumbral filaments of sunspots in terms of azimuthal variations around the sunspot axis (Hu, 1983). The model suggests the dark penumbral filaments are associated with local strong magnetic fields. Going on to fully nonlinear treatments, large classes of equilibria were found (Low, 1982a, 1984a; Low *et al.*, 1985). Among the more interesting equilibrium states constructed are three-dimensional inverted-U plasma loops, vertical filamentary fine-structures of quiescent prominences, prominences as conglomerates of density enhancements and depletions in sheared magnetic fields, and the coronal helmet-streamer structure as a three-dimensional object. In these constructions, the key task is to satisfy Equation (6). As in previous magnetostatic studies, the energy equation is not treated self-consistently. The main interest is to gain familiarity with what states are admissible under force equilibrium. The development has been facilitated by the applicability of analytic methods in the construction. Analytic solutions of equilibrium states have the advantage of being amenable to stability analysis by method of perturbations. Not only are models of three-dimensional bipolar plasma loops available, examples have also been found that are shown to be stable (Low, 1982a). An exact solution is shown in Figure 1 which displays the magnetic field lines of an isothermal plasma loop embedded in a field-free atmosphere.

Although the equilibria described above are interesting in their own right, they are very restricted by the special mathematical assumptions that generate them. For example, one class of solutions correspond to demanding $(\mathbf{B} \cdot \nabla)B_x = 0$, $(\mathbf{B} \cdot \nabla)B_y = 0$, which render Equation (6) trivially satisfied. These magnetic fields have vertically oriented tension forces. With Maxwell's equation $\nabla \cdot \mathbf{B} = 0$, we have three equations in three unknowns. Assuming an ignorable coordinate is unnecessary and a large class of solutions with three-dimensional variations become available. A limitation of these solutions is that a standard prescription of an arbitrary normal magnetic flux at the boundary cannot be imposed. This limitation has been removed in a new class of equilibria, thus bringing magnetostatic equilibria as three-dimensional models to a level of sophistication previously restricted to potential and linear force-free fields (Low, 1985; Bogdan and Low, 1986).

It is instructive to look at the new class of equilibria admitted under Equation (1). Let us assume the electric current density flows in closed circuits in the atmosphere and, in particular, it flows in horizontal planes. Setting the vertical component of the electric

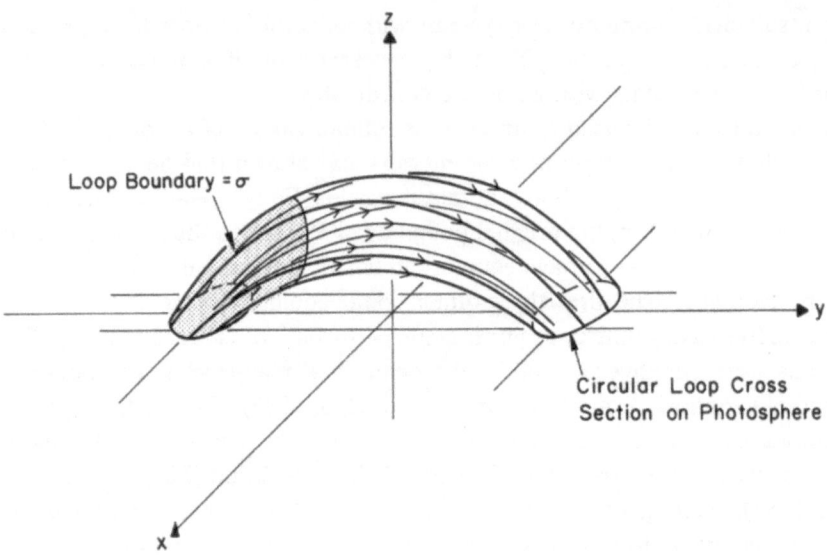

Fig. 1.   A perceptive view of the magnetic field lines of a three-dimensional plasma loop. In this special solution, there is no external magnetic field, the internal magnetic field is potential and an electric current flows in the loop boundary σ. The magnetic field lines on the loop boundary are drawn with thick curves. The plasmas inside and outside are in isothermal hydrostatic equilibrium.

current density to zero, the magnetic field takes the form

$$\mathbf{B} = \frac{\partial \phi}{\partial x}\,\hat{x} + \frac{\partial \phi}{\partial y}\,\hat{y} + \psi \hat{z}, \tag{7}$$

where ϕ and ψ are arbitrary functions. The compatibility equation (6) reduces to a simple form that requires ϕ and ψ to be related by

$$\psi(x, y, z) = \Psi\left(z, \frac{\partial \phi}{\partial z}\right), \tag{8}$$

where Ψ is an arbitrary function of two variables. For a given Ψ, Maxwell's equation $\nabla \cdot \mathbf{B} = 0$ leads to the equation

$$\frac{\partial^2 \phi}{\partial x^2} + \frac{\partial^2 \phi}{\partial y^2} + \frac{\partial}{\partial z}\,\Psi\left(z, \frac{\partial \phi}{\partial z}\right) = 0. \tag{9}$$

If we prescribe the boundary normal flux, Equations (7) and (8) allow us to translate it into a boundary condition on the derivatives of ϕ. Equation (9) then poses a boundary value problem for ϕ, given Ψ. To each solution so constructed, one can go on to construct the magnetic field $\mathbf{B}$ and the equilibrium pressure and density (Low, 1985). Notice that this problem includes the potential field problem, corresponding to $\Psi = \partial \phi / \partial z$, as a special case. Choosing the functional form of Ψ is a difficult task. Its complete arbitrariness stems from not having imposed an energy equation. Energy processes in the solar atmosphere is so complex that a limited approach based on a formal prescription of Ψ seems a reasonable first step. For a fixed form of Ψ, an input of magnetograph data gives the boundary conditions which determine the magnetic field and plasma equilibrium pressure and density. The latter can be compared with spectroscopic observation of plasma conditions.

Despite a richness of magnetic topology exhibited by the solutions to Equation (9), these solutions are still specialized, being restricted by having their electric current density flowing in horizontal planes. For example, magnetic fields with twisted field lines do not belong to this class of solutions. It seems likely that further work with Equation (6) may uncover more complicated equilibria. There is nothing remarkable about a magnetic field not satisfying Equation (6). An arbitrary deformation of a magnetic field, initially in equilibrium, readily violates Equation (6). What is interesting is that, to satisfy Equation (6) for equilibrium, it is not necessary to have a system with an ignorable coordinate. This conforms with the reality of a variety of solar structures which are quasi-static and yet do not possess a high degree of symmetry. On the other hand, Equation (6) for compatibility requires the equilibrium magnetic field to have a symmetry as postulated by Parker. The required symmetry need not be as strong as those associated with ignorable coordinates, but is in any case quite definite in the following sense (Low, 1980b). We can think of any magnetic line of force in each localized region to be the intersection between a pair of surfaces associated with the Euler potential U and V (Stern, 1966):

$$\mathbf{B} = \nabla U \times \nabla V. \tag{10}$$

The condition $\nabla \cdot \mathbf{B} = 0$ is satisfied and we say all magnetic fields are generated by a pair of independent Euler potentials. Compatibility relation (6), which acts on the equilibrium magnetic field alone, relates U and V in a definite manner so that only one generating function is free. In geometric terms, the relationship between U and V gives the U and V surfaces a certain symmetry in a generalized sense. Introducing an ignorable coordinate is a simple way of imposing that symmetry. For example, with axisymmetry, $\partial/\partial\varphi \equiv 0$, we have the magnetic field given by equation (2) where we identify $U = \varphi$ and $V = H$. In this case the magnetic field has just one generating function, namely, H.

3. Nonequilibrium

The term nonequilibrium describes circumstances under which a given set of physical laws forbids a system to reach equilibrium. Nonequilibrium occurs commonly but, in many cases, it is not physically interesting and not much thought is given to it. That a stone unsupported in gravity cannot be in equilibrium is not saying something interesting. In contrast, other cases of nonequilibrium can be subtle and have profound implications. For instance, the virial theorem states that just pressure and Lorentz force cannot self-confine a plasma in equilibrium. In nature, gravity can provide a means of confinement but not so if the pressure is too large. A highly heated, thermally conducting corona cannot be in hydrostatic equilibrium with a tenuous interstellar medium, and this realization led Parker to the theory of the solar wind (Parker, 1963). In astrophysical environments, the magnetic Reynolds number is extremely large, typically about 10^9–10^{15} for the solar corona, so that magnetic fields are frozen into the plasma to a very good approximation. Hydromagnetic evolution must proceed in a manner so as to preserve the global magnetic-field topology. This strong constraint on how a plasma can move has far reaching consequences. It can provide stability to a given static equilibrium by denying the system of destabilizing plasma displacements which are otherwise admissible, see Section 4. The same constraint can also prove so prohibitive as to prevent a magnetic field from establishing equilibrium. Parker (1972) has been first to recognize this hydromagnetic property and its implications in the astrophysical environment.

3.1. NONEQUILIBRIUM IN SOLAR MAGNETIC FIELDS

The following instructive problem was treated by Parker (1972). An initially uniform magnetic field in equilibrium with a uniform pressure is anchored at a pair of end-plates separated at distance L. To keep the basic physical point simple, gravity is neglected. The lines of force are interwoven arbitrarily over one another by shuffling the magnetic footpoints on one of the end-plates and then re-anchored. To allow the use of an expansion technique, let us take the footpoint shuffling to be confined to cells of characteristic dimension l much smaller than L. The interesting question arising is whether the deformed magnetic field can establish an equilibrium irrespective of what complex topology it has acquired. To be sure, the stressed magnetic field can drive a great deal of motion between the end plates and in the absence of dissipation, kinetic motion and perhaps hydromagnetic shock-steepening can go on forever. Let there be sufficient kinetic viscosity to damp out these distracting effects. The system must then settle eventually to a minimum-energy static state with its endowed magnetic-field topology. Parker showed that this terminal state exists only if its magnetic field is translationally invariant in the direction of the initial uniform magnetic field. An expansion of the terminal state about the initial state in the parameter $\varepsilon = l/L$ revealed that all terms in the expansion must possess that invariance. Only special interweaving of the lines of force can meet this stringent demand. Hence, no terminal equilibrium state exists, in general. When no equilibrium is available, thin sheets of electric currents form. Equilibrium prevails everywhere outside while the current sheets collapse ceaselessly, squeezing out the sandwiched plasma along magnetic field lines. However small the electrical resisitivity, Ohmic reconnection of lines of force eventually sets in at the current sheets and proceeds to change the magnetic-field topology until, presumably, a simpler topology compatible with an equilibrium state is attained. The system can then find rest. Parker's problem has generated considerable interest (Yu, 1973; Parker, 1979, 1982, 1983c, 1985a, b, d; Low, 1980b; Rosner and Knobloch, 1982). Tsinganos (1982b) considered a variation of the problem, in which a steady fluid flow and magnetic field are present in the initial state. He showed that, in general, an arbitrary interweaving of the lines of forces cannot lead to an equilibrium involving steady hydromagnetic flows.

Tsinganos *et al.* (1984) gave a different treatment of Parker's problem without the use of the expansion technique. Drawing the mathematical parallel between the equations describing magnetic field lines in equilibrium and the equations describing particle motions in phase space, they showed that Parker's nonequilibrium follows from the Kolmogoroff–Arnold–Moser Theorem in Hamiltonian mechanics. Van Ballegooijen (1985) considered Parker's problem without the plasma pressure, looking for force-free magnetic fields. Let z denote the Cartesian coordinate in the direction of the initial uniform magnetic field. Employing an unconventional expansion scheme, he found an interesting class of force-free fields that interweave with a pattern varying with z. The winding pattern represented by some functions α and h varies with z not arbitrarily but according to

$$\frac{\partial \alpha}{\partial z} = \frac{\partial(h, \alpha)}{\partial(x, y)}, \qquad \alpha = \left(\frac{\partial^2 h}{\partial x^2} + \frac{\partial^2 h}{\partial y^2} \right), \tag{11}$$

which we recognize to be the equation describing the time evolution of a two-dimensional vorticity if we identify z with time. The vorticity equation is nonlinear and no general treatment of it is available. There is a question whether, in general, solutions to this equation can be found for a given situation without the singularities that correspond to electric current sheets in Van Ballegooijen's treatment. If singularities can be avoided, we would have equilibrium states that somehow have eluded the mathematical analyses of Parker (1972) and Tsinganos *et al.* (1984). Finding such an equilibrium would be an interesting development, but it does not change the original conclusion of Parker. These force-free fields, if they exist, require careful interweaving of the lines of force to give the peculiar variation of the winding pattern with z as described by the vorticity equation. They cannot, therefore, be considered to be very general.

Parker (1985) had re-examined the problem of the force-free fields. A detailed analysis shows that the interweaving of lines of force can be described by two scalar functions. The two scalar functions are analogs of the two independent Euler potentials U and V in Equation (10). Equilibrium requires full freedom in these two functions to satisfy two independent equations but one of the functions is predetermined by field-line winding. Equilibrium is thus, in general, impossible. To appreciate the basic point, it helpts to note the interplay between the two principal effects in the making of a hydromagnetic nonequilibrium. Firstly, it takes a certain compatibility for the anisotropic Lorentz force to be balanced by scalar pressures in equilibrium, a point already touched upon in Section 2. Secondly, the highly electrically conducting medium can adjust only by displacements that preserve the magnetic-field topology. The former is a local condition while the latter is a global constraint. No difficulty arises for adjusting a local magnetic field to the right symmetry in order to achieve local equilibrium. It is the global constraint of the magnetic topology that precludes, in general, the possibility of so adjusting at all localities to achieve equilibrium everywhere (Low, 1980b). In the construction of three-dimensional equilibria described in Section 2, as well as van Ballegooijen's force-free fields, the approach to the question is different from that of Parker. Instead of dealing with a fully random magnetic field, classes of magnetic fields are identified by direct construction with the demand that they have topologies compatible with equilibrium. Each of these equilibrium magnetic fields is restricted by having only one free generating function.

A fraction of the coronal magnetic flux is open into the interplanetary space, presumably through the coronal holes. Elsewhere, especially over the active regions, the magnetic fields are closed. In the conventional picture, the active region magnetic fields are mainly bipolar arches, with footpoints anchored in the dense photosphere. At the photosphere, the plasma dominates over the magnetic field and its small-scale random motion shuffles the magnetic footpoints as in Parker's idealized problem. Higher up, in the chromosphere and corona, the magnetic field dominates and adjusts dynamically in response to the footpoint motion (Gold, 1964). An important implication of Parker's calculation is that the magnetic field cannot find equilibrium since there is no reason to expect the photospheric motion to take any specialized form. In spite of the large magnetic Reynolds number, electric current sheets form and dissipate all the time and everywhere in the corona. It has been suggested that this ubiquitous nonequilibrium dissipates magnetic energy to heat the corona (Parker, 1972, 1979, 1981a, b, 1985a, c; Sakurai and Levine, 1981; Sturrock and Uchida, 1981). This idea explains, in a natural way, the direct connection between magnetic field and coronal heating as indicated by detailed observation and modelling of the X-ray corona (Rosner *et al.*, 1978) and the fact that most stars have X-ray coronae. The shuffling of magnetic footpoints and the current sheets it creates are expected to be on too small a scale observable by presently available instruments (Parker, 1985c). Future instruments with resolution significantly below one arc second can provide useful information on this mechanism for coronal heating. Estimating theoretically the energy input into the corona from the continual dissipation of current sheets is difficult because the dynamical non-equilibrium is not amenable to theoretical treatment and our knowledge of resistive reconnection of magnetic fields remains qualitative (Parker, 1973; Vasyliunas, 1975). Parker (1985c) gave rough calculations to suggest that the process can meet the demand of about 10^7 erg cm^{-2} s^{-1} needed to heat the corona in the active region. In the coronal holes, where the magnetic field is believed to be open, the mechanism does not work well. Any twist in the lines of force simply propagates, largely as Alfvén waves, out along the open ends. The Alfvén waves can dissipate viscously and heat up the corona. Several calculations have been presented to illustrate this important process (e.g. Hollweg, 1984; Heyvaerts and Priest, 1984; Leer *et al.*, 1982; Parker, 1985c).

The photosphere executes large-scale mean motion in addition to the ever-present random component. For example, there is the solar differential rotation. While the small-scale interweaving of lines of force are continuously created and dissipated, the larger scale distortion may either also dissipate smoothly or relax to some metastable equilibrium whenever the larger scale magnetic topology so permits. Parker (1985c) points to observations that imply gradual relaxations of large scale coronal magnetic fields to potential states (Sheeley *et al.*, 1975). If a metastable equilibrium state is available, free energy can be stored to be liberated impulsively, after further evolutions, in such spectacular events as flares, prominence eruptions, coronal transients, etc. (Low, 1982b; Spicer, 1982). The basic cause of this energy release can also be attributed to nonequilibrium. Here, an ordered magnetic field is involved. Before we discuss this kind of nonequilibrium, let us digress to some developments in the study of contained laboratory plasmas, of relevance to our discussion up to now.

3.2. Nonequilibrium in contained plasmas

In the approximation that the dense photosphere is a lower boundary to the atmosphere above it, an important feature of solar magnetic fields is that they generally thread through that boundary. In contrast, the laboratory plasma has a magnetic field wholly contained within rigid, electrically highly conducting walls. In the latter, lines of force can close upon themselves or be ergodic in a volume, in the sense of a single, infinitely-long, winding line of force filling up a volume (Grad, 1967). Ergodicity obtains in most situations except when there is special symmetry. With axisymmetry, ergodicity in a volume may be avoided. At most, a line of force may be ergodic on a surface. Whereas, a constant-α force-free field with three-dimensional variations may have smoothly behaving spatial distribution of the field vector and yet its lines of force are ergodic in a volume (Hénon, 1966; Dombre et al., 1985). Moreover, introducing a small component of random magnetic field to a magnetic field readily creates ergodicity in a volume where previously no such ergodicity exists. Magnetic fields with such chaotic field lines can be in equilibrium only with a uniform pressure. Since $(\mathbf{B} \cdot \nabla)p = 0$, spatially varying equilibrium pressures are necessarily magnetic surfaces made up of field lines, the existence of which is not compatible with ergodic field lines that fill up a space. It has been conjectured by Grad (1967) and studied by Arnold (1974) and Moffat (1985) that a contained magnetic field with an arbitrary topology cannot find equilibrium without the formation of electric current sheets.

The topology of wholly contained magnetic fields can be described in terms of the magnetic helicity defined to be the scalar product of the magnetic field with its vector potential. In each element of volume bounded by magnetic surfaces, we can integrate for the total helicity. That the volume be bounded by magnetic surfaces is a crucial requirement for the total helicity to be gauge invariant and physically meaningful. If the magnetic field is frozen into the plasma, the helicity of each individual magnetic volume element is conserved (Woltjer, 1958; Moffat, 1969). This conservation law is a neat way of expressing the 'frozen-in' condition as applied to contained plasma. Taylor (1974) hypothesized that in a plasma relaxing by dissipative processes, the total helicity of a closed system may be approximately conserved although the helicity of each magnetic volume element may not. In other words, the dissipation of total helicity is a higher order effect than the dissipation of total energy. Berger (1984) provided support for this hypothesis by deriving a bound on the rate of change of total helicity compared with rate of change of total energy. In that case, a simple variational calculation shows that, neglecting plasma pressure, a state of minimum energy with a fixed total helicity is a constant-α force-free magnetic field (Woltjer, 1958). Taylor's hypothesis has been successful in explaining the phenomenology of the so-called reversed pinch in the laboratory. Although a quantitative theory for Taylor's hypothesis is still lacking, it is plausible that nonequilibrium is at the root of the process (Parker, 1985a). The dissipative processes and initial perturbations probably introduce a random component to the magnetic field in the subsequent development of a turbulent state. Since the pressure is neglected, the admissible terminal equilibrium state is a force-free magnetic field. A variable α force-free state cannot survive in the ergodic field because the α surfaces are magnetic surfaces, i.e., $\mathbf{B} \cdot \nabla \alpha = 0$. The terminal state must, therefore, have a constant α. The possibility of a similar relaxation in the solar atmosphere was explored by Norman and Heyvaerts (1983). They suggested that flares are the results of Taylor's relaxation taking a solar magnetic field to a constant-α force-free field. There are two difficulties with the idea, namely, solar magnetic fields are not closed systems and the lines of force penetrate the lower boundary. We will have occasion to address the first difficulty in the next section. The second difficulty vitiates the use of total helicity in the model since the total helicity is no longer gauge invariant. Remedy is provided by the concept of gauge-invariant relative helicity, applicable to magnetic fields whose lines of force penetrate the boundary, invented recently by Berger and Field (1984). Relaxation of solar magnetic fields subject to an invariant total relative helicity has been investigated by Berger (1985).

3.3. Nonequilibrium in symmetric systems

Symmetry in a magnetic field is a necessary but not sufficient condition for equilibrium. Even in systems invariant in a given direction, nonequilibrium can arise in other interesting ways. Wherever there is an X-type magnetic neutral point, unless pressure is artificially controlled, there is a tendency for equilibrium to fail, resulting in the neutral point being ironed out into a neutral sheet terminated at the ends by a pair of three-fold vertices (Parker, 1957; Sweet, 1958; Syrovatskii, 1981; Bobroba and Syrovatskii, 1979; Priest and Raadu, 1976; Hu and Low, 1982). This effect is particularly interesting in closely packed twisted magnetic flux tubes (Parker, 1982, 1983c, 1985a, d; Vainshtein and Parker, 1985). Close packing and the tendency of flux tubes to expand against their neighbours lead naturally to hexagonal packing in order to minimize the total energy. For flux tubes with a similar sense of twist, at least one of the three boundaries at every vertex unavoidably has oppositely oriented magnetic fields aligned along it. Without artificially controlling the pressure in that boundary to keep the opposite magnetic fields apart, reconnection sets in. Most interesting is that the reconnection between flux tubes of like twists creates pairs of flux tubes of opposite twists (Parker, 1982). So the bundle of flux tubes must reconnect until an end state of a pair of two oppositely twisted flux tubes remain. In this end state, the boundary between the two flux tubes automatically have magnetic fields of like orientations on either side. The dynamics of flux tubes as a two dimensional problem is a basic ingredient of two-dimensional hydromagnetic turbulence (Montgomery and Vahalas, 1979; Montgomery *et al.*, 1979; Bogdan, 1984; Parker, 1985d).

Finally, we turn to the nonequilibrium encountered in the shearing of force-free magnetic fields. There is a popular notion that the large-scale photospheric motion of magnetic footpoints can shear a quasi-static solar magnetic field to a point beyond which no equilibrium is available. The highly stressed magnetic field then breaks abruptly into a dynamical state, liberating a large amount of energy by various dissipative mechanisms that are presumed to set in. This model is attractive as an explanation of the onset of flares and other similar violent eruption on the Sun. The point in question is that flares do not take place all the time but occur at sporadic times to punctuate quasi-steady evolutions of solar magnetic fields. Moreover, flares are likely to occur when the magnetic geometry is most complex, say, as indicated by $H\alpha$ fibrils (Zirin, 1974). To put the idea into a model, a system invariant in a direction, say, the x-coordinate, has been considered by various people. Writing the magnetic field in the form $\mathbf{B} = Q\hat{x} + (\partial P/\partial z)\hat{y} - (\partial P/\partial y)\hat{z}$, the force-free equation

$$(\nabla \times \mathbf{B}) \times \mathbf{B} = 0 \qquad (12)$$

requires that Q is a strict function of P. Let us set $Q = \lambda F(P)$, where λ is a constant parameter. Then, Equation (12) reduces to

$$\nabla^2 P + \lambda^2 F \frac{\mathrm{d}F}{\mathrm{d}P} = 0. \qquad (13)$$

Let us consider a bipolar magnetic field in the half space $z > 0$ with footpoints on $z = 0$, identified with the photosphere. The problem is to compute the force-free magnetic field given the magnetic flux distribution on $z = 0$ and the precise topology with which bipolar field lines leave and re-enter $z = 0$ (Sturrock and Woodbury, 1967). This problem is intractable and we treat a simpler problem instead. A magnetic field with footpoints progressively displaced with shear in the x-direction must lengthen its lines of force in that direction. This manifests in an increasing x-component of the field. By picking a fixed functional form of $F(P)$, we can increase $B_x = \lambda F(P)$ through the parameter λ. The normal flux at $z = 0$ is given by the value of P at $z = 0$. Let us demand that $\mathbf{B}$ vanish rapidly at infinity in $z > 0$ for an isolated magnetic field. It follows that Equation (13) for a fixed $F(P)$ and increasing values of λ^2 poses a Dirichlet boundary value problem whose solutions describe a quasi-static sequence of force-free states associated with a slow shearing displacement of the footpoints. Solutions to this problem have been reviewed by Birn and Schindler (1981) and Low (1982b). Explicit solutions show that if λ is increased beyond a critical value λ_c, physically acceptable solutions do not exist. An evolutionary path of a quasi-static magnetic field through the λ-sequence of solutions would terminate at the $\lambda = \lambda_c$ critical solution. The interpretation is that some form of nonequilibrium sets in at $\lambda = \lambda_c$ (Low, 1977a). To give generality to this intriguing behavior of force-free magnetic fields, Birn et al. (1978) pointed out that the general absence of solutions to boundary value problems involving the nonlinear Equation (13) for large values of λ^2 has been established in general theorems in the theory of partial differential equations.

What is the physical nature of the $\lambda = \lambda_c$ critical point? Low (1977b) presented an example to identify the $\lambda = \lambda_c$ state to be the first marginally stable state in a sequence of $\lambda < \lambda_c$ absolutely stable states. In reality, solar magnetic fields are never exactly force-free. Pressure gradients are always present and have been neglected in the force-free approximation. If $\lambda = \lambda_c$ corresponds to a marginally stable state, the true stability of that force-free field is contingent upon higher order effects due to the small background pressure which has been neglected. However, the claim that $\lambda = \lambda_c$ corresponds to a marginally stable state in Low's model was untenable. The correction to an error in Low's analysis leads to all $\lambda \leq \lambda_c$ states being absolutely stable (Cargill *et al.*, 1985). In another demonstration, Low (1978) showed that if the small background pressure were to be accounted for, all $\lambda < \lambda_c$ force-free fields can readily accommodate for its presence in equilibrium by a slight adjustment of the lines of force. However, the $\lambda = \lambda_c$ force-free field is pathological in that no first order adjustment from the force-free state can accommodate the pressure introduced, no matter how small that pressure is. Thus, we have nonequilibrium.

Recently, Aly (1984) considered the problem with a novel application of the virial theorem. Among the several interesting results he obtained, is one that explains in a direct physical manner why the boundary value problem posed by Equation (13) has no solutions if λ^2 is too large. Without restricting to any symmetry, the total Lorentz force acting on a volume is the sum of Maxwell stresses on the boundary. If we take the volume to be the half space $z > 0$ and assume that the Maxwell stress vanishes at infinity in $z > 0$, the total upward force L_z per unit length in the x-direction is

$$L_z = \int\limits_{z=0} [B_z^2 - (B_x^2 + B_y^2)] \, dy. \tag{14}$$

For a force-free field, $L_z = 0$ and we have

$$\int\limits_{z=0} (B_x^2 + B_y^2) \, dy = \int\limits_{z=0} B_z^2 \, dy. \tag{15}$$

In the problem with invariance in the x-direction, we are moving the footpoints in the x-direction. Hence, the distribution of B_z on $z = 0$ is unchanged in the process and the right-hand side of Equation (15) is independent of the footpoint displacements. It then becomes clear that increasing $B_x = \lambda F(P)$ by holding $F(P)$ fixed and increasing λ cannot proceed beyond a critical value $\lambda = \lambda_c$ obtained when $B_y \equiv 0$ on $z = 0$. Therefore, a transverse magnetic component at $z = 0$ which is too large, is not compatible with force-free equilibrium. Aly (1984) has gone on to conjecture that displacing the footpoints of the magnetic field beyond $\lambda = \lambda_c$ is possible by relaxing the fixed form of $F(P)$ so that Equation (15) is still satisfied. To demonstrate this possibility requires treating the footpoint connectivity explicitly, leaving the function $F(P)$ free to be determined consistently, a problem that continues to challenge us. It is probable that some special cases may behave as Aly conjectured. But, in general, nonequilibrium may set in at the $\lambda = \lambda_c$ critical point such as demonstrated by Low (1978). It is also intriguing to note that violating Equation (15) is the same as saying that we have a non-vanishing total Lorentz force L_z in Equation (14). This suggests the presence of an upward or a downward acting Lorentz force appearing in the region $z > 0$.

4. Linear Hydromagnetic Stability

Given an equilibrium state, its stability or instability against linear perturbations is an important property to consider. As reported in Section 2, a wealth of magnetostatic states is available for stability analysis. The perturbation equations are difficult to treat so that only the simplest equilibria is amenable to a fruitful study of stability. In any case, studying the simplest equilibria has the advantage of allowing us to see the essential physics clearly. The energy principle of Bernstein *et al.* (1958) for ideal hydromagnetic stability has been used commonly in recent work. This principle determines stability or instability without the need to know the precise nature of the perturbation and its frequency, and is relatively easy to apply as compared to treating the perturbations explicitly in an eigenvalue analysis, or in an initial value problem (Friedberg, 1982). The essence of the energy principle is the conservation of energy, obtaining for a given set of appropriate boundary conditions. If we take all magnetic fieldlines in a solar situation to be rooted to the photosphere treated as a rigid perfect electrical conductor, the boundary condition, at $z = 0$, for plasma displacement ξ is

$$z = 0, \qquad \xi = 0. \tag{16}$$

This condition comes from the continuity of the tangential electric field and the non-vanishing normal magnetic field at the boundary. If the equilibrium magnetic field does not penetrate the boundary, as in a contained laboratory plasma, a different set of boundary conditions apply (Bernstein *et al.*, 1958). In addition to Equation (16), let us consider a spatially isolated equilibrium so that we set $\xi = 0$ at infinity in $z > 0$. We then have a conservative system. The energy principle states that the equilibrium is unstable if and only if a static displacement ξ can be found, subject to the boundary conditions, such that the change in potential energy,

$$\delta W = -\tfrac{1}{2} \int_{z>0} \xi \cdot F(\xi) \, dV, \tag{17}$$

is negative. The second-order differential operator F is defined in terms of the unperturbed equilibrium state. First principles show that F is self-adjoint in consequence of energy conservation. Technically, it is much easier to prove instability as it merely calls for constructing a destabilizing displacement. For stability, a minimization of δW over all admissible displacements is required to show a positive minimum. The Euler–Lagrangian equations are second-order partial differential equations posing an eigenvalue problem with the above boundary conditions. Sometimes, a clever manipulation of the integral defining δW shows it is manifestly positive and the problem then becomes mathematically trivial.

A multitude of hydromagnetic instabilities are known in the laboratory. Yet, observations show the corona populated with all kinds of long-lived plasma loops (Vaiana and Rosner, 1978). These structures are probably not in strict equilibrium. Theory suggests that small-scale interweaving of magnetic field lines and dissipation of current sheets take place continually. Over larger scales, volume electric currents must be present to give rise to the observed plasma structures. The long lives of these structures suggest they are in approximate equilibrium with a remarkable stability. Both the energetics and force balance go into maintaining the approximate equilibrium and its stability. The magnetic field plays a central role in both the large-scale ordering and the small-scale dissipative heating. The study of mechanical stability of ideal hydromagnetic equilibria tends to suppress the role of thermal stability and is only half the story. But, it captures a great part of the physics because ideal hydromagnetic instabilities have rapid rise times. Results of stability analyses tend to be clothed in tedious mathematics. The following is an attempt to give a qualitative physical picture of the interplay between various stabilizing and destabilizing effects in the solar atmosphere.

Hydromagnetic instabilities are classified by tracing their mathematical origins to terms in the integrand in Equation (17) that render δW negative, and by the specific forms of the associated destabilizing displacements (Friedberg, 1982). Current-driven instabilities are those arising from terms involving the electric current parallel to the magnetic field in the equilibrium state. When this current is large, the system is unstable to a kink displacement. Pressure-driven instabilities may arise from terms involving the electric currents transverse to the magnetic field in the equilibrium state, whose existence necessitates the presence of pressure gradients for force balance. Here, the interchange displacement is relevant. If the positions of two adjacent magnetic flux tubes are physically interchanged, the change in the potential energy of the system is given by the net work done in stretching the individual flux tubes and in expanding or compressing the plasma. In the classical case of a plasma bordering a region of strong magnetic field, the interface is stable (or unstable) if the plasma is concave (or convex) to the magnetic field.

A wealth of stability results on the contained laboratory plasma exist and has been brought to bear on similar concerns in solar physics. To the extent stability of equilibrium states are relevant to understanding the long-lived solar plasma structures, two agents serve to suppress the instabilities found with the laboratory plasma. The first is the so-called line-tying effect arising from the magnetic field lines in the corona being anchored in the dense photosphere. Under the frozen-in condition, the anchoring restricts the plasma displacements. In contrast, a magnetic field fully detached from the boundary allows freer plasma displacements. The second effect comes from the solar gravity. Depending on the circumstance, the presence of gravity can be either stabilizing or destabilizing. For example, gravity packs the photosphere so densely as to provide line-tying of the magnetic field.

4.1. Equilibria without Gravity

By definition, the instabilities of force-free magnetic fields are current-driven. Parker (1966) showed that a force-free flux tube tends to buckle if twisted severely. In an application of the energy principle, Anzer (1968) showed that an infinitely long flux tube with no confining boundary is kink unstable for all rates of twist. Raadu (1972) was first to recognize the importance of line-tying in flux tubes of finite lengths. Introducing endplates to Anzer's flux tube, the kink instability can be suppressed if the twist between the endplates is moderate, typically less than one full twist. This class of problems and the one involving a translationally invariant field in the half-space $z > 0$ that is anchored at $z = 0$ have been investigated extensively. It has been conjectured that all line-tied force-free field in the latter class are stable (e.g., Hood and Priest, 1981; Cargill et al., 1985).

Molodensky (1975) showed that a general constant force-free magnetic field is stable to perturbations with wavelength less than $1/\alpha$ where α is the usual proportionality between electric current and magnetic field. A contained plasma can admit perturbations of wavelength no larger than the distance between the containing walls. A force-free field having a constant α is stable in this plasma if that distance is smaller than $1/\alpha$ (Voslamber and Callebaut, 1962; Berger, 1985). Molodensky's result has bearing on the hypothesis of Norman and Heyvaerts (1983) that the Taylor relaxation may operate in the solar atmosphere giving a flare. We already pointed out in Section 3 that in place of the total helicity used by these authors, the gauge-invariant relative total helicity of Berger and Fields (1984) should be employed. Another difficulty with the idea is that the solar-flare magnetic field is not a closed system confined by rigid walls and the analogy of the Taylor relaxation in the laboratory does not carry through in an obvious way. The question is what volume defines the total relative helicity. It is unavoidable that the flaring region is not isolated but must interact with its surrounding. Can there be a basis to speculate that this interaction determines a maximum wavelength for the disturbances, given by the typical dimension of about 4×10^4 km for a large flaring region? If so, Taylor relaxation can take a magnetic field to a constant α force-free field with $1/\alpha$ being larger than that dimension. Molodensky's result then ordains this as a stable terminal state.

Pressure-driven instabilities in the absence of gravity have been investigated extensively in connection with solar loop plasmas (van Hoven, 1982). A plasma enhancement confined by concave magnetic field lines is unstable to interchange displacement. Numerous calculations demonstrate how line-tying and magnetic shear can act to suppress this instability by interfering with the interchange of elemental flux tubes. Magnetic shear orients adjacent flux-tube elements at different angles so that the straight forward interchanging of two parallel adjacent flux-tube elements to lower the potential energy of the system is thwarted. However, shearing is not always stabilizing since it introduces an electric current parallel to the equilibrium magnetic field, with potential for current-driven instabilities. The competition between pressure-driven instabilities and line-tying is most apparent in a laminar magnetic field, without the complication of the shear and having an ignorable coordinate perpendicular to the magnetic field (Bernstein et al., 1958; Gilman, 1970; Zweibel, 1981; Low, 1984; Bogdan, 1985). For such a system, the most unstable displacements are those with large wave numbers associated with the ignorable coordinate. Note that if a shear is present, and a field component in the direction of the ignorable coordinate exists, such displacements are stabilizing because they would tangle up the magnetic field. Without shear, the energy integral δW can be reduced to a sum of independent path integrals along individual magnetic field lines from one footpoint to the other. Instability, if present, then manifests in the form of negative contributions to one of these path integrals. An interesting feature is that to maximise the negative contributions, the plasma displacements ξ must vary as little as possible along the field line. This conflicts with the rigid boundary condition which demands that ξ vanishes at the footpoints. The longer a field line is, the weaker is the line-tying effect but it does not mean that short field lines are always stable. Much depends on the magnitude of the negative contribution to the path integral associated with pressure-driven instability along a given field line.

The use of rigid boundary conditions is an over simplification. The photosphere is not rigid despite its high density. It can also introduce flows into the atmosphere above it. In a more realistic model, the base of the atmosphere is allowed to move and the energy of the atmosphere is no longer conserved. Alternative boundary conditions to allow for displacements at the boundary have been proposed and used extensively (e.g. van Hoven *et al.*, 1981; Ray and van Hoven, 1982; Migliuolo and Cargill, 1983). The displacements at the boundary in these new models are artificially constrained to be parallel to the boundary magnetic field and of special forms to conserve energy in the region above the photosphere. The stability problem reduces once again to an energy principle and its associated eigenvalue problem. However, the physical meaning of these new models is not clear (see the discussion is Rosner *et al.*, 1985). Although flows at the boundary are allowed, the boundary remains fixed in space. Artificially restricting the displacements at the boundary to regain energy conservation seems to defeat the purpose of relaxing the rigid boundary conditions, which is to allow the photosphere to respond to and influence disturbances above it. The artificiality of the restriction is seen in revised boundary conditions that demand flows at the two footpoints of each bipolar magnetic field line to be so well correlated as to conserve total energy. This raises the question of why general perturbations, which are in principle naturally occurring in an arbitrary fashion, should correlate across the large distances between pairs of footpoints. A proper treatment calls for allowing an exchange of flows and energy between the corona and the photosphere through the transition layer but that takes the model out of reach. Zweibel (1985) recently considered a simple model where the photosphere is taken to be a thin layer, sandwiched between a rigid boundary below and a free-moving contact surface with the corona above.

4.2. THE EFFECT OF GRAVITY

In the pressure of gravity, the equilibrium pressure must vary along a magnetic line of force rather than being constant; see Equation (4). The Lorentz force acts perpendicularly to the magnetic field along which the pressure gradient alone must support the plasma weight. The magnetic flux tubes may be regarded to be separate hydrostatic atmospheres stacked together in equilibrium. The hydrostatic stratification gives rise to buoyancy as an added effect in determining stability. Buoyancy is often destabilizing. The magnetic field tends to behave buoyantly like a hot, weightless fluid, such as in the instability suggested by Parker (1966) to operate in the condensation of interstellar clouds. A similar process probably takes place in the corona (Sweet, 1971). The effect of gravity can also be stabilizing. It can pack plasma to a high density, such as in the photosphere and provides anchors for line-tying. Many equilibria involving gravity are unstable, of course, but despite the added complication of gravity, several equilibria are among the few that can be technically proved to be stable. For example, Zweibel (1982) proved the stability of the classical prominence model of Kippenhahn and Schlüter (1957), some twenty-five years after that model was written down. The equilibrium state in that case consists of a vertical prominence sheet cradled in hammock-shaped magnetic field lines. The magnetic tension force is everywhere vertical and balanced by gravity so that all displacements from equilibrium result in a restoring force. The other forces in the system arise from the plasma and magnetic pressures which sum into a total pressure which is uniform in space. Any perturbation of this uniform total pressure also generates a restorative force. Another example of stable equilibrium is that of a three-dimensional plasma loop in an isothermal atmosphere. The special solution shown in Figure 1 has no magnetic field outside the plasma loop. In more general solutions, an external magnetic field is present (Low, 1982a). For these more general solutions, the interior of the plasma loop and its exterior contain potential magnetic fields in stably stratified atmospheres. The two regions are absolutely stable. Separating them is an electric current surface, the boundary of the three-dimensional loop, across which the pressure and magnetic field jump discontinuously. Depending on the sign of the pressure jump, the loop interior may have higher or lower density, with a corresponding lower or higher magnetic field, compared to its exterior. The stability of this structure hinges on the stability of the current boundary surface. Take the case of a high-density loop and consider the apex of the inverted-U loop. Both at the topside and the underside of the loop, stability depends on the action of buoyancy and magnetic field-line curvature. At the topside, the heavy fluid in the interior sits beneath the light fluid in the exterior and the configuration is stabilizing. On the other hand, the strong magnetic field in the exterior is concave to the strong pressure in the loop interior and that is unstable to an interchange displacement. The stability of the topside of the loop, therefore, obtains only if the stabilizing effect of the local stratification dominates. In contrast, on the underside of the loop, there is the heavy fluid of the loop interior sitting on the light fluid of the exterior, whereas the strong magnetic field of the exterior is convex to the strong pressure in the loop interior. The former is destabilizing whereas the latter stabilizing to interchange displacements. Local stability depends on the latter being the dominant of the two. At the current surface, the total equilibrium pressure is continuous but its derivative is discontinuous, in general. The stability of the surface then depends strictly on the sign of the jump of the derivative of the total pressure normal to the surface (Bernstein et al., 1958). In the plasma-loop model of Low (1982a), both the total pressure and its derivative are continuous across the current surface of the loop. The current surface is neutrally stable. So, the destabilizing and stabilizing effects, as discussed above, compensate exactly everywhere on the current surface. The plasma loop is thus stable as a whole.

5. Discussion

The hydromagnetic problems reviewed have been inspired and motivated by solar phenomena, but they hold basic physical interests in their own rights. A great deal of effort has actually been directed at these problems at a basic level, for there is need to educate ourselves in the basic principles and to discover novel properties. The study of magnetostatic equilibrium has progressed beyond symmetric systems, with ignorable coordinates, to fully three-dimensional states which are geometrically realistic enough for comparison with observation. Our intuition for the hydromagnetic stability of equilibrium states can now be based upon a large variety of calculations showing the interplay of current- and pressure-driven instabilities, magnetic shear, magnetic field line-tying and gravity. Only a few years ago, any equilibrium state amenable to theoretical treatment was found to be dismayingly unstable. Now examples of stable equilibrium states exist, including a model for a three-dimensional plasma loop. Three-dimensionality does not merely make constructing a magnetostatic equilibrium mathematically difficult. New physics is encountered in the realization that special symmetry in the magnetic field is a pre-requisite for equilibrium. The astrophysical magnetic field is usually embedded in a plasma of extremely high-electrical conductivity. The constant agitation and small scale motions of the plasma readily interweave the 'frozen-in' magnetic field lines in an arbitrary manner. In general, the magnetic field does not have a topology compatible with equilibrium. Hence, nonequilibrium is the natural state of affairs. Electric current sheets must form and dissipate, a process by which the magnetic field dissipates its topology, as Parker (1972) puts it, in its attempt to seek an equilibrium. The quasi-static structures on the Sun can be long-lived, suggesting that on the large scale, equilibrium obtains approximately in these structure. On the other hand, the ubiquity of small-scale chaotic motions in the photosphere implies a co-existing small-scale nonequilibrium arising from local interweaving of lines of force. The small-scale nonequilibrium can give rise to a general coronal heating. The electric currents associated with the large-scale equilibrium stores energy, built up by more ordered photospheric motions, for release in violent events such as the flare.

To what extent and in what manner the small-scale nonequilibrium is coupled to the large-scale equilibrium with its apparent stability is an important question we should next address. What we have, in a strict sense, is a dynamical steady state of which the magnetostatic state with its hydromagnetic stability properties is only a lowest order approximation. Of this approximation, we have gone far in understanding the basic physics, as can be seen in Sections 2 and 4. There is a lot more to explore theoretically within this crude approximation, but we should think of the next step. Perhaps theory alone is insufficient to push us on to the next step of development. A shove from the observer helps. It will be interesting to see how well, if at all, three-dimensional magnetostatic models compare with observation. Future observations with spatial resolutions below one arc second will test the predictions of nonequilibrium and relate detailed equilibrium structures with their stability properties. The results of these and other observational tests will not only push us on but also point out directions for future theoretical developments.

Acknowledgements

I thank Peter Gilman and Gene Parker for helpful comments.

References

Aly, J. J.: 1984, *Astrophys. J.* **283**, 349.
Altschuler, M. A. and Newkirk, Jr., G.: 1969, *Solar Phys.* **9**, 131.
Anzer, U.: 1968, *Solar Phys.* **3**, 298.
Arnold, V.: 1974, 'The Asymptotic Hopf Invariant and Its Application' (in Russian), Proc. Summer School in Differential Equations, Erevan, Armenian S.S.R. Acad. Sci.
Berger, M. A.: 1984, *Geophys. Astrophys. Fluid Dyn.* **30**, 79.
Berger, M. A.: 1985, *Astrophys. J.* (in press).
Berger, M. A. and Field, G. B.: 1984, *J. Fluid Mech.* **147**, 133.
Bernstein, I. B., Frieman, E. A., Kruskal, M. D., and Kulsrud, R. M.: 1958, *Proc. Roy. Soc. London* **A244**, 17.
Biermann, L.: 1941, *Vierteljahrsschr. Astron. Ges.* **76**, 194.
Birn, J. and Schindler, K.: 1981, in E. R. Priest (ed.), *Solar Flare Magnetohydrodynamics*, Gordon and Breach, New York, Ch. 6.
Birn, J., Goldstein, H., and Schindler, K.: 1978, *Solar Phys.* **57**, 81.
Bobrova, N. A. and Syrovatskii, S. I.: 1979, *Solar Phys.* **61**, 379.
Bogdan, T. J.: 1984, *Phys. Fluids* **27**, 994.
Bogdan, T. J. and Low, B. C.: 1986, *Astrophys. J.* (submitted).
Cargill, P. J., Hood, A. W., and Migliuolo, S.: 1985, *Astrophys. J.* (in press).
Cowling, T. G.: 1953, in G. P. Kuiper (ed.), *The Sun*, Univ. of Chicago Press, Chicago, p. 532.
Dombre, T., Frish, U., Greene, J. M., Hénon, M., Mehr, A., and Soward, A. M.: 1985, *Chaotic Streamlines and Lagrangian Turbulence: the ABC Flows* (preprint).
Flå, T., Osherovich, V. A., and Skumanich, A.: 1982, *Astrophys. J.* **261**, 700.
Friedberg, J. P.: 1982, *Rev. Mod. Phys.* **54**, 801.
Gilman, P. A.: 1970, *Astrophys. J.* **162**, 109.
Gold, T.: 1964, in W. N. Hess (ed.), *Physics of Solar Flares*, AAS-NASA Symposium, p. 389.
Grad, H.: 1967, *Phys. Fluids* **10**, 137.
Hénon, M.: 1966, *Compt. Rend. Acad. Sci. Paris* **262**, 312.
Heyvaerts, J. and Priest, E. R.: 1984, *Astron. Astrophys.* **137**, 63.
Hollweg, J. V.: 1984, *Astrophys. J.* **277**, 392.
Hood, A. W. and Priest, E. R.: 1981, in R. M. Bonnet and A. K. Dupree (eds.), *Solar Phenomena in Stars and Stellar Systems*, D. Reidel Publ. Co., Dordrecht, Holland, p. 509.
Hu, Y. Q.: 1983, *Chinese J. Space Sci.* **3**, 261.
Hu, Y. Q. and Low, B. C.: 1982, *Solar Phys.* **81**, 107.
Hu, W. R., Hu, Y. Q., and Low, B. C.: 1983, *Solar Phys.* **83**, 195.
Hundhausen, J. R., Hundhausen, A. J., and Zweibel, E. G.: 1981, *J. Geophys. Res.* **86**, 11117.
Kippenhahn, R. and Schlüter, A.: 1957, *Z. Astrophys.* **43**, 36.
Landman, D. A. and Finn, G. D.: 1979, *Solar Phys.* **63**, 221.
Leer, E., Holzer, T. E., and Flå, T.: 1982, *Space Sci Rev.* **33**, 161.
Lerche, I. and Low, B. C.: 1980, *Solar Phys.* **67**, 229.
Low, B. C.: 1975, *Astrophys. J.* **197**, 251.
Low, B. C.: 1977a, *Astrophys. J.* **212**, 234.
Low, B. C.: 1977b, *Astrophys. J.* **217**, 988.
Low, B. C.: 1978, *Astrophys. J.* **239**, 377.

Low, B. C.: 1980a, *Solar Phys.* **65**, 147.

Low, B. C.: 1980b, *Solar Phys.* **67**, 57.

Low, B. C.: 1981, *Astrophys. J.* **246**, 538.

Low, B. C.: 1982a, *Astrophys. J.* **363**, 952.

Low, B. C.: 1982b, *Rev. Geophys. Space Sci.* **20**, 145.

Low, B. C.: 1984a, *Astrophys. J.* **277**, 415.

Low, B. C.: 1984b, *Astrophys. J.* **286**, 772.

Low, B. C.: 1985, *Astrophys. J.* **293**, 31.

Low, B. C. and Wu, S. T.: 1981, *Astrophys. J.* **248**, 335.

Low, B. C., Hundhausen, A. J., Hu, Y. Q., and Kane, C.: 1985, *J. Geophys. Res.* (in press).

Migliuolo, S. and Cargill, P. J.: 1983, *Astrophys. J.* **271**, 820.

Migliuolo, S. and Cargill, P. J., and Hood, A. W.: 1984, *Astrophys. J.* **281**, 413.

Milne, A. M., Priest, E. R., and Roberts, B.: 1979, *Astrophys. J.* **232**, 304.

Moffatt, H. K.: 1969, *J. Fluid Mech.* **35**, 117.

Moffatt, H. K.: 1985, *J. Fluid Mech.* (in press).

Molodensky, M. M.: 1975, *Solar Phys.* **43**, 311.

Montgomery, D. and Vahalas, G.: 1979, *J. Plasma Phys.* **21**, 71.

Montgomery, D., Turner, L., and Vahala, G.: 1979, *J. Plasma Phys.* **21**, 239.

Morozov, A. I. and Solov'ev, L. S.: 1980, *Rev. Plasma Phys.* **8**, 1.

Nakagawa, Y. and Raadu, M. A.: 1972, *Solar Phys.* **25**, 127.

Norman, C. A. and Heyvaerts, J.: 1983, *Astron. Astrophys. J.* **124**, L1.

Osherovich, V. A.: 1982, *Solar Phys.* **77**, 63.

Parker, E. N.: 1963, *Interplanetary Dynamical Processes*, Interscience, New York.

Parker, E. N.: 1966, *Astrophys. J.* **145**, 811.

Parker, E. N.: 1972, *Astrophys. J.* **174**, 499.

Parker, E. N.: 1973, *Astrophys. J.* **180**, 247.

Parker, E. N.: 1974, *Solar Phys.* **36**, 249.

Parker, E. N.: 1979, *Cosmical Magnetic Fields*, Oxford University Press, Oxford.

Parker, E. N.: 1981a, *Astrophys. J.* **244**, 631.

Parker, E. N.: 1981b, *Astrophys. J.* **244**, 644.

Parker, E. N.: 1982, *Geophys. Astrophys. Fluid Dyn.* **22**, 195.

Parker, E. N.: 1983a, *Astrophys. J.* **264**, 635.

Parker, E. N.: 1983b, *Astrophys. J.* **264**, 642.

Parker, E. N.: 1983c, *Geophys. Astrophys. Fluid Dyn.* **2**, 85.

Parker, E. N.: 1985a, *Magnetic Topology, Nonequilibrium, and Dissipation* (preprint).

Parker, E. N.: 1985b, *Dynamical Nonequilibrium of Magnetic Fields with Arbitrary Interweaving of the Lines of Force* (preprint).

Parker, E. N.: 1985c, *Heating the Corona of the Sun* (preprint).

Parker, E. N.: 1985d, *The Hydrodynamics of Magnetic Nonequilibrium* (preprint).

Pizzo, V.: 1985, *Astrophys. J.* (in press).

Priest, E. R. and Raadu, M. A.: 1976, *Solar Phys.* **47**, 41.

Raadu, M. A.: 1972, *Solar Phys.* **22**, 425.

Ray, A. and van Hoven, G.: 1982, *Solar Phys.* **79**, 353.

Roberts, B.: 1976, *Astrophys. J.* **204**, 268.

Roberts, B. and Webb, A. R.: 1978, *Solar Phys.* **56**, 5.

Rosner, R. and Knobloch, E.: 1982, *Astrophys. J.* **262**, 349.

Rosner, R., Tucker, W. H., and Vaiana, G. S.: 1978, *Astrophys. J.* **220**, 643.

Rosner, R., Low, B. C., and Holzer, T. E.: 1985, in P. A. Sturrock, T. E. Holzer, D. Mihalas, and R. Ulrich (eds.), *Physics of the Sun*.

Sakurai, T. and Levine, R. H.: 1981, *Astrophys. J.* **248**, 817.

Schlüter, A. and Temesvary, S.: 1958, *IAU Symp.* **6**, 263.

Sheeley, Jr., N. R., Bohlin, J. D., Brueckner, G. E., Purcell, J. D., Scherrer, V., and Tousey, R.: 1975, *Solar Phys.* **40**, 103.

Spicer, D. S.: 1982, *Space Sci. Res.* **31**, 351.

Spruit, H. C.: 1977, *Solar Phys.* **55**, 3.

Stern, D. P.: 1966, *Space Sci. Rev.* **6**, 147.

Sturrock, P. A. and Uchida, Y.: 1981, *Astrophys. J.* **246**, 331.

Sturrock, P. A. and Woodbury, E. T.: 1967, in P. A. Sturrock (ed.), *Plasma Astrophysics*, Academic Press, London.

Sweet, P. A.: 1958, *IAU Symp.* **6**, 123.

Sweet, P. A.: 1971, *IAU Symp.* **43**, 457.

Syrovatskii, S. I.: 1981, *Ann. Rev. Astron. Ap.* **19**, 163.

Taylor, J. B.: 1974, *Phys. Rev. Letters* **42**, 1277.

Tsinganos, K. C.: 1982a, *Astrophys. J.* **252**, 775.

Tsinganos, K. C.: 1982b, *Astrophys. J.* **259**, 832.

Tsinganos, K. C., Distler, J., and Rosner, R.: 1984, *Astrophys. J.* **278**, 409.

Vaiana, G. S. and Rosner, R.: 1978, *Ann. Rev. Astron. Astrophys.* **16**, 393.

Vainshtein, S. I. and Parker, E. N.: 1985, *Astrophys. J.* (in press).

Van Ballegooijen, A.: 1985, *Astrophys. J.* (in press).

Van Hoven, G.: 1981, in E. R. Priest (ed.), *Solar Flare Magnetohydrodynamics*, Gordon and Breach, New York, Ch. 4.

Van Hoven, G., Ma, S. S., and Einaudi, G.: 1981, *Astron. Astrophys.* **97**, 232.

Vasyliunas, V. M.: 1975, *Rev. Geophys. Space Sci.* **13**, 303.

Voslamber, D. and Callebaut, D. C.: 1962, *Phys. Rev.* **128**, 2016.

Woltjer, L.: 1958, *Proc. Nat. Acad. Sci. USA* **44**, 489.

Yu, G.: 1973, *Astrophys. J.* **181**, 1003.

Zirin, H.: 1974, *Vistas Astron.* **6**, 1.

Zweibel, E. G.: 1981, *Astrophys. J.* **249**, 731.

Zweibel, E. G.: 1982, *Astrophys. J.* **258**, L53.

Zweibel, E. G.: 1985, *Geophys. Astrophys. Fluid Dyn.* (in press).

NONRADIATIVE ACTIVITY ACROSS THE H–R DIAGRAM: WHICH TYPES OF STARS ARE SOLAR-LIKE?

JEFFREY L. LINSKY*

Joint Institute for Laboratory Astrophysics, National Bureau of Standards and University of Colorado, Boulder, Colorado 80309, U.S.A.

Abstract. Major advances in our understanding of nonradiatively heated outer atmospheric layers (coronae, transition regions, and chromospheres) and other solar-like activity in stars has occurred in the past few years primarily as a result of ultraviolet spectroscopy from IUE, X-ray imaging from the *Einstein* Observatory, microwave detections by the VLA, and new optical observing techniques. I critically review the observational evidence and comment upon the trends with spectral type, gravity, age, and rotational velocity that are now becoming apparent. I define a solar-like star as one which has a turbulent magnetic field sufficiently strong to control the dynamics and energetics in its outer atmospheric regions. The best indicator of a solar-like star is the direct measurement of a strong, variable magnetic field and such data are now becoming available, but good indirect indicators include photometric variability on a rotational time scale indicating dark starspots and nonthermal microwave emission. X-rays and ultraviolet emission lines produced by plasma hotter than 10^4 K imply nonradiative heating processes that are likely magnetic in character, except for the hot stars where the heating is likely by shocks in the wind resulting from radiative instabilities. I conclude that dwarf stars of spectral type G–M and rapidly rotating subgiants and giants of spectral type F–K in spectroscopic binary systems are definitely solar-like. Dwarf stars of spectral type A7–F7 are almost certainly solar-like, and T Tauri and other pre-Main-Sequence stars are probably solar-like. Slowly rotating single giants of spectral type F to early K are also probably solar-like, and the helium-strong hottest Bp stars are interesting candidates for being solar-like. The O and B stars exhibit some aspects of activity but probably have weak fields and are not solar-like. Finally, the A dwarfs and the cool giants and supergiants show no evidence of being solar-like.

1. Introduction

To the best of our knowledge today, the only unique property of the Sun is its proximity. Until less than a decade ago, one presumed that the broad range of phenomena that are seen across the surface of the Sun on many spatial and temporal scales at all wavelengths must also occur on stars in some then-undetermined region of the Hertzsprung–Russell (H–R) diagram. In the absence of sufficient sensitivity at X-ray, ultraviolet, and microwave wavelengths and of indirect ways of achieving high-spatial resolution, however, this presumption could not be tested. In that era, solar physics and stellar astronomy addressed different sets of problems and were deemed by many to be wholly different and even unrelated fields that were developing in splendid isolation of each other.

Traditionally, the astronomy of stars near solar type has concentrated on understanding such questions as evolution, surface chemical composition, and atmospheric structure within the constraints of hydrostatic equilibrium, radiative equilibrium, and LTE spectral line formation, and has tended to ignore the effects of rotation, magnetic fields, mass loss, and nonradiative heating and momentum deposition processes. While recently certain of these assumptions, such as LTE line formation and no mass loss have

---

* Staff Member, Quantum Physics Division, National Bureau of Standards.

Solar Physics **100** (1985) 333–362.   0038–0938/85.15.
© 1985 *by D. Reidel Publishing Company*

been relaxed, for the most part solar-type stars are studied assuming radiative equilibrium and spatial and temporal homogeneity (a consequence of the above assumptions).

On the other hand, solar physics has concentrated on understanding the extreme inhomogeneity and time variability observed in the solar atmosphere and how the associated phenomena depend on magnetic fields, nonradiative heating processes, and hydrodynamic phenomena. The very existence of outer atmospheric layers hotter than predicted by radiative equilibrium (i.e. chromosphere, transition region, and corona) implies local nonradiative heating processes. I have proposed definitions from these different regions (Linsky, 1980, 1985) applicable to stars in terms of the local non-radiative heating rates, temperature gradients, and dominant terms in the local energy balance. The existence of observable structures within these 'layers' is a consequence of the inhomogeneous solar magnetic field that provides additional local heating in both quasi-continuous and flare modes, thermal isolation from the environment, stimulation of flows along the field lines, and inhibition of flows across the field lines. The solar wind also is a consequence of nonradiative activity as the momentum deposition presumably is both thermal (a consequence of the hot corona) and magnetic in character. The study of such phenomena in stars is what is now called the solar-stellar connection. Thus what I call solar-like phenomena are those which are nonradiative in character, are fundamentally magnetic in origin, and are almost certainly a consequence of dynamo-regenerated magnetic fields produced in or at the base of a stellar convection zone. In this review I will summarize what we now know about solar-like phenomena in stars and, in particular, which regions of the H–R diagram contain stars exhibiting these phenomena. Beginning in 1978, four developments have revolutionized our recognition and understanding of nonradiative activity and structure in cool stars. The International Ultraviolet Explorer (IUE) satellite launched in 1978 has obtained ultraviolet spectra of many stars of almost all types, providing evidence for 10^4–10^5 K plasma in the chromospheres and transition regions (TRs) of many of these stars. The *Einstein* (HEAO-2) satellite, also launched in 1978, has detected soft X-ray fluxes from many stars of almost all types, providing evidence for 10^6–10^8 K plasma in the coronae of many of these stars. The Very Large Array (VLA), dedicated in 1980, is observing microwave emission from an increasing number of cool stars and binary systems. In the future even more powerful instruments will be available for ultraviolet [Hubble Space Telescope (HST), Far Ultraviolet Spectrograph Explorer (FUSE)], X-ray (EXOSAT now in orbit, ROSAT, AXAF), and microwave (VLBA and space VLBI) observations. Finally, there has been a revolution in ground-based observations that has permitted us to study magnetic fields, active regions, starspots, rotation, and even differential rotation using the techniques of high-resolution spectroscopy, Ca II line monitoring, precise photometry, and 'Doppler imaging'. These four events radically transformed the data base concerning nonradiative activity from famine to feast. So much has been learned in the past few years that a comprehensive review of the topic no longer is feasible. Instead, I will concentrate on the major achievements and the important unsolved problems.

There are several themes in this review. First, the types of atmospheric regions and phenomena that have been investigated in detail on the Sun are indeed present on most cool stars. Second, we must presume *a priori* that all stars of late spectral type are just as inhomogeneous and complex as the Sun even though we presently lack the spatial resolution to verify this except for eclipsing systems or by rotational modulation observations. Thus, we should strive to incorporate inhomogeneity into our models or at least be suitably apologetic when we cannot take inhomogeneity into account. Third, we now have ample evidence that nonradiative heating and probably also nonradiative momentum deposition are fundamental phenomena in the atmospheres and winds of cool stars. Fourth, the evidence is accumulating rapidly that magnetic fields lie at the heart of much of the rich phenomenology of 'activity' in cool stars. This is not to say that magnetic fields control all phenomena, but rather that magnetic fields usually determine the geometry, time variability, nonradiative heating rates, inhomogeneity, and ultimately the global energy balance in stars located in a wide range of the cool half of the H–R diagram.

Before proceeding, I should mention a number of reviews that provide more detailed treatments of different aspects of this topic. Broad surveys of X-ray emission from stellar coronae include those of Vaiana (1981, 1983), Golub (1983), Stern (1983), Linsky (1981a, b), and Rosner *et al.* (1985). Gibson (1981, 1985), Gary (1985), Mullan (1985), and Dulk (1985) have summarized the microwave observations of cool star coronae; and Linsky (1981c, 1982), Brown (1983), and Baliunas (1983) have summarized the ultraviolet observations of stellar chromospheres and transition regions. The extensive evidence for mass loss in cool stars has been reviewed by Cassinelli (1979), Dupree (1982), and Hartmann (1981, 1983); and the direct and indirect evidence for magnetic fields in these stars has been summarized by Vogt (1983), Marcy (1983), Zwaan (1983), and Linsky (1983a). Linsky (1982) has discussed the energy balance in the outer atmospheres of cool stars. The X-ray, ultraviolet, and microwave emissions from the M dwarf flare stars have been reviewed by Gibson (1983), Johnson (1983), Worden (1983), Giampapa (1983a), and Linsky (1983b); and Giampapa (1983b) and Feigelson (1983) have reviewed similar data for the pre-Main-Sequence stars. Finally, reviews of the emission from RS CVn binaries include those of Bopp (1983), Charles (1983), Linsky (1983a); and Dupree (1983) and Rucinski (1985) have reviewed this topic for the contact binary systems.

2. Types of Evidence for Nonradiative Activity in Stars

I will define the term *nonradiative activity* to include those phenomena and physical properties that occur when the energy balance in a stellar atmosphere departs greatly from pure radiative equilibrium. Nonradiative heating produces the hot atmospheric layers that we call chromospheres, transition regions (TRs), and coronae. These layers are presumed to be inhomogeneous and variable in time. In addition, momentum can be imparted to the outer layers of a star by a number of possible mechanisms to produce mass loss by a stellar wind. Except perhaps for the dusty M supergiants, the deposition of momentum in the winds of cool stars does not come directly from the stellar radiation field and, therefore, could be considered an aspect of nonradiative activity. Thus in general terms the evidence for nonradiative activity consists of the following:

(1) Thermal radiation from plasmas substantially hotter than can be explained by an atmosphere in radiative equilibrium. The prime spectral diagnostics are X-ray and ultraviolet emission lines and continua, as well as thermal microwave emission. In addition, a few spectral features in the visible and near infrared, including the H and K and infrared triplet lines of Ca II, Hα, and He I 10830 Å and 5876 Å, are useful.

(2) Nonthermal radiation from relativistic particles in magnetic fields. Such radiation is detected during flares in the microwave and perhaps also in hard X-rays. Indeed, some portion of the 'quiescent' microwave emission from M dwarfs could be nonthermal in character.

(3) Stochastic emission variations indicating flaring or rapid heating of atmospheric structures like magnetic flux tubes or active regions. An atmosphere in radiative equilibrium should be a steady emitter, since the stellar luminosity changes only on very long, evolutionary time-scales, except near the endpoints of a star's life. Radial and nonradial pulsations, on the other hand, can occur even for an atmosphere in radiative equilibrium and can be maintained by purely radiative processes.

(4) Large scale atmospheric inhomogeneities indicated by cyclic variations of the stellar spectrum on a rotational time-scale; for example, rotational modulation of the emission from bright active regions in the ultraviolet, X-ray, and microwave; or modulation of the optical continuum due to an inhomogeneous surface distribution of cool, dark starspots.

(5) Mass loss produced by such nonradiative acceleration processes as waves and an outwardly decreasing thermal pressure gradient.

In this review I will discuss the evidence for the first four points primarily by considering in turn the different spectral regions. As we proceed through this topic, it is important to keep in mind that the diagnostics of the heated or accelerated plasma may not be reliable for several reasons:

(1) Very small contrast between the desired emission line and the background stellar photosphere – a problem especially important in the ultraviolet for early F and A-type stars.

(2) Non-LTE effects. In some cases, non-LTE effects can produce a spurious emission feature that would seem to indicate nonradiatively heated plasma.

(3) Most stars are members of binary systems, and quite often the duplicity or multiplicity is not readily apparent either from optical imaging, composite colors, or variable radial velocities. Since the vast majority of stars are cool dwarfs that are faint in the optical but intrinsically bright in X-rays, ultraviolet emission lines, and microwave emission, one can easily be fooled into ascribing the evidence for hot plasma to the optically dominant primary star when, in fact, an unsuspected secondary star may be the source of much or all of the high-temperature emission.

(4) Close companions can alter the adjacent star by tidally-induced rapid rotation, mass exchange, or X-ray illumination and heating. Furthermore, as may occur in the

case of the RS CVn-type binaries, magnetic fields of the two stars might interact and heat plasma between the two stars.

(5) Interstellar and circumstellar absorption can decrease or totally eliminate measurable X-ray and UV radiation from a star. This is especially important for distant stars in the galactic plane such as pre-Main-Sequence stars.

(6) Instrumental problems can be very important. For example, sensitivity limitations can lead to biased samples or the inability to observe entire classes of objects. The failure to detect high-excitation emission features does not imply that a given star lacks hot plasma, but merely that the emission measure of the plasma must be less than an instrumental upper limit. Furthermore, subtle imperfections in the instruments themselves can lead to false conclusions. An example is the UV light leak in the *Einstein* High Resolution Imager (HRI) that falsely implied that Sirius A and Vega are X-ray sources.

3. Evidence for Hot (10^6–10^8 K) Coronal Gas: X-Ray Emission

X-ray emission in the continuum and discrete lines can be produced by thermal processes (free-free, free-bound, and bound-bound) and, in principle, by nonthermal processes involving high-energy particles in magnetic fields. Both mechanisms provide a direct means to identify a coronal plasma heated nonradiatively. While pioneering soft X-ray experiments on rockets and the ANS, SAS-3, and HEAO-1 satellites were able to detect the very brightest X-ray sources among the nearby stars, major progress required more sensitive imaging instruments, in particular the Imaging Proportional Counter (IPC) and HRI focal plane instruments on the *Einstein* X-Ray Observatory (HEAO-2). These instruments detected X-ray emission from nearly every type of star except the luminous cool giants and supergiants (Vaiana, 1981; Vaiana *et al.*, 1981; Ayres *et al.*, 1981b; Helfand and Caillault, 1982; Linsky, 1981b; Johnson, 1983), and soundly contradicted the previously held theory of acoustic wave heating of stellar coronae. The Solid State Spectrometer (SSS) instrument on *Einstein* obtained low-resolution soft X-ray spectra of RS CVn systems, Algol, and two dMe stars (Swank *et al.*, 1981; Swank and Johnson, 1982), while the higher resolution Crystal Spectrometer had the sensitivity to observe only the bright source Capella among the cool stars (Vedder and Canizares, 1983). The EXOSAT spacecraft is now observing many targets, and future missions will include ROSAT, which will undertake an all sky survey, and the AXAF, which is planned have very high-resolution imaging capability.

Linsky (1981b) has summarized the physical quantities that may be inferred from these data. The imaging instruments (primarily the IPC and HRI) are useful for identifying X-ray sources, studying their time variability, and measuring their broad band (0.25–4 keV) fluxes. The IPC also provided a rough estimate of the plasma temperature and emission measure for coronae hotter than about 1×10^6 K. Low-resolution spectroscopy (e.g., the *Einstein* SSS) permits one to begin to distinguish the contributions of multi-temperature plasmas.

The *Einstein* X-Ray Observatory detected X-ray emission from Main-Sequence (dwarf) stars of essentially all spectral types except the A stars. The hot stars (spectral type O and early B) turned out unexpectedly to be very strong X-ray emitters with $L_x/L_{bol} \approx 10^{-7}$ and $\log L_x$ in the range 31–34 (Cassinelli *et al.*, 1981; Pallavicini *et al.*, 1981). By comparison, the quiet Sun value is typically $\log L_x = 26.6$, but $L_x/L_{bol} \approx 10^{-7}$. Cassinelli and Olson (1979) proposed that X-rays from the hot stars are produced in a thin corona at the base of the cool wind, but the presence of soft X-rays that should be absorbed by the overlying wind led Cassinelli *et al.* (1981) to propose instead that the X-ray emitting material is distributed throughout the wind. This picture is consistent with X-ray emission from bow shocks produced by instabilities in the massive radiatively-driven winds (e.g., Lucy and White, 1980; Owocki and Rybicki, 1984). Waldron (1984) has argued, however, that the soft X-ray spectra and $L_x/L_{bol} \approx 10^{-7}$ relation are consistent with a hot corona at the base of these winds when one includes the ionization of the wind by the coronal X-rays. Cassinelli (1985) recently summarized the evidence for nonradiative activity in these stars, including SSS spectra that suggest the presence of $T > 15 \times 10^6$ K plasma, presumably at the base of the wind, that is too hot to be confined by gravity. This suggests magnetic confinement; nonthermal radio emission (see next section) from some OB stars also indicates hot plasma confined by magnetic fields. Also, Mangeney and Praderie (1984) found that the late O and B-type stars in common with dwarf stars of later spectral types appear to obey a simple power law dependence of L_x on the effective Rossby (dynamo) number, suggesting a common heating mechanism by dynamo-generated magnetic fields. On the other hand, Rosner *et al.* (1985) argue against the solar analogy on the grounds that it does not explain the nearly constant value of $L_x/L_{bol} \approx 10^{-7}$ for these stars. Thus the hot stars could have solar-like coronae at the base of their winds, although most of the observed X-ray emission is probably produced in the wind by shocks.

The A-type stars have proved puzzling, because both Vega (A0 V) and Sirius A (A1 V) were detected by the HRI (but not the IPC) at values of L_x/L_{bol} an order of magnitude below the 10^{-7} relation that characterizes the O and B stars. Golub *et al.* (1983) argued that these detections were real, but Schmitt *et al.* (1985a) showed that the HRI signals apparently were due to a spurious UV light leak. These authors also concluded that the X-ray emission detected from the other normal A-type stars at a level of $\log L_x = 29$ probably is from known or suspected K and M dwarf companions, although two detected Ap stars in their sample exhibit no obvious evidence of duplicity (cf. Cash and Snow, 1982). We, therefore, have no evidence as yet that stars of spectral type B5–A5 have 10^6 K coronae; we can only say that if such coronae exist, they must be of low luminosity, $\log L_x < 27$.

Schmitt *et al.* (1985a) have completed a comprehensive study of the 125 visually bright late A and F-type stars observed by *Einstein*, and have carefully considered the possible confusion introduced by X-ray bright cool star companions (known and unknown). They found that the hottest star in their sample with a detected 10^6 K corona is α Aql (Altair, A7 IV–V, $T_{eff} = 7650$ K), and that there is a significant increase in L_x between the A stars and the F stars. They argued that solar-type dynamos begin at spectral type A7, even though convective zones are very thin in these stars, on the basis of the X-ray emission observed from Altair and a correlation of L_x with the Rossby number. In an earlier study based on 14 F stars with detected X-ray emission, Walter (1983) argued that the onset of solar-like magnetic dynamos begins in cooler stars (spectral type F5 V, $B - V \approx 0.45$) on the basis that at this spectral type L_x and stellar rotational velocity ($v \sin i$) are correlated (as expected for a dynamo) and these velocities are becoming small (as expected for stars with winds and magnetic fields). The location in the H–R diagram where stars with coronal properties similar to the Sun first appear may be a question of definition: a corona may in principle be heated by a variety of mechanisms and magnetic fields may be generated by a variety of possible dynamos.

The *Einstein* Observatory detected X-ray emission from essentially all Main-Sequence (dwarf) stars of spectral type early F to late M that are sufficiently nearby. These data led Rosner *et al.* (1985) to conclude that all of these stars have hot coronae and are thus 'solar-like', although the quiescent X-ray luminosities cover an enormous range ($\log L_x = 26$–31). The coolest Main-Sequence stars (M dwarfs) are particularly interesting because a very large fraction of their bolometric luminosity is emitted in the X-ray region (L_x/L_{bol} larger than 10^{-3} in some cases), these stars often exhibit extremely luminous X-ray flares (Haisch, 1983), and there is evidence that the mean quiescent levels of the X-ray emission drops for stars cooler than spectral type M5. This latter point is not yet proven as Golub (1983) showed on the basis of *Einstein* observations of about 40 cool dwarfs that typical values of L_x drop rapidly at spectral type M5, whereas Rucinski (1984) has argued that the fraction of the stellar luminosity emitted as X-rays (L_x/L_{bol}) does not vary with spectral type among the M dwarfs. Recently Bookbinder *et al.* (1984) have shown that these two results are not really inconsistent; a significant decrease in L_x does occur at spectral type M5 if one takes into account X-ray upper limits for a volume-limited sample. If confirmed by more sensitive surveys, this result is very interesting because many authors have proposed that the dynamo operates in a thin boundary between the convective envelope and the radiative core, and different interior model calculations disagree on whether stars cooler than M5 are fully convective (e.g., Grossman *et al.*, 1974) or retain their radiative cores (e.g. Cox *et al.*, 1981). Thus the turn-on of solar-like coronal behavior near spectral type A7 and the decrease in X-ray emission near spectral type M5 appears to be correlated, respectively, with the beginning of a convective zone and its extension (perhaps) all the way to the center of the star in a way consistent with the generation of magnetic fields by dynamo action.

As single stars evolve with age off the Main-Sequence to lower surface gravity and higher luminosity, their X-ray emitting coronae may or may not change radically in character. For the hot stars (spectral types O and early B) the $L_x/L_{bol} \approx 10^{-7}$ behaviour is unchanged (e.g., Cassinelli *et al.*, 1981; Pallavicini *et al.*, 1981). It is interesting that the two somewhat cooler luminous stars detected by *Einstein* [α Oph (A5 III) and α Car = Canopus (F0 Ib)] are also consistent with the $L_x/L_{bol} \approx 10^{-7}$ relation, and thus may not have solar-like coronae. On the other hand, luminous stars of spectral type mid F to early K with detected X-ray emission exhibit a similar dependence of L_x on rotational velocity as Main-Sequence stars and thus these luminous stars likely have solar-like coronae heated by magnetic fields which in turn are generated by dynamo processes.

Luminous K and M stars have very different X-ray emission from dwarf stars of similar spectral type. Ayres *et al.* (1981b) and Haisch and Simon (1982) were unable to detect single giant stars of spectral type cooler than about K2 III and supergiant stars of spectral type cooler than about G2 Ib at very low flux levels. For example, the upper limit on the L_x/L_{bol} ratio for α Boo (K2 III) is a factor of $> 10^3$ smaller than for the Sun, which itself is only a weak X-ray source. These observations led to the concept of a 'dividing line' in the H–R diagram separating stars with significant coronal emission (single giants with spectral types earlier than about K1 III and Main-Sequence stars) from stars without detected₎emission (giants later than spectral type K2 III and supergiants later than about G2 Ib). Although the absence of detections of solar-like coronae to the right of the 'dividing line' among single stars is no proof that such stars do not have hot coronae, it does indicate the rapid decrease in the emission measure of hot plasma as one proceeds from left to right across this 'dividing line' in the H–R diagram (see Figure 1). The X-ray dividing line is essentially coincident with a similar boundary for the disappearance of C IV emission (see below) and the onset of large mass loss rates detected in such cool plasma ions as Mg II (e.g., Cassinelli, 1979; Dupree, 1982; Hartmann, 1981, 1983). Thus it is likely that stars to the right of the boundary lines do not have 'solar-like' outer atmospheres, although the physical reason for a transition from nonsolar-like atmospheres is not clear at this time.

Figure 1 summarizes which types of stars are solar-like on the basis of X-ray emission from coronae and other activity indicators. Important stars mentioned in this review are included in the figure. A great deal has been learned about the properties of the hot plasma in stellar coronae and their dependence on stellar parameters, in particular:

(1) There is a monotonic decrease in L_x with increasing stellar age (see Stern, 1983). This is based on systematic studies of the Hyades (age 4×10^8 yr, Stern *et al.*, 1981), Ursa Major (age 1.6×10^8 yr, Walter *et al.*, 1984), Pleiades (age 6×10^7 yr, Caillault and Helfand, 1985; Micela *et al.*, 1985), and pre-Main-Sequence stars (e.g., Ku and Chanan, 1979; Feigelson and Kriss, 1981; Montmerle *et al.*, 1983; Simon *et al.*, 1985a). The pre-Main-Sequence stars reviewed by Feigelson (1983) and Rosner *et al.* (1985) have values of L_x up to 10^3 times that of Main-Sequence stars of similar spectral type, rapid variability, and intrinsically hard spectra, properties that could be explained by continuous flaring. However, all of these stars show absorption at low energies by circumstellar matter, and the stars with the most massive circumstellar envelopes exhibit no observable X-ray emission, as if their coronae are smothered (Walter and Kuhi, 1981).

(2) L_x generally increases with increasing rotational velocity for stars redder (cooler) than colour index $B - V = 0.45$, corresponding to spectral type F5 V (e.g., Walter, 1983). This result may explain the age effect (point 1), because such stars spin down with age, and it provides further indirect evidence that the coronal heating process is magnetic in character with the magnetic fields regenerated by a dynamo mechanism. However, the functional dependence of L_x on age (Caillault and Helfand, 1985) is different from the $t^{-1/2}$ law initially proposed by Skumanich (1972). Indeed, the early F stars have large L_x/L_{bol} ratios which are uncorrelated with rotational velocity

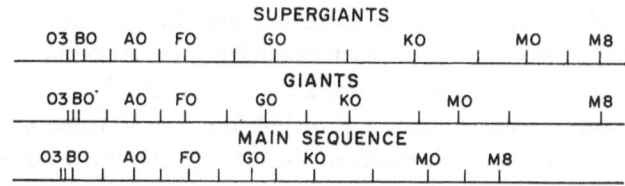

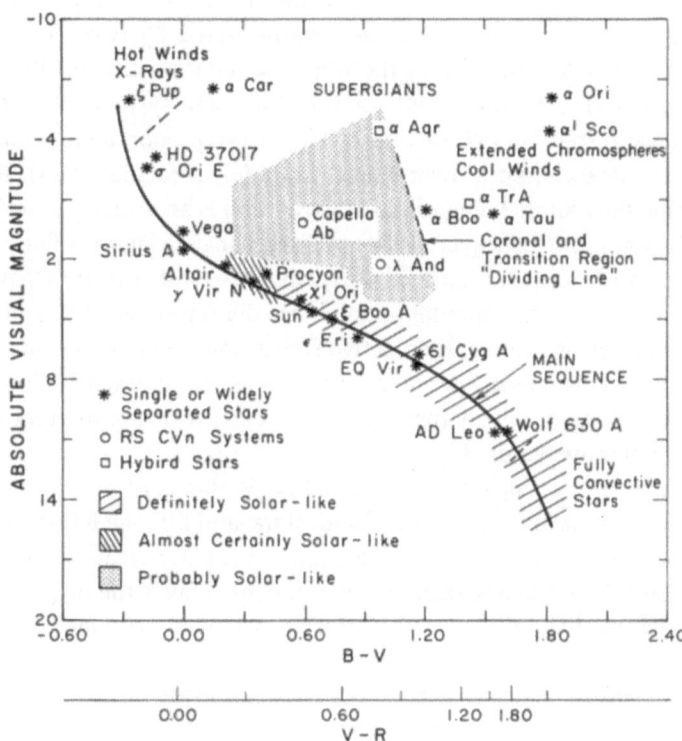

Fig. 1. An H–R diagram showing schematically which types of stars are definitely or probably solar-like on the basis of direct or indirect indicators of strong, turbulent magnetic fields. Also indicated are the regions of the H–R diagram where massive winds occur and hot plasma is apparently absent. Important stars mentioned in the text are identified in the diagram.

(Pallavicini *et al.*, 1981; Walter, 1983). One possible explanation for this behaviour is that the dynamo which occurs in stars having very thin convective zones is qualitatively different from that occurring in the cooler stars. There remains a disagreement, however, whether the functional dependence of X-ray emission on rotation is of the form $L_x \sim (v \sin i)^2$ as proposed by Pallavicini *et al.* (1981) for stars later than spectral type F5, or whether there is a more complex relation in which L_x/L_{bol} depends linearly on rotational velocity for rotational periods less than 12 days and then decays exponentially for stars with longer periods (Walter, 1981, 1982; Caillault and Helfand, 1985). Furthermore, there is evidence for saturation at high-rotational velocities and young ages (Vilhu and Rucinski, 1983; Caillault and Helfand, 1985).

The relation of L_x to rotation is more complex for the rapidly rotating, tidally synchronous spectroscopic binaries, such as the RS CVn, Algol, and W UMa systems, which are very strong X-ray emitters (log L_x = 30–31). As discussed by Rosner *et al.* (1985), the relation between L_x/L_{bol} and period (orbital and rotational periods are the same) is biased by the rough proportionality between the radius of the cooler (more active) stars and stellar separation.

(3) On the basis of *Einstein* SSS data, Swank *et al.* (1981) have found that the coronae of RS CVn and Algol systems are characterized by two temperatures (one component at roughly 5×10^6 K and the other hotter than 2×10^7 K). Swank and Johnson (1982) found a similar result for the dMe star systems Wolf 630 AB and AD Leo. The lower resolution IPC data also suggest two predominant temperatures for the coronae of many stars, both dwarfs and giants (e.g., Vaiana, 1983). Rosner *et al.* (1985) argued that the bimodal temperature distribution is an artifact of the low spectral resolution of the data, but Schrijver (1984) pointed out that high-resolution spectra of three stars (Procyon*, Capella, and σ^2 CrB) obtained with the EXOSAT objective gratings are well fitted by the bimodal temperature distributions. If this latter result is confirmed by future observations, then active stellar coronae may be characterized by two different types of structures, perhaps different classes of magnetic loops. Alternatively, the hot components might represent a population of nonthermal electrons as suggested by the microwave data.

(4) In their study of the eclipsing system AR Lac, Walter *et al.* (1983) found evidence for discrete active regions in the coronae of both stars, and that the K0 IV star possesses an extended component to its corona (extending one stellar radius above its surface) that might be hotter than the nonextended component. This is the only star other than the Sun for which we have direct evidence of an extended coronal geometry seen in X-rays.

(5) Coronal magnetic fields are needed both to confine the hot plasma and probably also to heat it.

4. Evidence for Hot (10^6–10^9 K) Coronal Gas: Microwave Emission

The known or suspected mechanisms for microwave emission from the Sun (and stars) include thermal bremsstrahlung, gyroresonance emission by thermal electrons spiraling in coronal magnetic fields, gyrosynchrotron emission from nonthermal electrons, and coherent processes. Prior to the commencement of VLA observations, the types of late-type stars detectable by interferometers such as the NRAO three-element interferometer or single dish radio telescopes were severely limited by sensitivity and source confusion. The only detected sources consisted of dMe and RS CVn systems while flaring, interacting binary systems, and two M supergiants having massive winds (α Ori

---

* However, Jordan *et al.* (1985) pointed out that the EXOSAT high-resolution spectrum of Procyon can just as well be explained by a one-temperature corona ($T \approx 1.5 \times 10^6$ K) in agreement with the *Einstein* IPC pulse height spectrum (Schmitt *et al.*, 1985b).

and α Sco). The factor of 100 increase in sensitivity (a 3σ noise level of 0.1 mJy is achievable at 6 cm) and the factor of 400 improvement in angular resolution of the VLA compared to a 100 m single disk telescope revolutionized the field of stellar radio astronomy, just as *Einstein* revolutionized the field of stellar X-ray astronomy. There now are at least five classes of nonthermal microwave sources (dMe stars, pre-Main-Sequence stars, RS CVn binaries, O stars, and magnetic B stars) and three classes of thermal wind sources (cool giants, O stars, and pre-Main-Sequence stars). As yet, however, no dwarf stars similar in spectral type to the Sun have been detected.

The VLA has now observed about a dozen dMe stars as quiescent and flaring radio sources (Gary and Linsky, 1981; Linsky and Gary, 1983; Topka and Marsh, 1982; Fisher and Gibson, 1983; Gibson, 1983, 1985). These sources all appear to be variable. Because the dMe stars all are detected at levels far above the bremsstrahlung fluxes predicted using emission measures consistent with the X-ray luminosities, and because the microwave emission often is circularly polarized, these authors have argued that the quiescent flux is likely due to gyroresonant or synchrotron emission. Linsky and Gary (1983) suggested that a nonthermal tail of electrons with $E > 10$ keV might be responsible for most of the synchrotron emission. The 6 cm luminosities for the quiescent emission lie in the range 1×10^{13}–5×10^{14} ergs s^{-1} Hz^{-1}. The microwave emission can be highly variable on both short time-scales ($\lesssim 0.2$ s flaring) (e.g., Lang *et al.*, 1983) and long time-scales of hours to days. The probable emission mechanisms (masering for the flares and gyrosynchrotron emission for the longer term variability) imply highly structured coronae similar to that of the Sun in which the local magnetic field confines and accelerates the emitting electrons.

Except for the Sun, no single dwarf stars of spectral types F, G, and K have been detected positively despite several searches (e.g., Linsky and Gary, 1983). Luminosity upper limits for nearby stars as low as 3×10^{12} ergs s^{-1} Hz^{-1} at 6 cm have been achieved for such stars as ε Eri (K2 V) and 61 Cyg AB (K5 V + K7 V), but these limits are still above those predicted on the basis of bremsstrahlung emission alone from their X-ray coronae. Gary and Linsky (1981) originally detected the young star χ^1 Ori (G0 V), but repeated observations have revealed that the source is highly variable and the original detection could be explained by a flare on its M dwarf companion. Furthermore, no single A-type dwarfs have yet been detected. For comparison purposes, the Sun placed at a distance of d parsecs would have a typical quiet flux density of $0.01 \, d^{-2}$ mJy, whereas the sensitivity limit of the VLA is now 0.1–0.2 mJy. Thus, only stars that are considerably more luminous than the Sun at radio wavelengths presently can be detected as radio sources by the VLA.

VLBI mapping of RS CVn spectroscopic binary systems (Mutel *et al.*, 1984, 1985; Lestrade *et al.*, 1985) is becoming a powerful tool for inferring the nature of the emission processes and structure of the magnetic field in these systems. The observations to date typically show at low flux density (quiescent emission) a single circularly polarized source with brightness temperature of roughly 10^9 K, or at high flux density a compact unpolarized core source smaller than a stellar diameter and a halo comparable in size to the binary system. These authors have interpreted the data by means of an expanding coronal loop model in which the core is gyrosynchrotron emission from a power law energy distribution of electrons trapped in a compact loop at the early stage of a solar-like flare, while the halo emission is from loops that have expanded to the dimensions of the binary system and are now optically thin gyrosynchrotron emitters. Kuipers and van der Hulst (1985) have proposed a model in which the unpolarized gyrosynchrotron emission is from electrons in the high-energy tail of the hot coronal thermal gas seen in the X-rays rather than a power law (nonthermal) distribution.

An unexpected recent discovery is nonthermal microwave emission from hot O-type stars. Thermal bremsstrahlung microwave emission from the winds of many hot stars has been detected by the VLA and is useful in deriving the stellar mass loss rates. However, Abbott *et al.* (1984, 1985), now find that five of the detected hot stars are emitting by a nonthermal process on the basis that (1) the flux is ten or more times larger than can be explained by thermal bremsstrahlung from a wind with mass loss rates inferred from ultraviolet line profiles and infrared excess emission; (2) the microwave spectral index is inconsistent with thermal bremsstrahlung emission; and (3) the sources are unresolved with the 'A' array. White (1985) has proposed a model in which electrons are accelerated to relativistic energies by shocks in the wind and radiate by the synchrotron process, and Underhill (1984) has proposed that a portion of the microwave flux is due to gyroresonant emission. For either model the atmosphere must be highly inhomogeneous for the nonthermal emission not to be completely covered by bremsstrahlung absorption (Cassinelli, 1985).

Another class of nonthermal radio emitters is the helium strong Bp stars with strong dipole magnetic fields. These stars are believed to be synchrotron emitters on the basis of their large microwave fluxes, flat spectral indices, and very small mass loss rates (Drake *et al.*, 1985). Two T Tauri stars are also nonthermal sources (probably gyro-synchrotron emission) and many pre-Main-Sequence stars are thermal sources due to bremsstrahlung from their winds (Dulk, 1985; Brown, 1985).

Among the late-type giants and supergiants, the single (or widely separated binaries) detected so far are α Ori (M2 Iab), α^1 Sco (Ml Ib), α^1 Her (M5 II), α Boo (K2 III), β UMi (K4 III), α Tau (K5 III), and μ Gem (M3 III) (cf. Newell and Hjellming, 1982; Drake and Linsky, 1983a, 1985). The emission mechanism for these stars likely is bremsstrahlung from a cool (6000–8000 K) chromosphere and wind. Also, flares have been reported from α Ori, π Aur (M3 II), and R Aql (gM5e–8e). None of the X-ray emitting single F and G giants has yet been detected as a microwave source, whereas the RS CVn and Algol systems are readily detected as flaring and quiescent sources with single dish antennae and the Green Bank interferometer (Gibson, 1981; Feldman, 1983) and now also as quiescent sources with the VLA. With increasing binary separation (longer period), the RS CVn systems often are less luminous (Mutel and Lestrade, 1985). An example is the nearest long period system, Capella (G6 III + F9 III), with a period of 104 days that has now been detected at 2 and 6 cm (Drake and Linsky, 1985). The observed 0.2 mJy flux at 6 cm is consistent with the flux expected on the basis of bremsstrahlung from the X-ray corona alone, without a nonthermal component, and the Capella system thus is the only purely coronal bremsstrahlung source known except for the quiet Sun. Linsky (1983c) argued that the decrease in nonthermal radio emission, flaring, and other nonthermal properties with increasing separation in RS CVn binary systems is evidence that interacting magnetic fields of the two stars are the source of the nonradiative activity.

5. Evidence for 10^5 K (Transition Region) Gas and 10^4 K (Chromospheric) Gas: Ultraviolet and Optical Data

The available evidence for the existence of chromospheric (10^4 K) and transition region (TR, 10^5 K) plasma in the outer atmospheres of late-type stars consists primarily of emission lines observed in the ultraviolet. These lines are formed either by collisional excitation, recombination and subsequent cascade, or fluorescence. The first mechanism is generally thermal in character, although excitation by nonthermal electrons streaming down loops from the corona may contribute; the second and third mechanisms also are thermal but nonlocal in the sense that the ionizing or stimulating radiation originates elsewhere, often in an overlying higher temperature plasma. For example, the He I 10830 Å and He II 1640 Å lines likely are formed at relatively cool temperatures following multiple ionization of neutral helium by coronal X-rays and subsequent recombination. The Ca II H and K lines and hydrogen Hα lines are among the few lines in the visible which are useful indicators of chromospheric plasma. In addition to spectral lines, chromospheric plasma can be observed by microwave emission and ultraviolet continuum emission – the former has been detected so far only from K and M supergiants, while the latter is readily observed only in dMe stars during flares. However, Haisch and Basri (1985) and Simon *et al.* (1985b) have proposed that the 1600–1700 Å continuum can be used to infer the temperature at the base of the chromosphere.

Prior to observations from space, we could observe only relatively cool chromospheric plasma using the Ca II H and K lines and could obtain indirect evidence for hot coronae from the He I 10830 Å line (Zirin, 1982; Zarro and Zirin, 1985). Linsky and Avrett (1970) and Linsky (1977) have summarized these data and the usefulness of various spectroscopic diagnostics. The first few space observations by rockets, balloons, and *Copernicus*, were superseded by IUE, which has observed hundreds of late-type stars in the 1200–3200 Å spectral region at both low and high resolution. These data in turn will be superseded by the more sensitive and versatile instruments on Space Telescope and the proposed Far Ultraviolet Spectrograph Explorer (FUSE) mission.

It is important to recognize that instrumental spectral range and sensitivity limit the plasma temperatures that can be observed. For example, *Copernicus* was capable of observing only the Lα and Mg II lines in late-type stars (except for the very brightest stars like Capella), and thus could only observe plasma as hot as 10^4 K. IUE, on the other hand, can observe emission features of C IV at 1550 Å and N V at 1240 Å formed in plasma as hot as 150 000 K. The 912–1216 Å spectral range of the proposed FUSE contains the strong resonance lines of O VI formed at 300 000 K, and the 100–912 Å spectral range, also observable by FUSE, contains lines emitted by coronal plasma as hot as 3×10^7 K (Fe XXIV).

Emission lines formed in the chromospheres and TRs of Main-Sequence stars later than spectral type F0 are readily detected by IUE and the Ca II H and K lines are easily observed by telescopes on the ground. The important lines from plasmas at 5000–10 000 K are Ca II (3933, 3968 Å), Mg II (2796, 2803 Å), H I (1216 Å), C I (1657 Å), Si II (1808, 1817 Å), and O I (1305 Å multiplet). At higher temperatures $(3 \times 10^4 - 2 \times 10^5$ K) the strongest available emission lines are of C II (1335 Å), Si III (1892 Å), C III (1909 Å), Si IV (1394, 1403 Å), C IV (1548, 1550 Å), and N V (1238, 1242 Å).

The search for hot plasma in the early F and A-type stars in the optical region from the ground and in the ultraviolet from IUE is severely hampered by the poor contrast between ultraviolet emission lines and the very bright photospheric continuum in these stars. The existence of chromospheres and TRs on these stars is, therefore, an unanswered question at this time. This topic has been reviewed by Linsky (1981c) and most recently by Wolff (1983). The hottest stars exhibiting emission in the Ca II H and K lines are the F0 dwarf γ Vir N ($B - V = 0.36$, Warner, 1968) and the F0 supergiant, α Car ($B - V = 0.15$, Warner, 1966). Occasionally, Ca II emission has been reported in the A7 III, possible δ Scuti star, γ Boo ($B - V = 0.19$, LeContel et al., 1970; Auvergne et al., 1979). Dravins et al. (1977) demonstrated that transient emission occurs in the δ Scuti stars from shock waves formed when the photosphere has maximum outward acceleration. Careful studies of the Ca II lines at high dispersion in the early A-type stars (e.g., Freire et al., 1978) and in A dwarfs in young clusters (Dravins, 1981) show no evidence for emission.

Extending the search for emission features in the A-type stars to shorter wavelengths offers some prospect for improvement, because the photospheric continuum becomes fainter toward shorter wavelengths. In their extensive survey of Mg II emission, Böhm-Vitense and Dettmann (1980) detected stars as early as α Cae ($B - V = 0.34$), but the hottest dwarf star that might exhibit Mg II emission features is α Aql ($B - V = 0.22$), observed by Blanco et al. (1982). Because this rapidly rotating A7 IV–V star also has been detected at Lα (Blanco et al., 1980) and in X-rays (Golub et al., 1983), it must contain nonradiatively heated plasma with temperatures of 10^4 K to in excess of 10^6 K. Accordingly, α Aql appears to be the earliest dwarf star that exhibits the nonradiatively heated atmospheric layers typical of the late-type stars.

Many investigators have searched the 1175–2000 Å region for evidence of emission in the C II, Si IV, and C IV lines. Among the earliest type stars detected are HD 127739 ($B - V = 0.35$, Saxner, 1981), and the Ursa Major Stream star α Cr v ($B - V = 0.32$, Walter et al., 1984). Attempts to detect emission from A-type stars (e.g., Crivellari and Praderie, 1982) have been unsuccessful so far. In particular, the detection of C IV emission in one spectrum of HD 21389 (A0 Ia) by Underhill (1980) probably is spurious in view of the narrow line widths and unusual flux ratio. The disappearance of the strongest emission features into the photospheric continuum 'noise', together with the absence of verifiable continuum emission in excess of that expected on the basis of radiative equilibrium, He I 10830 Å features, or line variability (Wolff, 1983), prevent us from determining whether the A-type stars (except for the very coolest) have nonradiatively heated atmospheres. On the other hand, many Ap stars have strong dipole magnetic fields so that nonradiative heating could potentially occur by a number of possible mechanisms in such stars.

As a result of extensive observations of ultraviolet spectra and the Ca II H and K lines, we have learned a great deal about the nonradiatively heated chromospheres and TRs of late-type stars. These results have been reviewed in detail by Brown (1983), Dupree (1982), and Linsky (1981c, 1982, 1983c); I list here only some of the highlights of this work:

(1) Emission lines characteristic of chromospheric plasmas generally are observed in all stars later than early F spectral type and in all luminosity classes. Evidence for TRs (10^5 K plasma) generally is present in dwarf stars cooler than about F0 V, and in the giants and supergiants of spectral types F and G. Linsky and Haisch (1979) proposed and Simon et al. (1982) confirmed the existence of a 'dividing line' in the H–R diagram near spectral type K1 III such that TR lines are generally not observed in single stars to the right (cooler) of this boundary (see Figure 1). Whether the existence of the dividing line is due to the true absence of any plasma at 10^5 K in these stars or merely to the rapid decrease in the emission measure of such plasma with decreasing effective temperature, cannot be determined at this time. However, the upper limit to the C IV surface flux in α Boo (K2 III) is only 1% that of the quiet Sun (Ayres et al., 1982). The location of the boundary in the H–R diagram is the same as the X-ray boundary proposed by Ayres et al. (1981b). The absence, or small amount, of 10^5 K plasma in the cooler giants could be a result of the rapid decrease in rotational velocity as giants evolve across these 'boundaries' (Ayres et al., 1981b; Gray, 1981), leading to weakened dynamo generation of magnetic fields and thus decreasing heating and open magnetic field configurations conducive to strong winds.

(2) Stars cooler and more luminous than these boundaries typically have large mass loss rates (Cassinelli, 1979) as inferred from circumstellar absorption features and infrared emission from dust among the M supergiants or asymmetric Ca II and Mg II emission lines in the K stars (Stencel, 1978; Stencel and Mullan, 1980). Stencel et al. (1981) and Carpenter et al. (1985) have used line ratios within the C II 2325 Å multiplet to estimate electron densities in the chromospheres of late-type giants and supergiants. They find that stars hotter than the boundary have high-density, geometrically thin chromospheres, whereas stars cooler than the boundary have low density chromospheres that are geometrically extended (1–5 times the photospheric radius). These size scales are rough estimates since constant electron density was assumed. Brown and Carpenter (1984) have derived chromospheric temperatures of 7000–9000 K for these stars from C II 1335 Å/2325 Å flux ratios. Hartmann and Avrett (1984) have derived a chromospheric model for α Ori (M2 Iab) which is geometrically extended with temperatures and densities in this range. Drake and Linsky (1983b) have developed a co-moving PRD radiative transfer code to derive chromospheric properties and mass loss rates from asymmetric emission lines.

(3) Hartmann *et al.* (1980, 1981) proposed a third class of stars, the hybrid stars, which show evidence for both strong mass loss and 10^5 K TR plasma. Prototypes of stars in this class are α Aqr (G2 Ib) and α TrA (K4 II). Proposed explanations for the hybrid nature of these stars include an Alfvén wave heated and accelerated wind (Hartmann *et al.*, 1981), isolated hot flux tubes imbedded in a cool wind (Linsky, 1982), shocks in an inhomogeneous wind (Mullan, 1984), or (in the particular case of α TrA) a previously unknown F dwarf companion (Ayres, 1985). We do not yet know whether any of these proposed explanations is correct. There is also some evidence for rotational modulation of the Mg II emission lines (Brosius *et al.*, 1985), perhaps indicating an inhomogeneous distribution of emission across the surface of these stars.

(4) Surface fluxes for both TR and chromospheric emission features vary greatly from one star to another. Stars with very large surface fluxes include the dMe stars (Linsky *et al.*, 1982), RS CVn systems (Simon and Linsky, 1980), young stars like those in the Hyades (Zolcinski *et al.*, 1982), and pre-Main-Sequence stars (e.g., Giampapa, 1983b; Calvet *et al.*, 1985). There is a clear decrease in surface fluxes, and thus nonradiative heating rates, with increasing age on the main sequence (e.g., Simon and Boesgaard, 1983; Barry *et al.*, 1984; Haisch and Simon, 1985), but the (age)$^{-1/2}$ dependence originally proposed by Skumanich (1972) to describe the behavior of Ca II fluxes is not valid for stars younger than the Hyades (Duncan, 1983). In particular, Simon *et al.* (1985b) found that the decay with age is better fit by an exponential than a power law and that the *e*-folding time is shorter for high excitation TR lines than for chromospheric lines. The surface fluxes also increase with increasing rotational velocity, and Vilhu (1984), Hartmann *et al.* (1984), and Simon *et al.* (1985b) found a functional dependence of chromospheric and TR line emission on the Rossby number, further strenghtening the association of magnetic fields with the nonradiative heating process. The functional dependencies of emission features on rotational period depend on the temperature of formation (Marilli and Catalano, 1984) in such a way as to explain the correlations of coronal, TR, and chromospheric features with respect to each other (Ayres *et al.*, 1981).

(5) The importance of magnetic fields in determining the geometric structure and energy balance in the chromospheres and TRs of late-type stars has been summarized by Linsky (1983a). The evidence is of several types. First, the existence of individual solar-like active regions on stars is suggested by the cyclical modulation of UV emission lines at the rotational period, in phase with the dark starspots deduced from optical light curves (Baliunas and Dupree, 1982; Marstad *et al.*, 1982; Linsky, 1983c). Second, the large dispersion in heating rates for stars of the same spectral type, which can be explained readily only by different magnetic field strengths and geometries on these stars, is contrary to the predictions of purely acoustic wave heating (Linsky and Ayres, 1978; Basri and Linsky, 1979). Third, the empirical functional dependence of chromospheric heating rates on gravity and effective temperature strongly suggests heating by slow mode MHD waves (Stein, 1981; Ulmschneider and Stein, 1982). Fourth, the existence of flaring in dMe stars (e.g., Haisch, 1983), RS CVn systems (e.g., Linsky, 1983c), and T Tauri stars (e.g., Walter and Kuhi, 1984) implies rapid conversion of

stored (presumably magnetic) energy to heat as occurs in solar flares. Fifth, there is evidence for differential rotation with latitude for dMe stars (Vogt, 1983) and G dwarfs (Baliunas *et al.*, 1985), which is consistent with dynamo generation of fields. Finally, the existence of systematic redshifts of 10^4-10^5 K emission lines (Brown *et al.*, 1984; Ayres *et al.*, 1983; Ayres, 1984) likely is analogous to the downflows of hot plasma observed over solar magnetic active regions.

There is evidence that the large scale distribution of magnetic fields across the surface of a star is significantly different in the F stars compared to cooler stars. Rotational modulation of surface brightness inhomogeneities due to isolated starspots (Radick *et al.*, 1983; Campbell, 1984) and plages (Vaughan *et al.*, 1981; Baliunas *et al.*, 1983) are not observed in Main-Sequence stars earlier than about F6 V or in the F9 III active component of Capella (Ayres *et al.*, 1983; Ayres, 1984). Since many of these stars have very large X-ray and ultraviolet emission line fluxes, the absence of rotational modulation implies a nearly uniform distribution of magnetic active regions across the surface of these stars, whereas the cooler active stars generally have only one or two dominant active regions. Giampapa and Rosner (1984) interpreted this different behaviour as a consequence of dynamo activity in thin convective zones in the F stars producing many small, rather than a few large, active regions (see also Ayres, 1984).

6. Evidence for Nonradiatively Heated Photospheres

Evidence for nonradiative heating in stellar photospheres could consist of a derived temperature structure that is hotter than predicted on the basis of radiative equilibrium alone. Alternatively, one could consider as evidence photospheric emission in a spectral interval that is brighter than predicted on the basis of a radiative equilibrium photospheric model. Either type of evidence requires an accurate radiative equilibrium model. This is a difficult requirement for several reasons: (1) a reasonably complete description of line blanketing including molecules and taking nonLTE effects into account is needed; (2) an accurate theory of convection, including overshoot, must be incorporated; (3) the solar photosphere is highly inhomogeneous owing to the presence of nonradiatively heated, small scale flux tubes; and (4) efficient cooling by CO in the nonmagnetic regions is a likely basis for thermal instability (Ayres, 1981). Thus one-component radiative equilibrium models are not realistic physically for the Sun and, therefore, for late-type stars in general. We are thus left with the following quandary: against what are we to compare an empirical photospheric temperature distribution in order to infer the existence of nonradiative heating at photospheric levels?

For the Sun, Chapman (1981) derived empirical temperature structures for spatially averaged plages (active regions) and estimated temperature structures for isolated flux tubes by the analysis of the cores and wings of the Ca II, Mg II, and Lα lines. These models are compared with quiet Sun models in Figure 2. Other models indicating photospheric temperature enhancements in magnetic regions have been computed by Vernazza *et al.* (1981), Morrison and Linsky (1978), and others. While such models are not directly compared to radiative equilibrium models, the systematic enhancement of

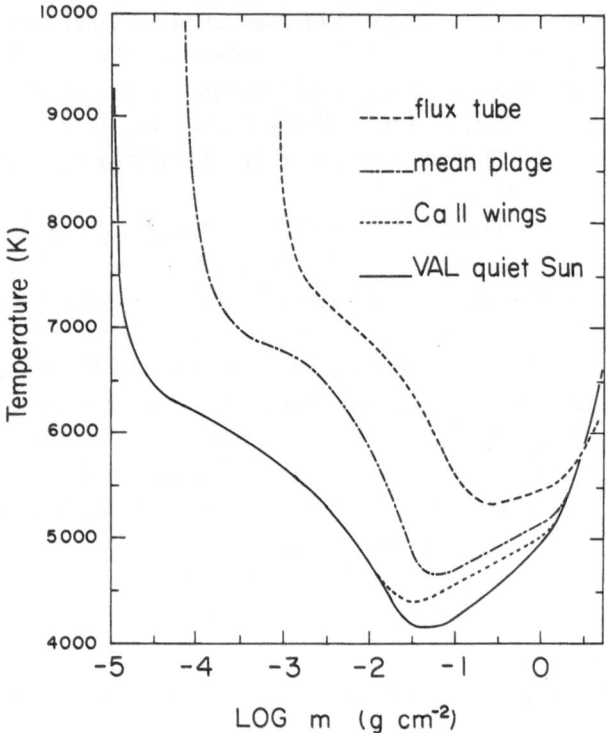

Fig. 2. Plage, flux tube, and quiet Sun models. The solid line is the VAL quiet Sun model (Vernazza *et al.*, 1981). The short dashed lines (Ca II wings) represent a modification of the VAL designed to reproduce the Ca II H and K damping wings. The dash-dot curve is a plage model based on H I Lα, Ca II K, and Mg II k data obtained by the LPSP experiment on OSO-8. The long dashed (higher) curve represents a flux tube model with a chromospheric portion matching the OSO-8 plage profiles with a 20% filling factor. The photospheric portion ($m > 0.3$ g cm^{-2}) is similar to the class of flux tube models advocated by Chapman (1977). From Chapman (1981), courtesy of Colorado Associated University Press.

the photospheric temperature structure in magnetic regions clearly suggests the presence of nonradiative heating, at least in the magnetic regions.

The extension of such arguments to late-type stars should be viewed with caution because of the difficulty of computing accurate radiative equilibrium models that properly incorporate atmospheric inhomogeneities. Nevertheless, the models of dMe stars computed by Giampapa *et al.* (1982) on the basis of the Ca II and Hα lines have hotter temperatures in the temperature minimum region than do dM stars of similar effective temperatures. An analogous argument can be made for the active F–K dwarfs compared to the less active dwarfs (Kelch *et al.*, 1979) and the active subgiants in RS CVn systems compared to nonactive stars of similar spectral type (Simon and Linsky, 1980; Baliunas *et al.*, 1979). Recently, Haisch and Basri (1985) have found that the brightness of the 1600 Å continuum in solar type dwarfs is enhanced in young, active stars, indicating enhanced nonradiative heating in the temperature minimum region. For example, the star HD 1835, which has the same spectral type (G2 V) as the Sun but is much younger (0.9×10^9 yr vs 4.6×10^9 yr) has a 1600 Å continuum brightness corresponding to nearly complete coverage by a solar-like plage. From their analysis of the 1700 Å continuum in F7 V–G2 V stars, Simon *et al.* (1985b) concluded that the upper photospheres of some young stars are nonradiatively heated by up to 300 K. This result is important because the far ultraviolet photospheric continuum is an efficient radiator and thus a small increase in temperature can correspond to local nonradiative heating far in excess of that needed to heat the entire chromosphere and corona (Athay, 1976).

The ultraviolet continua of A-stars have been examined for evidence of temperatures in excess of those predicted by radiative equilibrium models. Praderie *et al.* (1975) proposed that emission in the short wavelength wing of the Lyman alpha line of Vega (A0 V) implies nonradiative heating, but Snijders (1977) and Hubeny (1981) have shown that this spectral feature is consistent with non-LTE radiative equilibrium models. Again, further observational and theoretical work is needed.

7. Direct Measurements of Stellar Magnetic Fields

An important theme of this review of the solar-stellar connection is that magnetic fields and their interplay with fluid motions lie at the heart of phenomena associated with solar and stellar 'activity'. While the importance of this theme has been recognized for some time, it must be considered speculative without accurate and believable direct measurements of magnetic fields on stars. Fortunately, great strides have been made recently in this formidable task (see reviews by Marcy, 1983, 1984; Giampapa, 1984), but the data are still few and the inadequacies of the commonly used interpretation techniques are only now beginning to be appreciated and addressed.

The classical approach to determining stellar magnetic fields involves measuring circular polarization in absorption line wings produced by the Zeeman effect – the so-called 'Babcock technique'. While this approach has been successful for magnetic A and B stars, presumably owing to their simple bipolar field geometries, it has not yet led to fully credible field measurements in solar-type stars despite many attempts (e.g., Boesgaard, 1974; Boesgaard *et al.*, 1975; Vogt, 1980; Bonsack and Simon, 1983). Pushing this technique to its limit, Borra *et al.* (1984) (cf. Brown and Landstreet, 1981), used a Griffin-type mask spectrometer to measure the polarization signal from 1500 spectral lines simultaneously. They were able to push the net field measurement sensitivity to as low as $\pm$ 5–10 G. They obtained upper limits for many late-type stars, but they did claim (as a 3.9σ result) to have measured a net longitudinal magnetic field of 25.0 ± 6.4 G on ξ Boo A (G8 V) on one occasion and possibly also a field on UX Ari.

Unfortunately, the measurement of a net longitudinal magnetic field, even if of high quality, is not particularly useful: if the Sun is a reliable guide, stellar photospheric fields will be strong, but spatially concentrated, and will almost completely cancel because of the complex geometry. Most important, it is the strong fields that dominate the energetics and dynamics of stellar activity, and we must measure these fields rather than the globally averaged weak fields. Apart from sunspots and pores, the solar field is generally about 1500 G and is concentrated into thin flux tubes with an area filling factor of roughly 1%. To measure such fields on stars requires a very different type of technique. Most recent work has followed an approach described by Robinson (1980) in which one searches in unpolarized light for enhanced line broadening in the high Landé g-component of a pair of nearly identical lines, but with very different g-values. This technique is difficult to apply in practice because one needs high S/N and high-spectral resolution profiles of lines with minimal blending. Nevertheless, this approach does measure the absolute strength (B) of the strong fields, if they are predominantly of one value, independent of cancellation effects, and does provide an estimate of the area filling factor (f) of such fields.

The first use of this technique by Robinson *et al.* (1980) led to an estimated field strength of 2550 ± 390 G covering 20–45% of the surface of the active star ξ Boo A (G8 V) and a smaller field for 70 Oph A (K0 V). The discovery of a stellar field so much stronger and more widespread than on the Sun led several groups to apply the Robinson technique to other stars in a systematic way. The largest amount of such data so far has been obtained by Marcy (1984) who observed 29 G and K dwarf stars at the Lick Observatory. He detected fields in 19 of these stars with strengths in the range 600–3000 G and filling factors in the range 20–90%. Marcy and Bruning (1984) applied this technique to G and K type active giants but obtained no positive detections. Gray (1984) applied a Fourier deconvolution technique to analyze many lines per star, and measured fields in 7 out of 18 stars. The positive results were exclusively of spectral types G0 V–K2 V; none of the F2 V–F9 V stars exhibited measurable fields. Giampapa *et al.* (1983) reported measurements of a magnetic field on the RS CVn-type binary system λ And using infrared lines near 1.6 μm, taking advantage of the λ^2 dependence of the Zeeman pattern.

Before commenting on trends these authors found in their data, I must list the problems inherent in the Robinson technique that cast some doubt concerning the validity of some of these magnetic field measurements:

(1) Kurucz and Hartmann (1984) and others have called attention to the perverse role that weak line blends can play in producing false magnetic signals, since the primary signature of the field with this technique is excess broadening of the high g line profile in the wings very close to the continuum. While the commonly used Fe I 6173.34 Å ($g = 2.5$) line is relatively blend-free for G and early K-type stars, the blending problem can be important for this and most other lines in the late K and M stars. One method of minimizing the effects of line blending, as shown by Gray (1984), is to use many lines in the analysis, since each line will be blended differently. A second approach, employed by Giampapa *et al.* (1983) and Saar *et al.* (1985), is to compare the line profiles of active stars to the same line profiles of nonactive stars of similar spectral type, which have been corrected for different rotational velocities by matching the profiles of low g lines in both stars. This method presumes that the line blending and atmospheric models are similar for both stars. Kurucz and Hartmann (1984) and Saar *et al.* (1985) both pointed out that Marcy's (1984) estimate of a magnetic field ($B = 2990$ G, $f = 0.28$) on the very inactive star 61 Cyg A (K5 V) may be spurious because of line blends.

(2) Gray (1984), Kurucz and Hartmann (1984), and others have pointed out that the

conventional analysis techniques then in use assumed that all components (magnetic and nonmagnetic) of the absorption line profile are optically thin, which is generally an inadequate assumption for most of the lines studied. Since the central core of the line profile (consisting of the π and nonmagnetic components) is more opaque than the Zeeman split σ components, the ratio of line depths of the σ components to the central core (which is proportional to the fractional coverage of the magnetic field when all components are optically thin) will instead overestimate the fractional coverage. Saar (1985) has developed a code to take line saturation effects into account and has applied it to the analysis of profiles for two flare stars − EQ Vir (dK5e, Saar $et\ al.$, 1985), and AD Leo (dM3.5e, Saar and Linsky, 1985).

(3) The atmospheric structures in the magnetic and nonmagnetic regions may differ considerably. If, for example, stellar fields are confined to dark starspot umbrae, then the filling factor estimated assuming similar atmospheric structures will be under-estimated roughly by the ratio of the actual continuum fluxes. This ratio depends on the stellar T_{eff} and wavelength region studied, but is so large (about 5 at 6000 Å for the Sun) that the derived magnetic filling factors would exceed unity and be unrealistic. Thus the filling factors derived to date do not refer to the (presumably larger) fields in starspot umbrae, although Sun $et\ al.$ (1985) showed that in the infrared the spot umbral fields will dominate the derived spatially-averaged magnetic parameters if the spot umbrae cover more than 10% of the surface. If the observed fields are located in spot penumbrae then the filling factors are underestimates, and if they are located in bright faculae (as is likely for EQ Vir) they are overestimates. In any case, the derived field strengths will refer to the atmospheric region that is the dominant contributor to the observed profile. All analysis techniques used to date assume the same Doppler broadening parameters for the magnetic and nonmagnetic regions, and they also assume no systematic flows in both regions. Any, or all, of these assumptions could be in error.

(4) Zeeman patterns are generally not simple triplets as is commonly assumed and the magnetic fields on a star may have a distribution of strengths about a mean value. Saar's (1985) code can accommodate the exact Zeeman pattern, and Gray (1984) has discussed the effects of a distribution of field strengths on the derived values of B and f.

These major uncertainties apply more or less to most of the stellar magnetic field estimates to date and must be overcome before we can believe any conclusions concerning the pattern of magnetic fields in solar-like stars. Nevertheless, I believe that there are some interesting trends that will probably be supported by future work.

(1) Magnetic fields have been detected in main sequence stars between spectral type G0 V and dM3.5e. It will be difficult to extend the spectral range to hotter stars because rotational velocities increase rapidly toward earlier spectral type and will mask the Zeeman components. I expect, however, that the F dwarfs are solar-like because of the observed X-ray and transition region emission. Magnetic fields are also hard to measure on the M dwarfs because of severe line blending and faintness in the visual, but both problems may be reduced by observing in the infrared. Saar and Linsky (1985) measured the field on AD Leo (dM3.5e) using the Kitt Peak 4 m FTS to observe lines near 2.2 μm,

and other lines are available even further to the infrared where the λ^2-dependence of Zeeman splitting enhances field detection and simplifies the analysis. In particular, the 12 μm lines recently discovered in the solar spectrum (Brault and Noyes, 1983; Chang and Noyes, 1983) appear to be simple Zeeman triplets and are obvious candidates for study when appropriate instruments become available.

(2) Magnetic fields have not yet been detected in evolved stars with the exception of the marginal detection of fields in λ And (G8 III–IV) by Giampapa *et al.* (1983). Marcy and Bruning (1984) argue that the nondetection of fields in giant stars is not due to the absence of such fields in these stars, which have strong X-ray and C IV emission, but rather that the photospheric fields have systematically lower strengths than in the main sequence stars. Again observations in the infrared are required.

(3) Although Marcy (1983, 1984) determined field strengths for different stars in the range 600–3000 G, the mean fields derived for the G dwarfs and the K dwarfs are the same (roughly 1100 G) and similar to nonspot solar fields, about 1000–1500 G (cf. Tarbell and Title, 1977; Sun *et al.*, 1985). On the other hand, Saar *et al.* (1985) derived a field of 2500 $\pm$ 300 G for EQ Vir (dK5e) and Saar and Linsky (1985) derived a field $\approx$ 3800 G for AD Leo (dM3.5e), the coolest dwarf stars with reliably measured fields. They argued that there is a real increase in B toward the cooler stars, which can be simply explained by the equipartition of magnetic and thermal pressure in the photosphere (scaled relative to the Sun), since the photospheric thermal pressure near continuum optical depth unity increases with decreasing effective temperature (and increasing surface gravity) along the Main-Sequence. Since photospheric pressures are lower in giant stars owing to lower gravity, Marcy and Bruning (1984) argued that the low equipartition field strengths are responsible for the nondetections of the giant stars.

(4) Marcy (1983, 1984) called attention to a systematic increase in area filling factor from the early G to the late K dwarfs. His mean value for the G dwarfs is 0.16 and for the K0–K5 dwarfs is 0.54. This result is intriguing, but it could be spurious as the effect of increasing line blending and line saturation would be to produce systematically larger inferred filling factors for the cooler stars. More reliable estimates of filling factors are needed to determine whether this trend is real. Using the same data set, Gray (1985) concluded that the product Bf is constant ($\approx$ 620 G) for all G and K dwarfs studied: In particular, the number of magnetic field lines per unit area averaged over a star is a 'universal magnetic constant'. Saar and Linsky (1985) have argued that this result cannot be a universal relation as proposed because the Sun lies far below the trend ($Bf \approx$ 15 G), and both EQ Vir ($Bf \approx$ 2000 G) and AD Leo ($Bf \approx$ 2700 G) lie far above the trend.

(5) Timothy *et al.* (1981) and Marcy (1984) have reported large changes in the magnetic parameters for ε Eri (K2 V) from night to night. Since this star has an 11-day rotation period, this rapid variability cannot be explained by rotation of active regions on and off the disk. Instead something more drastic is happening to the large-scale field. More observations are needed to confirm this unsolar-like behavior.

(6) Marcy (1983, 1984) showed that the empirical functional dependence of the chromospheric heating rate on T_{eff} and B is most consistent with the magnetic slow

mode mechanism. He also concluded that the X-ray flux increases with the magnetic area filling factor and that the total magnetic flux is proportional to $V_{rot}^{0.55}$, where V_{rot} is the stellar rotational velocity, but both relations are based on only a few data. If confirmed by subsequent work, however, all three relations would constitute powerful evidence that activity in solar-like stars is primarily magnetic in origin.

8. Conclusion: Which Types of Stars are Solar-Like?

Until now we have surveyed which types of stars exhibit nonradiative activity that appears to be analogous to activity seen in the Sun, and we have called attention to the roles that magnetic fields and turbulent fluid motions likely play in controlling this activity on the Sun and stars. We can now attempt to answer the question posed in the title of this review: 'What types of stars are solar-like?'

This question should not be answered in phenomenological terms, that is by the appearance of solar-like phenomena, because phenomena may have many different origins. A good case in point is X-ray emission from O-type stars which could, in principle, be due to a magnetically heated corona or shocks in a radiatively unstable wind. Instead, the term 'solar-like' should be defined in physical terms. Since the nonradiative activity of the Sun is magnetic in origin and in most solar atmospheric regions of interest (e.g., the corona, TR, chromosphere, flares, photospheric faculae, sunspots) the magnetic pressure exceeds or equals the thermal and turbulent pressures, *I will define a solar-like star as: a star which has a turbulent magnetic field sufficiently strong to control the dynamics and energetics in its outer atmospheric regions.* This particular definition was chosen because it is the variability in space and time of the solar magnetic fields, driven by sub-photospheric conventive motions, that produces the observed time variable and spatially inhomogeneous phenomena that we call solar activity. The mere presence of a magnetic field is not sufficient to produce activity, because the local magnetic pressure ($B^2/8\pi$) must exceed the thermal pressure (nkT) and the turbulent pressure ($\frac{1}{2}\rho v_{turb}^2$) before the field can control the fluid motions and thus define the geometry (closed loops) and energy balance by restricting the transfer of heat and flows across the field lines. Also in the Sun the strongest nonradiative heating occurs in flux tubes where the magnetic pressure dominates.

If we adopt this definition, then the most reliable method for identifying solar-like stars is by direct measurement of their magnetic fields, in particular, the strong facular or spot fields measured in unpolarized light, rather than the weak net fields that may be measured in polarized light by the Babcock technique. To date strong fields have been measured reliably in only dwarf stars between spectral type G0 and M3.5. These fields are generally close to equipartition values in the deep photosphere and thus, by solar analogy should have pressures large compared to thermal and turbulent in closed flux tubes in the upper photosphere and higher layers. All of these stars are very active as expected. Thus main sequence stars of spectral type G, K, and M are solar-like. It is important to extend the detections of strong fields to other types of stars. Also, systematic studies of the correlation of variations in the field parameters with different signatures of activity would be extremely valuable in determining how the spatial organization, strength, and filling factor of the field and the activity phenomena depend upon the depth of the convective zone, effective temperature, gravity, rotation rate, age, proximity of a binary companion, and other parameters.

In the absence of a measured strong field, we can less reliably determine whether a star is solar-like by observing the various aspects of stellar activity. The best indirect indicator of strong fields is photometric variability indicative of an inhomogeneous distribution of dark starspots. By this criterion we can add the RS CVn and W UMa systems to the list of solar-type stars and thus extend the range of such stars into the G and early K subgiants and giants. Photometric variability has been detected as early as spectral type F6 V. The pre-Main-Sequence T Tauri stars are also variable photometrically, but the origin of this variability could be changing circumstellar obscuration.

Nonthermal microwave emission, if produced by the gyroresonance or gyro-synchrotron process, indicates the presence of magnetic fields in hot plasma, either in a corona or a hot wind. The important question is whether the inferred field is strong enough to control the local dynamics or whether it is a minor contributor. In the solar corona the magnetic field corresponding to equal magnetic and gas pressure ($P \sim 10$ dynes cm^{-2}) in coronal loops is only 15 G, whereas fields in the range 100–600 G are commonly inferred from the observed gyroresonance emission. It is likely that the magnetic pressure exceeds the gas pressure also in the coronae of dwarf stars of spectral types F–M. On the other hand, the fields required to produce equal magnetic and kinetic (ram) pressure in O and B stars are roughly 35 G at the critical point. This estimate is based on parameters for ζ Pup (O4f): mass loss rate $5 \times 10^{-6} M_\odot$ yr^{-1}, critical point radius = $1.5 R_* = 30 R_\odot$, and critical point expansion velocity = 200 km s^{-1}. By comparison, White (1985) derived a photospheric magnetic field of 5 G and much smaller coronal fields to explain the synchrotron emission from hot stars. Furthermore, if photospheric fields were even as large as 100 G, they could lead to spin down rates faster than observed (Friend, 1985). Thus magnetic fields in O stars exist, but they only play a minor role in the dynamics and energy balance. The hot stars by my definition are thus not solar-like notwithstanding proposed correlations of L_x with the Rossby (dynamo) number.

Flaring is generally thought to be an excellent indicator of strong magnetic fields in a stellar corona as the rapid heating and large energies presumably result from the conversion of magnetic energy to nonthermal particles and eventually to heat. Stars known to flare include the M dwarfs, RS CVn binaries, and pre-Main-Sequence stars.

Finally, X-ray emission and TR emission lines are indicative of 10^6–10^8 K and 10^5 K plasma, respectively, that is heated by nonradiative processes. I have previously cited arguments why the heating process is magnetic in character for most types of stars, except the hot stars for which the heating process could be by shocks in the wind resulting from radiative instabilities.

In conclusion, the present evidence concerning which types of stars are solar-like is as shown schematically in Figure 1 and summarized below:

– Dwarf stars of spectra types G, K, and M are definitely solar-like on the basis of measured strong magnetic fields and many indirect indicators. Among the coolest M dwarfs there is evidence for decreased X-ray luminosity consistent with decreased dynamo generation of fields in stars that are probably fully convective.

– Dwarf stars of spectral type A7–F7 are almost certainly solar-like, although this statement cannot be definitive without measurements of strong magnetic fields. The solar-like indirect indicators are photometric and Ca II flux variability on rotation time-scales, indicating isolated dark starspots and chromospheric plages in the cooler F dwarfs, and strong X-ray, TR, and chromospheric emission lines in stars as hot as α Aql (A7 IV–V). Flares and microwave emission have not yet been detected from these stars. The correlation of coronal and chromospheric emission with the Rossby number is consistent with the magnetic character of the heating process, and as noted above, the fields may be more uniformly distributed across the stellar surface than in the cooler stars because of the thin convective zones.

– There is as yet no direct or indirect evidence of turbulent magnetic fields or activity in the dwarf A stars, and on this basis I presume that they are not solar-like.

– The hot O and B stars are X-ray sources and are in some cases microwave synchrotron sources, but in general the magnetic fields appear to be weak and thus the stars are not solar-like. An exception may be the helium strong Bp stars. Two of these stars (σ Ori E and HD 37017) have kilogauss fields, X-ray emission, and nonthermal microwave emission (Drake *et al.*, 1985) and are thus candidates for being called solar-like. These two stars could be unusual as they are the hottest magnetic Bp stars and lie at the boundary where detectable mass loss begins. The cooler magnetic B and A stars are not X-ray or microwave sources.

– Single giants of spectral type F, G, and early K have not yet been detected as microwave sources, show no evidence for starspots or flares and have no detectable magnetic fields. On the other hand, many of these stars are luminous X-ray and TR emission line sources. Thus they have hot coronae that cannot be confined without magnetic fields, and in the absence of viable alternative explanations are probably heated by magnetic processes like the dwarf stars and rapidly-rotating binary stars of similar effective temperature. On the basis of this indirect evidence, I suspect that these stars are solar-like.

– Rapidly rotating subgiants of spectral type F, G, and early K in spectroscopic binary sustems (RS CVn, W UMa, and RS CVn types) exhibit flares, nonthermal radio emission, strong X-ray and TR emission features, and evidence for dark starspots and plages. In addition a strong field has been reported for one system, λ And. Stars in these systems are clearly solar-like.

– Giant stars cooler than about spectral-type K2 III and supergiant stars of spectral types G, K, and M generally show none of the indicators of strong magnetic fields, but instead have extended chromospheres and strong winds. Even the so-called hybrid stars show only weak TR lines and winds that may be accelerated by Alfvén waves due to weak fields. These stars are also not solar-like.

– Pre-Main Sequence stars exhibit flaring, photometric variability, nonthermal microwave emission, and strong X-ray and TR emission. Although strong magnetic fields have not yet been measured in these stars, they are likely present and probably even stronger than the young Main Sequence stars already observed. These stars are probably solar-like with activity indicators orders of magnitude larger than for the Sun.

Acknowledgements

This work was supported in part by NASA grants NAG5-82 and NGL-06–003–057 to the University of Colorado. I would like to thank my colleagues, D. C. Abbott, T. R. Ayres, A. Brown, K. G. Carpenter, S. Drake, D. M. Gibson, S. H. Saar, and F. M. Walter for stimulating discussions and for their comments on the text.

References

Abbott, D. C., Bieging, J. H., and Churchwell, E.: 1984, *Astrophys. J.* **280**, 671.
Abbott, D. C., Bieging, J. H., and Churchwell, E.: 1985, in A. B. Underhill and A. G. Michalitsianos (eds.), *The Origin of Nonradiative Heating/Momentum in Hot Stars*, NASA Conf. Publ. 2358, p. 47.
Athay, R. G.: 1976, *The Solar Chromosphere and Corona: Quiet Sun*, D. Reidel Publ. Co., Dordrecht, Holland, p. 282.
Auvergne, M., Le Contel, J.-M., and Baglin, A.: 1979, *Astron. Astrophys.* **87**, 15.
Ayres, T. R.: 1981, *Astrophys. J.* **244**, 1064.
Ayres, T. R.: 1984, *Astrophys. J.* **284**, 784.
Ayres, T. R.: 1985, *Astrophys. J.* **291**, L7.
Ayres, T. R., Marstad, N. C., and Linsky, J. L.: 1981a, *Astrophys. J.* **247**, 545.
Ayres, T. R., Linsky, J. L., Vaiana, G. S., Golub, L., and Rosner, R.: 1981b, *Astrophys. J.* **250**, 293.
Ayres, T. R., Simon, T., and Linsky, J. L.: 1982, *Astrophys. J.* **263**, 791.
Ayres, T. R., Schiffer, F. H. III, and Linsky, J. L.: 1983, *Astrophys. J.* **272**, 223.
Ayres, T. R., Stencel, R. E., Linsky, J. L., Simon, T., Jordan, C., Brown, A., and Engvold, O.: 1983, *Astrophys. J.* **274**, 801.
Baliunas, S. L.: 1983, *Publ. Astron. Soc. Pacific* **95**, 532.
Baliunas, S. L., Horne, J. H., Porter, A., Duncan, D. K., Frazier, J., Lanning, H., Misch, A., Mueller, J., Noyes, R. W., Soyumer, D., Vaughan, A. H., and Woodard, L.: 1985, *Astrophys. J.* **294**, 310.
Baliunas, S. L., Vaughan, A. H., Hartmann, L., Middlekoop, F., Mihalas, D., Noyes, R. W., Preston, G. W., Frazier, J., and Lanning, H.: 1983, *Astrophys. J.* **275**, 752.
Baliunas, S. L., Avrett, E. H., Hartmann, L., and Dupree, A. K.: 1979, *Astrophys. J.* **233**, L129.
Baliunas, S. L. and Dupree, A. K.: 1982, *Astrophys. J.* **252**, 668.
Barry, D. C., Hege, K., and Cromwell, R. H.: 1984, *Astrophys. J.* **277**, L65.
Basri, G. S. and Linsky, J. L.: 1979, *Astrophys. J.* **234**, 1023.
Blanco, C., Catalano, S., and Marilli, E.: 1980, *Proc. Second European IUE Conf. Tübingen*, ESA SP-157, p. 63.
Blanco, C., Bruca, L., Catalano, S., and Marilli, E.: 1982, *Astron. Astrophys.* **115**, 280.
Boesgaard, A. M.: 1974, *Astrophys. J.* **188**, 567.
Boesgaard, A. M., Chesley, D., and Preston, G. W.: 1975, *Publ. Astron. Soc. Pacific* **87**, 353.
Böhm-Vitense, E. and Dettmann, T.: 1980, *Astrophys. J.* **236**, 560.
Bonsack, W. K. and Simon, T.: 1983, in J. O. Stenflo (ed.), *Solar and Stellar Magnetic Fields: Origins and Coronal Effects*, D. Reidel Publ. Co., Dordrecht, Holland, p. 35.
Bookbinder, J., Golub, L., Schmitt, J. H. M. M., and Rosner, R.: 1984, *Bull. Am. Astron. Soc.* **16**, 515.
Bopp, B. W.: 1983, in P. B. Bryne and M. Rodono (eds.), 'Activity in Red Dwarf Stars', *IAU Colloq.* **71**, 363.
Borra, E. F., Edwards, G., and Mayor, M.: 1984, *Astrophys. J.* **284**, 211.
Brault, J. and Noyes, R.: 1983, *Astrophys. J.* **269**, L61.

Brosius, J. W., Mullan, D. J., and Stencel, R. E.: 1985, *Astrophys. J.* **288**, 310.

Brown, A.: 1983, in S. L. Baliunas and L. Hartmann (eds.), *Cool Stars, Stellar Systems, and the Sun*, Springer-Verlag, Berlin, p. 282.

Brown, A.: 1985, in F. D. Kahn (ed.), *Cosmical Gas Dynamics*, VNU Science Press, Utrecht (in press).

Brown, A. and Carpenter, K. G.: 1984, *Astrophys. J.* **287** L43.

Brown, A., Jordan, C., Stencel, R. E., Linsky, J. L. and Ayres, T. R.: 1984, *Astrophys. J.* **283**, 731.

Brown, D. N. and Landstreet, J. D.: 1981, *Astrophys. J.* **246**, 899.

Caillault, J. P. and Helfand, D. J.: 1985, *Astrophys. J.* **289**, 279.

Calvet, N., Basri, G., Imhoff, C. L., and Giampapa, M. S.: 1985, *Astrophys. J.* **293**, 575.

Campbell, B.: 1984, *Astrophys. J.* **283**, 209.

Carpenter, K. G., Brown, A., and Stencel, R. E.: 1985, *Astrophys. J.* **289**, 676.

Cash, W. and Snow, T.: 1982, *Astrophys. J.* **263**, L59.

Cassinelli, J. P.: 1979, *Ann. Rev. Astron. Astrophys.* **17**, 275.

Cassinelli, J. P.: 1985, in A. B. Underhill and A. G. Michalitsianos (eds.), *The Origin of Nonradiative Heating/Momentum in Hot Stars*, NASA Conference Publ. 2358, p. 2.

Cassinelli, J. P. and Olson, G. L.: 1979, *Astrophys. J.* **229**, 304.

Cassinelli, J. P., Waldran, W. L., Sanders, W. T., Harnden, F. R., Jr., Rosner, R., and Vaiana, G. S.: 1981, *Astrophys. J.* **250**, 677.

Chang, E. S. and Noyes, R. W.: 1983, *Astrophys. J.* **275**, L11.

Chapman, G. A.: 1977, *Astrophys. J. Suppl.* **33**, 35.

Chapman, G. A.: 1981, in F. Q. Orrall (ed.), *Solar Active Regions*, Colorado Assoc. Univ. Press, Boulder, p. 43.

Charles, P. A.: 1983, in P. B. Byrne and M. Rodono (eds.), 'Activity in Red Dwarf Stars', *IAU Colloq.* **71**, 415.

Cox, A. N., Shaviv, G., and Hodson, S. W.: 1981, *Astrophys. J.***245**, L37.

Crivellari, L. and Praderie, F.: 1982, *Astron. Astrophys.* **107**, 75.

Drake, S. A. and Linsky, J. L.: 1983a, *Astrophys. J.* **274**, L77.

Drake, S. A. and Linsky, J. L.: 1983b, *Astrophys. J.* **273**, 299.

Drake, S. A. and Linsky, J. L.: 1985, *Astrophys. J.* (in press).

Drake, S. A., Abbott, D. C., Bieging, J. H., Churchwell, E., and Linsky, J. L.: 1985, in R. Hjellming and D. Gibson (eds.), *Radio Stars*, D. Reidel Publ. Co., Dordrecht, Holland, p. 247.

Dravins, D.: 1981, *Astron. Astrophys.* **96**, 64.

Dravins, D., Lind, H., and Särg, K.: 1977, *Astron. Astrophys.* **54**, 381.

Duncan, D. K.: 1983, in S. L. Baliunas and L. Hartmann (eds.), *Cool Stars, Stellar Systems, and the Sun*, Springer-Verlag, Berlin, p. 128.

Dulk, G. A.: 1985, *Ann. Rev. Astron. Astrophys.* **23**, 169.

Dupree, A. K.: 1982, in *Advances in Ultraviolet Astronomy: Four Years of IUE Research*, NASA Conference Publication 2238, p. 3.

Dupree, A. K.: 1983, in P. B. Byrne and M. Rodono (eds.), 'Activity in Red Dwarf Stars', *IAU Colloq.* **71**, 447.

Feigelson, E. D.: 1983, in S. L. Baliunas and L. Hartmann (eds.), *Cool Stars, Stellar Systems, and the Sun*, Springer-Verlag, Berlin, p. 27.

Feigelson, E. D. and Kriss, G. A.: 1981, *Astrophys. J.* **248**, L35.

Feldman, P. A.: 1983, in P. B. Byrne and M. Rodono (eds.), 'Activity in Red Dwarf Stars', *IAU Colloq.* **71**, 429.

Feldman, U., Doschek, G. A., and Mariska, J. T.: 1979, *Astrophys. J.* **229**, 369.

Fisher, P. L. and Gibson, D. M.: 1983, in M. S. Giampapa and L. Golub (eds.), *Second Cambridge Workshop on Cool Stars, Stellar Systems, and the Sun*, SAO Special Report 392, Vol. II, p. 109.

Freire, R., Czarny, J., Felenbok, P., and Praderie, F.: 1978, *Astron. Astrophys.* **68**, 89.

Friend, D. B.: 1985, in A. B. Underhill and A. G. Michalitsianos (eds.), *The Origin of Nonradiative Heating/Momentum in Hot Stars*, NASA Conf. Publ. 2358, p. 142.

Gary, D.: 1985, in R. Hjellming and D. Gibson (eds.), *Radio Stars*, D. Reidel Publ. Co., Dordrecht, Holland, p. 185.

Gary, D. E. and Linsky, J. L.: 1981, *Astrophys. J.* **250**, 284.

Giampapa, M. S.: 1983a, in P. B. Byrne and M. Rodono (eds.), 'Activity in Red Dwarf Stars', *IAU Colloq.* **71**, 223.

Giampapa, M. S.: 1983b, in S. L. Baliunas and L. Hartmann (eds.), *Cool Stars, Stellar Systems, and the Sun*, Springer-Verlag, Berlin, p. 14.

Giampapa, M. S.: 1984, in A. Mangeney and F. Praderie (eds.), *Space Research Prospects in Stellar Activity and Variability*, Observatoire de Paris-Meudon, p. 309.

Giampapa, M. S. and Rosner, R.: 1984, *Astrophys. J.* **286**, L19.

Giampapa, M. S., Worden, S. P., and Linsky, J. L.: 1982, *Astrophys. J.* **258**, 740.

Giampapa, M. S., Golub, L., and Worden, S. P.: 1983, *Astrophys. J.* **268**, L121.

Gibson, D. M.: 1981, in R. M. Bonnet and A. K. Dupree (eds.), *Solar Phenomena in Stars and Stellar Systems*, D. Reidel Publ. Co., Dordrecht, Holland, p. 545.

Gibson, D. M.: 1983, in P. B. Byrne and M. Rodono (eds.), 'Activity in Red Dwarf Stars', *IAU Colloq.* **71**, 273.

Gibson, D. M.: 1985, in R. Hjellming and D. Gibson (eds.), *Radio Stars*, D. Reidel Publ. Co., Dordrecht, Holland, p. 213.

Golub, L.: 1983, in P. B. Byrne and M. Rodono (eds.), 'Activity in Red Dwarf Stars', *IAU Colloq.* **71**, 83.

Golub, L., Harnden, F. R., Jr., Maxson, C. W., Rosner, R., Vaiana, G. S., Cash, W., Jr., and Snow, T. P., Jr.: 1983, *Astrophys. J.* **271**, 264.

Gray, D. F.: 1981, *Astrophys. J.* **251**, 155.

Gray, D. F.: 1984, *Astrophys. J.* **277**, 640.

Gray, D. F.: 1985, *Publ. Astron. Soc. Pacific* **97**, 719.

Grossman, A. S., Hays, D., and Groboske, H. C., Jr.: 1974, *Astron. Astrophys.* **30**, 95.

Haisch, B. M.: 1983, in P. B. Byrne and M. Rodono (eds.), 'Activity in Red Dwarf Stars', *IAU Colloq.* **71**, 255.

Haisch, B. M. and Basri, G.: 1985, *Astrophys. J. Suppl.* **58**, 179.

Haisch, B. M. and Simon, T.: 1982, *Astrophys. J.* **263**, 252.

Hartmann, L.: 1981, in R. M. Bonnet and A. K. Dupree (eds.), *Solar Phenomena in Stars and Stellar Systems*, D. Reidel Publ. Co., Dordrecht, Holland, p. 331.

Hartmann, L.: 1983, in J. O. Stenflo (ed.), 'Solar and Stellar Magnetic Fields: Origins and Coronal Effects', *IAU Symp.* **102**, 419.

Hartmann, L. and Avrett, E. H.: 1984, *Astrophys. J.* **284**, 238.

Hartmann, L., Dupree, A. K., and Raymond, J. C.: 1980, *Astrophys. J.* **236**, L143.

Hartmann, L., Dupree, A. K., and Raymond, J. C.: 1981, *Astrophys. J.* **246**, 193.

Hartmann, L., Baliunas, S. L., Duncan, D. K., and Noyes, R. W.: 1984, *Astrophys. J.* **279**, 778.

Helfand, D. J. and Caillault, J. P.: 1982, *Astrophys. J.* **253**, 760.

Hubeny, I.: 1981, *Astron. Astrophys.* **98**, 96.

Johnson, H. M.: 1983, in P. B. Byrne and M. Rodono (eds.), 'Activity in Red Dwarf Stars', *IAU Colloq.* **71**, 109.

Jordan, C., Brown, A., Walter, F. M., and Linsky, J. L.: 1985, *Monthly Notices Roy. Astron. Soc.* (in press).

Kelch, W. L., Linsky, J. L., and Worden, S. P.: 1979, *Astrophys. J.* **229**, 700.

Ku, W. H.-M. and Chanan, G. A.: 1979, *Astrophys. J.* **234**, L59.

Kuipers, J. and van der Hulst, J. M.: 1985, *Astron. Astrophys.* **149**, 343.

Kurucz, R. L. and Hartmann, L.: 1984, SAO Preprint, No. 2015.

Lang, K. R., Bookbinder, J., Golub, L., and Davis, M. M.: 1983, *Astrophys. J.* **272**, L15.

LeContel, J. M., Praderie, F., Bijaoui, A., Dantel, M., and Sareyan, J. P.: 1970, *Astron. Astrophys.* **8**, 159.

Lestrade, J. F., Mutel, R. L., Preston, R. A., and Phillips, R. B.: 1985, in R. Hjellming and D. Gibson (eds.), *Radio Stars*, D. Reidel Publ. Co., Dordrecht, Holland, p. 275.

Linsky, J. L.: 1977, in O. R. White (ed.), *The Solar Output and its Variations*, Colorado Associated University Press, Boulder, p. 477.

Linsky, J. L.: 1980, *Ann. Rev. Astron. Astrophys.* **18**, 439.

Linsky, J. L.: 1981a, in R. Giacconi (ed.), *X-ray Astronomy with the Einstein Satellite*, D. Reidel Publ. Co., Dordrecht, Holland, p. 19.

Linsky, J. L.: 1981b, in S. S. Holt (ed.), *X-ray Astronomy in the 1980's*, NASA TM 83848, p. 13.

Linsky, J. L.: 1981c, in R. M. Bonnet and A. K. Dupree (eds.), *Solar Phenomena in Stars and Stellar Systems*, D. Reidel Publ. Co., Dordrecht, Holland, p. 99.

Linsky, J. L.: 1982, in *Advances in Ultraviolet Astronomy: Four Years of IUE Research*, NASA Conference Publication 2238, p. 17.

Linsky, J. L.: 1983a, in J. O. Stenflo (ed.), 'Solar and Stellar Magnetic Fields: Origins and Coronal Effects', *IAU Symp.* **102**, 313.
Linsky, J. L.: 1983b, in P. B. Byrne and M. Rodono (eds.), 'Activity in Red Dwarf Stars', *IAU Colloq.* **71**, 39.
Linsky, J. L.: 1983c, in S. L. Baliunas and L. Hartmann (eds.), *Cool Stars, Stellar Systems, and the Sun*, Springer-Verlag, Berlin, p. 244.
Linsky, J. L.: 1985, in R. Stalio and J. Zirker (eds.), *Third Trieste Workshop* (in press).
Linsky, J. L. and Avrett, E. H.: 1970, *Publ. Astron. Soc. Pacific.* **82**, 169.
Linsky, J. L. and Ayres, T. R.: 1978, *Astrophys. J.* **220**, 619.
Linsky, J. L. and Gary, D. E.: 1983, *Astrophys. J.* **274**, 776.
Linsky, J. L. and Haisch, B. M.: 1979, *Astrophys. J.* **229**, L27.
Linsky, J. L., Bornmann, P. L., Carpenter, K. G., Wing, R. F., Giampapa, M. S., and Worden, S. P.: 1982, *Astrophys. J.* **260**, 670.
Lucy, L. B. and White, R. L.: 1980, *Astrophys. J.* **241**, 300.
Mangeney, A. and Praderie, F.: 1984, *Astron. Astrophys.* **130**, 143.
Marcy, G. W.: 1983, in J. O. Stenflo (ed.), 'Solar and Stellar Magnetic Fields: Origins and Coronal Effects', *IAU Symp.* **102**, 3.
Marcy, G. W.: 1984, *Astrophys. J.* **276**, 286.
Marcy, G. W. and Bruning, D. H.: 1984, *Astrophys. J.* **281**, 286.
Marilli, E. and Catalano, S.: 1984, *Astron. Astrophys.* **133**, 57.
Marstad, N., Linsky, J. L., Simon, T., Rodono, M., Blanco, C., Catalano, S., Marilli, E., Andrews, A. D., Butler, C. J., and Byrne, P. B.: 1982, in *Advances in Ultraviolet Astronomy: Four Years of IUE Research*, NASA Conf. Publ. No. 2238, p. 554.
Micela, G., Sciortino, S., Serio, S., Vaiana, G. S., Golub, L.: 1985, *Astrophys. J.* **292**, 172.
Montmerle, T., Koch-Miramond, L., Folgarone, E., and Grindlay, J.: 1983, *Astrophys. J.* **269**, 182.
Morrison, N. D. and Linsky, J. L.: 1978, *Astrophys. J.* **222**, 723.
Mullan, D. J.: 1984, *Astrophys. J.* **283**, 303.
Mullan, D. J.: 1985, in R. Hjellming and D. Gibson (eds.), *Radio Stars*, D. Reidel Publ. Co., Dordrecht, Holland, p. 173.
Mutel, R. L. and Lestrade, J. F.: 1985, *Astron. J.* **90**, 493.
Mutel, R. L., Doiron, D. J., Lestrade, J. F., and Phillips, R. B.: 1984, *Astrophys. J.* **278**, 220.
Mutel, R. L., Lestrade, J. F., Preston, R. A., and Phillips, R. B.: 1985a, *Astrophys. J.* **289**, 262.
Newell, R. T. and Hjellming, R. M.: 1982, *Astrophys. J.* **263**, L85.
Owocki, S. P. and Rybicki, G. B.: 1984, *Astrophys. J.* **284**, 337.
Pallavicini, R., Golub, L., Rosner, R., Vaiana, G. S., Ayres, T. R., and Linsky, J. L.: 1981, *Astrophys. J.* **248**, 279.
Praderie, F., Simonneau, E., and Snow, T. P.: 1975, *Astrophys. Space Sci.* **38**, 337.
Radick, R. R. Lockwood, C. W., Thompson, D. T., Warnock, Q. III, Hartmann, L. W., Mihalas, D., Worden, S. P., Henry, G. W., and Sherlin, J. M.: 1983, *Publ. Astron. Soc. Pacific* **95**, 621.
Robinson, R. D.: 1980, *Astrophys. J.* **239**, 961.
Robinson, R. D., Worden, S. P., and Harvey, J. W.: 1980, *Astrophys. J.* **236**, L155.
Rosner, R., Golub, L., and Vaiana, G. S.: 1985, *Ann. Rev. Astron. Astrophys.* **23**, 413.
Rucinski, S. M.: 1985, in P. P. Eggleston and J. E. Pringle (eds.), *Interacting Binaries*, D. Reidel Publ. Co., Dordrecht, Holland, p. 13.
Rucinski, S. M.: 1984, *Astron. Astrophys.* **132**, L9.
Saar, S. H.: 1985, in preparation.
Saar, S. H. and Linsky, J. L.: 1985, *Astrophys. J.* (in press).
Saar, S. H., Linsky, J. L., and Beckers, J. M.: 1985, *Astrophys. J.* (in press).
Saxner, M.: 1981, *Astron. Astrophys.* **104**, 240.
Schmitt, J. H. M. M., Golub, L., Harnden, F. R., Jr., Maxson, C. W., Rosner, R., and Vaiana, G. S.: 1985a, *Astrophys. J.* **290**, 307.
Schmitt, J. H. M. M., Harnden, F. R., Jr., Peres, G., Rosner, R., and Serio, S.: 1985b, *Astrophys. J.* **288**, 751.
Schrijver, C. J.: 1984, in *Eighteenth ESLAB Symposium, X-Ray Astronomy*, Scheveningen, The Netherlands (in press).
Simon, T. and Boesgaard, A. M.: 1983, in J. O. Stenflo (ed.), 'Solar and Stellar Magnetic Fields: Origins and Coronal Effects', *IAU Symp.* **102**, 161.

Simon, T. and Linsky, J. L.: 1980, *Astrophys. J.* **241**, 759.

Simon, T., Linsky, J. L., and Stencel, R. E.: 1982, *Astrophys. J.* **257**, 225.

Simon, T., Cash, W., and Snow, T. P., Jr.: 1985a, *Astrophys. J.* **293**, 542.

Simon, T., Herbig, G., and Boesgaard, A. M.: 1985b, *Astrophys. J.* **293**, 551.

Skumanich, A.: 1972, *Astrophys. J.* **171**, 565.

Snijders, M. A. J.: 1977, *Astrophys. J.* **214**, L35.

Stein, R. F.: 1981, *Astrophys. J.* **246**, 966.

Stencel, R. E.: 1978, *Astrophys. J.* **223**, L37.

Stencel, R. E. and Mullan, D. J.: 1980, *Astrophys. J.* **238**, 221.

Stencel, R. E., Linsky, J. L., Brown, A., Jordan, C., Carpenter, K. G., Wing, R. F., and Czyzak, S.: 1981, *Monthly Notices Roy. Astron. Soc.* **196**, 47P.

Stern, R. A.: 1983, in S. L. Baliunas and L. Hartmann (eds.), *Cool Stars, Stellar Systems, and the Sun*, Springer-Verlag, Berlin, p. 150.

Stern, R., Zolcinski, M. C., Antiochos, S. K., and Underwood, J. H.: 1981, *Astrophys. J.* **249**, 647.

Sun, W. H., Giampapa, M. S., and Worden, S. P.: 1985, *Astrophys. J.* (in press).

Swank, J. H. and Johnson, H. M.: 1982, *Astrophys. J.* **259**, L67.

Swank, J. H., White, N. E., Holt, S. S., and Becker, R. H.: 1981, *Astrophys. J.* **246**, 208.

Tarbell, T. D. and Title, A. M.: 1977, *Solar Phys.* **52**, 31.

Timothy, J. D., Joseph, C. L., and Linsky, J. L.: 1981, *Bull. Am. Astron. Soc.* **13**, 828.

Topka, K. and Marsh, K. A.: 1982, *Astrophys. J.* **254**, 641.

Ulmschneider, P. and Stein, R. F.: 1982, *Astron. Astrophys.* **106**, 9.

Underhill, A. B.: 1980, *Astrophys. J.* **235**, L149.

Underhill, A. B.: 1984, *Astrophys. J.* **276**, 583.

Vaiana, G. S.: 1981, in R. Giacconi (ed.), *X-ray Astronomy with the Einstein Satellite*, D. Reidel Publ. Co., Dordrecht, Holland, p. 1.

Vaiana, G. S.: 1983, in J. O. Stenflo (ed.), 'Solar and Stellar Magnetic Fields: Origins and Coronal Effects', *IAU Symp.* **102**, 165.

Vaiana, G. S., Cassinelli, J. P., Fabbiano, G., Giacconi, R., Golub, L., Gorenstein, P., Haisch, B. M., Harnden, F. R. Jr., Johnson, H. M., Linsky, J. L., Maxson, C. W., Mewe, R., Rosner, R., Seward, F., Topka, K., and Zwaan, C.: 1981, *Astrophys. J.* **245**, 163.

Vaughan, A. H., Baliunas, S. L., Middelkoop, F., Hartmann, L. W., Mihalas, D., Noyes, R. W., and Preston, G. W.: 1981, *Astrophys. J.* **250**, 276.

Vedder, P. W. and Canizares, C. R.: 1983, *Astrophys. J.* **270**, 666.

Vernazza, J. E., Avrett, E. H., and Loeser, R.: 1981, *Astrophys. J. Suppl.* **45**, 635.

Vilhu, O.: 1984, *Astron. Astrophys.* **133**, 117.

Vilhu, O. and Rucinski, S. M.: 1983, *Astron. Astrophys.* **127**, 5.

Vogt, S. S.: 1980, *Astrophys. J.* **240**, 567.

Vogt, S. S.: 1983, in P. B. Byrne and M. Rodono (eds.), 'Activity in Red Dwarf Stars', *IAU Colloq.* **71**, 137.

Waldran, W. L.: 1984, *Astrophys. J.* **282**, 256.

Walter, F. M.: 1981, *Astrophys. J.* **245**, 677.

Walter, F. M.: 1982, *Astrophys. J.* **253**, 745.

Walter, F. M.: 1983, *Astrophys. J.* **274**, 794.

Walter, F. M. and Kuhi, L. V.: 1981, *Astrophys. J.* **250**, 254.

Walter, F. M., Gibson, D. M., and Basri, G. S.: 1983, *Astrophys. J.* **267**, 665.

Walter, F. M. and Kuhi, L. V.: 1984, *Astrophys. J.* **284**, 194.

Walter, F. M., Linsky, J. L., Simon, T., Golub, L., and Vaiana, G. S.: 1984, *Astrophys. J.* **281**, 815.

Warner, B.: 1966, *Observatory* **86**, 82.

Warner, B.: 1968, *Observatory* **88**, 217.

White, R. L.: 1985, *Astrophys. J.* **289**, 698.

Wolff, S. C.: 1983, *The A-Stars: Problems and Perspectives*, NASA SP-463.

Worden, S. P.: 1983, in P. B. Byrne and M. Rodono (eds.), 'Activity in Red Dwarf Stars', *IAU Colloq.* **71**, 207.

Zarro, D. M. and Zirin, H.: 1985, *Astrophys. J.* (in press).

Zirin, H.: 1982, *Astrophys. J.* **260**, 655.

Zolcinski, M. C., Antiochos, S. K., Stern, R. A., and Walker, A. B. C., Jr.: 1982, *Astrophys. J.* **258**, 177.

Zwaan, C.: 1983, in J. O. Stenflo (ed.), 'Solar and Stellar Magnetic Fields: Origins and Coronal Effects', *IAU Symp.* **102**, 85.

SOLAR AND STELLAR ACTIVITY: THE THEORETICAL APPROACH

GAETANO BELVEDERE

Istituto di Astronomia, Universita' di Catania, Italy

Abstract. The unified sight of solar and stellar activity has revealed a worthwhile concept under several aspects, gaining in the last decade the increasing favour of observers and theorists, and the term solar-stellar connection has recently been introduced to point out the complementarity of solar and stellar observations in the background of the basic role played by the magnetic field.

The great development of stellar activity observations suggests a much wider scenario than it were possible to imagine even a.few years ago and stimulates theoretical work, most of which is in the framework of the $\alpha-\omega$ dynamo theory.

Although dynamo theory seems to be plausible and successful in capturing the fundamental mechanism of solar and stellar activity, several uncertainties and intrinsic limits do still exist and are discussed together with alternative or complementary suggestions.

Further, it is stressed the relevance of nonlinear problems in dynamo theory – as magnetoconvection, growth and stability of flux tubes against magnetic buoyancy, hydromagnetic global dynamos – to improve our understanding of both small and large scale interaction of rotation, turbulent convection and magnetic field, and of the transition from linear to nonlinear regime. Finally, recent dynamo models of stellar activity are critically reviewed, as to the dependence of activity indexes and cycles on rotation rate and spectral type.

Open problems to be solved by future work are outlined, pointing out the role of ever increasing stellar data in widening out our comprehension of the dynamo operation modes, which seem to depend on stellar structure, rotation and age.

1. The Unified View of Solar and Stellar Activity

Nothing is sure in our present knowledge of stellar activity but the basic role played by the magnetic field. Thanks to the fact that our nearest star, the Sun, is an active star, we have learned much on the phenomenology of magnetic activity and recognized the relevance of the analogy to the Sun in the interpretation of similar phenomena occurring in other stars. Should the Sun have not been an active star, probably things would have been more puzzling and, may be, the problem less stimulating. We know that the magnetic field plays a dominant role in the Sun's activity and we expect the same to be true for other stars. Although the physics of the interaction between rotation and convective modes – which seems to be relevant in interpreting stellar activity – is very complex and poorly understood despite of the ever more refined attempts made in the recent years, the prominent role of the magnetic field seems to be beyond discussion. However, we have learned from stellar observations that the mechanism giving rise to stellar activity may operate in a multimodal way. Here is the fundamental point: the complementarity of solar and stellar observations to increase our knowledge and understanding of the basic phenomena and mechanisms of activity*.

* To my knowledge, this concept is clearly stated for the first time in Godoli (1967). We recall, however, that the idea of stellar activity in analogy to the Sun is already present in Hale (1905; quoted by Goldberg (1983)) and in Eberhard and Schwarzschild (1913).

Solar Physics **100** (1985) 363–383. 0038–0938/85.15.

In a way solar observations give the possibility of a detailed study of activity phenomena in a close astrophysical laboratory and provide a guideline to explore stellar activity on the basis of what is learned from the Sun; on the other hand stellar observations may offer a large sample of phenomenology on a multiplicity of physical situations and time-scales, suggesting a more general context and widening out our ideas about the meaning itself and the operation modes of stellar activity. The analogy to the Sun should, therefore, be used with caution: even if dynamo theory has revealed promising for understanding the Sun's magnetic activity, and its extension to the more general stellar case seems to be valuable, nevertheless differences in the dynamo working mode are expected as a consequence of different stellar characteristics, essentially structure, efficiency of convection, depth of the convection zone (c.z.), rate of rotation and age. So that solar activity is to be regarded as a particular regime, for a star of solar parameters and age, in the broader context of stellar activity.

Unity and multiplicity are, therefore, the two complementary aspects of stellar activity: unity as for the basic mechanism, multiplicity as for the dependence of activity modes on stellar parameters.

The unified view of solar and stellar activity has revealed a worthwhile concept under several aspects, gaining in the last decade the increasing favour of observers and theorists and the term solar-stellar connection has recently been introduced to point out the complementarity of solar and stellar observations in the study of stellar activity. In the most recent years the data collected by IUE and EINSTEIN have sensibly widened out our knowledge of stellar activity phenomenology in different bands of the electro-magnetic spectrum and at different levels in stellar atmospheres. In particular, X-ray observations have pointed out the basic relevance of magnetic fields in sustaining stellar coronae and the close link among X-ray emission, rotation and convection (Vaiana, 1981, 1983; Rosner, 1980, 1983a; Linsky, 1981, 1983; Pallavicini et al., 1981; Stern et al., 1981; Walter, 1981, 1982; Walter and Bowyer, 1981; Stern and Skumanich, 1983; Vilhu, 1984; Mangeney and Praderie, 1984).

Nevertheless the more classical Ca II emission flux observations have still confirmed to be a powerful method of investigation on chromospheric activity, rotation rate and stellar cycles (Wilson, 1978; Vaughan and Preston, 1980; Vaughan, 1980; Vaughan et al., 1981; Skumanich and Eddy, 1981; Middelkoop, 1982; Catalano and Marilli, 1983; Noyes, 1983; Baliunas et al., 1983; Noyes et al., 1984a; Marilli and Catalano, 1984; Catalano, 1984; Skumanich et al., 1984).

In the same context, recent Mg II emission flux chromospheric observations have added to our knowledge of rotation, convection and magnetic activity in late-type stars (Hartmann et al., 1984), as well as Hα-line investigations (Young et al., 1984; Skumanich and Young, 1984).

Further, photometric variations in RS CVn, BY Dra, and red-dwarf stars seem to be more than a promising tool to give further insight on surface phenomena as starspots, flares, rotation rates (short term modulation) and stellar cycles (long term modulation) (Catalano et al., 1980; Hall, 1981; Hartmann, 1981; Rodonó, 1981, 1983, 1985a; Vogt, 1983; Gershberg, 1983; Mullan, 1985). Flares observed in lower Main Sequence, red-dwarfs and pre-Main-Sequence objects are of extreme importance to our understanding of the energetics of such violent magnetic field instabilities (Gershberg, 1978; Kodaira, 1983; Byrne, 1983; Rodonó, 1985b).

Last but not least, magnetic field measurements are now sensibly improved by the Robinson-Preston technique (Robinson, 1980; Preston, 1971) that allows to measure (not-averaged) field intensity and filling factors by comparison of magnetically sensitive and insensitive lines. Results for 29 G–K stars are reviewed by Marcy (1983) and range in the 800–1500 G interval, with filling factors 0.5–0.8, the latter much larger than in the Sun.

The great development of stellar activity observations suggests a much wider scenario than it were possible to imagine even a few years ago. This has enhanced the theoretical effort to predict, interpret and reproduce the observed features, giving rise to several models, most of which are in the framework of dynamo theory. The refinement of the methods and the improvement of the results of the theoretical analysis should have in turn a feedback on the observational programs, suggesting new aspects of stellar activity to be investigated in the future.

2. Basic Points of Dynamo Theory

As is well known, most current theoretical research in stellar activity is done in the framework of dynamo theory, which attempts to explain the generation and evolution of cosmical magnetic fields in terms of induction effects in conducting fluid masses. In the author's opinion dynamo theory is a plausible and somewhat successful theoretical background to understand the basic mechanism of solar and stellar activity, even if several uncertainties and intrinsic limits do still exist. For the development of the main concepts and exhaustive detailed information we address to the following authors: Moffatt (1978, 1979, 1984); Parker (1979); Krause and Rädler (1980); Vainshtein et al. (1980). Dynamos may be separated into two classes: kinematic, linear dynamos (KD) in which the velocity field u is assigned externally, without considering the feedback of the magnetic field B on the motion; hydromagnetic, nonlinear dynamos (HD), in which the back-reaction of the magnetic field is taken into account and the whole system of the magnetohydrodynamical (MHD) equations is to be solved simultaneously, assuring the internal consistency.

An adequate model should, of course, include equations describing the physics of the system (generally a rotating spherical shell in which convection is developed), in particular the energy transport equation and the equation of state.

The reliability of the kinematic approximation depends on the ratio of magnetic energy density to kinematic energy density. If this ratio is small compared to the unity, the kinematic approach is sufficient. For instance, if we want to describe the solar cycle evolution of the poloidal and toroidal magnetic fields averaged on large spatial scales ($B < 10$ G), the kinematic approximation is reasonable. But things are radically different if we want to describe the intermittent field behaviour in convective cells (granules and supergranules), where the observed field intensity is of the order 10^3 G, or the flux tube accumulation at the bottom of the convection zone, where a field intensity as large as 10^4 G is realistically assumed. In the latter circumstances the hydromagnetic approach is to be dealt with.

2.1. The α–ω dynamo in the mean field electrodynamics

In the kinematic approach, the dynamo problem may be reduced to an eigenvalue problem for the magnetic field **B**, governed by the induction equation:

$$\partial \mathbf{B}/\partial t = \mathrm{curl}\,(\mathbf{u} \times \mathbf{B}) - \mathrm{curl}\,(\eta \,\mathrm{curl}\,\mathbf{B}), \tag{1}$$

where **u** is the velocity field and $\eta = 1/\mu\sigma$ is the ohmic diffusivity, with μ = magnetic permeability and σ = electrical conductivity. For a given **u**, solutions of the form $B \sim \exp(a + ib)t$ are searched to determine the growth rate a and the oscillatory frequency b of the eigenmodes, consistently with some physical conditions for the dynamo maintenance of the field (see, e.g., Belvedere *et al.*, 1980b).

In the framework of the Mean Field Electrodynamics (MFE) (Steenbeck *et al.*, 1966; Steenbeck and Rädler, 1969a, b; Krause and Rädler, 1980), the vector fields **u** and **B** are expressed as sums of *mean* (large scale, slowly varying in time) fields and *fluctuating* (small scale, rapidly varying in time) fields:

$$\mathbf{u} = \langle \mathbf{u} \rangle + \mathbf{u}', \quad \mathbf{B} = \langle \mathbf{B} \rangle + \mathbf{B}' .$$

This leads to the following equations, for the mean and the fluctuating magnetic fields respectively:

$$\partial \langle \mathbf{B} \rangle /\partial t = \mathrm{curl}(\langle \mathbf{u} \rangle \times \langle \mathbf{B} \rangle + \langle \mathbf{E} \rangle) - \mathrm{curl}(\eta \,\mathrm{curl}\,\langle \mathbf{B} \rangle), \tag{2}$$

$$\partial \mathbf{B}'/\partial t = \mathrm{curl}(\langle \mathbf{u} \rangle \times \mathbf{B}' + \mathbf{u}' \times \langle \mathbf{B} \rangle + \mathbf{G}) - \mathrm{curl}(\eta \,\mathrm{curl}\,\mathbf{B}'), \tag{3}$$

where $\mathbf{E} = \langle \mathbf{u}' \times \mathbf{B}' \rangle$ represents an additional mean electromotive force generated by the turbulent interaction between the fluctuating velocity and magnetic fields and

$$\mathbf{G} = \mathbf{u}' \times \mathbf{B}' - \langle \mathbf{u}' \times \mathbf{B}' \rangle .$$

If we assume homogenous turbulence ($\langle \mathbf{u} \rangle = 0$) and introduce the first order smoothing or quasi-linear approximation, which consists in neglecting **G** in (3), we get the following simplified equation for **B**':

$$\partial \mathbf{B}'/\partial t = \mathrm{curl}(\mathbf{u}' \times \langle \mathbf{B} \rangle) - \mathrm{curl}(\eta \,\mathrm{curl}\,\mathbf{B}') . \tag{3'}$$

This implicitly means to assume **B**' small compared to **B** and is consistent with Equation (3) in two cases:

(a) Magnetic Reynolds number $R_m = Ul/\eta \ll 1$, where U is a typical turbulent velocity and l a typical eddy size.

(b) Stroughal number $U\tau/l \ll 1$, where τ is the correlation time.

These conditions correspond, respectively, to the *high resistivity* case (in which the advection term curl ($\mathbf{u}' \times \langle \mathbf{B} \rangle$) is balanced by the dissipative one) and to the *rapid fluctuation* case (in which the advection term is balanced by the time variation of the fluctuating field **B**').

Integrating Equation (3′) over a time interval which is short enough for $\mathbf{u}'$ and $\langle \mathbf{B} \rangle$ to be considered time independent, linearity of $\mathbf{B}'$ in $\langle \mathbf{B} \rangle$ and its space derivates allows us to express $\mathbf{E} = \langle \mathbf{u}' \times \mathbf{B}' \rangle$, in the case of isotropic turbulence, as

$$\mathbf{E} = \alpha \langle \mathbf{B} \rangle - \beta \operatorname{curl} \langle \mathbf{B} \rangle ,$$

so that we get the MFE induction equation in the form:

$$\partial \langle \mathbf{B} \rangle / \partial t = \operatorname{curl}(\langle \mathbf{u} \rangle \times \langle \mathbf{B} \rangle + \alpha \langle \mathbf{B} \rangle) - \operatorname{curl}[\eta + \beta) \operatorname{curl} \langle \mathbf{B} \rangle] . \qquad (4)$$

Here $\alpha \langle \mathbf{B} \rangle$ represents a mean electromotive force – parallel or antiparallel to the mean magnetic field $\langle \mathbf{B} \rangle$ – generated by turbulence and β is the turbulent magnetic diffusivity.

If turbulence lacks mirror symmetry, it can be shown (e.g. Krause and Rädler, 1980) that α is proportional to the mean helicity $\langle \mathbf{u} \cdot \operatorname{curl} \mathbf{u}' \rangle$ of the fluctuating velocity field $\mathbf{u}'$: $\alpha \sim - \tau \langle \mathbf{u}' \cdot \operatorname{curl} \mathbf{u}' \rangle$. A detailed and accurate analysis of the basic principles of the mean field approach to spherical dynamo models including the α^2, $\alpha\omega$, $\beta\omega$ mechanisms and precise statements on the α-effect can be found in Rädler (1983a, b).

In the α–ω dynamos, the advection term $\langle \mathbf{u} \rangle \times \langle \mathbf{B} \rangle$ generates toroidal fields form poloidal fields by differential rotation (ω-effect), while the $\alpha \langle \mathbf{B} \rangle$ electromotive force regenerates poloidal fields from toroidal fields by cyclonic turbulence (Parker, 1955a, b, 1979) through the twisting action of the Coriolis force on magnetic field loops in the convective cells (α-effect). Turbulent diffusion, parametrized by β, mixes large-scale magnetic fields, whilst does not affect small scale fields ultimately smoothed out by ohmic diffusion. Since $\beta \gg \eta$, the turbulent diffusion time R^2/β is much smaller than the ohmic diffusion time R^2/η. Therefore, a reasonable fit to the solar cycle period can easily be attained, with $R =$ solar radius and $\beta \sim 5 \times 10^{12}$ cm^2 s^{-1}, a typical value of the solar c.z. The role of turbulent diffusion is clearly of fundamental importance in the α–ω dynamos, which should better be called α–β–ω dynamos.

The turbulent diffusion time R^2/β and the period of the oscillatory field $(R/\alpha\Delta\omega)^{1/2}$, where $\Delta\omega$ is the differential rotation, are the dominant time-scales involved in the dynamo process. For marginal dynamo instability* these two time-scales are expected of the same order of magnitude. In the solar case, the probably too large estimated α-value, leads to a theoretical period too short by an order of magnitude.

The relative strength of the poloidal to the toroidal field is given by $(\alpha/\Delta\omega R)^{1/2}$. In the solar case this ratio is 10^{-2}, depending, however, on the magnitude of α, the estimates of which seem to be in excess.

The mean field (dynamo wave) propagates along the surfaces of isorotation (Parker, 1955; Yoshimura, 1975) in the direction of $\alpha \nabla \omega \times \mathbf{i}_\varphi$, where $\mathbf{i}_\varphi$ is the azimuthal unit vector. This implies, with $\alpha \gtrless 0$ in the northern and southern emispheres respectively (Stix, 1976), $\partial \omega/\partial r < 0$ (angular velocity increasing inward), since the observational constraint of the butterfly diagram (equatorward migration) must be taken into account. However, the recent measurements by Duvall and Harvey (1984) and Duvall *et al.*

---

* Marginal dynamo instability arises when the dynamo number $N_D = R_\alpha R_\omega$ is slightly larger than a critical value. For fixed $R_\omega = \Delta\omega R^2/\beta$, this implies $R_\alpha > R_{\alpha c}$, a critical value.

(1984), based on the analysis of solar oscillations, seem to indicate a nearly constant angular velocity in the c.z., at least at the equator. This may imply serious difficulties to the $\alpha-\omega$ approach with radial shear driven dynamos, thus suggesting to shift the attention to the latitudinal shear. To this regard a very recent model of solar differential rotation by Pidatella *et al.* (1985), which includes rotational perturbation of the radiative transport in a realistically described convection zone and overshooting layer, indicates that the anysotropy of viscosity seems to be the dominant mechanism in generating differential rotation, and that the angular velocity should slightly decrease inwards.

The parity of the mean field with respect to the equator can be even or odd, depending on which modes are excited at lower $R_{\alpha c}$. For instance, Belvedere *et al.* (1980b) find no apparent preference for the observed odd parity modes. Yoshimura *et al.* (1984a, b, c), in a series of detailed papers, put in evidence that, for latitudinal gradient dominated differential rotation, the selection rule for parity depends on the anisotropy of the turbulent magnetic diffusivity. When diffusivity for poloidal fields is larger, odd parity is selected; when diffusivity for toroidal fields is larger even parity is selected. In the case of radial gradient dominated differential rotation, growth rates of odd (even) parity prevail, provided that the radial gradient increases inward (outward), corresponding to equatorward (poleward) migration of the wings of the butterfly diagram.

2.2. LIMITS AND PROBLEMS OF THE $\alpha-\omega$ THEORY

Criticism against the $\alpha-\omega$ dynamo theory in the MFE concerns especially two points: the rather crude method of closure – strictly justified only if $\mathbf{B}'$ is small compared to $\langle \mathbf{B} \rangle$ – involved in the first-order smoothing approximation (for the Sun, indeed, $R_m \gg 1$ and $U\tau/l \sim 1$), and the role of turbulent diffusion.

A possibility of overcoming the former difficulty is in Cowling's (1981) argument that '$\mathbf{B}$' varying rapidly in space is no longer large compared with $\langle \mathbf{B} \rangle$, being rapidly smoothed out by ohmic diffusion at small length scales' (see also Cowling's (1981) discussion of the induction equation for $\mathbf{B}''$, the part of $\mathbf{B}'$ correlated with $\mathbf{u}'$). The second remark is in part due to the misunderstanding of the role of turbulent diffusion, which applies only to the mean field (which is the average field, *not* the field in between the flux concentrations). A detailed discussion about these points can be found in Stix (1981) and Belvedere (1983).

Nonetheless, the MFE $\alpha-\omega$ theory is far from being satisfactory, despite of its success in reproducing the main characteristics of solar cycle (see, e.g. Köhler, 1973; Stix, 1976a, b; Yoshimura, 1975, 1978a, b, 1983; Parker, 1979; Belvedere *et al.*, 1980b), namely:
– equatorward migration of the active regions (butterfly diagram);
– polarity reversals of the polar fields;
– polarity rules in the two hemispheres;
– ratio and phase relationship of the poloidal field to the toroidal field;
– cycle period;
– rigid rotation of preferential longitudes and some chromospheric-coronal structures.

This seems to confirm the capability of the MF dynamo equation (4) of capturing in a simple way the essential mechanism that maintains the solar cycle (Weiss, 1981), but several questions remain still open and should be investigated deeply in a more consistent and detailed non-linear approach:

– The role of helicity and turbulent diffusion should be clarified on both large and small scales, leading to more plausible intrinsic determinations of α and β in the context of the theory of turbulence*.

In particular the present estimates of α seem to be too large. Note, however, that in the case of strong toroidal flux concentrations, a strong α-effect is needed against the no more negligible Lorentz force (Gilman and Miller, 1981).

– The depth in the c.z. at which dynamo operates is still matter of discussion: spatial separation of the ω-effect is not plausible in so far as it would imply problems of upward and downward field transport which are not easily overcome by means of turbulent diffusion. On the other hand, since stability of flux ropes against magnetic buoyancy (see Section 3.2) suggests that the ω-effect is located at the bottom of the c.z. or in the overshooting layer, also the α-effect is expected to operate mainly in the deep levels. This is also supported by the argument that the α-effect on flux tubes rapidly rising towards the surface would be ineffective in the top half of the c.z. (Golub *et al.*, 1981).

– The feedback of the magnetic field on the velocity field through the Lorentz force is expected to be relevant in the case of toroidal flux ropes wound by the ω-effect at the bottom of the c.z. or in the observed case of strong intermittent fields at the edges of cellular patterns.

– The problem of the stability of flux tubes under magnetic buoyancy forces, hydrodynamical drag and Coroilis force is of basic importance to get a reasonable estimate of the float-up times, which should be comparable to the amplification time and to the diffusion time, both expected of the order of the cycle period.

At the present all these points are investigated separately, leading to a gradual but promising advancement (see Sections 3.1 and 3.2).

2.3. ALTERNATIVES TO THE $\alpha-\omega$ THEORY AND THE STATISTICS OF THE SOLAR CYCLE

Piddington (see, e.g., 1981 and earlier references therein) does not agree with the concept of turbulent diffusion of the magnetic field and claims that eddy diffusivity leads to shear amplification of fields within the eddies. In his opinion no merging of fields can be accomplished by turbulence. Piddington emphasizes also the fact that convective motions and buoyancy would tend to transport upwards the 'newly' generated poloidal field, so that it could not efficiently be operated by differential rotation, the strongest action of which occurs in the lower part of the c.z.

The alternative scenario proposed by Piddington consists of a primordial non-reversing dipole – buried in the radiative region to avoid turbulence – whose field lines oscillate in the meridian planes with a period of 22 yr and are acted by the ω-effect, generating toroidal fields of opposite signs in two consecutive cycles. However, neither

---

* For recent developments in the theory of turbulence, see Moffatt (1981, 1983). Topological studies on helicity and magnetic helicity are in Moffatt (1984) and Moffatt and Proctor (1985).

clear fundament is given to the energy source and the mechanism of such a dipole oscillation, nor a satisfactory quantitative treatment is developed.

Layzer *et al.* (1979) propose an alternative framework consisting of an original field generated by the Biermann (1950) mechanism, amplified during the fully convective phase and giving rise ultimately to a large scale tangled field in the uniformly rotating radiative core. Differential rotation in the overshooting layer generates a toroidal component which is wound and unwound alternatively. This torsional magnetic field oscillation would explain the solar cycle, the observed phenomenology of which would be due to flux tubes floating upwards to the surface. Once again, this alternative scenario appears to be too speculative, no well-developed physical arguments being given.

Incidentally, we recall that the 11 yr period torsional oscillation discovered by Howard and LaBonte (1980) has been proposed by the same authors (LaBonte and Howard, 1982) to sustain the magnetic cycle. However, even if this idea is appealing since derived by a clear observational evidence, it seems a bit difficult to accept it, on the ground of an energetic argument: it is unlikely that a rather weak oscillating field can compete with the much stronger differential rotation or turbulent convection fields. It seems more reasonable that the toroidal oscillation be driven by the longitudinal component of the Lorentz force associated to the dynamo waves which generate the solar cycle itself (Yoshimura, 1981).

In this context we recall that the mean velocity derived from the proper motions of sunspot groups seems to show a 11 yr oscillation, both in longitude and latitude (Tuominen J. *et al.*, 1983; Tuominen I. and Virtanen, 1984). According to Tuominen I. *et al.* (1984), these cyclic mean flows are interpreted as a natural consequence of the feedback of the oscillating mean magnetic field. Indeed, solving the mean-field $\alpha-\omega$ dynamo equations gives the observed components of the mean flows, thus confirming the validity of the dynamo approach.

Perhaps, the most exciting alternative to dynamo theory comes from the regular magnetic torsional oscillations in the core, the 'chronometer hidden deep in the Sun', proposed by Dicke (1978, 1979, 1982a, b). He suggests that a sort of magnetohydrodynamic oscillator acts deeply in the solar core. This regular oscillation of the core causes it to periodically shed magnetic field, floating upwards to the surface, and controls rigidly the solar cycle phase maintenance and modulates the solar luminosity.

The fundamental difference between torsional oscillations in the core and turbulent dynamos is in the fact that the former predict a strict phase maintenance, whilst the latter do not. So that, these alternative theories may be checked by the degree of the solar cycle phase wandering from a perfect clock. This matter is still controversial, since, for instance, Dicke (1978) finds no statistical indication of a random walk in the phase of the solar cycle, whilst Gough (1983) outlines that 'there is some indication from the statistical analysis that, contrary to Dicke's conclusion, phase is not maintained by the cycle. This adds weight to theories of a cycle with little or no memory, though it hardly proves tham. Moreover, it does not even necessarily suggest that a turbulent dynamo is operating in the body of the convection zone'.

Now we come to a radical question: how significant is to speak of *cyclic* activity of the Sun? Indeed large time-scale records of solar activity seem to indicate a real aperiodicity rather than a quasi-periodicity (Williams, 1981; Wallenhorst, 1982). Further, from the study of anomalies in C-abundances, Stuiver (1980) found a series of grand-minima, the most recent of which are the Spörer and the Maunder minima. So that, one could be led to conclude that chaos 'governs' solar activity, at least on large time-scales. On the other hand, this is not so surprising if we consider the interesting simulation by Barnes *et al.* (1980), who reproduced the variance in the sunspot number at maximum, starting from Gaussian white noise. Moreover, this simulation exihibited intervals of 50 yr or more very low output, once every 500 yr on average, reminiscent of grand-minima.

As we shall see later, nonlinear dynamo models are able to describe chaotic behaviour as well as cyclic behaviour, the transition from each to other being related to some characteristic parameters.

3. Nonlinear Problems in Dynamo Theory

Essentially this class of theoretical attempts arises from introducing the feedback Lorentz force $(1/4\pi)$ (curl $\mathbf{B}$) $\times$ $\mathbf{B}$ into the dynamo equations. This is strictly necessary when the ratio of the magnetic energy density to the kinetic energy density is expected to be of order unity or larger. The main topics so far considered are:

(i) Growth of the mean magnetic field limited by the Lorentz force; magnetoconvection.

(ii) Growth and stability of flux tubes in the convection zone against magnetic buoyancy.

(iii) Hydromagnetic global dynamos, which solve at once the complete magneto-hydrodynamic problem, assuring internal consistency.

3.1. MAGNETO-CONVECTION

Nonlinear magnetoconvection simulations have been worked out to explain the presence of intense intermittent fields (~ 1500 G) observed at the solar surface, through mechanisms able to describe the formation of isolated flux tubes in the convection zone.

Galloway and Weiss (1981) have studied Boussinesq convection in the presence of magnetic fields and found that magnetic flux is rapidly concentrated into sheets at the lateral boundaries of the convective cells, while flux expulsion from the cell interior takes a longer time, of the order of a few turnover times. Turbulent convection concentrates magnetic flux until the equipartition value $\mathbf{B}_e$ (generally $\ll \mathbf{B}_p$, the pressure equlibrium value, except that at the top of the c.z.) is attained. Fast evacuation of matter from the flux tube ('collapse') is then expected, to get the pressure equilibrium value $\mathbf{B}_p$. Weaker flux ropes are shredded and dispersed, giving rise to smaller size activity features. The observed total flux at the Sun's surface should be compatible with the toroidal flux contained in a shallow layer located at the bottom of the c.z. This agrees with the observational evidence of the coronal hole pattern corotating nearly uniformly, as it were anchored to a deep level in the c.z. (Golub *et al.*, 1981). In the final state the cell edge field is amplified by a factor $R_m^{1/2}$, where R_m is the magnetic Reynolds number, while the flux expulsion time goes as $R_m^{1/3}$.

The latter result is also found analytically in a simple and elegant model by Moffatt and Kamkar (1983), and agrees with the results of Weiss (1966).

Nonlinear three-dimensional magnetoconvection and the magnetic field spectrum have been studied by Knobloch (1981a, b), who finds that nonlinear concentration of magnetic flux by turbulent motions occurs at the cell edges, in agreement with the previous authors (see also Peckover and Weiss, 1978), and that different scales of flux tubes arise as a result of different scales of motions. Also a theoretical prediction of size and spatial distribution of flux tubes is given, as well as the field strength as a function of the tube radius, on the assumption that flux concentrations are formed from an initial uniform field in the presence of a given turbulence spectrum.

The kinematics of exagonal magnetoconvection has recently been studied by Galloway and Proctor (1983) in order to investigate the evolution of an imposed magnetic field subject to three-dimensional motions. This analysis does not include the Lorentz force feedback, but may be of some relevance to the problem of building up magnetic flux deeply in the c.z., where kinematic effects should be important.

Numerical three-dimensional simulations in the nonlinear fully compressible case, including radiative transfer and ionization effects, have been performed by Nordlund (1983). The results show that there exist a strong tendency for the magnetic field to be swept up and concentrated in the intergranular spaces, attaining intensities as high as 1 kilogauss.

For recent reviews on magnetoconvection we address to Galloway (1984) and Nordlund (1984).

3.2. FLUX TUBE DYNAMICS

The problem of growth and stability of flux tubes in the convection zone under the action of turbulent convective motions, magnetic buoyancy, hydrodynamical drag, and rotational forces is one of the most debated in the recent years. For a general overview we address to Parker (1979, Chapters 8, 10, 13), Spruit (1981a, b), and Schüssler (1984). Of course, this problem is strictly connected with the time-scale of the magnetic flux rise to the solar surface to form active regions.

Parker (1975) suggested that toroidal flux generation should occur deeply in the c.z., where the interaction between convection and rotation is strong, and in order the rise time of tubes to be comparable with the time scale of the solar cycle. However, there are some difficulties: tubes in thermal equilibrium with the surroundings would float upwards due to magnetic buoyancy. But the constraint that they should remain in the c.z. for some years leads to an upper limit of $\sim 10^2$ G for the field intensity, which is too small compared to the equipartition value. On the other hand, if the tubes were neutrally buoyant, their internal temperature should be lower than the surroundings, but it is not very clear which mechanism could maintain the temperature difference. However, an interesting suggestion has been proposed to this regard by Yoshimura (1985), namely the cooling of flux tubes by inhibition of convective heat transport due to magnetic fields inside the tubes. It is supposed that the energy transport inside the tube is mainly due to radiation with a steeper radiative temperature gradient than the surroundings. Since extremely small temperature differences are able to suppress magnetic buoyancy, this cooling mechanism should easily operate as a stabilizing mechanism against floating.

Recently, Parker (1984a) has advanced new arguments to overcome the difficulties raised by the magnetic buoyancy: a flux ejection dynamo effect could operate in the deep convection zone, where the cyclonic rotation of the convective cells may be as large as 180°. If this is the case, the initially direct horizontal field at the bottom of a convective cell becomes reverse when conveyed to the top of the cell. According to this mechanism, we expect ejection of reverse field at the stellar surface, providing a net gain in direct field which accumulates at the bottom of the convection zone.

Another interesting remark by Parker (1984b) is that only magnetic flux which reconnects can escape from stellar surfaces. For the Sun, due to the wide longitude differences between strong bipolar regions, only some percent of the generated flux should escape freely into space. Of course, this should have severe quantitative implications on models of flux concentration in the deep convective layers.

The suggestion of the lower part of the c.z. and the boundary overshooting layers as a site for magnetic flux storage is also founded on stability requirements. Indeed, the strict polarity rules of active regions and the well definite laws which surface activity does obey, seem to imply that the subsurface magnetic field is highly organized and not subject to strong deformation by convective motions (Van Ballegooijen, 1982b).

The stability of adiabatic flux tubes at the bottom of the c.z. has been studied in detail by Van Ballegooijen (1982a, b, 1983) and Spruit and Van Ballegooijen (1982). Several mechanisms operating against magnetic buoyancy and buoyant instabilities such as small upward displacements or wave-like disturbances have been examined, in particular hydrodynamical drag and the Coriolis force resulting from retrograde mass flow in the toroidal tubes, induced by angular momentum conservation in the equatorward motion. An interesting conclusion is that stability can be achieved in the boundary layers, with field intensities of the order 10^4–10^5 G.

The extent and thermal stratification of the convective overshoot region, as a suitable site for magnetic flux storage, has been investigated by Schmitt et al. (1984), adopting a semitheoretical approach borrowed from studies of buoyancy in the Earth's atmosphere. The results are clearly in favour of the existence of a stabilizing boundary layer of about 500 km thickness between adiabatically and non-adiabatically stratified regions. This should be of some relevance to the recent paper by Van Ballegooijen and Title (1985), according to whom the existence of a magnetic boundary layer at the bottom of the c.z. is essential to understand the solar cycle. They suggest indeed that active regions are formed by interaction of convective cells with the magnetic layer with a small chance of about 1%, while the mechanism of flux disappearance is rather effective in restoring the horizontal field at the base of the c.z. on a short time-scale after each emergence of an active region. Further, the magnetic flux storage in a thin stable boundary layer has two important implications on the α–ω dynamo: the radial gradient would not be important in winding up the toroidal field; the magnitude of the α-effect would be lowered – as required to fit the 22 yr period – since the interaction between convection and flux tubes would be greatly reduced.

3.3. HYDROMAGNETIC GLOBAL DYNAMOS

So far, most theoretical work in stellar activity has been done in the framework of the $\alpha-\omega$ linear theory and we have to admit that the results are not so bad, indicating at least a trend which seems to be independent of the single approachs.

There is no doubt that linear theory is more tractable analytically and numerically, and more suitable to an easier interpretation of the results. Further, it may have captured the essentials of dynamo action. However, linear theory is basically limited by three reasons: (i) the magnetic field amplitude cannot be predicted, (ii) the velocity field does not 'feel' the magnetic field; (iii) multi-modal dynamos cannot be described.

On the other hand, there is no doubt that the nonlinear approach is more rigorous and self-consistent and can in principle describe a large variety of dynamo operation modes. This latter property has gained a particular relevance in the recent years, due to the discovery of the solar grand-minima and the evidence for the Vaughan-Preston gap (the latter should be taken with some caution, see Section 4.3.).

In the general context of nonlinear global convectively driven stellar dynamos, Yoshimura (1978a) has examined the feedback of the induced magnetic field on the motion, suggesting an oscillatory modulation of the differential rotation-global convection pattern associated to the solar cycle, the phase relation of which should give insight on the dynamo process in the c.z. Moreover, a drastic weakening of global convection by Lorentz force is proposed to explain the long-period absence of observed activity.

Global hydromagnetic dynamos in a rotating spherical shell have been worked out by Gilman (1982, 1983a, b, and references therein), Gilman and Miller (1981) in a large series of numerical experiments in the Boussinesq approximation. This approach is well correct in principle, but solutions depend on the adjustment of a large number of physical quantities and parameters defining the regime (viscosity, thermal diffusivity, Reynolds, Prandtl, Taylor numbers, etc.) and sometime it is a bit difficult to interpret the results clearly. Difficulties are found when applying the models to the Sun, in particular chaotic magnetic field behaviour, no polar reversal, no cycles, no preferred equatorial symmetry, poleward field migration. Nevertheless this does not exclude in principle applicability to active stars, since we expect that dynamo operates in different modes. To this regard it is interesting Gilman's (1983a, b) 'regime diagram', which attempts to predict the dynamo operation mode as a function of the electrical conductivity (related to the dynamo number) and a parameter which measures the influence of rotation on the dynamics of the system, namely the ratio of the differential rotation kinetic energy to the total kinetic energy. This way, three basic regimes are identified: no dynamos, dynamos with cycles, dynamos without cycles (chaotic), as a consequence of the combined influence of the magnetic diffusivity, the efficiency of convection, differential rotation and α-effect, and the Lorentz force.

Glatzmaier (1984, 1985a, b) presents the first numerical simulations of global stellar dynamos made in the compressible case. They reproduce the equatorial acceleration and are in agreement with the angular velocity slight decrease with depth through most of the c.z. (Duvall and Harvey, 1984; Duvall *et al.*, 1984). A magnetic field poleward latitude drift is found, if the dynamo operates in the turbulent convective layers, whilst, if the dynamo is located in the transition region at the bottom, a tendency to equatorward migration comes out, this depending on the sign of helicity. Theoretical giant cell velocities at the surface are larger than observed, but surface turbulence may greatly reduce the amplitude of the motions, as pointed out by Snodgrass and Howard (1984).

The transition from periodic to aperiodic dynamo waves has recently been investigated by Weiss *et al.* (1984) in a nonlinear model of oscillatory stellar dynamo. They analyze the stability of the hydromagnetic regime as a function of the dynamo number and show that periodic solutions (Parker's waves) may become unstable, leading to aperiodic (chaotic) behaviour through successive transitions. This paper improves Yoshimura's (1978b) first attempt to demonstrate that aperiodic behaviour may be produced by nonlinear dynamos, insofar as it avoids introducing arbitrary time lags, and shows that aperiodic magnetic cycles with grand-minima may occur naturally. According to these authors the solar cycle is consistent with deterministic chaos in a nonlinear dynamo, in contrast with the picture of stochastic irregularities of a periodic oscillator. This conclusion indeed agrees with the evidence that the record of solar activity seems to be aperiodic rather than quasi-periodic and with the observations of chaostic oscillations in the activity cycles of late-type stars (see the end of Section 2.3).

4. Theory of Stellar Activity

It is well known that Main-Sequence stars later than F7 and giants later than G0 have outer convective envelopes, the thickness of which increases with the decreasing effective temperature. The interaction between rotation, convection and magnetic field, leading in the Sun to the observed differential rotation and dynamo driven magnetic activity, should in principle do the same job in other stars with convective envelopes, although different characteristics are expected, depending on basic parameters as luminosity class, spectral type, rotation rate and age.

In principle, models of stellar activity should at least predict magnetic field strengths, filling factors, butterfly diagrams and cycle periods. Note that poleward migration, non-cyclic or chaotic activity, absence of polar reversal *cannot* be excluded in the general statement of the problem, since we have learned both from recent observations and nonlinear theory that dynamo can operate in a multi-modal way.

At the present time, however, theoretical research on magnetic activity in stars other than the Sun is mostly driven by the analogy to solar activity models and by the increasing amount of data provided by stellar observations, in the general framework of the solar-stellar connection (see Section 1).

4.1. THE THEORETICAL TREND OF ACTIVITY IN LATE-TYPE STARS

The first attempt to compute models of differential rotation, magnetic activity and X-ray coronal emission for Main Sequence and giants has been made by Belvedere *et al.* (BPS, 1980c, d; BCP, 1981, 1982) following the analogy to the Sun (Belvedere and Paternó, 1977; Belvedere *et al.*, 1980a). The results show that differential rotation, magnetic field strength, latitude extension of the activity belts (according to the computed butterfly diagrams) do increase with the increasing angular velocity of rotation and the advancing spectral type (later than $\sim G$).

It has pointed out in these papers that the ratio of the global convection turn-over time-scale to the rotational time-scale – namely $\omega d/U \sim \omega d^2/\nu$ where d is the thickness of the convection zone and ν is the kinematic viscosity – which is a measure of the strength of the interaction of rotation and convection, does increase with the advancing spectral type, leading to increasing differential rotation. This in turn implies a larger R_ω, thus a smaller $R_{\alpha c}$, that means a more favoured dynamo action for faster rotators and later spectral types. In particular the toroidal magnetic field strength, estimated under the hypothesis of energy equipartition between the velocity and magnetic fields, can be expressed in the form ω (an increasing function of the spectral type).

Durney and Robinson (1982, hereafter referred to as DR), assuming that the rise time of flux tubes is of the order of the amplication time, have estimated the magnetic field strength along the main sequence and found that both magnetic field strength along the Main Sequence and area coverage increase with ω and $(B - V)$. These results are essentially confirmed by Robinson and Durney (1982), where a simplified local system of equations, including the magnetic buoyancy term, is solved in the lower part of the convection zone.

Knobloch *et al.* (1981), under the assumption that the rate of production of toroidal field is comparable to the rate of escape by magnetic buoyancy, find that the field strength increases with the decreasing stellar mass.

Paternó and Zuccarello (1983), adopting DR's assumption, confirm this theoretical trend, their results being supported by reasonable comparison between the observed X-ray fluxes for F–G spectral types and those computed in the framework of the model of X-ray emission sustained by magnetic energy dissipation in coronal loops (Golub *et al.*, 1980; BCP, 1981, 1982).

Rosner (1983b), comparing observed stellar X-ray luminosities with a theoretical expression deduced from solar corona scaling laws, predicts an increase of the filling factor with ω and $(B - V)$.

These theoretical results, even quantitatively different, clearly show that present $\alpha-\omega$ dynamo models, although subject to various assumptions, do converge in predicting that stellar activity should increase with ω and $(B - V)$ and, in particular, that $B \sim \omega$ (an increasing function of $(B - V)$), in qualitative agreement with the main trend shown by almost all existing observational evidence (either direct or indirect; see Section 1).

The theoretical prediction is basically founded on the following argument, characteristic of the $\alpha-\omega$ mean field dynamos. The dynamo number N_D, which is a measure of the dynamo excitation level, is expressed by $N_D = R_\alpha R_\omega = \alpha\Delta\omega d^3/\beta^2$. Now, $\Delta\omega \sim \omega$ (BPS, 1980c) and $\alpha \sim -\tau_c \langle \mathbf{u}' \cdot \mathrm{curl}\,\mathbf{u}' \rangle \sim \omega d$, since the mean helicity, which depends on the influence of the Coriolis force, scales as ωU and $\tau_c \sim d/U$. So that, taking into account that $\beta \sim Ud$ it follows $N_D \sim (\omega d/U)^2 = R_0^{-2} =$ (turnover time/rotational time)^2, where Ro is the Rossby number. Therefore, we expect the level of magnetic activity, which increases with N_D as confirmed by all nonlinear calculations so far performed (e.g., Gilman, 1983b), to be an increasing function of the angular velocity ω and the convective turnover time d/U. This theoretical trend has clearly been confirmed by recent measurements of the CaII emission flux in late type stars. Indeed the ratio $F_{\mathrm{Ca\,II}}/F_{\mathrm{bol}}$ correlates fairly well with P_{rot}/τ_c, where P_{rot} is the rotational period (Noyes, 1983; Noyes et al., 1984a), in agreement with the general predictions of dynamo theory. Similar results are found measuring the MgII$_{\mathrm{h,\,k}}$ emission flux (Hartmann et al., 1984) and using chromospheric, transition region and coronal activity indicators (see e.g., Vilhu, 1984). We recall, however, that the correlation between integrated CaII emission luminosity L_K and rotation period seems to be independent of the spectral type (Marilli and Catalano, 1984).

4.2. STELLAR ACTIVITY CYCLES

We know from observations (Wilson, 1978; Vaughan and Preston, 1980; Vaughan, 1980; Vaughan et al., 1981) and begin to understand (e.g., Gilman, 1983a, b; Weiss et al., 1984) that dynamo can operate in a multimodal way, giving rise, in particular, both to chaotic (non-cyclic) and cyclic activity, In other words, stellar activity does not necessarily imply cycles, although stellar cycles do exist. The cycle period should again depend in principle on rotational rate, spectral type and age, so that comparison between predicted and observed stellar cycle periods should add to our understanding of the dynamo working modes. Unfortunately, very scarce work has been carried out on this topic and some disagreement exists between theoretical predictions and observations. However, the matter is still controversial also on the observational point of view, so that more data should be available before attempting to draw even preliminary conclusions.

According to Belvedere et al. (1980d), the cycle period should increase with $(B - V)$, this result being obtained in the case of marginal dynamo instability for which the dynamo wave period $(R/\alpha\Delta\omega)^{1/2}$ is of the order of the diffusion time in the convection zone d^2/β. Following Durney and Robinson (1982), who assume that the cycle period is of the order of the amplification time, the cycle period should decrease with $(B - V)$. Both models disagree with the fact that at least the up-to-now observed stars (mainly K–M) seem to show no cycle dependence on spectral type (see, however, Noyes et al., 1984b). As for the ω-dependence, dynamo theory predicts that the cycle period goes inversely proportional to the angular velocity of rotation, since the dynamo wave period $(R/\alpha\Delta\omega)^{1/2} \sim (1/\omega)$, if $\Delta\omega \sim \omega$ and $\alpha \sim \omega d$.

Most recently, Noyes et al. (1984b) have found an empirical relation between cycle period P_{cycl}, rotational period P_{rot} and spectral type, namely $P_{\mathrm{cycl}} \sim (P_{\mathrm{rot}}/\tau_c)^n$ with $n \sim 1.25$, and discussed it in the framework of simple nonlinear models. This empirical relation is consistent with models in which dynamo action is limited by losses due to magnetic buoyancy, but is based on a relatively small sample of objects, so that further data are required to clarify the matter both on the observational and theoretical points of view. To this regard, we mention that Kleeorin *et al.* (1983) give non-monotonic dependences $P_{cycl}(\omega)$ for a given spectral type, different for rapid rotating and slowly rotating objects. For the former the cycle period should be ω-independent, but the present (scarce) observational evidence shows that it depends sensibly on the rotational velocity for large rotation rates (Vogt, 1983). A detailed analysis of some observational and theoretical aspects of stellar cycles can be found in Paternó (1984).

4.3. OPEN PROBLEMS TO BE SOLVED BY FUTURE MODELS

Although stellar activity theory may have captured some basic concepts in understanding the dynamo operation mode as a function of the rotation rate and the spectral type, several open problems do still remain and should be solved in the future work. For instance:

– We know from observations that the magnetic flux increases with ω and $(B - V)$, while filling factors are inversely proportional to the magnetic field strength (Marcy, 1983; Gray, 1984). The reason for that is not very clear. Perhaps it may depend on the magnetic field topology (open or closed m.f. lines), as suggested by Giampapa (1985), or the extent of the c.z. (Giampapa and Rosner, 1984). Further, the theoretical field strengths so far suggested (see Section 4.1) are to be taken just as indicative estimates. However, as pointed out by Mullan (1984), fields as high as 10 kilogauss may be present in spots on cool dwarfs without violating observational constraints. Clearly, nonlinear models are expected to give more consistent estimates.

– The dependence of filling factors on the rotational velocity is not evident: indeed slowly rotating dM stars show high filling factors as well as rapidly rotating dMe stars (Giampapa, 1985). Dynamo theory should account for. However, we have to remark that the present data are not sufficient to draw safe conclusions. In addition line blends may lead to wrong measurements.

– The interpretation of the Vaughan–Preston gap in terms of a discontinuity in dynamo action (bimodal behaviour) may suffer from the fact that the dependence of the mean chromospheric emission on rotation and spectral type is essentially the same for stars above and below the gap (Noyens *et al.*, 1984a). Incidentally, we recall that the gap could be fictitious, just the result of the tendency toward an upper limit of chromospheric Ca II emission, combined with the lower limit from photospheric flux (Hartmann *et al.*, 1984)

– Poleward migration of active regions (spot-like features resembling solar coronal holes) has been observed by a sofisticated technique for imaging stellar surfaces (Vogt and Penrod, 1983). This is predicted by dynamo models (see Gilman, 1983b), but we need further investigation to clearly understand which parameters discriminate between equatorward and poleward migration (see Section 2.1).

– Stable longitudes (lasting hundreds of rotations) have been detected on stars other than the Sun (Skumanich and Young, 1984). No convincing explanation of such features is yet available, since in most models axisymmetry is assumed.

– Solar (and stellar, see Vogt *et al.* (1983) for the quiescent phase of AD Leo) grand-minima are of extreme interest to understand the dynamo operation modes. Non-linear theory can account for (Ruzmaikin, 1981; Weiss *et al.*, 1984), but the physical picture is still unsatisfactory.

– The age-dependence of stellar activity needs a better theoretical understanding, in connection with the mass and angular momentum loss by stellar wind and the magnetic braking (Gray, 1982a, b, c; Hartmann, 1983).

We conclude stressing the outstanding importance of the information that surface oscillations can give us on the internal structure and dynamics of stars. Indeed, they offer a unique tool to probe stellar interiors and therefore to test the theory of stellar structure and evolution, which is one of the basic supports of our knowledge of the Universe. In the framework of stellar activity, the analysis of solar (and stellar) oscillations can provide accurate data on basic ingredients of dynamo, namely the depth of the convection zone and the radial dependence of the angular velocity in the c.z. and in the radiative core. Here we do not discuss the matter in detail; for this we address to Gough (1981, 1983, 1984), Christensen-Dalsgaard (1984) and Belvedere and Paternó (1984). We just want to recall two points:

– The determination of the solar internal rotation from first order frequency splitting imposes serious constraints to the $\alpha-\omega$ theory, casting doubt on radial shear driven solar dynamos.

– The presence of a non-axisymmetric magnetic field in the core could in principle be revealed by the second order frequency splitting. If so, this should have strong implications on theories of stellar activity, and would give basic support to torsional oscillation driven cyclic activity.

Acknowledgements

I am very grateful to the colleagues who kindly sent me published or forthcoming papers or bibliography: J. Christensen-Dalsgaard, B. R. Durney, D. J. Galloway, G. A. Glatzmaier, R. E. Gershberg, M. S. Giampapa, P. A. Gilman, G. Godoli, D. O. Gough, D. F. Gray, J. Heyvaerts, R. Howard, K. Kodaira, P. Maltby, H. K. Moffatt, D. J. Mullan, A. Nordlund, R. W. Noyes, E. N. Parker, F. Praderie, E. Priest, K. H. Rädler, R. Rosner, A. A. Ruzmaikin, A. Skumanich, I. Tuominen, A. A. Van Ballegooijen, S. S. Vogt, J. C. Vial, N. O. Weiss, H. Wöhl, H. Yoshimura. I sincerely apologize to some authors whose papers I did not quote since they may have a better emphasis in other reviews of this special issue of Solar Physics.

I want to express my hearty thanks to Prof. T. G. Cowling for his letter of encouragement and to Keith Moffatt for encouragement and stimulating discussions during his stay in Catania.

I dedicate this work to Donatella and Alessandro.

References

Baliunas, S. L., Vaughan, A. H., Hartmann, L. W., Middelkoop, F., Mihalas, D., Noyes, R. W., Preston, G. W., Frazier, J., and Lanning, H.: 1983, *Astrophys. J.* **275**, 752.

Barnes, J. A., Sargent, H. H., and Tryon, P. V.: 1980, in R. O. Pepin, J. A. Eddy, and R. B. Merrill (eds.), *The Ancient Sun*, Pergamon Press, New York.

Belvedere, G.: 1983, in P. B. Byrne and M. Rodonó (eds.), 'Activity in Red Dwarf Stars', *IAU Colloq.* **71**, 579.

Belvedere, G. and Paternó, L.: 1977, *Solar Phys.* **54**, 289.

Belvedere, G. and Paternó, L. (eds.): 1984, *Mem. Soc. Astron. Ital.* **54**, 1.

Belvedere, G., Chiuderi C., and Paternó, L.: 1981, *Astron. Astrophys.* **96**, 133.

Belvedere, G., Chiuderi C., and Paternó, L.: 1982, *Astron. Astrophys.* **105**, 289.

Belvedere, G., Paternó, L., and Stix, M.: 1980a, *Geophys. Astrophys. Fluid. Dyn.* **14**, 209.

Belvedere, G., Paternó, L., and Stix, M.: 1980b, *Geophys. Astrophys.* **86**, 40.

Belvedere, G., Paternó, L., and Stix, M.: 1980c, *Geophys. Astrophys.* **88**, 240.

Belvedere, G., Paternó, L., and Stix, M.: 1980d, *Geophys. Astrophys.* **91**, 328.

Byrne, P. B.: 1983, in P. B. Byrne and M. Rodono (eds.) 'Activity in Red Dwarf Stars', *IAU Colloq.* **71**, 157.

Catalano, S.: 1984, in A. Mangeney and F. Praderie (eds.), *Space Research Prospects in Stellar Activity and Variability*, Obs. Meudon, p. 243.

Catalano, S., Frisina, A., and Rodonó, M.: 1980, in M. J. Pavlec, D. M. Popper, and R. K. Ulrich (eds.) *Close Binary Stars: Observations and Interpretations*, D. Reidel Publ. Co., Dordrecht, Holland, p. 405.

Christensen-Dalsgaard, J.: 1984, in A. Mangeney and F. Praderie (eds.), *Space Research Prospects in Stellar Activity and Variability*, Obs. Meudon, p. 11.

Cowling, T. G.: 1981, *Ann. Rev. Astron. Astrophys.* **19**, 115.

Dicke, R. H.: 1978, *Nature* **276**, 676.

Dicke, R. H.: 1979, *Nature* **280**, 24.

Dicke, R. H.: 1982a, *Nature* **300**, 693.

Dicke, R. H.: 1982b, *Solar Phys.* **78**, 3.

Durney, B. R. and Robinson, R. D.: 1982, *Astrophys. J.* **253**, 290.

Duvall, T. J. and Harvey, J. W.: 1984, *Nature* **310**, 19.

Duvall, T. J., Dziembowski, W. A., Goode, P. R., Gough, D. O., Harvey, J. W., and Leibacher, J. W.: 1984, *Nature* **310**, 22.

Eberhard, G. and Schwarzschild, K.: 1913, *Astrophys. J.* **38**, 292.

Galloway, D. J.: 1984, in *Nonlinear Plasma Astrophysics*, EPS Meet. Trends in Physics, Prague, MPA 155.

Galloway, D. J. and Proctor, M. R. E.: 1983, *Geophys. Astrophys. Fluid Dyn.* **24**, 109.

Galloway, D. J. and Weiss, N. O.: 1981, *Astrophys. J.* **243**, 945.

Gershberg, R. E.: 1983, in P. B. Byrne and M. Rodonó (eds.), 'Activity in Red Dwarf Stars', *IAU Colloq.* **71**, 487.

Giampapa, M. S.: 1985, *Astrophys. J. Letters* (in press).

Giampapa, M. S. and Rosner, R.: 1984, *Astrophys. J. Letters* **286**, L19.

Gilman, P. A.: 1982, in M. S. Giampapa and L. Golub (eds.), *Smithsonian Astrophys. Obs.*, SP 392, Vol. I, p. 165.

Gilman, P. A.: 1983a, in J. D. Stenflo (ed.) 'Solar and Stellar Magnetic Fields: Origin and Coronal Effects', *IAU Symp.* **102**, 247.

Gilman, P. A.: 1983b, *Astrophys. J. Suppl.* **53**, 243.

Gilman, P. A. and Miller, J.: 1981, *Astrophys. J. Suppl.* **46**, 211.

Glatzmaier, G. A.: 1984, *J. Comput. Phys.* **55**, 461.

Glatzmaier, G. A.: 1985a, *Astrophys. J.* (in press).

Glatzmaier, G. A.: 1985b, *Geophys. Astrophys. Fluid Dyn.* **31**, 137.

Godoli, G.: 1967, *Oss. Astrof. Catania Pubbl.* **115**, 224.

Goldberg, L.: 1983, in P. B. Byrne and M. Rodonó (eds.), 'Activity in Red Dwarf Stars', *IAU Colloq.* **71**, 653.

Golub, L., Maxson, C., Rosner, R., Serio, S., and Vaiana, G. S.: 1980, *Astrophys. J.* **238**, 343.

Golub, L., Rosner, R., Vaiana, G. S., and Weiss, N. O.: 1981, *Astrophys. J.* **243**, 309.

Gough, D. O.: 1981, in S. Sofia (ed.), *Variations of the Solar Constant*, NASA Conf. Publ. 2191, p. 185.

Gough, D. O.: 1983, *ESA Journal*, **7**, 325.

Gough, D. O.: 1984, *Mem. Soc. Astron. Ital.* **54**, 13.

Gray, D. F.: 1982a, *Astrophys. J.* **261**, 259.

Gray, D. F.: 1982b, *Astrophys. J.* **262**, 682.

Gray, D. F.: 1982c, *Mem. Soc. Astron. Ital.* **53**, 931.

Gray, D. F.: 1984, *Astrophys. J.* **277**, 640.

Hall, D. S.: 1981, in R. M. Bonnet and A. K. Dupree (eds.), *Solar Phenomena in Stars and Stellar Systems*, D. Reidel, Publ. Co., Dordrecht, Holland, p. 431.

Hartmann, L. W.: 1981, in R. B. Dunn (ed.), *Solar Instrumentation: What's Next?* Sac. Peak Nat. Obs., Sunspot NM, p. 170.

Hartmann, L. W.: 1983, in J. D. Stenflo (ed.), 'Solar and Stellar Magnetic Fields: Origin and Coronal Effects' *IAU Symp.* **102**, 419.

Hartmann, L. W., Baliunas, S. L., Duncan, D. K., and Noyes, R. W.: 1984, *Astrophys. J.* **279**, 778.

Howard, R. and LaBonte, B. J.: 1980, *Astrophys. J.* **239**, L33.

Kleeorin, N. I., Ruzmaikin, A. A., and Sokoloff, D. D.: 1983, *Astrophys. Space Sci.* **95**, 131.

Knobloch, E.: 1981a, *Astrophys. J. Letters* **247**, L93.

Knobloch, E.: 1981b, *Astrophys. J.* **248**, 1126.

Knobloch, E., Rosner, R., and Weiss, N. O.: 1981, *Monthly Notices Roy. Astron. Soc.* **197**, 45 p.

Kodaira, K.: 1983, in P. B. Byrne and M. Rodonó (eds.), 'Activity in Red Dwarf Stars', *IAU Colloq.* **71**, 561.

Köhler, H.: 1973, *Astron. Astrophys.* **25**, 467.

Krause, F.: 1984, *Astron. Nachr.* **305**, 281.

Krause, F. and Rädler, K. H.: 1980, *Mean Field Magneto-Hydrodynamics and Dynamo Theory*, Pergamon Press, Oxford.

LaBonte, B. J. and Howard, R.: 1982, *Solar Phys.* **75**, 161.

Layzer, D., Rosner, R., Doyle, H. T.: 1979, *Astrophys. J.* **229**, 1126.

Linsky, J. L.: 1981, in R. B. Dunn (ed.), *Solar Instrumentation: What's Next?* Sac. Peak Nat. Obs. Sunspot NM, p. 180.

Linsky, J. L.: 1983, in P. B. Byrne and M. Rodonó (eds.), 'Activity in Red Dwarf Stars', *IAU Colloq.* **71**, 39.

Mangeney, A. and Praderie, F.: 1984, *Astron. Astrophys.* **130**, 143.

Marcy, G. W.: in J. O. Stenflo (ed.), 'Solar and Stellar Magnetic Fields: Origin and Coronal Effects', *IAU Symp.* **102**, 3.

Marilli, E. and Catalano, S.: 1984, *Astron. Astrophys.* **133**, 57.

Middelkoop, F.: 1982, *Astron. Astrophys.* **107**, 31.

Moffatt, H. K.: 1978, *Magnetic Field Generation in Electrically Conducting Fluids*, Cambridge Univ. Press, Cambridge

Moffatt, H. K.: 1979, *Geophys. Astrophys. Fluid. Dyn.* **14**, 147.

Moffatt, H. K.: 1981, *J. Fluid. Mech.* **106**, 27.

Moffatt, H. K.: 1983, *Reports Progr. Phys.* **46**, 621.

Moffatt, H. K.: 1984a in T. Tatsumi (ed.), *Turbulence and Chaotic Phenomena in Fluids*, Elsevier Sci. Publ., Amsterdam, p. 223.

Moffatt, H. K.: 1984b, *Proc. of L'Ecole d'Astrophysique*, Goutelas, France.

Moffatt, H. K. and Kamkar, H.: 1983, in A. M. Soward (ed.) *Stellar and Planetary Magnetism*, Gordon and Breach, New York, p. 91.

Moffatt, H. K. and Proctor, M. R. E.: 1985, *Geophys. Astrophys. Fluid Dyn.* (to appear).

Mullan, D. J.: 1984, *Astrophys. J.* **279**, 746.

Mullan, D. T.: 1985, in H. R. Johnson and F. Querci (eds.), 'The Atmospheres of M. Stars. II. Theoretical Work: *M Stars Monograph Series on Non-Thermal Phenomena in Stellar Atmospheres*, (in press).

Nordlund, Å.: 1983, in J. O. Stenflo (ed.), 'Solar and Stellar Magnetic Fields: Origin, and Coronal Effects', *IAU Symp.* **102**, 79.

Nordlund, Å.: 1984, in T. D. Guyenne and J. J. Hunt (eds.), *Proc. 4th European Meeting on Solar Physics*, ESA SP-220, p. 37.

Noyes, R. W.: 1983, in J. O. Stenflo (ed.), 'Solar and Stellar Magnetic Fields: Origin and Coronal Effects', *IAU Symp.* **102**, 133.

Noyes, R. W., Hartmann, L. W., Baliunas, S. L., Duncan, D. K., and Vaughan, A. H.: 1984a, *Astrophys. J.* **279**, 763.

Noyes, R. W., Weiss, N. O., and Vaughan, A. H.: 1984b, *Astrophys. J.* **287**, 769.

Pallavicini, R., Golub, L., Rosner, R., Vaiana, G. S., Ayres, T., and Linsky, J. L.: 1981, *Astrophys. J.* **248**, 279.

Parker, E. N.: 1955a, *Astrophys. J.* **121**, 491.

Parker, E. N.: 1955b, *Astrophys. J.* **122**, 293.

Parker, E. N.: 1975, *Astrophys. J.* **198**, 205.

Parker, E. N.: 1979, *Cosmical Magnetic Fields*, Clarendon Press, Oxford.

Parker, E. N.: 1984a, *Astrophys. J.* **276**, 341.

Parker, E. N.: 1984b, *Astrophys. J.* **281**, 839.

Paternó, L.: 1984, in J. C. Pecker and Y. Uchida (eds.), *Active Phenomena in the Outer Atmosphere of the Sun and Stars*, CNRS-Obs., Paris, p. 343.

Paternó, L. and Zuccarello, F.: 1983, *Astrophys. J. Letters* **275**, L1.

Peckover, R. S. and Weiss, N. O.: 1978, *Monthly Notices Roy. Astron. Soc.* **182**, 189.

Pidatella, R. M., Stix, M., Belvedere, G., and Paternó, L.: 1985, 'The Role of Inhomogeneous Heat Transport and Anisotropic Momentum Exchange in the Dynamics of Stellar Convection Zones: Application to Models of the Sun's Differential Rotation', *Astron. Astrophys.* (in press).

Piddington, J. H.: 1981, *Astrophys. J.* **247**, 293.

Preston, G. W.: 1971, *Astrophys. J.* **164**, 309.

Rädler, K. H.: 1983a, in A. M. Soward (ed.), *Stellar and Planetary Magnetism*, Gordon and Breach, New York, p. 17.

Rädler, K. H.: 1983b, in A. M. Soward (ed.), *Stellar and Planetary Magnetism*, Gordon and Breach, New York, p. 37.

Robinson, R. D.: 1980, *Astrophys. J.* **239**, 261.

Robinson, R. D. and Durney, B. R.: 1982, *Astron. Astrophys.* **108**, 322.

Rodonó, M.: 1981, in E. B. Carling and Z. Kopal (eds.), *Photometric and Spectroscopic Binary Systems*, D. Reidel Publ. Co., Dordrecht, Holland, p. 285.

Rodonó, M.: 1983, *Adv. Space Res.* **2**, 225.

Rodonó, M.: 1985a, in L. V. Mirzoyan (ed.), 'Coordinated Ground-Based and Space Observations of UV Cet-Type Flare Stars', *Byurakan Symp. on Flare Stars and Related Objects*, Armenian Acad. Sci. (in press).

Rodonó, M.: 1985b, in H. R. Johnson and F. Querci (eds.), 'The Atmospheres of M Stars. I. Observations', *M Stars Monograph Series on Non-Thermal Phenomena in Stellar Atmospheres*, (in press).

Rosner, R.: 1980, in A. K. Dupree (ed.), 'Cool Stars, Stellar Systems and the Sun', *Smithsonian Astrophys. Obs.*, SP 389, 79.

Rosner, R.: 1983a, in J. O. Stenflo (ed.), 'Solar and Stellar Magnetic Fields: Origin and Coronal Effects', *IAU Symp.* **102**, 279.

Rosner, R.: 1983b, *Adv. Space Res.* **2**, 3.

Ruzmaikin, A. A.: 1981, *Comments Astrophys.* **9**, 85.

Schmitt, J. H. M. M., Rosner, R., and Bohn, H. U.: 1984, *Astrophys. J.* **282**, 316.

Schüssler, M.: 1984, in T. D. Guyenne and J. J. Hunt (eds.), *Proc. 4th European Meeting on Solar Physics*, ESA SP-220, p. 67.

Skumanich, A. and Eddy, J. A.: 1981, in R. M. Bonnet and A. K. Dupree (eds.), *Solar Phenomena in Stars and Stellar Systems*, D. Reidel, Publ. Co., Dordrecht, Holland, p. 349.

Skumanich, A. and Young, A.: 1984, in B. J. LaBonte, G. A. Chapman, H. S. Hudson, and R. C. Willson (eds.), *Solar Irradiance Variation on Active Region Time-Scales*, NASA Conf. Publ. 2310.

Skumanich, A., Lean, J. L., White, O. R., and Livingston, W. C.: 1984, *Astrophys. J.* **282**, 776.

Snodgrass, H. B. and Howard, R.: 1984, *Astrophys. J.* **284**, 848.

Spruit, H. C.: 1981a, in R. M. Bonnet and A. K. Dupree (eds.), *Solar Phenomena in Stars and Stellar Systems*, D. Reidel Publ. Co., Dordrecht, Holland, p. 289.

Spruit, H. C.: 1981b, in S. D. Jordan (ed.), *The Sun as a Star*, NASA SP-450, p. 385.

Spruit, H. C. and Van Ballegooijen, A. A.: 1982, *Astron. Astrophys.* **106**, 58.

Steenbeck, M. and Krause, F.: 1969a, *Astron. Nachr.* **291**, 49.

Steenbeck, M. and Krause, F.: 1969b, *Astron. Nachr.* **291**, 271.

Steenbeck, M., Krause, F., and Rädler, K. H.: 1966, *Z. Naturforsch.* **21a**, 369.

Stern, R. A. and Skumanich, A.: 1983, *Astrophys. J.* **267**, 232.

Stern, R. A., Zolcinsky, M. R., Antiochos, S. K., and Underwood, J. M.: 1981, *Astrophys. J.* **249**, 647.

Stix, M.: 1976a, in V. Bumba and J. Kleczek (eds.), 'Basic Mechanisms of Solar Activity', *IAU Symp.* **71**, 367.

Stix, M.: 1976b, *Astron. Astrophys.* **47**, 243.

Stix, M.: 1977, *Astron. Astrophys.* **59**, 73.

Stix, M.: 1981, *Solar Phys.* **74**, 79.

Stuiver, M.: 1980, *Nature* **286**, 868.

Tuominen, I. and Virtanen, H.: 1984, *Astron. Nachr.* **305**, 225.

Tuominen, I., Krause, F., Rüdiger, G., and Virtanen, H.: 1984, *Proc. Nordic Astron. Meeting*, Helsinki.

Tuominen, J., Tuominen, I., and Kyröläinen, J.: 1983, *Monthly Notices Roy. Astron. Soc.* **205**, 691.

Vaiana, G. S.: 1981, *Space Sci. Rev.* **30**, 151.

Vaiana, G. S.: 1983, in P. B. Byrne and M. Rodonó (eds.) 'Activity in Red Dwarf Stars', *IAU Colloq.* **71**, 651.

Vainshtein, S. I., Zeldovich, Ya. B., and Ruzmaikin, A. A.: 1980, *Turbulent Dynamo in Astrophysics*, Moscow (in Russian).

Van Ballegooijen, A. A.: 1982a, *Astron. Astrophys.* **106**, 43.

Van Ballegooijen, A. A.: 1982b, *Astron. Astrophys.* **113**, 99.

Van Ballegooijen, A. A.: 1983, *Astron. Astrophys.* **118**, 275.

Van Ballegooijen, A. A. and Title, A. M.: 1985, 'A Boundary-Layer Model of the Solar Dynamo', preprint.

Vaughan, A. H.: 1980, *Publ. Astron. Soc. Pacific* **92**, 392.

Vaughan, A. H. and Preston, G. W.: 1980, *Publ. Astron. Soc. Pacific* **92**, 385.

Vaughan, A. H., Baliunas, S. L., Middelkoop, F., Hartmann, L. W., Mihalas, D., Noyes, R. W., and Preston, G. W.: 1982, *Astrophys. J.* **250**, 276.

Vilhu, O.: 1984, *Astron. Astrophys.* **133**, 117.

Vogt, S. S.: 1983, in P. B. Byrne and M. Rodonó (eds.), 'Activity in Red Dwarf Stars', *IAU Colloq.* **71**, 137.

Vogt, S. S. and Penrod, G. D.: 1983, *Publ. Astron. Soc. Pacific* **95**, 565.

Vogt, S. S., Soderblom, D. R., and Penrod, G. D.: 1983, *Astrophys. J.* **269**, 250.

Wallenhorst, S. G.: 1982, *Solar Phys.* **80**, 379.

Walter, F.: 1981, *Astrophys. J.* **245**, 677.

Walter, F.: 1982, *Astrophys. J.* **253**, 741.

Walter, F. and Bowyer, S.: 1981, *Astrophys. J.* **245**, 671.

Weiss, N. O.: 1966, *Proc. Roy. Soc.* **A293**, 310.

Weiss, N. O.: 1981, in R. M. Bonnet and A. K. Dupree (eds.), *Solar Phenomena in Stars and Stellar Systems*, D. Reidel Publ. Co., Dordrecht, Holland, p. 499.

Weiss, N. O., Cattaneo, F., and Jones, C. .: 1984, *Geophys. Astrophys. Fluid Dyn.* **30**, 305.

Williams, G. E.: 1981, *Nature* **291**, 624.

Wilson, O. C.: 1978, *Astrophys. J.* **226**, 379.

Yoshimura, H.: 1975a, *Astrophys. J. Suppl.* **29**, 467.

Yoshimura, H.: 1975b, *Astrophys. J.* **201**, 740.

Yoshimura, H.: 1978a, *Astrophys. J.* **220**, 692.

Yoshimura, H.: 1978b, *Astrophys. J.* **226**, 706.

Yoshimura, H.: 1981, *Astrophys. J.* **247**, 1102.

Yoshimura, H.: 1983, *Astrophys. J. Suppl.* **52**, 363.

Yoshimura, H.: 1985, *Publ. Astron. Soc. Japan* **37**, 171.

Yoshimura, H., Wang, Z., and Wu, F.: 1984a, *Astrophys. J.* **280**, 865.

Yoshimura, H., Wang, Z., and Wu, F.: 1984b, *Astrophys. J.* **283**, 870.

Yoshimura, H., Wang, Z., and Wu, F.: 1984c, *Astrophys. J.* **285**, 325.

Young, A., Skumanich, A., and Harlan, E.: 1984, *Astrophys. J.* **282**, 683.

STELLAR ANALOGS OF SOLAR MAGNETIC ACTIVITY

ROBERT W. NOYES

Harvard–Smithsonian Center for Astrophysics, Cambridge, Mass., U.S.A.

Abstract. The techniques and principal results of observational studies of stellar activity are summarized. Both chromospheric and coronal emission clearly track surface magnetic field properties, but it is not well known how the detailed relation between the emission and surface magnetic fields varies with spectral type. For lower Main-Sequence stars of the same spectral type, there is clear evidence of a close relationship between mean activity level and rotation period P_{rot}. There is also less definitive evidence for a similar dependence on convective overturn time τ_c, such that activity depends on the single parameter Ro = P_{rot}/τ_c. For single stars, stellar rotation, and magnetic activity both decline smoothly with age. This implies a feedback between angular momentum loss rate and activity level. Temporal variations in mean stellar activity level mimic the solar cycle only for old stars like the Sun, being much more irregular for younger stars. The characteristic timescale of the variations (the 'cycle period') appears to depend on Ro for old stars, but shows no clear dependence on either rotation rate or spectral type for younger stars. Further data on mean activity and its variation for a large number of lower Main-Sequence stars should contribute significantly to our understanding of the causes of stellar magnetic activity.

1. Introduction

Observational research into the 'solar-stellar connection' has blossomed in recent years, as solar phenomena have been increasingly more thoroughly studied on other stars, and as this study has produced fresh insights into solar physics. Nowhere has progress been more evident than in the study of stellar magnetic activity. Freed of the restriction of studying only the activity of our own star, with its particular rotation rate and degree of convection, we have been able to explore in some detail just how magnetic activity depends on rotation and convection. This is leading to a firmer empirical foundation for theories of dynamo generation of solar and stellar magnetic fields. At the same time, stellar observations now show rather clearly that the Sun's rotation and activity are not substantially different from that of other G2V stars of the same age. This suggests that activity on solar mass stars of different ages can provide a rather reliable indication of the past and future history of solar activity.

It is generally believed that solar and stellar activity originates in a magnetic dynamo powered by the interaction of solar rotation and convection (see Belvedere, 1985). This interaction produces differential rotation within the solar convection zone; the differential rotation in turn, together with the convective motions creates the internal field amplification, transport, and dissipation responsible for surface magnetic fields and the 22-year magnetic cycle. Similar processes should occur in other rotating stars with outer convection zones. Stellar interior theory predicts that, like the Sun, all Main-Sequence stars later than about F0 have outer convection zones, whose thickness as a fraction of their radius increases steadily with advancing spectral type, until stars later than about mid-M are fully convective (e.g., Copeland *et al.*, 1970). Thus these stars are obvious candidates to display the same surface magnetic phenomena that delight and challenge solar physicists.

Solar Physics **100** (1985) 385–396. 0038–0938/85.15.
© 1985 *by D. Reidel Publishing Company*

Other late-type stars in addition to those on the Main-Sequence have both convection zones and significant rotation, and therefore might be expected to exhibit magnetic activity. These include the pre-Main-Sequence T Tauri stars, as well as post-Main-Sequence subgiant and giant stars. In this context, it is interesting that the same relation seen between equatorial rotation velocity and chromospheric or coronal emission seems to hold for all these stellar classes alike (Pallavicini *et al.*, 1981). This provides a suggestive, albeit circumstantial, indication that the underlying physical processes generating stellar magnetic fields are similar for all late-type stars with convection and rotation.

It has been known for many years that the overall level of stellar magnetic activity declines with age in Main-Sequence stars (for a review, see Skumanich and Eddy. 1981). However, the decline of activity is not simply a result of the aging of a star. This is nicely shown by the binary RS CVn stars, whose rotation rates, being tidally locked to the orbital period of the system, do not decline with age as rapidly as those of single stars: the activity level of RS CVn's is similar to that of younger single stars with the same rotation rate and, hence, larger than that of single stars of the same age. Thus rotation, and not age, appears to be the fundamental parameter controlling stellar activity. The decline of rotation with age for single stars is generally thought to be caused by stellar winds, which carry off angular momentum despin the star, and thus cause the magnetic activity to decrease. Feedback can occur because the stellar wind torques themselves are dependent upon magnetic activity. This leads to a regulation of the rotation speed, so that stars of a given mass settle down to a rotation rate that depends only on their age.

This general scenario of the evolution of stellar magnetic activity has been broadly discussed in recent years, and is now rather widely accepted, based on a growing body of supporting evidence. In Section 2 of this review we discuss the measurements that have led to this scenario, and are now leading to a rather more detailed picture. Some aspects of this more detailed picture are discussed in Section 3.

2. The Observational Study of Stellar Magnetic Activity

Observationally, the study of stellar magnetic activity involves both the characterization of that activity and also the measurement of the rotation and convection upon which activity appears to depend. In this section we discuss these two aspects in turn.

2.1. MAGNETIC ACTIVITY INDICATORS AND THEIR VARIATION

Probably the most widely used indicator of stellar magnetic activity is the flux in the Ca II H and K line cores (hereafter denoted 'H and K'). On the Sun the spatial location and the strength of H and K emission is well known to correlate closely with surface magnetic flux (Leighton, 1959; Skumanich *et al.*, 1975). The H and K emission in other lower Main-Sequence stars also appears to be tied to surface magnetic activity, as is qualitatively shown by the self-consistency of the rotation-activity-age scenario when H and K are used as a measure of activity. More direct support is found in the correlations seen between H and K emission enhancements and photometric or spectroscopic indications of starspots (e.g., Vogt, 1983), and of area covered by photospheric fields (Marcy, 1983). The correlations are consistent with the picture, based on the solar analogy, that stellar activity is concentrated in localized active regions, containing enhanced chromospheric emission, starspots, and large magnetic flux; the correlated variations in the stellar data are produced as stellar rotation brings active regions on or off the facing hemisphere. Also, the remarkably solar-like long-term variations of H and K emission found by Wilson (1978) for many slowly-rotating lower Main-Sequence stars show rather convincingly that these stars undergo magnetic activity cycles similar to the Sun's, with H and K providing a good indicator of mean surface fields.

However, to relate H and K emission to direct measures of magnetic activity, such as the total magnetic flux Φ (independent of sign) through a stellar photosphere, requires quantitative knowledge of the relation of H and K emission to Φ for stars differing from the Sun in spectral type, rotation rate, or age. Here there is little guidance from solar physics, and theoretical predictions, relating chromospheric heating to Φ and convective properties (e.g., Ulmschneider and Stein, 1982) are not definitive. As a result different empirical approaches have been tried, leading to somewhat different conclusions. For example, Catalano and Marilli (1983) have noted a rather tight relation between the surface rotation period P_{rot} and F_{HK} for lower Main-Sequence stars, where F_{HK} is the flux in the H and K line cores normalized to the stellar surface area. This relation is independent of spectral type. On the other hand Noyes *et al.* (1984a) have found a different rotation-activity relation that does depend on spectral type. They find that R_{HK}, the ratio of F_{HK} to the bolometric luminosity per unit area σT^4, varies tightly with the dimensionless ratio P_{rot}/τ_c. Here τ_c is a characteristic convective turnover time near the bottom of the convection zone. It increases rapidly with advancing spectral type for stars with relatively thin convective envelopes, but more slowly for later types with thicker envelopes; its detailed dependence on spectral type is a function of convective efficiency (Gilman, 1981). The ratio P_{rot}/τ_c, known as the Rossby number, measures the influence of coliolis forces upon convection. It is an attractive idea that chromospheric activity depends on the Rossby number, since that ratio is a measure of the efficiency of dynamo generation of magnetic fields (e.g., Durney and Latour, 1978; Noyes *et al.*, 1984).

However, because the observational correlations found using the two different normalizations described above are both reasonably tight, its is difficult to choose either F_{HK} or R_{HK} as the more 'fundamental' magnetic activity indicator purely on empirical grounds. Also, dynamo theory gives little reason to prefer one measure of activity over the other. Nevertheless the physical interpretations (that is, either independence of magnetic activity on spectral type or an explicit dependence on spectral type through the convection overturn time τ_c) are critically different in their implications for the causes of magnetic activity.

The question is complicated further when we consider whether H and K emission provides the same measure of chromospheric heating in other late-type stars as it does for the Sun. In the Sun H and K emission makes up (together with the similar Mg II h and k emission) the major part of the non-thermal losses from the chromosphere. In later-type stars, however, Hα produces an increasingly large fraction of the emission due to non-radiative heating. If non-radiative heating is taken as the fundamental indicator of stellar magnetic activity, then Hα and other major sources of energy loss must be included. Studies by Herbig (1985), Simon *et al.* (1985), and others show that Hα does indeed share most of the correlations with activity shown by H and K. Clearly, for quantitative estimates of non-radiative heating, one must measure all significant chromospheric emissions, with due consideration of the variation of their relative importance with spectral type.

Even more fundamentally, there is no *a priori* justification for treating chromospheric heating as a direct measure of magnetic activity level. The relation between total chromospheric losses and surface magnetic field distribution is poorly known for stars other than the Sun because our empirical knowledge of stellar magnetic fields (e.g., Marcy, 1983) is very primitive. Furthermore, the relation between the surface field distribution and the deep-seated magnetic dynamo is still more obscure. The steps from dynamo generation of magnetic fields deep in the interior to the emission of chromospheric radiation are many and complex, probably including: formation of concentrated flux tubes near the bottom of the convection zone, their rise to the surface, their buffeting by motions in the convection zone, upward propagation of the magnetic energy thus produced, disipation of that energy in the chromosphere, and conversion of the deposited heat into non-LTE emission from various atomic species. In practice, the justification for the use of chromospheric loss rates to indicate magnetic dynamo activity is an *a posteriori* one based on detailed solar correlations and less detailed stellar ones. In this situation, the researcher must be on guard against allowing pre-conceptions to influence conclusions.

A different indicator of stellar magnetic activity is the soft X-ray flux, about which there now exists much data from *Einstein* and other satellites. The X-ray flux from lower main sequence stars is also found to be tightly-dependent on rotation (e.g., Pallavicini *et al.*, 1981); this is scarcely surprising given the well-known dependence of solar X-ray emission on underlying magnetic structures. It is more surprising, perhaps, that the surface-averaged X-ray flux F_X is well-correlated with rotation, and with F_{HK}, over four orders of magnitude in F_X, and for both giants and dwarfs (Schrijver and Rutten, 1985, Pallavicini *et al.*, 1981).

In the Sun, soft X-ray emission, like H and K, is controlled by the photospheric magnetic field pattern, but in a rather different way. Whereas the spatial pattern of chromospheric emission is nearly identical to that of the photospheric field (at least on scales of an arc second and larger), that of coronal X-ray emission depends not only on the location of photospheric footpoints, but also on the geometry of the field lines emerging from the photosphere; this geometry determines the size scale of coronal loops and their total emission measure (Rosner *et al.*, 1978). From stellar X-ray data, then, we may in principle obtain information about surface magnetic fields complementing and extending that from chromospheric lines. For example, in the Sun the ratio EM_x/EM_{ch}, or the ratio of emission measure in a coronal loop to that from its chromospheric footpoints, depends on the size of the loop or equivalently the separation of its footpoints. If as suspected active regions in younger more rapidly rotating stars have higher filling factors and higher chromospheric surface brightness, then the ratio EM_x/EM_{ch} should be larger for these stars. This in fact is found to be the case (Schrijver, 1983; Schrijver and Rutten, 1985).

In principle direct measurement of stellar photospheric magnetic fields can provide a direct probe of stellar magnetic activity, and at the same time should allow a better calibration of chromospheric indicators of magnetic activity. Unfortunately, the mean field of a Sun-like star is only on the order of one gauss, because fields of opposite sign in the photosphere nearly cancel. The resulting Zeeman shift of spectral lines is not detectable by current techniques. However, as shown by Robinson et al. (1980), the broadening of Zeeman-sensitive lines, which is independent of the sign of the field, is measurable if the area occupied by magnetic fields is a sufficient fraction of the stellar surface. Using this measure, Marcy (1983) and other investigators have attempted to determine both the field strength and the area coverage of active regions on the surface of a number of late-type dwarfs. This technique appears promising, but only for stars with high magnetic activity, so that strong fields cover about 5% or more of the surface area–some five times the value for the Sun.

An unambiguous indicator of stellar magnetic activity is provided by starspots, which are now well-enough studied that their physical similarity to sunspots is very certain. Starspots have been studied most intensively in late dMe stars and rapidly rotating (for their spectra class) RS CVn stars, both of which show spot coverage very much larger than the Sun. Recently, other stars as early as the Sun but younger and more rapidly rotating have also shown the presence of spots, again with coverage considerably higher than the Sun. This dependence on rotation and spectral type appears qualitatively to be an association of large spot coverage with small Rossby number Ro $= P_{rot}/\tau_c$, at least for stars of spectral class G or later. To exemplify this, we note that spots on solar-type stars rotating more rapidly than the Sun (smaller P_{rot}, implying smaller Ro) may cover several percent of the surface area (Lockwood et al., 1984) compared to a coverage of about 0.1% for the Sun at maximum. Spots on slowly-rotating stars but of later-type (larger τ_c, also implying smaller Ro) such as 61 Cyg A (K5, $P_{rot} = 38$ days) also may have coverage of several percent (Dorren and Guinan, 1982). Finally, spots on rapidly rotating late type stars (smaller P_{rot} and larger τ_c, implying very much smaller Ro) such as BY Dra (MO, $P_{rot} \sim 4$ days) may have coverage of several tens of percent (Vogt, 1983).

Vogt and Penrod (1983a, b) have implemented a very useful technique for studying starspots known as Doppler imaging. This technique, which is best applied to rapidly rotating stars, utilizes the fact that stellar rotation causes the contribution from photospheric regions at different projected distances from the star's rotation axis to be located at correspondingly Doppler-shifted locations in the line profile. Doppler imaging permits mapping of starspot size, location in longitude and latitude (for a star with suitably inclined rotation axis), and in some cases even shape. Applying this technique to the RS CVn star HR 1099, they found spot characteristics for the K1 IV primary star that are very different from sunspots; a large dark area was deduced to lie essentially at the rotation pole, with a long extension toward the equator at nearly constant longitude. The geometry in this case was more like a solar coronal hole than a sunspot. Vogt and Penrod have suggested that such a dark area may be more like a solar complex of activity, within which individual spots may grow and decay.

Similar ideas have been suggested by Uchida (1984), who noted that the drift in latitude and longitude of the large dark areas on RS CVn's over many years is not at all like sunspots (which grow and decay *in situ*) but rather similar to the solar drift of complexes of activity in both longitude (cf. Gaizauskas *et al.*, 1983) and latitude (as is seen in the Sun in the well-known butterfly diagram). This picture also helps to explain the near-rigid rotation rate of the giant dark areas deduced by modelling of their latitude and longitude variation with time, which shows much less than the expected stellar differential rotation rate but is in accord with the rotation of solar activity complexes (and coronal holes).

The apparently discrepant behaviour of giant starspots and sunspots is perhaps not surprising, given the enormously greater area coverage of such starspots. Nevertheless, the correlations between strong chromospheric and coronal emission, large magnetic field coverage, and the presence of large spots supports the conclusion that we are dealing with qualitatively similar processes of surface magnetic activity and atmospheric heating in the Sun and other rotating stars with outer convection zones.

2.2. EMPIRICAL STUDIES OF STELLAR CONVECTION AND ROTATION

If, as theory suggests and observations appear to confirm, magnetic activity depends strongly on rotation and convection, it is important to obtain the fullest information possible on these properties for other stars of various ages, spectral types, and levels of chromospheric or coronal activity.

The measurement of stellar rotation is relatively straightforward in concept, especially if one restricts the investigation to the rotation rate averaged over the visible surface of the star. Two well-proven methods are: (a) determination of projected equatorial velocity $v \sin i$ (where i is the inclination of the rotation axis to the line-of-sight), through measurement of rotational line broadening, and (b) direct measurement of rotation periods from rotational modulation of light curves by localized surface features, such as starspots or chromospheric active regions. These methods have their individual advantages: the first is relatively fast (requiring only a single observation), and works even if there is little inhomogeneity over the stellar surface; the second is more precise, especially for velocities less than a few km s^{-1}, and avoids the ambiguity of sini. Together the two complementary methods have provided the basic rotation data underlying the rotation-activity connection.

However, differential rotation with latitude and depth, in addition to rotation itself is important for the stellar dynamo mechanism. Here the stellar measurements become much more difficult. The latitude dependence of differential rotation is well-explored on the Sun, and helioseismology now is providing the first information on its depth dependence (e.g., Gough, 1985). Comparable data for other stars would be of enormous value. Some information about surface differential rotation has been obtained from modelling of starspot light curves (see, e.g., Vogt, 1981, Dorren and Guinan, 1984), in which it is inferred that starspots at greater latitudes have slower rotation rates – that is, the angular velocity is found to decrease toward the poles, as for the Sun. In addition, there is evidence from H and K rotational modulation curves of some lower Main-Sequence stars that two or more active regions are simultaneously present and rotating at different rates (Baliunas *et al.*, 1985); however, the extraction of a latitude-dependence of differential rotation from such data is very difficult (Gilliland, 1984; Gilliland and Fisher, 1985). This is because the H and K emission, if like that on the Sun, is widely distributed over the stellar surface (in quiet as well as active areas), so that a very high signal-to-noise ratio in the data is required for successful inversion to yield latitude information. The difficulties are compounded by evolution of surface activity over several rotation periods.

It should in principle be possible to extract differential rotation data simply from variation of rotational modulation periods as stars undergo their activity cycles. For example the mean rotation period from rotational modulation of solar activity indicators like spots, plages seen in H and K, radio emission, etc., might be expected to decrease late in the activity cycle when active regions are closer to the more rapidly rotating equator. Unfortunately, however, the effect on the Sun is too small to rise above the noise introduced by growth and decay of activity (Labonte, 1984).

Finally, one may hope to measure the depth-dependence of differential rotation through stellar oscillations, just as solar oscillation data are yielding the internal rotation of the Sun (see Christensen-Dalsgaard, 1984). The fact that stellar data can give oscillation frequencies, and splittings, at most only for the lowest-degree oscillation modes severely constraint this possibility. It may not be too much to hope, however, that in the future at least such data may be obtainable; when compared with surface rotation periods and latitude profiles deduced as described above, some knowledge of the depth-dependence of rotation may indeed be obtained.

Convection in stars is the second property whose direct study is important for understanding stellar activity. In the Sun, convection is directly manifested by the readily visible granulation and supergranulation. In the Sun seen as a star, the influence of granulation on absorption line profile asymmetries is detectable (Dravins and Linds, 1984), so that granulation should be indirectly observable on other stars as well. Dravins and Linds find that on α Cen A, a near-twin of the Sun, the asymmetry of photospheric spectral lines is very similar to that for the Sun, and that other F and K Main-Sequence stars also show rather similar behaviour. However, F and G giants show very different patterns of line asymmetry, perhaps due to larger surface convection cells with different radial temperature or velocity profiles. In addition, small blue shifts of high-excitation lines relative to low-excitation lines are seen in the solar spectrum (Dravins and Larsson, 1984) due to the preferential strengthening of high-excitation lines in rising hot granules; this effect should be visible in other stars and could give further information on stellar granulation. Gray (1982; see also Gray and Toner, 1985) finds from observed stellar line asymmetries that stellar granular velocities increase with increasing effective temperature for F and G Main-Sequence stars, and that the velocity dispersion increases as well.

The question of the stellar analogues to supergranulation is an important one, in view of the major role played by supergranulation in structuring photospheric magnetic fields and the overlying chromosphere and corona. Unfortunately, we do not understand solar supergranulation well enough even to explain its characteristic size scale, so any attempt to predict the properties of stellar supergranulation seems premature. It would be even more difficult to observe stellar analogs of the solar giant cells thought to play the fundamental role of convection in dynamo activity, given that their detection in the Sun is preliminary at best. Thus our knowledge of stellar convective properties relevant to magnetic activity generation remains principally theoretical.

One attribute of convection particularly important for dynamo generation of magnetic fields is the convective turnover time τ_c near the bottom of the convection zone; its theoretically-calculated value depends sensitively upon the assumed depth of the convection zone, or equivalently upon the convective efficiency that determines that depth (see, for example, Gilman, 1981). This efficiency is frequently parameterized by the ratio $\alpha = L/H$, where L is the mixing length and H the pressure scale height. While τ_c or α cannot be measured directly, they can be inferred, in the Sun at least, from observational data pertaining to interior structure. In particular (see, e.g., Gough, 1985), helioseismology data have been used to determine the depth of the solar convection zone to be $D_c \sim 200\,000$ km, or $D_c/R_0 \sim 0.3$, consistent with a value of α near 1.5 (Berthomieu et al., 1980, Lubow et al., 1980). This value is in keeping with independent estimates of α based on comparison of stellar evolution calculations and the known parameters of the present-day Sun (Gehren, 1982).

In principle, similar information can be obtained for the convective envelopes of other stars from measurement of their global oscillations (Christensen-Dalsgaard, 1984), but just as in the case for seismic probing of the internal rotation of stars, the problems are formidable. Interpretation of measured frequencies in terms of interior structure will be greatly aided by independent information on the stellar radius, mass, and age, but this information is difficult to obtain for most stars.

3. Current Understanding of Stellar Magnetic Activity

From observations of the sort described above, we are led to a picture of stellar magnetic activity that is in rough accord with the general concepts of dynamo mtheory – namely, a mean level of activity that depends on rotation and convection through the ratio of their characteristic time-scales P_{rot} and τ_c. Of course it would be remarkable (and unlikely) that such a complex process as stellar activity could be accurately parameterized by such a simple ratio, but this parameterization may be a useful first approximation.

At a somewhat deeper level, the stellar data reveal numerous complexities, which may in time lead to deeper understanding of the processes involved. One of these is the Vaughan–Preston 'gap' in chromospheric emission (Vaughan and Preston, 1980) which appears at a specific activity level, or equivalently at a specific (mass-dependent) rotation rate. The gap might be caused by a sudden decrease in the efficiency of dynamo production of magnetic fields at a certain critical rotation rate (Durney *et al.*, 1981; Knobloch *et al.*, 1981); if so, the implied difference in dynamo activity above and below the gap places an important constraint on any successful dynamo theory. On the other hand the gap could be simply a reflection of some past stellar birthrate fluctuation in the solar neighbourhood, or even a statistical fluctuation within the relatively small sample of stars studied (Hartmann *et al.*, 1984). There is no evidence that the dependence of overall chromospheric activity level on Ro is different for stars above and below the gap (Noyes *et al.*, 1984a), although more data are necessary for a definitive determination on this point.

There does, however, appear to be a significant difference in the long-term variation of chromospheric activity between stars above and below the gap – that is, relatively rapidly and slowly rotating stars. Over time-scales measured in years, stars above the gap tend to show irregularly varying activity, while those below the gap tend to show smoothly varying, repetitive variability, much like the solar cycle (Vaughan, 1980). This finding is based on observations of a relatively small sample of stars, extending over about fifteen years – an impressive time span in comparison with most astronomical observations but rather short for the study of Sun-like cycles. Thus, whether the gap represents a true demarcation line is still not clear.

Differing long-term activity behaviour for rapidly rotating and slowly rotating Main-Sequence stars, however, might be expected from dynamo theory. For example, Parker (1971) noted that a star with rapid rotation, that is, a star with large dynamo number or equivalently small Rossby number, could generate magnetic fields in higher modes than the lowest-degree mode characteristic of rotation just above the threshold for dynamo activity. The interference of several dynamo modes, with different characteristic periods and surface geometries, would produce activity that varies relatively irregularly in time, as is observed for the more rapid rotating stars in Wilson's sample.

Interestingly, for most stars there do not appear to be striking correlations between activity cycle period or amplitude, and stellar rotation rates (Balianas and Vaughan, 1985). For slowly-rotating stars only, however, Noyes *et al.* (1984b) have noted that those with the most clearly-defined cycles show a correlation between cycle period and Rossby number, so that more rapidly rotating stars of the same mass tend to have shorter cycle periods. This result is consistent with dynamo models in which magnetic flux build-up is limited by losses due to magnetic buoyancy. However, more data, extending over a number of years, are needed to firmly establish this condition, or to determine how cycle properties depend on rotation for more rapidly rotating stars (see also Belvedere, 1985).

In addition to the general decline of activity level as a star ages and spins down, the surface properties of individual magnetic activity regions appear to change as a star ages. For example, Simon *et al.* (1985) find from analysis ;of extreme ultraviolet stellar spectra not only that the fractional area covered by plages declines as a star ages, but that their surface brightness declines as well; for very young stars the surface brightness in C IV is more than three times solar. Also, the ratio of 1700 Å continuum flux to the bolometric flux indicates substantial non-thermal heating of the upper photosphere, by some 300 K. Presumably these effects reflect the changing configuration of the surface magnetic field (characterized for example by the filling factor) as the magnetic dynamo behaviour evolves.

All the manifestations of stellar activity discussed above appear to be controlled by the interaction of stellar rotation and convection. Another effect of solar and stellar magnetic activity, discussed in this volume by Hartmann (1985), is mass and angular momentum loss by stellar winds. This effect produces feedback upon the activity itself, for the angular momentum loss causes the rotation gradually to decrease. In principle, observations of rotation rate as a function of age for stars of different mass yield the stellar angular momentum loss rate as a function of stellar mass and rotation. This of course requires that angular momentum loss *is* a unique function of these tow parameters. Some evidence that this is actually the case is found in measured rotation periods for stars of the same age, such as the (presumably coeval) members of open clusters. Stars in the Hyades cluster (age about 7×10^8 yr) are found to have a well-defined mass-dependent rotation period with relatively small scatter (Lockwood *et al.*, 1984; see also Noyes, 1983; Duncan *et al.*, 1984); the mean rotation period increases smoothly from about 2 days for F5 stars to about 12 days for K0 stars. This suggests that some regulating mechanism is at work early in the lifetime of a star, perhaps even in its pre-Main-Sequence stage, which erases any effects of initial conditions in setting the star's initial rotation rate.

If the rotation period of lower main sequence stars is in fact a function only of their mass and age, it is an important challenge to discover that function, for it would yield extremely important information about the evolution of the tellar winds that appear to control angular momentum loss. At the moment, we have only crude knowledge of the rate of stellar spindown. The now-classic work of Skumanich (1972) which yielded stellar rotation rates decreasing with age as $T^{-1/2}$ appears to be in need of some correction as new data accumulate on rotation rates in young clusters. Barry *et al.* (1984) infer and exponential decay of Ca II emission with age; similar results are found for extreme ultraviolet chromospheric and transition zone lines by Simon *et al.* (1985). The latter authors, however, found that the characteristic braking time, $\tau_0 = -\omega/\dot{\omega}$, increases as a star ages from about 1×10^9 yr for young (less than the age of the Hyades) to at least several times 10^9 yr for stars of solar age. Future measurements of rotation rates for stars of different ages and masses should be of considerable importance for exploring the evolution of stellar winds and angular momentum loss.

4. Summary and Conclusions

The subject of solar and stellar magnetic activity and its evolution is a rapidly growing one. It is profiting from the convergence of several lines of observational and theoretical reasoning, including: (a) stellar activity monitoring, from both ground (broadband photometry, chromospheric photometry), and space (IUE, Einstein); (b) helioseismology and potentially astroseismology, which present new opportunities to study the structure and dynamics of stellar interiors; and (c) theoretical advances in our understanding of convection, rotation, and dynamos. The expected availability of more powerful space instruments will be a major stimulus for future progress in the field. Prospects for ground-based research are equally bright, but will probably require dedicated long-term observing programs at relatively large telescopes; this represents something of a departure rom the typical usage of large telescopes. However, the case for such intensive and focused study seems strong, largely because the subject area, and the questions to be asked, are well-defined. The nature of future progress can be rather clearly anticipated from our present vantage point.

One interesting benefit of studying the evolution of solar and stellar magnetic activity, of course, is increased knowledge of the evolution of activity in our own Sun. For example, we may already infer that, somewhat after its arrival on the main sequence, the Sun had a rotation period of several days, and very substantial magnetic activity, characterized by very much larger spot coverage, and chromospheric and coronal activity, than the present-day Sun. This must have had important consequences for the evolution of primordial planetary atmospheres and in particular the prebiological terrestrial atmosphere (Zahnle and Walker, 1982; Canuto *et al.*, 1982). Probably the young Sun did not have a regular activity cycle as it does now; that regularity may have appeared only when the Sun's rotation period slowed to about 20 days, at an age of perhaps 2×10^9 yr. Possibly the mean activity cycle period has gradually lengthened since then to its present value, as the Sun has continued to lose angular momentum (Noyes *et al.*, 1984b). The apparent long-term intermittency of activity, as exemplified by the Maunder minimum, may be verifiable through statistical studies of the fraction of solar-like stars that show clear activity cycles. To carry out such studies, however, requires a long-term data set for many stars. The continuing acquisition of such data on stellar activity and its variability is an important requirement for progress in our knowledge of solar and stellar activity.

References

Baliunas, S. and Hartmann, L. (eds.): 1984, *Cool Stars, Stellar Systems, and the Sun*, Springer, New York.
Baliunas, S. and Vaughan, A.: 1985, *Ann. Rev. Astron. Astrophys.* **23**, 379.
Baliunas, S., Horne, J., Porter, A., Duncan, D., Frazer, J., Lanning, H., Misch, A., Mueller, J., Noyes, R., Soyumer, D., Vaughan, A., and Woodard, L.: *Astrophys. J.* (in press).
Barry, D., Hege, K., and Cromwell, R.: 1984, *Astrophys. J.* **277**, L65.
Belvedere, G.: 1985, *Solar Phys.* **100**, 363 (this volume).
Berthomieu, G., Cooper, A. J., Gough, D. O., Osaki, Y., Provost, J., and Rocca, A.: 1980, in H. A. Hill and W. A. Dziembowski (eds.), *Nonradial and Nonlinear Stellar Pulsations*, p. 307.
Bonnet, R. M. and Dupree, A. K. (eds.): 1981, *Solar Phenomena in Stars and Stellar Systems*, D. Reidel Publ. Co., Dordrecht, Holland.
Canuto, V., Levine, J., Augustsson, T., and Imhoff, C.: 1982, *Nature* **296**, 816.
Catalano, S. and Marilli, E.: 1983, *Astron. Astrophys.* **121**, 190.
Christensen-Dalsgaard, J.: 1984, in A. Mangenay and F. Praderie (eds.), *Space Research Prospects in Stellar Activity and Variability*, Obs. de Paris, Meudon, p. 11.
Copeland, H., Jensen, J. O., and Jorgensen, H. E.: 1970. *Astron. Astrophys.* **5**, 12.
Dorren, J. D. and Guinan, E. F.: 1982, in M. Giampapa and L. Golub (eds.), *Second Cambridge Workshop on Cool Stars*, SAO Spec. Rept. 392, Cambridge, Mass., p. 49.
Dorren, J. D. and Guinan, E. F.: 1984, in S. Baliunas and L. Hartmann (eds.), *Cool Stars, Stellar Systems, and the Sun*, p. 259.

Dravins, D. and Larsson, B.: 1984, in S. Keil (ed.), *Small-Scale Dynamical Processes in Quiet Stellar Atmospheres*, p. 306.

Dravins, D. and Linds, J.: 1984, in S. Keil (ed.), *Small-Scale Dynamical Processes in Quiet Stellar Atmospheres*, p. 414.

Duncan, D., Baliunas, S., Noyes, R., Vaughan, A., Frazier, J., and Lanning, H.: 1984, *Publ. Astron. Soc. Pacific* **96**, 707.

Durney, B. R. and Latour, J.: 1978, *Geophys. Astrophys. Fluid Dyn.* **9**, 241.

Durney, B. R., Mihalas, D., and Robinson, R.: 1981 *Publ. Astron. Soc. Pacific* **93**, 537.

Gaizauskas, V., Harvey, K. L., Harvey, J., and Zwaan, C.: 1983 *Astrophys. J.* **265**, 1056.

Gehren, T.: 1982, *Astron. Astrophys.* **109**, 187.

Gilliland, R.L.: 1984, in C. Baliunas and L. Hartmann (eds.), *Cool Stars, Stellar Systems, and the Sun*, p. 146.

Gilliland, R. L. and Fisher, R.: 1986 *Publ. Astron. Soc. Pacific* **97**, 285.

Gilman, P. A.: 1981, in D. F. Gray and J. L. Linsky (eds.), *IAU Colloq.* **51**, 19.

Gough, D. O.: 1985, *Solar Phys.* **100**, 65 (this volume).

Gray, D. F.: 1982, *Astrophys. J.* **255**, 200.

Gray, D. F. and Toner, C. G.: 1985, preprint.

Hartmann, L. W.: 1985, *Solar Phys.* **100**, 587 (this volume).

Hartmann, L. W., Soderblom, D., Noyes, R., Burnham, N., and Vaughan, A.: 1984, *Astrophys. J.* **276**, 254.

Herbig, G.: 1985, *Astrophys. J.* **289**, 269.

Hill, H. A. and Dziembowski, W. A. (eds.): 1980, *Nonradial and Nonlinear Stellar Pulsation*, Springer, Berlin.

Keil, S. A. (ed.): 1984, *Small-Scale Dynamical Processes in Quiet Stellar Atmospheres*, National Solar Observatory, Sunspot.

Knobloch, E., Rosner, R., and Weiss, N.: 1981, *Monthly Notices Roy. Astron. Soc.* **197**, 45 p.

Labonte, B.: 1984, *Astrophys. J.* **276**, 335.

Leighton, R. B.: 1959, *Astrophys. J.* **130**, 366.

Lockwood, G. W., Thompson, D. T., Radick, R. R., Osborn, W. H., Baggett, W. E., Duncan, D. K., and Hartmann, L. W.: 1984. *Publ. Astron. Soc. Pacific* **96**, 714.

Lubow, S. H., Rhodes, E. J., Jr., and Ulrich, R. K.: 1980, in H. A. Hill and W. A. Dziembowski (eds.), *Nonradial and Nonlinear Stellar Pulsations*, p. 300.

Marcy, G. W.: 1983, in J. O. Stenflo (ed.), *IAU Symp.* **102**, p. 3.

Noyes, R. W.: 1983, in J. O. Stenflo (ed.), *IAU Symp.* **102**, p. 133.

Noyes, R. W., Hartmann, L. W., Baliunas, S. L., Duncan, D. K., and Vaughan, A. H.: 1984a, *Astrophys. J.* **279**, 763.

Noyes, R. W., Weiss, N. O., and Vaughan, A. H.: 1984b, *Astrophys. J.* **287**, 709.

Pallavicini, R., Golub, L., Rosner, R., Vaiana, G. S., Ayres, T., and Linsky, J. L.: 1981, *Astrophys. J.* **248**, 279.

Parker, E. N.: 1971, *Astrophys. J.* **165**, 139.

Robinson, R. D., Worden, S. P., and Harvey, J. W.: 1980, *Astrophys. J.* **236**, L155.

Rodonó, M. and Byrne, P. (eds.): 1983, *Activity in Red-Dwarf Stars*, D. Reidel Publ. Co., Dordrecht, Holland.

Rosner, R., Tucker, W. H., and Vaiana, G. S.: 1978, *Astrophys. J.* **220**, 643.

Schrijver, C. J.: 1983, *Astron. Astrophys.* **127**, 289.

Schrijver, C. J. and Rutten, R. G. M.: 1985, *Astron. Astrophys.* (preprint).

Simon, T., Herbig, G., and Boesgaard, A.: 1985, *Astrophys. J.* (in press).

Skumanich, A.: 1972, *Astrophys. J.* **171**, 565.

Skumanich, A. and Eddy, J. A.: 1981, in R. M. Bonnet and A. K. Dupree (eds.), *Solar Phenomena in Stars and Stellar Systems*, D. Reidel Publ. Co., Dordrecht, Holland, p. 349.

Skumanich, A., Smythe, C., and Frazier, E.: 1975, *Astrophys. J.* **200**, 747.

Stenflo, J. O. (ed.): 1983, 'Solar and Stellar Magnetic Fields: Origins and Coronal Effects', *IAU Symp.* **102**.

Uchida, Y.: 1984, in J. C. Pecker, Y. Uchida (eds.), *Japan–France Seminar on Active Phenomena in the Outer Atmosphere of the Sun and Stars*, CNRS, Paris, p. 300.

Ulmschneider, P. and Stein, R. F.: 1982, *Astron. Astrophys.* **106**, 9.

Vaughan, A. H.: 1980, *Publ. Astron. Soc. Pacific* **92**, 392.

Vaughan, A. H. and Preston, G. W.: 1980, *Publ. Astron. Soc. Pacific* **92**, 385.

Vogt, S. S.: 1981, *Astrophys. J.* **250**, 327.

Vogt, S. S.: 1983, in M. Rodonó and P. Byrne (eds.), *Activity in Red-Dwarf Stars*, D. Reidel Publ. Co., Dordrecht, Holland, p. 137.

Vogt, S.S. and Penrod, G. D.: 1983a, in M. Rodonó and P. Byrne (eds.), *Activity in Red-Dwarf Stars*, D. Reidel Publ. Co., Dordrecht, Holland, p. 379.

Vogt, S. S. and Penrod, G. D.: 1983b, *Publ. Astron. Soc. Pacific* **95**, 565.

Wilson, O. C.: 1978, *Astrophys. J.* **226**, 379.

Zahnle, K. and Walker, J.: 1982, *Rev. Geophys. Space Phys.* **20**, 280.

THE EMERGENCE OF MAGNETIC FLUX

CORNELIS ZWAAN

Sterrewacht 'Sonnenborgh', Zonnenburg 2, 3512 NL Utrecht, The Netherlands

Abstract. This paper first summarizes the morphology and dynamics of emerging flux regions and arch filament systems and then discusses detailed observations of a particular active region with emerging magnetic flux.

The central part of the growing active region shows abnormal granulation and a weak magnetic field that, locally, is transverse. In the border zone, strong downward flows occur in the chromopshere and photosphere (small features with strong magnetic fields (faculae, pores) are formed here.) Near the leading and following edge, sunspots are formed by the coalescence of such small magnetic elements.

The observational data are interpreted by means of a heuristic model of an emergent magnetic loop-shaped bundle consisting of many flux tubes. In this model we incorporate the theory of convective collapse and the buoyancy of flux tubes. The observed complexity in the structure and dynamics, including strong transverse fields and velocity shear, is attributed to the emergence of several flux regions within the active region at different orientations.

1. The Birth of an Active Region

The first manifestation of a new active region in the solar atmosphere is the appearance of a small, compact and very bright bipolar plage (Fox, 1908; Waldmeier, 1937; Sheeley, 1969). Soon the *arch filament system* (AFS) is observed, which is seen in the line core of Hα as a set of roughly parallel dark fibrils connecting faculae of opposite polarity. These arch filament systems have been so named and studied in detail by Bruzek (1967, 1969).

The faculae of opposite polarity move apart. The rate of expansion exceeds 2 km s^{-1} during the first half hour (Harvey and Martin, 1973), then drops to values between 1.3 and 0.7 km s^{-1} during the next six hours (Harvey and Martin, 1973), or to 0.5 km s^{-1} during the first 10 hr. (0.2 km s^{-1} is typical for the next few days (Mosher, 1977).) New magnetic flux emerges near the line $B_{\parallel} = 0$, which separates the opposite magnetic polarities in the plage, in between the diverging faculae that are made up from the magnetic flux which emerged earlier. If sufficient flux is available, pores and, eventually, sunspots are formed near the leading and following edges of the expanding plages (Zirin, 1972, 1974).

An arch filament system lasts as long as magnetic flux emerges, but individual fibrils live for about half an hour. Hence, after some time the AFS connects only faculae in the inner part of the growing active region. The loops are between 20 and 30 Mm long and the maximum heights are about 5 Mm (Bruzek, 1967, 1969). The top of a loop ascends with a speed of up to 10 km s^{-1}. In both legs, the matter flows down at speeds of up to about 50 km s^{-1}, as measured in Hα (Bruzek, 1969).

Initially, the AFS's are oriented almost at random (Weart, 1970), but the systems rotate, and those living longer than one day assume the 'correct' orientation of active

Solar Physics **100** (1985) 397–414. 0038–0938/85.15.

regions: nearly parallel to the equator (Weart, 1972; Frazier, 1972). These properties indicate that the emergence and initial growth of a new active region follow distinct patterns. For this, Zirin (1972) coined the term Emerging Flux Region (EFR). Such an EFR is bipolar; it stands out as an intruding feature that, at first, does not disturb the surrounding old magnetic structure, except for some pushing aside. EFR's range from small bipolar units, with a magnetic flux $\Phi \lesssim 10^{20}$ Mx, that do not develop beyond ephemeral active regions, to large bipoles $\Phi > 5 \times 10^{21}$ Mx in which sunspots develop.

The course of events suggests that an emerging flux region is caused by the emergence of the top of a loop bundle, shaped as a peaked arch, consisting of many magnetic flux tubes (Figure 1). In this way, magnetic flux may be inserted into the atmosphere at the

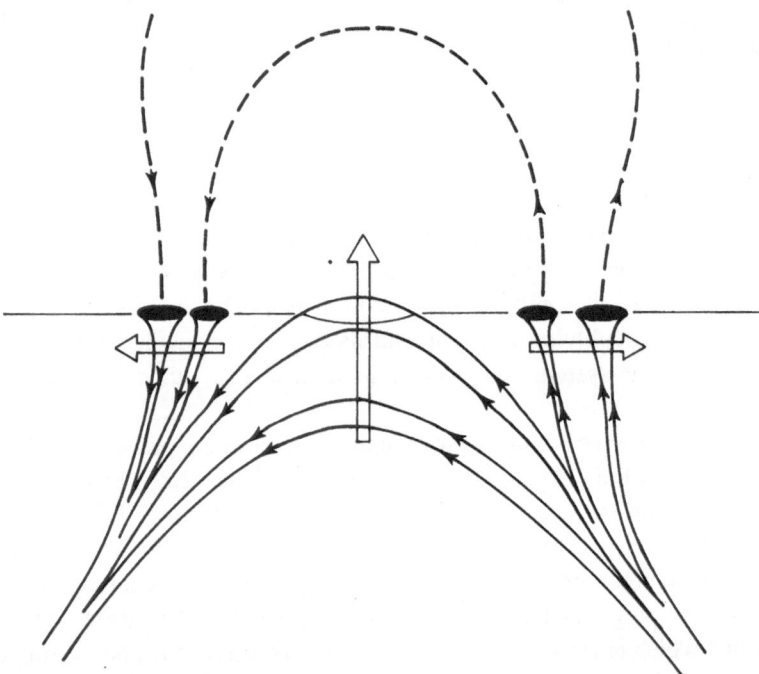

Fig. 1.   A model for the emergence of magnetic flux, separation of polarities and coalescence of sunspots. Broad arrows indicate local displacements of flux tubes. (Real emerging flux regions consist of many more separate flux loops.)

observed large rates between about 4×10^{15} and 3×10^{16} Mx s^{-1} (Mosher, 1977) through a small area. Zwaan (1978) pointed out that the observed properties of growing active regions refute the notion that some initially weak magnetic field already present in the atmosphere, may be concentrated into a strong field by convection in the top of the convection zone. The latter scenario has been advocated by Schmidt (1968) and Meyer et al. (1974), and was recently recapitulated by Priest (1984, Section 8.6.1).

When examined closely, a large active region appears to be formed by contributions from several emerging flux regions that occur, simultaneously or in quick succession, within a few days (Weart, 1970, 1972; Schoolman, 1973; Mosher, 1977). The EFR's constituting an active region are related: they frequently occur in closer proximity and quicker succession than chance would permit. Moreover, in many cases, the poles of several EFR's which have appeared as separate bipoles at different orientations and at different times, merge in such a way that a single simply bipolar active region results (example: McMath region 11972, 24–28 July, 1972, shown in Mosher (1977, Figures 1 and 2)). Occasionally, the contributing EFR's do not immediately combine so nicely, and a region is formed with a crooked $B_{\parallel} = 0$ line dividing the polarities, or even a magnetically complex region.

If sufficient magnetic flux emerges, sunspots are produced as part of the whole process. Presumably all sunspots are formed by coalescence of recently emerged flux tubes. Once a sunspot is formed and is still growing, magnetic features are seen to be streaming into that spot. Vrabec (1974) distinguishes two cases which he describes by examples. The first case concerns a narrow pair of sunspots of opposite polarity being formed by an emerging flux region, from which the magnetic features move straight into the rapidly growing spots on either side, with velocities between 0.25 and 1.0 km s^{-1}. In this case, time series of magnetograms are needed to see the inflow. The second case refers to the inflow of small umbrae, pores and faculae of leading polarity into a leader sunspot. These magnetic elements are arranged in a pronged tail trailing behind the large spot and they stream along the prongs as tracks, with an average speed of 0.25 km s^{-1}. Because there are small spots among the streaming features, this type of inflow is visible in white-light pictures. McIntosh (1981, Figure 15) shows an example of a string of four almost equally large sunspots of leading polarity, coalescing within one day into one large sunspot. Spots coalescing into a larger one are usually formed in different EFR's. In some of these regions the emergence may have terminated some time ago. Flux emergence and coalescence occur in one continous process, however; usually that whole process takes no more than a few days.

There are conflicting statements as to where an EFR appears with respect to the pre-existing chromospheric network. According to Bumba and Howard (1965), the new faculae occur within the Ca II network structure (which the authors translate into 'between supergranules') throughout the emergence phase. Also Born (1974) maintained that an AFS remains rooted in the chromospheric network. Bruzek (1969) noticed a coincidence in the length of the Hα loops in arch filament systems and diameters of supergranules, and he suggested that an AFS may cover a supergranule. Frazier (1972) inferred that supergranules bring the flux tubes to the surface and transport them laterally to the vertices of the cells. However, Harvey and Martin (1973) found that EFR's occur adjacent to, but not in the network. According to Zirin (1974), there is no connection whatsoever between EFR's and the elements of the network.

Definite statements concerning the appearance of EFR's with respect to the *pre-existing* network should be considered with skeptism. The comparison between the position of an EFR and the – interpolated – position of the previous network is not simple. Moreover, the evolution of an EFR suggests control by the dynamics of the field itself, and not by the local environment of the photosphere (Schoolman, 1973). Finally, there is the consideration that, even if a flux loop is not set rising by convection somewhere deep in the convection zone, a rising loop, as indicated in Figure 1, must create and therupon maintain a convective bubble traveling with the top of the loop.

During the flux emergence, the plage is amorphous and very bright. Once emergence stops, the brightness decreases sharply, and the plage takes the well-known appearance of the filigree of fine facular points (see Zirin, 1974).

Within EFR's the photosphere is disturbed as well: locally the intergranular lanes are darker than normal, and aligned in the inferred direction of the horizontal component of the magnetic field (Bray and Loughhead, 1964; Section 3.3.4.). Bray and Loughhead attribute the alignments to the action of the horizontal field in the top of a magnetic flux loop rising through the photosphere.

In addition to the strong downflows ($v \simeq 50$ km s^{-1}) in the chromospheric feet of arch filament systems, downward flows with velocities of up to about 1 km s^{-1} are found in the photosphere of emerging flux regions. These downflows occur over large parts of the EFR. Maximum velocities are found over and close to rapidly growing sunspots (Gopasyuk, 1967, 1969; Bachmann, 1978). Kawaguchi und Kitai (1976) found that the downflows occur shortly before the birth of sunspots and throughout the phase of rapid sunspot growth. In addition to the downflows, Howard (1971) found upward flow over small portions of young active regions.

This summary of data shows that emerging flux regions are important events. Two aspects are of specific interest: first, the information they carry on magnetic structure in the convection zone shortly before emergence and, secondly, the adjustment of the magnetic field to atmospheric conditions upon its breakout. The remainder of this review is a discussion of recent research on a particualr emerging flux region. Much of this work is presented in the thesis by Brants (1985c), which is reviewed here as a basis for further interpretation.

2. Emerging Magnetic Flux in McMath Region 16164, 24 July, 1979

On 24 July, 1979, Zwaan *et al.* (1985) observed a rapidly growing active region using the Vacuum Tower Telescope and Echelle Spectrograph at Sacramento Peak Observatory. The new magnetic flux emerged in an old active region, McMath 16164, consisting of enhanced network only. On KPNO magnetograms, the area of interest showed on 22 July enhanced network of negative polarity (the leading polarity on the S hemisphere), while on 23 July (15 : 00 UT) there were one or two tiny emerging flux regions. The *Solar-Geophysical Data* lists a small sunspot group consisting of two pores on 23 July (13 : 40 UT). During 24 July, at least after the start of the observations at 14 : 00 UT, magnetic flux was emerging vigorously, there were very bright plages and a conspicuous arch filament system (Figure 4), a compact, nearly bipolar magnetic structure (Figure 3). and a sunspot group consisting of several pores (Figure 2). By 25 July, a medium-large active region had formed, with a full-fledged, single, leader sunspot and

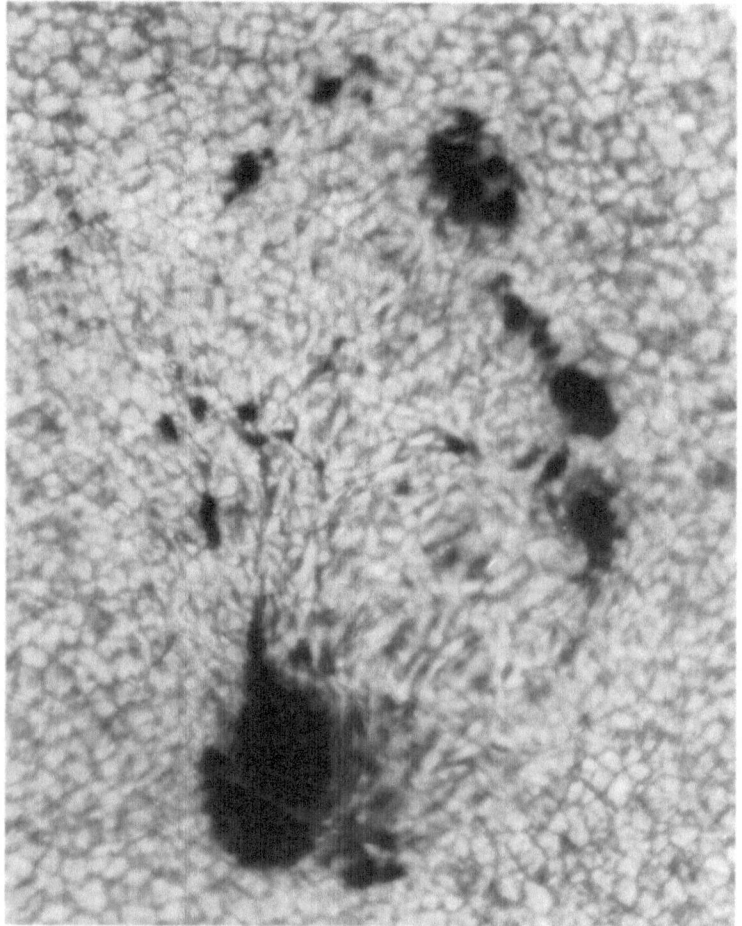

Fig. 2a.   The new active region in McMath region 16164, 24 July, 1979, 17:35:39 UT, in white light.

two clusters of following spots (Figure 5); the region was still growing in magnetic flux and in size.

The data reviewed in this paper were obtained during 24 July, the day of vigorous flux emergence.

2.1. Observational program and analysis

The observational procedure of Zwaan *et al.* (1985) was designed for the study of fine structures in active regions. The investigation of the rapidly growing active region in McMath 16164 is based mainly on white-light slitjaw pictures, and on line profiles in opposite directions, of circular polarization: a standard line for magnetic-field measurement Fe I $\lambda 3602.5$ Å $(g = 2.5)$, the umbral line Ti I $\lambda 6064.5$ Å $(g = 2.0)$ and the Zeeman-insensitive line Fe I $\lambda 5691.5$ Å.

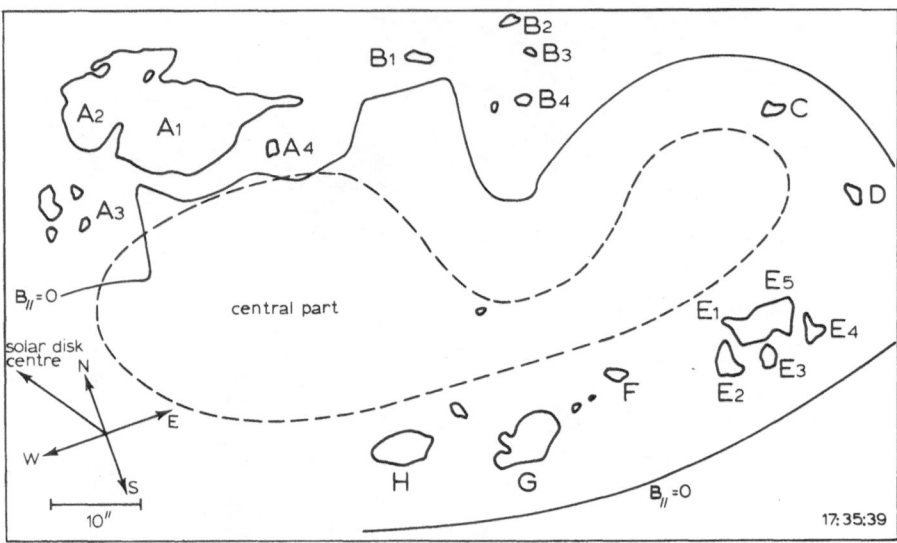

Fig. 2b. Contour map identifying the sunspot pores in Figure 2a. The approximate position of the line $B_\| = 0$ separating the polarities is indicated by the full line. The dashed contour encloses the central part, which shows abnormal granulation with dark alignments and protopores; the filigree is absent there.

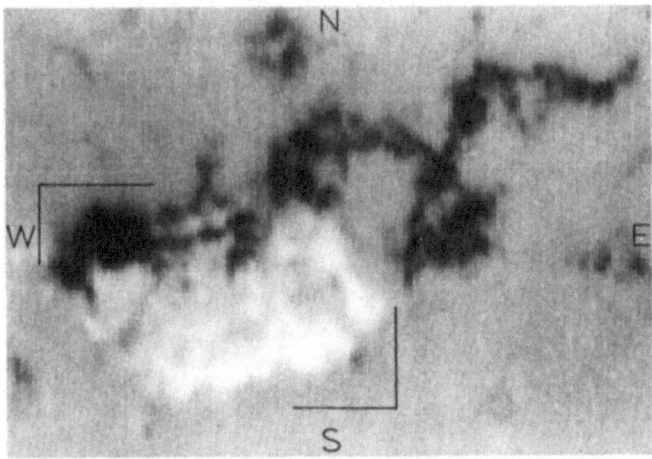

Fig. 3. The magnetogram of McMath region 16164, 24 July, 1979, 15 : 28 : 38, showing the rapidly growing new active region (within the frame) and the ambient enhanced network of negative polarity (black), the leading polarity on the southern hemisphere (by courtesy J. W. Harvey, National Solar Observatory, Tucson).

   Brants and Steenbeek (1985) studied the structure and evolution of the photospheric features in the emerging flux region using white-light slitjaw pictures and one direct photograph of superior quality (Figure 2a). Selected frames were digitized in order to determine areas, positions and approximate intensities of sunspot pores and other photospheric features.

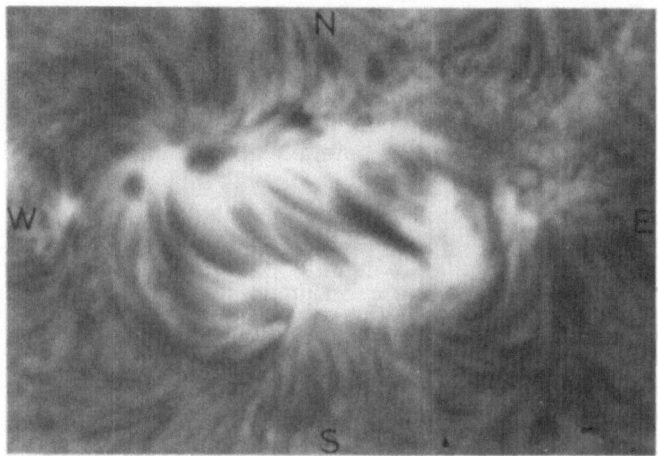

Fig. 4.   The arch filament system and faculae in McMath region 16164, 24 July, 1979, 18 : 48 : 42, obtained in the center of the Hα line (by courtesy of V. Gaizauskas, Ottawa River Solar Observatory, Ottawa).

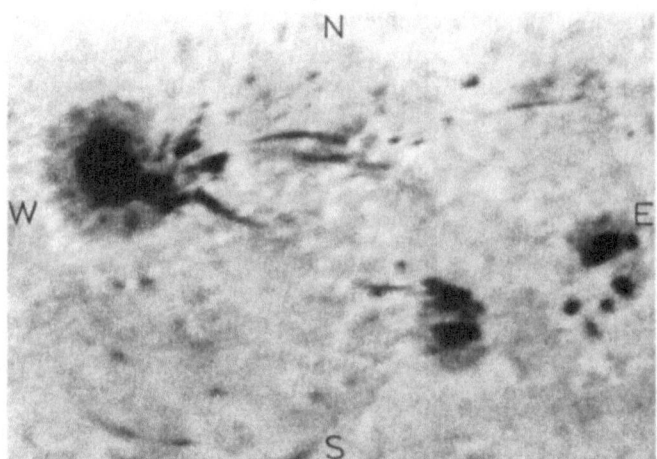

Fig. 5.   The growing active region in McMath region 16164 on 25 July, 1979, 14 : 42 : 12, obtained in Hα − 1.0 Å. Some dark streaks are caused by Doppler shifts in parts of the AFS (by courtesy of V. Gaizauskas, Ottawa River Solar Observatory, Ottawa).

Brants (1985a, c) described the analysis of the spectrograms. Parameters characterizing the observed line profiles are interpreted in terms of magnetic structure paramaters B, ψ, $f = 1 - s$, and Δv, assuming that the atmospheric structure within the resolution element consists of two components: (i) magnetic elements with a magnetic field strength B and an inclination angle ψ with respect to the line-of-sight, yielding a fraction of $f = 1 - s$ of the continuum intensity, and (ii) non-magnetic photosphere yielding a fraction s of the continuum intensity. The upward velocity of the magnetic plasma with respect to the non-magnetic photosphere is measured by Δv. Brants demonstrates that, although the inclinations ψ and the filling factor f are very uncertain, these uncertainties largely cancel out in the determination of the mean magnetic flux density $\Phi/A = B f \cos \psi$ over the resolution element. The field strength B may be estimated fairly accurately from the separation between the extrema of the Stokes's V-profile, provided that the mean flux density Φ/A is sufficiently high. If both the field strength B and the filling factor f are sufficiently high, B may be estimated from the magnetic broadening of the line profile in the total intensity I, even if the field is strongly inclined. A comparison between the magnetic line-width enhancement and the amplitude of the Stokes's V-profile permits estimates for B, ψ, and f.

2.2. TYPES OF MAGNETIC STRUCTURE

Brants (1985b, c) presents an analysis of the 900 picture elements of 0.8×0.6 (arc sec)^2 within eleven settings of the spectrograph slit across the emerging flux region. In about 20% of these elements, strong fields are found with $B > 1000$ G and filling factor $f > 0.6$. There are two distinct ranges in the inclination: the strong fields are either slightly to moderately inclined, or nearly transverse. The magnetic field covered by the eleven slit settings consists of three catagories:

Slightly to moderately inclined, strong fields occur in about 16% of the resolution elements; the inclination angle ψ is definitely less than 60°. The pores contain about 45% of these fields; the field strengths are between 1900 and 2600 G (Brants and Zwaan, 1982). A small fraction (about 10%) of these strong fields occur in faculae, which Brants (1985b) has defined as 'line gaps' in the Zeeman-insensitive line Fe I $\lambda 5691.5$ Å. The remaining 45% of the elements with slightly inclined and strong fields are neither conspicuous in continuum brightness nor in $\lambda 5691.5$ Å line-core brightness.

Nearly transverse, strong magnetic fields are found in 4% of the elements. The inclination is large ($\psi \gtrsim 80°$), the field strength B is between 1000 and 2000 G, or even larger if the filling factor f is smaller than unity.

Weak field is found in 80% of the elements: there the mean magnetic flux density is low (less than 120 G), and there are no indications for transverse strong fields. Hence, the field strength is low, or the filling factor is low, or both. Although the flux densities for individual elements are not reliable because of the noise, the result that most of the weak field is of following (positive) polarity is significant.

Brants (1985b, c) describes the line-of-sight velocities in terms of a large-scale component, and small-scale flows extending over no more than a few picture elements. These flows are discussed below.

3. The Large-Scale Structure of the Emerging Flux Region

Comparison between the white-light pictures (Figure 2), the Hα filtergram (Figure 4), Ca II K filtergrams and the magnetogram (Figure 3) shows that on 24 July the sunspot pores are, indeed, found close to the edges of the expanding region. The boundary of the region at the following (S and E) edge is particulary well visible on the magnetogram because there the polarity is opposite to the polarity of the ambient old magnetic field.

Different parts may be distinguished in the growing active region. The flux of leading (negative, black) polarity is mainly confined to a relatively small area in and around the rapidly growing sunspot pores indicated in Figure 2b by A1 through A4. The line $B_{\parallel} = 0$ separating the polarities passes quite close to the S side of the pore mass A1 (see Figure 2b and 3) – the largest fraction of the region is covered by the magnetic flux of following polarity. This marked asymetry is abnormal – most EFR's are nearly symmetric in the polarities.

The strong fields at slight to moderate inclination are found in the periphery, within the pores, and close to the pores at the side towards the center of the region. Some of these patches of strong fields are visible as faculae (line gaps), most of which are found near the rapidly growing pores G and H (see Brants, 1985b, his Figure 1).

The central part of the region, indicated by the dashed contour in Figure 2b, shows a disturbed granulation. It looks fuzzy, there are alignments in dark intergranular lanes, and protopores. There is no filigree. The magnetic flux is of following polarity but for some inclusions of leading polarity that are found in the KPNO magnetic data (Figure 3), and in some of the spectrograms. The mean magnetic flux density in the central part is relatively low and there are very few picture elements with strong fields, among which is the only facula in the central part (Brants 1985a, b, c).

The area around pores B2, B3, and B4 shows little activity: the granulation looks normal and the filigree is clearly visible (Figure 2a). Moreover, pores B2, B3, and B4 are already decaying (see Section 4.3).

The strip around the polarity dividing line $B_{\parallel} = 0$, where it passes close to the spots indicated by A and B in Figure 2b, is characterized by: (i) a nearly transverse, very strong magnetic field (all the elements with nearly transverse fields and $1000 < B \lesssim 2000$ G, mentioned in Section 2.2, are found there); (ii) a large-scale upward flow, that exceeds 0.5 km s^{-1} in most of the strip.

Large-scale downward flow exceeding 0.5 km s^{-1} is found in the vicinity of the rapidly growing pores G and H, at the side toward the center of the region.

The arch filament system is well visible in Hα filtergram (see Figure 4) which is the only high-quality Hα filtergram found. It has been recorded more than an hour after the observations at SPO had been stopped, and more than three hours after the KPNO magnetogram had been recorded. Some information on the arch filament system(s) during the observations could be obtained from Ca II K slitjaw pictures (although they show AFS's much less clearly than Hα filtergrams), and from Ca II H spectrograms (although they refer to a tiny fraction of the disk). From the data at hand, it appears that from 14:00 until 19:00 UT on 24 July, 1979 conspicuous loops were located between A1/A4 and the E cluster of pores, between A3 and G, and the strongest loops between A3 and H (see Figure 4). Hence during the observations, at least two EFR's with AFS's were in operation: one between the spots A3 and H, oriented SSE-NNW, and one between A1 and C, D, and E, oriented about E–W.

4. Dynamics of Emerging Flux in McMath 16164

4.1. EXPANSION, ROTATION, AND SHEAR

The large-scale dynamics in the region confirmed facts and figures summarized in
Section 1: the active region expanded chiefly in the direction of the long axis, which is
approximately oriented E–W. Brants and Steenbeek (1985, Figure 6) found that during
the observing run on 24 July, the E–W expansion amounted to 0.7 km s^{-1}. The long
axis rotated by about 2° per hour in a direction towards the 'proper' orientation for a
mature active region.

Comparison between the magnetograms of 24 and 25 July shows that the expansion
of the region continued. Blinking the two magnetograms reveals that the expansion
was caused by the rapid westward drift of the leading sunspot A, the following part of
the region remaining nearly stationary with respect to the ambient magnetic structure.
This westward motion of leading sunspots in young active regions is well known (see
Waldmeier, 1955, Section 45; Bray and Loughhead, 1964, Section 6.2.5), but in this
particular case, the average speed between 24 and 25 July is quite high: 0.5 km s^{-1}.

Brants and Steenbeek (1985) found a strong shearing motion in the leading pore
cluster. The pores in the leading edge indicated A2 and A3 in Figure 6 moved with
respect to the pore cores indicated by A1 with speeds of up to 0.6 km s^{-1} to the north,
hence, in a direction about perpendicular to the westward motion of the pore cluster
A1. The light bridge between the parts indicated by A1 and A2 was squeezed by the
shear, but it remained visible during the three hours of white-light observations. The
velocity shear was probably related to, and perhaps caused by, the two EFR's operating
in different orientations during the observations (see Section 3).

4.2. DOWNWARD FLOWS AND THE FORMATION OF STRONG MAGNETIC FIELDS

The data obtained at SPO confirms the presence of large-scale downward flows over
parts of the photosphere in emerging flux regions. In the vicinity of rapidly growing
pores, these downflows range from 0.5 to 2.0 km s^{-1} (Brants, 1985b, c). In addition,
Brants found small-scale downflows in areas not exceeding a few arc sec in diameter.
The velocities range up to about 1.0 km s^{-1}. Many, but not all, of these small-scale
downflows are superimposed on large-scale downward flows exceeding 0.5 km s^{-1}.

Zwaan et al. (1985, Figure 4) happened to cover with one setting of the spectrograph
slit more than one half of an arch loop. In Ca II H3, the top of the loop was rising with
7 km s^{-1}; along the leg covered by the slit the mass flow was downwards, with the speed
increasing to at least 40 km s^{-1} in the footpoint. The foot of the loop was very bright
in the wing of Ca II H, but the continuum was slightly darker than in the normal
granulation. The photospheric spectral lines indicate an enhanced magnetic field
strength and a downward flow of about 2.0 km s^{-1} (Brants, 1985a). On the white-light
slitjaw pictures this downflow area developed through the stage of a protopore into a
pore indicated by G3 in Figure 7 (Brants and Steenbeek, 1985). This small pore G3
coalesced with other to form the large single pore G shown in Figure 2. This example

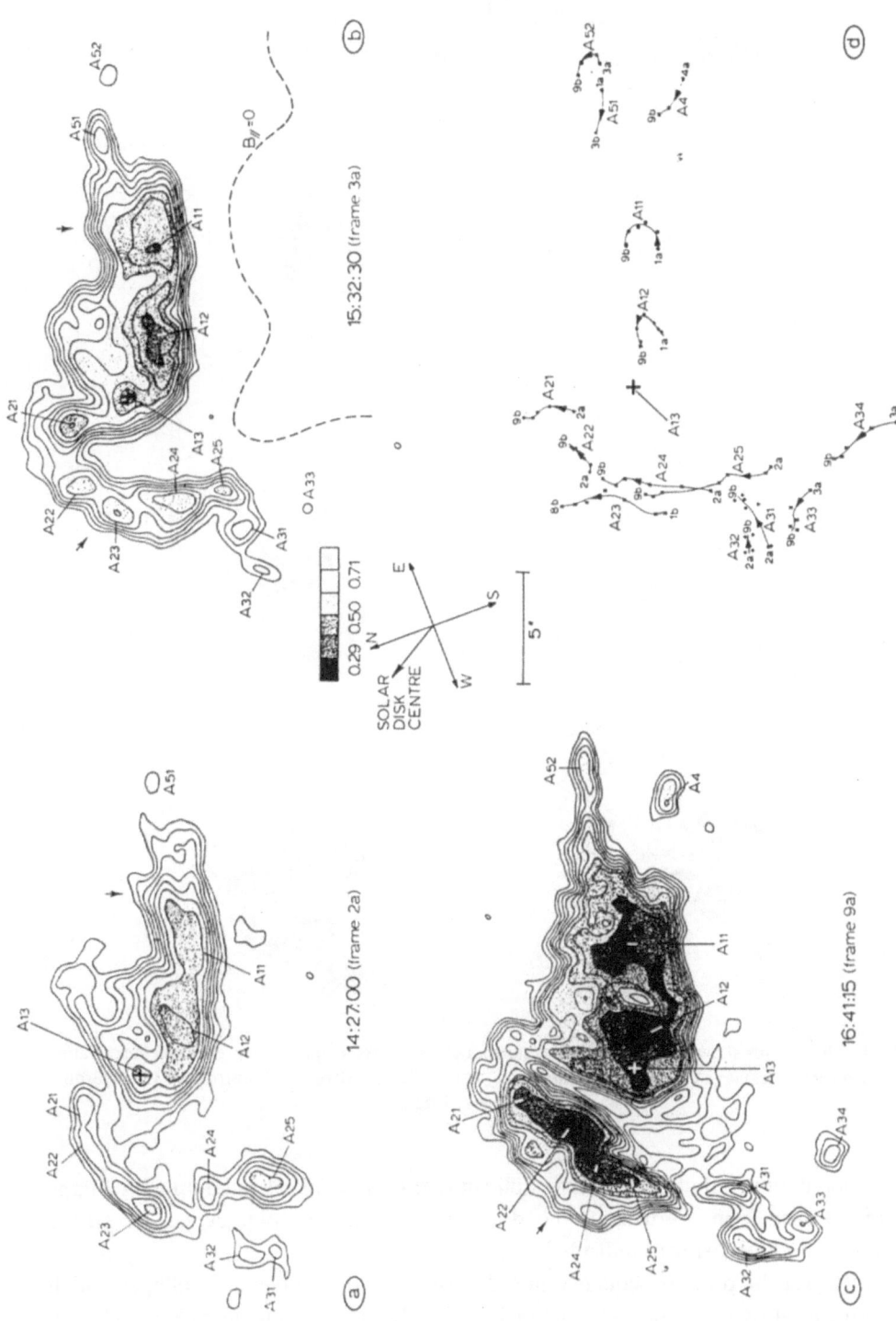

Fig. 6. Frames (a)–(c): isophote maps showing the evolution of the leading sunspot. The arrows point to rapidly expanding boundaries without inflows of distinct elements visible in white light. Frame (d): trajectories of the dark cores relative to the core A 13; 1a and 9b refer to the first and the last frame in the time series (from Brants and Steenbeek, 1985).

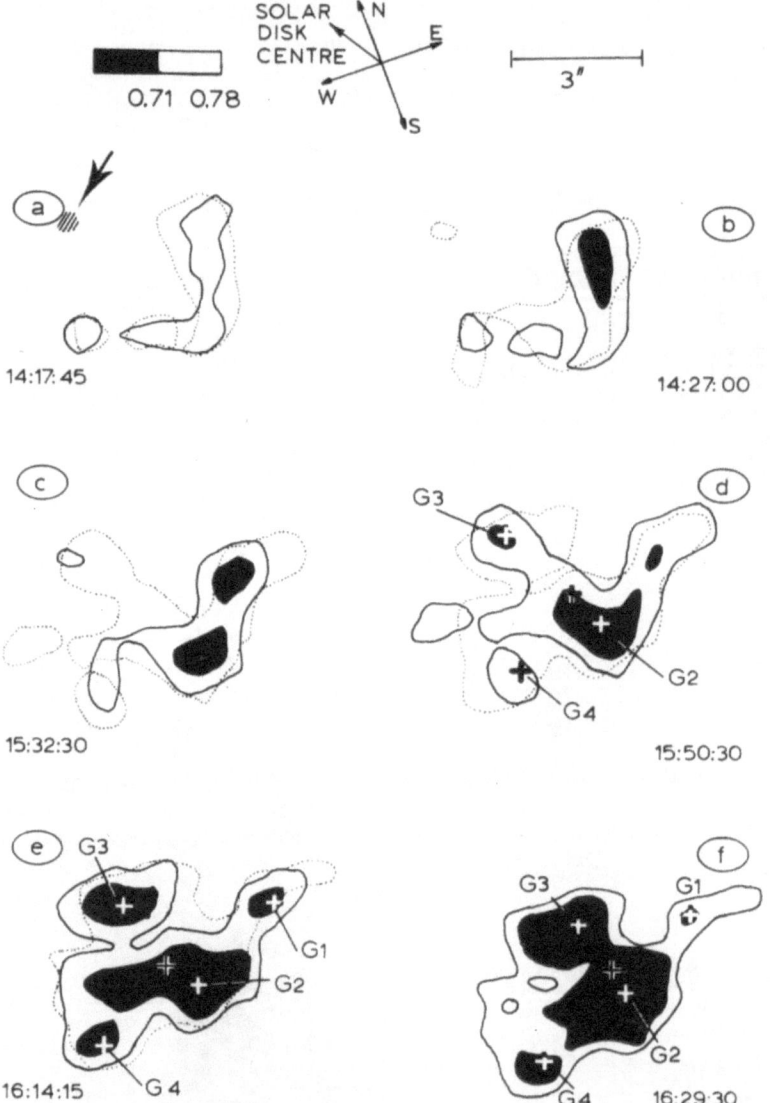

Fig. 7.   Isophote maps showing the growth of a sunspot pore. Dotted isophotes represent the next frame. The shaded area in frame (a), indicated by the arrow, is the site of a strong downward flow (from Brants and Steenbeek, 1985).

is exceptional only in that the spectrograph slit had been so nicely lined up with a large part of an arch loop. Perusal of the data indicates many coincidences of strong downflows in chromosphere and photosphere.

We interpret the process sequence just described as the convective collapse which transforms a rather weak magnetic field ($B \lesssim 500$ G) into a strong field ($B > 1000$ G). The idea (Parker, 1978; Zwaan, 1978) is that the initial field is strong enough to impede convective heating, hence, the photospheric part of the flux tube cools by radiative losses. This cooling starts a downdraft which is then enhanced in the convectively unstable medium. The downflow helps the cooling and carries the cooling down deeper into the tube. With decreasing temperature, the gas pressure decreases and, hence, the flux tube becomes compressed to a higher field strength. Spruit (1979) has calculated the structure of slender flux tubes in the collapsed state – nearly independent of the initial strength the field strength becomes about 1500 G.

In the rising-loop model (Figure 1), the coalescence of magnetic elements of the same polarity into sunspots is readily explained by the buoyancy of the tops of the collapsed flux tubes: provided that some force keeps the tubes contained deep in the convection zone, the tops of the tubes coalesce while the tubes adjust to the vertical position.

Once a dark pore is formed, the field strength has attained a value $1900 < B < 2600$ G (Brants and Zwaan, 1982), and the downward flow has stopped. This suggests that a pore is the end product of the convective collapse. Presumably, flux tubes of smaller magnetic flux become faculae after collapse. In any case, there are no significant small-scale downward flows within faculae (Brants, 1985b).

4.3. EVOLUTION OF SUNSPOTS IN EMERGING MAGNETIC FLUX

Most of the sunspot pores and pore clusters in the region show two aspects of positive growth: an increase in area and a decrease in continuum brightness (see Brants and Steenbeek (1985, Table II)). One pore cluster (E in Figure 2) decreased in area within the selected $I = 0.78$ isophote but it became more compact and much darker during the observations.

Sunspot growth consists of two phases: (1) the convective collapse of the flux tubes and (2) the coalescence of flux tubes. Apparently only faculae and small pores may be formed in just one collapse process. The formation of larger pores and sunspots requires coalescence. In some cases, the coalescence may be followed in time series of white-light pictures, if some of the constituting elements are large enough to show up as pores – the formation of pore G Figure 7) is such a case. However, the pore indicated by H in Figures 2 and 8 is seen to grow and to darken steadily without an influx of distinct photospheric features. Presumably, the flux tubes streaming into the existing pore H are not visible as pores because they are either too small or in an early stage of convective collapse. The idea that pore H grows through an influx of many small magnetic elements is supported by the large area of downward flow and several faculae in the vicinity of pore H (see Brants 1985b, Figure 1).

The study of the evolution of the leading spot, indicated by A in Figures 2 and 6, is facilitated by the persistence of the dark cores after coalescence. Protopores and pores are seen to flow into the spot A along two streams (see Figure 6): from the East (A4, A51, A52) and from the south (A24, A25, A31, A32,...). These two streams are probably caused by the two emerging flux regions mentioned in Section 3, which are comprised in the shearing velocity field discussed in Section 4.1. Brants and Steenbeek point at growth occurring at the NE and NW boundaries without an influx of conspicuous photospheric features.

By the end of the observing period on 24 July, the leading spot A was large enough (area $\simeq 70$ Mm$^2 = 25 \times 10^{-6}$ hemisphere) to develop a penumbra, but apparently such a process requires more time then a few hours. The next day, however, when the area had become about 100×10^{-6} hemisphere, the spot had developed a complete penumbra (Figure 5).

While the majority of the pores are growing, some are decaying. Brants and Steenbeek give examples of protopores and pores that form and disappear again during the duration of the observing run. The pore marked F in Figure 2 decayed by a decrease in area and a spectacular loss in contrast (see Brants and Steenbeek, 1958; Figure 13). These cases concern pores of following polarity, and the posibility of an early decay of spots of following polarity is well known. More striking is the decay of pores of leading polarity, viz. the pores indicated by B2, B3, and B4 in Figure 2. These pores had the same proper motion, which differed markedly from the motion of the growing spots of leading polarity A1 and B1 (see Figure 6 in Brants and Steenbeek, 1985). This may indicate that the pores B2, B3, and B4, possibly together with the decaying pore F of opposite polarity, had been formed earlier than the other spots from an EFR very early during 24 July or late during 23 July.

4.4. Upward flows, dark alignments and emergence of magnetic flux

Large-scale upward flows with velocities larger than 0.5 km s^{-1} occurred in the zone around the main polarity dividing line $B_{\parallel} = 0$ within the region close to the leading spots (Brants, 1985b, Figure 1). These upward-moving photospheric areas coincided with, or were close to nearly transverse, strong magnetic fields ($1000 < B \lesssim 2000$ G, see Section 2.2). Moreover, many of the resolution elements with a strong and nearly transverse field showed an additional small-scale upward flow and many of these elements were *darker* than elements in the normal granulation. These observational data suggest a picture of the tops of magnetic loops being pushed upwards through the top of the convection zone into the photosphere, thereby impeding the turbulent heat exchange. It is not likely, however, that the high field strengths $B > 1000$ G are typical for the tops of flux loops during emergence. The fields were found in a strong velocity shear (Section 4.1). They are part of, or pushed against, a strong magnetic flux concentration in the leading sunspot pore that started emerging quite early in the development of the new active region.

Elsewhere in the region, strongly inclined but weak fields were indicated by polarity reversals. Brants (1985b) estimates the local field strength B = 500 $\pm$ 300 G. These elements showed a tendency towards upward flow but they were not located in large-scale upward flows. These elements may correspond to emerging tops of unstrained flux loops.

During the observations, probably most of the magnetic flux was emerging in the central part of the region. One indication is that this part was spanned by an arch filament system. Although the following polarity dominated over the central part, there were inclusions of leading polarity. Moreover, the granulation was disturbed: the alignments of abnormally dark intergranular lanes (Figure 2a and 8) were parallel to the loops in the arch filament system and to the inferred direction of the horizontal component of the magnetic field. Hence, the interpretation of the alignments by the effect of tops of magnetic flux loops on the granulation is plausible.

The system of alignments was maintained by a succession of alignments, one alignment being visible during about 10 minutes (Brants and Steenbeek, 1985). From the diameter ($\simeq 1500$ km, see Figure 8) and the lifetime, the speed of rise is estimated at $v \simeq 3$ km s^{-1}. If this speed equals the Alfvén speed, the field strength is about 600 G, and the flux about 1×10^{19} Mx. The latter figure agrees with estimates of the magnetic flux per loop in arch filament systems (Born, 1974). So the present data suggest that the magnetic field emerges with approximately the equipartition strength in the top of the convection zone.

5. Discussion

This section attempts to fit pieces of information together, and to identify some of the open questions.

The magnetic flux emerges in the central part of the growing region. It is estimated that the field strength during emergence is about equal to the equipartition value for the top of the convection zone, that is $B_{eq} \simeq 500$ G (Section 4.4). A better determination of the field strength at emergence is urgently needed. During emergence, the magnetic field and the granular convection are in a transient state: the interaction between the magnetic field and convection produces a system of alignments of particularly dark intergranular matter. This dark matter is observed moving upward. One alignment lasts for about 10 min, which probably corresponds to the time it takes for one flux loop to cross the photosphere.

After the top of a flux loop has risen to a few thousand km height above the photosphere, it becomes visible in Hα as a loop in the arch filament system. The draining of the chromospheric loop and the convective collapse of the photospheric feet produce strong downward flows. During this phase, the chromospheric faculae are particularly bright. The collapse produces the normal small-scale magnetic elements – faculae of normal brightness and small pores, at the well-known field strengths between about 1500 and 2500 G, and without noticeable vertical mass flow. The partitioning between the magnetic field and the granular convection is then established: the faculae are organized in the filigree structure which consists of bright elements that are between neat granules.

Large pores and sunspots are formed by coalescence of thin magnetic tubes. This coalescence may happen in a continuous process so that flux tubes flow into a growing sunspot while collapsing (the growth of pore H is probably an example, see Section 4.3). Sunspots and pores may also coalescence many hours after flux emergence and convective collapse have stopped. Even pores and spots of the same polarity but formed by different EFR's may fuse into a single sunspot.

The heuristic model of the emergent loop consisting of many magnetic flux tubes

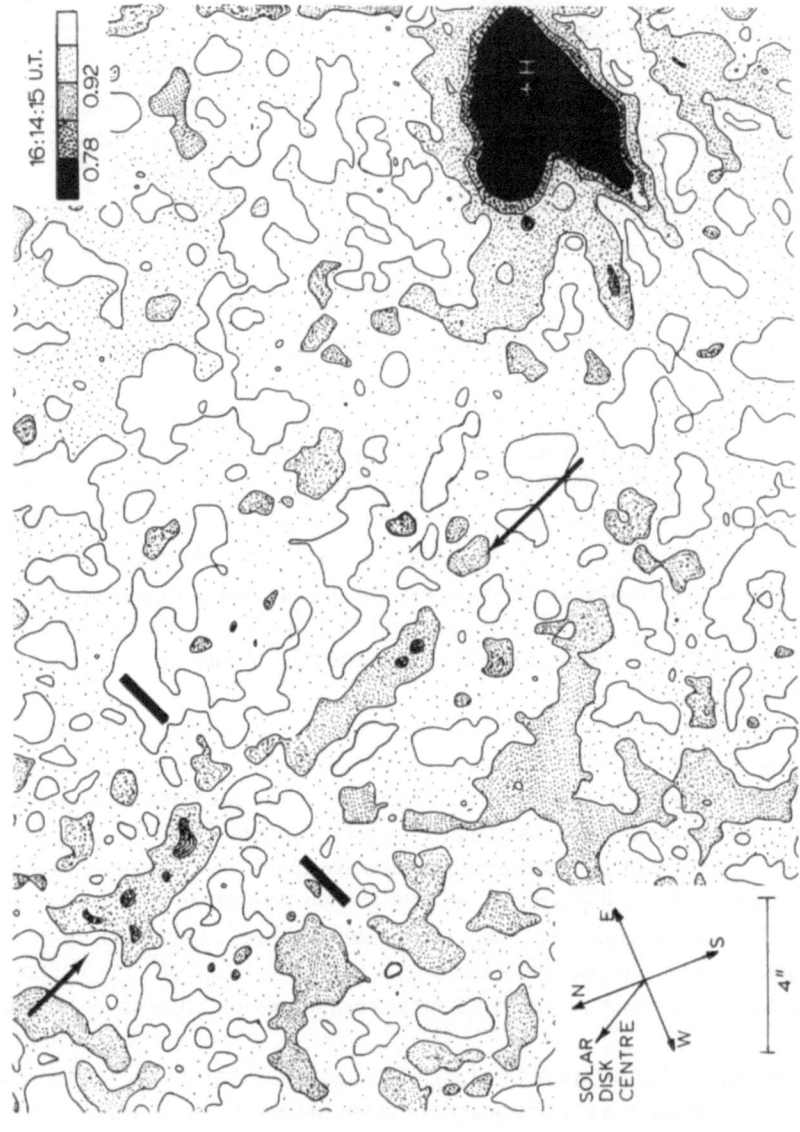

Fig. 8. Isophote map showing the environment of the rapidly growing pore H and a dark alignment (between arrows) in the granulation in the central part of the growing active region (from Brants and Steenbeek, 1985).

(simplified to a few in Figure 1), combined with the theory of the convective collapse, explains the observed structure and dynamics of emerging flux regions: their central part where the magnetic and convective structures are in a transient state, the border zones where the arch filament system is rooted and the strong downflows are found, and the outer leading and following edges where sunspots are formed by coalescence. The dynamics within an emerging flux region, in particular the separtion of the opposite polarities and the coalescence of patches of the same polarity, are largely determined by the buoyancy of the (collapsed) flux tubes. The formation of a single large sunspot requires a force in the convection zone to contain the bundle of flux tubes constituting the spot.

Complexity in the magnetic structure and the velocity field of a growing active region can be produced by contributing EFR's occurring at different orientations.

Acknowledgements

The observations obtained during 1979 at Sacramento Peak Observatory were prepared and carried out in collaboration with L. E. Cram who, with H. A. Mauter, made the setup work. J. W. Harvey supplied magnetograms covering 22–28 July 1979, and he showed us the evolution of the magnetic field on the data display in Tuscon. V. Gaizauskas provided the $H\alpha$ filtergrams.

This paper reflects many discussions with J. J. Brants while he prepared his PhD thesis. He and R. J. Rutten read a draft of this paper.

References

Bachmann, G.: 1978, *Bull. Astron. Inst. Czechosl.* **29**, 180.
Born, R.: 1974, *Solar Phys.* **38**, 127.
Brants, J. J.: 1985a, *Solar Phys.* **95**, 15.
Brants, J. J.: 1985b, *Solar Phys.* **98**, 197.
Brants, J. J.: 1985c, Thesis Utrecht.
Brants, J. J. and Steenbeek, J. C. M.: 1985, *Solar Phys.* **96**, 229.
Brants J. J. and Zwaan, C.: 1982, *Solar Phys.* **80**, 251.
Bray, R. J. and Loughhead, R. E.: 1964, *Sunspots*, Chapman and Hall, London.
Bruzek, A.: 1967, *Solar Phys.* **2**, 451.
Bruzek, A.: 1969, *Solar Phys.* **8**, 29.
Bumba, V. and Howard, R.: 1965, *Astrophys. J.* **141**, 1492.
Fox, Ph.: 1908, *Astrophys. J.* **28**, 253.
Frazier, E. N.: 1972, *Solar Phys.* **26**, 130.
Gopasyuk, S. I.: 1967, *Izv. Krymsk. Astrofiz. Obs.* **37**, 29.
Gopasyuk, S. I.: 1969, *Izv. Krymsk. Astrofiz. Obs.* **40**, 111.
Harvey, K. L. and Martin, S. F.: 1973, *Solar Phys.* **32**, 389.
Howard, R.: 1971, *Solar Phys.* **16**, 21.
Kawaguchi, I. and Kitai, R.: 1976, *Solar Phys.* **46**, 125.
Kiepenheuer, K. O.: 1953, in G. P. Kuiper (ed.), *The Sun*, University of Chigago Press, Chigago, Chapter 6.
McIntosh, P. S.: 1981, in L. E. Cram and J. H. Thomas (eds.), *The Physics of Sunspots*, Sunspot, NM, p. 7.
Meyer, F., Schmidt, H. U., Weiss, N. O., and Wilson, P. R.: 1974, *Monthly Notices Roy. Astron. Soc.* **169**, 35.

Mosher, J. M.: 1977, Thesis Caltech.

Parker, E. N.: 1978, *Astrophys. J.* **221**, 368.

Priest, E. R.: 1984, *Solar Magnetohydrodynamics*, D. Reidel, Dordrecht, Holland.

Schmidt, H. U.: 1968, in K. O. Kiepenheuer (ed.), 'Structure and Development of Solar Active Regions',
   IAU Symp. **35**, 95.

Schoolman, S. A.: 1973, *Solar Phys.* **32**, 379.

Sheeley, N. R.: 1969, *Solar Phys.* **9**, 347.

Spruit, H. C.: 1979, *Solar Phys.* **61**, 363.

Vrabec, D.: 1974, in R. G. Athay (ed.), 'Chromospheric Fine Structure', *IAU Symp.* **56**, 201.

Waldmeier, M.: 1937, *Z. Astrophys.* **14**, 91.

Waldmeier, M.: 1955, *Ergebnisse und Probleme der Sonnenforschung*, Geest und Portig, Leipzig.

Weart, S. R.: 1970, *Astrophys. J.* **162**, 987.

Weart, S. R.: 1972, *Astrophys. J.* **177**, 271.

Zirin, H.: 1972, *Solar Phys.* **22**, 34.

Zirin, H.: 1974, in R. G. Athay (ed.), 'Chromospheric Fine Structure', *IAU Symp.* **56**, 161.

Zwaan, C.: 1978, *Solar Phys.* **60**, 213.

Zwaan, C., Brants, J. J., and Cram, L. E.: 1985, *Solar Phys.* **95**, 3.

MODERN OBSERVATIONS OF SOLAR PROMINENCES

TADASHI HIRAYAMA

Tokyo Astronomical Observatory, University of Tokyo, Mitaka, Tokyo 181, Japan

Abstract. We review observational studies of solar prominences with some reference to theoretical under-standings. We lay emphasis on the following findings: (1) An important discovery was made by Leroy, Bommier, and Sahal-Bréchot concerning the direction of the magnetic field inside some high-altitude, high-latitude prominences, where the field vector points in the *opposite* direction from the one which would be expected from the potential field calculated from the observed photospheric magnetic field. (2) Landman suggests the possibility of a high total density of $\approx 10^{-11}\,\mathrm{g\,cm^{-3}}$ for the main body of quiescent prominences, ≈ 50 times higher than the value hitherto believed. (3) Flow patterns, nearly parallel to the magnetic neutral lines, were detected in the 10^5 K plasma near and in prominences. (4) Coronal loop structures were found overlying prominences as viewed from X-ray photographs. We propose also an evolutionary scheme by taking the magnetic field topologies into account.

The fundamental question why a prominence is present remains basically unanswered.

1. Introduction

Prominences are fascinating objects, abundant in variety, beautiful, and above all mysterious. Prominence study in recent years has become intermingled with other branches of solar physics: the formation of prominences with coronal heating, eruptive prominences with flares, and the evolution of quiescent prominences with that of active regions and complexes of activity. Measurements of magnetic fields, spectroscopic diagnoses, and Hα, X-ray, and EUV morphologies are, I believe, still in their infancy, because there is not yet enough information to solve the fundamental problems of prominences (formation, stability, and eruption).

First, I briefly summarize some important historical developments: (1) Studies of prominence developments in Ca II K and Hα begun with the newly invented heliographs (and helioscopes) by Hale with Ellerman in 1903 and Deslandres in 1910, and culminated in a monumental document of d'Azambuja and d'Azambuja (1948). (2) Apart from the discovery of helium in 1868, spectroscopic studies revealed that prominence spectra are similar to that of the chromosphere; e.g., Schwarzschild in 1906. (3) Lyot's coronagraph and his filter, invented in the 1930's, have made a great impact on prominence study. (4) Prominences were found to lie on the neutral line of the photospheric longitudinal (line-of-sight) magnetic field (Babcocks in 1955); the actual detection of a magnetic field in prominences (Zirin and Severny, 1961) followed. (5) Magnetostatic models have appeared (Menzel; Dungey; Kippenhahn and Schlüter, 1957). (6) The study of the prominence-corona interface begun using EUV emissions from space observations and also at radio wavelengths together with the study of the influence of EUV radiations.

Many people including Pettit, Menzel and Evans, de Jager (1959), and Zirin (1979) have classified prominences. Here the following scheme is adopted for the purpose of

Solar Physics **100** (1985) 415–434.   0038–0938/85.15.

definition and presenting this review. (1) Quiescent prominences (quiescents or QP): the word 'quiescent' has been used as opposed to 'activated', but here it means that they occur outside active regions (in quiescent regions). QPs are *long* in shape, consist of cold and dense plasma suspended in the corona, are invariably found on the line of zero magnetic field in the line-of-sight component (zero longitudinal field). They are found in the outskirts of active regions, between active regions, and in the polar belts (polar crowns). (2) Active region prominences (more often called active region filaments, ARF): these are found in active regions. The only difference from QPs is that they are low, and composed of horizontally sheared fibrils (see Figure 1). (3) Post-flare loops or loop prominences (prominence systems): formerly these were often called sunspot prominences. These are now understood as being the result of cooling processes of the 10^7 K plasma in solar flares. (4) Surges: they consist of cool gas of high speed, probably ejected along the magnetic field lines.

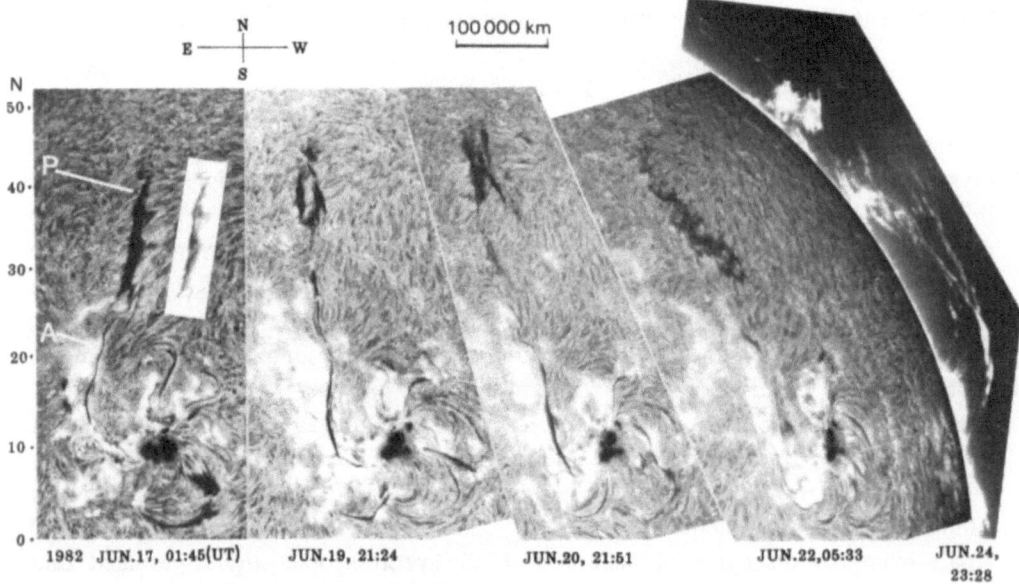

Fig. 1. Quiescent prominences (*P*) and active region prominences (*A*). Note the chromospheric fine structures near prominences. Ordinate: latitude. Photos taken at Norikura Solar Observatory by H. Morishita. This Hale region 14422 (CMP = 18.8^d) was newly born, but in an older region. It is hence not very typical. Particularly the limb picture shows unexpected high-rising streams, unusually high as compared with the two ARFs below.

Erupting or rising prominences (or eruptives) can be placed in Category (1) or (2). Sprays are mostly high-speed eruptive prominences from active regions (e.g., Tandberg-Hanssen *et al.*, 1980) and, hence, belong to Category (2).

The following sections review recent studies, concentrating mainly on quiescent and active-region prominences, since erupting prominences, surges, and post-flare loops were reviewed extensively in Sturrock (1980) (earlier references are in Tandberg-Hanssen, 1974; and Jensen *et al.*, 1979).

2. Spectroscopic Characteristics of Quiescent Prominences and
Their Energy Balance

Since the study of thermodynamic structures, including the energy balance inside a prominence, has been reviewed (Hirayama, 1979), I will describe only very recent results on quiescent prominences. (We still lack information on low-lying active region filaments.) First, the temperature can be reliably determined by comparing line widths of optically thin hydrogen, helium, and metallic lines. We can also use the colour temperature of hydrogen Lyman continuum, which, however, gives only temperatures of the outer parts because of the high opacity (Kanno et $al.$, 1981). For example, Orrall and Schmahl (1980) gave 7520 K. A similar value of 7770 K came from observations of the He I 504 Å continuum (see Heasley and Milkey, 1983). And a general tendency of increasing temperature towards periphery is seen in many instances. The lowest (yet reliable) temperature is 4300 K, compatible with the widths of the helium lines. The average is 6500 K (5000–8000 K) in the main part. A part of quiescents shows $T \gtrsim 20\,000$ K, which is inferred from emission-line wings. If an one-digit value is needed, 7000 K is recommended.

The electron density can best be determined from the Stark effect of high Balmer lines. A variety of values can be found beyond observational errors by the same method (e.g., Hirayama, 1972, Figure 8). Using the same observations, I newly determined the average electron density of $10^{11.02}$ cm^{-3} for five hedgerow quiescent prominences at 57 different positions, and $10^{10.48}$ cm^{-3} for two curtain-like old quiescent prominences at six different positions. Here the maximum value is $10^{11.4}$ cm^{-3}, and if $n_e \leq 10^{10.3}$, the determination becomes difficult. Sahal-Bréshot (1984) has derived an electron density of $\approx 7 \times 10^9$ cm^{-3} from the Hanle effect, in which depolarization of the Hβ line caused by collisions as compared to He D$_3$ is utilized (see also Leroy, 1985). $n_e \leq 10^{10.0}$ may well be expected at the fainter part of quiescents. The $electron$ pressure in the main body of quiescents can be taken at 0.10 dyn cm^{-2}.

New results that have essential bearing on the whole physics of quiescent prominences are emerging from non-LTE line intensity calculations (Landman, 1983; 1984; 1985a; see also Nikaido and Kawaguchi, 1983); namely, Landman derives the ionization of hydrogen of $n_{\mathrm{H\,II}}/n_{\mathrm{H\,I}} \approx 0.07$ and consequently, with the inferred electron density of $n_e \approx 10^{11.3}$, a total gas pressure of 3–6 dyn cm^{-2}. (The total number density of hydrogen is $n_{\mathrm{H}} \approx 5 \times 10^{12}$ cm^{-3}.) The value of $n_{\mathrm{H\,II}}/n_{\mathrm{H\,I}}$ and the gas pressure are more than one order of magnitude smaller and larger, respectively, than previous values (Hirayama, 1979, p. 14). Landman obtained these values by comparing the intensity ratios of the Balmer lines, lines of Sr II and Ba II, and of Mg I and Na I, with non-LTE calculations using observations by Ye (1961), Yakovkin and Zel'dina (1964), and himself. The problem is that while I find $n_e < 10^{10.3}$ from the Stark effect for a prominence in my data, the metallic intensity ratio still gives $n_e \approx 10^{11.3}$ (from Figure 3 of Foukal and Landman, 1984) for the same object, and I suspect that similar discrepancies might be found in the hydrogen ionization and in the total gas pressure. Summarizing, $0.05 \leq n_{\mathrm{H\,II}}/n_{\mathrm{H\,I}} \leq 1$ will be the best range, though again at the periphery of prominences $n_{\mathrm{H\,II}}/n_{\mathrm{H\,I}} > 1$ may be expected.

The non-LTE transfer equations are beginning to be treated with the use of a cylindrical (Heasley, 1977) and a two-dimensional geometry (Vial, 1982) instead of an infinite slab model, or by incorporating a partical frequency redistribution in place of complete redistribution (Milkey *et al.*, 1979; Heinzel, 1983). The former authors found that the observed intensity ratio of hydrogen Hα to Lα ($\approx$ 1–2) can be made consistent with the partial redistribution approach giving a factor of two smaller intensity in Lα than the complete redistribution, because of the photon trapping due to coherent scattering.

The non-thermal velocity, often called turbulent velocity, is known to be 2–10 km s^{-1} in the main body and generally increases towards the periphery or towards fainter part of quiescents (e.g., Hirayama, 1964; 1979). A rather peculiar behaviour of the widths of metallic lines was detected by Landman (1985b), who found that the Fe II b$_3$ line is systematically narrower (20%) than the Mg I b$_2$ lines, and he suggests selective inhibition of ionic motions in the prominence magnetic field. Mariska *et al.* (1979) found from a detailed study of a prominence at the prominence-corona interface that the non-thermal motions increase from $\approx$ 10 km s^{-1} at 2×10^4 K to 12–30 km s^{-1} at 2×10^5 K, and at higher temperature it rather decreases towards higher altitude. The cause of the general increase of the non-thermal velocities towards the periphery is not known, though amplification of travelling wave amplitudes due to the decreasing density at the periphery could be conceived.

Radiative equilibrium models of quiescents were presented by Heasley and Mihalas (1976), where the incident radiation from the corona, chromosphere, and photosphere determines the temperature and its variation inside the prominence. The observed temperature of $\approx$ 7000 K is reproduced as a result of incident Lyman continuum radiation if a thin slab model is considered. But for a thicker model, which is closer to the observation, either a small temperature, of 4500 K, is expected, or to get 7000 K they needed an extra heating source, or diffuse penetration of Lyc radiation in the filamentary structure. Further work in this area is certainly necessary and should incorporate Landman's gas pressure and the prominence-corona interface.

The determination of the helium abundance is an important application of prominence spectroscopy: Heasley and Milkey (1978) obtained a number ratio to hydrogen of 10%, which however is doubted because too large a thickness (of 10 000 km) and too small a gas pressure (of 0.01 dyn cm^{-2}) were used. This implies that the true value may be larger than 10%. On the other hand Hirayama *et al.* (1979) obtained $\geq$ 14% and possibly 16%, which is not believed because it differs from the *cosmic abundance* of 10% (see also Heasley and Milkey, 1983). There might be a difference between the main body of the Sun and prominences, as is the case with flare-associated solar wind plasmas.

3. The Magnetic Field and Its Topology

Since the magnetic field must play an essential role for the formation, support and eruptions of prominences, the measurement of the field is of vital importance. Measurements of the longitudinal field by the Zeeman effect were begun by Zirin and Severny

(1961), and are summarized by Tandberg-Hanssen (1974) and Leroy (1979). Good agreement in the field strengths was obtained from different lines such as $H\alpha$, $H\beta$, $He\,D_3$, and metallic lines. An alternative method which has recently been used extensively is based on the Hanle effect (first application by Hyder, 1965; latest summary by Leroy, 1985). The Pic du Midi polarimeter uses broad wavelength bands at D_3, and also $H\alpha$ and $H\beta$, while the Sac Peak–HAO polarimeter uses the full Stokes parameters along the entire line profile of D_3 (Landi degl' Innocenti, 1982). Note that though the latter can occasionally secure the circular components also, both are essentially measuring the linear polarization due to the Hanle effect. (Linear polarization due to the Zeeman effect is too small in quiescent prominences).

In the following a very brief account of the observation from the Hanle effect method is given (Leroy, 1985; Sahal-Bréchot, 1984). A strong line emitted in prominences scatters the photospheric light coherently and gives a linear polarization ($p_{max} \gtrsim 6\%$ near the limb in $He\,D_3$) directed horizontally to the solar surface if there is no magnetic field (Figure 2(b), $B = 0$). If the magnetic field is present as shown in Figure 2(a), not only the polarization degree decreases ($p \approx 2\%$ in D_3), but also the direction will tilt ($\varphi \neq 0$ in Figure 2(b)). This results from the precession of an atom around the magnetic field lines if interpreted in terms of a classical oscillator. In quantum mechanical words the interference effects between the wave functions of the partially overlapping Zeeman sublevels cause the tilt and decrease of polarization. If the field is too strong, only the field orientation is inferred due to the complete separation of the sublevels (the critical field strength is $\approx 10 B_c$, where B_c is the field strength yielding a Larmor period equal to the radiative lifetime). Quantum mechanical calculation shows that the three unknowns ($|\mathbf{B}|$, Ψ, θ) can be expressed as functions of two observable quantities, p and φ. If one uses two lines having different functional forms such as (1) the main (blue) component and the faint red component of $He\,D_3$ or (2) $He\,D_3$ and $H\beta$, it is possible to know the three unknowns. However, still only the absolute value of the azimuthal angle θ (Figure 2(a), (c), (d)) can be derived, as is the case with the Zeeman effect (see

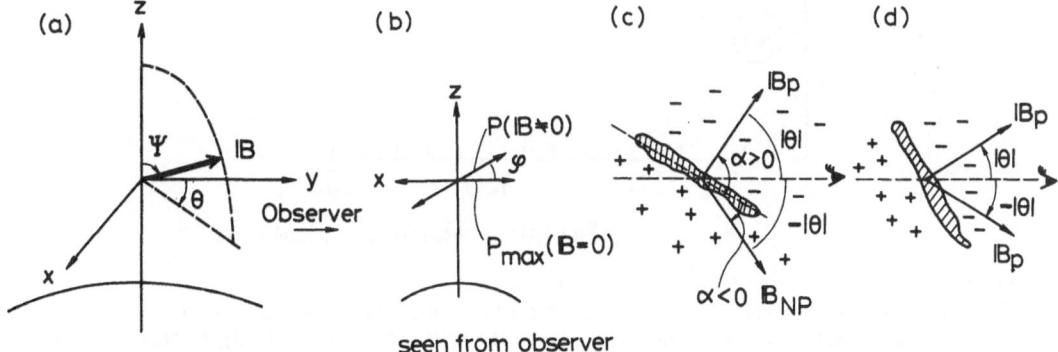

Fig. 2.   The Hanle effect and the relation between the long axis of a filament and the magnetic field direction B. B_P is the potential-like field. B_{NP}, a non-potential-like field.

below). The accuracy is $\Delta B \approx 0.5$ G (for $B \approx 10$ G), $\Delta|\theta| \approx 5°$, and $\Delta\Psi \approx 10°$, while for the longitudinal Zeeman instrument $\Delta B \approx 1$ G.

We can summarize the results from recent magnetic field measurements together with earlier ones as follows: first the absolute magnetic field strength for quiescents is, on the average,

$$|B| = 8 \text{ G}.$$

It depends upon the solar activity as shown in Figure 3. There may also be another preferred strength, viz. 20 G, in active region filaments. On the other hand, active prominences such as surges, post-flare-loops, and some active filaments show 20–150 G and reach to 200 G (longitudinal field: Harvey, 1969; Rust, 1972; Tandberg-Hanssen, 1974). A remarkable fact is that the magnetic field strength does not generally decrease with height inside the prominence as far as observations reach. A slow increase with height, of 0.5–1×10^{-4} G km^{-1}, is seen (Rust, 1972; Leroy *et al.*, 1983), and this increase has been considered to be consistent with a stability criterion for the 'V'-shaped part of the magnetic field model of quiescents, though Athay *et al.* (1983) did not confirm the increase from the detailed inspection of a few prominences.

Next we come to the question whether the magnetic field is horizontal at the place

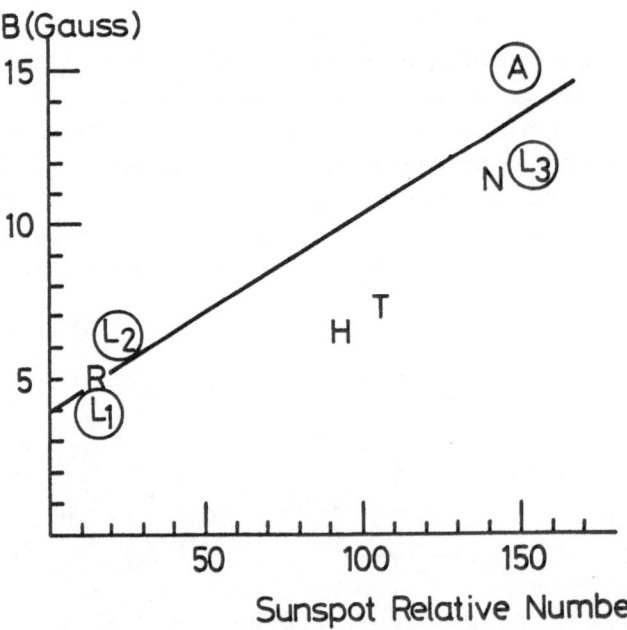

Fig. 3.   Magnetic field strength versus relative spot number. The symbols in circles give the total field strength from the Hanle-effect. Those without circle are from the Zeeman longitudinal field measurement. *A*, Athay *et al.* (1983); *H*, Harvey (1969); $L_{1,2}$, Leroy (1979); L_3, Leroy *et al.* (1983); *N*, Nikolsky *et al.* (1984); *R*, Rust (1972); *T*, Tandberg-Hanssen (1974). The straight line was obtained from the weighted total field measurements. The average sunspot number over 22 years (1961–1983) is 69.3, hence, the average field is $B = 8$ G.

where cool emission originates. A definitive conclusion came from the analysis by Athay *et al.* (1983), whose averaged data show that the median value of the deviations from the horizontal field ($\Psi = 90°$) is only $|\Delta\Psi| = 3°$. A similar result was obtained by the Pic du Midi observations: $|\Delta\Psi| \approx 15°$ from $H\beta$ and D_3 (Leroy *et al.*, 1984). Early observations by Leroy (1979) had excluded small Ψ's because the observed depolarization was considerable. Thus we are forced to accept that not only the vertical curtain-like filamentary structures in hedgerow prominences but also funning tree-like structures are not situated along the field direction (Athay *et al.*, 1983, prominences *G* and *D*).

We now treat one of the most important problems of the magnetic field measurements; what is the angle between the magnetic vector and the long axis of prominences (the angle α in Figure 2(c)). So far four methods have been used to discriminate one of the two values of α (or θ).

(1) If two field vectors (hereafter assumed horizontal) are found both at either side of a prominence, one can conclude as with the case of Figure 2(d) that the magnetic field vector points into the same direction as photospheric magnetic field vector, in which case we call a potential-like prominence *P* (often called the Kippenhahn–Schlüter (1957) type, $\alpha > 0$). The opposite case is called a non-potential-like prominence *NP* (called the Kuperus–Raadu (1974) type, $\alpha < 0$). See Figure 4 (A–a and A–d) and also Anzer

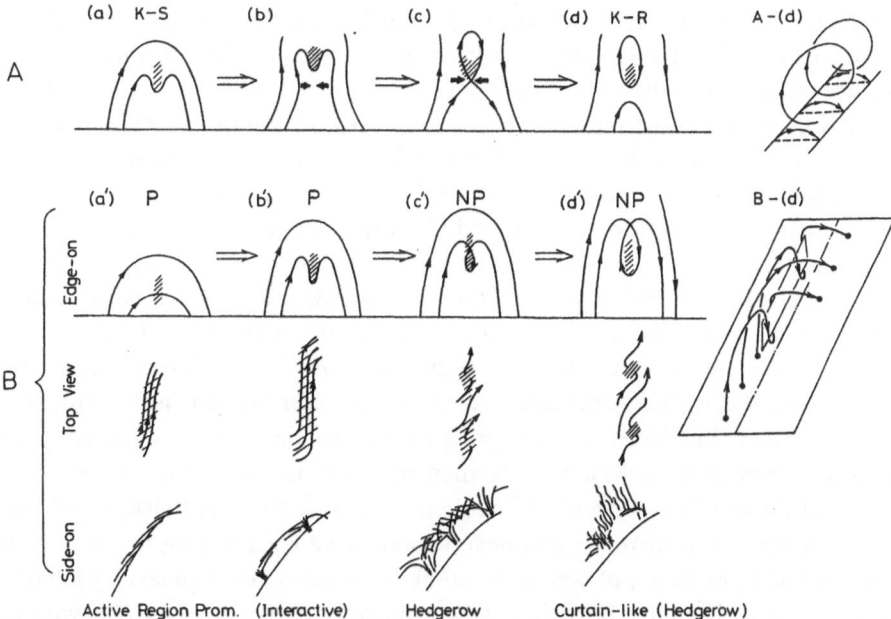

Fig. 4. Magnetic field configurations and prominence development. The thinly shaded area is the prominence. K–S: the Kippenhahn-Schlüter model. K–R: the Kuperus–Raadu model. *P*: potential-like. *NP*: non-potential-like. The 'A' series is from Pneuman (1983). Prominence drawings at the bottom do not necessarily correspond to the 'A' series. The 'B' series is from the present study. In B–c' and d', the magnetic field in between each vertical thread would not have a pit, hence is inhomogeneous along the prominence axis. The morphological development is described in Section 6.

(1979). Leroy *et al.* (1984) identified eight *NP*-type (mostly hedgerow type) and five *P*-type prominences by this method. Also two *NP* prominences can be selected from 12 prominences observed by Athay *et al.* (1983, Table III). By this method the value of the tilt angle, α, cannot be determined.

(2) If one observes the linear polarization in Hα together with that in He D$_3$, and if the optical depth of Hα is known, say, from the total intensity of Hα, it is possible to determine the tilt angle since the observed p and ϕ become different for a positive azimuthal angle $|\theta|$ and a corresponding negative value $-|\theta|$ in an optically thick line. About ten high-altitude prominences have been classified by this method to be all of the *NP*-type (Bommier *et al.*, 1985a).

(3) If the long axis of a prominence lies parallel to the line-of-sight, one can unambiguously determine the value of $|\alpha|$. Statistically $|\alpha|$ clusters around 20–30°. On the other hand for non edge-on prominences one can again see clustering for a number of prominences around $|\alpha| = 20$–30°, if one assumes that these non edge-on prominences are all of the *NP*-type. However if a *P*-type is assumed, such clustering cannot be seen. Hence statistically one can say that the prominences under study are more likely to be of the *NP*-type (Leroy *et al.*, 1983, the first report of the *NP*-type).

(4) If prominences are supposed to have a preferential value of the tilt angle α then the other (fictitious) solution of α will become a linear function of the angle between the long axis of prominences and the line-of-sight. By searching for a best fit one can deduce this preferential tilt angle; the results from Pic du Midi and Sac Peak observations agree with each other and give $\alpha = -20 \sim -30°$ on the average. This means that a large part of the prominences observed are of the *NP* (Leroy *et al.*, 1984; Bommier *et al.*, 1985b).

The first method appears rather secure and the results from the fourth method seem reliable. Altogether the conclusions are the following: magnetic structures of prominences with maximum heights higher than $\approx 30\,000$ km are consistent with the *NP*-type of models, and the mean field strength and tilt of the field direction from the long axis of prominences are $\alpha \approx -25°$ and $B \approx 5$ to 10 G. These prominences often show curtain-like filamentary structures. On the other hand, prominences with maximum heights lower than $\approx 30\,000$ km are consistent with the *P*-type models (Kippenhahn–Schlüter), and give $\alpha \approx +20°$ and $B \approx 20$ G (Leroy *et al.*, 1984; Bommier *et al.*, 1985b). These prominences are bright, often sharp-edged in He I D$_3$ and occur essentially at low latitudes. And polar crown prominences are apparently *not* of the *P*-type (Leroy *et al.*, 1983). The Zeeman magnetic measurements show also the smallness of the absolute value of α (Tandberg-Hanssen, 1974; Nikolsky *et al.*, 1984). And since independent methods and observations have been employed, there may be no doubt about the existence of non potential-like structures. It would be very interesting if one prominence is found to be of the *P*-type in its lower parts and of the *NP*-type higher up as has been searched for by Leroy, and also if one would find in a long filament that one end is a typical active region low lying filament of the *P*-type and the other end is a high quiescent prominence of the *NP*-type.

The magnetic field orientation which is here taken to be roughly *parallel* to the long axis of prominences seems to be changing during the solar cycle (Leroy *et al.*, 1984,

Figure 13). The orientation of the field in the polar crown is the same as the photospheric one which would result from the distortion by differential rotation of an initially meridian line of force.

Measurements of active region filaments (ARF) are scarce because the photopheric scattered light entering into the coronagraph makes them very difficult. However Doppler shifts have been seen in many cases, suggesting that, in contrast to hedgerow vertical-curtains, the magnetic field is nearly parallel to the long filamentary structures of ARFs. I could not find any positive evidence of a pit, essential to the Kippenhahn–Schlüter type model, from a number of good Hα pictures taken at various observatories, particularly near the limb. Since the top height of ARFs (a few thousands to $\approx 10^4$ km) is much higher than the pressure scale height of ≈ 200 km or so, there is no way of *statically* supporting the heavy material. In conformity with the back and forth motions along filaments seen in Hα movies with Doppler measurements (Martin, 1973), there should be supersonic or near-sonic motions of gas filling up magnetic flux tubes (Meyer and Schmidt, 1968). Long fibrils of ARFs are often seen twisted, but I do know of any examples showing one full turn of, say, two fibrils in an ARF; the twist angle seems to reach at most 180° (Morishita, 1985).

In the following I will briefly describe the developments of the magnetic topology of ARFs to QPs, taking into account new observational facts concerning potential versus non potentiallike field directions. Figure 4–A depicts one possible way of development: a classical Kippenhahn–Schlüter type model (A–a) would change, say by shearing motion due to differential rotation, to a high-altitude prominence (A–b), and finally by slow reconnection (c) to another stable Kuperus–Raadu type model (d) (Pneuman, 1983). Another case which I propose is shown as (b′) in Figure 4: here a potential-like ARF is the starting model. And again due to differential rotation and/or large scale motion, shear would develop in the neutral line. The prominence, passing the K–S type in (b′) evolves to a non-potential-like (c′), and finally to *NP*-type (d′). Thus (d′) would look like as though it were of the Malville-type as reviewed by Anzer (1979), but the essential difference is that there is no neutral point in (d′), as is seen in the top view: there has been no topology change from (a′) to (d′), only a slight twist with deep pits (see Section 6).

4. Prominence-Corona Interface and Relations to the Corona

The prominence-corona interface (PCI) can be studied best with EUV lines as well as with radio emissions (see a review by Schmahl, 1979). Much of the recent work in this area has been done with data from OSO VII and OSO VIII, Skylab, SMM, and from rocket observations.

Using the intensity ratio of density-sensitive lines (e.g., CIII 1176/977 emitted at 5×10^4 K), the electron pressure P_e in quiescent prominences is found to be $\frac{1}{5}$ to $\frac{1}{2}$ (Schmahl, 1979), or $\frac{1}{2}$ to 1 (Mariska *et al.*, 1979) of the quiet chromosphere-corona transition zone, which is around 0.1 dyn cm^{-2}. An exceptional case was reported by Poland and Tandberg-Hanssen (1983), who obtained $P_e \approx 1$ dyn cm^{-2} ($n_e \approx 10^{11}$ cm^{-3}). The lower than normal pressure value could partly be understood in terms of a lower hydrostatic pressure at the higher altitude if a magnetic flux tube containing 10^6 K plasma connects the base of the chromosphere and the prominence material.

If we adopt 0.5 at 1.0×10^5 K for the intensity ratio of the PCI to the quiet Sun, being the average over nine prominences (Schmahl, 1979), the thickness of the PCI, L, becomes several times larger than the quiet value, since the line intensity is proportional to $P_e^2 L$ at a fixed temperature. The thermal conductive flux to the cool prominence matter will then be smaller than the quiet value (Withbroe and Noyes, 1977) of 2×10^5 erg cm^{-2} s^{-1}. However, since Orrall and Schmahl (1980) estimated that there may be on the average about ten interfaecs in the line-of-sight, the thickness decreases, and on the other hand the conduction flux increases by this factor. Schmahl concludes that at a single interface the conductive flux may be balanced by the radiation flux at the interface (the hydrogen Lα radiates 2×10^5 erg cm^{-2} s^{-1} and the interface lines lose about $\frac{1}{7}$ of that).

Observations on the disk show that quiescents are seen dark in every EUV lines observed, even in the lines emitted longward of the hydrogen Lyc (Schmahl, 1979, Figure 8; Schmieder *et al.*, 1984, for C IV 1548; Hirayama *et al.*, 1985, for C III 977). Filaments observed in H I Lα (Bonnet *et al.*, 1980) and He II 304 are rather faint like in Ca II K, though they are still darker than the average atmosphere.

The interpretation for a high-altitude prominence seen on the disk should be similar as with those seen on the limb, i.e., a small pressure and/or small sheath thickness causes smaller intensity, hence darkness. However, in the case of low-lying active region filaments, the thickness of PCI should be extremely small, since the magnetic field, probably running parallel to the fibrils, will prohibit lateral heat conduction. And I would suspect that the chromosphere just below and at the periphery of the filament also contributes to the small intensity, because the small fibrils often found on both sides of a filament are nearly horizontal (prominence corridor) and, hence, leave little space for the conventional transition region. This is not inconsistent with the results of Athay *et al.* (1985) that the 10^5 K plasmas are in the form of loops which are presumably overlying the filaments.

Since the results from radio observations will be presented in this volume by Kundu, only a brief summary is given. Raoult *et al.* (1979) summarize the values found for the contrast between the radiation temperature of filaments T_f and the quiet atmosphere, T, as a function of wavelengths: $C = (T_f - T)/T$. The C-value is about -0.4 at 5 cm, increases towards shorter wavelengths and reaches -0.04 at 3.1 cm. If the spatial resolution becomes as good as in the EUV, the C-value will decrease. Thus independent inference can be obtained for the PCI. Although Rao and Kundu (1980) demonstrate the similarity of the size of radio depressions and Hα filaments with 15" resolution, the prominence corridor in the chromosphere may be contributing to the radio depression, as is the case with C IV (10^5 K) in active region filaments. Full disk pictures at 36 GHz with a 48" resolution (Kosugi *et al.*, 1985) show that the radio depressions are clearly not confined to Hα filaments, but are much longer, which suggests that they are better tracers of the magnetic neutral line. Since the effect of the overlying corona at this wavelength of 8.3 mm is small, the coronal cavities with or without filaments will not affect the depressions as suggested by Schmahl *et al.* (1980, early references therein).

It has long been known from eclipse photos that quiescents are situated below a coronal cavity over which two to three coronal loops stride and a coronal streamer extends. Serio *et al.* (1979), among others, have clarified the following characteristics of coronal cavities from Skylab X-ray photographs. (1) Coronal cavities are dark long strips along magnetic neutral lines where, however, only a fractional length of 0.2–0.5 contains filaments or segments of a filament. An average coronal cavity has a length of 60×10^4 km (a range of 20–300×10^4 km), a width of $\approx 6 \times 10^4$ km and a height of $> 5 \times 10^4$ km, and an electron density of 3×10^8 cm^{-3} in areas without a filament (n_e is half of this where a filament is found, $P_g \approx 0.04$ dyn cm^{-2}). (2) The fractional length depends on the lifetime of the neutral line region. At average 'middle ages' of four to seven solar rotations, it becomes largest. Moreover each segment of a filament ranging from a few times 10^4 km to tens of 10^4 km survives longer (≈ 5–8 days) in the 'middle age' regions than in younger and older regions ($\approx$ a few days). (3) The structure of the cavities is roughly unchanged, apart from luminosity changes, when prominences have temporarily disappeared. Whether the structure of helmet loops is sheared or not is not known, however, because only the 'leg' portions are occasionaly seen.

In the proximity of active regions, Davis and Krieger (1982) have shown that coronal loops are seen striding over quiescent prominences (not necessarily in the cavity), where several loops of $\approx 2 \times 10^4$ km width are seen to cross the long axis of filaments with large angles. Footpoints are found near the boundary of supergranulations. However, where the X-ray loops are well visible, prominences in Hα are inconspicuous (and vice versa), and they suggest that these loops are later cooled to prominence matter.

The active region filaments show a different behavior: Webb and Zirin (1981) found that the arcades of stable, but sometimes highly sheared X-ray loops form over ARFs and, hence, almost parallel to the neutral lines. However, we should notice that X-ray loops cross the neutral line rather at right angles in ARFs situated farthest from active regions (top part of their Figure 5(a) and (b)). The footpoints of X-ray loops are again in strong magnetic regions, but Webb and Zirin (1981) did not find any example where footpoints extend into sunspot umbrae. See the summary on field directions in Section 6.

5. Systematic Flows and Oscillatory Motions

The study of systematic motions is important for understanding the formation of prominences. In quiescents it is still inconclusive whether the cool material is moving upward or downward as a whole: Hα prominence movies show often downward motions of 5 km s^{-1} or less (Engvold, 1976), and spectroscopic studies on the disk show on the average 0.7 km s^{-1} downward motion (Kubota, 1980), while upward motions of 0.5 km s^{-1} or so are reported by other observers (e.g., Martres *et al.*, 1981; Schmieder *et al.* 1984). In the UV line of C IV at 1548 Å emitted at 10^5 K, Schmieder *et al.* (1984) found upward flows of 5.6 km s^{-1} for three quiescents. However, if the general quiet region down-draft of about 4 km s^{-1} is subtracted, the up-flow becomes only 1.6 km s^{-1}, which is rather marginal. There is an indication that the flow patterns of 10^5 K plasmas are unidirectional, coming up from one footpoint, then presumably moving into the prominence, and finally down to the other (Engvold *et al.*, 1985). However, the quiescent filaments they treated were at high altitudes, hence, projection effects might change the conclusion. Photospheric motions (< 0.3 km s^{-1}) are known to be depressed in the regions below quiescents (Martres *et al.*, 1981, and earlier papers therein). However it is not clear whether this is related to diminished oscillations, or because the locus of zero velocity comes close to the quiescent.

Next we turn to active region filaments. The material flow seen in Hα movies goes evidently along the long axes of filaments. Off-band Hα pictures also indicate downflow to the sunspot at the ends of the filaments (Martin, 1973). Steady flows around ARFs, persisting over several days, were discovered from C IV 1548 Å line Dopplergrams (10^5 K) with the UVSP–SMM experiments (Athay *et al.*, 1985, and references therein). The striking fact is that the locus of zero velocity (V_0-line) almost coincides with the neutral line of the longitudinal magnetic field (B_0-line). This is well seen near the limb, indicating that the horizontal flows (≈ 20 km s^{-1}) are predominant. (In quiescents the V_0-line is also found parallel to the B_0-line; Engvold *et al.*, 1985.) For the majority of cases downflows from the top of a loop to both footpoints are suggested, although the intensity maps do not always show loops. The 10^5 K loops overlying ARFs seem to be also low lying, because otherwise the V_0-line would not coincide with the B_0-line. Near the limb one can see a portion of a filament perpendicular to as well as parallel to the limb, and yet the V_0-line is distinctively visible in both cases. Hence, it is not clear if all the loops are making small angles to the B_0-line (sheared loop). Also there are cases indicating unidirectional flow patterns from one end of a loop to the other. At the photospheric level, in spite of the fact that there are some reports in literature on the V_0-line coinciding with the B_0-line (e.g. Harvey and Harvey, 1976), the correspondence with ARFs is not clear.

A new type of dramatic motions in a part of a quiescents was reported by Ligett and Zirin (1984): material seems to go up in a semi-circular trajectory projected against the sky-plane, then comes down again along a semi-circular trajectory with velocities of 15–75 km s^{-1}. Sometimes the whole prominence begins to show rotation. This phenomenon might be related to the change from A–b to A–d in Figure 4, or from B–b' to B–c' and B–d' (small eruptions in between the threads).

Oscillatory motions of prominences have been known for over half a century (see Bashkirtsev and Mashnich, 1984). The latter authors have made extensive observations using Hβ Doppler shifts (relative measurements at two points in a prominence); a median period of 50 min (ranging from 42 to 82 min) with an amplitude of 0.7 km s^{-1} was obtained from 15 prominences. See interpretations in Tandberg–Hanssen (1974).

Malherbe *et al.* (1981) found from Hα-Dopplergrams of a quiescent filament that the 3–5 min oscillations are almost invisible unlike the adjacent chromosphere, while Tsubaki and Takeuchi (1986) detected 3.5–4 min Doppler oscillations in a prominence. Further work in the period range below 5 min is evidently necessary.

6. Prominence Evolution and Comments on the Theory of Formation

The evolution of prominences has been studied extensively by the d'Azambuja and d'Azambuja (1948), and reviewed by Kiepenheuer (1953), de Jager (1959), and Martin (1973). In the following description I will add also our findings obtained from a study of Hα pictures obtained by Dr Dunn (Sac Peak), and Mr Morishita (Norikura). At birth a typical active region prominence is known to be almost directed north–south, as follows from extrapolation to the birth date of the region. While generally extending their length poleward, the direction of the long axis tends to be more and more tilted following the diffential rotation. Gradually the low latitude portion will disappear and the high latitude portion appears in a prominence; hence, as a whole it appears to have migrated poleward. (In addition there are of course pole and equatorial ward migrations of *prominence belts* within the solar cycle). A prominence born between two active regions behaves similarly. According to Martin, the time needed to develop to its maximum (stable) length is inversely proportional to the magnetic field strength of the adjacent regions.

Prominences are formed and stay along the neutral lines, except for some active region filaments which start from penumbrae. There is at least one condition for the formation of prominences, as suggested by Martin (1973): prior to the formation of filaments, Hα fibrils along the neutral line become aligned nearly parallel to the neutral line. In filaments seen on weaker field regions, their alignment becomes less frequent, and even in strong field regions there are exceptions. After the formation of the prominence, there may be two cases in *nearly parallel* fibrils: One is as would be obtained by horizontal shearing motion in between two regions of the oppositite polarity (Foukal, 1971), and the other is an arrowhead-like fibril orientation, often shifted away from prominences (Figure 1, left of P; Martin, 1973, Figure 6). Another point I see from Hα pictures is that, more often than not, one end of an active region filament stems from the bay-like portion of a bright plage area of one polarity. It may be interpreted as material being collimated easily because of converging (horizontal) field lines towards the entrance of a bay from its inner portion. Incidentally X-ray loops seem also to emerge often from the one polarity bay on iso-gauss maps (see maps in Howard and Švestka, 1977). Martin further notes that in a filament between active regions, a necessary condition for the formation is that two opposite polarity regions approch each other. However, it is also clear from the magnetograms that the existence of strong fields or steep gradients between opposite polarity regions is not a sufficient condition for the *existence* of prominences.

In the following I will try to assign various morphological types of prominences to a single evolutionary sequence of magnetic regions (Figure 4B). The ages of the filaments are not fundamental because (1) a segment of quiescents lasts at most a week and (2) even a very young prominence that re-appeared after its sudden disappearance looks morphologically similar to the old erupted one. We start with low-lying active region filaments (as in Figures 1 and 4B (a′)). When they grow in length to extend outside active regions, there appears a tree-like structure on one end of a filament or on both. And between the trees, large horizontal motions can be seen (Pettit's interactive one; Figure 4B (b')). Then gradually the trees develop their boughs. When we have obtained several trees, we have a fully developed hedgerow prominence, generally in latitudes of $\lesssim 30$–$50°$, (c') – drawing. This is the most stable of all kinds; d'Azambuja and d'Azambuja suggested that possibly all prominences would evolve into this form if they lived long enough. Motions are seen parallel to the long axis at high altitudes and downwards in the tree trunks (see Martin).

Prominences having vertical curtain-like structures, presumably in deeper pits (Figure 4B (d') – drawing), seem to belong to the faintest magnetic fields, or the latest in their development. Here boughs in a typical hedgerow have lost their horizontal structures; a general continual trend from horizontal structures in ARFs to the present vertical ones. They are often seen in high latitude, and occasionally in equatorial regions as judged from Dunn's pictures mostly taken in 1956.

We can summarize the field configuration as follows, taking what was observed in Hα, EUV and X-ray pictures and from the Hanle and Zeeman effect at their face value. In active region filaments the magnetic field seems to be nearly parallel to the long axis. Generally it is highly sheared, and in potential-like configurations in the prominence itself or in the corona as well as in 10^5 K plasmas. However, in some cases chromospheric fibrils below the filament make large angles with the axis. On the other hand in low latitude, low altitude quiescent prominences the field orientations may still have a potential-like configurations ($\alpha = +20°$), but at coronal temperatures they seem to make large angles. In the base of the chromosphere fibril structures are found either parallel or highly oblique to the long axis (Figure 1). Finally high latitude, high-altitude prominences show non-potential-like field directions of $\alpha = -20 \sim -30°$, but we have no informations on coronal and 10^5 K plasmas, except for helmet-streamer structure.

Before leaving this section, let me pass to a discussion of some basic *unsolved problems* of the theory of prominence formation, without referring to many recent papers.

Low-lying active region filaments: magnetically channeled stationary syphon flows due to the pressure difference between the two base points have been discussed (Mayer and Schmidt, 1968). However, a more fundamental question yet to be answered may be how to create the highly sheared fields in the neutral line region within one to two rotations, particularly at the pole-ward side.

Quiescent prominences: a basic problem is how to bring enough material to the prominence site, in order to eventually form the cool condensations through radiation losses (Kiepenheuer, 1953). If the cavity does not provide enough gas (Saito and Tandberg-Hanssen, 1973), the chromospheric gas should be sucked up in an 'evaporation processes' or by syphon flows due to (enhanced) coronal heating (Pikel'ner, 1971). Unfortunately we have almost no knowledge of the (enhanced) coronal heating rate, nor of the radiation losses below, say, 3×10^4 K, which is essential for understanding the final formation of cool gas, nor whether the material is continuously renewed or not. If the 'evaporation' occurs in a single event like in flares, the draining of mass through the pit may not be essential. The diffusion time across the magnetic field in fully ionized plasma is very long: a year for a scale length of 200 km.

In the case of partially ionized material, I find from Cowling (1957, p. 111, Equation

(6.28)) that the effective diffusion time becomes $t_D = (B^2/8\pi)/[2^{-1}(\rho_n/\rho_i)\rho_n g^2 \tau_{ni}]$, where I used $|\mathbf{j} \times \mathbf{B}| = (\rho_i + \rho_n)g$. Here ρ_n, ρ_i, g, and τ_{ni} are the density of neutral hydrogen, that of ionized hydrogen, the surface gravity, and the mean collision time of neutral hydrogen with ions (protons), respectively. With $B = 8$ G, a density-independent quantity of $\rho_n \tau_{ni} = 7 \times 10^{-16}$ at 7000 K, and $\rho_n/\rho_i = 1$, one obtains $t_D = 28$ days, and if $\rho_n/\rho_i = 10$, $t_D = 2.8$ days. Therefore, over a height distance of, say, 2.4×10^4 km with $t_D = 2.8$ days, a mean down-flow velocity of 0.1 km s^{-1} is derived. And we find again that it is crucial to determine precisely the ionization degree of hydrogen in order to settle whether the formation is continuous or not.

Raadu (1979) proposed another way of bringing material into prominence (see also Kuperus and Raadu, 1974). He assumes that the photospheric gas motion underlying the neutral line is downward, driven by e.g., a giant convective cell. And magnetic fluxes and, hence, coronal gas together with the photospheric gas are brought horizontally to the neutral line from both sides. In the sense that the Sun as a whole would somehow needs submergence of magnetic fields, possibly in the neutral lines, this idea is attractive. The most important future task here may be rather an observational one (e.g., Harvey and Harvey, 1976): (1) To find systematic photospheric motions in the neutral line, and (2) if we find any motions, to look for the difference in motions and their time history between the place below the prominence (or its segment) and the prominence-free regions.

In short, a future successful theory *must* then explain why the (segment of a) prominence is there and not in another place where the photospheric magnetic field distribution would look similar.

7. Erupting Prominences, Post-Flare Loops, and Surges

Studies of erupting prominences, post-flare loops, and surges have been reviewed recently in Martin (1980), Moore (1980), and Rust and Hildner (1980) as well as by Jensen *et al.* (1979), Švestka (1976), and Tandberg-Hanssen (1974); also coronal mass ejections are treated in this latter volume. Hence, I will only present some comments concerning, particularly, unsolved theoretical problems.

7.1. ERUPTING PROMINENCES

Prominence eruptions are now considered to be a part of the flare phenomena, where MHD explosions and energy releases occur: if erupting fields happen to include a prominence, we observe an erupting priminence at the early stage of the flare. And if not, we see only disturbing motions in a prominence at the flare site, and presumably the upper corona (X-ray corona) erupts. Kahler (1981) has shown that when the prominence does not erupt, X-ray (weak) flares tend to appear as loops perpendicular to the prominence axis, and when it erupts they are parallel to the axis. Before the instability starts, 10^5–10^6 K arches appear striding over ARFs with large angles (Schmahl *et al.*, 1982). These do not tell whether in the absence of the prominence eruption the corona has erupted or not, but they give certainly important information. As to the cause of the eruptions, sunspot movements, flux emergences, and photospheric motions have been investigated. However a casual relationship was established only for a limited number of cases (Martin, 1980).

At the theoretical side there is no doubt that the MHD instability, the kink instability in particular, is a key factor for interpreting the erupting prominence. After an exploratory work using a slender electric current rope (Hirayama, in Bruzek and Kuperus, 1972, p. 11), Sakurai (1976) has made three-dimensional nonlinear MHD simulations of the kink instability of a magnetic flux rope, and has shown that the smaller the twist of the magnetic field, the higher the final height attained and the smaller the overall twisting motions, in accordance with observations. The following effects should be included in order to proceed beyond Sakurai's treatments; external fields, the field tied with the photosphere, the dynamical effect of reconnections below the prominence, and the cool heavy mass. Approximate methods using, say, electric current ropes as have been done by Anzer, Pneuman, and others could be also helpful.

It seems at present fashionable to visualize *spiralling fields* in and around erupting prominences, which are assumed to result from reconnections under the prominence (Hirayama, 1974, Figure 1; Moore, 1980). The outward expansions of two-ribbon flares could also be explained by successive reconnections through sucking-up motions of the rising prominence, and/or of 10^6 K X-ray ropes. However, I myself have some doubts on these scenarios in two respects: first, good photographs of erupting prominences in big events like 19 December 1973 observed by Skylab in He II 304 Å (Rust and Hildner *et al.*, 1980, p. 276) or the one in Kiepenheuer (1953, p. 413) do not show concrete evidence of many turns; if any, one turn occurs in a whole prominence. Secondly, in flares occurring on both sides of a highly sheared erupting ARF, the topology of the magnetic fields seems to be much more complicated than the above simple picture (e.g. Tanaka, 1976; Morishita, 1985).

7.2. POST-FLARE LOOPS

As mentioned in Section 1, these are understood as being the results of a cooling process (and consequent falling) from a 10^7 K-flare plasma with the high density of 10^{10}–10^{12} cm^{-3} (e.g., Moore, 1980). Here, once the energy release in the flare is *assumed*, one can reproduce the *basic* observed features from numerical simulations. One such calculation is shown in Figure 5 (Hirayama and Endler, 1975), where, starting from a vertically standing column of a stationary active-region corona and chromosphere, the one-dimensional equations of motion, energy, and mass have been solved, they give rise to a 10^7 K flare and upward flow (evaporation) by the extra thermal energy input into the corona. The figure shows the temperatue variation as viewed at a fixed point in the column, after the instant when the enhanced energy release is cut off and reduced to the original coronal value. After the moment when the plasma temperature becomes 10^4 K in model E, where the cool plasma is flowing down, the corona resumes again its temperature of $10^{6.3}$ K. Apart from the crucial energy release problem, the interpretation and/or cause of the very small thickness of 2-100 km (e.g. Hirayama,

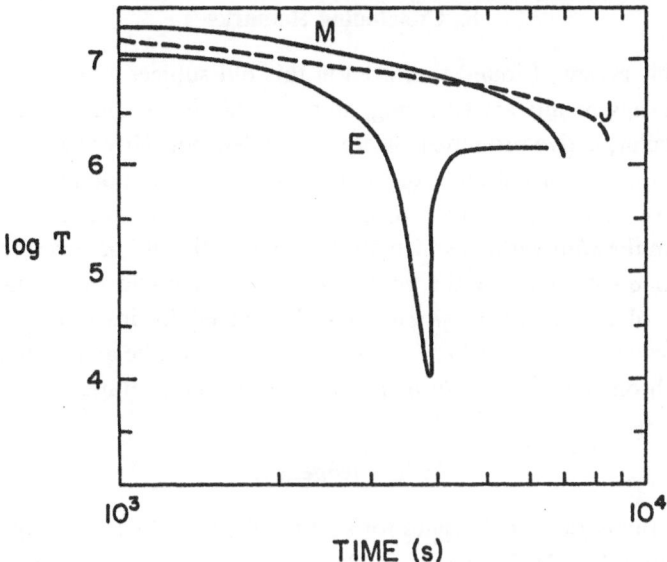

Fig. 5. Formation of a post-flare loop. Model 'E' shows cool falling material, and the recovery to coronal temperature. 'E' is a slow flare: column height = $10^{4.5}$ km; linearly increasing energy input till 10^3 s (max. 10^9 erg cm^{-2} s^{-1}); max. electron density = 6×10^9 cm^{-3}, initial and final density = 3×10^8 cm^{-3} (at $10^{4.3}$ km).

1979; Foukal and Landman, 1984; an interesting case in Jefferies and Orrall, 1965) should be investigated in detail.

7.3. SURGES

A surge is the collimated ejection of chromospheric material, probably along the magnetic field, with velocities of 50–250 km s^{-1}, thus attaining maximum heights of 2–20 × 10^4 km. There are reccurrence tendencies of the order of one hour. Sometimes they occur near the neutral point of the longitudinal field, but at least in one case it started between umbrae of the same polarity (Rust, 1972, Figure 9). Their shape in their bases looks occasionally roundish, or curved (e.g., Foukal, 1971, Figure 1(b)), but in other cases rather straight. Disturbances are seen to travel from the flare site to the roots of surges at their onset as 'puffs' (Giovanelli and McCabe, 1958), and evolving magnetic structures, including the appearance of satellite sunspots, appear to be related to surge occurrences (Roy, 1973).

Steinolfson et al. (1979) have successfully reproduced the observational characteristics of surges in their one-dimensional simulations by the pressure gradient force. However, they needed an initial pressure in the chromosphere as high as five to even thirty times that of the normal chromosphere. Uchida and Shibata (1985) explained matter ejections by the Lorentz force where the energy was assumed to be originally stored in a twisted magnetic flux rope like a pressed spiral spring. In both cases the crucial future problem seems to be how to realize the initial conditions such as the very high pressure.

8. Concluding Remarks

In preparing this review, I found that even in this old subject a very large quantity of new and interesting results are emerging, and I could discuss only a small fraction of them. In particular, active prominences are largely left out. However, I cannot escape the conclusion that important questions concerning the formation and stability remain basically unanswered, and I hope that these fundamental questions will be attacked intensively from the *observations*, which after all render the ultimate verdict, and I hope that this be done especially by the young generation. I could even imagine that the heating theory of *the corona in general* may be solved by inspecting a part of the prominence formation. And further, who knows if there will be an international society for the research on *stellar and galactic prominences* in the centuries ahead of us!

Acknowledgements

I would like to thank many colleagues for sending their works at my request, and I am particularly thankful to Drs U. Anzer, E. Hiei, J. L. Leroy, D. A. Landman, B. C. Low, M. Makita, K. Tanaka, E. Tandberg-Hanssen, Y. Uchida, and H. Zirin for giving comments on specific questions I raised. Prof. C. de Jager and Dr J. M. Pasachoff kindly improved the manuscript. Special thanks go to Dr R. B. Dunn and Mr H. Morishita whose high-resolution pictures were of great help.

Note added in Proof: J. M. Fontenla and M. Rovira (1985, *Solar Phys.* **96**, 53) have started to construct the radiative equilibrium models of prominence threads where the equations of hydrogen radiative transfer and heat conduction were simultaneously solved, and the results were compared with the observation of H I Lα and Lβ lines (Vial, J. C.: 1982, *Astrophys. J.* **253**, 330) by adding a number of thread models (≈ 50) where each thread model has a thickness of less than ten km. The observed thread diameter is a few hundred km (Engvold, 1976), which is, however, an upper limit. This line of investigation seems the only way towards the full understanding of the energy balance problem of quiescent prominences.

References

Anzer, U.: 1979, *IAU Colloq.* **44**, p. 322.

Athay, R. G., Querfeld, C. W., Smartt, R. N., Landi degl'Innocenti, E., and Bommier, V.: 1983, *Solar Phys.* **89**, 3.

Athay, R. G., Jones, H. P., and Zirin, H.: 1985, *Astrophys. J.* **288**, 363.

Bommier, V., Landi degl'Innocenti, E., Leroy, J. L., and Sahal-Bréchot, S.: 1985a, in M. J. Hagyard (ed.), *Measurements of Solar Magnetic Fields*, NASA Conf. Publ. CP-2374, p. 335.

Bommier, V., Leroy, J. L., and Sahal-Bréchot, S.: 1985b, in M. J. Hagyard (ed.), *Measurements of Solar Magnetic Fields*, NASA Conf. Publ. CP-2374, p. 375.

Bonnet, R. M., Bruner, E. C., Acton, L. W., and Brown, W. A.: 1980, *Astrophys. J. Letters* **237**, L47.

Bruzek, A. and Kuperus, M.: 1972, *Solar Phys.* **23**, 3.

Bashkirtsev, V. S. and Mashnich, G. P.: 1984, *Solar Phys.* **91**, 93.

Cowling, T. G.: 1957, *Magnetohydrodynamics*, Interscience Publ. Inc., New York.

Davis, J. M. and Krieger, S.: 1982, *Solar Phys.* **80**, 295, and **81**, 325.
d'Azambuja, L. and d'Azambuja, M.: 1948, *Ann. Obs. Paris-Meudon*, **6**, No. 7.
de Jager, C.: 1959, in S. Flügge (ed.) *Handbuch der Physik* **52**, Springer Publ. Co., Berlin, p. 80.
Engvold, O.: 1976, *Solar Phys.* **49**, 283.
Engvold, O., Tandberg-Hanssen, E., and Reichman, E.: 1985, *Solar Phys.* **96**, 35.
Foukal, P.: 1971, *Solar Phys.* **19**, 59.
Foukal, P. and Landman, D. A.: 1984, *IAU Colloq.* **86**, 25.
Giovanelli, R. G. and McCabe, M.: 1958, *Australian J. Phys.* **11**, 353.
Harvey, J. W.: 1969, Thesis, Univ. Colorado, Boulder, Colorado.
Harvey, K. L. and Harvey, J. W.: 1976, *Solar Phys.* **47**, 233.
Heasley, J. N.: 1977, *J. Quant. Spectrosc. Radiat. Transfer* **18**, 541.
Heasley, J. N. and Mihalas, D.: 1976, *Astrophys. J.* **205**, 273.
Heasley, J. N. and Milkey, R. W.: 1978, *Astrophys. J.* **221**, 677.
Heasley, J. N. and Milkey, R. W.: 1983, *Astrophys. J.* **268**, 398.
Heinzel, P.: 1983, *Bull. Astron. Inst. Czech.* **34**, 1.
Hirayama, T.: 1964, *Publ. Astron. Soc. Japan* **16**, 104.
Hirayama, T.: 1972, *Solar Phys.* **24**, 310.
Hirayama, T.: 1974, *Solar Phys.* **34**, 323.
Hirayama, T.: 1979, *IAU Colloq.* **44**, 4.
Hirayama, T. and Endler, F.: 1975, *Bull. Am. Astron. Soc.* **7**, 352.
Hirayama, T., Nakagomi, Y., and Okamoto, T.: 1979, *IAU Colloq.* **44**, 48.
Hirayama, T., Tanaka, K., Watanabe, T., Akita, K., Sakurai, T., and Nishi, K.: 1985, *Solar Phys.* **95**, 281.
Howard, R. and Švestka, Z.: 1977, *Solar Phys.* **54**, 65.
Hyder, C. L.: 1965, *Astrophys. J.* **141**, 1374.
Jefferies, J. T. and Orrall, F. Q.: 1965, *Astrophys. J.* **141**, 505.
Jensen, E., Maltby, P., and Orrall, F. Q. (eds.): 1979, 'Physics of Solar Prominences', *IAU Colloq.* **44**.
Kahler, S. W.: 1981, *Solar Phys.* **71**, 337.
Kanno, M., Withbroe, G. L., and Noyes, R. W.: 1981, *Solar Phys.* **69**, 313.
Kiepenheuer, K. O.: 1953, in G. Kuiper (ed.) *The Sun*, Chicago Univ. Press, Chicago, Ch. 4.
Kippenhahn, R. and Schlüter, A.: 1957, *Z. Astrophys.* **43**, 36.
Kosugi, T., Ishiguro, M., and Shibasaki, K.: 1986, *Publ. Astron. Soc. Japan* **38**, 1.
Kubota, J.: 1980, in F. Moriyama and J. C. Henoux (eds.) *Proc. Japan–France Seminar on Solar Physics*, p. 178.
Kuperus, M. and Raadu, M. A.: 1974, *Astron. Astrophys.* **31**, 189.
Landi degl'Innocenti, E.: 1982, *Solar Phys.* **79**, 291.
Landman, D. A.: 1983, *Astrophys. J.* **270**, 265.
Landman, D. A.: 1984, *Astrophys. J.* **279**, 438.
Landman, D. A.: 1985a, *Astrophys. J.* **290**, 369.
Landman, D. A.: 1985b, *Astrophys. J.* **295**, 220.
Leroy, J. L.: 1979, *IAU Colloq.* **44**, p. 56.
Leroy, J. L.: 1985, in M. J. Hagyard (ed.), *Measurements of Solar Magnetic Fields*, NASA Conf. Publ. CP-2374, p. 121.
Leroy, J. L., Bommier, V., and Sahal-Bréchot, S.: 1983, *Solar Phys.* **83**, 135.
Leroy, J. L., Bommier, V., and Sahal-Bréchot, S.: 1984, *Astron. Astrophys.* **131**, 33.
Liggett, M. and Zirin, H.: 1984, *Solar Phys.* **91**, 259.
Malherbe, J. M., Schmieder, B., and Mein, P.: 1981, *Astron Astrophys.* **102**, 124.
Mariska, J. T., Doschek, G. A., and Feldman, U.: 1979, *Astrophys. J.* **232**, 929.
Martin, S. F.: 1973, *Solar Phys.* **31**, 3.
Martin, S. F.: 1980, *Solar Phys.* **68**, 217.
Martres, M.-J., Mein, P., Schmieder, B., and Soru-Escaut, I.: 1981, *Solar Phys.* **69**, 301.
Meyer, F. and Schmidt, H. U.: 1968, *Z. Angew. Math. Mech.* **48**, 218.
Milkey, R. W., Heasley, R. N., Schmahl, E. J., and Engvold, O.: 1979, *IAU Colloq.* **44**, 53.
Moore, R. L., McKenzie, D. L., Švestka, Z., Widing, K. G., and 12 co-authors: 1980, in P. A. Sturrock (ed.), *Solar Flares, Skylab Workshop II*, Colorado Assoc. Univ. Press, Boulder, Colorado, p. 341.
Morishita, H.: 1985, *Tokyo Astron. Bull. 2nd Ser.*, No. 272, 3123.
Nikaido, Y. and Kawaguchi, I.: 1983, *Solar Phys.* **84**, 49.

Nikolsky, G. M., Kim, I. S., Koutchmy, S., and Stellemacher, G.: 1984, *Astron. Astrophys.* **140**, 112.
Orrall, F. Q. and Schmahl, E. J.: 1980, *Astrophys. J.* **240**, 908.
Pikel'ner, S. B.: 1971, *Solar Phys.* **17**, 44.
Pneuman, G. W.: 1983, *Solar Phys.* **88**, 219.
Poland, A. I. and Tandberg-Hanssen, E.: 1983, *Solar Phys.* **84**, 63.
Raadu, M. A.: 1979, *IAU Colloq.* **44**, p. 167.
Rao, A. P. and Kundu, M. R.: 1980, *Astron Astrophys.* **80**, 373.
Raoult, A., Lantos, P., and Fürst, E.: 1979, *Solar Phys.* **61**, 335.
Roy, J. R.: 1973, *Solar Phys.* **28**, 95.
Rust, D. M.: 1972, AFCRL-72-0048 (Sacramento Peak Observatory Project 7649).
Rust, D. M., Hildner, E., and 11 co-authors: 1980, in P. A. Sturrock (ed.), *Solar Flares, Skylab Workshop II*, Colorado Assoc. Univ. Press, Boulder, Colorado, p. 273.
Saito, K. and Tandberg-Hanssen, E.: 1973, *Solar Phys.* **31**, 105.
Sahal-Bréchot, S.: 1984, *Ann. Phys. France* **9**, 705.
Sakurai, T.: 1976, *Publ. Astron. Soc. Japan* **28**, 177.
Schmahl, E. J.: 1979, *IAU Colloq.* **44**, 102.
Schmahl, E. J., Bobrowsky, M., and Kundu, M. R.: 1981, *Solar Phys.* **71**, 311.
Schmahl, E. J., Mouradian, Z., Martres, M. J., and Soru-Escaut, I.: 1981, *Solar Phys.* **81**, 91.
Schmieder, B., Vial, J.-C., Mein, P., and Tandberg-Hanssen, E.: 1984, *Astron. Astrophys.* **127**, 337.
Serio, S., Vaiana, G. S., Godoli, G., Motta, S., Pirronello, V., and Zappala, R. A.: 1978, *Solar Phys.* **59**, 65.
Steinolfson, R. S., Schmahl, E. J., and Wu, S. T.: 1979, *Solar Phys.* **63**, 187.
Sturrock, P. A. (ed.): 1980, *Solar Flares, Skylab Workshop II*, Colorado Assoc. Univ. Press, Boulder, Colorado.
Švestka, Z.: 1976, *Solar Flares*, D. Reidel Publ. Co., Dordrecht, Holland.
Tanaka, K.: 1976, *Solar Phys.* **47**, 247.
Tandberg-Hanssen, E.: 1974, *Solar Prominences*, D. Reidel Publ. Co., Dordrecht, Holland.
Tandberg-Hanssen, E., Martin, S. F., and Hansen, R. T.: 1980, *Solar Phys.* **65**, 357.
Tsubaki, T. and Takeuchi, A.: 1986, *Solar Phys.* (submitted).
Uchida, U. and Shibata, K.: 1985, *Solar Phys.* (submitted).
Vial, J. C.: 1982, *Astrophys. J.* **254**, 780.
Webb, D. F. and Zirin, H.: 1981, *Solar Phys.* **69**, 99.
Withbroe, G. L. and Noyes, R. W.: 1977, *Ann. Rev. Astron. Astrophys.* **15**, 363.
Yakovkin, N. A. and Zel'dina, M. Yu: 1964, *Soviet Astron. – AJ* **7**, 643.
Ye, S.-H.: 1961, *Izv. Krymsk. Astrofiz. Obs.* **25**, 180.
Zirin, H.: 1979, *IAU Colloq.* **44**, p. 193.
Zirin, H. and Severny, A. B.: 1961, *Observatory* **81**, 55.

21 MAY 1980 FLARE REVIEW

CORNELIS DE JAGER and ZDENĚK ŠVESTKA

Laboratory for Space Research, Utrecht, The Netherlands

Abstract. A review is given of observations and theories relevant to the solar flare of 21 May, 1980, 20 : 50 UT, the best studied flare on record. For more than 30 hr before the flare there was filament activation and plasma heating to above 10 MK. A flare precursor was present ≥ 6 min before the flare onset. The flare started with filament activation (20 : 50 UT), followed by thick-target heating of two footpoints and subsequent ablation and convective evaporation involving energies of 1 to 2×10^{31} erg. Coronal explosions occurred at 20 : 57 UT (possibly associated with a type-II burst) and at 21 : 04 UT (associated with an Hα spray?). Post-flare loops were first seen at 20 : 57 UT, and their upward motion is interpreted as a manifestation of successive field-line reconnections. A type-IV radio burst which later changed into a type-I noise storm was related to a giant coronal arch located just below the radio noise storm region. Some implications and difficulties these observations present to current flare theories are mentioned.

1. Introduction

We have selected the major solar flare of 21 May, 1980 for a review, because it was one of the best observed, and best analyzed flares in the whole history of solar research. Its observations on board the Solar Maximum Mission (SMM) spacecraft led to the first study of images of hard X-ray brightenings at footpoints of loops during the impulsive phase (Hoyng *et al.*, 1981) and to the discovery of giant post-flare coronal arches following two-ribbon flares (Švestka *et al.*, 1982a). Antonucci *et al.* (1985) found high-velocity upward motions during the impulsive phase which yield information about the early phase of chromospheric evaporation, and Lemmens and De Jager (1985) found related lateral motions pointing to a chromospheric explosion. During the growth of post-flare loops, Švestka an Poletto (1985) gave a direct evidence for a late release of energy high in the corona which might represent hard X-ray images of the reconnection of field lines. The mass ejection associated with this flare was imaged both by the NRL–Solwind coronagraph on board the P78–1 spacecraft and by the Helios A zodiacal-light photometer (McCabe *et al.*, 1985). Several other papers have analyzed the flare observations and contain attemps to interpret them.

In addition to these observations of the flare itself, we also have good data about the active region in which the flare occurred, because coordinated observations of the region were made for several days as a part of the Flare Build-up Study Program (Harvey, 1983; Schadee *et al.*, 1983). We are thus in possession of an extraordinary set of data on pre-flare variations, the flare build-up, flare onset, impulsive phase, decay phase with post-flare loops, the associated mass ejection, and the post-flare corona. We have tried to summarize these various aspects of this flare event in the following review.

2. Pre-Flare Variations in the Active Region

The major flare (importance 2B/X1) started at 20 : 50 UT on 21 May, 1980 (Hoyng *et al.*, 1981) in Hale region 16850 at 13 S 15 W. The SMM spacecraft pointed at this

Solar Physics **100** (1985) 435–463. 0038–0938/85.15.

active region (with a few short gaps) from 16 UT on 17 May. Schadee *et al.* (1983, further abbreviated as SDJS) analyzed the observations made by the Hard X-Ray Imaging Spectrometer (HXIS) from 12:50 UT on 19 May and detected an almost permanent X-ray emission along the filament channel of the active region.

This emission was extremely weak, about 10^{-4} times the 3.5–5.5 keV X-ray flux from an average flare event (cf. SDJS Figure 2 and Table I) so that only long integrations (for 20 min or more) revealed statistically significant X-ray images. The activations along the filament channel lasted apparently long enough, or repeated themselves so often, that most integrations yielded distinct images (cf. SDJS Figure 2). However, in spite of the low countrate, the temperature in these phenomena was found to be close to 11 million degrees, without any involvement of flares or detectable flare-like brightenings (cf. SDJS Table IV).

This high temperature was determined from the (5.5–8.0 keV)/(3.5–5.5 keV) flux ratio under the assumption of a Maxwellian distribution of velocities (thermal plasma, cf. Mewe *et al.*, 1985). Thus it may either represent a real temperature in heated loops embedding the filament or, as Mandelstam and his co-workers (private communication) believe, it may be a fictitious 'temperature' of a non-thermal plasma produced by intermittent, but very frequent accelerations along the filament channel. They base their suggestion on the fact that in some other, possibly analogous, events the ion temperature, deduced from linewidths of X-ray lines, was substantially lower.

As from 14 UT on 20 May this semi-permanent weak X-ray emission along the filament channel began to appear close to the site of the most intense enhancement later seen in the onset phase of the flare (A in Figure 1; cf. SDJS Figure 4). The emission

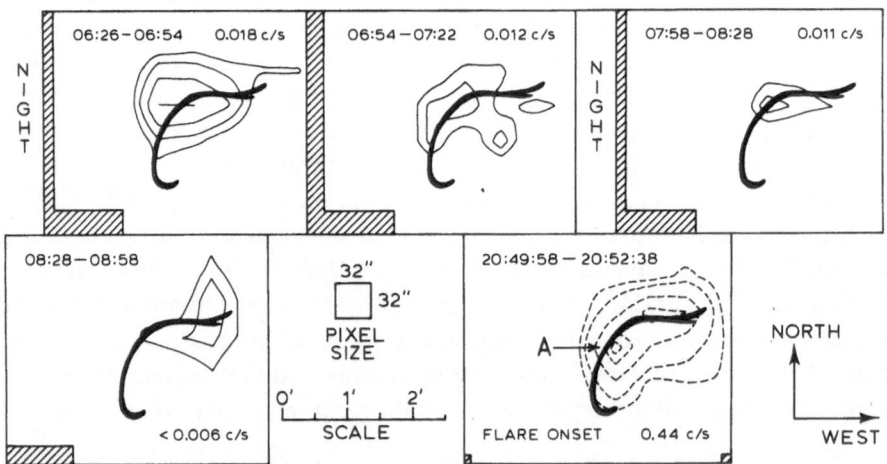

Fig. 1.   Last HXIS images on 21 May, 1980 of the long-lived X-ray brightening which coincided in position with the brightest patch (A) of the later coming flare (cf. SDJS, Figure 2) and subsequent shift of the emission to the western end of the filament. Energy range 3.5–5.5 keV in HXIS coarse field of view. The eastern border of the FOV is hatched and the filament is indicated. Data at the top of each frame: UT beginning and end of count integration; mean countrate in the pixel with maximum count. The last frame (with dashed contours) shows the onset of the flare. (After Schadee *et al.*, 1983.)

stayed in this position until ~07 UT on 21 May (see the last images of this period in Figure 1; note that 'close to' means 'within the spatial resolution of 32 arc sec'). It is of interest that at 18 UT on 20 May the zodiacal-light photometer aboard Helios A began to record a weak mass ejection at a distance above 20 solar radii from the Sun. It became most pronounced at 6 UT on 21 May and decayed thereafter until the mass ejection from the major flare succeeded it in about the same position (cf. Figure 2, flare ejection

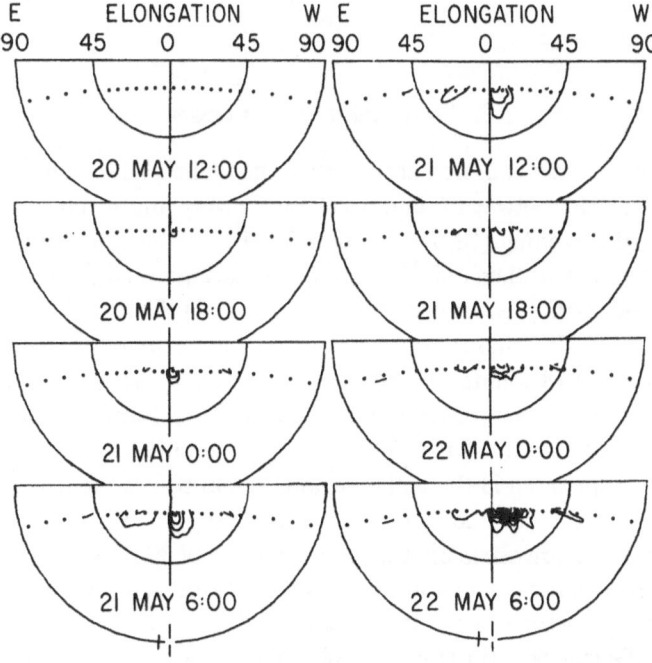

Fig. 2.   Helios A photometer images of excess electron density along the line-of-sight in contour steps of 3×10^{14} cm^{-2}. The photometer reading nearest to the Sun is at 16° S from the ecliptic plane; see paper by Jackson in this issue for more information about the plots. (Courtesy of B. Jackson, UCSD, La Jolla.)

appears at 22 May 0:00 and 6:00 UT; see paper by Jackson in this issue for an explanation of the plottings). Thus, from the viewing position of Helios A (30° west from the Sun–Earth line) there was a weak mass-flow into interplanetary space at the position angle of AR 16850 for more than one day prior to the major flare, without any obvious source other than the long-lived brightenings along the filament channel detected close to the position of the later flare by SDJS. This observation tends to support the interpretation by Mandestam's group that the weak X-ray enhancements are images of intermittent acceleration processes along the filament channel.

Also Hα images of the filament showed remarkable changes for at least a day before the flare (Harvey, 1983), though the flare activity was very low (only four Hα subflares were reported during two days preceding the major flare). According to Gaizauskas (in

Harvey, 1983) the main filament changes were seen near its west end, but parts of the filament showed vertical motions which could be interpreted as rotational motions of the filament along its own axis. The western-end activity might have been provoked by the rapidly moving magnetic flux in this area: a nearby spot moved with a mean velocity of 130 m s^{-1} on 20 May and 260 m s^{-1} on 21 May during the five last hours preceding the major flare. Large electric currents flowing along the filament might be the cause of the vertical motions (Harvey, 1983) and of the intermittent instabilities seen in X-rays by HXIS.

3. Flare Build-up and Onset

As Figure 1 shows, X-ray emission at the site of the main brightening of the major flare (site A) ceased to exist between 6 and 8 UT on 21 May and it did not reappear until $\sim$19:20 UT in the evening (SDJS, Section 3.1.4.). Within the coarse resolution of 32 arc sec its position was then quite identical with the later brightening in the flare. This brightening stayed slightly enhanced (0.014 counts s^{-1}) till 19:53 UT (onset of SAA extending into the SMM night), and reappeared in the same position at 20:44 UT (SMM sunrise) with 0.18 counts s^{-1}, slowly brightening. Faster enhacement (which can be considered to be the flare onset) started at 20:50 UT (imaged in Figure 1) and reached flare intensity at 20:55 UT. Thus one may tentatively assume that a pre-flare X-ray brightening appeared in the active region 90 min before the flare onset, and that with certainty an X-ray precursor was present at $\leq$20:44 UT, i.e., $\geq$6 min before the flare onset. For a comparison: in another (smaller) flare of 30 April, 1980, with the much better resolution of 8 arc sec in the HXIS fine field of view, De Jager *et al.* (1983) observed a flare precursor exactly at the flare site for $\sim$30 minutes prior to the flare.

According to Gaizauskas (in Harvey, 1983), strong downward motions appeared at the west end of the filament about 2 hr before the flare. At the same time, a kink developed in the middle of the filament, very close to the main site of the brightening in the flare, and these changes may have been associated with the pre-flare X-ray enhancement mentioned above. Rust (in Hoyng *et al.*, 1981) suggested that the kink was due to the emergence of new flux exactly below it, manifested by the appearance of a new tiny spot just beneath that position. However, Harvey's (1983) analysis of magnetograms suggests that this spot did not form by emergence but by the compression of existing flux at the surface. This compression was ultimately caused by newly erupting flux, evidenced by the magnetograms, but this flux was located to the west of the spot and the kink. The net flux at a location directly beneath the activated filament actually decreased. Nevertheless, as Gaizauskas (1985) suggests, these changes may have destroyed the equilibrium between the filament and its surroundings, so that emerging flux could have been indirectly the source of the filament disruption and associated flare emission at the site of the kink.

According to Harvey (1983) these magnetic field changes were first detected about 80 min prior to the flare and they increased steadily in strength. This has striking similarity with the HXIS observations mentioned before: an X-ray enhancement at the position of the brightest flare patch first appeared 90 min prior to the flare and its intensity was continuously growing.

Both Harvey (1983) and Gaizauskas (1985) believe that the described flare build-up and flare onset are consistent with the model of filament eruption of Kuperus and Van Tend (1981). However, according to McCabe *et al.* (1985) the filament did never rise and erupt (cf. Section 8). As the flare developed at the kink, a set of bright Hα loops formed across the neutral line and along it towards the south obscuring this part of the filament for a while. After the flare the filament was fragmented, but it reoccurred essentially at the same position as before the flare. The growing system of post-flare loops indicates that the field above the active region opened and subsequently reconnected (following the model by Kopp and Pneuman, 1976), but the field opening must have been accomplished only above the filament. This is also confirmed by the fact that the separation of the two bright Hα flare ribbons (at the footpoints of supposedly reconnected loops) was quite large since the flare onset, with ~ 25 000 km distance between the outer edges (Švestka and Poletto, 1985).

The onset of the flare was described by Hoyng *et al.* (1981). The filament broadened and became diffuse near the site of the main flare emission at 20 : 45 UT and parted at 20 : 48 UT when the 8–20 Å X-ray emission (measured aboard the GOES-2 satellite) began to grow. By 20 : 50 : 38 UT the filament has broken into two distinct sections and at 20 : 54 : 50 UT several Hα patches reached kernel brightness. The flare entered its impulsive phase; this is described in the following section.

4. The Impulsive Phase

The study of many flares, mainly during the Solar Maximum Year 1979–1981, and chiefly by the Solar Maximum Mission, has shown that at least during 1980 virtually all flares of some importance had an impulsive phase. That phase is characterized by the following five main properties:

– The occurrence of impulsive bursts, both in the hard X-ray energy range, $E \gtrsim 20$ keV, and in the GHz microwave rage, $f \gtrsim 10$ GHz.

– The location of the hard X-ray burst emission in small, well defined areas, with characteristic diameters of the order of 3000 to 10^4 km. These areas, called flare *footpoints* or flare *kernels*, are best visible in medium and high energies ($E \gtrsim 15$ keV), but are also detectable in lower energies, after a subtle procedure of 'gradual background substraction' (De Jager and Boelee, 1984; De Jager *et al.*, 1984).

These first two properties indicate thick-target interactions of beams of electrons incident on the chromosphere (Duijveman, 1983).

Violent upward motions appear during the first few minutes of the impulsive phase; they are observed through shortward displayed spectral line components, mostly in the Ca XIX lines around 3.19 Å (Antonucci *et al.*, 1984). These velocity components range between 150 and 400 km s⁻¹; they are usually large in the first half minute of the impulsive phase and decrease to smaller values in a few minutes time. On the average only one fifth of the Ca XIX emitting gas shows this upward motion component.

These phenomena are ascribed to convective upward motion of chromospheric gas heated to tens of MK.

– There is a gradual increase of the thermal energy content of the flare, reaching maximum value at the end of the impulsive phase (De Jager and Boelee, 1984).

This is interpreted as being the consequence of energy injection during the impulsive phase, principally by the electrons producing the hard X-ray bursts. Their energy is fed into the gradual phase component of the flare plasma by the convective motions described just earlier.

– There exists a high-temperature flare-plasma component during the very first part of the impulsive phase which is clearly indicated, both from Doppler widths of spectral lines (Antonucci *et al.*, 1984, 1985) and from spectral broad-band intensity ratios in the medium and high-energy spectral range (15–30 keV; Hoyng, 1982; Duijveman *et al.*, 1982; De Jager, 1985b). Hence the hot flare plasma has at that time a bi-thermal or non-thermal character. The temperature differences between the high-temperature and the 'low'-temperature components approach zero in a few minutes time, quasi-exponentially, with an *e*-folding time of ca. 1.5 min.

This behaviour is ascribed to kernel heating by beams of energetic particles. This heating, directly associated with the high-energy bursts, causes chromospheric ablation and results in the convective motions, just described (De Jager, 1985b). This upward moving gas, spreading around and above the flare footpoints, is proper to the gradual phase.

There are more properties of the impulsive phase; some of them have a fairly general character, but we think that the five listed above are the most essential ones. Let us examine these properties now for the 21 May, 1980 flare.

4.1. IMPULSIVE BURSTS AND FOOTPOINTS

Figure 3 shows the time-development of the hard X-ray emission of the flare, as observed by the Hard X-Ray Burst Spectrometer (HXRBS) aboard SMM (Hoyng *et al.*, 1981). Only the first two peaks marked 1 and 2 are also clearly visible in the high-energy channel around 300 keV. These are the most energetic bursts. The later ones apparently involve electrons with smaller energies.

As shown first by Hoyng *et al.* (1981), these bursts are primarily associated with flare kernels, as appears from Figure 4, which is taken from a later, similar study, by Antonucci *et al.* (1985). There, the flare images are shown in low (3.5–8 keV) and high (16–30 keV) energies. The first high-energy image (20 : 55 : 15–20 : 55 : 35 UT) shows that there are two kernels, marked *A* and *B*. Hoyng *et al.* (1981) have shown that these kernels are situated at either side of, but very close to, the magnetic inversion line, which suggests that the kernels are loop footpoints. It is important to note that these kernels are *also* visible in the earliest low-energy images, being indicated even in images taken as early as 20 : 50 UT, see Figure 1. Hence, as it seems, the footpoints emission was already there before the onset of the hard X-ray bursts (Figure 3). The footpoints are

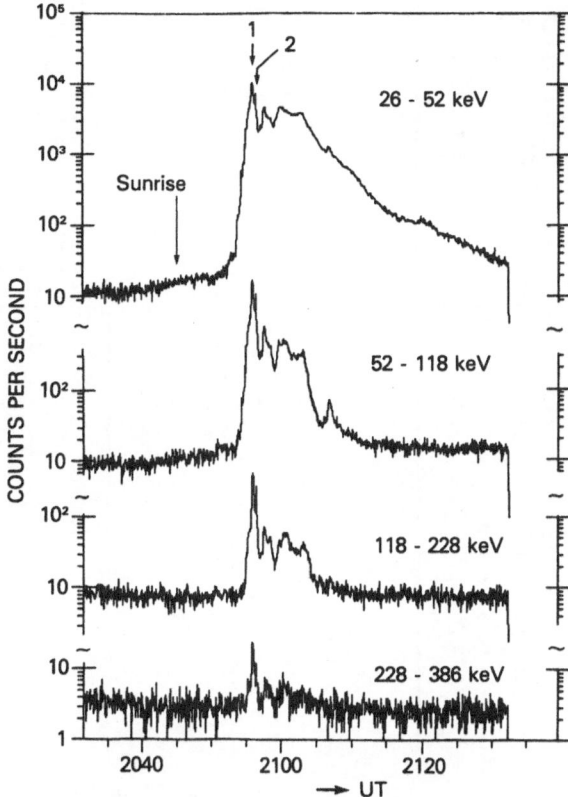

Fig. 3.   Countrates during the 21 May, 1980 flare, observed in various energy channels of the Hard X-Ray Burst Spectrometer aboard SMM (From Hoyng *et al.*, 1981.)

seen particularly well in the image of 20 : 55 : 15 UT at burst maximum, an observation that applies to more flares seen by HXIS; it appears more clearly for the 21 May flare because of the high countrates of the flare. It is also important to note that at 20 : 55 : 35 UT, hence only half a minute later, the maximum countrate in low energy shifted to the central part of the loop: already after that short time interval the gradual phase component overruled the impulsive one, in low-energy images. After 21 : 57 UT the footpoints had disappeared also in high-energy pictures (Hoyng *et al.*, 1981). The existence of a third footpoint, initially mentioned by Hoyng *et al.* (1981), and called footpoint *C* by them, was not confirmed in the later research (Antonucci *et al.*, 1985), where calibration corrections were applied in a more refined way. Note, however (cf. Section 5), that a source of a coronal explosion occurs at the place of footpoint *C*.

The observations show therefore the existence of two footpoints in the flare, prior to and during the most energetic bursts, between 20 : 50 and 20 : 57 UT. Thereafter the gradual phase component was the most important one in the whole range of HXIS energies (3.5–30 keV).

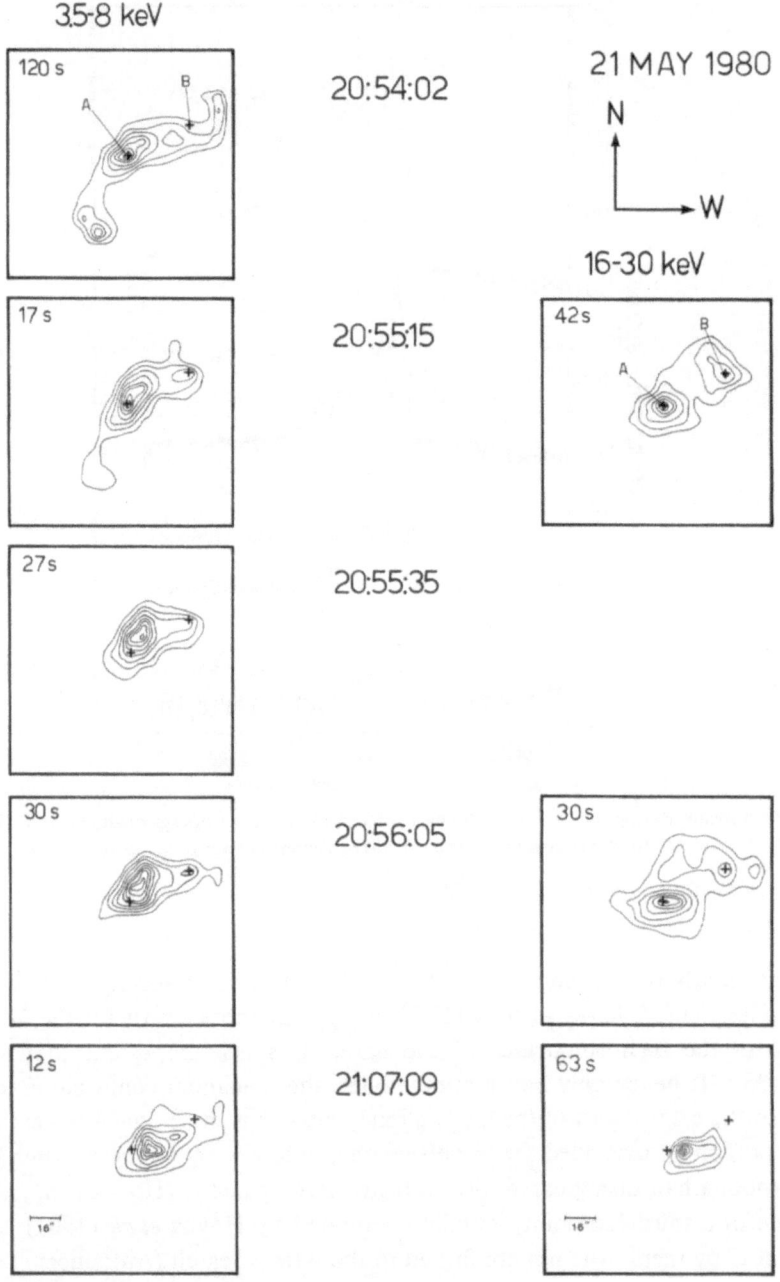

Fig. 4.   X-ray images of the 21 May flare in the low-energy (3.5–8 keV) and high-energy (16–30 keV) ranges. The crosses A and B mark the two footpoints. The first high-energy image is averaged over the period covering the low-energy images at 20 : 55 : 15 and 20 : 55 : 35 UT. (From Antonucci *et al.*, 1985.)

4.2. THE HIGH-TEMPERATURE FLARE COMPONENT; UPWARD MOTIONS

Duijveman (1983) found from an analysis of HXIS observations in different energy channels that immediately after the occurrence of the first impulsive hard X-ray spikes (cf. Figure 3) the plasma showed evidence for a low-temperature component (20 MK) and a high-temperature one (40 MK). This has since been confirmed in various ways.

Figure 5, taken from Antonucci *et al.* (1985), shows the Ca XIX spectrum between 20 : 54 and 21 : 07 UT. Remarkable is the strong broadening of the lines in the initial phase, indicating ionic Doppler temperatures T_D of 130 MK at 20 : 53 : 59 UT, decreasing to 40 MK around 20 : 55 : 35 UT. In the later, fully gradual phase of the flare, at 21 : 07 UT, the ionic Doppler temperature (14 MK) is virtually equal to the electron temperature T_e (16 MK) derived from the intensity ratios and degrees of ionisation of the spectral lines. From a fit to line profiles, Bely-Dubau *et al.* (1982) found electron temperatures from 15 MK (21 : 00 UT) to 12 MK (21 : 20 UT).

These observations give rise to the following two comments. First, the ion-electron exchange time in typical flare conditions is short, of the order of a second or less (Duijveman and Hoyng, 1983; De Jager, 1985b) so that the thermalisation time is very short as compared to the time intervals that are relevant for the present discussion. Therefore, the ionic Doppler temperatures must be either electron temperatures of a hot flare plasma component (which could indicate that the flare plasma should be bi-thermal, or multi-thermal) or else the Doppler broadening indicates the presence of turbulent motions of the order of 220 ± 30 km s^{-1} at 20 : 54 UT. Antonucci *et al.* (1985) favour the latter possibility, while one of us (De Jager, 1985b) has given arguments for the bi-thermal interpretation. That this may be the case is supported by the broad-band HXIS intensity measurements (Hoyng, 1982). But only observations of a large number of lines, including those of other ions, with other molecular weights, should give the decisive answer.

Secondly, although the time dependence of the temperature difference $\Delta \log T$ ($= \log T_D - \log T_e$) is badly known, with only a few data points available, they do not contradict the results obtained for the 8 April, 1980 and 5 November, 1980 flares (De Jager, 1985b) that $\Delta \log T$ goes exponentially to zero with an e-folding time of about 1.5 min.

The time development of the line-of-sight convective motion component in the flare plasma is given in Figure 6 (marked v'), together with the emission measures (EM) for the undisplaced component (full-drawn line) and the upward moving component (broken line). This shows that the upward motions were 370 km s^{-1} at 20 : 55 : 13 UT, remained > 300 km s^{-1} until 20 : 57 UT, and were ~ 150 km s^{-1} after 21 : 03 UT. The upward moving component apparently did not go to zero even as late as 21 : 05 UT, from which one infers that the footpoint heating continued till at least 21 : 05 UT. We will show later (Section 5) that the observations of chromospheric explosions in this flare support this inference. At 20 : 57 UT the emission measure of the displaced component was above one fourth of the undisplaced one. At 21 : 05 UT it had decreased to about one tenth, but it was not zero.

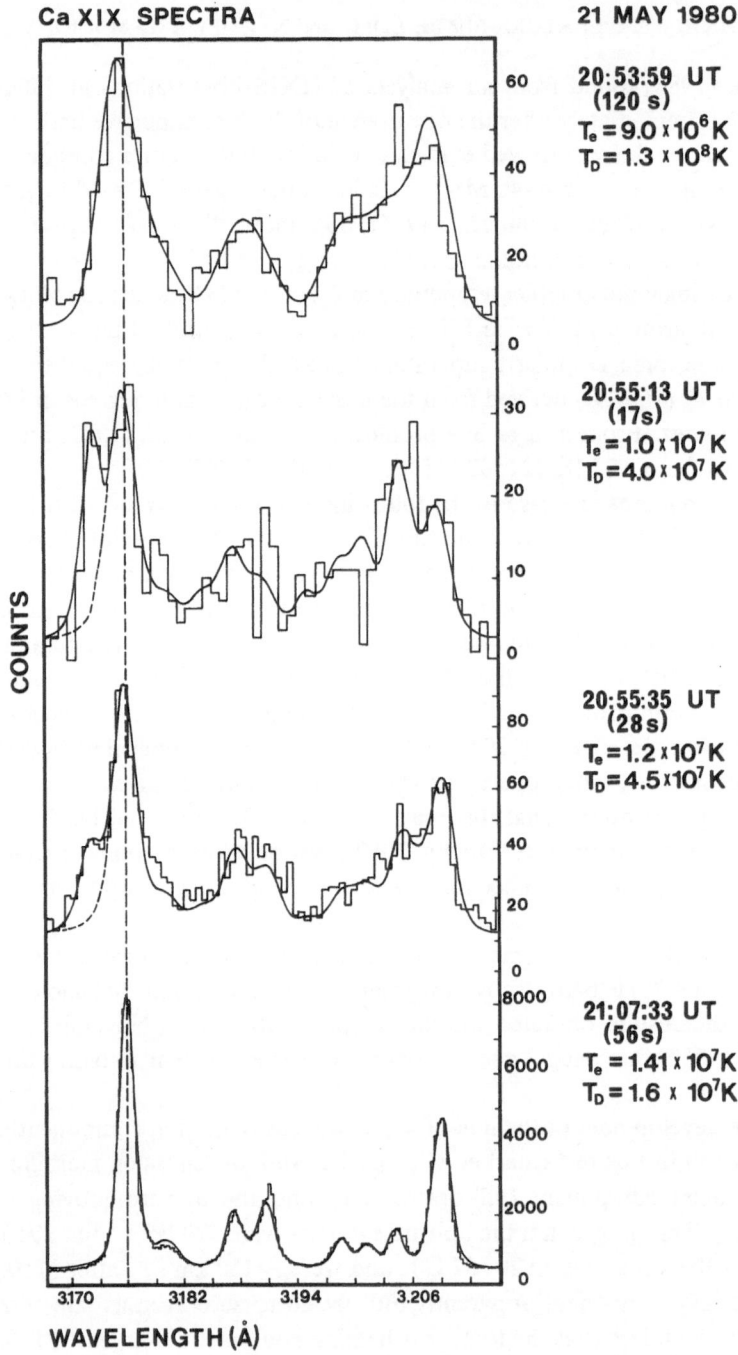

Fig. 5.   Four Ca XIX spectra of the flare of 21 May, 1980, the first three being taken in the impulsive phase.
The broadening and the shortward-shifted components are clearly visible. (From Antonucci *et al.*, 1985.)

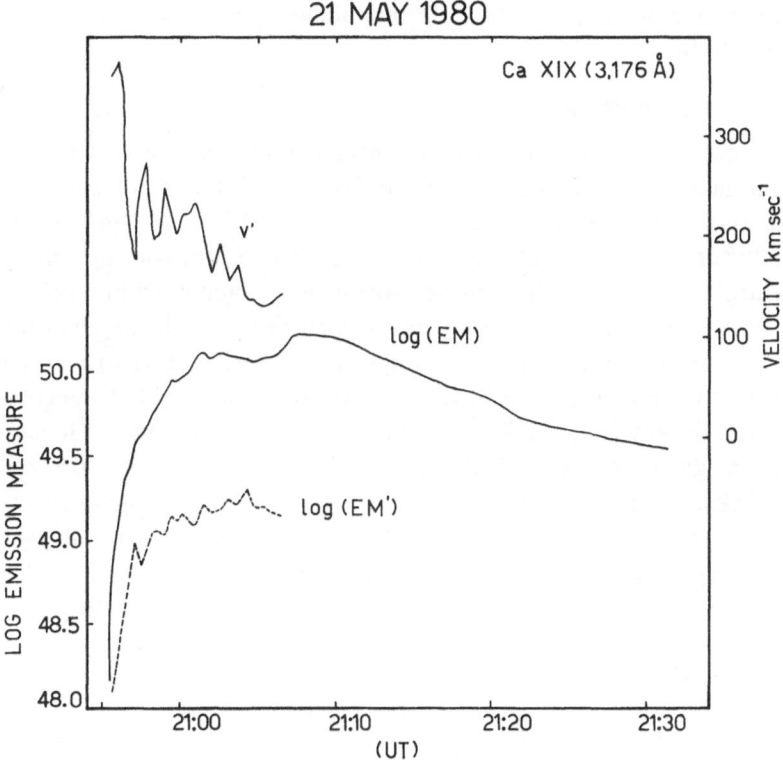

Fig. 6. Temporal evolution of the upward velocity component v', the emission measure for the kernel gas (EM) and for its displaced component (EM'). (From Antonucci *et al.*, 1985.)

4.3. ENERGIES INVOLVED IN THE IMPULSIVE PHASE PROCESSES

In order to examine the energy balance of the flare, we have to compare three quantities: injection energy, convection energy, and thermal energy of the gradual phase.

4.3.1. *The Injection Energy*

The energy injected by an (assumed) beam of energetic electrons is determined from the energy spectrum of the hard X-ray burst, assuming thick-target interaction and taking for the spectrum of the incident electrons a power law (Duijveman *et al.*, 1983; Dennis *et al.*, 1985):

$$I(E) = AE^{-\gamma} \text{ photons cm}^{-2}\,\text{s}^{-1}\,\text{keV}^{-1}.$$

From $I(E)$ the energy spectrum is derived, and one can compute the energy contained in that beam for $E \geq E_0$, the cut-off energy. The choice of E_0 is often an arbitrary act.

Thus Duijveman *et al.* (1982), taking a cut-off of 16 keV, found for the spike at 20:56 UT (marked 1, 2 in Figure 3) an integrated energy of 3×10^{29} erg. However, as shown before, electron beam incidence continued after 20:57 UT, albeit to a lesser degree. Dennis *et al.* (1985) calculated for this flare over the whole impulsive phase an integrated energy input of $(1 \pm 0.1) \times 10^{31}$ erg above $E_0 = 25$ keV, a value that seems nearer to the truth.

4.3.2. *The Convection Energy*

The second energy quantity one has to consider is that of the convected gas. We calculate this in the following way (cf. also De Jager, 1985b): according to Dennis *et al.* (1985) the kernels of the flare had a total area $A = 2.2 \times 10^{18}$ cm^2 and a volume $V = 25 \times 10^{26}$ cm^3. Since one fifth of the gas in this volume is moving upward, as shown by the spectral observations, the upward moving gas is contained in a column with a height $\frac{1}{5} \times V/A = 2600$ km, a value that may be identified with the depth of the ablated chromospheric 'hole'. Taking an average upward velocity (Figure 6) $v = 220$ km s^{-1} this means that the kernel gas needs 12 s to move over the height of the column. The kernel electron density is $n_e = 1.7 \times 10^{11}$ cm^{-3} (Dennis *et al.*, 1985). Hence, the production of hot gas by chromospheric ablation is $\dot{N} = \frac{1}{12} \times n_e \times \frac{1}{5}v = 6.9 \times 10^{36}$ electrons s^{-1}. Taking an exponential decrease of the production of high-temperature gas with an *e*-folding time of $\tau = 1.5$ min ($\simeq 100$ s), the total number of hot gas particles produced during the first, most violent, part of the impulsive phase is $N\tau = 6.9 \times 10^{38}$. The amount of thermal energy carried upward is hence $3N\tau kT$, and with $T = 4 \times 10^7$ K this yields 1.1×10^{31} erg, a quantity equal to that of the energy input by electron beams (with $E_0 = 25$ keV), as derived in Section 4.3.1.

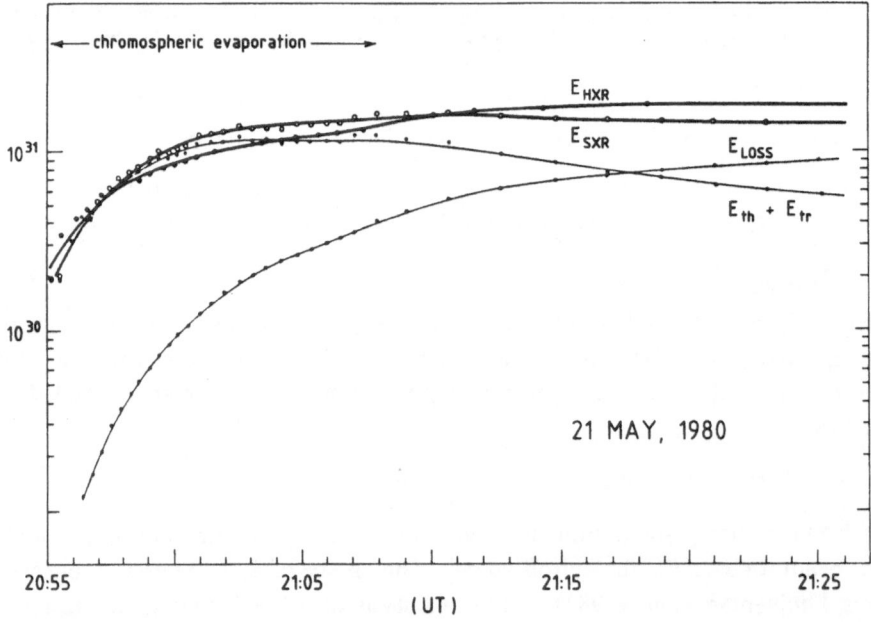

Fig. 7.   Temporal variation, for the 21 May, 1980 flare, of the energy content of the low-energy X-ray emitting plasma E_{SXR}, the high-energy X-ray plasma ($E_0 = 25$ keV, E_{HXR}), the energy lost by radiation and conduction (E_{LOSS}), and the thermal and turbulent energies ($E_{\text{th}}, E_{\text{tr}}$). (From Antonucci *et al.*, 1984.)

4.3.3. *The Thermal Energy of the Gradual Phase*

The total thermal energy content of the flare in its gradual phase is easily derived from the emission measure Y, the area A (taking the volume $V = A^{3/2}$) and the kinetic temperature. However, energy losses should be taken into account, in a way first described by Antonucci *et al.* (1982). From detailed study of some flares, Antonucci *et al.* (1984, 1985) derived maximum values for E_{th} of 1 to 2×10^{31} erg, depending on the model assumed.

Figure 7 is a diagram due to Antonucci *et al.* (1984) showing for the 21 May flare the time development of E_{SXR}, the energy of the soft X-ray plasma; it is the sum of the thermal and turbulent energies, with radiative and conductive losses substracted. E_{HXR} is the energy contained in the high-energy flare plasma (> 25 keV). It is of the order of 2×10^{31} erg at 21:10 UT.

We conclude that in spite of some slight inconsistencies in the models used by various authors and in their basic assumptions, there is good order-of-magnitude equality between the electron beam energy input, the thermal energy convected upward, and the eventual thermal energy of the gradual phase plasma. All values are $\simeq 1$ to 2×10^{31} erg.

5. Transition to the Flare's Gradual Phase, Coronal Explosions

Let us take the values of Dennis *et al.* (1985; cf. also Wu *et al.*, 1986) for the parameters of the impulsive phase at the peak time of the soft (~ 5 keV) X-ray emission, 21:05 UT. The flare had at that time an area $A = 4 \times 10^{18}$ cm^2, a volume $V = 4 \times 10^{27}$ cm^3, a total thermal energy content of 7×10^{30} erg (slightly less than the value given in Figure 7), and an electron density of 8×10^{10} cm^{-3}. The energy content of the flare in the gradual phase has been acquired by footpoint heating and consequent ablation and convection, the latter with velocities $\simeq 200$–300 km s^{-1} in the first part of the impulsive phase and about 150 km s^{-1} at 21:05 UT.

What happens further with the convected plasma can be examined by studying the lateral motions. The coronal explosions for this flare were investigated by Lemmens and De Jager (1985). This is done (De Jager, 1985a) by determining for each pixel the time at which local intensity maximum is reached and by drawing isochrones connecting these times of local maximum.

There appear to be *two* explosions in this flare. The first one started at or before 20:58:30 UT, just after the first impulsive hard X-ray spikes (Figure 8) and was progressing outward: initially with a speed of about 100 km s^{-1}, then accelerating to ~ 350 km s^{-1} and, thereafter, decreasing in speed again, first to about 100 km s^{-1} and later to lesser values. The explosion started simultaneously in two small areas almost coinciding with the two footpoints A and C of the flare as defined by Hoyng *et al.* (1981). It is remarkable that footpoint B (see Figure 4) did not coincide with an explosion source.

Also the second explosion (Figure 6 of Lemmens and De Jager, 1985) started in the footpoints A and C, but it was less violent, with velocities rather of the order of tens

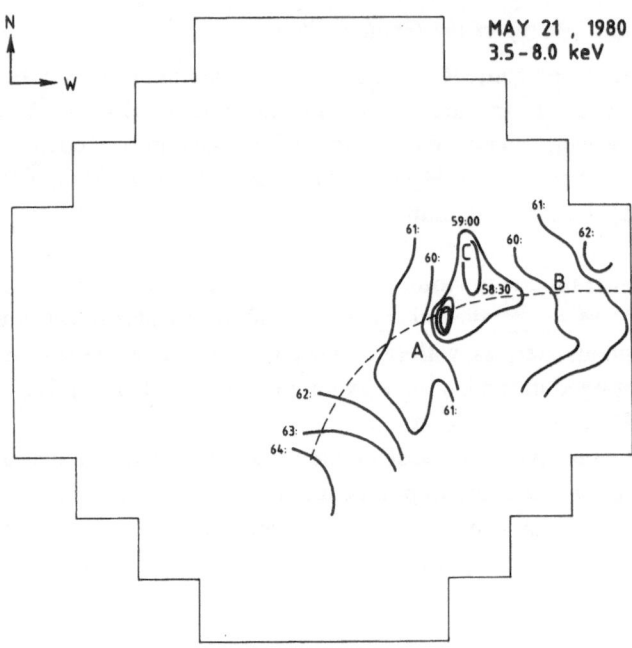

Fig. 8. Coronal explosion in the 21 May flare, observed in the low-energy HXIS channels (3.5–8.0 keV), after the impulsive spikes at 20:56–20:57 UT. The lines are isochrones labeled with the time in minutes after 20:00 UT. (From Lemmens and De Jager, 1985.)

of km s^{-1}. This second explosion was probably related to the burst complex at 21:00–21:03 UT, but it may also be associated with the spray that was seen in Hα since 21:50 UT (cf. Section 8).

The following interpretation has been proposed for the coronal explosions. Lemmens and De Jager suggest that they reflect the lateral outstreaming of gas initially convected upwards in the flare kernels. These authors showed for the 1980 November 12, 02:50 UT flare that the velocity variation can be mimiqued best by assuming flare plasma streaming, with conservation of mass, in cylindrical tubes which occupy a volume situated over the kernel and being larger than it. The fact that this kind of motion seems to be the case for any explosion observed so far (in contrast to the often assumed restricted streaming into and inside one loop connecting the footpoints) indicates that any flaring region must be a most complicated system of magnetic loops and should certainly not contain only one or a few distinct loops: a spaghetti-bundle model or a fanning-out fountain model look more probable. Also, reconnection between any of such loops in the course of a flare must be a normal process.

This interpretation may be supported by the observation of Antonucci et al. (1985) – cf. Figure 4 – that immediately after first kernel heating the center of gravity of the gradual emission component tended to move upward. The pictures at 21:07 UT in Figure 4 show a southward offset of the flare intensity maximum. But already at 20:57 UT the centroid of the emission had shifted southward by 5 arc sec (cf. Figure 9).

This indicates upward growth of the whole loop system, first described by Švestka *et al.* (1982a) and discussed in detail in the next section.

A possible consequence of this model had earlier been proposed by one of us (De Jager, 1985a): the upward moving component of the explosion becomes a magneto-hydrodynamical shock that may accelerate in less dense plasma, and then show up as a type II radioburst. A type II burst was indeed observed at Culgoora (Švestka *et al.*, 1982a) starting at 20:57 UT, close to the onset time of the first coronal explosion.

6. Growth of Post-Flare Loops

The shift of the maximum of brightness to the SW in Figure 4 shows the beginning of the growth of (post-) flare loops. Since the flare was at 13 S and 15 W, the shift to the SW reflects a growth in altitude in projection on the solar disk.

Figure 9, in its top, presents several images of the flare during more than two hours of its development. It is well-known that the post-flare loops are brightest at their tops (cf. e.g., Nolte *et al.*, 1979) so that there is little doubt that we image here the growth of the post-flare loops in the solar corona.

The central part of Figure 9 illustrates the time variation of temperature and emission measure in the brightest part of the flare imaged above. The electron temperature, determined from the flux ratio of HXIS bands 3 (8.0–11.5 keV) and 1 (3.5–5.5 keV) results here in much higher values than the temperature derived from line intensities and line profiles in Section 4.2. That is because the temperature here was deduced for the brightest part of the flare, whereas the line profiles and temperatures in Section 4.2. (Figure 5) represent averages over the whole flaring region.

The bottom part of Figure 9 shows the time variation of the distance d of the brightest portion of the loops from the $H_{\parallel} = 0$ line (Švestka and Poletto, 1985). For radial loops the real altitude would be 3.0 d. One can see that the growth of the flare loops in altitude was not continuous: the tops of the loops stayed for minutes at a given altitude before, quite abruptly, other loop tops began to appear above them. One may wonder if this was related to the second coronal explosion which lasted from 21:04 until 21:10 UT (Lemmens and De Jager 1985, Figure 6). A similar observation was made by Hinotori on 13 May, 1981, when the tops of flare loops stayed at the same altitude for 12 min (Tsuneta *et al.*, 1983: cf. Figure 14 in the paper by Dennis in this issue). Also in movies of post-flare loops in the Hα line one can sometimes see such abrupt jumps. Whereas, however, the jumps seen in Hα might have been due to effects of differences in the cooling of plasmas at different initial densities, the X-ray jumps are clearly related to the loops' excitation or formation. It is of interest that Forbes and Priest (1983) theoretically predicted the existence of multiple X-type neutral lines at several different altitudes above two-ribbon flares, where distended field lines may subsequently recon-nect and thus produce new sets of loops at different higher altitudes.

The first 'jump', at 20:57 UT, apparently indicates the first formation of (post-) flare loops, probably through reconnection of earlier opened (distended) field lines (Kopp and Pneuman, 1976). The temperature at the top of these loops was close to 24 MK

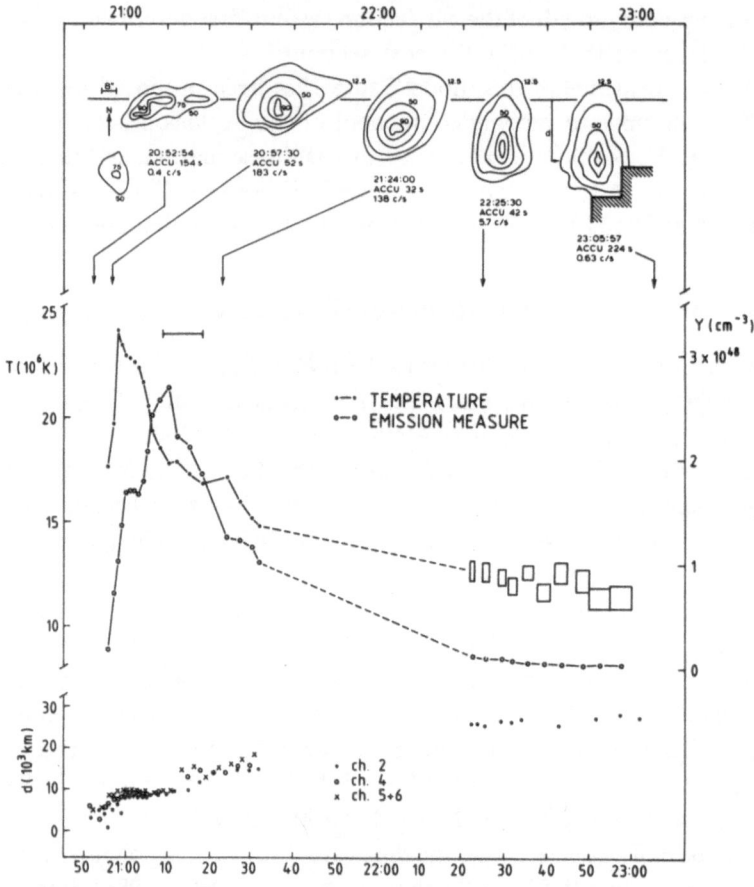

Fig. 9.   Top: HXIS images of the flare of 21 May, 1980 in 3.5–8.0 keV X-rays (after Švestka *et al.*, 1982a).
Center: temperature T and emision measure Y (per pixel of HXIS fine field of view) determined for the four
brightest pixels from the ratio of HXIS bands 3 (8.0–11.5 keV) and 1 (3.5–5.5 keV). *Bottom*: the distance
of the brightest part of the flare (tops of the loops) from the $H_\parallel = 0$ line, d, in projection on the disk, as
determined from HXIS bands 2, 4, and 5 + 6. (After Švestka and Poletto, 1985.)

(Figure 9), but at least some of them must have cooled very fast, since the first Hα loops
were seen at 21:00 UT. The cooling time of 3 min from $T \simeq 24$ MK to $T \simeq 10$ kK
shows that the electron density might have been as high as 1.5×10^{12} cm^{-3} in the early
formed loops.

Figure 10 shows an enlargement of the second abrupt rise in altitude that is evident
at 21:12 UT in Figure 9 (Švestka and Poletto, 1985). It demonstrates clearly that the
rise was first seen in the highest energy channel 6 (22–30 keV) with a later follow-up
at lower energies. This can be interpreted as a new release of energy in a very small
volume of plasma at extremely high temperature (a rise from 20 to 50 MK enhances the
flux in HXIS channel 6 by a factor 2200, but only by a factor 3 in channel 1). A likely
candidate for the process we observe here is the reconnection of field lines, in apparent

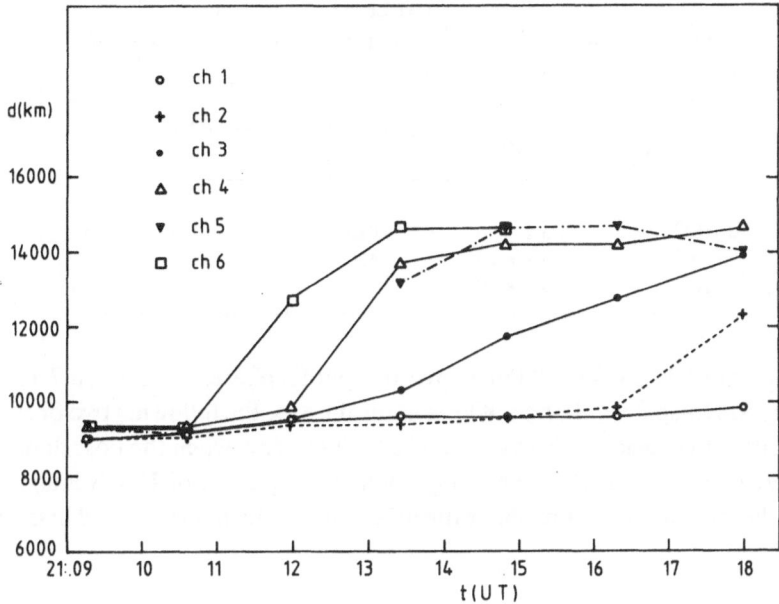

Fig. 10.   The variation of *d* in HXIS channels 1 (3.5–5.5 keV) – 6 (22–30 keV) during the jump in altitude which occurred at 21 : 12 UT (cf. the bottom part of Figure 9; from Švestka and Poletto, 1985).

agreement with the Kopp and Pneuman (1976) model of the formation of post-flare loops (in the basic configuration suggested by Sturrock, 1968; and with the multiple neutral lines deduced by Forbes and Priest, 1983). Kopp and Poletto (1985) compared the observations with model predictions and have concluded that there is good evidence for magnetic reconnection to be directly witnessed in the corona during this flare.

The available data do not allow us to deduce a unique set of parameters for the new source of the energy release. However, a reasonable agreement with the observation is found for a source with a temperature of 50 MK from 21 : 08 through 21 : 13 UT, growing in size (supposedly as an arcade of loops was formed) and reaching an emission measure of 3×10^{45} cm^{-3} at the end of its development. Then the source cooled, and its emission measure was fast increasing (possibly through evaporation of chromospheric material into the new loops) to 3×10^{47} cm^{-3} at 21:17 UT, i.e., by two orders of magnitude during 4 min. Table I gives the estimated size of the new source of energy release at 21:13 UT for various (unknown) electron densities.

The new brightening at a higher altitude (43 000 km if the loops extended radially, but probably at 21 000 km as we will see in Section 7.1) occurred 23 min after the flare onset and 17 min after the maximum in the hard X-ray burst. Thus it represents an energy release late in the post-maximum phase of the flare, which gives direct evidence that, in two-ribbon flares, energy is still being released after the end of the impulsive phase.

The post-flare loops could be clearly seen in X-rays for two satellite orbits, up to 23 : 07 UT (2 hr and 15 min after the flare onset). At that time the loop temperature decreased from 24 MK to 12 MK and their altitude increased to $d = 28\,000$ km in

TABLE I

Size of the new source at 21:13:19 UT for various assumed electron
densities

n_e (cm^{-3})	Volume (cm^3)	Radius (if source spherical) (km)
10^9	3.4×10^{27}	9330
10^{10}	3.4×10^{25}	2010
10^{11}	3.4×10^{23}	430
10^{12}	3.4×10^{21}	90

projection (most likely 41 000 km above the photosphere, cf. Section 7.1). Unfortunately, the SMM spacecraft did not look at the flare for the following two orbits. It came
back to this active region only at 03 : 08 UT the next day, when the post-flare loops had
completely disappeared. However, long enough integration of HXIS images revealed
then another coronal structure above the flare region, which is discussed in the following
section.

7. Post-Flare Coronal Arch

The very low background of HXIS made it possible to detect in X-rays large arch-like
structures in the solar corona, which are formed, or become enhanced, after two-ribbon
flares characterized by post-flare loops. Flares of other kinds apparently do not produce
or affect these structures (Švestka, 1984a, b). The flare of 21 May 1980 was the first
event, after which this new post-flare coronal phenomenon was detected (Švestka et al.,
1982a).

Figure 11 shows a series of images of the arch which became visible when the SMM
returned to the flaring active region at 03 : 08 UT on May 22. All these images are
half-an-hour integrations of an extremely weak structure: the peak countrate of the last
image, at 06 : 36 UT, only slightly exceeded 1 count min^{-1} per pixel; yet this is still four
times more than the average background noise in the 3.5–5.5 keV energy band of HXIS.

One can see that the brightest part (apparently the top) of the arch stayed in the same
position from 03 : 20 UT (6.5 hr after the flare onset) till the end of the HXIS imaging
period. Last images prior to the first one in Figure 11 are from 23 : 07 UT when the
post-flare loops still fully dominated the field so that it was extremely difficult to
distinguish the much weaker coronal arch. Nevertheless, a careful analysis made it
possible to separate the brightest part (top) of the arch from the loops and the results
are shown in Figure 12 (Hick and Švestka, 1985): since 23 : 02 UT the top of the arch
was in the same position as in the much later images of Figure 11 (the righthand image).
Prior to that, from the beginning of the orbit at 22 : 28 UT, the maximum was in a
neighbouring pixel, but still at about the same altitude above the flare. If one follows
the time development of the X-ray flux in these two pixels, one finds that the flux in the
first pixel peaked at ∼ 22 : 56 UT, whereas the other pixel peaked some 10 min later and
since then remained the brightest pixel in the field.

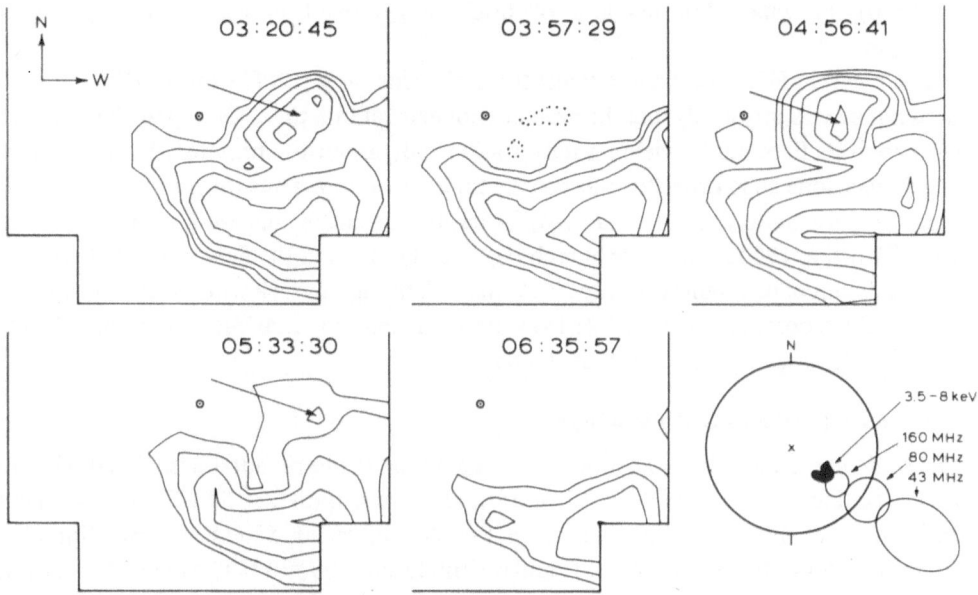

Fig. 11. X-ray contours of integrated images of the post-flare coronal arch observed on 22 May, 1980. Energy range 3.5–5.5 keV; HXIS coarse field of view, resolution 32 arc sec. The given times are mean times of 1525 s integrations. The position of the parent flare at 20 : 52 UT on 21 May (6.5 hr prior to the first image of the arch) is marked by dotted contours in the upper central image. Arrows point to a variable region below the arch discussed by Švestka *et al.* (1983; the reader is referred to this paper which analyses similar variations in another flare). The insert shows the relative position of the X-ray arch (black) and three radio images of the stationary type-I noise storm as seen at Culgoora at 03 : 20 UT on 22 May. (After Švestka, 1984b.)

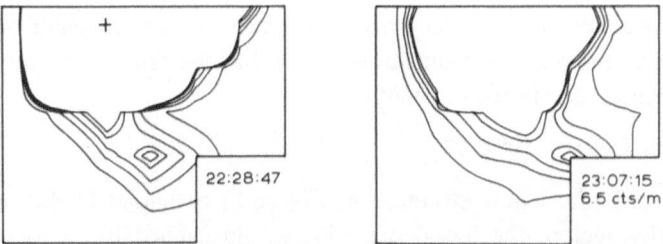

Fig. 12. X-ray contours during the SMM orbit next to the flare occurrence. HXIS coarse field of view, energy band 3.5–5.5 keV. The two images were recorded at respectively the beginning and end of the SMM orbit. (After Hick and Švestka, 1985.)

Thus the arch apparently consisted of at least two different structures which peaked in brightness at different times but stayed at a projected altitude of $\sim 100\,000$ km above the $H_\parallel = 0$ line for more than 8 hr. Taking into account the spatial resolution of the HXIS coarse field of view (32 arc sec), the extreme upper speed limit for any motion of this arch is 1 km s^{-1} so that the arch was clearly a stationary structure. From this point of view it differs from other events of this kind, discovered later (Švestka,

1984a, b), in which the maxima of brightness moved upwards with speeds of 8–12 km s⁻¹.

The fact that HXIS could not detect the arch prior to 22 : 28 UT during the flare orbit does not necessarily imply that the arch did not exist at that time. During the flare, HXIS worked in its flare mode, yielding information only about the brightest X-ray features, which was done so in order to enhance the time resolution. Thus the arch, much weaker than the flare, simply was not imaged. In another event of this kind, on 6 November, 1980 (Švestka, 1984a) the coronal arch began to brighten at the flare onset, but reached the maximum of its brightness only $2^h 15^m$ later. This agrees well with the data presented here: the two components of the 21 May arch reached their maxima $2^h 06^m$ and $2^h 16^m$, respectively, after the beginning of the flare.

7.1. ASSOCIATED RADIO EMISSION

The radio spectrograph at Culgoora observed a type II metric burst associated with the flare, starting at 20 : 57 UT on May 21 and lasting until 21 : 57 UT. The possible association of this burst with the coronal explosion of 20 : 58 UT was mentioned in Section 5. Possibly since 21 : 10 UT, and definitely since 21 : 30 UT, a type IV burst was observed at 20–200 Mhz. Stewart, in Švestka et al. (1982a), assumes that the latter burst was stationary, but no position data is available until 23 : 16 UT when the helio-graph started its observations. It was stationary for at least six hours after that. The burst was at first a mixture of a continuum and a noise storm, but it could be classified as a pure type I noise storm after 23 : 50 UT. The Culgoora heliograph could follow the noise storm until the end of the observing period at 05 : 10 UT on May 22.

As one can see in the insert in Figure 11, the type I noise storm radio emission was observed directly above the X-ray arch, so that the arch clearly images in X-rays the lowest part of the radio noise storm region. We can estimate the real altitudes of the radio images at various frequencies, since the radio emission is visible only at levels in the corona where the plasma frequency is lower that the radio frequency v, i.e., above the level for which the electron density is

$$n_e = (v/9)^2 \times 10^6 \, \text{cm}^{-3} \, ,$$

with v in MHz. If the whole structure in Figure 11 extended in the radial direction above the active region, the found densities would exceed by factors 30 to 100 the electron densities n_0 in the quiet corona. One has to incline the whole structure by 20° to 30° to the south–west to get $n_e/n_0 \simeq 10$ which is more reasonable for an active region (Hick and Švestka, 1985). This inclination is also suggested by coronagraphic observa-tions of the active region on the western limb (Wagner, in Švestka et al., 1982a): the coronagraph showed a (projected) southward tilt of roughly 20° from the vertical corresponding to a real south-west tilt of ∼25°.

Švestka and Poletto (1985) have demonstrated (in their Figure 2) that the post-flare loops grew exactly in the direction (PA = 225°) where the Culgoora images of the radio storm and the HXIS X-ray images of the arch were projected. Thus, since it is reasonable to believe that the arch is an upper product of the same reconnection processes that lead to the formation of the post-flare loops (cf. Section 4.7), we may assume that the post-flare loops, the X-ray arch, and the radio noise storm all extended in the same spatial direction, tilted by $\sim 25°$ to the south-west. In that case the brightest part of the arch was at an altitude of 145 000 km above the photosphere; the height of the flare loops was 21 000 km after the abrupt rise at 21:12 UT and 41 000 km at the end of HXIS observations at 23:07 UT.

7.2. Physical parameters of the arch

Hick and Švestka (1985) have determined the temperature T and emission measure Y at the top of the arch by comparing X-ray fluxes in HXIS bands 2 (5.5–8.0 keV) and 1 (3.5–5.5 keV). At the time of its maximum brightness ($\sim 23:00$ UT) the arch temperature was found to be 6.3 (± 1.3) MK. In the same phase of development the temperature T in the well-observed second arch of November 6 (Švestka, 1984a) was ~ 13.4 MK. In that arch, T peaked 1 hr 20 min earlier reaching ~ 14.5 MK and decreased 4 hr later to ~ 8.5 MK. In contrast to it, T remained remarkably constant in the May arch: $T \approx 7.1$ (± 1.8) MK at 03:40 UT and $T \approx 6.4$ (± 1.3) MK at 05:20 UT on May 22. This has been also confirmed by temperature determinations from the Ca XIX line complex observed by the Bent Crystal Spectrometer on board of the SMM. Acton, Gabriel and Rapley (in Švestka *et al.*, 1982a) found $T = 7.0$ (± 0.5) MK at 03:25 UT, 6.0 (± 0.5) MK at 03:53 UT, and 6.4 (± 0.5) MK at 05:19 UT.

Taking $T = 6.3$ MK, the emission measure at 23:00 UT was $Y \approx 1.0 \times 10^{47}$ cm^{-3} per pixel (of the coarse field of view). Assuming that the linear thickness of the arch at the location of the maximum of brightness was of the same order as its altitude, $l \simeq h \simeq 145\,000$ km, we find an electron density $n_e \simeq 1.1 \times 10^9$ cm^{-3} and the corresponding energy density $\varepsilon = 3n_e kT \simeq 2.9$ erg cm^{-3}. With the emitting volume estimated to $V \simeq 5.0 \times 10^{29}$ cm^{-3} the total energy content was $E = V\varepsilon \simeq 1.4 \times 10^{30}$ erg (which is about one tenth of the initial energy input into the flare – Section 4.3), and the total mass $M \simeq 8.9 \times 10^{14}$ g.

In comparison with the second arch of November 6, the post-flare coronal structure of 21/22 May was a weak event. The corresponding values for the (much brighter) November arch were (Švestka, 1984a) $E \simeq 1.2 \times 10^{31}$ erg and $M \simeq 4.4 \times 10^{15}$ g. Yet, in both cases the arch energy is still a significant fraction of the energy of the parent flare and the mass of the arch is comparable with the mass of an active region filament. The values of E and M are also of the order of magnitude of the energy and mass values found for coronal transients.

7.3. Proposed interpretation

Švestka *et al.* (1982a) have tried to explain the arches as a byproduct of the Kopp and Pneuman (1976) model for two-ribbon flares, modified for sheared field lines by Anzer and Pneuman (1982). Figure 13 illustrates this idea.

Prior to the flare (Figure 13(a)) a system of sheared active region loops bridges the $H_{\parallel} = 0$ line. When the flare starts, these loops are extended high into the corona. At a certain level a reconnection process starts (Figure 13(b)) due to the resulting

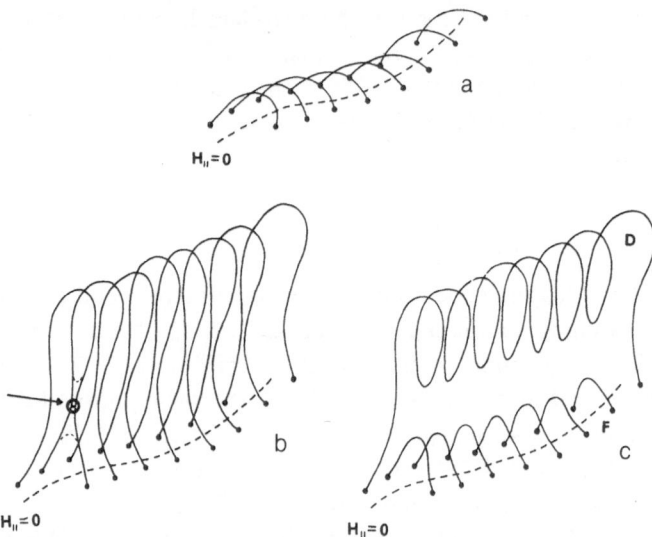

Fig. 13.   If the pre-flare magnetic field is sheared (a) and is extended upwards at the onset of the flare (b), reconnection occurs between neighboring field lines (arrow in (b)). As a result we get then less sheared post-flare loops below (F in (c)), while the upper disconnected loops (D in (c)) become interconnected along the $H_{\parallel} = 0$ line. This may schematically represent the post-flare arch. (From Švestka *et al.*, 1982a.)

unbalance between the magnetic pressure and gas pressure, leading to the formation of less sheared post-flare loops in the low corona and another system of reconnected fieldlines above.

In the original Kopp and Pneuman model the field lines open. This, however, can never be achieved if the field is sheared, as has been pointed out by Aly (1984). In that case the dynamical situation of the system cannot lead to an open field for the simple reason that such a configuration is a state of maximal energy for the force-free field. However, the field opening can be closely approached as we schematically indicated in Figure 13(b); then the reconnection gives rise to systems of coil-like magnetic fieldlines extending along the $H_{\parallel} = 0$ line of which one is schematically shown in Figure 13(c).

There is a long sequence of such reconnecting coils that merge, mix, partly reconnect, and eventually give rise to a very complex magnetic field formation above the flare site in which plasma, excited in the reconnection process, is confined and seen in X-rays as the post-flare arch.

The fact that solely the two-ribbon flares, with their growing systems of loops, give rise to these arches, greatly supports this interpretation. Another strong supporting fact is the observed propagation of the growing loops and the extension of the arch and radio type I noise storm on 21/22 May along the same direction of $PA = 225°$ (cf. Section 7.2). The main objection against this model is the stationarity of the arch. In the original picture by Anzer and Pneuman (1982) the sequentially reconnected coils move steadily upwards, driven by magnetic pressure from below. To keep the whole structure fixed at a constant altitude for eight or more hours needs some force that compensates for this drive. It has been suggested that this braking agent is related to the magnetic complexity of the arch which orginates through reconnections between the sequential coils and eventually creates a self-supporting post-flare structure above the active region.

Švestka *et al.* (1982a) assumed that the plasma heated during the reconnection process, and particles accelerated in the reconnection, propagate into the complex field of the arch, are trapped there, and produce the observed X-ray emission. Achterberg and Kuipers (1984), on the other hand, have suggested that the diffusion of electrons with energies in excess of 100 keV out of the flaring loops in the solar corona are the energy source of the observed X-ray and radio emission. The diffusion agent is low-frequency magnetic turbulence in the lower corona which appears to play a more important role than regular drifts for the cross-field electron transport.

These authors did not try to explain the origin of the loops. The sequential reconnection process suggested by Kopp and Pneuman seems necessary, because the growth of post-flare loops can last up to 11 hr (cf. Nolte *et al.*, 1979). Therefore, heating and acceleration processes during the reconnection have to be taken into account. Nevertheless, the interpretation proposed by Achterberg and Kuijpers may perhaps help to understand the propagation of accelerated electrons across magnetic field lines into the post-flare arch.

7.5. ORIGIN OF THE ARCH

In the November event (Švestka, 1984a) the whole arch began to brighten at the very onset of the flare, reaching maximum brightness about 2 hr 15 min later, but at the same time the maximum of brightness was steadily moving upwards. This has been interpreted as the filling-up of a pre-existing arch with heated plasma while a new arch-like structure was being formed from below. It took about six hours before the moving structure reached an altitude of $\sim 150\,000$ km, the altitude of the stationary March arch.

The May arch apparently reached this altitude in less than 1.5 hr. Thus the average rise velocity must have been higher than 17 km s^{-1} if starting at $< 50\,000$ km altitude (as on November 6). Since we have no data about the arch for the first 1.5 hr after the flare onset, such a rise can neither be verified nor excluded.

It is, of course, also possible that the arch was built up very fast at the very onset of the flare as a counterpart of a coronal transient. We will see in Section 8 that there was no filament eruption in the May 21 flare: still a part of the magnetic field above the filament might have risen and create the arch near the time of the flare onset. This elevated structure would then be further enhanced by the long-lasting reconnection process schematically shown in Figure 13.

An alternative possibility is that the arch was a pre-existing structure in the corona, which just brightened when the flare appeared below it, as was the case in all three arches on November 6/7. But in all these three events there had been an earlier two-ribbon flare which could be held responsible for the creation of the pre-existing arch. No such flare preceded the flare event of May 21 (no significant activity at all preceded the May 21 flare, cf. Schadee *et al.*, 1983).

Nevertheless, as we saw in Figure 2, there was a leakage of plasma from the active region into interplanetary space prior to the flare, possibly associated with the semi-permanent weak activity along the filament channel (cf. Section 1). Since we know very little about the nature of the arch, one cannot exclude that its basic magnetic structure was also built up during this pre-flare period.

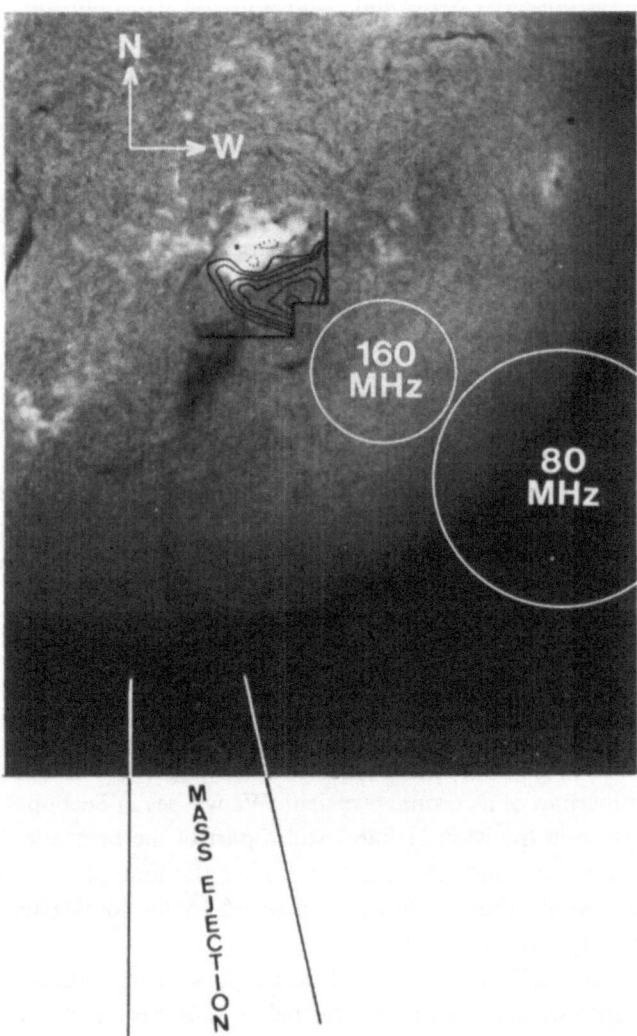

Fig. 14.   Hα image of the Sun (Haleakala) at 21:20 UT on 21 May, 1980, showing the two-ribbon flare (white) and the associated dark spray (black). Projected on this image is the X-ray post-flare arch from Figure 12 (HXIS) and sites of the type I noise storm at 160 and 80 MHz (Culgoora) at 03 : 20 UT on 22 May. The cone of the mass ejection (Solwind, NRL) is shown at the bottom. (From McCabe *et al.*, 1985.)

8. Associated Mass Ejection

The NRL-Solwind coronagraph aboard the P78–1 satellite revealed a very bright and narrow feature in the corona propagating southward ($PA \simeq 180$–$192°$; Howard and Sheeley in McCabe *et al.*, 1985). The feature was first seen at $21:43$ UT (53 min after the flare onset) and its leading edge reached $6.5\,R_\odot$ at $23:19$ UT. The average speed of its propagation in projection on the plane of sky was 420 km s^{-1}.

Further information can be obtained about this mass ejection from the viewpoint of the zodiacal light photometers on board the Helios A spacecraft (Jackson in McCabe *et al.*, 1985). From both the Solwind viewpoint (at Earth) and the Helios viewpoint (about $30°$ east of the Sun–Earth line), the primary mass of the ejection was limited to a narrow angular extent moving southward of the Sun. The real speed of propagation in space, deduced from a comparison of the Solwind and Helios data was $\gtrsim 670$ km s^{-1}.

Under the assumption that the mass ejection originated in the flare, the direction disagrees by $\sim 40°$ from the direction expected if the flare filament erupted and produced the observed transient. The mass ejection should then extend approximately along the same direction as the X-ray arch and the radio noise storm (cf. Figure 14). However, as McCabe *et al.* (1985) have emphasized, the active region filament never erupted, and did not disappear. Instead, Hα-emitting material was ejected from the south–eastern end of the filament channel in the form of a spray at $21:05:20$ UT (cf. Figure 14). The observed mass ejection was seen in the prolongation of the path of this spray, which apparently was the real source of the recorded coronal transient in this particular case.

The coexistence of stationary post-flare coronal structures (X-ray arches) and fast moving mass ejecta (coronal transients) in the post-flare corona poses a serious problem. In this event the coexistence can be easily understood: the mass ejection went through a narrow cone in another direction than where the stationary X-ray and radio structures were seen. But some other events are more complex: on 6 November, 1980, e.g., at $03:30$ UT, Helios A (Jackson, private communication) and Culgoora (Stewart, private communication) clearly evidenced the presence of a very extensive transient. This has also been confirmed in later images by Solwind of the post-flare corona (Howard and Sheeley in McCabe *et al.*, 1985). Still, this flare also produced (or revived) a giant semi-stationary arch staying above the active region for more than 11 hr (Švestka, 1984a). We do not understand yet what really happens in the flaring and post-flare corona in such a case, but this problem is actually an old one: as early as in 1961, Pick distinguished between the first and second part of a radio type IV burst, of which part one corresponds to a moving type IV and part two to a stationary component of the same burst. An excellent example was imaged by Wild (1969) on 22 November, 1968. What we see here, appear to be X-ray images of the 'part two' which has developed into a noise storm.

The flare was a source of high-energy particles recorded in space. Helios A, at 0.35 AU from the Sun, yielded the following maximum fluxes (Wibberenz, private communication):

$$0.8\text{–}2.0 \text{ MeV electrons:} \quad 5 \times 10^3 \text{ m}^{-2} \text{ s}^{-1} \text{ sr}^{-1} \text{ MeV}^{-1};$$

$$3.8\text{–}12.6 \text{ MeV protons:} \quad 1 \times 10^4 \text{ m}^{-2} \text{ s}^{-1} \text{ sr}^{-1} \text{ MeV}^{-1};$$

$$27\text{–}37 \text{ MeV protons:} \quad 2 \times 10^2 \text{ m}^{-2} \text{ s}^{-1} \text{ sr}^{-1} \text{ MeV}^{-1};$$

27–37 MeV helium nuclei: no recognizable flux .

A minor shock wave from the flare hit Helios A at ~ 21 UT on 22 May; the time difference of ~ 24 hr corresponds to the mean speed of propagation of ~ 590 km s^{-1} (Sheeley *et al.*, 1984).

9. Final Scenario

From all these data we may suggest the following scenario for the flare of 21 May, 1980.

For more than 30 hr prior to the flare, the active region filament was greatly activated and X-ray images gave evidence for the existence of plasma with temperature in excess of 10 MK along the filament channel; or, alternatively, for intermittent, but very frequent acceleration processes near the $H_{\parallel} = 0$ line. It is likely that the weak mass-flow imaged by Helios A above $16 R_{\odot}$ over the active region at that time was linked to this filament activity (Section 2). No important flares, however, did occur during that period.

A possible pre-flare X-ray brightening appeared at the flare site 90 min before the flare, with the X-ray flux growing by an order of magnitude during this period. An intense X-ray flare precursor was definitely present for $\gtrsim 6$ min before the flare onset (Section 3).

Strong downward motions appeared at the west end of the filament about 2 hr before the flare. At the same time, a kink developed in the middle of the filament, very close to the main site of the brightening in the flare. These destabilisations of the filament seemed to be linked to changes in the magnetic field first detected 80 min prior to the flare: a newly emerging flux compressed the existing flux below the kink, where a small new sunspot began to be seen (Section 3).

By 20 : 50 UT the filament had broken into two distinct parts at the site of the kink and the flare began to develop, supposedly through excitation of some coronal loops at this site and by opening (extending) of magnetic fieldlines at higher altitudes. Since the filament did not erupt (Section 8), the field opening must have been accomplished only above the filament, leaving the lower loop system, containing the filament, closed (Section 3). At that time already, the first faint indications of footpoint heating appeared (Figure 1; Section 4.1).

At the peak of the flare-associated hard X-ray burst, at 20 : 55 UT, hard X-ray images showed two distinct kernels of emission at opposite sides of the $H_{\parallel} = 0$ line: obviously the footpoints of an excited flare loop in which accelerated electrons produced thick-target X-ray emission (Section 4.1). Shortly before that, profiles of the Ca XIX lines showed remarkable broadening which can be interpreted either by turbulent motions (with $\bar{v} \simeq 220\ \mathrm{km\ s^{-1}}$ at $20:54$ UT) or by a multi-thermal structure (with the hot component as hot as $130\ \mathrm{MK}$ at the very beginning). Chromospheric evaporation manifested itself through convective upward motions since the time of the hard X-ray peak ($370\ \mathrm{km\ s^{-1}}$ at $20:55:13$ UT), and the motions were still present ten minutes later which implies that even at that time the footpoint heating still continued (Section 4.2). However, the thermal gradual phase component dominated the X-ray images since $20:57$ UT (Section 4.1).

The energy involved in the impulsive phase processes can be estimated to 1 to 2×10^{31} erg (Section 4.3).

Shortly after the peak of the impulsive phase, at or before $20:58:30$ UT, a coronal explosion was detected in X-rays, with initial lateral speed of $100\ \mathrm{km\ s^{-1}}$ accelerating to $\sim 350\ \mathrm{km\ s^{-1}}$ and decreasing thereafter. It might have been the lateral component of a wave of which the upward moving component transformed into a shock wave that manifested itself as a type II radioburst high in the solar corona (starting at $20:57$ UT) (Section 5).

A second coronal explosion started at about $21:04$ UT and could perhaps be associated with the spray seen in Hα since $21:05$ UT (Sections 5 and 8).

At $20:57$ UT X-ray images gave the first evidence of the existence of (post-) flare loops. In Hα, these loops were first seen at $21:00$ UT which indicates that the electron density in the first (lowest) loops was $\leq 10^{12}\ \mathrm{cm^{-3}}$. The loops grew upwards, but the growth was not continuous: after staying for a while at a constant height, new loops began to appear abruptly at a higher altitude. One of these jumps in altitude, at $21:12$ UT, gives evidence for a late release of energy high in the corona in a very small and extremely hot volume which may be most likely a product of field-line reconnection. This tends to support strongly the model of Kopp and Pneuman (1976) in which successive reconnections of earlier opened field lines produce the post-flare loops (Section 6).

About 1 hr after the flare onset the Solwind coronagraph began to image the mass ejection coming from the flare. Later on, its images were also obtained by Helios A from a different viewpoint. These observations show that the mass ejection was in the prolongation of the path of a powerful spray that was ejected at $21:05$ UT from the south–eastern end of the filament channel, while the filament itself did not erupt (Section 8).

From $21:30$ UT, Culgoora observed a stationary type IV burst associated with the flare, which changed into a type I noise storm at about $23:50$ UT. From 1.5 hr after the flare, X-ray images revealed the existence of a giant coronal arch just below the radio noise-storm region. The arch extended along the filament channel over the active region to an altitude of $\sim 150\,000$ km. As the radio noise storm above it, the arch was completely stationary and stayed at the same altitude for more than 8 hr, slowly decaying in brightness (Section 7.1). Its energy content is estimated to be 1.4×10^{30} erg and its total mass 9×10^{14} g (Section 7.2). It is suggested that the arch was the upper product of the reconnection process that gave rise to the post-flare loops below (Section 7.3).

We believe that the results obtained for this particularly well observed flare have extended our knowledge of the flare phenomenon in several important points. However, on the other hand, the observations also show how complicated the flare process actually is. Let us give here just two very brief examples.

In another paper in this issue, Bryan Dennis describes and discusses the characteristics of the Hinotori flares of classes B and C. The present flare, without any doubt, was both a class B (with emission at footpoints) and a class C (with emission high in the corona). Thus it belonged to the 'controversial events' mentioned by Dennis and we have some doubts whether oversimplified classifications can really help to understand flares.

Another believe is that two-ribbon flares and coronal mass ejections follow the eruption of a filament. Well, clearly, as in this flare, the eruptive field changes can occur entirely above the filament channel, leaving the filament in its preflare position. Still, the filament did not stay intact: it became fragmented and a powerful spray erupted near its eastern end some five minutes after the maximum of the flare impulsive phase.

One can see that variations in the altitude at which the field opens (erupts) can clearly produce a great variety of consequent flare configurations in the corona. And that is a variation of only one parameter in the flare process, and only in one kind of the flare phenomenon: the two-ribbon (dynamic) flares. We do not dare to go deeper into this problem here, but theoreticians should give some thought to it when modelling 'a flare'.

Acknowledgements

The extraordinarily rich set of space and ground-based data on this flare was obtained during an interval of coordinated observations organized by the Flare Build-up Study Project of the Solar Maximum Year. Thanks are due to all who participated in it and to the FBS coordinator Dr. Paul Simon in Meudon.

We are obliged to Hans Braun in Utrecht for his skilfull help when preparing the illustrations.

References

Achterberg, A. and Kuijpers, J.: 1984, *Astron. Astrophys.* **130**, 111.

Aly, J. J.: 1984, preprint.

Antonucci, E.: 1982, *Mem. Soc. Astron. Ital.* **53**, 495.

Antonucci, E. and Dennis, B. R.: 1983, *Solar Phys.* **86**, 67.

Antonucci, E., Gabriel, A. H., Acton, L. W., Culhane, J. L., Doyle, J. G., Leibacher, J. W., Machado, M. E., and Rapley, C. G.: 1982, *Solar Phys.* **78**, 107.

Antonucci, E., Gabriel, A. H., and Dennis, B. R.: 1984, *Astrophys. J.* **287**, 917.

Antonucci, E., Dennis, B. R., Gabriel, A. H., and Simnett, G. M.: 1985, *Solar Phys.* **96**, 129.

Anzer, U. and Pneuman, G. W.: 1982, *Solar Phys.* **79**, 129.

Bely-Dubau, F., Dubau, J., Faucher, F., Gabriel, A. H., Loulegrue, M., Steenman-Clark, L., Volonté, S., Antonucci, E., Rapley, C. G.: 1982, *Monthly Notices Roy. Astron. Soc.* **201**, 1155.

De Jager, C.: 1985a, *Solar Phys.* **96**, 143.

De Jager, C.: 1985b, *Solar Phys.* **98**, 267.

De Jager, C. and Boelee, A.: 1984, *Solar Phys.* **92**, 227.

De Jager, C., Boelee, A., and Rust, D. M.: 1984, *Solar Phys.* **92**, 245.

De Jager, C., Machado, M. E., Schadee, A., Strong, K. T., Švestka, Z., Woodgate, B. E., and van Tend,W.: 1983, *Solar Phys.* **84**, 205.

Dennis, B. R., Lemen, J. and Simnett, G.: 1985, *The Relationship between Hard and Soft X-Rays in Flares Observed with the Solar Maximum Mission during 1980*, preprint, also reproduced in Chapter V of SMM Workshop Proceedings.

Duijveman, A.: 1983, *Solar Phys.* **84**, 189.

Duijveman, A. and Hoyng, P.: 1983, *Solar Phys.* **86**, 279.

Duijveman, A., Hoyng, P., and Machado, M. E.: 1982, *Solar Phys.* **81**, 137.

Forbes, T. G. and Priest, E. R.: 1983, *Solar Phys.* **84**, 169.

Gaizauskas, V.: 1985, in *Proceedings of the NASA-SMM Workshop on Solar Flares,* Section 5, Chapter 3.

Harvey, J. W.: 1983, *Adv. Space Res.* **2**, No. 11, 31.

Hick, P. and Švestka, Z.: 1985, *Solar Phys.* (in press).

Hoyng, P.: 1982, *The Observatory* **102**, 119.

Hoyng, P., Duijveman, A., Machado, M. E., Rust, D. M., Švestka, Z., Boelee, A., de Jager, C., Frost, K. J., Lafleur, H., Simnett, G. M., van Beek, H. F., and Woodgate, B. E.: 1981, *Astrophys. J.* **246**, L155.

Jackson, B. V., Schadee, A., and Švestka, Z.: 1985 (in preparation).

Kopp, R. A. and Pneuman, G. W.: 1976, *Solar Phys.* **50**, 85.

Kopp, R. A. and Poletto, G.: 1985, in D. Neidig (ed.), *Proc. of Sac. Peak Summer Meeting on Low-Temperature Flare Plasma*, in press.

Kuperus, M. and van Tend, W.: 1981, *Solar Phys.* **71**, 125.

Lemmens, A. and de Jager, C.: 1985, *Solar Phys.* (submitted).

McCabe, M. K., Švestka, Z, Howard, R. A., Jackson, B. V., and Sheeley, N. L., Jr.: 1985, *Solar Phys.* (submitted).

Mewe, R., Gronenschild, E. H. B. M., and van den Oord, G. H. J.: 1985, Astron. *Astrophys. Suppl.* (in press).

Nolte, J. T., Gerassimenko, M., Krieger, A. S., Petrasso, R. D., and Švestka, Z.: 1979, *Solar Phys* **62**, 123.

Pick, M.: 1961, *Ann. Astrophys.* **24**, 183.

Schadee, A., De Jager, C., and Švestka, Z: 1983, *Solar Phys.* **89**, 287. (Abbreviated as SDJS).

Sheeley, N. R., Howard, R. A., Koomen, M. J., Michels, D. J., Schwenn, R., Muhlhauser, K. H., and Rosenbauer, H.: 1984, *J. Geophys. Res.* **90**, 163.

Sturrock, P. A.: 1968, *IAU Symp.* **35**, 471.

Švestka, Z.: 1984a, *Solar Phys.* **94**, 171.

Švestka, Z.: 1984b, *Mem. Soc. Astron. Ital.* **55**, 725.

Švestka, Z. and Poletto, G.: 1985, *Solar Phys.* **97**, 113.

Švestka, Z., Stewart, R. T., Hoyng, P., van Tend, W., Acton, L. W., Gabriel, A. H., Rapley, C. G., and 8 co-authors: 1982, *Solar Phys.* **75**, 305.

Švestka, Z, Schrijver, J., Somov, B., Dennis, B. R., Woodgate, B. E., Fürst, E., Hirth, W., Klein, L., and Raoult, A.: 1983, *Solar Phys.* **85**, 313.

Tsuneta, S., Takakura, T., Nitta, N., Ohki, K., and 4 co-authors: 1983, *Solar. Phys.* **86**, 313.

Wild, J. P.: 1969, *Solar Phys.* **9**, 260.

Wu, S. T., de Jager, C., Dennis, B. R., Hudson, H. S., Simnett, G. M., Strong, K. T., and 15 co-authors: 1985, in M. R. Kundu and B. E. Woodgate (eds.), *Proc. SMM Workshop on Solar Flares*, (in preparation).

SOLAR HARD X-RAY BURSTS

BRIAN R. DENNIS

*Laboratory for Astronomy and Solar Physics, NASA Goddard Space Flight Center,
Greenbelt, MD 20771, U.S.A.*

Abstract. The major results from SMM are presented as they relate to our understanding of the energy release and particle transportation processes that lead to the high-energy X-ray aspects of solar flares. Evidence is reviewed for a 152–158 day periodicity in various aspects of solar activity including the rate of occurrence of hard X-ray and gamma-ray flares. The statistical properties of over 7000 hard X-ray flares detected with the Hard X-Ray Burst Spectrometer are presented including the spectrum of peak rates and the distribution of the photon number spectrum. A flare classification scheme introduced by Tanaka is used to divide flares into three different types. Type A flares have purely thermal, compact sources with very steep hard X-ray spectra. Type B flares are impulsive bursts which show double footpoints in hard X-rays, and soft-hard-soft spectral evolution. Type C flares have gradually varying hard X-ray and microwave fluxes from high altitudes and show hardening of the X-ray spectrum through the peak and on the decay. SMM data are presented for examples of type B and type C events. New results are presented showing coincident hard X-rays, O v, and UV continuum observations in type B events with a time resolution of ≤ 128 ms. The subsecond variations in the hard X-ray flux during ≲ 10% of the stronger events are discussed and the fastest observed variation in a time of 20 ms is presented. The properties of type C flares are presented as determined primarily from the non-imaged hard X-ray and microwave spectral data. A model based on the association of type C flares and coronal mass ejections is presented to explain many of the characteristics of these gradual flares.

1. Introduction

This paper is a review of some of the more recent results on solar hard X-ray bursts primarily for the Solar Maximum Mission (SMM). I have used the flare classification scheme introduced by Tanaka (1983) to lend some order to the apparently random nature of hard X-ray bursts.

It is instructive to first reflect on the state of our knowledge of hard X-ray bursts prior to the immense number of observations during this solar maximum, now determined to have been in December, 1979. Prior to that time thare were no hard X-ray images of solar flares apart from observations of over-the-limb events, no observations of variations on time-scales much less than 1 s, no spectral observations that could resolve the steepest spectra, and no definitive polarization measurements at all. We now have X-ray images for many flares with better than 10″ resolution at energies up to 30 keV from SMM and 40 keV from Hinotori. The Hard X-Ray Burst Spectrometer (HXRBS) has provided observations of over 7000 flares with a time-resolution of 128 ms and many of these with 10 ms resolution. Significant advances in spectral resolution and in polarization measurements have also been made but only in relatively short duration observations. Lin *et al.* (1981) observed one flare and several microflares with sub-keV resolution on a balloon flight and were able to resolve the X-ray spectrum which became as steep as E^{-11} at photon energies E up to 35 keV. Tramiel *et al.* (1984) made

Solar Physics **100** (1985) 465–490. 0038-0938/85.15.

polarization measurements in the energy range from 5 to 20 keV on a flight of the Space Shuttle and obtained upper limits between 2.5 and 12.7% that are consistent with the more sophisticated polarization models of Leach and Petrosian (1983).

What have we learned from the new observations? In this paper I will concentrate on the answers to this question that concern the improvements in our knowledge of the flare processes that lead to the production of hard X-rays. In particular I will show that we have begun to be able to differentiate, in some specific flares, between the three basic models for the production of hard X-rays: thick and thin-target interactions, and thermal (Brown 1971; Lin and Hudson, 1976; Tucker, 1975; Crannell *et al.*, 1978).

Although it is almost universally accepted that the hard X-rays are electron-ion bremsstrahlung, the determination of the spectrum of the emitting electrons still depends on knowledge of the temperature and density of the plasma with which the electrons are interacting. It is necessary to know, in any particular flare, which model (or combination of models) is correct since the determination of the electron spectrum from the observed X-ray spectrum depends on the model assumed. More importantly, the correct model must be known in order to determine the role of the fast electrons in the overall flare energetics, and ultimately to determine the fundamental energy release machanism or mechanisms of the flare. It now appears clear, as indicated below, that different models apply for different flares and at different times during individual flares.

This paper begins in Section 2 with a discussion of the results which have been obtained from a statistical analysis of the flares detected with HXRBS including evidence for a 158 day periodicity, the spectrum of peak counting rates, and the distribution of spectra. The flare classification scheme is presented in Section 3 and properties of types B and C flares are presented in Sections 4 and 5, respectively.

2. Statistical Analysis

One of the first things that can be done with the large number of events recorded with instruments on SMM is to determine the global properties of flares including the rate of occurrence, the spectrum of sizes, rise times, durations, spectral parameters, etc. Several catalogues of SMM events have already been prepared and distributed. These included the SMM Event Listing for 1980 (Pryor *et al.*, 1981) and the HXRBS Events Listing for 1980–82 (Dennis *et al.*, 1983). A comprehensive catalogue of data on all SMM events is available on the VAX computer at the SMM Data Analysis Center and plans exist to issue, in the near future, a revised version of the 1980 listing, a complete listing of all HXRBS events, and a new listing of all SMM events recorded since the repair.

2.1. THE 152–158 DAY PERIODICITY

The most notable result from this statistical analysis is the discovery of a 152–158 day periodicity in the rate of flare occurrence during this solar cycle, cycle 21. This periodicity was first noticed by Rieger *et al.* (1985) in the tendency of 139 solar flares detected with the Gamma Ray Spectrometer (GRS) above 300 keV to occur in groups with a mean spacing of 154 days.

A much larger sample of 6775 hard X-ray flares detected with HXRBS above ~ 30 keV was used by Kiplinger *et al.* (1985) to reveal a similar periodicity illustrated in Figure 1. The number of flares detected per day by HXRBS is plotted vs time from launch in 1980 until the middle of 1984 with corrections made for the SMM duty cycle. A power spectrum analysis of the data plotted in Figure 1 was carried out using a Fourier transform routine that allows for the data gaps (Deeming, 1975) and this confirms the existance of a strong Fourier component having a period of 158 days and a semi-amplitude of approximately 40% of the mean flare rate. The sine wave superposed on the data in Figure 1 represents this Fourier component plus the first-order trend. Eight peaks can be seen in the flare rate at the expected times given by the sine wave up to June, 1983. The expected peak in the fall of 1983 was not present but a ninth strong peak was observed in April/May, 1984. Subsequently, no increase in flare rate was observed in the fall of 1984 but some weak activity was observed in January/February, 1985.

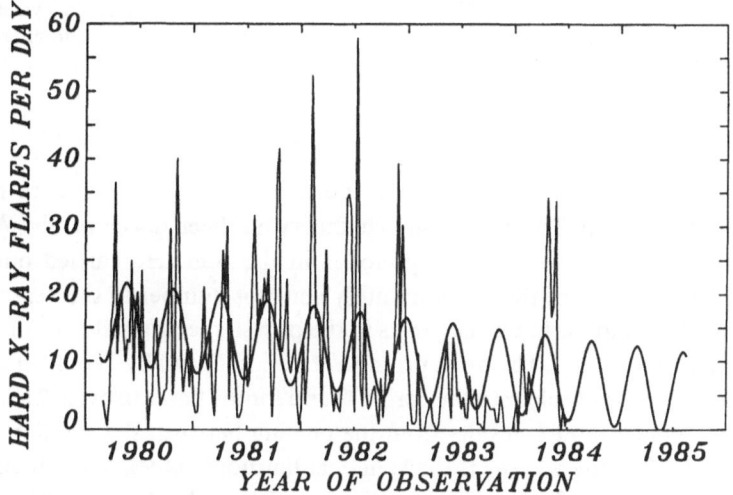

Fig. 1.   Plot of the weekly averaged rate of flares detected with HXRBS (corrected for the duty cycle) as a function of time from launch in February 1980 until after the repair in April 1984. The sine wave superimposed on the data is the Fourier component with a period of 158 days determined from the power spectrum analysis. A total of 6675 flares were included in the analysis. The energy threshold varied from 25 keV at launch to 33 keV in 1984.

A similar periodic component is present in an analysis of 6102 events with X-ray emission that does not exceed 150 keV. The power spectrum of this data set shown in Figure 2 also shows a strong peak at 0.0633 days^{-1} corresponding to a period of the 139 gamma-ray flares used by Rieger *et al.* (1985). A periodicity of 152 days is also present in the rate of GOES events with a classification above M2.5 (Rieger *et al.*, 1985), and Bogart and Bai (1985) report a 157 day periodicity of flare occurrence in the microwave data.

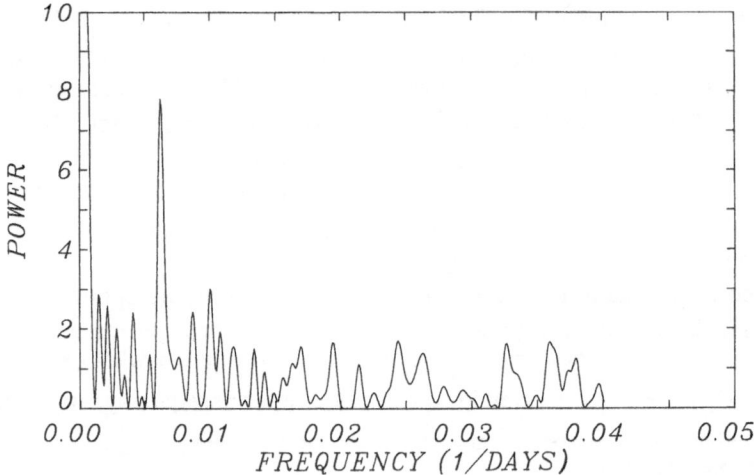

Fig. 2. Power spectrum of the rate versus time of HXRBS low energy events. Only the HXRBS events detected at energies not exceeding 150 keV were used in the analysis so that the data set would not include the gamma ray flares used by Rieger *et al.* (1985). A total of 6102 events detected from launch through July 31, 1984 were used in the analysis. The peak in the power spectrum at 0.0633 days^{-1} confirms the existence of the 158 day periodicity in this independent data set.

The interpretation of this 152–158 day periodicity in the solar flare rate during cycle 21 is still uncertain. The most complete theory has been developed by Wolff (1983) based on the rotational spectrum of *g*-modes in the Sun. He carried out a Fourier analysis of the variation of the mean monthly sunspot number in the years from 1749 to 1979 and found narrow but weak peaks in the power spectrum at a set of frequencies consistent with this model. The most prominent period found by Wolff below 200 days is 155.4 days, remarkably close to the period found for the flares in cycle 21. This period, according to Wolff's model, results from the beating between the rotation of an $l = 2$ mode and an $l = 3$ mode where l is the spherical harmonic index. An attempt has been made to match the beat frequencies of other *l*-modes to the other peaks in the power spectrum obtained from the HXRBS event rate but without success (Kiplinger, private communication).

In concluding this section, it is clear that there is compelling evidence for a periodicity in the rate of occurrence of solar flares during cycle 21 with a period on 152–158 days. It is impossible to say, at present, if this is a manifestation of the interacting *g*-modes as suggested by Wolff (1983). However, the possibility that this may be the case and the unique information concerning the solar interior that could then be determined lends great importance to this result and to the continued long-term accumulation of data.

2.2. THE SPECTRUM OF PEAK RATES

A second result which has been obtained from the large number of flares recorded by SMM is an accurate spectrum of peak rates involving over 6000 flares recorded with HXRBS. This differential spectrum shown in Figure 3 was obtained from all complete

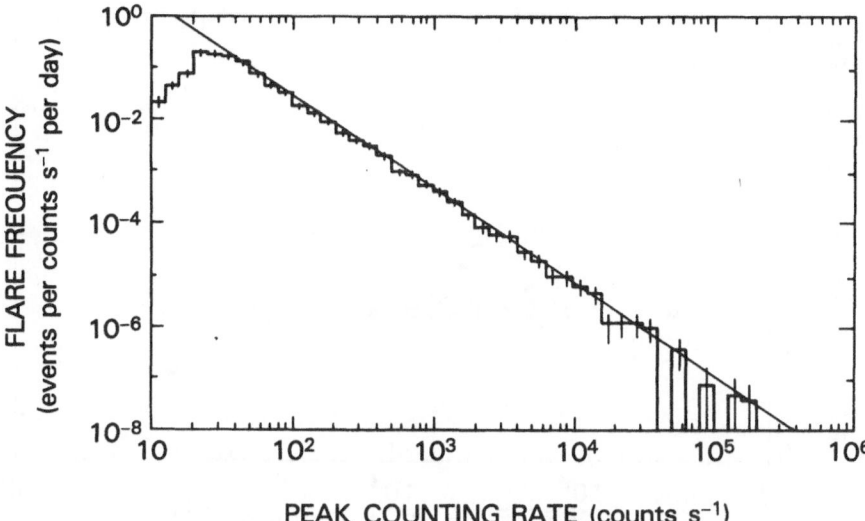

PEAK COUNTING RATE (counts s⁻¹)

Fig. 3. Peak rate spectrum of all complete events detected with HXRBS from launch to February 1985. The straight line through the data corresponds to the power-law expression given in the text with a spectral index of -1.8.

events recorded with HXRBS above ~ 30 keV between launch and the beginning of 1985 and it can be well represented by an expression of the form

$$N(P) = 110\, P^{-1.8} \text{ flares (counts s}^{-1} \text{ day)}^{-1},$$

where $N(P)$ is the rate of flares detected with a peak rate P above background measured with HXRBS as the sum of the rate in all 15 channels. This expression fits the data over more than three orders of magnitude; the turnover below a peak rate of ~ 40 counts s^{-1} almost certainly results from the reduced efficiency of finding events in the HXRBS data that result in a less than doubling of the HXRBS background rate.

This differential spectrum of peak rates shown in Figure 3 is not directly comparable to the spectrum reported by Datlowe *et al.* (1974) and renormalized by Lin *et al.* (1984). Their integral spectrum is for the peak photon flux at 20 keV. It can be related to Figure 3 by noting that for a typical power-law photon spectrum with an index of 5, a peak HXRBS rate above background of 1000 counts s^{-1} corresponds to a photon flux at 20 keV of ~ 30 photons cm^{-2} s^{-1} keV^{-1}. This is the highest flux included in Datlowe *et al.*'s spectrum. Thus, Figure 3 extends the spectrum some two orders of magnitude to larger events. This is possible because the effects of pulse pile-up were less severe in HXRBS than in the OSO-7 detector (Datlowe, 1975) as a result of the ~ 0.06 cm thick aluminum window and the ~ 0.01 cm thick dead-layer on the HXRBS CsI(Na) crystal. In spite of these relatively thick absorbers, pulse pile-up is still a problem for HXRBS when the total counting rate greatly exceeds 10^4 counts s^{-1}, especially when the photon spectrum is steeper than $\sim E^{-5}$.

There is no indication that the spectrum of peak rates changes shape as a function

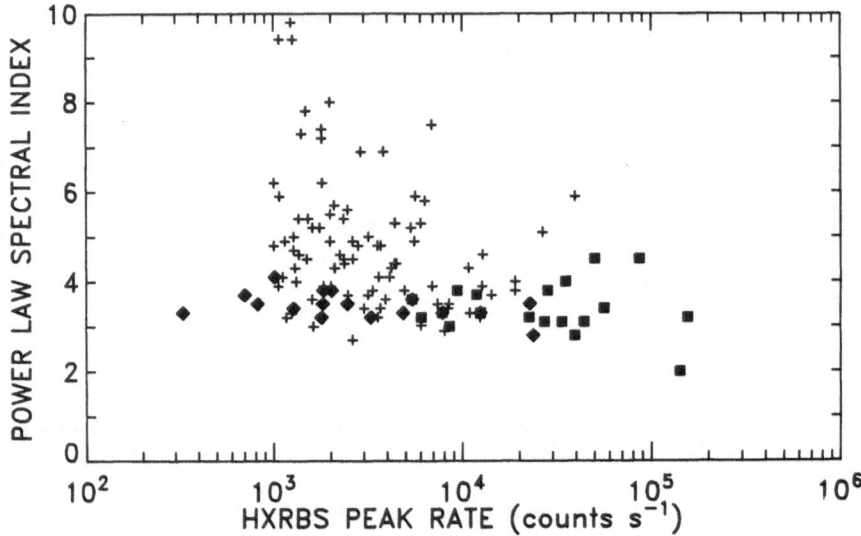

Fig. 4. The hard X-ray spectral index γ obtained from HXRBS data at the time of peak counting rate versus the peak counting rate summed over all 15 channels. The plus signs are for all flares in 1980 for which HXRBS recorded the true peak of the event and the peak rate was > 1000 counts s^{-1}. The filled squares are for the 1980 and 1981 flares listed by Bai *et al.* (1985) that showed evidence for gamma-ray line emission in the Gamma-Ray Spectrometer (GRS) on SMM. The filled diamonds are for the gradual microwave-rich flares recorded in 1980, 1981, and 1982 and also listed by Bai *et al.* (1985).

of time. When separate spectra are obtained for each of the five years from 1980–1984, all are consistent with the same slope as that shown in Figure 4. Furthermore, the power-law slope of -1.8 is consistent with the slope of -0.8 obtained for the integral spectrum in 1971/1972 by Datlowe *et al.* (1974) and -0.9 obtained for the integral spectrum in 1969/1971 using the detector very similar to HXRBS on OSO-5 (Dennis, unpublished).

2.3. Distribution of spectral index

The HXRBS observations allow a counting rate spectrum to be obtained every 128 ms. From each such spectrum or from spectra accumulated over longer intervals, the spectrum of the incident photon flux can be determined as a function of time throughout the flares of interest. Unfortunately, as with all scintillation spectrometers such as HXRBS, the relatively poor energy resolution and the other factors contributing to the non-diagonal elements of the instrument response matrix, make it necessary to assume a form for the incident spectrum before the instrument deconvolution can be carried out. A power-law spectral form is commonly used and this can be expressed as follows:

$$N(E) = AE^{-\gamma} \text{ photons cm}^{-2} \text{ s}^{-1} \text{ keV}^{-1},$$

where $N(E)$ is the photon flux (strictly flux density) in units of photons cm^{-2} s^{-1} keV^{-1} at an energy E in keV that is incident on the detector. The parameters A and γ are determined from the observations using the technique described by Batchelor (1984).

The power-law spectral analysis has been carried out for many HXRBS events and the results have been used to determine the spectrum and energy content of the electrons producing the bremsstrahlung X-rays (Wu *et al.*, 1985). In Figure 4 we show the distribution in the value of γ determined at the time of peak HXRBS counting rate for all complete flares recorded in 1980 where that peak rate was greater than 1000 counts s^{-1}. The value of γ is seen to range from slightly less than 3 to as high as 10. It must be realized that, again because of the poor energy resolution of scintillation spectrometers, spectra steeper than E^{-7} cannot be accurately measured, and consequently, any value of γ greater than ~ 7 in Figure 4 should be considered as a lower limit at $\gamma = 7$. Corrections were applied for the effects of pulse pile-up using the technique described by Datlowe (1975, 1977) so that we believe that the variation in the mean value of γ with the peak rate is real. This may be a further manifestation of the Big Flare Syndrome (Kahler, 1982) such that bigger flares tend to have harder spectra. Figure 4 also shows that flares that are relatively microwave rich and those that produce gamma ray lines also have harder X-ray spectra (Bai *et al.*, 1985).

Figure 4 can be compared to the distribution of γ given by Datlowe *et al.* (1974) with the understanding that they were looking at smaller flares with observations extending down to lower energies: 10 keV as compared to 25 keV for HXRBS. The values of γ obtained by Datlowe *et al.* (1974) extend from 2 to 7 with a median value of 4.0. This is clearly at least 1.0 lower than the median value of the smallest events in Figure 4. A median value of ~ 5.0 was obtained by Kane (1973) using OGO-5 data. Thus, it is possible that the generally steeper spectra found here is a result of the average spectrum being flatter from 10 to 25 keV and steeper at higher energies. This would agree with the sometimes better fit obtained at the peak of some flares to an exponential function expected for thermal bremsstrahlung (Dennis *et al.*, 1981; Kiplinger *et al.*, 1983b).

3. Flare Classification Schemes

It has long been known that not all flares are alike, and various classification schemes have been proposed based on the appearance of different flares in certain wavelength or energy bands and different phases. These schemes have suffered from the difficulty of relating them to one another and in determining if the different classes were merely phenomenological or actually indicated the occurrence of different physical processes. It is vitally important to determine if different processes are occurring to produce the different types of flares. It has been one of the frustrations of solar flare physics that any generalization about flares that one tries to make is immediately met with many counter examples. If the counter examples can be shown to be different types of flares with a different process producing the particle acceleration, for example, then that would be an important step forward in understanding flares.

Pallavicini *et al.* (1977) developed a useful classification scheme based on soft X-ray images of limb flares made from Skylab. They identified three principal groupings: (a) flares characterized by compact loops, (b) flares with point-like appearance, and (c) flares characterized by large and diffuse systems of loops.

Considerable advances have been made in classifying flares as a result of more recent observations. This is best exemplified by the scheme introduced by Tanaka (1983) and used by Tsuneta (1983), Ohki *et al.* (1983), and Tanaka *et al.* (1983). This scheme is based on the hard X-ray imaging results from Hinotori but has found widespread value in coordinating other observations. Tanaka (1983) also divided flares into three classes referred to as type A, B, and C, but it appears that the Pallavicini *et al.*'s A and B groups may correspond to Tanaka's type B and A flares, respectively. Some flares, particularly the larger ones, can show characteristics of more than one type during their different phases. The three different flare types have the following distinct properties:

Type A flares: hot thermal flares with $T \sim 3 - 5 \times 10^7$ K, compact, < 5000 km, low altitude.

Type B Flares: typical impulsive bursts, double footpoints seen in hard X-rays, highly sheared loop with lengths $\gtrsim 2 \times 10^4$ km, possible existence of electron beams producing hard X-rays in thick-target interactions.

Type C Flares: high altitude X-ray sources, $\sim 5 \times 10^4$ km, gradually varying X-ray and microwave fluxes, X-ray spectral hardening and microwave delay suggesting trapped non-thermal electrons and/or continuous acceleration.

In this review I shall concentrate on discussions of SMM observations of type B flares obtained mainly during 1980 and on type C flares observed mainly after 1980. I will not discuss any observations of type A flares since no good examples have been reported in the HXIS data set. The reader is referred to papers by Tsuneta *et al.* (1984a) and Tsuneta (1983) for discussion of Hinotori observations of this type of flares. There is some statistical evidence that type C flares tend to occur preferentially during the decreasing phase of the solar cycle, and this may explain the difference between the 1980 results from the SMM Hard X-Ray Imaging Spectrometer (HXIS) in 1980 and the initial 1981 results from Hinotori. It appears that primarily type B flares were detected with HXIS whereas the Hinotori team initially reported observations of type C flares.

4. Properties of Type B Flares Observed with SMM

4.1. THE 1980, NOVEMBER 5 FLARE AT 22 : 32 UT

The best example of a type B flare in the SMM data set is the flare on 1980, November 5 starting at 22 : 32 UT. The soft and hard X-ray time profiles are shown in Figure 5, where the earlier flare at 22 : 26 UT can also be seen. Although the first flare was observed at 15 GHz with the Very Large Array (Hoyng *et al.*, 1983), it was a factor of ten less intense in X-rays than the second flare and the statistical significance of the HXIS images in the highest energy bands (16–30 keV) is poor. Hence, the second flare has been the subject of more extensive study than the first flare (Duijveman *et al.*, 1982; Duijveman and Hoyng, 1983; Rust *et al.*, 1985; MacKinnon *et al.*, 1984; Wu *et al.*, 1985; Dennis *et al.*, 1985).

As shown in Figure 5, the hard X-ray time profile of this second flare was relatively

5 NOVEMBER 1980

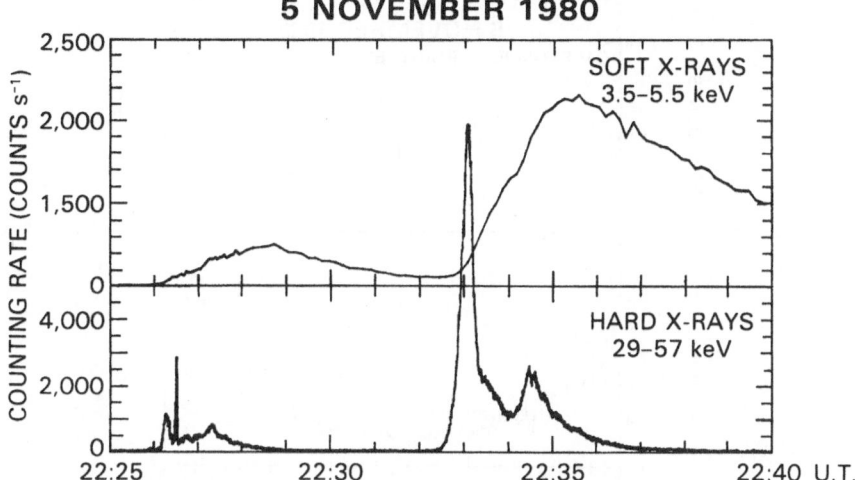

Fig. 5.   Soft and hard X-ray emissions as functions of time for the two flares on 1980, November 5. The soft X-ray counting rate is the sun of the counts in the HXIS band 1 in selected pixels of the coarse field of view. The hard X-ray rate is the sun of counts in HXRBS channel 1 and 2.

simple, with a single, relatively smooth and intense spike lasting for 17 s (FWHM) followed some 90 s later by a second, less intense peak. The detailed spectral evolution of the first peak can be seen in Figure 6 where the HXRBS counting rates in three different energy ranges and the value of the power-law spectral index γ determined from the 34–405 keV HXRBS data are plotted on an expanded time-scale. The value of γ decreases from 6 near the beginning of the event to 3.5 at the time of the peak rate and increases back to ≥ 6 on the decay. This soft-hard-soft spectral evolution was first reported by Kane and Anderson (1970) and is considered typical for impulsive flares. Some impulsive flares, however, show high energy tails on the spectrum above 50 to 100 keV that remain hard or become harder after the peak (Crannell *et al.*, 1978; Dennis *et al.*, 1981). Indeed this is also true for the November 5 peak above ~ 100 keV. It is not clear if this early indication of spectral hardening after an impulsive spike is the beginnings of a type C flare as discussed below. It is possible that there is a continuous progression from purely impulsive type B events with no spectral hardening after the peak to purely gradual type C events with continuous spectral hardening through the peak. There are some events observed with HXRBS that clearly show type B and type C characteristics on different peaks within the same flare. Hence, events classified as type B or type C may just represent the extreme cases; impulsive and gradual phases may occur in all flares at some intensity and/or energy levels. It would seem, however, that the relative importances of these two phases must differ by several orders of magnitude for flares now classified as type B and type C.

The HXIS 16–30 keV images show three well resolved bright patches during the first impulsive peak in the second flare on 1980, November 5 at 22 : 33 UT (Figure 7). Two patches labelled A and B following Duijveman *et al.* (1982) are separated by

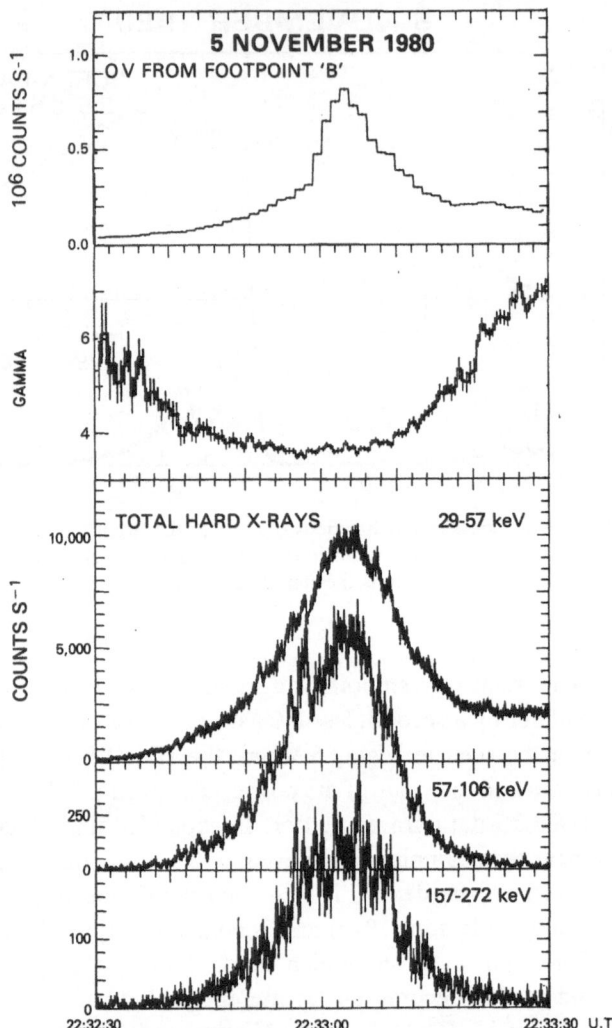

Fig. 6.   An expanded plot of the major hard X-ray peak shown in Figure 5 for the 1980, November 5 flare at 22:33 UT. The bottom three graphs show the HXRBS counting rates as functions of time in three different energy ranges. The middle plot shows the variation of the power-law spectral index γ obtained from the 34–405 keV HXRBS counting rate data (channels 2–14) on the same time-scale. The top plot shows the time variation of the UVSP counting rate in O v summed over all 9 pixels shown in Figure 7. More than 90% of the counts came from the three southern-most pixels (Dennis *et al.*, 1985). The error bars on all the plots represent $\pm 1\sigma$ uncertainties based on the Poisson fluctuations alone.

1.6×10^4 km and a third patch labelled C is separated from B by 7×10^4 km. Comparisons with Hα images and magnetograms suggest that the bright patches were at the footpoints of two magnetic loops joining A and B, and B and C. Duijveman *et al.* (1982) claimed that the footpoints B and C started to brighten simultaneously to within ~ 5 s implying a velocity of $\geq 2 \times 10^9$ cm s^{-1} along the assumed semi-circular loop. They showed that for this velocity to be interpreted as an Alfvén velocity, the magnetic field

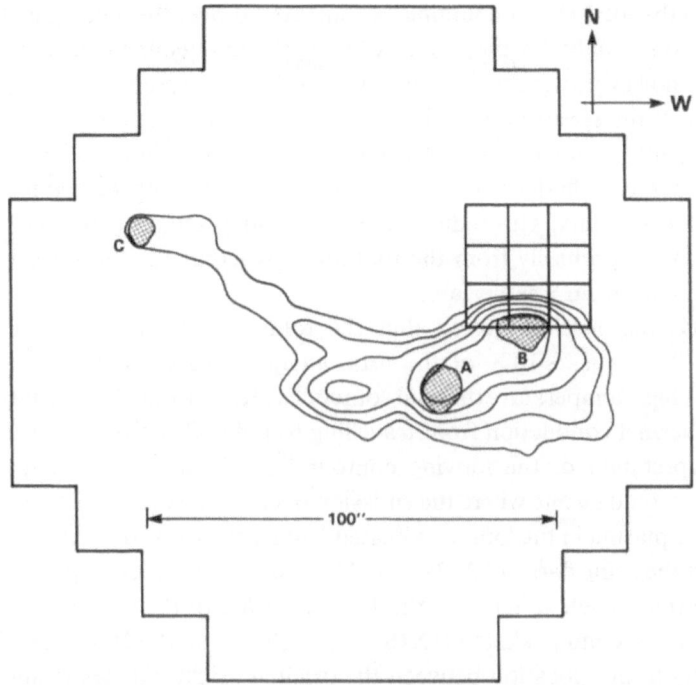

Fig. 7. Contour plot obtained from the sum of three HXIS images showing the location of the soft and hard X-ray emission at the time of the most intense hard X-ray peak on 1980, November 5. Accumulation of the first image began at 22:32:53 UT. Each image had an accumulation time of 4.5 s with a 4.5 s gap between images. The contour lines were obtained from the 3.5–8 keV data with the collimator response deconvolved according to the method given by Švestka *et al.* (1983). The contour levels are at the following counts pixel^{-1}: 25, 50, 100, 200 and 400; the deconvolved peak rate was 902 counts pixel^{-1}. The cross-hatched areas labelled A, B, and C were obtained from the 16–30 keV data similarly deconvonved. The outer edge of these areas corresponds to a contour level of 40 counts pixel^{-1} for A and B and 20 counts pixel^{-1} for C with a deconvolved peak rate at 53 counts pixel^{-1}. The 3 × 3 array of 10″ × 10″ squares represents the 9 UVSP pixels used for the O v observations shown in Figure 6.

in the loop must have been $\gtrsim 900$ G, implausibly high for such a large loop. They concluded that the most likely explanation for the simultaneous brightening was a beam of fast electrons travelling along the loop and producing X-rays in thick-target inter-actions at the footpoints.

MacKinnon *et al.* (1984) have shown that only a small fraction of the hard X-rays in the HXIS 16–30 keV images come from the pixels that define the bright patches A, B, and C. They claim that the remainder come from a diffuse region around A and B although the interpretation of the two or three images at the time of the first hard X-ray burst is complicated by the presence of a diffuse background resulting from the leakage of high energy photons through the walls of the instrument collimator. This conclusion that only a small fraction of the photons come from the footpoints is supported, however, by an extrapolation of the HXRBS spectrum above 30 keV down to the 16–30 keV HXIS energy range. Good agreement is found between this extrapolated spectrum and the total HXIS counting rate integrated over the whole fine field of view. When the counts in only the fine field of view pixels that define the three bright patches are summed, however, the resulting flux is a order of magnitude below the HXRBS extrapolation. If the spectrum of only the impulsive component is used, i.e., the more gradually varying component is subtracted from the fluxes in each energy band, then the spectrum becomes flatter at energies below 50 keV, E^{-3} compared to $E^{-3.9}$ for the spectrum of the total flux. This reduces the discrepancy and suggests that the impulsive component comes primarily from the footpoints whereas the more gradually varying component comes from a larger area.

Rust (1984) has also suggested that the total energy in electron beams in the November 5 flare was $\leq 10\%$ of the total energy released and that the remainder appeared as a high-temperature thermal source. Evidence from HXIS images in bands 1 and 2 for a thermal conduction front travelling to point C is given in Rust et al. (1985) but the interpretation of the moving contour lines is ambiguous. An equally valid interpretation would be one where the emission from loop BC increased uniformly along the loop as the plasma in the loop was heated by fast electrons travelling along the loop.

Later on in the same flare at 22 : 34 : 30 UT, a second hard X-ray peak occurred with an E^{-7} spectrum, much softer than the E^{-4} spectrum of the first peak. Furthermore, at the time of the second peak, the HXIS images show that the 16–30 keV X-rays came predominantly from a location between the original bright patches A and B. The soft X-ray flux shows a second increase at about this time, as indicated in Figure 5, suggesting that almost the same amount of energy was released as during the first impulsive peak. The simplest explanation of these observations is that this second energy release served mainly to heat the plasma injected into the loop as a result of the first non-thermal energy release. Such a transition from an essentially non-thermal model to a thermal model during the impulsive phase of many flares has been proposed by Smith (1985) and by Tsuneta (1985). Smith (1985) proposes a dissipative thermal model, a term originally suggested by Emslie and Vlahos (1980), and he suggests that an increase in the ratio of plasma to magnetic pressure (the plasma β) at the energy release site later in the flare results in this transition from a non-thermal to a thermal model. It must be pointed out, however, that the blue shifts in the Ca XIX lines observed with the Bent Crystal Spectrometer (Acton et al., 1980) persist into the later stages of the impulsive phase of the 1980, November 5 flare indicating continuing chromospheric evaporation with upward velocities of ~ 200 km s^{-1} even during the second hard X-ray peak (Antonucci et al., 1984).

4.2. SIMULTANEOUS UV AND HARD X-RAY EMISSION

Another way to differentiate between the different flare models is to look at the timing relationship between UV and hard X-ray emission during impulsive flares. The simultaneity to within 1 s of UV and hard X-ray emission has been known since the observations of sudden ionospheric disturbances (SIDs) in conjunction with hard X-ray bursts (Donnelly and Kane, 1978; Kane et al., 1979). SIDs are caused by bursts of solar UV radiation between 10 and 1030 Å coming primarily from the transition region and low corona. The combined UVSP, HXIS, and HXRBS observations have now provided the required spatial information to allow the simultaneity of the UV and hard X-ray emissions to be used to significantly constrain the flare models.

Poland *et al.* (1982) and Cheng *et al.* (1981) have shown that the transition region line radiations – O v at 1371 Å, Si iv at 1402 Å, and O iv at 1401 Å – are strongly concentrated at the footpoints of magnetic loops. Woodgate *et al.* (1983) have shown the simultaneity of the O v emission and the 25–300 keV X-ray flux to within 1 s for several impulsive flares observed on 1980, November 8. Here we show in Figure 6 the UV and hard X-ray observations for the 1980, November 5 flare at 22 : 33 UT from Dennis *et al.* (1985). The total O v emission summed over all nine UVSP pixels near footpoint B shown in Figure 7 is plotted in Figure 6 on the same time-scale as the hard X-ray counting rate recorded by HXRBS in three energy ranges. This is a particularly clear example of the simultaneity to $\lesssim 1$ s of a hard X-ray peak and the increase in the O v emission from a footpoint. It supports the conclusion reached by Woodgate *et al.* (1983) that the lack of any detectable time delay is inconsistent with flare models in which the hard X-rays are initially produed at the loop top followed by the formation of thermal conduction fronts which travel to the footpoints where the UV burst is produced by heating. Models in which both X-rays and UV radiation are both produced at the footpoints, or an electron beam transmits energy between the loop top and the footpoints in less than 1 s, are allowed by these observations.

Attemts to quantify the expected O v flux for a given electron beam intensity have met with limited success. The theoretical problem is complicated by the fact that there are two competing effects of an electron beam incident from the corona. One is the depression of the transition zone to a lower altitude and higher density. This tends to increase the amount of material at the O v emitting temperature around 2.5×10^5 K suggesting that the O v flux should increase rapidly. The other effect suggests the opposite, however, and that is the steepening of the temperature gradient in the transition zone. This tends to reduce the amount of 2.5×10^5 K material and suggests that the O v flux should decrease. Thus, it is important to accurately model the transition zone in any theoretical calculation of the expected O v flux and this typically is difficult to do in the computer simulations because of its extremely small scale height. Poland *et al.* (1984), Emslie and Nagai (1984), and Mariska and Poland (1985) have carried out some exploratory model calculations to study the relationship between the energy emitted in hard X-rays and in the O v line assuming that both result from an electron beam. Poland *et al.* (1984) find from the observations of individual flares that there is a definite relation between hard X-ray and O v emission throughout a given flare but that the flux ratio is different in different flares. They attribute these differences to the initial conditions in the flaring loops and their model calculations support this conclusion.

More recent observations made after the SMM repair have shown simultaneity of UV continuum and hard X-ray features in an impulsive flare to within 0.25 s (Woodgate, 1984). In a flare on 1984, May 20 at 02 : 59 UT, UVSP was recording at around 1600 Å with a time-resolution of 75 ms in one pixel and HXRBS was recording the counting rate above 24 keV with 10 ms time resolution. (The UVSP wavelength drive was not working at this time so that the exact wavelength of the UV observations is not known nor is it known for certain if it was recording line or continuum emission. However, the two detectors separated by 17 Å recorded very similar rates and consequently it is believed that they were both in the continuum with little, if any, contribution from line emission.) The overall time profile of the impulsive phase of the event in UV and hard X-rays is shown in Figure 8 and the first feature is shown on an expanded time-scale

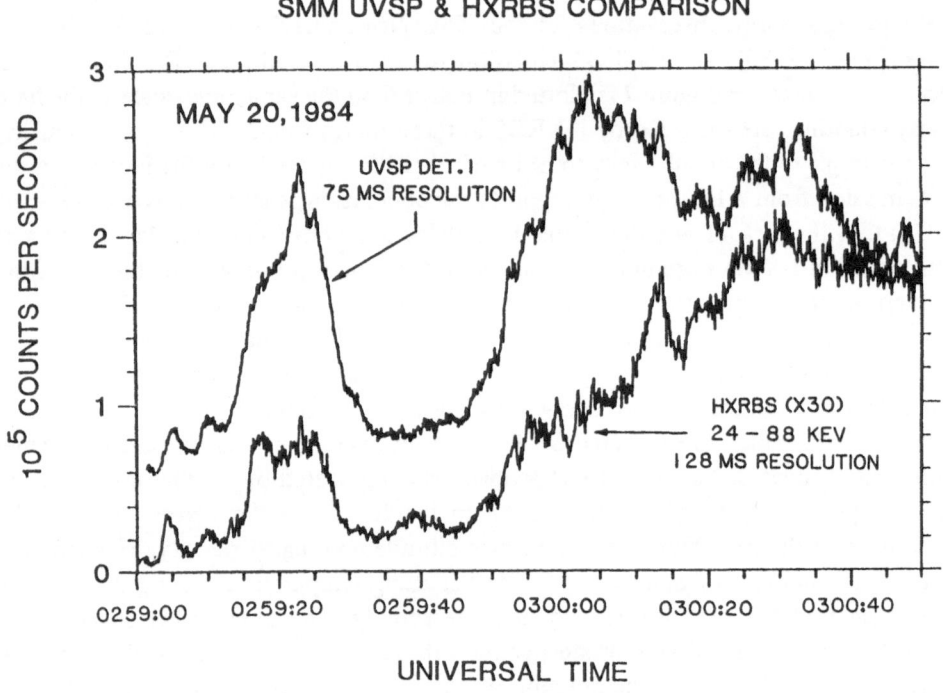

Fig. 8.   Impulsive bursts during the flare on 1984, May 20 beginning at 02:59 UT showing the similarity between the hard X-ray and UV continuum time profiles.

in Figure 9. There are clearly significant differences between the overall UV and hard X-ray time profiles but the similarities are striking. In particular, the sharp feature at 02:59:23 UT shows up clearly in both time profiles with the start and peak times coincident to within 0.1 s.

If we assume that the detected UV emission for this event was continuum at ~ 1600 Å, then it was probably emitted low in the chromosphere at a temperature of ~ 5000 K according to Cook and Brueckner (1979). Thus, it is difficult to see how an electron beam incident on a footpoint from the corona could result in the measured 0.1 s simultaneity since electrons cannot reach deep in the chromosphere (Woodgate, private communication). The possibility of the energy release itself occuring in the chromosphere would of course, alleviate the problem.

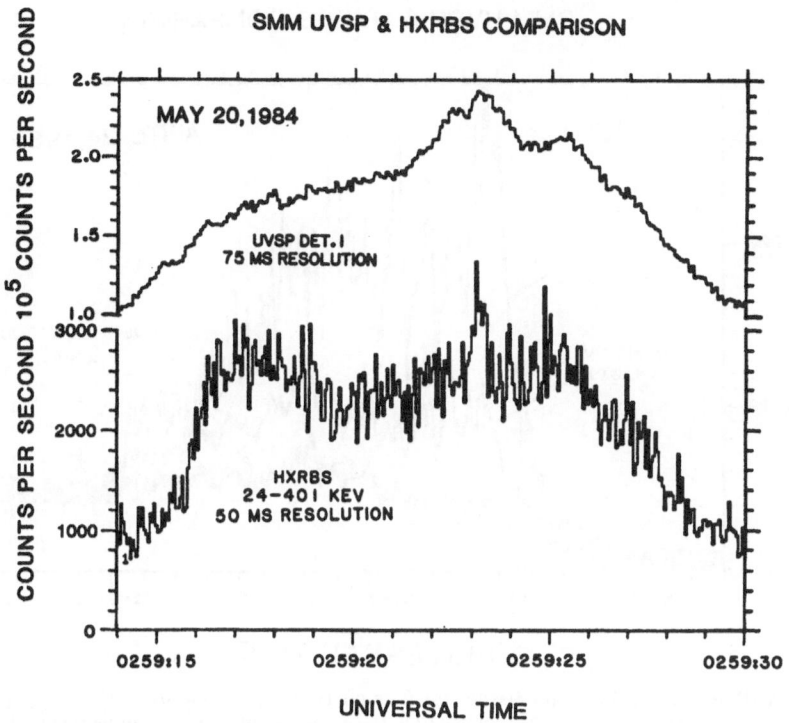

Fig. 9. The first major feature of the flare on 1984, May 20 shown in Figure 8 plotted on an expanded time-scale. The UVSP detector 1 time profile is believed to be for continuum emission at approximately 1600 Å and has a time-resolution of 75 ms. The large scatter in the hard X-ray counting rate reflects the statistical uncertainty resulting from the mean rate at 02:59:20 UT of 115 counts per 50 ms interval. The feature at 02:59:23 UT is, however, clearly significant in both the hard X-ray and UV time profiles.

Even more recent coincident UV and hard X-ray observations with 128 ms time resolution should ultimately help to differentiate between the different possible models. An impulsive flare on 1985, April 24 at 01:48 UT was recorded by both UVSP and HXRBS. UVSP was recording separately the emission in both of the O v line and the continuum 17 Å from the line with a time-resolution of 128 ms. Thus, the time profiles of the emission from the transition zone and from the lower chromosphere were obtained at the same time as the hard X-ray profile above 25 keV. The three time profiles are shown on the same time-scale in Figure 10, where the rich detail is evident in this flare. It is too early in the analysis of this data to make any definitive statement about the relative timing, but it does appear as if the first peak in hard X-rays and in the continuum emission are coincident to within 1 s. The O v emission, however, seems to reach its first peak a second or so later. This surprising result, if it can be verified, would support the idea that at least the initial energy release occurs in the chromosphere and not in the corona.

BRIAN R. DENNIS

SMM UVSP & HXRBS COMPARISON

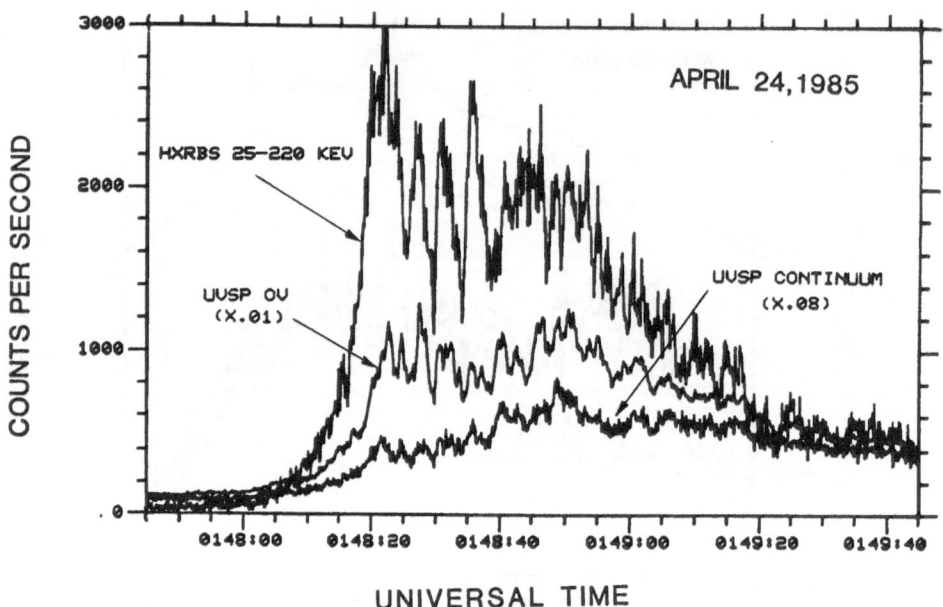

Fig. 10. UVSP and HXRBS time profiles for the flare on 1985, April 24 at 01:48 UT. The top profile is the hard X-ray counting rate between 25 and 220 keV, the center profile is the counting rate in the Ov line, and the bottom profile is the counting rate in the continuum 17 Å from the Ov line. The time-resolution for all three profiles is 128 ms,

4.3 RAPID FLUCTUATIONS IN THE HARD X-RAY FLUX

The impulsive phase of solar flares is characterized by rapid variations in the hard X-ray and microwave flux. This must reflect the variations in the rate of energy release in the flare and the production of energetic electrons but the propagation of the electrons from the site of energy release to the location is also a factor in determining the observed variations. A detailed analysis of the most rapid variations that are observed has proved to be a useful tool to place limits on the models for the production and propagation of fast electrons.

Before 1980, observations of hard X-rays had revealed variations on time-scales as short as 1 s but as soon as better time-resolutions become available, variations on time-scales as short as a few tens of ms were observed (Orwig *et al.*, 1981; Kiplinger *et al.*, 1983a; Hurley *et al.*, 1983).

One of the fastest variations observed with HXRBS is shown in Figure 11, where two seconds of the counting rate are plotted with a time-resolution of 20 ms for an impulsive event that occurred on 1980, June 6. The most dramatic variation is the large, unresolved rise from 20 to 53 counts per 20 ms at 23:34:45.7 UT. Numerous other fast variations (not shown) occurring within 10s of the plotted interval support the reality of this rapid fluctuation.

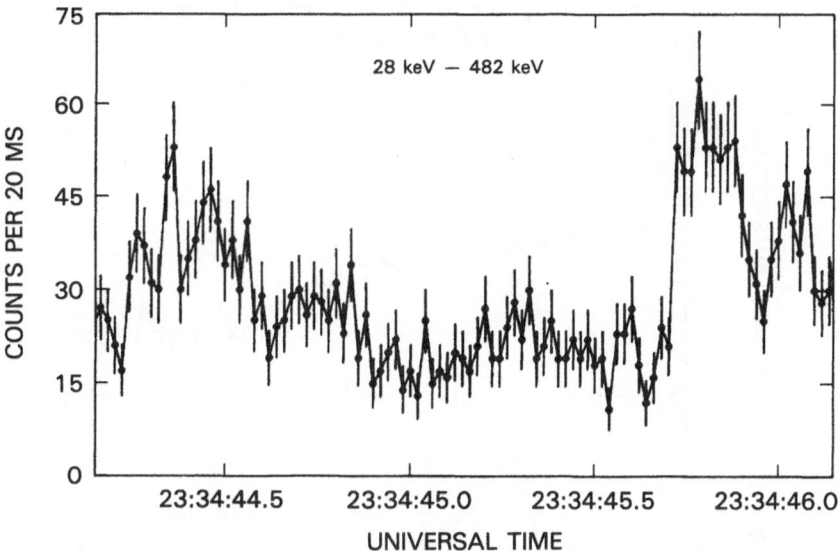

Fig. 11.   Two seconds of HXRBS memory data showing very rapid X-ray variations in a solar flare which occured on 1980, June 6. The counting rate is plotted at a time resolution of 20 ms per point with $\pm 1\sigma$ statistical error bars.

It must be remembered that only about 10 % of all flares detected with HXRBS with a sufficiently high counting rate to see variations on a sub-second time-scale, in fact, showed such rapid variations (Kiplinger *et al.*, 1983a). Thus, sub-second variations are not a common phenomenon. Nevertheless, they can be used to constrain models for energy release and propagation. The short time-scales involved are upper limits to the time-scales of the acceleration process itself. Although both thermal and non-thermal interpretations of sub-second variations are viable, Kiplinger *et al.* (1983a) have shown that a thermal interpretation has no energetic advantage over a non-thermal inter- pretation for the production of the hard X-rays. Furthermore, Kiplinger *et al.* (1984) have shown for a particularly well resolved sub-second spike, that the observations allow constraints to be placed on the loop length and on the electron pitch-angle distribution assuming a non-thermal interpretation.

5. Properties of Type C Flares

The canonical type C flare identified in the Hinotori data is the flare on 1981, May 13 (Tsuneta *et al.*, 1984b). A part of the time profile of this event as determined with HXRBS is shown in Figure 12 together with the evolution of the power-law spectral index γ. We see immediately two differences between this event and the 1980, November 5 event shown in Figures 5 and 6. First of all, the 1981, May 13 flare is a much more gradually varying event with the hard X-ray peak lasting for ~ 8 min (FWHM) compared to 17 s for the major peak on November 5. The beginning of the event was not seen with HXRBS since SMM did not emerge from spacecraft night until

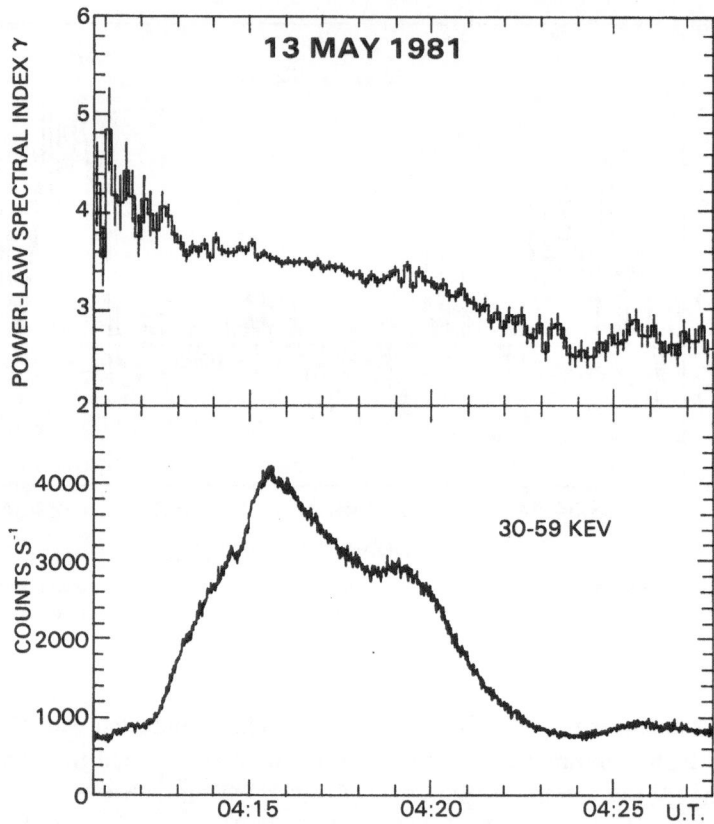

Fig. 12.   The bottom trace shows the variation of the HXRBS 30–59 keV (channels 1 and 2) counting rate vs time for the major peak of the 1980, May 13 event. The upper trace shows the value of the best fit power-law spectral index γ as a function of time. The spectral fits were made using the HXRBS counting rates in channels 3–14 (59–419 keV) to show the spectral hardening through the peak and on the decay. Individual spectra at different times during the event are shown in Figure 13.

04:08 UT but Tsuneta *et al.* (1984b) report that this event showed no impulsive component in either hard X-rays or microwaves. The second difference is that for the May 13 event the hard X-ray spectrum gets progressively harder (γ decreases) through the peak in contrast to the soft-hard-soft spectral evolution of the November 5 event shown in Figure 6. The soft-hard-harder spectral evolution is also evident in Figure 13 where individual spectra are plotted for specific times on the rise and decay of the hard X-ray peak. The photon flux density was determined as a function of photon energy from the 15-channel HXRBS counting rate data using the technique described by Batchelor (1984) and a power-law fit was made to the points corresponding to channels 3–14. The lower two channels were excluded because there is a steep, low-energy component to the spectrum near the beginning of the rise and towards the end of the decay of the event as shown in Figure 13. Clearly, it is the spectrum above ~ 60 keV that participates in the spectral hardening with time. The steep spectrum at lower

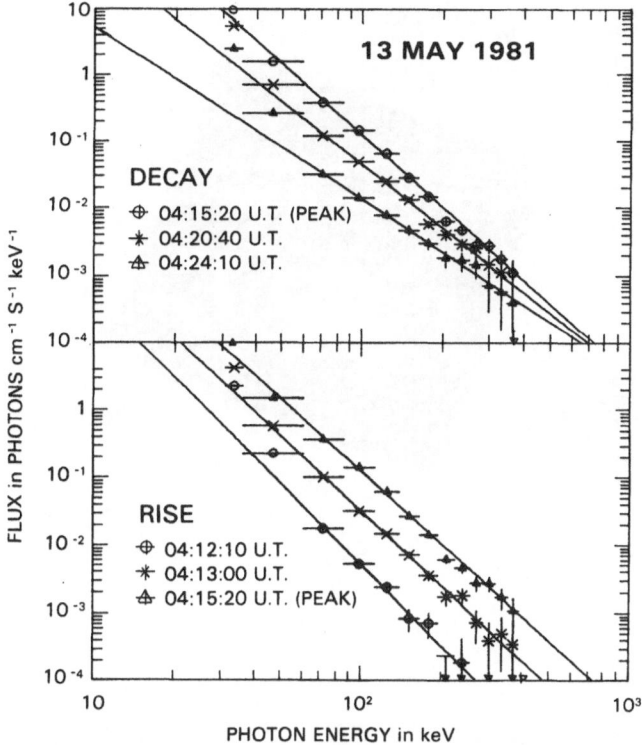

Fig. 13. Hard X-ray spectra at five different times during the major feature of the 1981, May 13 event shown in Figure 12. The data points were derived from the HXRBS counting rates assuming the indicated least-squares power-law fit through the points for channels 3–14. The vertival lines through the data points are ± 1 σ error bars based on the statistical uncertainties alone. The horizontal lines through the points are not error bars but indicate the widths of the HXRBS channels.

energies is most probably a high temperature thermal component similar to that reported by Lin *et al.* (1981).

The most striking result concerning the May 13 flare, however, is that the altitude of the X-ray and microwave source was $\sim 4 \times 10^4$ km above the photosphere (Tsuneta *et al.*, 1984b). This result was obtained from the 14–38 keV image in Figure 14, which shows that the X-ray source was displaced by ~ 1 arc min towards the limb from the two-ribbon Hα flare. Considering the location of the Hα source at N 9–13 E, 54–48 (*Solar Geophysical Data*), this displacement corresponds to the altitude indicated above. This is again in marked contrast to the type B flares, where at least a large fraction of the hard X-rays are believed to come from low altitudes at the footpoints of magnetic loops.

Since hard X-ray imaging has not been available for many flares, attempts have been made to select type C flares based on their more readily available properties. Cliver *et al.* (1985) have selected ten gradual hard X-ray bursts (GHB's) including the 1981, May 13 event that all show the following similar characteristics:

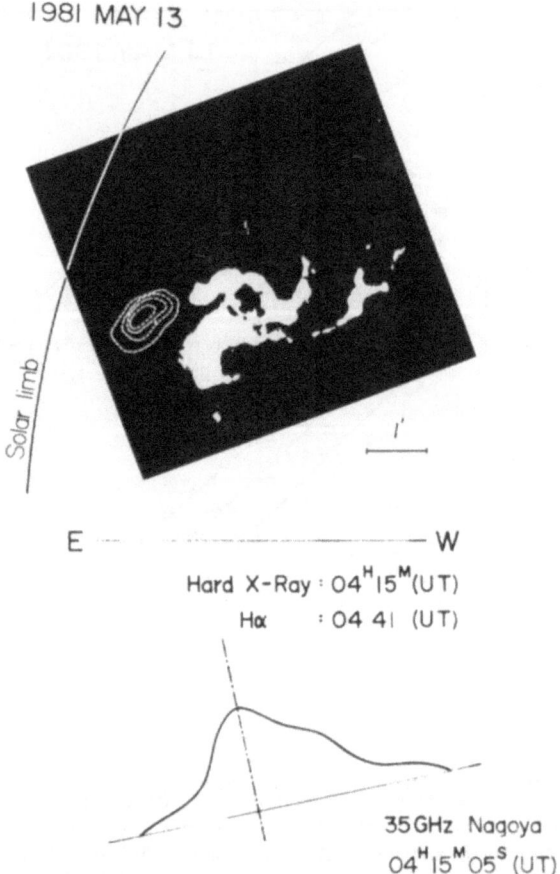

Fig. 14.   Overlay from Tsuneta (1983) of the Hinotori hard X-ray image of the 1981, May 13 type C flare with an Hα photograph taken 26 min later. The one-dimensional microwave brightness distribution at 35 GHz is oriented to show that the bulk of this emission was from the same high altitude location as the X-ray source (Kawabata *et al.*, 1983).

(1) the X-ray spectrum is harder than average (see Figure 4) with a typical value for γ of between 3 and 4;

(2) the hard X-ray spectrum systematically hardens (or at least does not soften) through the peaks and on the decay;

(3) the GHB's are often preceded, by as much as 60 min, by an impulsive phase although sometimes, as in the case of the 1981, May 13 event, no impulsive phase may be seen at all;

(4) the GHB's occur in the later, parallel-ribbon phase of major flares;

(5) a coronal mass ejection was observed, or inferred, in association with at least nine of the ten GHB's;

(6) the associated microwave bursts were also very gradual and peaked sometimes many minutes after the hard X-ray peak as shown for the 1981, April 26 flare in Figure 15;

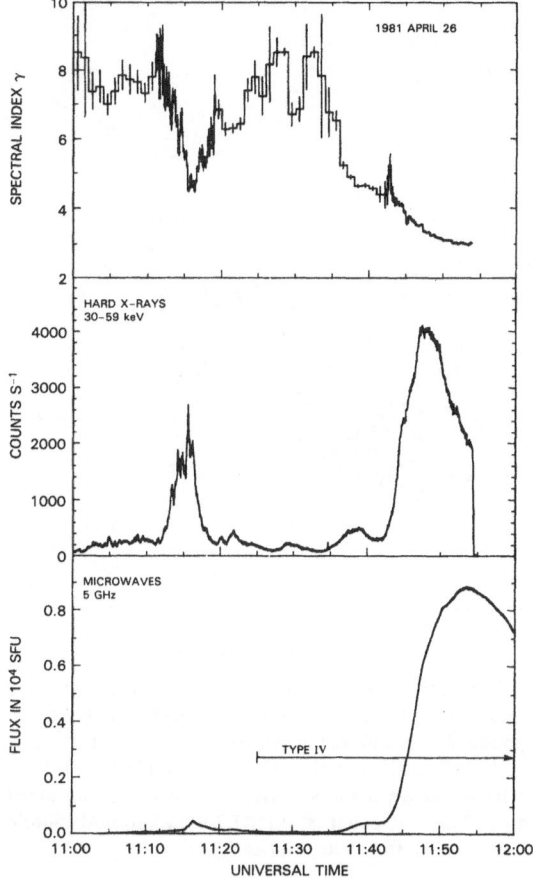

Fig. 15.   Hard X-ray and microwave time-intensity profiles of the 1981, April 26 solar flare. The upper panel shows the variation of the X-ray power-law spectral index γ with time.

(7) the GHB's were typically microwave rich in comparison with the hard X-ray flux as shown in Figure 16 and had relatively low peak frequencies suggesting a low density source and weak magnetic fields;

(8) the GHB's are associated with long duration soft X-ray events suggesting continuous energy release for up to several hours.

These properties of GHB's lead Cliver *et al.* (1985) to propose that these events result from particle acceleration by magnetic reconnection following a coronal mass ejection. This simplified picture of such an event shown in Figure 17 was first proposed by Cliver (1983). In this model, the magnetic field lines are envisioned as being stretched as the coronal mass ejection moves out. Magnetic reconnection takes place below the ejected material and it is at this point that the electrons are accelerated. Electrons moving upwards produce the associated radio type IV emission; electrons moving downwards are trapped in the low density loops below the reconnection point and produce the hard X-rays and microwaves.

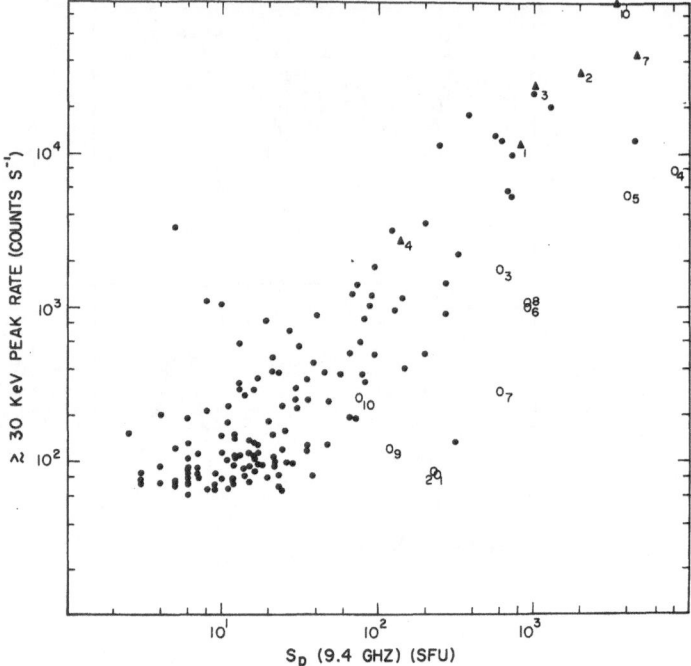

Fig. 16.   Scatter plot of HXRBS > 30 keV peak hard X-ray count-rate vs the peak 9.4 GHz flux-density for impulsive (△) and gradual (○) components of the GHBs analysed by Cliver *et al.* (1985). For comparison, the data points for ~130 impulsive events observed both by SMM and Toyokawa Observatory during April, August, and December 1981 are also plotted (●). For the comparison sample, we considered events to be impulsive if the 9.4 GHz emission peak occurred within ± 0.5 min of the HXRBS > 30 keV maximum. The ~40 counts s^{-1} HXRBS background counting rate was not subtracted from the > 30 keV peak rates.

It is not certain if the observed spectral hardening can be attributed to the decrease in energy loss rate with increasing electron energy in the trap or to the time dependence of the electron acceleration process itself. The density of the X-ray emitting source for the 1981, May 13 flare has been estimated by Tsuneta *et al.* (1984b) to be 3×10^{10} cm^3 from the soft X-ray observations. The decay time for 100 keV electrons in a trap with this density would be ~ 1 s assuming only Coulomb losses. Bai and Dennis (1985) give this argument and others in favor of the so-called second-step acceleration interpretation (Bai and Ramaty, 1979), in which protons are accelerated to > 1 MeV and electrons to relativistic energies using particles accelerated by the primary (or first step) mechanism as injection particles. For other flares where density information is not generally available, a combination of both continuous acceleration and energy losses may well be important.

6. Conclusions

We have reviewed some of the more important results from SMM that have contributed greatly to our understanding of the processes that lead to the high-energy X-ray aspects

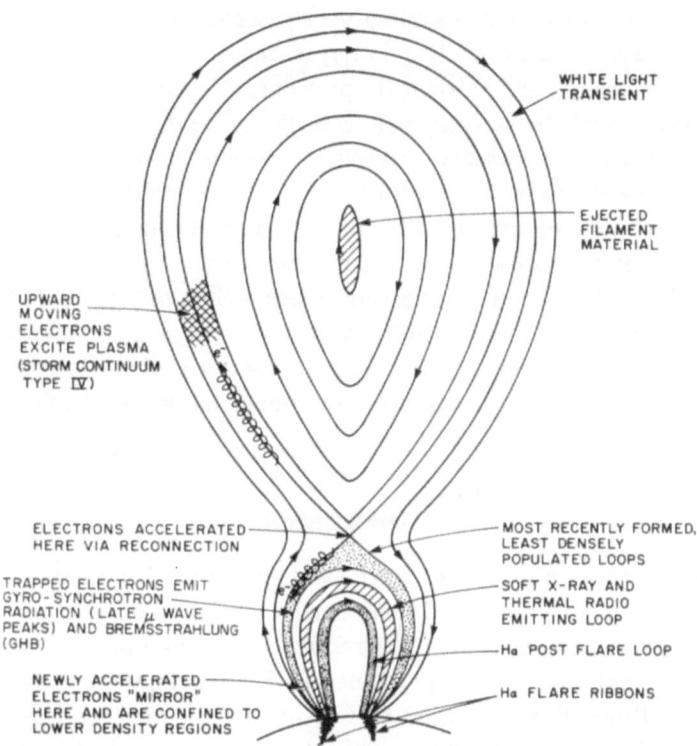

Fig. 17. Proposed geometry in which electrons accelerated via neutral 'point' reconnection are trapped and give rise to gradual hard X-ray bursts and secondary microwave peaks. The figure also indicates how concomitant type IV storm continuum might occur. The drawing is not to scale. Characteristic heights of the soft X-ray loops and the leading edge of the transient at a time ~ 30 min after the flare flash phase are $\sim 3 \times 10^4$ km, respectively.

of solar flares. The 152–158 day periodicity in the rate of occurence of hard X-ray flares and other manifestations of solar activity may prove to be important in determing the dynamics of solar rotation as Wolff (1983) believes but only further analysis and more extensive data sets will resolve that issue. The classification of flares into three different types suggests that we are beginning to be able to distinguish between the three basic models for X-ray production: the purely thermal emission from type A flares, the thick-target interactions at the footpoints in the impulsive type B flares, and the coronal thin-target interactions in the gradual type C flares. This interpretation is still controversial and many flares exhibit characteristics of all three types. Nevertheless, the fact that in certain flares we have begun to distinguish between the different models for X-ray production has allowed us to refine the constraints on the basic flare energy release mechanism or mechanisms.

Higher-resolution hard X-ray images with better time resolution and extending to higher photon energies will allow us to improve the identification of the correct X-ray production models in any specific flare. The simultaneity of footpoint brightening on time scales of fractions of a second is a very powerful indicator of the existence of electron beams. The detection of the thin-target emission from such beams in the loop would also give a clear picture of where the particles are accelerated in the loop and with what pitch-angle distributions. Imaging to energies above the 30–40 keV achieved with SMM and Hinotori is necessary to clearly separate the thermal emission of the $10–30 \times 10^6$ K plasma from the higher energy X-ray component. Furthermore, higher spectral resolution than is possible with scintillation spectrometers is required to clearly resolve the very steep spectra of this thermal emission.

These requirements for future observations are well within the capabilities of new instruments being proposed for flights during the next solar maximum. Thus, the opportunity exists to build on our successes during this cycle 21 of solar activity and to achieve even more dramatic advances in our understanding of the fundamental flare processes during cycle 22.

Acknowledgements

I am grateful to Larry Orwig, Alan Kiplinger, and the HXRBS and SMM computer staff for their assistance in data analysis. I have benefitted greatly from the freely shared ideas of many people and the open access to the SMM and ground-based observations from many groups around the world. In particular, I wish to acknowledge the use of UVSP data thanks to Bruce Woodgate, the SOON microwave observations thanks to Ed Cliver, and the Hinotori, Hα, and microwave results shown in Figure 14 thanks to S. Tsuneta. Finally, I much appreciate the careful reading of an early manuscript by Alan Kiplinger and Takeo Kosugi, and the rapid typing of Gloria Wharen.

References

Acton, L. W., Culhane, J. L., Gabriel, A. H., and 21 co-authors: 1980, *Solar Phys.* **65**, 53.
Antonucci, E., Gabriel, A. H., and Dennis, B. R.: 1984, *Astrophys. J.* **287**, 179.
Bai, T.: 1982, in R. E. Lingenfelter, H. S. Hudson, and D. M. Worrall (eds.), *Gamma Ray Transients and Related Astrophysical Phenomena*, American Institute of Physics, New York, p. 409.
Bai, T. and Dennis, B. R.: 1985, *Astrophys. J.* **292**, 699.
Bai, T. and Ramaty, R.: 1979, *Astrophys. J.* **227**, 1072.
Bai, T., Hudson, H. S., Pelling, R. M., Lin, R. P., Schwartz, R. A., and von Rosenvinge, T. T.: 1983, *Astrophys. J.* **267**, 433.
Bai, T., Kiplinger, A. L., and Dennis, B. R.: 1985, *Astrophys. J.* (in preparation).
Batchelor, D. A.: 1984, Ph.D. Thesis, University of North Carolina at Chapel Hill; also NASA TM86102.
Bogert, R. S. and Bai, T.: 1985, presented at the Solar Physics Division meeting Tucson, May 13–15, 1985.
Brown, J. C.: 1971, *Solar Phys.* **18**, 489.
Cheng, C.-C., Tandberg-Hanssen, E., Bruner, E. C., Orwig, L. E., Frost, K. J., Woodgate, B. E., and Shine, R. A.: 1981, *Astrophys. J.* **248**, L39.
Cliver, E. W.: 1983, *Solar Phys.* **84**, 347.
Cliver, E. W., Dennis, B. R., Kane, S. R., Neidig, D. F., Sheeley, N. R., Jr., and Koomen, M. J.: 1985, *Astrophys. J.* (submitted).
Cook, J. W. and Brueckner, G. E.: 1979, *Astrophys. J.* **227**, 645.
Crannell, C. J., Frost, K. J., Mätzler, C., Ohki, K., and Saba, J. L.: 1978, *Astrophys. J.* **223**, 620.
Datlowe, D. W.: 1975, *Space Sci. Instr.* **1**, 389.
Datlowe, D. W.: 1977, *Nucl. Inst. Methods* **145**, 365.

Datlowe, D. W., Elcan M. J., and Hudson, H. S.: 1974, *Solar Phys.* **39**, 155.

Deeming, T. J.: 1975, *Astrophys. Space Sci.* **36**, 137.

Dennis, B. R., Frost, K. J., and Orwig, L. E.: 1981, *Astrophys. J.* **244**, L167.

Dennis, B. R., Frost, K. J., Orwig, L. E., Kiplinger, A. L., Dennis, H. E., Gibson, B. R., Kennard, G. S., and Tolbert, A. K.: 1983, NASA TM84998.

Dennis, B. R., Kiplinger, A. L., Orwig, L. E., and Frost, K. J.: 1985, *Proc. 2nd India–U.S. Workshop on Solar Terrestrial Physics, January 1984*, New Delhi, India; also NASA TM86187.

Donnelly, R. F. and Kane, S. R.: 1978, *Astrophys. J.* **222**, 1043.

Duijveman, A. and Hoyng, P.: 1983, *Solar Phys.* **86**, 289.

Duijveman, A., Hoyng, P., and Machado, M. E.: 1982, *Solar Phys.* **81**, 137.

Emslie, A. G. and Nagai, F.: 1984, *Astrophys. J.* **279**, 896.

Emslie, A. G. and Vlahos, L.: 1980, *Astrophys. J.* **242**, 359.

Hoyng, P., Marsh, K. A., Zirin, H., and Dennis, B. R.: 1983, *Astrophys. J.* **268**, 865.

Hurley, K., Niel, M., Talon, R., Estulin, I. V., and Dolidge, V. Ch.: 1983, *Solar Phys.* **86**, 367.

Kahler, S. W.: 1982, *J. Geophys. Rev.* **87**, 3439.

Kane, S. R.: 1973, in R. Ramaty and R. G. Stone (eds.), *High Energy Phenomena in the Sun*, NASA SP-342, p. 55.

Kane, S. R. and Anderson, K. A.: 1970, *Astrophys. J.* **162**, 1003.

Kane, S. R., Frost, K. J., and Donnelly, R. F.: 1979, *Astrophys. J.* **234**, 669.

Kawabata, K., Ogawa, H., and Suzuki, I.: 1983, *Solar Phys.* **86**, 247.

Kiplinger, A. L., Dennis, B. R., Emslie, A. G., Frost, K. J., and Orwig, L. E.: 1983a, *Astrophys. J.* **265**, L99.

Kiplinger, A. L., Dennis, B. R., Frost, K. J., and Orwig, L. E.: 1983b, *Astrophys. J.* **273**, 783.

Kiplinger, A. L., Dennis, B. R., Frost, K. J., and Orwig, L. E.: 1984, *Astrophys. J.* **287**, L105.

Kiplinger, A. L., Dennis, B. R., and Orwig, L. E.: 1985, *Bull. Am. Astron. Soc.* **16**, 891.

Leach, J. and Petrosian, V.: 1983, *Astrophys. J.* **269**, 715.

Lin, R. P. and Hudson, H. S.: 1976, *Solar Phys.* **50**, 153.

Lin, R. P., Schwartz, R. A., Kane, S. R., Pelling, R. M., and Hurley, K. C.: 1984, *Astrophys. J.* **283**, 421.

Lin, R. P., Schwartz, R. A., Pelling, R. M., and Hurley, K. C.: 1981, *Astrophys. J.* **251**, L109.

MacKinnon, A. L., Brown, J. C., and Hayward, J.: 1984, preprint.

Mariska, J. T. and Poland, A. I.: 1985, *Solar Phys.* **96**, 317.

Ohki, K., Takakura, T., Tsunata, S., and Nitta, N.: 1983, *Solar Phys.* **86**, 301.

Orwig, L. E., Frost, K. J., and Dennis, B. R.: 1981, *Astrophys. J.* **244**, L163.

Pallavicini, R., Serio, S., and Vaiana, G. S.: 1977, *Astrophys. J.* **216**, 108.

Poland, A. I., Machado, M. E., Wolfson, C. J., Frost, K. J., Woodgate, B. E., Shine, R. A., Kenny, P. J., Cheng, C.-C., Tandberg-Hanssen, E. A., Bruner, E. C., and Henze, W.: 1982, *Solar Phys.* **78**, 201.

Poland, A. I., Orwig, L. E., Mariska, J. T., Nakatsuka, R., and Auer, L. H.: 1985, *Astrophys. J.* **280**, 457.

Pryor, H., Pierce, M. J., Speich, D. M., Fesq, L. M., Spear, K. A., Nelson, J. J., and McGovern, J. G.: 1981, Solar Maximum Mission Event Listing, Internal Document of the SMM Data Analysis Center, NASA, GSFC.

Rieger, E., Share, G. H., Forrest, D. J., Kanback, G., Reppin, C., and Chupp, E. L.: 1985, *Nature* **312**, 623.

Rust, D. M.: 1984, *Adv. Space Res.* **4**, 191.

Rust, D. M., Simnett, G. M., and Smith, D. F.: 1985, *Astrophys. J.* **288**, 401.

Smith, D. F.: 1985, *Astrophys. J.* **288**, 801.

Švestka, Z., Schrijver, J., Somov, B., Dennis, B. R., Woodgate, B. E., Fürst, E., Hirth, W., Klein, L., and Raoult, A.: 1983, *Solar Phys.* **85**, 313.

Tanaka, K., 1983, in P. B. Byrne and M. Rodono (eds.), 'Activity in Red-Dwarf Stars', *IAU Colloq.* **71**, 307.

Tanaka, K., Nitta, N., Akita, K., and Watanabe, T.: 1983, *Solar Phys.* **86**, 91.

Tramiel, L. J., Charan, G. A., and Novick, R.: 1984, *Astrophys. J.* **280**, 440.

Tsuneta, S.: 1983, in J.-C. Pecker and Y. Uchida (eds.), *Proc. Japan–France Seminar on Active Phenomena in the Outer Atmosphere of the Sun and Stars, 3–7 October 1983*, CNRS and Observatoire de Paris, p. 243.

Tsuneta, S.: 1985, *Astrophys. J.* **290**, 353.

Tsuneta, S., Nitta, N., Ohki, K., Takakura, T., Tanaka, K., Makishima, K., Murakami, T., Oda, M., and Ogawara, Y.: 1984a, *Astrophys. J.* **284**, 827.

Tsuneta, S., Takakura, T., Nitta, N., Ohki, K., Tanaka, K., Makishima, K., Murakami, T., Oda, M., Ogawara, Y., and Kondo, I.: 1984b, *Astrophys. J.* **280**, 887.

Tucker, W. H.: 1975, *Radiation Processes in Astrophysics*, MIT Press, Cambridge.

Wolff, C. L.: 1983, *Astrophys. J.* **264**, 667.

Woodgate, B. E.: 1984, *Adv. Space Res.* **4**, 393.

Woodgate, B. E., Shine, R. A., Poland, A. I., and Orwig, L. E.: 1983, *Astrophys. J.* **265**, 530.

Wu, S. T., de Jager, C., Dennis, B. R., Hudson, H. S., Simnett, G. M., Strong, K. T., Bentley, R. D., Bornmann, P. L., Bruner, M. E., Cargill, P. J., Crannell, C. J., Doyle, J. G., Hyder, C. L., Kopp, R. A., Lemen, J. R., Martin, S. F., Pallavicini, R., Peres, G., Serio, S., Sylvester, J., and Veck, N. J.: 1985, in M. R. Kundu and B. E. Woodgate (eds.), *Proc. SMM Workshop on Solar Flares*, NASA, (in preparation).

HIGH SPATIAL RESOLUTION MICROWAVE OBSERVATIONS
OF THE SUN

M. R. KUNDU

Astronomy Program, University of Maryland, College Park, MD 20742, U.S.A.

Abstract. Over the past decade two large arrays – the Westerbork Synthesis Radio Telescope (WSRT) and the Very Large Array (VLA) built primarily for sidereal radio astronomy have been used for solar radio astronomical studies with spatial resolution of a few seconds of arc. In this review, we discuss some results obtained at Maryland using these instruments.

The quiet Sun observations made with the WSRT have premitted us to produce synthesized maps of supergranulation network at 6 cm wavelength. The brightness temperatures of typical network elements and cells are respectively $\sim 2.5 \times 10^4$ K and $\sim 1.5 \times 10^4$ K; thus the contrast is 1.7:1 which compares with 1.3:1 for Ca$^+$ K and 2:0 for Lα networks. Limb profiles in both equatorial and polar directions have been obtained; limb brightening is observed at both west and south limbs, peak limb temperature being about 40% higher than disk temperature. We have produced synthesized maps of disk filaments which correspond well to Hα disk filaments and regions of reduced emission in He I 10 830 Å spectroheliograms. Using the WSRT synthesized maps at 6 cm, we have compared the structure of a sunspot associated source with model computations. Using a new method of analysis we have been able to map the vertical as well as the horizontal component of the sunspot magnetic field at specific locations in the low corona. Using the VLA, we have mapped coronal loops at 20 cm; the radio emission is attributed to bremsstrahlung near the loop footpoints whereas gyroresonance process dominates near the loop top. Using the VLA, we have carried out simultaneous observations of a microwave burst at 2 and 6 cm. The 6 cm burst source is apparently located near the top of a flaring loop, while the 2 cm emission originates from the loop footpoints. The 6 cm emission is attributed to gyrosynchrotron radiation of thermal electons in the bulk heated plasma at $\sim 4 \times 10^7$ K, while the 2 cm emission is due to nonthermal particles released and accelerated during the flare process. A DC electric field flare model appears to explain the observed delay between the peaks at the two wavelengths. From the delay, the strength of the electric field in the flaring region is estimated.

1. Introduction

Over the past two decades, several large synthesis arrays have been built, primarily for studies in sidereal radio astronomy. Two of these arrays, namely the Westerbork Synthesis Radio Telescope (WSRT) in the Netherlands and the Very Large Array (VLA) in the U.S.A. have been used in a limited manner for solar radio astronomy. These arrays provide spatial resolution ranging from $\sim 1''$ to $10''$ at centimeter wavelengths. Since the radio emission at these wavelengths originates from the upper chromosphere, the transition region and the corona, it is a useful probe for the temperature and density in these regions. The radio emission from active and flaring regions at these wavelengths is strongly influenced by the magnetic fields and, therefore, can be used to estimate the magnetic fields in these regions. In particular, synthesized maps in total intensity and circular polarization obtained with a spatial resolution of $\gtrsim 1''$ arc offer a powerful tool for probing the magnetic field topology in the active and flaring regions. Since the flare energy release takes place at lower levels of the solar atmosphere, such maps taken at short time intervals should allow us to test different flare models. In this review, we present some results obtained at Maryland with the WSRT and VLA on the

Solar Physics **100** (1985) 491–514.   0038–0938/85.15.

quiet Sun (supergranulation network and limb profile), active regions (sunspot-associated regions and coronal loops) and flaring regions.

2. Quiet Sun Observations with the WSRT

High-resolution radio observations of the quiet Sun provide valuable information on the
structure of the solar atmosphere and on other quiet Sun phenomena such as the
supergranulation network and quiescent prominences. Kundu *et al.* (1979) observed the
Sun at 6 cm with the Westerbork Synthesis Radio Telescope (WSRT) during the period
1976, June 15–18 when the solar activity was near minimum. The two main objectives
of the observations were to study the supergranulation network and the limb brightening
at radio frequencies.

The supergranulation network which is associated with convective circulation at and
below the photosphere is clearly seen at visible and ultraviolet wavelengths. The
temperature and density of the network are believed to be determined primarily by the
magnetic fields which are concentrated at the edges of the cells by convective motions.
Since the radio emission is generated by free-free emission from the thermal plasma, it
is possible that, by mapping the supergranulation network at radio wavelengths, one can
study the temperature and density within the cell center and the network.

Radio observations of limb brightening have shown contradictory results. Interferometer observations of Christiansen and Warburton (1953) at 21 cm showed an
asymmetric distribution, with strong limb brightening in the E–W but no limb
brightening in the N–S. Single dish observations at shorter wavelengths, however, show
that the quiet Sun is circularly symmetric (e.g., Kundu and McCullough, 1972) and there
is no evidence for the asymmetry observed at 21 cm. However, accurate estimation of
limb brightening from these observations has proved difficult mainly because the
resolution of single dish telescopes is poor – of the order of $1'$ – and the exact beam
shape is not well known. Nonetheless, the general results indicate that at $\lambda \gtrsim 3$ cm there
is limb brightening, whereas at shorter wavelengths there is controversy as to whether
there is limb brightening or darkening (Kundu, 1965, 1982). Since the resolution with
the WSRT is $\sim 6''$ (E–W) $\times$ 15$''$ (N–S), one should be able to overcome many of the
resolution problems encountered with single dish observations.

2.1. SOLAR NETWORK

The supergranulation network was observed at 6 cm with the WSRT on June 16, 1976
by Kundu *et al.* (1979). Figure 1 shows the 'clean map' of the central $10' \times 10'$ field
with the radio emission from the plage region McMath 14271 subtracted from the
1.5×1.4 box at left center. The 'clean map' of Figure 1 reveals low-level structures.
The most intense feature on the 'clean map' has a 6 cm brightness temperature
$\sim 1.7 \times 10^4$ K above the mean solar disk temperature of 2.0×10^4 K. 'Typical' bright
radio network features in Figure 1 have brightness temperatures $\sim (5 \pm 1) \times 10^3$ K
above the mean solar disk temperature of 2.0×10^4 K. Similarly, the regions of low radio
brightness in Figure 1 typically have temperatures $\sim (5 \pm 1) \times 10^3$ K below the mean

beam

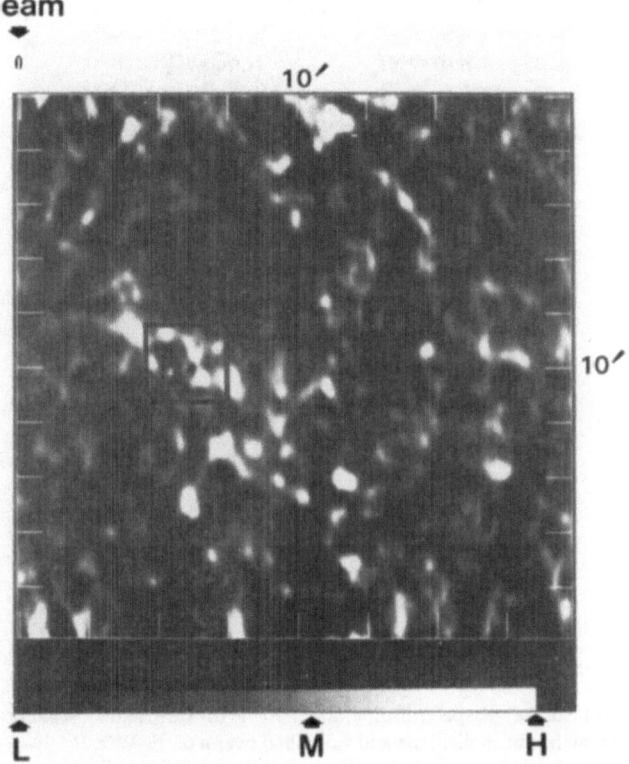

Fig. 1.   WSRT 6 cm 12 hr synthesis 'clean map' of 10′ × 10′ central disk field on 1976, June 16. The box indicates the 1.'5 × 1.'4 region from which radio plage MPR 14271 has been subtracted. The gray-scale intensities corresponding to the lowest, mean, and highest brightness temperatures of this map are indicated by L ($T \approx 11 \times 10^3$ K), M ($T \approx 2.2 \times 10^4$ K), and H ($T \approx 3.3 \times 10^4$ K).

solar disk temperature. The typical radio network contrast, defined as

$$\frac{\text{r.m.s. network temperature}}{\text{r.m.s. cell temperature}} \approx (1.7 \pm 0.2):1 ,$$

is consistent with the network contrast of ∼2 to 1 found by Reeves (1976) at levels where the mean temperature is $T \approx 2 \times 10^4$ K. Reeves (1976) found that the maximum network contrast (∼4:1) occurred at heights where the mean temperature was much greater ($T \approx 2 \times 10^5$ K).

2.1.1. *Comparison between 6 cm Map and Optical Photograph*

Figure 2 shows a digitized and smoothed (over the 6 cm beam) Ca⁺K photo for 16 June. A similarly digitized and smoothed 'control' photograph of the central disk region taken in the same Ca⁺K line at 14:03 UT on 1976, June 14 was also used. Inspection of Figures 1 and 2 indicates that there is a tendency for the radio and Ca⁺K features to be cospatial. The results of a two-dimensional numerical cross-correlation analysis between the radio map and the Ca⁺K maps for both June 16 and June 14 are given

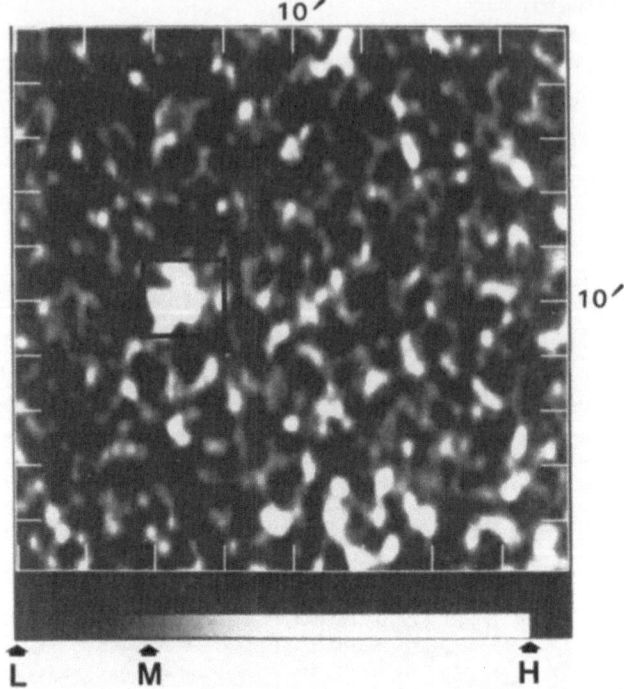

Fig. 2.   Sacramento Peak Ca$^+$K spectroheliogram of $10' \times 10'$ central disk region taken at 14:21 UT on 1976, June 16. This map has been digitized and smoothed over a $6''$ E–W $\times 15''$ N–S 'beam' to degrade its resolution to that of the 6 cm map of Figure 1. The box indicates $1.'5 \times 1.'4$ region surrounding MPR 14271 which was deleted during cross-correlation analysis. The gray scale intensities corresponding to the lowest, mean, and highest map intensities in this display are indicated by L, M, and H.

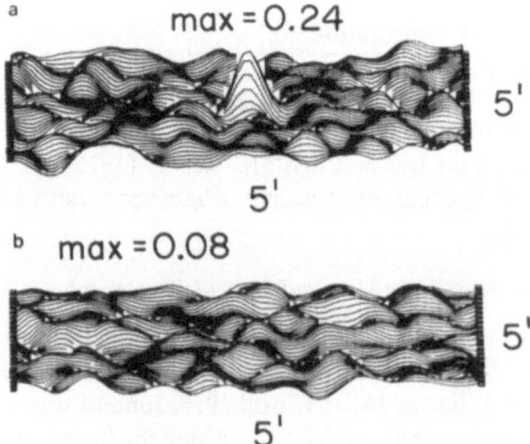

Fig. 3.   Two-dimensional cross-correlation of (a) WSRT 6 cm map with same-day Ca$^+$K and (b) WSRT 6 cm map with 2-day earlier Ca$^+$K.

in Figures 3(a) and 3(b). Figure 3(a) shows that there is a peak on the positive cross-correlation function (significant at the 3σ level) located within 5″ of the initial guess at the alignment between the 6 cm radio map (Figure 1) and the corresponding Ca^+K map for 1976 June 16 (Figure 2). For comparison, we have performed a two-dimensional cross-correlation between the 6 cm radio map of Figure 1 and the 'control' Ca^+K map of June 14. Figure 3(b) shows that there is no peak in the cross-correlation coefficient higher than the 1σ level for shifts $\lesssim 5'$. The radio versus same-day Ca^+K cross-correlation is likely to be higher than the 3σ level if the radio map could be synthesized in a shorter time. During the 12 h synthesis there is certainly some rearrangement of the supergranulation pattern which is known to have a lifetime of ~ 20 hr (Simon and Leighton, 1964).

2.1.2. *Spatial Distribution of 6 cm Quiet Sun Disk Emission*

The results of an autocorrelation analysis on the radio map of Figure 1, and on the June 16 Ca^+K map of Figure 2 and the June 14 Ca^+K 'control' map are shown in Figures 4(a)–(c). The three autocorrelation displays of Figures 4(a)–(c) all have central

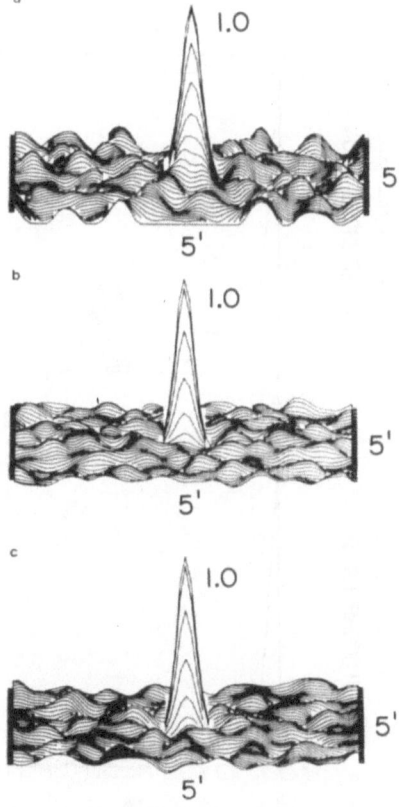

Fig. 4. Two-dimensional autocorrelation for (a) 1976, June 16 WSRT 6 cm map of Figure 1; (b) 1976, June 16 Sacramento Peak Ca^+K map of Figure 2; (c) 1976, June 14 Sacramento Peak Ca^+K 'control' map.

peaks with comparable full width at the 1σ level and secondary peaks within shifts $\lesssim 5'$ of no higher than 1.5σ. The azimuthal average autocorrelation functions about these central peaks are shown in Figures 5(a)–(c). The characteristic 'network width' W estimated from the full width at half-maximum (FWHM) of the central peak of the autocorrelation function $\approx 15''$ ($\sim 11\,000$ km) for the radio map, the Ca^+K map of June 16, and the control Ca^+K map of June 14. Note that the radio and Ca^+K widths

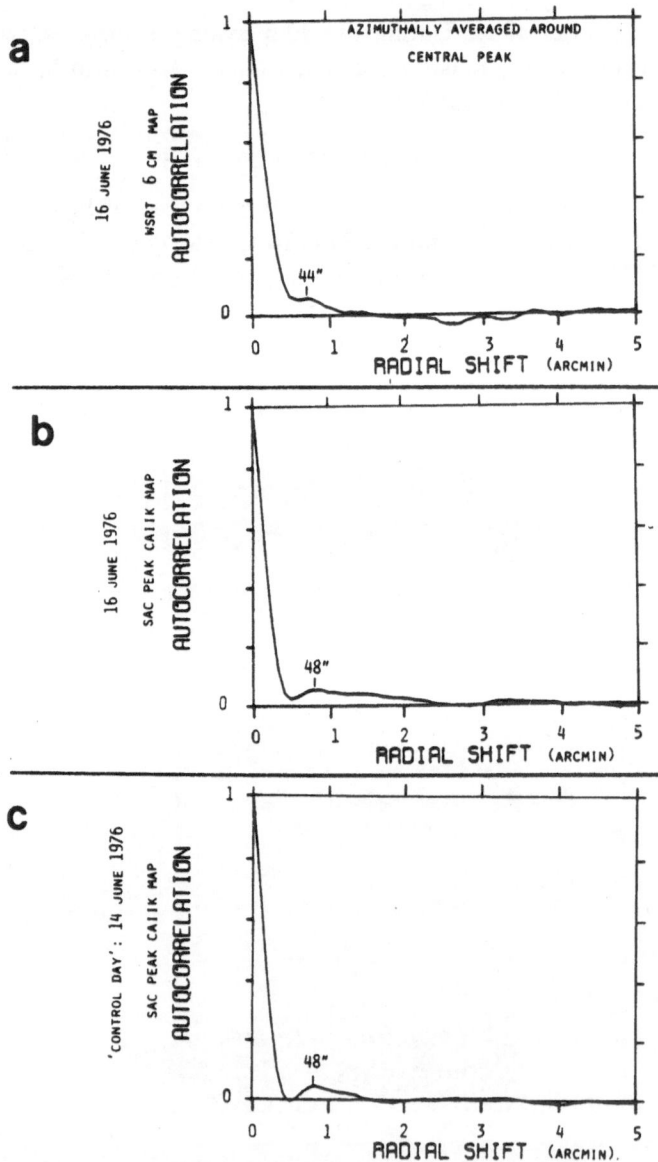

Fig. 5.   Autocorrelation of Figures 4(a)–(c) azimuthally averaged around central peak. (a) WSRT 6 cm map of 1976, June 16; (b) Ca^+K map of 1976, June 16; (c) Ca^+K 'control' map of 1976, June 14.

are larger than the width of 10″ found by Reeves (1976) in the EUV. The characteristic cell spacings L estimated from the positions of secondary peaks in Figures 5(a)–(c), are ~44″ for the 6 cm features, and ~48″ for the Ca⁺K features observed on June 16 and June 14.

The above study provides strong evidence that the supergranulation network observed at optical wavelengths can be observed at centimeter wavelengths. The emission at these wavelengths originates partly from the transition region and partly from the corona. Since the network elements do not appear in the coronal EUV lines, we believe that at centimeter wavelengths we are observing only the contribution from the transition region. Using the RATAN-600 radio telescope, at 2–4 cm with ~12″ × 72″ resolution, Pariiskii *et al.* (1976) had previously shown the existence of structures having spatial properties similar to the optical network; however, these results were based upon strip scans with a beam 72″ long. Thus, Kundu *et al.* (1979) produced the first synthesized two-dimensional map of the supergranulation network at radio wavelengths. The temperatures and densities of these structures at higher levels in the transition region and as observed at centimeter wavelengths are of extreme importance for studies of solar atmosphere models.

2.2. LIMB BRIGHTNESS DISTRIBUTION

The solar limb was observed in an attempt to synthesize the limb profile and study limb brightening at 6 cm. Four regions on the Sun whose heliocentric coordinates were N 70 E 00 (N limb), N 00 E 70 (E limb), N 00 W 70 (W limb), and S 70 E 00 (S limb) were synthesized on June 17, 1976, using the WSRT. To overcome the difficulty of synthesizing limb brightening due to grating responses and lack of zero and short baselines, Kundu *et al.* (1979) employed two different methods: least-squares fitting and deconvolution. For details see Kundu *et al.* (1979).

Figure 6 shows limb brightening of ~40–50% at the south limb. The peak of the limb brightening is at ~15″ outside the white-light limb, which has a radius of 15′46″ on

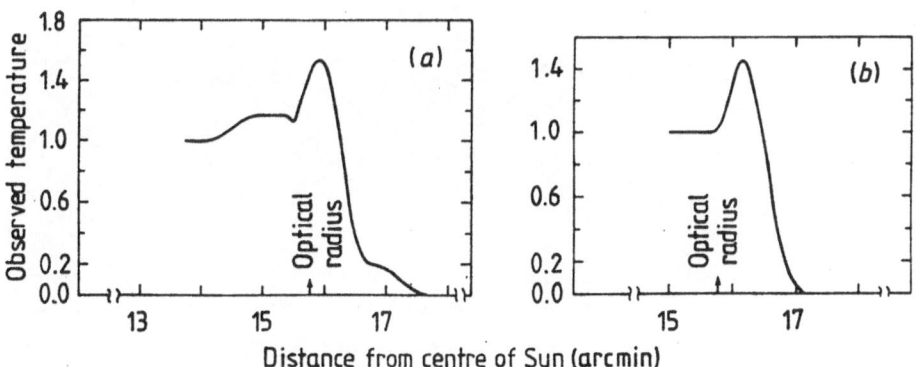

Fig. 6.   (a) Limb profile of the south limb, using the deconvolution method. The temperatures have been normalized by the temperature of the disk center; (b) limb profile of the west limb estimated from a two-component least-squares fit.

this day. The west limb also shows clear limb brightening and the limb brightening at
the west limb is at a radius larger than that at the south limb by $\sim 15''$.

The half-power radius of the Sun at 6 cm is larger than the white-light Sun by $\sim 40''$
at the south limb and by $\sim 50''$ at the west limb. The observations clearly show limb
brightening at both the west and south limbs. This is in contradiction with the earlier
results at 9 cm and longer wavelengths which indicate that the Sun shows limb
brightening at the east and west limbs, but not at the north and south limbs. A partial
explanation of the discrepancy could be that at these wavelengths the observations are
affected by the presence of coronal holes at the poles of the Sun which are well known
to evolve in both size and density on a time-scale of days to weeks.

3. Observations of Filaments

At centimeter wavelengths, the quiescent prominences seen on the disk appear as
depressions in brightness temperature that are well correlated with Hα dark filaments.

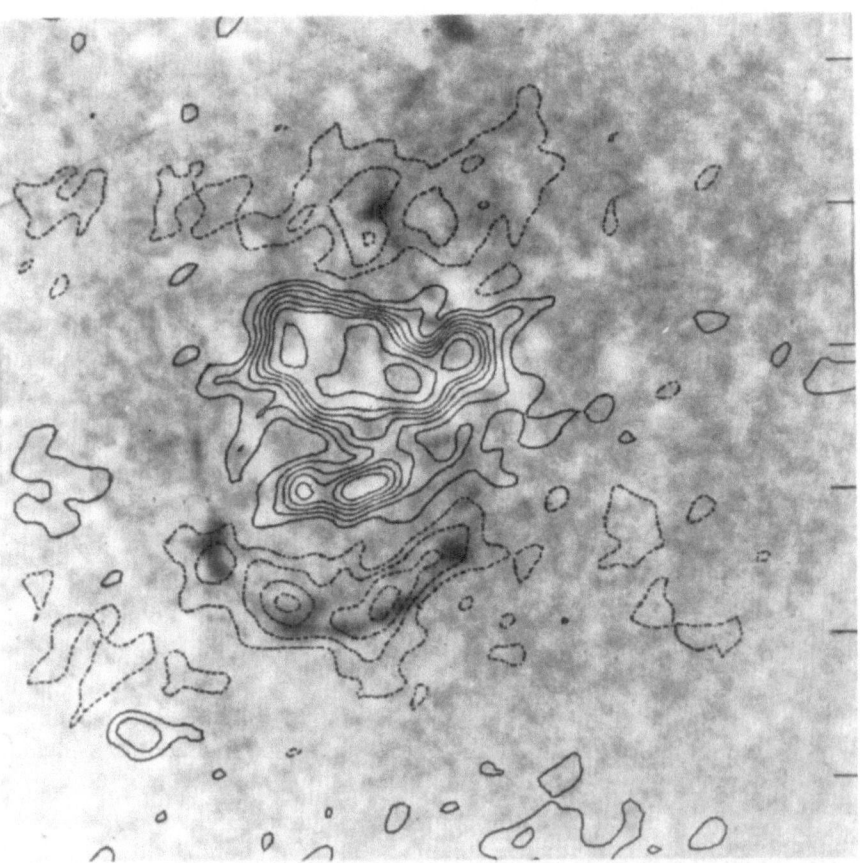

Fig. 7.   Radio filament at 6 cm superposed on the corresponding Hα picture (courtesy of Sacramento Peak
Observatory).

In a large number of cases, the sizes of the radio and optical filaments are similar (e.g., Schmahl *et al.*, 1981), although in some cases, the radio filaments are found to be broader than the corresponding Hα filaments (see, e.g., Kundu and McCullough, 1972; Kundu *et al.*, 1978). Since the filaments are optically thick they should have the same brightness temperature at radio wavelengths as the electron temperature at optical wavelengths. However, the radio observations indicate that the brightness temperature of a filament increases with wavelength. This has been interpreted as due to radiation from the filament transition sheath where the electron temperature increases from 6000 K at the filament to about 10^6 K, the temperature of the surrounding corona. Accurate determination of filament temperature as a function of wavelength is impor-

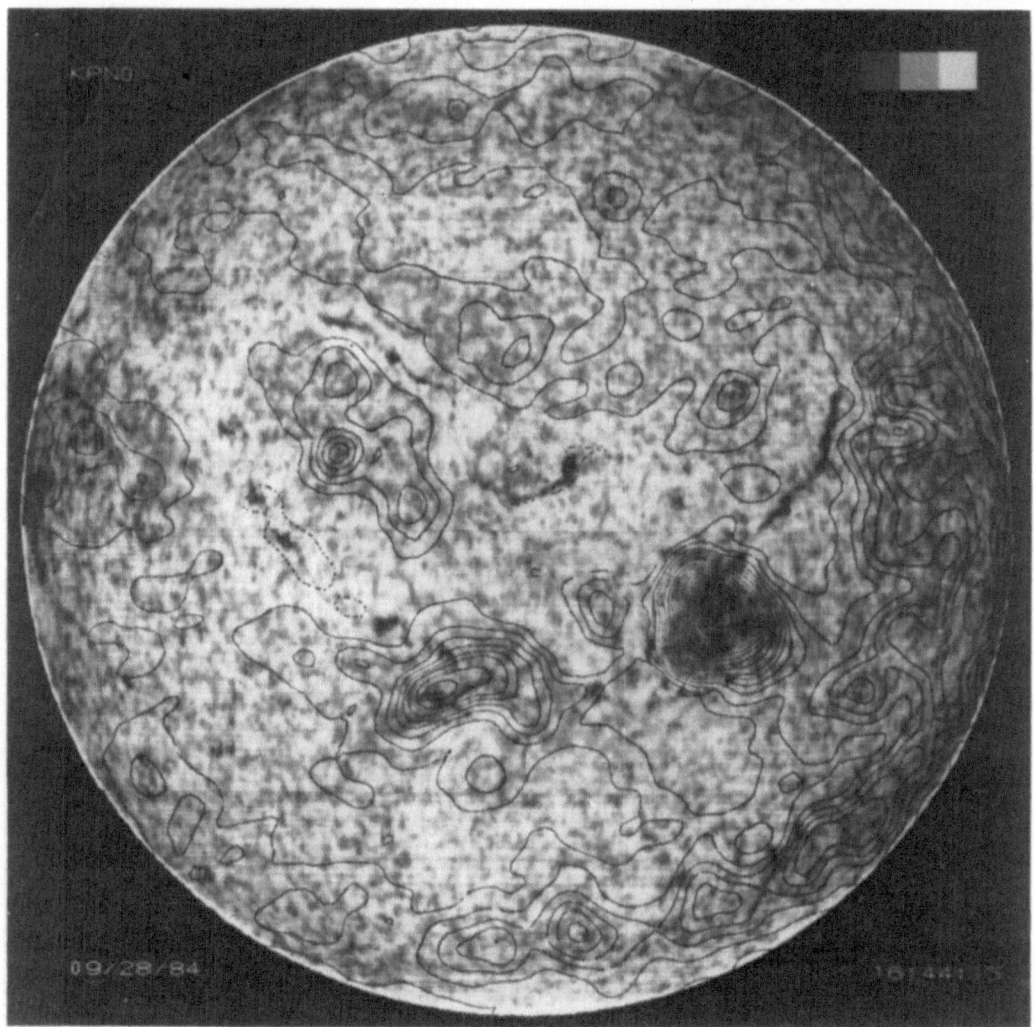

Fig. 8.   Radio filament at 20 cm superposed on the He I 10 830 Å spectroheliogram of KPNO.

tant, since it can provide information on the density and temperature structure of the filament transition sheath, which in turn determines the amount of thermal energy conducted into the filament from the corona (e.g., Rao and Kundu, 1977). Therefore, it is necessary to produce accurate two-dimensional maps of radio filaments with angular resolution as good as a few arc sec. Rao and Kundu (1980) produced the first synthesized map of a filament at 6 cm with angular resolution of $\sim 15''$, using the WSRT.

Figures 7 and 8 show a disk filament observed with the VLA at 6 and 20 cm, superposed on Hα and He I 10830 Å spectroheliograms, respectively. We see a good correspondence between the 6 cm depression features and Hα filaments, and also between the 20 cm depression features and filaments as well as the lack of emission in the He I picture of KPNO. Typical brightness temperatures of the radio filaments at 20 cm are $\sim 5.7 \times 10^4$ K + 10%. At 6 cm these values are $\sim 16\,000$ K $\pm$ 10% (Kundu *et al.*, 1985).

4. Microwave Active Regions

High-resolution microwave observations of solar active regions (e.g., Kundu *et al.*, 1977; Alissandrakis and Kundu, 1982; Lang and Willson, 1982; Chiuderi-Drago *et al.*, 1982) have shown that the slowly varying component of solar radio emission is associated with sunspots, plages and regions of transverse magnetic fields or neutral lines. The strong sunspot associated component at centimeter wavelengths with brightness temperatures of up to a few million degrees has been generally interpreted as arising from emission at low harmonics of the gyrofrequency. The weaker and extended low brightness temperature sources ($T_b \lesssim 10^5$ K) associated with chomospheric plages are usually interpreted as originating from thermal free-free emission from the transition region and corona above active regions. The moderately intense emission ($T_b \gtrsim 10^6$ K), sometimes associated with plages or with regions of transverse magnetic fields (as evidenced by neutral lines) is also attributed, at least in part, to low harmonic gyroradiation.

The total intensity structure of isolated spot associated sources as observed with a resolution of a few arc sec is fairly simple, with emission coming from an area about the size of the peunumbra (Kundu *et al.*, 1977), although some exceptions have been reported (Alissandrakis and Kundu, 1982; Kundu and Alissandrakis, 1984). The circular polarization structure is more complicated; a source with two polarization peaks was observed by Kundu *et al.* (1977), while observations with higher resolution (Alissandrakis and Kundu, 1982; Lang and Willson, 1982) showed that the circular polarization is higher at the edge of the total intensity source, forming a 'ring' or 'horseshoe' structure.

Two model computational studies of the slowly varying component have been made recently, one by Gel'freikh and Lubyshev (1979) and the other by Alissandrakis (1980). The two studies lead to basically the same conclusions: The structure of the source depends upon which of the low order (first to fourth) harmonics of the gyrofrequency are located in regions of high temperature (i.e., in the upper transition region or low corona). This, in turn, depends upon the wavelength and the local value of the magnetic field. Once a given harmonic satisfies the above condition, the intensity of the radiation is determined by its opacity, which has much stronger dependence on the angle θ between the field and the line-of-sight, than on any other physical parameter such as temperature and density.

There is always a region of small opacity, at the location where $\theta \simeq 0$. The extent of the region depends mainly on the harmonic number and the emission mode (ordinary or extraordinary); at 6 cm it is very small for the second harmonic in the ordinary mode and quite extended for the third harmonic in the extraordinary mode. The image of a given harmonic may look like a uniform disk, a disk with a depression corresponding to the low opacity region, or even a crescent. The location of the depression with respect to the rest of the image depends on the heliocentric distance of the source: as the source moves toward the limb, the depression moves from the center of the source toward the center of the disk.

If only the third harmonic is located in a high-temperature region (i.e., when the photospheric field is low and/or the observing frequency is high) there is little emission in the ordinary mode, while the structure of the extraordinary mode map has the form of a ring (if the source is located at the center of the disk) or a thin crescent. Due to the lack of ordinary mode radiation, the source is almost 100% polarized and the total intensity and circular polarization maps are almost identical. When both the third and second harmonics are located above the chromosphere, the resulting image in right or left circularly polarized radiation is a superposition of disks and crescents which gives a peculiar form to the source. The total intensity (I) image retains a form similar to the disk-crescent form of the right and left circularly polarized images, while the circular polarization (V) image has a different form. The circular polarization is low at the locations of high total intensity and shows maxima in regions of intermediate total intensity (Alissandrakis and Kundu, 1984).

The comparison of the observations of a source with the models gives the possibility of inferring which harmonics are responsible for the emission and thus estimating the magnetic field intensity. In this respect the cm wavelength maps can be considered as magnetograms; however, unlike Zeeman magnetograms they do not give the value of the magnetic field at a particular location, but rather the temperature of the region where the magnetic field has a given value. Observations at different frequencies can give the magnetic field as a function of temperature, and this in turn can give the field as a function of height if the variation of temperature with height is known.

5. Determination of Magnetic Fields in the Low Corona

Using the Westerbork Synthesis Radio Telescope (WSRT) at 6.16 cm wavelength Alissandrakis and Kundu (1984) studied a simple bipolar active region (McMath 16382) as it moved from longitude east 50° to west 16°. The region contained two large spots of opposite magnetic flux, with penumbral structure that appeared to rotate clockwise

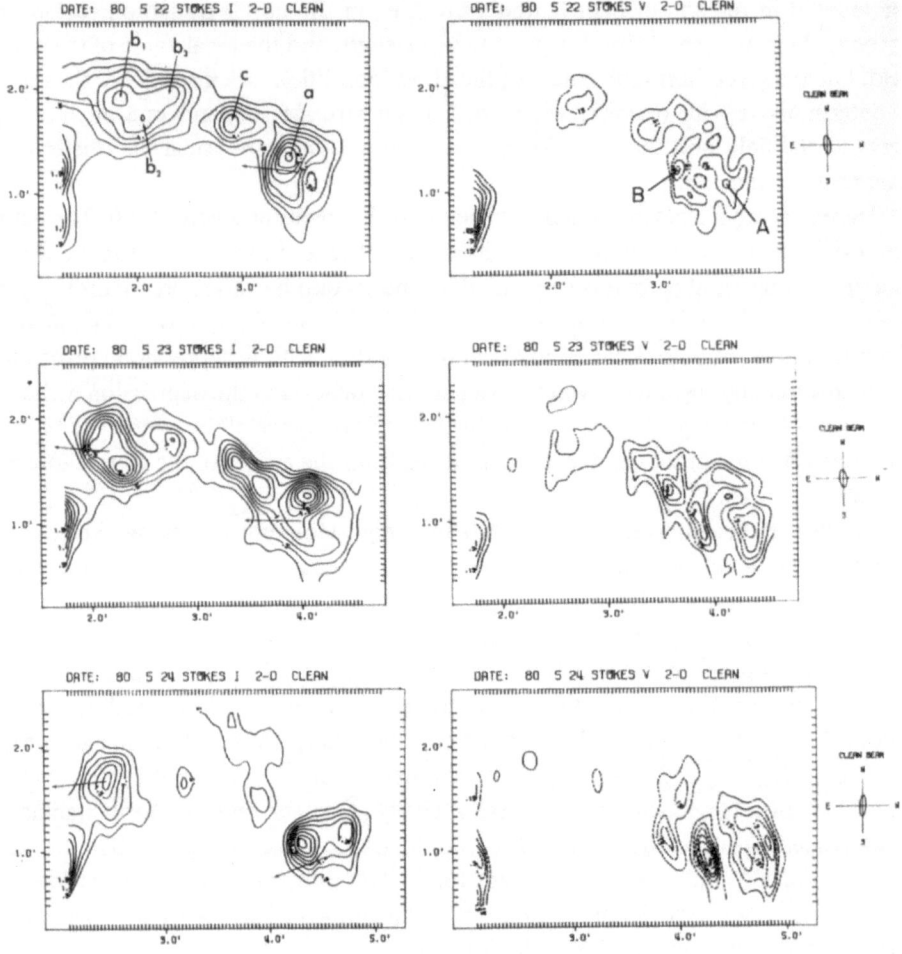

Fig. 9a.   22–24 May, 1980.

Fig. 9a-b.   Total intensity and circular polarization maps for McMath region 16862. Celestial north is up, celestial west to the right. The west part of region 16862 is visible near SE corner of the maps. Total intensity contours are in steps of 0.25×10^6 K for I-maps, 0.075×10^6 K for the V-maps; the contour at 0.25×10^6 K has not been drawn on the I-maps of May 24 and 25. The arrows show the direction of the limb.

around the leading spot and counterclockwise around the follower during the six days of observations (May 22–27, 1980). The microwave maps were analyzed in great detail to determine the strength and orientation of the coronal magnetic fields during this period.

The maps showed two prominent sources associated with the two spots and some weak emission between them. As defined by the lowest contours, the sources appeared to be displaced slightly limbward from the spots, indicating that the radio sources were located about 10^4 km (within a factor of 2) above the photosphere. The leading source

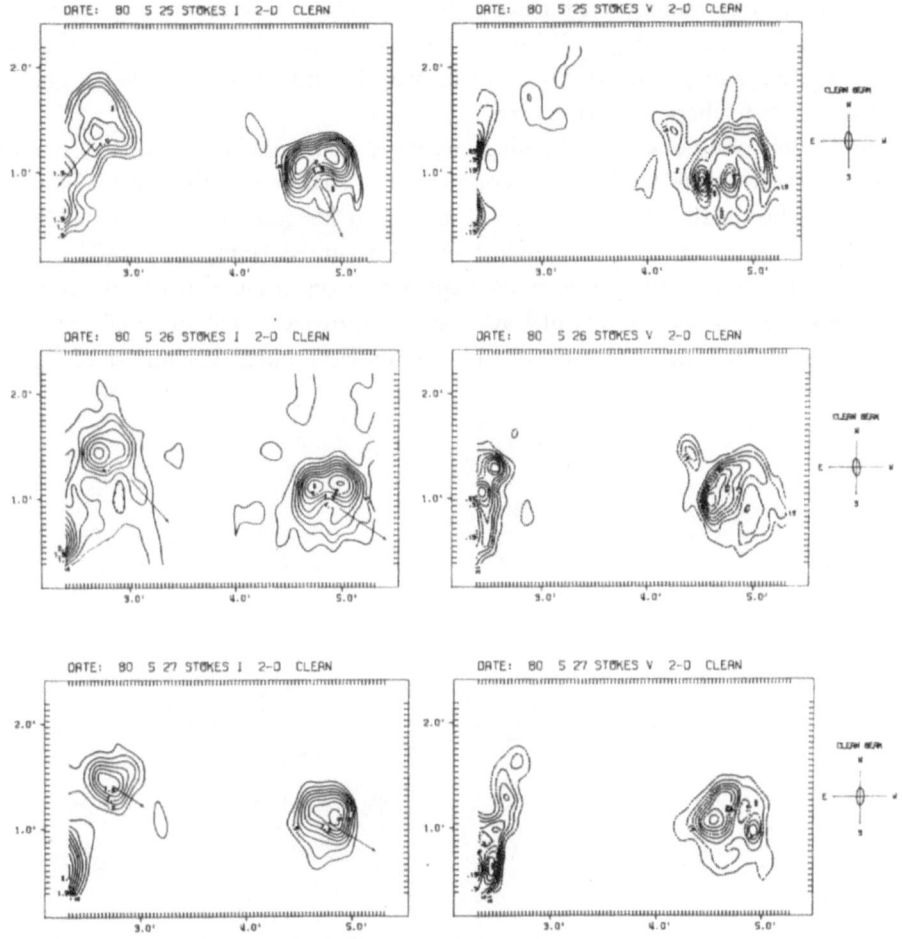

Fig. 9b.   25–27 May, 1980.

appeared roughly elliptical in the lowest contours, much like the leading spot, but the higher contours showed a crescent shape, with the intensity peak displaced in the limbward direction from the centroid of the lowest contours (see Figure 9). The intensity gradient of the convex part of the source, which overlay the penumbra, was greater than that of the concave part, which overlay the umbra. This gradient difference was larger when the spot was farthest from disk center. The peak brightness temperature of the source changed little, from a minimum of 2×10^6 to a maximum of 2.6×10^6 K, and the flux showed a maximum value of 4 s.f.u. near the central meridian.

An interesting question is whether the rotation of the spot has any effect on the shape and structure of the radio maps. Although structural changes give the impression that the source rotated in the clockwise sense throughout the observing period, this is not related to the rotation of the spot itself. It is rather a consequence of the fact that as the position of the source on the solar disk changes, so does the angle between the magnetic field and the linge-of-sight, and consequently the opacity of the harmonic layers. A careful inspection of the maps of May 24 and 25, shows that the source has almost the same shape and structure.

The leading source was left-handed circularly polarized (extraordinary mode) and was apparently unaffected by mode coupling. The structure of the I and V maps leads to the conclusion that both the second and third harmonics are responsible for the emission, otherwise the source would have been almost 100% polarized.

Model calculations of gyroresonance emission from a current-free dipole magnetic field embedded in a plane parallel atmosphere showed a striking similarity with the leading source of May 22 (Figure 10). In the model, the peak brightness of the L-map

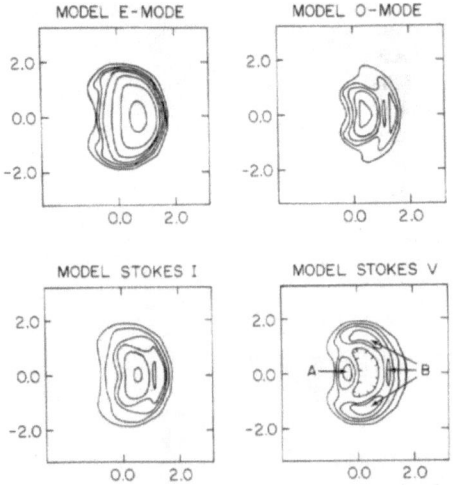

Fig. 10.   Computed maps of the extraordinary and ordinary mode radiation, and of total intensity (I) and circular polarization (V) of a sunspot-associated source at 6.16 cm. The magnetic field is that of a vertical dipole placed 2.7×10^9 cm below the base of the transition region with a dipole moment of 1.2×10^{31} c.g.s. units. The temperature structure is determined by a constant conductive flux of 6×10^6 erg cm^{-2} s^{-1} while the pressure is determined by hydrostatic equilibrium and a value of 10^{15} K cm^{-3} at the base of the transition region.

corresponds approximately to the electron temperature of the third harmonic level and the peak brightness of the R-map corresponds approximately to the electron temperature of the second harmonic level. Furthermore, the point where the magnetic field is parallel to the line-of-sight corresponds to the low opacity region where the maps show low intensity and high circular polarization. The angle l between the magnetic field vector and the vertical, determined from the maps (Figure 11) showed an almost linear increase as a function of distance from the source center. A plot of the location of the low-opacity region, with the direction and magnitude of the derived horizontal component (Figure 12) showed that the horizontal field had both an azimuthal and a radial component. However, the horizontal component in this model appeared to rotate

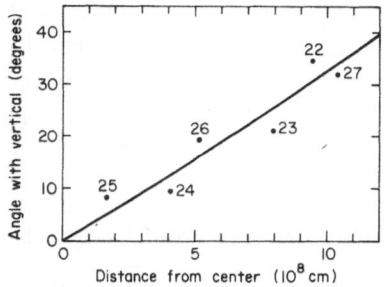

Fig. 11.   The angle of the magnetic field vector with the vertical as a function of distance from the center of the source. The solid line is a fit with a force-free model with $\gamma = 1.08 \times 10^{-8}$ cm. The numbers beside the data points are the dates of the observations.

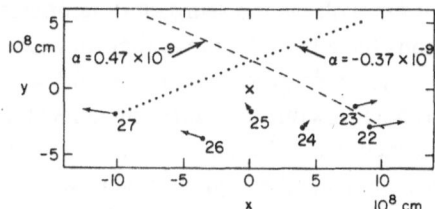

Fig. 12.   The position of the low intensity region projected on the horizontal plane. The arrows show the magnitude and direction of the horizontal component of the magnetic field. The $\times$ shows the center of the source. The dashed and dotted lines are loci of computed positions for two values of α.

counterclockwise, rather than clockwise, the direction of the rotation of the spot. In addition, the low opacity region was displaced limbward from spot center, rather than towards disk center, as current-free models predict. Alissandrakis and Kundu (1984) developed a force-free model to explain these discrepancies.

In the I-maps the axis of symmetry of the source is not directed toward the limb, which is an effect produced by a twisted magnetic field, further arguing for a force-free model, for which the analytical model of Schatzman (1961) was used. In this model, the projection of the magnetic field lines on the plane of the sky are logarithmic spirals forming an angle ϕ with the radial direction, and the angle l of the field vector to the vertical is a function of radius only. Both ϕ and l were determined from the maps. A least square fit to l (as estimated from the initial current-free model), combined with the best fit to ϕ yielded the value of the vertical magnetic scale height ($1/k$) and the ratio of the current to magnetic field strength (α).

For May 22–23, $\alpha = 0.47 \times 10^{-9}$ cm^{-1}; and $k = 0.98 \times 10^{-9}$ cm^{-1}. For May 27, the values obtained were not significantly different, but for May 24 and 25, the model was not suitable, implying almost 90° [180°] twists or negative radial components of the magnetic field. Alissandrakis and Kundu found that the structure of the maps could be reproduced by inclining the dipole axis of a current-free magnetic field by 45° from the vertical. They concluded that the magnetic field structure rotated by about 50°

counterclockwise between May 24 and 25, opposite to the sunspot rotation. The inclined dipole model also fit the other days of observation as did the force-free model.

These high spatial resolution microwave observations show the importance of radio maps for tracing the magnetic field of the low corona, particularly in combination with optical or X-ray observations.

6. Radio Observations of Coronal Loops

The observation of magnetic loops at radio wavelengths is important in view of the apparent abundance of such structures at X-ray wavelengths in the low corona (Rosner *et al.*, 1978). X-ray imaging of the Sun has shown that loops of hot plasma are a major structural component of the low corona. X-ray observations have been particularly useful in understanding the physics of active regions and coronal heating, and in enabling plasma parameters such as pressure and temperature to be determined. However, radio observations have the advantage that they can provide information on the loop magnetic fields, and their gradients along and across the loops, and the magnitude of possible toroidal fields due to electric currents along the loops. Kundu and Velusamy (1980) first reported the existence of a looplike structure connecting two sunspots of opposite polarity in an active region at 6 cm wavelength, and attributed the emission at the foot points ($T_b \sim 5 \times 10^6$ K) located near the sunspot umbra to the gyroresonance process, while the emission in the loop (outside of the foot points) was attributed to optically thin thermal bremsstrahlung ($T_b \sim 10^6$ K). On the other hand, Lang *et al.* (1982) observed a radio loop at 20 cm and attributed all the observed emission to thermal bremsstrahlung. Thus, there appears to be disagreement over the source of optical depth within the loop.

Figure 13 shows the total intensity map of a 20 cm active region (McConnell and Kundu, 1983); the 20 cm sources connect regions of opposite magnetic polarity and, therefore, they must be magnetic loops in the low corona. In order to derive a consistent

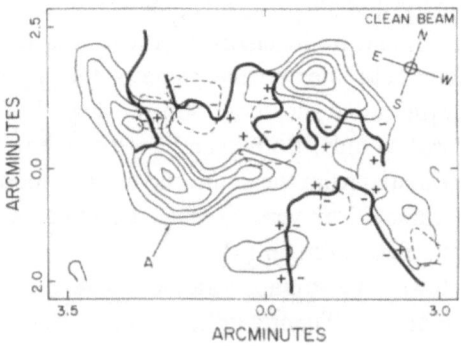

Fig. 13. Map of the 20 cm radio emission from an active region showing two loops. Note that each loop terminates in two regions of opposite polarity. The peak brightness temperature is 1.7×10^6 K, and the contour levels are 15, 30, 45, 60, 75, and 90% of the map maximum. The thick lines indicate neutral lines.

model of the radio emission in loops, McConnell and Kundu used models for active region loops and calculated the expected thermal bremsstrahlung and gyroresonance emissions. They showed that the brightness peak ($T_b \sim 1.7 \times 10^6$ K) of the observed 20 cm loops cannot be due to thermal bremsstrahlung, contrary to the conclusions of Lang *et al.* (1982). However, the secondary maxima near the loop foot points could be explained by thermal bremsstrahlung.

McConnell and Kundu (1983) considered thermal gyroresonance emission as the cause of the emission peak observed near the center of the loop. The magnetic field strength required to give resonance for 20 cm radiation is 170 G or 130 G, depending on whether the third or fourth harmonic dominates. Since the angle (θ) between the line-of-sight and the magnetic field is maximum near the top of the loop and the loop temperature is maximum at the top, thermal gyroresonance emission should peak at the top of the loop provided the magnetic field strength falls in the right range. This was as observed. Indeed, McConnell and Kundu showed that for realistic physical parameters of the loop ($N_e \simeq 5 \times 10^8$ cm^{-3}, $f = 1446$ MHz, $T \simeq 2 \times 10^6$ K and for large θ ($\theta \geqslant 65°$)), the required scale length of magnetic field variation, L_B would be $\gtrsim 2 \times 10^7$ cm and $\gtrsim 5 \times 10^9$ cm for $\tau_{gr} \gtrsim 1$ at the third and fourth harmonics respectively. Both these conditions are plausible but the former is probably more easily satisfied. Thus, the 20 cm loop emission can be explained by a model consisting of two components: a component due to bremsstrahlung which is important near the feet of the loop, and a gyroresonance component which peaks near the top of the loop.

7. Observations of Flares with the VLA

7.1. SPATIAL STRUCTURE AT 2 AND 6 CM WAVELENGTHS

In a flare the energy release is usually located in the lower corona and, therefore, the microwave observations can be used to derive the physical parameters in the flaring region such as the electron density, magnetic field strength, and number of gyrosynchrotron emitting electrons. With the high resolution of a few arc sec as presently available with the VLA, it is possible to locate the flare energy release site within the complex magnetic structure of the flaring active region. The observations conducted in the past (e.g., Alissandrakis and Kundu, 1978; Kundu *et al.*, 1982; Marsh and Hurford, 1980) indicate that the radio bursts are generally located over the magnetic neutral line whereas their Hα counterparts are located near the footpoints of the flaring loop. The close association of hard X-ray bursts with the microwave bursts (Kundu, 1961) suggests a common origin of energetic electrons radiating in the two spectral domains.

Simultaneous multifrequency observations in microwaves and hard X-rays are needed to properly understand the radiation mechanisms operating in different spectral domains and are, therefore, important in modelling a flare. Shevgaonkar and Kundu (1985) carried out simultaneous microwave burst observations at 6 and 2 cm with the VLA with spatial resolution of $5'' \times 3''$ at 6 cm and $2'' \times 1.6''$ at 2 cm, respectively. Figure 14 shows the time profiles of the burst at 6 and 2 cm and in hard X-rays. At 6

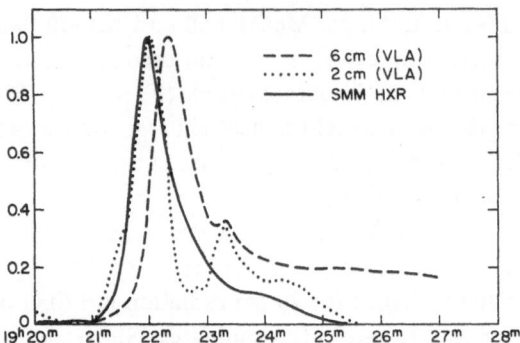

Fig. 14.   Flux versus time profiles of the burst. At 6 and 2 cm the shortest baseline of $\sim 2\,k\lambda$ is used. Hard X-ray profile is over an energy range 30–60 keV. The normalization factor is ~ 8.5 s.f.u. at 6 cm and ~ 0.9 s.f.u. at 2 cm. For hard X-ray profile maximum counts are 120.

and 2 cm, the shortest baseline of $\sim 2\,k\lambda$ was used to obtain the time profile. The burst consists of two peaks; the first peak is much stronger than the second one. The hard X-ray and the 2 cm emissions peak more or less simultaneously but the 6 cm peak is delayed by ~ 15 s. Figure 15 shows the total intensity maps at 6 and 2 cm around the peak of the burst. The observed degree of circular polarization was low.

From total intensity maps it is clear that the burst sources at 6 and 2 cm are not co-spatial. The source at 6 cm has its total intensity maximum lying between the two isolated bright components of the 2 cm source. Comparing the source locations with the

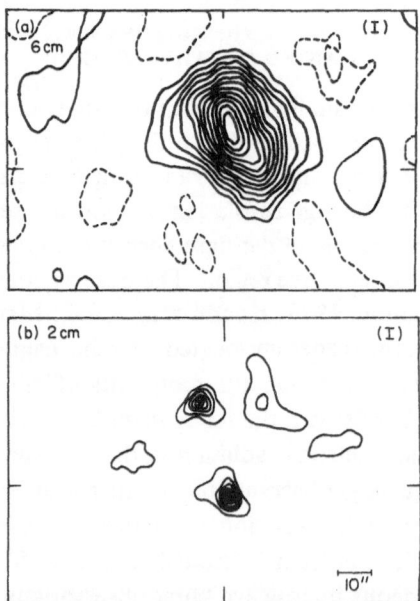

Fig. 15.   Total intensity (I) maps around the stronger peak of the burst at 6 and 2 cm. Contour interval is 2.6×10^6 K at 6 cm and 9.70×10^4 K at 2 cm.

magnetic structure of the quiescent active region it is apparent that the 6 cm source more or less overlies the magnetic neutral line whereas, the 2 cm sources coincide spatially with the strong magnetic regions. Moreover, the two components of the 2 cm source lie over the regions of opposite magnetic polarity.

From the spatial locations of the 6 and 2 cm sources within the magnetic structure of the active region it appears that the 6 cm emission originates from the top of the loop in which the flare energy is released while the 2 cm emission arises from the footpoints. The brightness temperatures at 6 and 2 cm are $\sim 4 \times 10^7$ K and $\sim 10^6$ K, respectively. These relatively low brightness temperatures and low degree of circular polarization may imply that the emission is thermal.

It is important to assess the relative contributions of the thermal and nonthermal emissions in the total burst radiation. First we estimate the brightness due to the nonthermal particles produced in the flare. At wavelengths $\lesssim 6$ cm and magnetic fields of a few hundred gauss at the site of flare energy release approximate formulae for gyrosynchrotron emission (Dulk and Marsh, 1982) can be used to compute the brightness temperature.

The computed brightness temperature T_B at 6 cm has been plotted against the magnetic field strength in Figure 16. The values chosen for the microwave source parameters are: density of nonthermal particles $N \sim 10^6$ cm^{-3}, power-law index of the electron energy distribution $\delta = 3$–5, angle between the line-of-sight and the direction of the magnetic field $\theta = 25$ and $70°$, and the strength of the magnetic field $B = 100$–600 G. From Figure 16, it is clear that the emission at 6 cm is optically thin

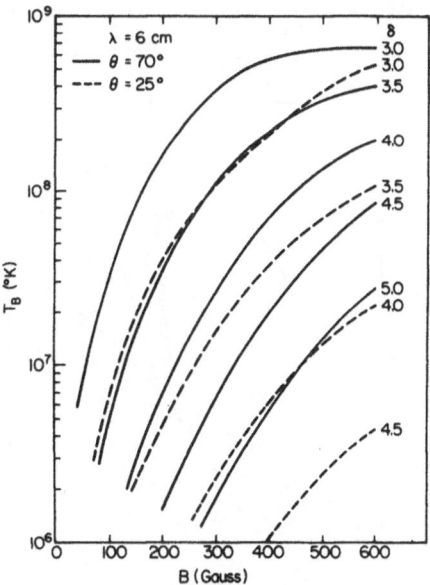

Fig. 16.   Brightness temperature due to gyrosynchrotron radiation of nonthermal electrons having a power-law energy distribution versus magnetic field. θ is the angle between the magnetic field and the line-of-sight. δ, power-law index of the energy distribution of the electrons. λ, wavelength used for computations (6 cm).

for the given magnetic field range and, therefore, the degree of polarization p should be $\gtrsim 30\%$ for a magnetic field greater than ~ 100 G. This means that the nonthermal particles, even if they are trapped near the top of the loop, cannot account for the observed low polarization of the 6 cm source. The other possibility is that the 6 cm region is heated to a high temperature of $\sim 4 \times 10^7$ K and the thermal particles produce gyrosynchrotron emission. Using the expression for the absorption coefficient for thermal mildly relativistic electrons (Dulk *et al.*, 1979; Zheleznyakov, 1970) it can be shown that for a density of $\sim 10^{10}$ cm^{-3} of the thermal electrons, the emission is optically thick for a magnetic field of $\gtrsim 200$ G, and the degree of polarization then should be zero. These computations clearly indicate that the observations presented here require that the 6 cm radiation should be dominated by a thermal component. Further, the bulk heated plasma should have a magnetic field of $\gtrsim 200$ G. A similar estimate of the magnetic field in a flaring region was obtained (Velusamy and Kundu, 1981) from the lifetime of electrons in a post-flare loop. If we assume an electron density $\sim 10^9$ cm^{-3}, the lower limit on the magnetic field is slightly increased but the conclusions remain the same.

Since the 6 cm emission originates from near the top of the loop, while the 2 cm emission comes from its footpoints, there should be beaming of nonthermal particles along the magnetic field lines to the footpoints. The possibility of a conduction front moving down the loop is ruled out because in that case the 2 cm emission will peak after the 6 cm emission and, therefore, will not be consistent with the observations. Although the nonthermal particles are beamed to the footpoints, it is not certain if the 2 cm

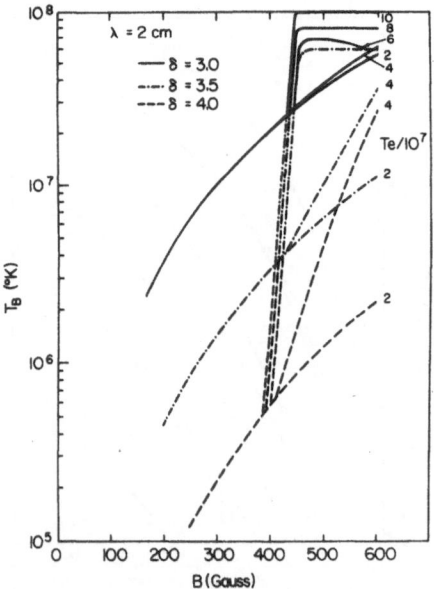

Fig. 17.   Brightness temperature at 2 cm from a plasma volume with both thermal and nonthermal particles versus magnetic field. T_e is the thermal temperature of the plasma and δ is the power-law index of the nonthermal particle energy distribution. Angle between the magnetic field and the line-of-sight is $70°$.

emission is directly due to gyrosynchrotron radiation of the beaming particles in the high magnetic field, because in that case, due to the small angle θ between the line-of-sight and the magnetic field, one would expect a high degree of circular polarization. A possibility to explain the absence of polarization in the 2 cm source is that the beam of electrons thermalizes in the dense plasma near the footpoints and that the 2 cm emission is due to optically thick gyroresonance radiation in the thermal plasma produced by thermalization of the beam of nonthermal particles. Since the location of the 2 cm sources relative to the 6 cm source supports the presence of nonthermal particles, albeit indirectly, the particles having initial high pitch angles must be present near the top of the loop. From their deficiency to emit at 2 cm in a magnetic field of $\gtrsim 200$ G, we can find from Figure 17 a lower limit on δ to be ~ 4. Even if the nonthermal particles do not produce sufficient emission at 2 cm if $\delta \gtrsim 4$, the thermal component will start contributing to the 2 cm radiation if the magnetic field is $\gtrsim 350$ G (Figure 17). Therefore, from the lack of co-spatiality between the 2 and 6 cm sources and the low degree of polarization in the 6 cm source we estimate a magnetic field of 200–350 G in the flaring region.

7.2. TIME DELAYS BETWEEN 2 AND 6 CM BURST EMISSION

From Figure 14 it is clear that the 2 cm and hard X-ray burst emissions rise almost simultaneously whereas, for ~ 10–20 s after the onset of 2 cm and hard X-ray emission, the 6 cm emission first rises slowly and then almost as impulsively as the 2 cm emission. A multiple kernel model has been proposed to explain large delays between the peaks of 5 and 15 GHz emission (Brown *et al.*, 1983). In this model the source is assumed to be composed of small kernels with different physical parameters. These kernels flare sequentially to produce mini bursts with much shorter duration whereas, the rise time of the ensemble of the mini bursts could be much longer. The different kernels have different physical parameters and, therefore, different spectral characteristics with the result that the spectral index of the burst changes with time. This change in the spectral index during the burst appears in the form of a pseudo delay.

Although this model is capable of explaining large delays, it requires that the physical conditions must change appreciably from one kernel to another. Further, the multiple kernel model could probably apply to this burst if the burst emission at two wavelengths starts simultaneously but their rise times are different. The initial slow rise and the subsequent impulsive rise at 6 cm cannot be explained by the multiple kernel model. Since we believe the emission at 6 cm to be due to the bulk heated plasma and that at 2 cm due to nonthermal particles, it may be possible to attribute the delay to the time difference between the heating of the plasma and the acceleration of the particles. Several solar flare models propose DC electric field to accelerate the particles (e.g., Holman, 1985). The high currents associated with the electric fields in turn produce Joule heating of the thermal plasma. Assuming the presence of electric fields in the flaring region, the Joule heating time (t_J = rise time at 6 cm) and the nonthermal particle production time (t_N = rise time at 2 cm) can be calculated from the formulae given by Holman (1985).

Since 6 cm emission is believed to be thermal and 2 cm emission due to nonthermal particles, t_J and t_N essentially represent the rise times of the 6 and 2 cm emission respectively. From the observed dimensions of the 6 cm source and a density of 10^{10} cm^{-3}, the total number of thermal particles $N_T \approx 10^{36}$. Taking the area of the current sheet to be equal to the source area and a magnetic field ~ 200 G, the thermal rise time is

$$t_J = 12 \times 10^5 \, v_e^{-1} \left(\frac{v_c}{v_e}\right)^2 \text{ s,}$$

where v_c = critical velocity = $6 \, (n_e/E)^{1/2}$ (cm s^{-1}); v_e is the electron thermal velocity (cm s^{-1}); v_e is the collision frequency in the current sheet $\approx 300 \, N_e \, T_e^{-3/2}$ (s^{-1}); n_e is the density in the current sheet (cm^{-3}); N_e is the density in the thermal plasma (cm^{-3}); E is the DC electric field (statvolt cm^{-1}). Also, during the initial slow rise phase at 6 cm the condition that $t_J > t_N$ requires $v_c/v_e \lesssim 5$ if $N_T/N_r \sim 10^4$; N_r is the total number of accelerated electrons.

Even if we assume $v_c/v_e \sim 2$, a high density of $\sim 2 \times 10^{13}$ cm^{-3} is needed in the current sheet to get the correct order of magnitude for t_J. To overcome this difficulty one can consider (a) the presence of plasma turbulence in the current sheet creating an anomalous resistivity which is many orders of magnitude higher than the classical value and/or (b) multiple current sheets, $\sim 10^4$ as suggested by Holman (1985) to generate nonthermal hard X-rays. However, multiple current sheets alone will not explain the initial slow rise phase. On the other hand, if anomalous resistivity is present at the burst onset, t_J and $t_N \ll 1$ s, there will be no delay and no initial slow rise phase. Therefore, at the burst onset the resisitivity should be classical. The generation of anomalous resistivity depends upon the instabilities which are excited if the drift velocity of the electrons (v_d) exceeds the threshold velocity v_{thr} ($= fC_s$); c_s = ion sound speed and f is a function of the ratio of electron to ion temperature (T_e/T_i). For $T_e \approx T_i$, $f = 35$; $T_e = 8 \, T_i$, $f = 8$; and $T_e \gg 10 \, T_i$, $f = 1$. Since $v_{\text{thr}} \sim T_e^{1/2}$ and $v_d \sim T_e^{3/2}$ (see Holman, 1985) the ratio $v_d/v_{\text{thr}} \sim T_e$, i.e., as the temperature increases the ratio v_d/v_{thr} increases and after a threshold temperature at which $v_d \approx v_{\text{thr}}$ the instabilities are induced, the resistivity increases greatly and the emission at both wavelengths increases impulsively. In the initial phase, from the relation $v_c/v_e \lesssim 5$ (condition for $t_J > t_N$) and $v_d/v_{\text{thr}} > 1$ (condition for anomalous resistivity) we have $10^{-11} \lesssim ET_e/n_e \lesssim 2 \times 10^{-10}$. For a density of 10^{11} cm^{-3} in the current sheet and $T_e \sim 5 \times 10^6$ K (assuming that at this temperature the anomalous resistivity is generated) we can estimate the electric field in the current sheet as $0.2 \lesssim E \lesssim 4\mu$ statvolt cm^{-1}.

8. Summary

The quiet Sun observations with the WSRT have shown that the 6 cm network has the same scale as the Ca$^+$K network ($\sim 15''$), somewhat larger than the width of the network in EUV. The cell sizes in radio and Ca$^+$K are also identical. The brightness temperature of typical network elements is $\sim 2.5 \times 10^4$ K, while that of the network cells is $\sim 1.5 \times 10^4$ K. The contrast between network elements and cells is, therefore, $\sim 1.7:1$, which compares with $\sim 1.3:1$ for Ca^+K network and $\sim 2:1$ for $L\alpha$ network. The radius of the 6 cm disk is $\sim 45''$ greater than the white-light Sun. The difference in size is larger than the synthesized beam of the telescope. There is evidence that the equatorial radius of the Sun is greater than the polar radius by $\sim 10''$, but the difference may not be too significant, since it is comparable to the synthesized beam. Limb brightening has been observed on both west and south limbs, and the peak limb temperature is about 40% greater than the temperature of the disk. The presence of limb brightening at the south limb which was not observed by earlier observers, suggests the absence or weakening of polar coronal holes. Using the VLA, we have produced synthesized maps of disk filaments at 6 and 20 cm which correspond well to $H\alpha$ disk filaments and regions of reduced emission in $He\,I$ 10830 Å spectroheliograms.

The WSRT observations of 6 cm active regions have premitted us to compare the structure of one sunspot associated source, as well as its center to limb variation with model computations. We have been able to identify features on the radio maps, such as the region where the magnetic field is directed along the line-of-sight. Using a new method of analysis we have been able to map the vertical as well as the horizontal component of the sunspot magnetic field at specific locations in the low corona. The vertical component decreases away from the center of the source, while the horizontal component has both a radial and an azimuthal part. Assuming a plane parallel model for the electron temperature above the spot we interpret these results in terms of a force-free magnetic field model as well as in terms of a dipole with its axis inclined to the vertical. Using the VLA, we have observed coronal loops at 20 cm; the radio emission is attributed to bremsstrahlung near the loop footpoints, whereas gyro-resonance process dominates near the loop top.

Simultaneous VLA observations of a microwave burst at 2 and 6 cm wavelengths show that the 6 cm burst source is located close to a magnetic neutral line, presumably near the top of a flaring loop, while the 2 cm emission originates from the footpoints of the loop. The 6 cm emission is dominated by gyrosynchrotron radiation of the thermal electrons in the bulk heated plasma at a temperature of $\sim 4 \times 10^7$ K, while the 2 cm emission is due to nonthermal particles released and accelerated during the flare process. From the observed low degree of polarization and the lack of the 2 cm source co-spatiality with the 6 cm source a magnetic field of 200–350 G and $\delta \gtrsim 4$ are estimated in the flare energy release site. A DC electric field flare model is invoked to explain the delay of 15 s between the peaks at the two wavelengths. From the delay the strength of the electric field is estimated to be 0.2–4 μ statvolt cm^{-1} in the flaring region.

References

Alissandrakis, C. E.: 1980, in M. R. Kundu and T. E. Gergely (eds.), 'Radio Physics of the Sun', *IAU Symp.* **86**, 101.
Alissandrakis, C. E. and Kundu, M. R.: 1978, *Astrophys. J.*, **222**, 342.
Alissandrakis, C. E. and Kundu, M. R.: 1982, *Astrophys. J.* **253**, L49.
Alissandrakis, C. E. and Kundu, M. R.: 1984, *Astron. Astrophys.* **139**, 271.

Brown, J. C., Mackinnon, A. L., Zodi, A. M., and Kaufmann, P.: 1983, *Astron. Astrophys.* **123**, 10.

Chiuderi-Drago, F., Bandiera, R., Falciani, R., Antonucci, E., Lang, K. R., Willson, R. F., Shibasaki, K., and Slottje, C.: 1982, *Solar Phys.* **80**, 71.

Christiansen, W. N. and Warburton, J. A.: 1953, *Australian J. Phys.* **6**, 190.

Dulk, G. A. and Marsh, K. A.: 1982, *Astrophys. J.* **259**, 350.

Dulk, G. A., Melrose, D. B., and White, S. M.: 1979, *Astrophys. J.* **234**, 1137.

Gel'freikh, G. B. and Lubyshev, B. I.: 1979, *Astron. Zh.* **56**, 562 (*Soviet Astron. – AJ* **23**, 316).

Holman, G. D.: 1985, *Astrophys. J.* **293**, 584.

Kundu, M. R.: 1961, *J. Geophys. Res.* **66**, 4308.

Kundu, M. R.: 1965, *Solar Radio Astronomy*, John Wiley Interscience, New York.

Kundu, M. R.: 1982, *Rep. Progr. Phys.* **45**, 1435.

Kundu, M. R. and Alissandrakis, C. E.: 1984, *Solar Phys.* **94**, 249.

Kundu, M. R. and McCullough, T. P.: 1972, *Solar Phys.* **24**, 133.

Kundu, M. R. and Velusamy, T.: 1980, *Astrophys. J.* **240**, L63.

Kundu, M. R., Alissandrakis, C. E., Bregman, J. D., and Hin, A. C.: 1977, *Astrophys. J.* **213**, 278.

Kundu, M. R., Fürst, E., Hirth, W., and Butz, M.: 1978, *Astron. Astrophys.* **62**, 431.

Kundu, M. R., Melozzi, M., and Shergoonkar, R. K.: 1985, *Astron. Astrophys.*, submitted.

Kundu, M. R., Rao, A. P., Erskine, F. T., and Bregman, J. D.: 1979, *Astrophys. J.* **234**, 1122.

Kundu, M. R., Schmahl, E. J., and Velusamy, T.: 1982, *Astrophys. J.* **253**, 963.

Lang, K. R. and Willson, R. F.: 1982, *Astrophys. J.* **255**, L111.

Lang, K. R., Willson, R. F., and Rayrole, J.: 1982, *Astrophys. J.* **258**, 384.

Marsh, K. A. and Hurford, G.: 1980, *Astrophys. J.* **240**, L111.

McConnell, D. and Kundu, M. R.: 1983, *Astrophys. J.* **269**, 698.

Pariiskii, Yu. N., Korolkov, D. V., Shivris, O. N., Kaidanovskii, N. L., Esepkina, N. A., Zverev, Yu. K., Stotskii, A. A., Akhmedov, Sh. B., Bogod, V. M., Boldyrev, S. I., Gel'freikh, G. B., Ipatova, I. A., Korzhavin, A. N., and Romantsov, V. V.: 1976, *Soviet Astron. – AJ* **20**, 577.

Rao, A. P. and Kundu, M. R.: 1977, *Solar Phys.* **55**, 161.

Rao, A. P. and Kundu, M. R.: 1980, *Astron. Astrophys.* **86**, 373.

Reeves, E. M.: 1976, *Solar Phys.* **46**, 53.

Rosner, R., Tucker, W., and Vaiana, G.: 1978, *Astrophys. J.* **220**, 643.

Schatzman, E.: 1961, *Ann. Astrophys.* **24**, 251.

Schmahl, E. J., Bobrowsky, M., and Kundu, M. R.: 1981, *Solar Phys.* **71**, 311.

Shevgaonkar, R. K. and Kundu, M. R.: 1985, *Astrophys. J.* **292**, 733.

Simon, G. W. and Leighton, R. B.: 1964, *Astrophys. J.* **140**, 1120.

Velusamy, T. and Kundu, M. R.: 1981, *Astrophys. J.* **243**, L103.

Zheleznyakov, V. V.: 1970, *Radio Emission of the Sun and Planets*, Pergamon Press, London.

ENERGETIC IONS IN SOLAR γ-RAY FLARES

HUGH S. HUDSON

CASS C-011 UCSD, La Jolla, CA 92093, U.S.A.

Abstract. Solar flares emit line and continuum γ-radiation as well as neutrons and charged particles. These high-energy emissions require the presence of energetic ions within the magnetic structures of the flare proper. We have already learned a great deal about the location and mode of particle acceleration. The observations have now become extensive enough so that we can begin to study the dynamics of the energetic ions within the flare structures themselves. This paper reviews the γ-ray and neutron observations and the theory of their emission, and discusses on this basis the presence of energetic ions deep within the flaring atmosphere.

1. Introduction

The field of solar γ-ray astronomy has provided one of the best examples – rare in astrophysics – of successful theoretical prediction. Dolan and Fazio (1965) and Lingenfelter and Ramaty (1967) presented extensive calculations of the γ-ray spectra expected from solar flares, and Chupp *et al.* (1973) then observed the solar γ-ray lines and continua by using the pioneering instrument on OSO-7. Similarly Lingenfelter *et al.* (1965) calculated solar neutron production, which shares much of the physics with the theory of γ-radiation, and Chupp *et al.* (1982) then provided the first observations from the Solar Maximum Mission satellite.

The γ-ray and neutron emissions result from the interactions of primary energetic ions with the material of the solar atmosphere. The existence of 'solar cosmic rays' and of solar radio emission had made it clear long before we could carry out successful γ-ray observations that particle acceleration accompanied solar flares. Giovanelli (1948) made perhaps the first suggestion of an important role for non-thermal particles – i.e., those forming a non-Maxwellian tail on the distribution function – in the energetics of phenomena in the flare proper, that part of the flare in the lower solar atmosphere. In recent years much research has centered around the observation of solar X-radiation and the physics of the bremsstrahlung-emitting electrons in solar flares. The ion component has received somewhat less attention, primarily because of the lack of relevant observations until recently. We do know now, however, that particle acceleration not only occurs commonly but may also represent a major fraction of the total energy of a solar flare (e.g., Lin and Hudson, 1976).

This review summarizes the theory and observations of the solar γ-ray and neutron emission (see also Ramaty *et al.*, 1975, 1983; and Chupp, 1984). We emphasize observations related to the primary high-energy ions in solar flares, basically meaning those from instruments sensitive enough to detect the nuclear emission-line spectrum, and do not discuss the literature on fast electrons in solar flares that began with the observations of Peterson and Winckler (1959). We also do not discuss the literature on interplanetary propagation and acceleration of charged particles, and because of the theoretical uncertainty in the relationship between the interplanetary particles and the

Solar Physics **100** (1985) 515–535.   0038–0938/85.15.
© 1985 *by D. Reidel Publishing Company*

particles in the flaring atmosphere itself, we try not to base conclusions upon the interplanetary observations.

2. Gamma-Ray and Neutron Observations

2.1. INSTRUMENTATION

The successful observation of γ-ray line emission from solar flares began with the OSO-7 observations (Chupp *et al.*, 1973) of the flares of August, 1972. The instrument used in these observations consisted of a sodium iodide scintillation counter of standard dimensions – 3″ diameter by 3″ length – in an active anticoincidence configuration to reduce background (see Peterson, 1973, for a discussion of instrumentation technique). The OSO-7 instrument successfully observed γ-ray emission lines and continuum from flares of 4 and 7 August, 1972, finding a fluence of $> 150 \ \gamma \, cm^{-2}$ in the 2.2 MeV line attributed to deuterium formation, plus evidence for others of the theoretically-predicted features: positron annihilation at 511 keV, inelastic-scattering lines of ^{12}C at 4.4 MeV and of ^{16}O at 6.1 MeV, plus high-energy continuum (see also Gruber *et al.*, 1973). With these data from the solar maximum of 1970, the field of solar γ-ray astronomy paused (observationally) for several years.

Further data then came from non-solar γ-ray instruments on board HEAO-1 (Hudson *et al.*, 1980) and HEAO-3 (Prince *et al.*, 1982). These instruments did not directly view the Sun, and the solar γ-radiation had to penetrate thick layers of anti-coincidence shielding designed to exclude it. Nevertheless, the HEAO-1 observations confirmed the existence of the solar γ-ray line emission and established that the high-energy particle acceleration occurred rapidly during the impulsive phase of a flare. The HEAO-1 instrument again consisted of Na I scintillation counters (Peterson, 1973), but the HEAO-3 instrument gave the much higher spectral resolution (a few keV) of a germanium solid-state spectrometer. There have been no further observations with this kind of instrumentation.

Systematic use of γ-ray lines as a tool for broader studies of solar flares really only began with the observation of many of them, using the spectrometer provided by the University of New Hampshire – Naval Research Laboratory – Max-Planck-Institut group on the Solar Maximum Mission (Chupp, 1984). This instrument, again a set of scintillation counters, contained approximately seven times the volume of detector as the earlier OSO-7 experiment, and viewed the Sun continuously rather than scanning across it. The successful observation of a large number of flares resulted, some thirty with γ-ray emission lines by the time of the Chupp (1984) review. The data have continued since the SMM launch in early 1980 up until at least the time of writing. Also during this maximum, a smaller scintillation counter experiment on board the *Hinotori* satellite has observed a number of γ-ray flares (Yoshimori *et al.*, 1983).

A representative γ-ray spectrum from a major solar flare appears in Figure 1, the total spectrum of the flare of 27 April, 1981, as observed by the Solar Maximum Mission (Chupp, 1984). It shows numerous emission lines (from C, N, O, Ne, Mg, Fe, plus the

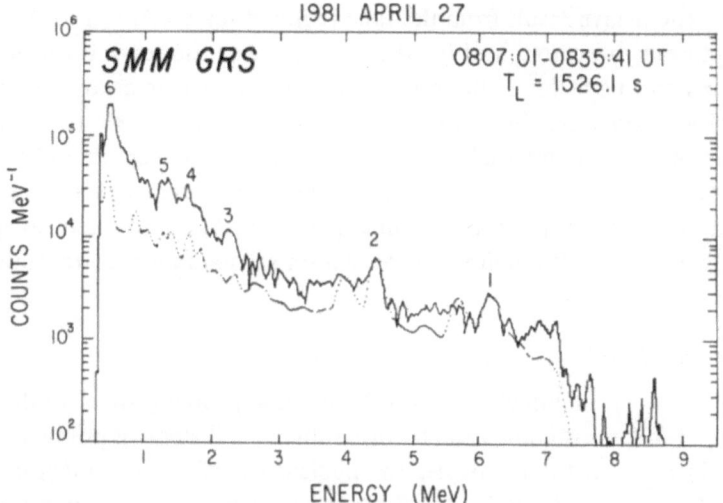

Fig. 1.   Integrated γ-ray spectrum for the flare of 27 April, 1981 (Chupp, 1984). The numbered features have the following identifications: (1) 6.129 MeV (^{16}O) (2) 4.439 MeV (^{12}C); (3) 2.313 MeV (^{14}N), 2.223 MeV (deuterium formation); (4) 1.634 MeV (^{20}Ne); (5) 1.369 MeV (^{24}Mg), 1.238 MeV (^{56}Fe); (6) 0.511 MeV (positron annihilation), ?0.478 MeV ($\alpha\alpha \rightarrow$ ^{7}Li), ?0.431 MeV ($\alpha\alpha \rightarrow$ ^{7}Be). Note that in this limb flare, the otherwise prominent 2.223 MeV line of deuterium formation does not appear strongly (see Figure 5).

delayed lines as described below) plus an intense continuum. The theoretical works cited above had predicted virtually all of the features of this spectrum, and we summarize this theoretical underpinning next before returning to a more detailed description of the data.

2.2. GAMMA-RAY EMISSION MECHANISMS

2.2.1. *Prompt Line Emission*

Prompt lines from target nuclei, prototypically the strong 4.43 MeV line of ^{12}C, result from direct collisional excitation (inelastic scattering) followed by spontaneous de-excitations of the target nucleus in the solar atmosphere. Typical cross-sections for these excitations have thresholds in the vicinity of several MeV primary kinetic energy (Ramaty *et al.*, 1975). The primary nuclei too may undergo excitation, in which case the resulting emission line, in the frame of an observer at rest, will have a Doppler shift corresponding to its motion. The so-called 'alpha-alpha' reactions result from the collision of a primary helium nucleus on a target helium nucleus, producing ^{7}Li and ^{7}Be lines in a process of non-thermal fusion. The de-excitation of the first excited states of the product nuclei have energies of 478 and 431 keV, respectively (Kozlovsky and Ramaty, 1974).

2.2.2. *Delayed Line Emission*

Two lines have appreciable delays (Lingenfelter and Ramaty, 1967): the 2.223 MeV line resulting from deuterium formation; and the 511 keV line resulting from positron annihilation. The delays result from the time required for a neutron to thermalize and undergo capture by a proton, and for the radioisotopes that emit positrons to decay. Both neutrons and positron-emitting radioisotopes appear as secondary products of primary nuclear interactions (Ramaty *et al.*, 1975). Emission in these lines may continue for hundreds of seconds, long after the prompt lines have become undetectable. The positron line and the alpha-alpha lines unfortunately lie too close together (in a portion of the spectrum with an intense continuum flare background) and so quantitative analysis of these lines will require high-resolution germanium spectroscopy.

2.2.3. *Bremsstrahlung*

Much of the γ-ray continuum above 300 keV, and also probably much of the continuum above about 7 MeV, must come from bremsstrahlung collisions of primary electrons on target ions. In between these energies the nuclear emission lines make a substantial contribution (Ramaty *et al.*, 1977). The same mechanism could extend to γ-ray energies if the primary electrons attain high enough energies. Suggestions that 'inner bremsstrahlung' caused by primary protons could explain the hard X-ray continuum (Boldt and Serlemitsos, 1969; Emslie and Brown, 1985) strongly disagree with the observed γ-ray lines fluxes (Hudson, 1973).

2.2.4. *Other Continua*

Continuum emission from secondary π mesons, either direct π^0 decay or as a result of bremsstrahlung from $\pi^\pm - \mu^\pm - e^\pm$ products, could contribute to the observed spectra, especially at higher energies. We have no direct means at present for distinguishing between these continua and direct bremsstrahlung from primary electrons, but theoretically a close relationship exists between the π production and the neutron and positron production.

When considering these mechanisms, one must always bear in mind the effects of the detector response and the 'geometry' of the observing circumstances. The ideal detector with truly proportional response for γ-radiation does not exist, so that the energy of the detected photon does not have a simple relationship to the amplitude of the resulting detector signal. The Na I scintillation counters produce considerable resolution spreading and some gain nonlinearity; in addition, a 'Compton continuum' of counts results if the detected photon only scatters out of the detector and does not meet a veto anticoincidence counter. A similar alteration of the energy of the photon can occur at the Sun if the photon scatters in the solar atmosphere before arriving at the detector.

The prompt γ-ray lines could come either from 'thick target' or 'thin target' interactions. In a thick target, the primary particles stop rapidly due to collision losses; in a thin target they have small energy losses and 'escape'. Thin target emission would result for example if the primary acceleration mechanism preferentially sent the energetic ions upwards into the corona into a low-density trap, or on open magnetic field lines into the solar wind.

2.3. Gamma-ray Fluxes

In the continuum, fluences range from 2–2000 γ cm^{-2} above 270 keV (Chupp, 1984). Rieger *et al.* (1983) observed eight events with spectra extending above 10 MeV, and found fluences in the range 0.8 to 51 γ cm^{-2}. This corresponds approximately to an E^{-1} integral spectrum for the strongest events, a spectrum as flat as the flattest spectra observed in the hard X-ray range (Hudson, 1978).

For the lines, no systematic tabulations of intensities as observed by the Solar Maximum Mission have yet appeared (but see Yoshimori *et al.*, 1983). Fluxes in the 4.43 MeV line range up to 0.18 ± 0.07 γ cm^{-2} s^{-1} (Hudson *et al.*, 1980) and down to the threshold of instrumental sensitivity, about 10^{-2} γ cm^{-2} s^{-1} for the SMM instrument – the limits strongly depend upon the continuum intensity and event duration. A broad 4–7 MeV band containing many emission lines provides a more practical observational mode for the prompt lines, and observed fluences range from about 1 to 100 γ cm^{-2} (Chupp, 1984); approximately one-third of this fluence comes from the 4.43 MeV line. A large fraction of the 4–7 MeV counting rate may also come from the underlying continuum; to determine the actual amount of continuum requires an uncertain interpolation between the observed continuum fluxes outside the region of the main nuclear γ-ray line emission.

Among the other lines, the deuterium-formation line stands out cleanly for disk flares because of its small line width, great strength – ranging up to 314 γ cm^{-2} for the

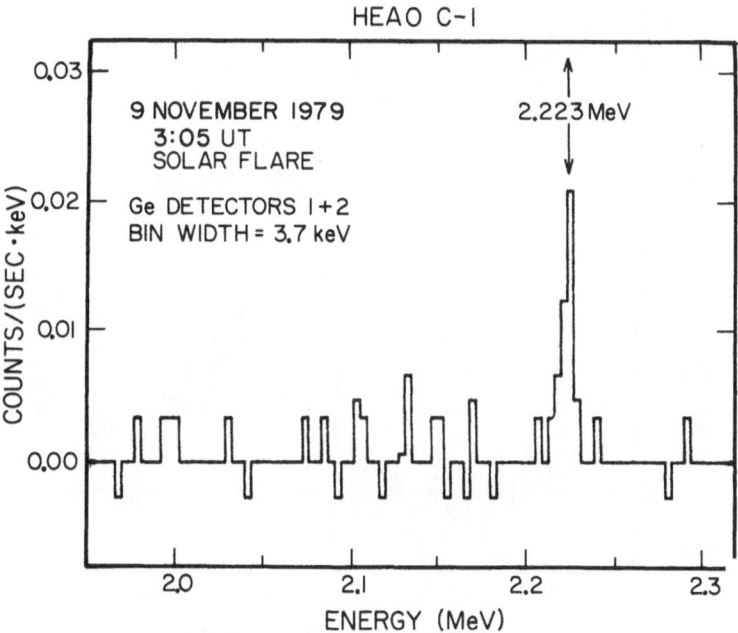

Fig. 2. Spectrum of the 2.2 MeV region obtained by the germanium spectrometer on board HEAO-3 (Prince *et al.*, 1982), for the period 3:04:20–3:06:30 UT, 9 November, 1979; subtraction of a background interval preceding the flare explains the negative counts.

strongest event recorded (Prince *et al.*, 1983) – and because of the lack of competition from the intense continuum found at lower energies. Figure 2 shows the observation of this line with the high-resolution Ge spectrometer on HEAO-3 (Prince *et al.*, 1982). These high-resolution observations give a line energy of 2.2248 ± 0.0010 MeV, in reasonable agreement with laboratory measurements.

The delayed 511 keV line may also produce a large number of γ-ray counts, sufficient to determine the time profile with some precision in major events (Share *et al.*, 1983). Share *et al.* found fluences of 14.6 ± 3.3 γ cm⁻² and 103 ± 8 γ cm⁻², respectively, for the two major flares of 21 June, 1980, and 3 June, 1982, noting that the limb location of the 21 June event may have reduced its 511 keV fluence. The time profile of emission at 511 keV generally resembles that predicted for radioactive decay (Ramaty *et al.*, 1975), but a mismatch at the detailed level exists for the 3 June, 1982 event (Share *et al.*, 1983; Murphy and Ramaty, 1985).

2.4. NEUTRON OBSERVATIONS

The initial reports of direct detection of solar neutrons (Chupp *et al.*, 1982), for the flare of 21 June, 1980, used data from the γ-ray spectrometer on board the Solar Maximum

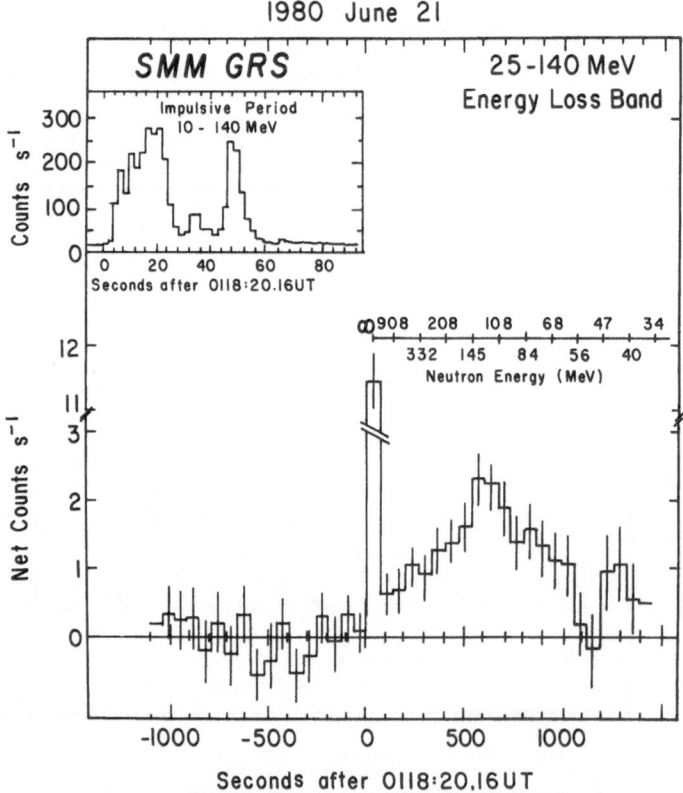

Fig. 3.   Arrival time history for the 25–140 MeV pulse-height range observed by the γ-ray spectrometer on SMM (Chupp *et al.*, 1982). The delayed peak after the impulsive emission (10–140 MeV shown in inset) corresponds to the time of flight of the neutron energies shown on the scale. Flare of 21 June, 1980.

Mission. Although this instrument does not give an unambiguous means for distinguishing counts due to neutrons from counts due to γ-rays, the timing of the counting rates for pulse heights > 25 MeV showed the existence of a delayed component as well as the prompt component associated with the flare itself. Lingenfelter *et al.* (1965) had predicted these delays, which can range up to a few hundred seconds as shown in Figure 3, as a result of the finite time of flight of neutrons from the Sun to the spacecraft.

In one of the most exciting chapters of flare research in the past solar maximum, this suggestion of the direct detection of neutrons at the Earth received confirmations from several sources: Evenson *et al.* (1983) found a flux of protons following the flare of 3 June, 1982, in observing conditions such that the protons must have originated in the decay of neutrons promptly emitted by the flare (Figure 4). Using the Roelof (1966)

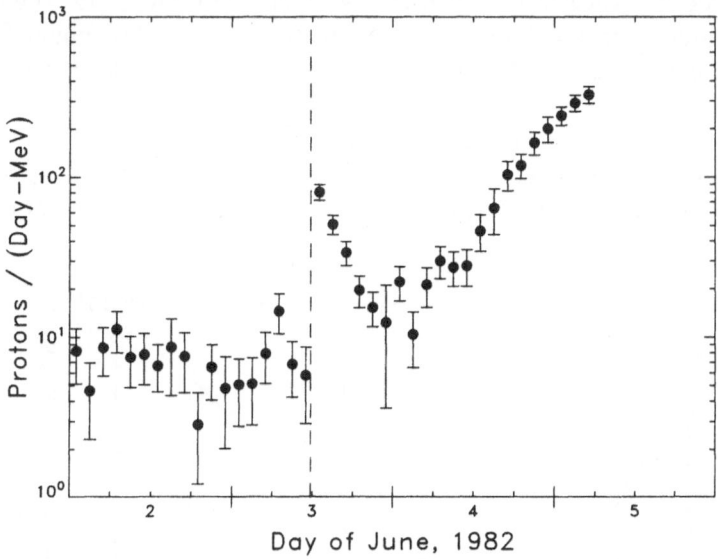

Fig. 4.   Observation by Evenson *et al.* (1983) of the decay protons resulting from neutron emission in the γ-ray flare of 3 June, 1982. The slower onset of direct protons (seen commencing on 4 June) results from the flare location at E82; the neutrons, cutting across the magnetic-field lines, had direct access to the field lines leading to the ISEE-3 instrument. The dashed line gives the epoch of γ-ray emission.

propagation theory, Evenson *et al.* found good agreement between the inferred neutron fluxes and the γ-ray emission level via the calculations of Lingenfelter and Ramaty (1967), concluding that the flare neutron emission occurred essentially isotropically. Further, the spectrum of the decay protons gives a good match to that expected (Lingenfelter and Ramaty, 1967) from a primary proton spectrum exponential in rigidity (see the discussion in the following section). Further striking confirmation of the solar neutron emission from this flare came from ground-based neutron-monitor observations at Jungfraujoch, Lomnický Štít, and probably Rome (Debrunner *et al.*, 1983). The three independent detections of the neutrons, using three very different techniques, give rather good agreement, as pointed out by Murphy and Ramaty (1985).

2.5. INFERENCE OF THE PRIMARY PARTICLE SPECTRUM

Since the 2.223 MeV line requires the production of neutrons from evaporation and spallation reactions, it essentially samples higher-energy primary particles than the inelastic-scattering lines. Thus the ratio of line fluences (total 4–7 MeV fluence to 2.223 MeV line fluence) gives a measure of the primary ion spectrum. Ramaty *et al.* (1983) show that this ratio varies approximately inversely with the αT product characterizing the Bessel-function spectra. Observationally, this line ratio turns out to vary only within observational errors from flare to flare. Ramaty *et al.* (1983) find the ratio to vary from 0.68 to 1.74, including a total of seven events from four different experiments but excluding the limb events. For the limb flares the deuterium-formation line shows a clear deficiency (pointed out by Wang and Ramaty, 1974) relative to the inelastic-scattering lines (Chupp, 1982; Yoshimori *et al.*, 1983). Figure 5 shows the quality of this correlation between the fluences of the prompt lines and the delayed 2.223 MeV line.

More generally, the flux in a given line or continuum represents an integral of the production cross-section over the particle spectrum and over the geometry of the source. In principle the observations of a number of components – each representing a different

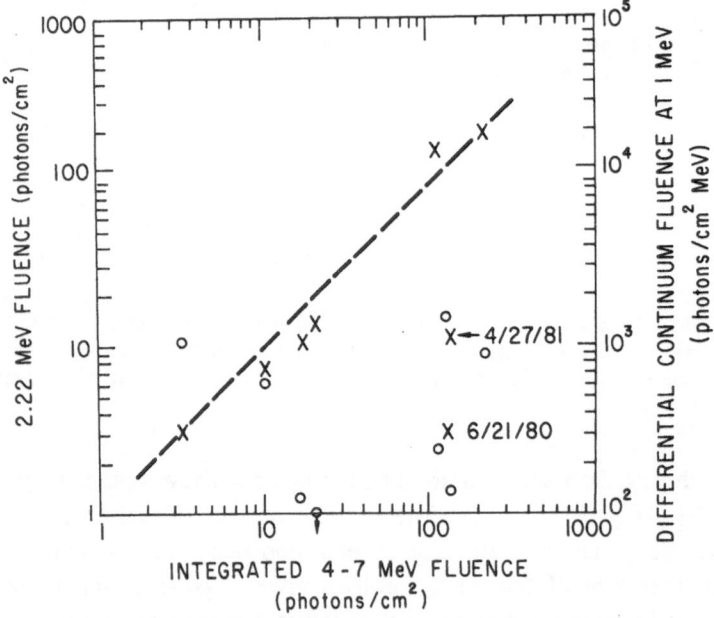

Fig. 5.   Correlation of 2.22 MeV fluence to 4–7 MeV fluence (× symbols), showing the 2.22 MeV deficit for the limb flares of 27 April, 1981, and 21 June, 1980. Also on this plot, the continuum fluence at one MeV (O symbols) appears to show no correlation with the 4–7 MeV fluence, suggesting an independent variability of the continuum and the nuclear emission lines (Chupp, 1982).

integral over the unknown spectrum and structure – constitute an 'inverse problem' for the inference of those unknowns. In practice Nature has given us such a restricted set of data that we cannot solve the inverse problem and must instead use a set of model calculations followed by comparison of the model data with the real data. If we consider the prompt lines alone, for example, to obtain the spectrum of the primary energetic particles we have an 'ill-posed' inverse problem because the γ-ray production cross-sections all have thresholds in roughly the same energy range. The 2.223 MeV line samples higher primary energies, as shown by the quantitative results of the theoretical calculations mentioned in Ramaty *et al.* (1983). With the addition of the 2.223 MeV line, however, one still has essentially only two points on the primary spectrum – one at about 10 MeV from the prompt lines, and one at about 50 MeV from the secondary neutron production resulting in the 2.223 MeV line. The constancy of the ratio (4–7 MeV)/2.22 MeV from flare to flare thus implies that a two-parameter spectrum can fit all of the spectra with the same values for the parameters, but does not help very much in determining which two parameters.

Murphy and Ramaty (1985) have added the neutron observations to such model calculations for inferring the primary spectrum of the flare of 3 June, 1982, for which the interplanetary and terrestrial observations mentioned above supplemented the direct neutron counting (Forrest, 1983) by the γ-ray spectrometer on board the Solar Maximum Mission. The data clearly prefer (Figure 6) a spectrum that falls off at higher energies more rapidly than a simple power law. This constitutes good evidence for a function that is exponential in rigidity:

$$N(E) \sim \exp(-P/P_0) \text{ MeV}^{-1},$$

as inferred also by Evenson *et al.* (1983) from the neutron spectrum. They found for the 3 June, 1982, flare, a best value of $P_0 = 125$ MV. The limits on π_0-decay γ-ray continuum also require the primary spectrum to turn downwards at high energies, as pointed out by Ramaty *et al.* (1975) and Rieger *et al.* (1983).

Theorists have frequently used a similar Bessel function spectral fit expected in a particular model of stochastic acceleration (Ramaty, 1979):

$$N(E) \sim K_2[12p/(Mc\alpha T)^{0.5}] \text{ MeV}^{-1},$$

where K_2 represents a modified Bessel function, with p and M the particle momentum and mass, respectively. Forman *et al.* (1982) give an approximation of this formula that one can easily calculate. If the theory predicting the Bessel-function form fails somehow, the product αT then becomes a parameter characterizing the spectrum above as an interpolation formula. For eight major disk flares, Murphy and Ramaty find αT values in the range 0.018–0.034, corresponding to best-fit power-law indices in the range 3.1–3.7 in the region of 30 MeV.

2.6. Number of primary particles and 'escaping' particles

For ten events, Bai *et al.* (1985) summarize estimates (Cliver *et al.*, 1983; Murphy and Ramaty, 1985) of the numbers of energetic protons (> 30 MeV), which we reproduce

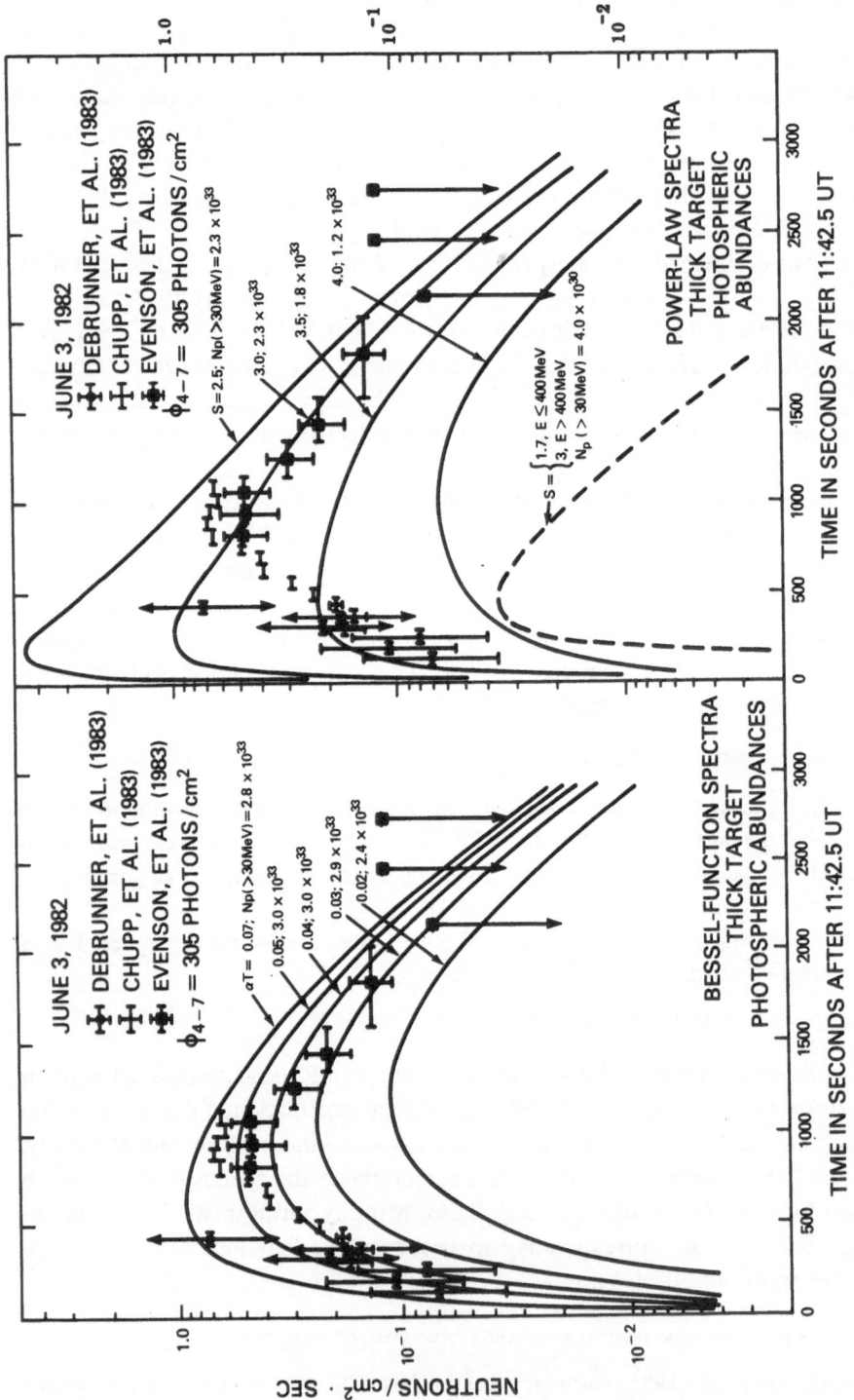

Fig. 6. Neutron time profiles for the 3 June, 1982 event, showing data from SMM (Chupp *et al.*, 1983), ISEE-3 proton observations (Evenson *et al.*, 1983), and ground-based neutron monitors (Debrunner *et al.*, 1983). The left profile shows a grid of model time profiles calculated for Bessel function primary spectra; the right panel shows power-law primary spectra. Murphy and Ramaty (1984) comment that for this event, a power-law spectrum fit the interplanetary data better, and this spectrum might also explain an apparent discrepancy noted in the time profile of the 511 keV line for this event.

TABLE I

Numbers of energetic protons (Bai *et al.*, 1985)

Flare Date	Location	Flare protons	IP protons	Impulsive or gradual?
4 Aug., 1972	N 14 E 08	1.0(33)	4.3(34)	Gradual
11 July, 1978	N 18 E 43	1.6(33)	?	Gradual
9 Nov., 1979	S 12 W 02	3.6(32)	none	Impulsive
7 June, 1980	N 12 W 74	9.3(31)	8.0(29)	Impulsive
1 July, 1980	S 12 W 38	2.8(31)	<4.0(28)	Impulsive
6 Nov., 1980	S 13 E 70	1.3(32)	3.0(29)	Impulsive
10 Apr., 1981	N 07 W 36	1.4(32)	7.0(31)	Gradual
3 June, 1982	S 09 E 72	2.9(33)	3.6(32)	see Bai *et al.* (1985)
21 June, 1980	N 19 W 90	7.3(32)	1.5(31)	see Bai *et al.* (1985)
9 Dec., 1981	N 12 W 16	<2.0(31)	1.0(32)	Gradual

as Table I. The data compare the inferred proton number at the Sun, assuming the Bessel-function spectral distribution, with the proton number in interplanetary space. As noted by others (e.g., Cliver *et al.*, 1983), little correlation exists between these numbers, whose ratio (interplanetary/solar) ranges from $< 1.4 \times 10^{-3}$ (for a W38 flare) to 43 (the 4 August, 1972 flare). Bai *et al.*, however, note that the large values of this ratio all correspond to *gradual* γ-ray line flares, a category associated with microwave richness, LDE soft X-ray events, type-II radio bursts, and coronal mass ejections (Kahler *et al.*, 1978). This distinction appears to clarify the relationship between the interplanetary particle population and the population in the flare itself: the impulsive γ-ray line flares do not allow protons to escape from the acceleration region near the Sun, whereas the gradual ones do.

2.7. INTERPRETATION OF THE TIMING DATA

The relative timing of the different components of the solar γ-ray emission tells us a great deal about the origin of these different components. The γ-ray emission lines, except for the delayed deuterium-formation and positron-annihilation lines, appear immediately during the impulsive phase. These 'prompt' lines, although present near the beginning of the flare, show a significant delay with respect to the 10–100 keV hard γ-rays (Bai, 1982b). This delay also appears in few hundred keV X-radiation (Hoyng *et al.*, 1976; Bai and Ramaty, 1979; Bai *et al.*, 1985), implying that the high-energy ions and relativistic electrons get accelerated more gradually, in a 'second step', than the primary impulsive-phase acceleration. The delays of hard X-rays range from 0.25 to 120 s (Bai, 1985). The second-step process occurs in conjunction with long-wavelength radio emission of type II and type IV, high microwave richness, large area of Hα emission (cf. Dwivedi *et al.*, 1984), and other characteristics distinguishing the 'extended' events (Hoyng *et al.*, 1976).

Figure 7 shows the delay in a well-known case (21 June, 1980) of impulsive γ-ray emission (Forrest and Chupp, 1983). Although Bai (1985) find no hard X-ray delay in this event, a pronounced delay in 4.1–6.4 MeV γ-radiation occurred, and Nakajima

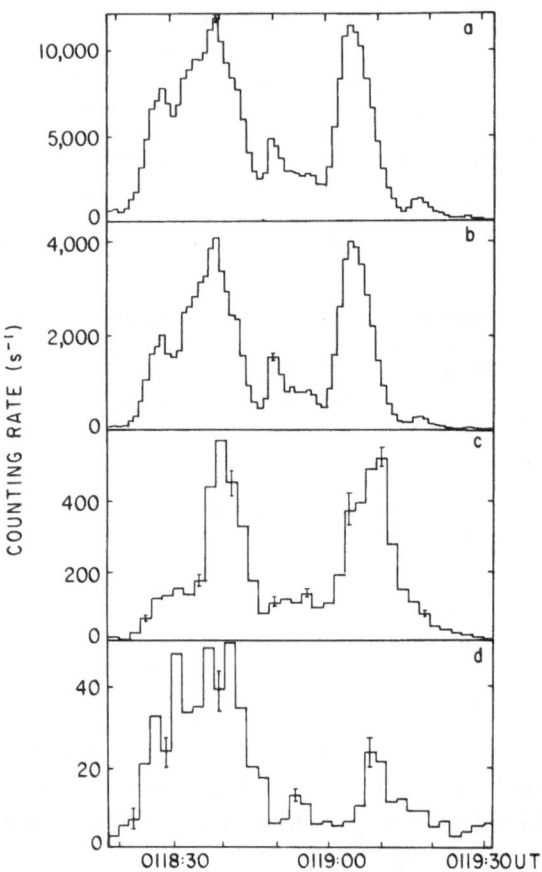

Fig. 7. Counting rates from the impulsive γ-ray flare of 21 June, 1980 (Forrest and Chupp, 1983), in (top to bottom) 40–80 keV, 80–140 keV, 4.1–6.4 MeV, and > 25 MeV channels of the Solar Maximum Mission γ-ray spectrometer. The 4.1–6.4 MeV channel clearly lags behind the hard X-ray time profile, leading Nakajima *et al.* (1983) to suggest the existence of successive accelerations within the impulsive phase of this event.

et al. (1983) describe this event as a good example of the successive acceleration of electrons and ions.

2.8. INTERPRETATION OF FLARE LOCATIONS

The distribution on the solar disk of flares seen at energies > 10 MeV shows a remarkable deviation from uniformity (Figure 8); all of the first fourteen cases occurred at solar longitudes > 60° from the central meridian. No agreed-upon explanation of this remarkable distribution has yet appeared, but clearly one possibility lies in exploiting the anisotropic bremsstrahlung emission of relativistic electrons in directional beams (Dermer and Ramaty, 1985; Petrosian, 1985). The nature of these beams remains obscure, however; quite possibly they result from the mirroring behaviour of the particles in the dense regions of the footpoints.

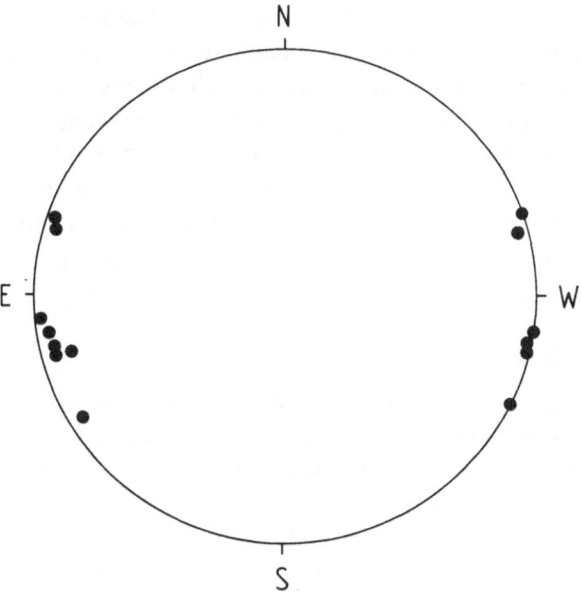

Fig. 8.   Solar locations of fourteen flares detected at > 10 MeV (Canfield *et al.*, 1985). This non-uniform distribution strongly suggests that the primary particles – probably relativistic electrons producing bremsstrahlung emission – have anisotropic distributions in the emission region.

3. Particle Transport in Flare Structures

The diagnostic utility of the γ-ray and neutron observations will increase if we can learn about the structure of the sources, or *vice versa*, the data may give us clues regarding the flare structure. Models assuming homogeneous and uniform sources may still work well enough for some purposes, but the need for thick-target calculations and the anomalous γ-ray spectra of limb flares, not to mention the apparent anisotropy for > 10 MeV, all motivate us to study the transport of particles within the flare structures. The solar flare structures that we can observe clearly consist of closed arch-like magnetic flux tubes, most beautifully seen in the Hα loop prominence systems following a major two-ribbon flare with its long-duration soft X-ray source (Kahler, 1977). When a flare occurs, the gas pressure on one of these magnetic flux tubes increases dramatically, rising perhaps from 0.1 dyne cm^{-2} to as much as 1000 dyne cm^{-2} (Machado and Linsky, 1975).

These closed-field structures have an uncertain relationship with open field lines that connect out into the solar wind and lead flare particles to Earth. We do not know to what extent we can identify these two particle populations – the one observed in interplanetary space and the other in the flare proper. We can liken the magnetic structures of an active region to a mound of slippery spaghetti on a plate – the stability of the heap requires uniformity of magnetic pressure on scales larger than those of the individual flux tubes that we see undergoing their dramatic pressure evolution during the flare energy release. How much magnetic flux passes near a flaring flux tube and then connects out into the solar wind (how many strands of spaghetti trail off the plate)? We do not yet know, but the energetic particles will provide tracers of these structures, as they do in the aurora, if we can understand their behaviour.

In the following sections we review the basic Coulomb scattering theory for energy losses of fast ions. Then a discussion of the theory of adiabatic motion of energetic ions follows. To make the discussion concrete, we consider the possibility of energy storage (Elliott, 1973) for post-flare loops as the result of trapped protons accelerated in the impulsive phase (Jefferies and Orrall, 1965; Hudson, 1985).

3.1. COULOMB SCATTERING

Coulomb scattering describes the basic physics of charged-particle interactions (e.g., Spitzer, 1962, and Ginzburg and Syrovatskii, 1964). The electrostatic force between two particles, one considered as a test particle and the other as a member of the background plasma, will change the energy and momentum of the test particle relative to the background. To apply the Coulomb collision theory, we must specify the velocity distribution functions of the test particles and field particles. In general, flare energy release should result in non-Maxwellian, anisotropic distributions. To avoid this complication, we assume here that the ion and electron components have separately relaxed to Maxwellian distributions characterized by temperatures T_p and T_e, respectively. We further assume that the ions are protons. From Spitzer,

$$\mathrm{d}T_p/\mathrm{d}t = (T_e - T_p)/\tau_{pe};$$

note that in the case of $T_p > T_e$ the test protons may actually have smaller velocities than the field electrons;

$$\tau_{pe} = 5.87 A_p A_e / n_e \ln \Lambda \, (T_p/A_p + T_e/A_e)^{3/2},$$

which for the assumed electron-proton plasma reduces to

$$\tau_{pe} = 12.6 T_e^{3/2}/n_e;$$

note that the equilibration depends solely upon the temperature and density of the electron distribution. This result holds for the assumption that 'the mean kinetic energies of electrons and protons are of the same order of magnitude' (Spitzer, 1962). For a typical soft X-ray temperature of 10^7 K and a density of 10^{10} cm^{-3}, we have $\tau = 39.8$ s.

For proton energies exceeding about $(M/m)kT$ (1.57 MeV at $T_e = 10^7$ K), the time-scale begins to increase approximately as

$$\tau_{pe} = 2.0 \times 10^{11} E_p^{3/2}/n_e,$$

with the proton energy E in MeV. For example, at 10 MeV in a density 10^{10} cm^{-3}, we have $\tau = 632$ s.

3.2. GEOMETRY, GUIDING-CENTER MOTION, AND TRAPPING

The geometry of the magnetic structures of solar flares governs the motion of the charged particles. To a first approximation, the ions gyrate in tight spirals around the magnetic field lines. The identification of a particle with a given field line, the guiding-center approximation (e.g., Northrop, 1963; and Hess, 1965) remains valid as long as the stresses such as field-line curvature and intensity gradients remain small. In the Earth's magnetosphere, for example, the field remains well-enough behaved so that three 'adiabatic invariants' exist: the gyration period of a particle around the field line, the bouncing motion of a particle between the magnetic mirrors at either end of the flux tube, and the azimuthal motion of the particle around the Earth in its essentially dipole field geometry. At present, unfortunately, we have little knowledge of the detailed structure of the coronal magnetic field, but clearly the third invariant has no place in solar magnetic structures because of their lack of azimuthal symmetry around the Sun. This means that particles will drift out of the flux tubes in which they initially appear, provided that they have not precipitated in the meanwhile.

To conserve the first adiabatic invariant of motion, the magnetic moment of the particle orbit, a particle will execute a bouncing motion between magnetic mirrors at the foot points of the flux tube, assuming the existence of a mirroring geometry. The bounce time has been estimated by Zweibel and Haber (1983) for a particular configuration as

$$\tau_b = 0.7 L_9 \, E_{\mathrm{MeV}}^{-0.5} [1/\mu_0 + f(\mu)],$$

where L_9 represents the half-length of the loop in units of 10^9 cm. The factor $f(\mu)$ allows for a further model dependence (magnetic geometry) on the pitch-angle cosine μ. As one can see from this estimation, bounce times for mirroring 1–100 MeV protons would typically be less than about one second.

As the particles bounce, they will drift across the field lines as their 'guiding centers' move in response to non-adiabatic forces. We do not know enough about the field distribution to estimate the gradient drift, but we can estimate the motion caused by macroscopic field-line curvature:

$$v_d = (mc/eB) \, v_{\parallel}^2 / R.$$

For parameters typical of soft X-ray loop structures of the gradual phase of a major two-ribbon flare, magnetic intensity $B = 100$ G; field-line radius of curvature $R = 3 \times 10^9$ cm; proton energy $E = 20$ MeV, we find $v_d = 10^4$ cm s^{-1}. For a flux-tube radius $0.1 R$, this gives a crossing time and, hence, an estimate of the 'escape' time, of 3×10^4 s, not far from the observed cooling time-scale of such post-flare loops. Note that if the loop lies in a vertical plane, the curvature drift leads the particle horizontally across the field lines. At present we know essentially nothing about the field structure outside the flaring loop, except that a comparable field intensity must exist there to maintain the mechanical equilibrium of the flaring loop.

3.3. COMPLICATIONS

The discussion above describes the basic minimum physics of the interactions between differentially heated electrons and protons in coronal magnetic flux tubes. Beyond this physics, what other processes could play roles important to the question of energy storage? We list several competing mechanisms:

(i) Scattering: the trapped protons may escape by pitch-angle scattering into the loss cone. This scattering could occur as a result of Coulomb collisions or 'anomalously' from interactions with waves (e.g., Zweibel and Haber, 1983; Meerson and Rogachevskii, 1983).

(ii) Charge exchange: low-energy protons can neutralize by picking up a free electron, thus permitting them to cross magnetic field lines freely – subject to photoionization – and thus to escape from a trap.

(iii) Filamentation: if the post-flare loops contain small-scale filaments, undetectable at present resolution, the higher density in the filaments would shorten the proton Coulomb lifetime.

These alternative processes would in each case reduce the storage time of the protons, making their energy less likely as an explanation for the observed late-phase heating in solar flares. From a theoretical standpoint it may seem extremely likely that pitch-angle scattering from ion-cyclotron waves, driven by the anisotropy of the loss-cone distribution of the mirroring trapped particles, would occur on time-scales much shorter than the Coulomb energy losses (Meerson and Rogachevskii, 1983), as we estimate below. Nevertheless the instability calculations are model-dependent and it is worthwhile to consider the Coulomb scattering alone, since this mechanism certainly does operate at all times (Bai, 1982b).

Meerson and Rogachevskii argue that the loss-cone distribution of trapped ions will lead to unstable growth of ion cyclotron waves, analogous to the Kennel–Petschek limitation for particle populations in the Van Allen belts. They find that this instability will occur for $\beta_h > \beta_*$, where β_h represents the ratio of energetic-particle isotropic pressure to magnetic pressure, and β_* represents a limiting value that depends upon the magnetic mirror ratio, particle energy, field intensity, and scale length of the trapping structure. Plausible values for post-flare loops lead to values on the order of 10^{-4} for β_*, lower than the value $\beta = 0.02$ inferred for a plasma with parameters as estimated by MacCombie and Rust (1979) in a 100 G field. Thus the Meerson and Rogachevskii analysis suggests that long-term trapping of an energetically significant population of ions cannot occur in post-flare loops.

3.4. ENERGY TRANSPORT AND SECONDARY EFFECTS

A proton accelerated to an energy of a few tens of MeV can penetrate from the corona deeply into the photosphere. This coincidence of the proton range and the column depth of the solar atmosphere led Švestka (1970) and Najita and Orrall (1970) to suggest that such protons could transport the energy needed to supply the intense visible continuum seen in the so-called white-light flares. Biswas and Radhakrishnan (1973) extended this discussion of flare heating by energetic protons, but no real test of the idea could occur until the γ-ray observations enabled us to tie down the actual particle fluxes more firmly. The γ-ray observations showed that, whereas ample energy existed in the energetic ions, their heating would predominantly excite the upper photosphere, rather than the vicinity of $\tau_{5000} = 1$ (Hudson and Dwivedi, 1982). This does not necessarily rule out ion transport for energizing the white-light flare, because we do not completely understand the physics of the white-light continuum formation. Canfield *et al.* (1985) provide a recent review of the observational and theoretical problems.

In addition to energy transport, a stream of energetic ions will transport momentum. Brown and Craig (1984) and McClymont and Canfield (1984) have discussed the relative importance of beam pressure, hydrostatic pressure, and magnetic pressure in the context of plausible particle beam intensities. One can imagine that accelerated particles could have important effects in heating and expansion of the flaring atmosphere, possibly related to the formation of the major large-scale and very energetic hydrodynamic effects of flares (e.g., Webb *et al.*, 1980). Most recently Tamres *et al.* (1985) conclude that the γ-ray producing ions, at a few tens of MeV, probably do not have important hydrodynamic effects in flares.

3.5. OTHER TECHNIQUES FOR REMOTELY SENSING HIGH-ENERGY IONS

One difficult point about understanding the role of protons in the vicinity of the flare itself has remained their essential undetectability: basically only those particles above minimum energies in the MeV range produce γ-ray spectral line emission, and even this has only generally become diagnostically useful for solar flares since the launch of the SMM in 1980. Other than theoretical inference (e.g., from the interplanetary particles) and from the γ-rays themselves, at present we have only limited ability for remotely sensing flare protons at low but non-thermal energies. Orrall and Zirker (1976) proposed the use of the wings of Lα, populated by charge-exchange collisions from low-energy non-thermal protons. The analogous use of Hα spectroscopy had led to the identification of proton aurora (Vegard, 1939). Recent theoretical work by Canfield and Chang (1985) suggests strongly that the spectroscopic technique will become very valuable as observational capability develops further. If this observational method materializes, it will provide a good opportunity to obtain high spatial resolution on the Sun by the techniques of optical and UV imaging. We should note that X-ray emission line profiles also give detailed information about ion velocities, especially for the bulk of the distribution.

4. Gamma-Ray Production in Flaring Magnetic Flux Tubes

The production of γ-rays and neutrons in closed magnetic flux tubes may follow the scenario described by Zweibel and Haber: the accelerated particles initially appear with some distribution in energy, pitch angle, and location. Under the reasonable hypothesis that the flux tube converges at its footpoints, magnetic mirrors will exist and at any given height and photon energy, a loss cone will, therefore, also exist. Particles at smaller pitch angles than the loss-cone opening angle will dump into the denser atmosphere at the footpoints, producing thick-target γ-rays and other secondaries. Particles outside the loss cone will tend to remain trapped, until they scatter into the loss cone, lose energy by collisions, drift onto field lines from which they can precipitate, or possibly escape into the outer corona or solar wind. In general, the escape fraction – defined as the ratio of the numbers of interplanetary and flare protons – remains small for impulsive γ-ray line events (e.g., Bai *et al.*, 1985). As long as the energetic ions can remain trapped in the flux tube, they will produce only thin-target γ-radiation and contribute to plasma heating via Coulomb collisions.

To compare the possible trapping time scales with the observations, we consider the flare of 21 August, 1973, for which MacCombie and Rust (1979) have given some of the best estimates of the physical conditions in the post-flare loops. Density 5×10^9 cm^{-3} and temperature decay time of 1.7×10^4 s lead to a proton energy of 43 MeV, well above the thresholds of the prompt γ-ray emission lines. If the total energy of these protons eventually converts into plasma heating, we can estimate the number of trapped protons from the energy of the soft X-ray source. Using the emission measure and density quoted by MacCombie and Rust, we find a total energy $W = 5 \times 10^{29}$ ergs. This requires aboqut 7×10^{33} protons at 43 MeV. For the 4.43 MeV γ-ray line of ^{12}C, Ramaty *et al.* (1979) give a cross-section of about 40 mb; assuming a carbon abundance of 1.6×10^{-4} relative to hydrogen, we would have a γ-ray flux of about 8×10^{-4} (cm^2 s)$^{-1}$. This lies below the faintest reported solar fluxes, so that the energy could be stored in protons and not detected via its γ-ray emission.

The gradual soft X-ray emission of a classical two-ribbon flare, such as the well-studied example of 29 July, 1973, may require replenishment of the plasma energy (e.g., Moore *et al.*, 1980). This requirement comes from the apparently anomalously long cooling time of the hot plasma. The argument is not absolutely compelling, but even if it were, the proton energy-storage mechanism could still work: for the example discussed above, only a fraction 2×10^{-5} of the total proton population needed to have been accelerated. A more serious constraint on ion energy storage comes from the threshold estimated by Meerson and Rogachevsky (1983) for the unstable growth of ion-cyclotron waves that would destroy the stability of the trapping, as discussed above. In addition, the mode of development of Hα spectral profiles in the expanding ribbons (Švestka *et al.*, 1980) argues against the transfer of energy via trapped energetic ions (Hudson, 1985).

5. Conclusions

High-energy particles in solar flares have become much better understood during the past solar maximum, thanks in large measure to the X-ray and γ-ray experiments on SMM and *Hinotori*. Due to the rapid increase in the number of flares observed at γ-ray energies, as well as the quality of the data, we can now make extensive comments about the behaviour of energetic ions in the actual lower-atmospheric structures of the flares. Since we have only very limited knowledge about coronal magnetic fields and about the connection between the flare particles and the interplanetary particles, any further insight gained from the propagation of non-thermal ions in this region will have great value. The outstanding problems involved in explaining and understanding the behaviour of energetic ions in solar flares center on the acceleration mechanisms, the problem of energy storage, the propagation of ions in the low corona, and the rather basic question of why some flares appear to accelerate high-energy ions, whereas others apparently do not.

To solve these problems will inevitably require new observations, since the observational investigation of high-energy phenomena in solar flares has only just begun: for γ-radiation, for example, essentially all of the data have come from whole-Sun scintillation-counter spectrometers, a technology devised in the 1950's. The achievement of sensitive spectroscopic observations with much better energy resolution, presumably with Ge spectrometers, and spatial resolution (source imaging) should rank as our first priorities. Imaging would be especially interesting if we could achieve angular resolutions of a few arc sec or better. The question raised by the non-uniform distribution of flares producing > 10 MeV radiation could be resolved by direct observations of anisotropic radiation, by comparing count-rates from different spacecraft (stereoscopic viewing) on the same event. Finally, a real exploitation of the possibility of using Doppler-shifted optical or EUV emission lines should be encouraged, since this appears to provide our only best avenue for observing the low-energy end of the distribution of energetic ions.

Acknowledgements

This work was supported by NASA under grant NSG-7161. It is a pleasure to thank Richard Lingenfelter for critical commentary, and his office for providing access to literally all of the literature in the field.

References

Bai, T.: 1982a, in R. E. Lingenfelter, H. S. Hudson, and D. M. Worrall (eds.), *Gamma-Ray Transients and Related Astrophysical Phenomena*, American Institute of Physics, New York, p. 409.
Bai, T.: 1982b, *Astrophys. J.* **259**, 341.
Bai, T. and Ramaty, R.: 1979, *Astrophys. J.* **227**, 1072.
Bai, T., Kiplinger, A. L., and Dennis, B. R.: 1985, preprint CSSA-ASTRO–85–15.
Biswas, S. B. and Radhakrishnan, S. V.: 1973, *Solar Phys.* **28**, 211.
Boldt, E. A. and Serlemitsos, P.: 1969, *Astrophys. J.* **157**, 557.
Brown, J. C. and Craig, I. J. D.: 1984, *Astron. Astrophys.* **130**, L5.
Canfield, R. C. and Chang, C.-R.: 1985, *Astrophys. J.* **295**, 215.
Canfield, R. C. and 15 co-authors: 1985, in *Proceedings of the SMM Workshop on Solar Flares* (to appear).
Chupp, E. A.: 1982, in R. E. Lingenfelter, H. S. Hudson, and D. M. Worrall (eds.), *Gamma-Ray Transients and Related Astrophysical Phenomena*, American Institute of Physics, New York, p. 363.
Chupp, E. A.: 1984, *Ann. Rev. Astron. Astrophys.* **22**, 359.
Chupp, E. A., Forrest, D. J., Heslin, J., Kanbach, G., Pinkau, K., Kanbach, G., Rieger, E., and Share, G. H.: 1982, *Astrophys. J.* **263**, L95.
Chupp, E. A., Forrest, D. J., Higbie, P. R., Suri, A. N., and Tsai, C.: 1973, *Nature* **241**, 333.
Cliver, E. W., Forrest, D. J., McGuire, R. E., and Von Rosenvinge, T. T.: 1983, *Proc. 18th Intern. Cosmic Ray Conf.* **10**, 342.
Debrunner, H., Fluckiger, E., Chupp, E. L., and Forrest, D. J.: 1983, *Proc. 18th Int. Cosmic Ray Conf.* **4**, 75.
Dermer, C. and Ramaty, R.: 1985, *Astrophys. J.* (submitted).
Dolan, J. F. and Fazio, G. G.: 1965, *Rev. Geophys.* **3**, 319.
Dwivedi, B. N., Hudson, H. S., Kane, S. R., and Švestka, Z.: 1984, *Solar Phys.* **90**, 331.
Elliott, H., 1973, in R. Ramaty and R. V. Stone (eds.), *High-Energy Phenomena in Solar Flares*, NASA SP-342, p. 12.
Emslie, A. G. and Brown, J. C.: 1985, *Astrophys. J.* **295**, 285.

Evenson, P., Meyer, P., and Pyle, K. R.: 1983, *Astrophys. J.* **274**, 875.

Forman, M. A., Ramaty, R., and Zweibel, E. G.: 1982, *The Acceleration and Propagation of Solar Flare Energetic Particles*, NASA TM-83989.

Forrest, D. J.: 1983, in M. L. Burns, A. K. Harding, and R. Ramaty (eds.), *Positron-Electron Pairs in Astrophysics*, American Institute of Physics, New York, p. 3.

Forrest, D. J. and Chupp, E. C.: 1983, *Nature* **305**, 291.

Ginzburg, V. C., and Syrovatskii, S. T., 1967, *The Origin of Cosmic Rays*. Macmillan, New York.

Giovanelli, R. G.: 1948, *Monthly Notices Roy. Astron. Soc.* **108**, 163.

Gruber, D. E., Peterson, L. E., and Vette, J. I.: 1973, in R. Ramaty and R. V. Stone (eds.), *High-Energy Phenomena in Solar Flares*, NASA SP-342, p. 147.

Hess, W. N.: 1965, in W. N. Hess (ed.), *Introduction to Space Science*, Gordon and Breach, New York, p. 165.

Hoyng, P., Brown, J. C., and van Beek, F.: 1976, *Solar Phys.* **48**, 197.

Hudson, H. S.: 1973, in R. Ramaty and R. G. Stone (eds.), *High-Energy Phenomena on the Sun*, NASA SP-342, p. 207.

Hudson, H. S.: 1978, *Astrophys. J.* **224**, 235.

Hudson, H. S.: 1985, *Proc. 19th Int. Cosmic Ray Conf.* **4**, 58.

Hudson, H. S. and Dwivedi, B.: 1982, *Solar Phys.* **76**, 45.

Hudson, H. S., Bai, T., Gruber, D., Matteson, J. L., and Nolan, P. L.: 1980, *Astrophys. J.* **236**, L91.

Jefferies, J. T. and Orrall, F. Q.: 1965, *Astrophys. J.* **141**, 519.

Kahler, S. W.: 1977, *Astrophys. J.* **214**, 891.

Kahler, S. W., Hildner, E., and van Hollebeke, M. A. I.: 1978, *Solar Phys.* **57**, 429.

Kozlovsky, B.-Z. and Ramaty, R.: 1974, *Astrophys. J.* **191**, L43.

Lin, R. P. and Hudson, H. S.: 1976, *Solar Phys.* **50**, 153.

Lingenfelter, R. E. and Ramaty, R.: 1967, in B. S. Shen (ed.), *High-Energy Nuclear Reactions in Astrophysics*. Benjamin, New York, p. 99.

Lingenfelter, R. E., Flamm, E. J., Canfield, E. H., and Kellman, S.: 1965, *J. Geophys. Res.* **70**, 4077 and 4087.

Lingenfelter, R. E., Ramaty, R., Murphy, R. J., and Kozlovsky, B.: 1985, in M. L. Cherry, K. Lande, and W. A. Fowler (eds.), *Solar Neutrinos and Neutrino Astronomy*. American Institute of Physics, New York, p. 121.

MacCombie, W. J. and Rust, D. M.: 1979, *Solar Phys.* **61**, 69.

Machado, M. and Linsky, J. L.: 1975, *Solar Phys.* **42**, 395.

McClymont, A. N. and Canfield, R. C.: 1984, *Astron. Astrophys.* **136**, L1.

Meerson, B. J. and Rogachevskii, I. V.: 1983, *Solar Phys.* **87**, 337.

Moore, R., McKenzie, D. L., Švestka, Z., Widing, K. G., and 12 co-authors: 1985, in P. A. Sturrock (ed.), *Solar Flares*, University of Colorado, Boulder, p. 341.

Najita, K. and Orrall, F. Q.: 1970, *Solar Phys.* **15**, 176.

Nakajima, H., Kosugi, T., Kai, K., and Enome, S.: 1983, *Nature* **305**, 292.

Murphy, R. J. and Ramaty, R.: 1985, *Adv. Space Res.* **4**, No. 7, p. 127.

Northrop, T. G.: 1963, *The Adiabatic Motion of Charged Particles*, Interscience, New York.

Orrall, F. B. and Zirker, J.: 1976, *Astrophys. J.* **208**, 618.

Peterson, L. E.: 1973, *Ann. Revs. Astron. Astrophys.* **13**, 423.

Peterson, L. E. and Winckler, J. R.: 1959, *J. Geophys. Res.* **68**, 979.

Petrosian, V.: 1985, *Astrophys. J.* **299** (to appear).

Prince, T. A., Ling, J. C., Mahoney, W. A., Riegler, G. R., and Jacobson, A. S.: 1982, *Astrophys. J.* **255**, L81.

Prince, T. A., Forrest, D. J., Chupp, E. L., Kanbach, G., and Share, G. H.: 1983, *Proc. 18th Int. Cosmic Ray Conf.* **4**, 79.

Ramaty, R.: 1979, in J. Arons, C. McKee, and C. Max (eds.), *Particle Acceleration Mechanisms in Astrophysics*, American Institute of Physics, New York, p. 135.

Ramaty, R., Kozlovsky, B., and Lingenfelter, R. E.: 1975, *Space Sci. Rev.* **18**, 341.

Ramaty, R., Kozlovsky, B., and Suri, A. N.: 1977, *Astrophys. J.* **214**, 617.

Ramaty, R., Kozlovsky, B., and Lingenfelter, R. E.: 1979, *Astrophys. J.* **400**, 487.

Ramaty, R., Murphy, R. J., Kozlovsky, B., and Lingenfelter, R. E.: 1983, *Solar Phys.* **86**, 395.

Rieger, E., Reppin, C., Kanbach, G., Forrest, D. J., Chupp, E. L., and Share, G. H.: 1983, *Proc. 18th Int. Cosmic Ray Conf.* **10**, 338.

Roelof, E. C.: 1966, *J. Geophys Res.* **71**, 1305.

Share, G. H., Chupp, E. L., Forrest, D. J., and Rieger, E.: 1983, in M. L. Burns, A. K. Harding, and R. Ramaty (eds.), *Positron-Electron Pairs in Astrophysics*, American Institute of Physics, New York, p. 15.

Spitzer, L. Jr.: 1962, *The Physics of Fully Ionized Gases*, Interscience, New York, Ch. 5.

Suri, A. N., Chupp, E. L., Forrest, D. J., and Reppin, C.: 1975, *Solar Phys.* **43**, 415.

Švestka, Z.: 1970, *Solar Phys.* **13**, 471.

Švestka, Z., Martin, S. F., and Kopp, R. A.: 1980, in M. Dryer and E. Tandberg-Hanssen (eds.), *Solar and Interplanetary Dynamics*, D. Reidel Publ. Co., Dordrecht, Holland, p. 217.

Tamres, D. H., Canfield, R. C., and McClymont, A. N.: 1985, *Astrophys. J.* (to appear).

Vegard, L.: 1939, *Nature* **144**, 1089.

Wang, H. T. and Ramaty, R.: 1974, *Astrophys. J.* **202**, 532.

Webb, D. F., Cheng, C.-C., Edberg, S. J., Martin, S. F., McKenna Lawlor, S., and McLean, D. J.: 1980, in P. A. Sturrock (ed.), *Solar Flares*, p. 471.

Yoshimori, M., Okudaira, K., Hirasima, Y., and Kondo, I., 1983, *Proc. 18th Int. Cosmic Ray Conf.* **4**, 89.

Zweibel, E. G. and Haber, D. A.: 1983, *Astrophys. J.* **264**, 648.

ENERGETIC SOLAR ELECTRONS IN THE INTERPLANETARY MEDIUM

R. P. LIN

Space Sciences Laboratory, University of California, Berkeley, CA 94720, U.S.A.

Abstract. ISEE-3 measurements extending down to 2 keV energy have provided a new perspective on energetic solar electrons in the interplanetary medium. Impulsive solar electron events are observed, on average, several times a day near solar maximum, with $\sim 40\%$ detected only below ~ 15 keV. The electron energy spectra have a nearly power-law shape extending smoothly down to 2 keV, indicating that the origin of these events is high in the corona. These coronal flare-like events often produced ^{3}He-rich particle events.

In large solar flares which accelerate electrons and ions to relativistic energies, the electron spectrum appears to be modified by a second acceleration which results in a double power-law shape above ~ 10 keV with a break near 100 keV and flattening from ~ 10–100 keV. Large flares result in long-lived (many days) streams of outflowing electrons which dominate the interplanetary fluxes at low energies. Even in the absence of solar activity, significant fluxes of low energy electrons flow out from the Sun.

Solar type-III radio bursts are produced by the escaping 2–10^2 keV electrons through a beam-plasma instability. The detailed ISEE-3 measurements show that electron plasma waves are generated by the bump-on-tail distribution created by the faster electrons running ahead of the slower ones. These plasma waves appear to be converted into radio emission by nonlinear wave-wave interactions.

1. Introduction

Non-relativistic electrons from the Sun were first observed about two decades ago (Van Allen and Krimigis, 1965; Anderson and Lin, 1966). Measurements since then have shown that these electrons are the particle type most frequently emitted by the Sun (see Lin, 1974 for review). The University of California, Berkeley experiment (Anderson *et al.*, 1978) on the ISEE-3 (International Sun Earth Explorer) spacecraft has provided significantly improved measurements of 20 keV to several hundred keV electrons in the interplanetary medium, and more importantly, has provided for the first time high-sensitivity measurements extending to 2 keV. Since the range of such low energy electrons is significantly less than or comparable to the integrated straight-line path through the corona, these measurements can provide information on the location of the source region for these particles. Furthermore, at these energies the propagation of the electrons appears to be nearly scatter free, so the observed temporal profiles tend to reflect the variations at the solar source of these particles.

The observations generally show that the solar corona is a dynamic and perhaps continuous source of electron acceleration. Impulsive acceleration of electrons occurs, on average, several times a day or more, with most of these events unaccompanied by Hα flares. A large fraction of the impulsive electron events is observed only at energies below ~ 15 keV, and almost all these events appear to originate high in the corona. These coronal flare-like events often also produce ^{3}He-rich particle events. On the other hand, the large flare events which produce > 10 MeV ion and relativistic electron acceleration, and which do not show ^{3}He enhancements, produce a significantly different shape of the electron spectrum. These large flares produce streams which last

Solar Physics **100** (1985) 537–561.   0038–0938/85.15.

for many days, during which the Sun appears to be a continuous source of low energy electrons. Interplanetary shocks and interplanetary type-III radio storms also produce long-lived streams.

Although the UCal particle experiment's period of operation in 1978–1979 was dominated by long-lived streams, there were a few periods of near constant quiet background. Even at these times, however, significant fluxes of low energy electrons were detected flowing out from the Sun.

Impulsive $2-10^2$ electron events produce solar type-III solar bursts. The detailed ISEE-3 measurements extending to 2 keV have provided new insight into the emission mechanism for solar type-III radio bursts. Here we review progress in the above areas, concentrating on the ISEE-3 measurements.

2. Impulsive $2-10^2$ keV Electron Events

Small non-relativistic solar electron events appear to be the most common type of impulsive particle emission from the Sun. Figure 1 shows an example of an impulsive solar electron event observed by ISEE-3. At the time of this event, the spacecraft was located in a halo orbit about the L1 Lagrange point about 10^6 km upstream of the Earth. This event is observed only at energies of 2–10 keV. ISEE-3 observations have shown that about 40% of the impulsive events (Table I) are observed only at energies below ~ 15 keV (Potter et al., 1980).

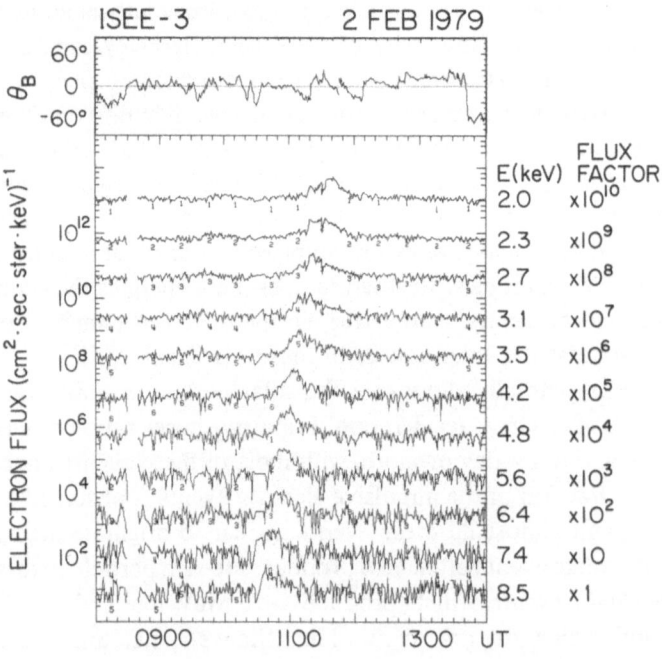

Fig. 1.   An impulsive solar electron event observed by the ISEE-3 spacecraft. The top panel shows the angle of the interplanetary magnetic field direction to the ecliptic plane. The detector views $\pm 25°$ from the ecliptic plane.

TABLE I

Impulsive electron events

	2–10 keV only	> 15 keV	Total
Number of events:	136	190	326
with interplanetary type-III radio burst (30 kHz to 2 MHz)[a]	73/94	154/163	227/257
with metric/decametric type-III radio burst[b]	13/68	77/113	90/181
with reported Hα flare[b]	3/62	59/93	62/155
with 10 cm microwave burst[c]	1/(90)	42/(148)	43/(238)
with hard X-ray burst[d]	4/62	41/91	45/153

[a] Comparison with ISEE-3 radio experiments, J. Fainberg and R. G. Stone, private communication.
[b] From *Solar Geophysical Data*.
[c] From *Solar Geophysical Data*; the coverage is uncertain.
[d] Comparison with ISEE-3 solar X-ray experiment, S. R. Kane, private communication.

Several important characteristics are illustrated by this electron event. There is clear velocity dispersion with the fastest electrons arriving earliest, consistent with the simultaneous injection of the electrons at the Sun and nearly scatter-free propagation along the interplanetary Archimedean spiral field line to 1 AU. The rapid rise and decay at all energies confirms the lack of significant scattering. The electron distributions are

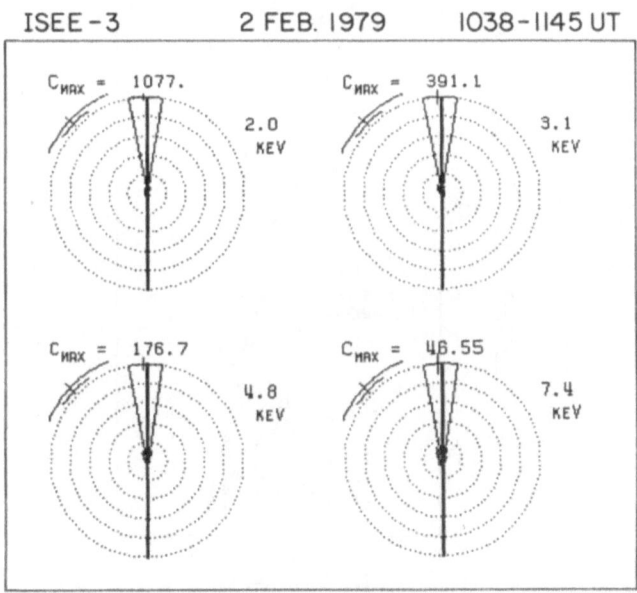

Fig. 2. 16-sector angular distribution of the electrons for the event of 2 February, 1979. Essentially all the counts fall into a single 22°.5 sector.

highly anisotropic and beamed along the magnetic field (Figure 2) so dips in the electron fluxes are observed when the magnetic field swings out of the field of view of the detector.

There appears to be a transition at ~ 15 keV in the way electrons propagate in the interplanetary medium. Below this energy, adiabatic effects, especially focusing due to the decrease in the interplanetary field magnitude with distance away from the Sun, essentially always dominate over scattering to give nearly scatter-free propagation and beam-like angular distributions (Anderson *et al.*, 1981). Above ~ 15 keV, the electrons often show the diffusive time profiles and more broad angular distributions characteristic of scattering-dominated propagation.

The electron energy spectra for this event and several others are shown in Figure 3. These electron spectra extend smoothly down to 2 keV, although there are some departures from a power law spectral shape. Since low-energy electrons have a short range in the lower corona, the extension of the spectra to low energies implies that the

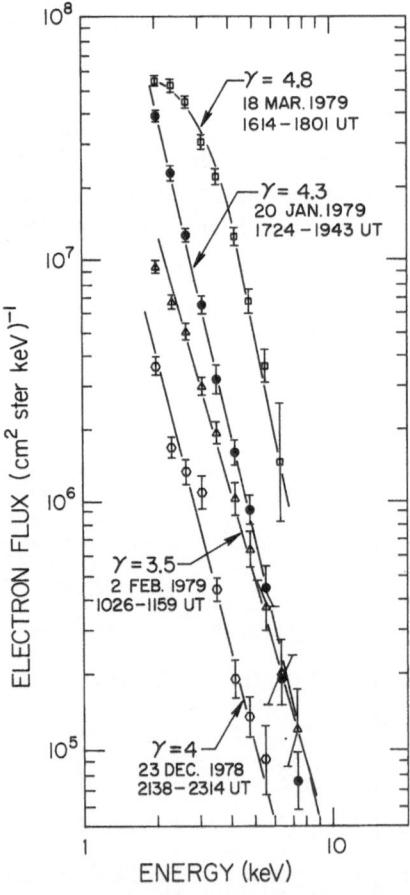

Fig. 3.   Time-integrated energy spectra for several electron events (from Potter *et al.*, 1980). The spectra are fit to a power law, $dJ/dE \approx E^{-\gamma}$.

acceleration occurs high in the corona. To compute the effect of traversing the corona, we have used a coronal density model derived from radio measurements (Fainberg and Stone, 1974), and Trubnikov's (1965) expression for the energy loss of electrons in a hydrogen plasma. Starting with a power-law electron spectrum $dJ/dE \approx E^{-\gamma}$ with $\gamma = 4$ for an electron source located at varying heights, and assuming straight radially outward paths, we obtain the resultant spectra at 1 AU (Figure 4). Comparison with the observed spectra indicates that the electrons most likely are accelerated at altitudes greater than $0.5 \, R_{\odot}$, although we cannot rule out the possibility that significant energy changes might have occurred for the electrons in their propagation from the Sun to 1 AU.

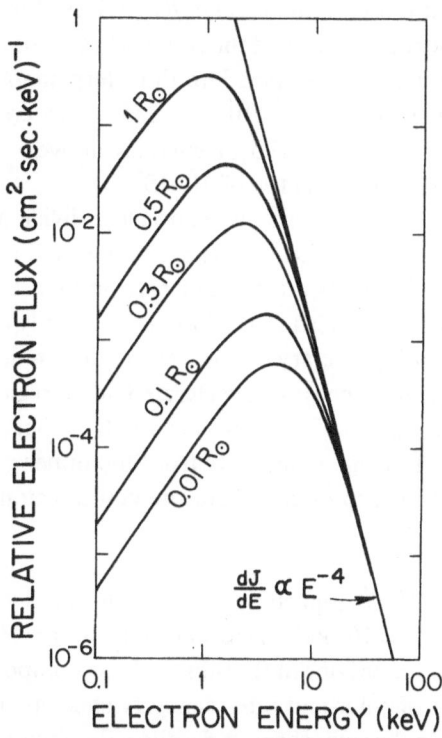

Fig. 4.   Computed electron spectra at 1 AU resulting from injection of a $dJ/dE \approx E^{-4}$ power-law spectrum at different heights in the solar atmosphere (see text for details).

A total of 326 impulsive electron events were observed by ISEE-3 in the ~ 15 month period from August, 1978 to November, 1979. These events appear to fill typically $\sim 60°$ in solar longitude in the interplanetary medium, so ~ 130 events/month must occur over the full Sun. Because the interplanetary medium is often filled with long-lived fluxes of electrons at these energies, small impulsive events will often not be detected. Thus the rate of occurrence of these events is probably significantly greater than estimated here.

About 40% of all the electron events (Table I) detected do not extend in energy above ~ 15 keV, but nearly all are observed to extend smoothly down to 2 keV, the limit of the measurement. Almost all impulsive electron events are accompanied by a solar type-III radio burst observed in the interplanetary medium, i.e., at frequencies of ~ 30 kHz to a few MHz. On the other hand, only $\sim 50\%$ are associated with metric/decametric type-III bursts and a minority (40%) appear to have an associated Hα flare or microwave bursts at ~ 3000 MHz (18%). Comparisons with solar hard X-ray observations show that only 29% were associated with a detectable hard X-ray burst. The associations with solar phenomena are even poorer for the electron events observed only at 2–10 keV energies. These poor associations are consistent with the picture of many of these impulsive electron events, particularly those limited to 2–10 keV events, originating at heights of ~ 0.5–$1\ R_\odot$, without associated chromospheric activity. We have, therefore, called these events *coronal flares*. Based on the observed FWHM duration at 1 AU, the electron injection at the Sun occurs on time-scales of on the order of a few minutes. The total energy released into the interplanetary medium in electrons above 2 keV is estimated to be $\sim 10^{25}$–10^{26} ergs for these coronal flares. For those events where the electrons are observed at energies above ~ 15 keV, these $\gtrsim 2$ keV energy figures could increase by a factor of 10–10^2.

From the interplanetary observations alone, no clear division can be made between the events seen only in the 2–10 keV energy range and events which extend to energies above 15 keV. The > 15 keV events are more intense and more closely associated with the chromospheric Hα flares and hard X-ray bursts. In a study of one event where both hard X-ray emission and interplanetary electrons were detected, Pan *et al.* (1984) found that a model with energetic electrons accelerated in a region extending from the chromosphere to high in the corona was consistent with the observations. In this model, the low-energy escaping electrons come only from the upper part of the region, while high-energy electrons are able to escape from the entire region.

2.1. ^{3}He-RICH EVENTS

Recently, Reames *et al.* (1985) reported that virtually all solar ^{3}He-rich events are associated with impulsive 2 to 10^2 keV electron events. Solar ^{3}He-rich events (reviewed by Ramaty *et al.*, 1980) represent one of the most striking composition anomalies among the observed populations of solar and interplanetary energetic particles, with ^{3}He/^{4}He ratios of order unity, while the typical ratios for the solar atmosphere or solar wind are a few times 10^{-4} (Geiss and Reeves, 1972). The solar wind ^{3}He/^{4}He ratio varies but always remains below 10^{-2}, and it is uncorrelated with the occurrence of solar ^{3}He-rich events (Coplan *et al.*, 1983). Other properties of ^{3}He-rich events include reduced ^{1}H/^{4}He ratios and enhanced abundances of Fe and other heavy elements, and a tendency to larger ^{3}He enhancements in smaller events.

With the high sensitivity of the ISEE-3 energetic particle telescopes of the Goddard Space Flight Center experiment, it is possible to obtain temporal profiles with good resolution ($\lesssim 1$ hr) for the ^{3}He fluxes at ~ 1 MeV nucl.$^{-1}$ energy for the small ^{3}He-rich events. Figure 5 shows the 17 May, 1979 event where velocity dispersion can be seen in the ^{3}He fluxes. In the lower panel the associated 2–10^2 keV impulsive electron event with its velocity dispersion can be seen. The timing of the particle onsets and maxima

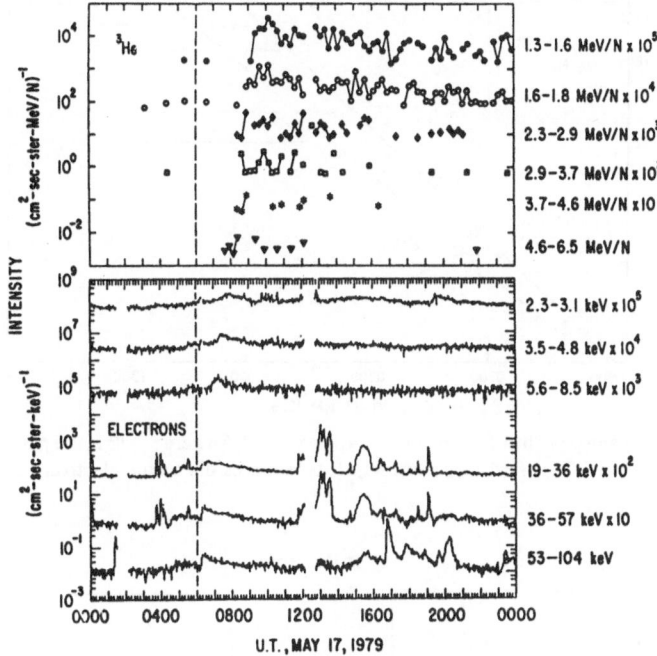

Fig. 5. Time histories of the intensities of ³He and electrons of the indicated energy during 17 May, 1979 (from Reames et al., 1985). The appearance of the plots differs partly because of the absence of continuous background fluxes for ³He. The lowest isolated points for each energy of ³He represent one particle per 15 min averaging period. The rapidly varying bursts (at ~04:00 UT and ~12:00 UT) observed in the higher, > 19 keV electron channels are due to particles from the Earth's bow shock. The 53–104 keV channel also responds to solar X-rays (~01:30, 16:30–20:30, and 23:30 UT). The dashed vertical line indicates the time of the type-III solar radio burst. Intensity scale factors are shown to the right of the energies.

are shown in Figure 6, where we have plotted those times versus $1/\beta$ (the particle velocity $v = \beta c$). If particles of all velocities were released at the same time t_0, and travelled the same distance L, their arrival times would be related by $t = t_0 + L/\beta c$, which would be a straight line on Figure 6. The observed onset time depends upon the relative instrument sensitivity, the rise time, and the event amplitude. The ³He onsets lag the time that would be expected from the electron measurements by no more than about 10–15 min. The traversal distance implied by the onset time of the lowest velocity ³He is ~1.3 AU. This compares with ~1.25 AU for the electrons. The times of maximum of ³He are also consistently later than those of electrons of the same velocity, implying a greater degree of scattering for ³He in the interplanetary medium. The mean path lengths implied by the ³He times of maximum are ~1.8–2.0 AU compared to ~1.45 AU for the electrons.

Energy spectra for the two species are shown in Figure 7. The spectral indices for the least-squares fits to a power-law are 2.7 ± 0.1 and 2.7 ± 0.3 for electrons and ³He, respectively. For electrons, only the region from 2.3 to 60 keV was fit because of

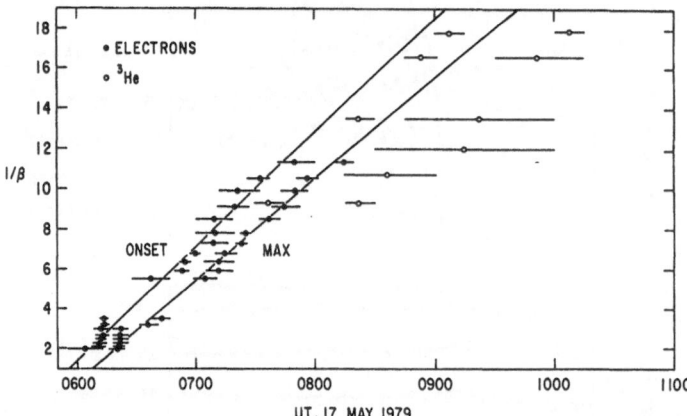

Fig. 6.   Time of onset and peak flux for particles plotted versus $1/\beta$ where $v = \beta c$ is the particle velocity (from Reames *et al.*, 1985). The lines represent least-squares fits to the electron data only.

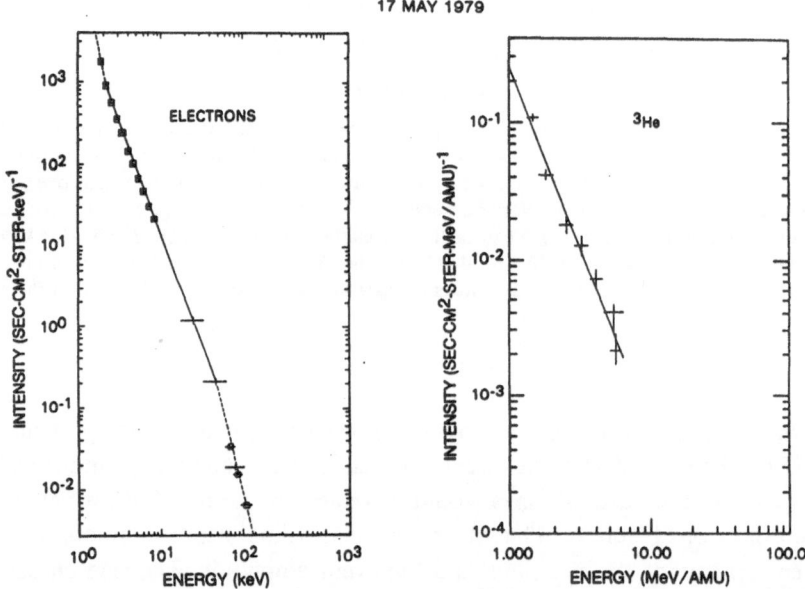

Fig. 7.   Event-averaged energy spectra for electrons and ^{3}He in the 17 May, 1979 event (from Reames *et al.*, 1985). Lines through the data are least-squares fit lines to power-law spectra with resulting spectral indices 2.7 ± 0.1 for electrons and 2.7 ± 0.3 for ^{3}He. The electron fit was confined to the 2.5 to 60 keV region.

apparent spectral changes at higher and lower energies. Note that ^{3}He and electrons have the same spectral slope over ranges of comparable particle velocities.

In the 15-month interval in 1978–1979, about a dozen ^{3}He-rich events were observed with sufficient statistics to provide good temporal profiles. All of these events were accompanied by an impulsive $2-10^{2}$ keV electrons with a temporal relationship similar to that shown in Figure 6. The solar associations for these ^{3}He-electron events are essentially the same as for electron events in general: close association with kilometric wavelength type-III bursts, small Hα subflares or no reported Hα association, very few hard X-ray bursts, and small or no soft X-ray bursts.

The close temporal associations and similar spectral slopes suggest that the same acceleration process may be responsible for both electron and ^{3}He acceleration. A preliminary search indicates that ions of a few hundred keV total energy may also be accelerated at the same time as the ~ 1 MeV/nucl. ^{3}He and the electrons. These low energy ion fluxes are usually difficult to associate with an impulsive injection at the Sun because the travel time from the Sun to 1 AU is long, $\gtrsim 6$ hr, and the flux increases for ^{3}He events are small. Furthermore, there are many variations in the low energy fluxes on these same time-scales, due to local interplanetary effects. Even so, for many ^{3}He-electron events, there are increases in the few hundred keV ion fluxes at the right time for releases simultaneous with the electrons. At these energies we have no information on the composition of the ions.

Thus, $2-10^2$ keV electrons, ~ 1 MeV/nucl. ^{3}He, and even lower energy ions may be accelerated together in coronal flare or flare-like events which occur at heights of $0.5-1 R_\odot$. This type of acceleration differs from large flare events, where ions are accelerated to > 10 MeV energies and electrons to relativistic energies, with no ^{3}He enrichment observed.

3. Long-Lived $2-10^2$ keV Electron Fluxes

The ISEE-3 spacecraft observes $2-10^2$ keV electron fluxes to be present in the interplanetary medium at all times. Figure 8 shows the daily minimum flux in three energy intervals: $2-2.3$ keV electrons, $19-45$ keV electrons and protons (but primarily electrons), and > 272 keV protons. This type of plot excludes the impulsive electron events discussed above since their time scales are typically less than a few hours. It can be seen that these particle fluxes are dominated by streams which last many days. Many of these streams arise from a large solar flare event or series of such events from a single active region (Anderson *et al.*, 1982). Others appear to be dominated by the passage of an interplanetary shock (Tsurutani and Lin, 1985). Still others, generally lower intensity ones, appear to be related to type-III radio burst storms observed at hectometric and kilometric wavelengths (Bougeret *et al.*, 1985).

3.1. LARGE SOLAR FLARES

Large solar flares are the most energetic natural particle accelerators found in the solar system, occasionally accelerating ions to tens of GeV and electrons to hundreds of MeV energies. Lin *et al.* (1982) presented IMP 6, 7, and 8 measurements of the energy spectrum of 20 keV to 20 MeV electrons observed from large solar flares (see Lin, 1974, and Simnett, 1974, for reviews of previous work). The aim of the study was to obtain information on the spectrum of the accelerated electrons at the Sun. To minimize propagation effects, only events from flares located at 30° to 90° W solar longitude were chosen for study. The energy spectra were constructed using the maximum flux observed

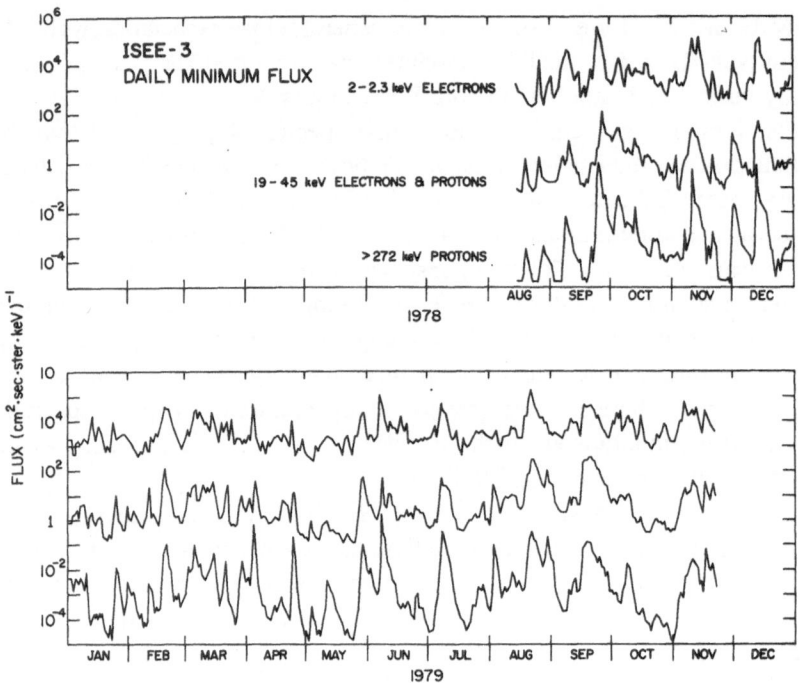

Fig. 8.   Plotted here at the minimum flux levels for each day for 2–2.3 keV electrons, 19–45 keV electrons and protons, and > 272 keV protons, for August, 1978 to November, 1979. The fluxes are dominated by long-lived streams, but even at the quietest times there are significant fluxes of electrons.

at each energy. These spectra were shown to be representative of the spectra of the electrons escaping from the Sun over this range of energies. Figure 9 shows the large solar flare event of 25 November, 1971. Terrestrial particle fluxes at energies $\lesssim 100$ keV are present from ~ 1030 onward but are not significant at the event maximum near 10:00 UT. Figure 10(a) shows the energy spectrum. A flat power law spectrum is observed below ~ 100 keV, changing to a much steeper power law at higher energies. Figure 10(b) shows the spectrum for the large event of 30 April, 1973, where the spectrum extends to ~ 20 MeV. In this case, the data at ≥ 10 MeV are a factor of more than 5 below an extrapolation of the 0.2 to 2.0 MeV spectrum, providing clear evidence for a steepening of the spectrum at very high energies. It was found that every event shows the same spectral shape: a double power law with a smooth transition around 100–200 keV and power law exponents of 0.6–2.0 below and 2.4–4.3 above. The more intense the event, the harder the spectrum observed. In some cases, the spectra are observed to steepen above 3 MeV.

Lin *et al.* (1982) also investigated whether the ions which are accelerated in the same large flares have a similar characteristic spectrum and whether that spectrum is related to that of the electrons. Van Hollebeke *et al.* (1975) found that proton spectra in the ~ 10–80 MeV range generally fit well to a power law. Figure 11 gives the best fit proton power-law spectral index in the > 10 MeV energy range versus the low energy

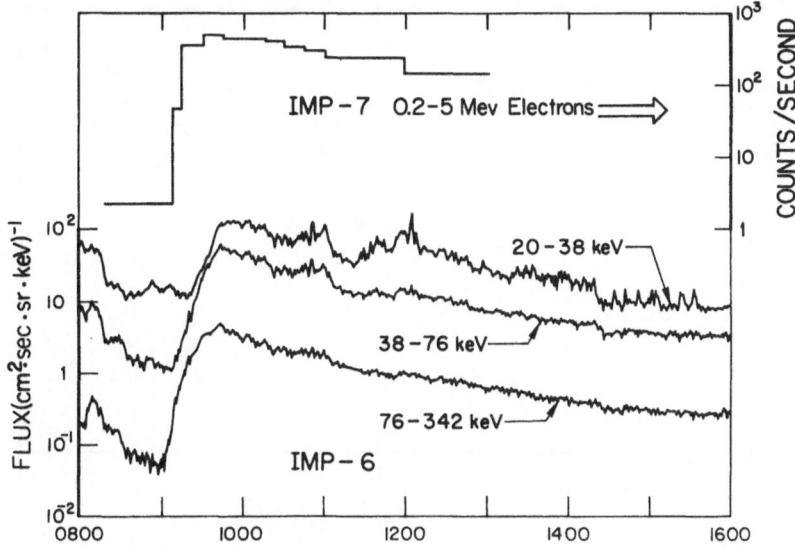

Fig. 9. The 25 November, 1977 solar electron event (from Lin *et al.*, 1982). Upstream terrestrial particles are present at IMP 6 prior to the event onset at ~09:00 and also sporadically between ~10:30 and 15:30 UT.

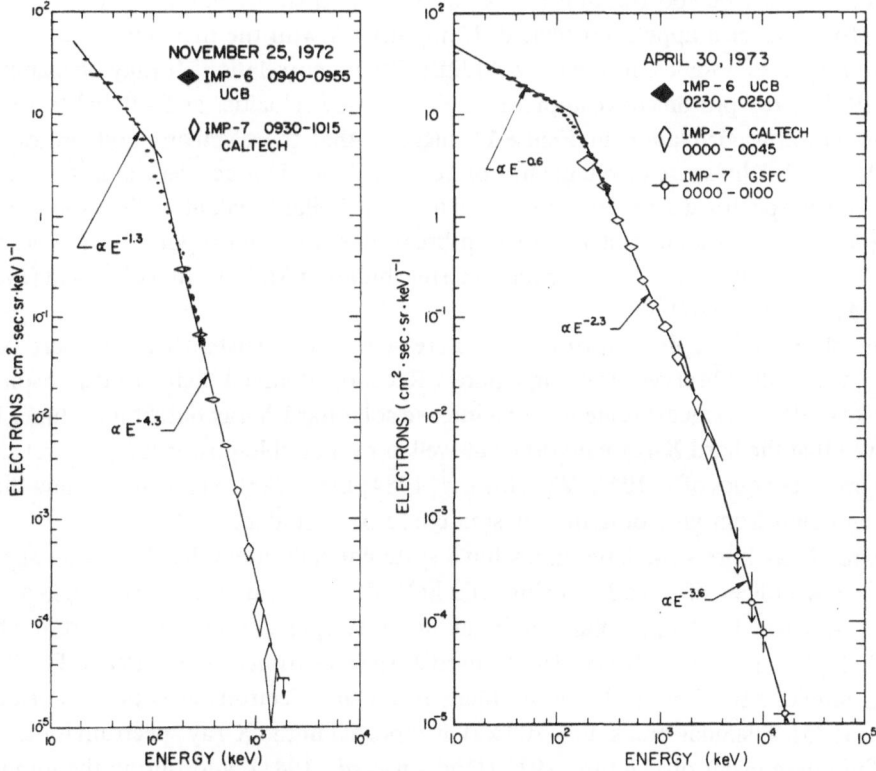

Fig. 10. The energy spectra for the 25 November, 1972 and 30 April, 1973 events (from Lin *et al.*, 1982). Both events show a smooth transition between ~100 and 200 keV. Some steepening is observed above ~3 MeV for the April event.

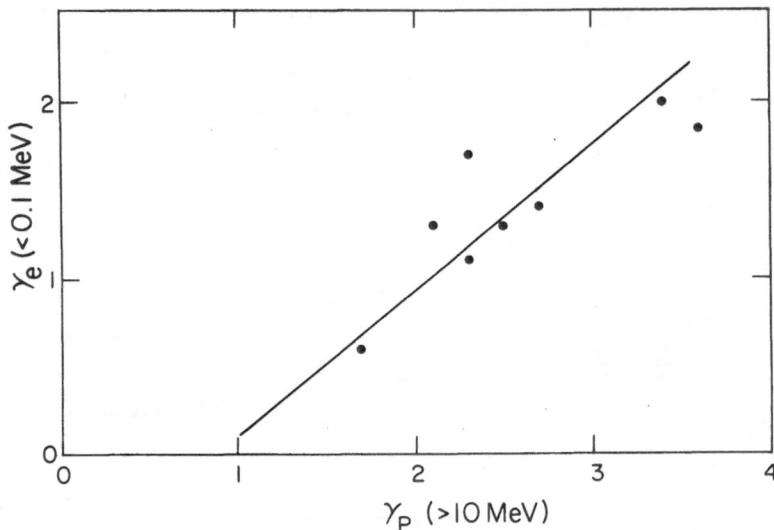

Fig. 11. Plotted here is the electron spectral exponent in the range 10–100 keV vs the proton spectral exponent in the 10–80 MeV range, for well connected large flare events observed by IMP 6, 7, and 8 (from Lin *et al.*, 1982).

< 0.1 MeV electron spectral index for eight of the events. As can be seen, the electron and proton spectra appear correlated. Comparisons with the high energy > 0.2 MeV electron spectral indices, however, yield little, if any, correlation. It may be significant that 10–80 MeV protons have approximately the same velocities as 5–40 keV electrons.

The spectral correlation in Figure 11 suggests that the electrons and protons are accelerated by the same mechanism. For velocity dependent acceleration, the break in the electron spectra at 100–200 keV would imply a similar break at ~ 200–400 MeV for protons. Although a characteristic sharp break has not been reported for protons at these energies, steepening of the spectra at a few hundred MeV is often observed (Fichtel and McDonald, 1967).

The observed escaping electron spectra are generally consistent with the hard X-ray spectra typically observed from large flares. Recently, detailed high spectral resolution ($\lesssim 1$ keV FWHM) measurements of an intense solar hard X-ray burst on 27 June, 1980 showed that the hard X-ray emission was well fit by a double-power-law spectral shape with break energies of $\sim 10^2$ keV (Schwartz, 1984; Lin, 1984). Single power law shapes and emission from an isothermal plasma could be excluded.

Hard X-ray bursts for large flares have systematically much harder X-ray spectra, power-law indices of $\gamma_x < 2$–3 below 100 keV, than typical small flares ($\gamma_x \gtrsim 4$). For the 4 August, 1972 large flare, a break to a steeper power-law spectrum above ~ 100 keV energies was observed in the hard X-ray spectrum (Hoyng, 1975). This X-ray break implies a break in the bremsstrahlung producing electrons at ~ 140 keV (Ramaty *et al.*, 1975). A similar break at ~ 100 keV is reported in the X-ray spectrum for an over the limb flare event on 22 July, 1972 (Hudson *et al.*, 1981), and during the impulsive phase of the 30 March, 1969 large flare (Frost and Dennis, 1971). The reported X-ray power-law indices correspond to electron spectral indices consistent with those obtained from interplanetary measurements.

ISEE-3 provides the first measurements of the electron spectrum below ~ 20 keV for large solar flares. Figure 12 shows the electron spectrum from 2 keV to ~ 1 MeV for the

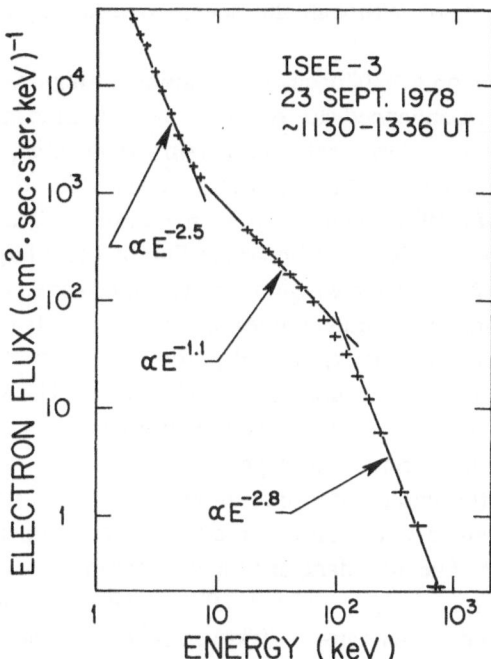

Fig. 12. The energy spectrum for the large solar flare event of 23 September, 1978 as observed by ISEE-3.

large solar flare of 23 September, 1978. The characteristic double-power-law shape with a break near $\sim 10^2$ keV and flattening below is observed. Below ~ 10 keV, however, the spectrum breaks upward again, with a power law index similar to that at energies above $\sim 10^2$ keV. Although the propagation of these lower energy $\lesssim 10$ keV electrons is nearly scatter-free while the $\gtrsim 20$ keV electrons propagate more diffusively, this difference is unlikely to produce a spectral change as large as observed here. We believe that this spectral break near ~ 10 keV reflects a real break in the injection spectrum at the Sun.

The electron spectra for small flare or coronal flare events tends to be smooth and free of breaks in the ~ 10 keV region. The spectra for large flare events suggests that a second acceleration occurs in large flares which accelerate both ions and electrons to MeV energies and above. This second acceleration appears to pick up electrons above ~ 10 keV in energy and accelerate them further. Solar flare gamma-ray observations from the Solar Maximum Mission (SMM) spacecraft indicate that if there is a physically distant second acceleration process in large flares, it must occur within seconds of the initial impulsive electron acceleration in some large flares (Chupp, 1983).

This second acceleration may be produced by the flare shock wave as it passes through the solar corona. The spectra observed here may be characteristic of the collisionless shock acceleration process. We note that flare shock waves in the interplanetary medium near 1 AU, corotating shock waves in the distant heliosphere, and bow shocks in front of planetary magnetospheres all accelerate particles to similar spectral shapes (Lin, 1980), although the energy ranges and flux levels are vastly different.

Large solar flare electron events decay away rapidly at high energies (see Figure 9). At low energies, however, enhanced electron fluxes persist in the interplanetary medium for many days following a large solar flare (Anderson *et al.*, 1982). The angular distribution for these long-lived electron fluxes still appears to be dominated by adiabatic effects (Anderson *et al.*, 1981) with a strong net outflow. Thus it appears that a long-lived source of particles is created near the Sun by these large flares and/or their associated active region. This source region supplies interplanetary magnetic field lines over a wide range of heliographic longitudes for many days. It is not known whether this long-lived source results from storage of particles from the large flare high in the corona or from continued acceleration in the high corona. The spectral shape of the low-energy electrons, however, stays a power law down to ~ 2 keV throughout the stream without significant variation in slope.

In some instances, the initial large flare impulsive event is not observed, but the flare-associated interplanetary shock is observed by ISEE-3, superimposed on a slowly rising long-lived stream. The interplanetary shock is commonly observed to accelerate electrons locally at energies below ~ 10 keV (Potter, 1981; Tsurutani and Lin, 1985), in contrast to the situation at energies > 20 keV, where shock acceleration is only very rarely observed to have any effect.

3.2. INTERPLANETARY TYPE-III STORMS

Besides the streams produced by large solar flares and interplanetary shocks there are many less intense long-lived streams which appear to be related to interplanetary type-III radio storms (Fainberg and Stone, 1970). These storms consist of thousands of type-III radio bursts emitted per day at frequencies of from a few MHz to tens of kHz, corresponding to a height range of ~ 10 solar radii to 1 AU. They last from 1 to 12 days and up to 3 storms are observed per solar rotation. An average of ~ 25 storms per year occurred in 1978–1982 (Fainberg *et al.*, 1982). The storms correlate with type-I and type-III radio storms observed at higher frequencies.

These interplanetary radio storms are usually associated with long-lived streams of electrons which show little, if any, impulsive features (Bougeret *et al.*, 1985). Apparently impulsive injections of electrons are so frequent in a storm that the fluxes from individual injections are washed out by the time the electrons reach 1 AU. The energy spectrum of these storm-related stream electrons appears similar to that observed for impulsive solar electron events, and the peak electron fluxes are usually significantly lower than for large flares or interplanetary shocks.

3.3 QUIET-TIME ELECTRONS

In 1978–1979, close to solar maximum, there are only a few short periods when the interplanetary medium appears to be free of long-lived streams associated with solar activity. At these times it may be possible to detect the quiet-Sun emission of supra-thermal electrons. Figure 13 shows the spectrum of electrons at one of the quietest times

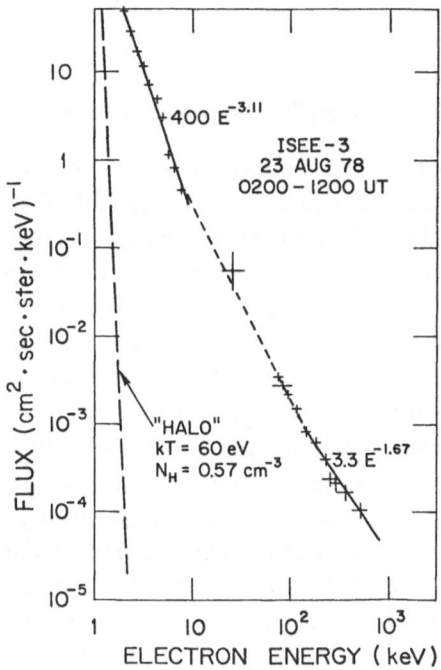

Fig. 13.   The electron spectrum at an apparent quiet time. An extrapolation of the typical solar wind halo electron spectrum (Feldman *et al.*, 1975) is shown for comparison.

in the 15-month period of observations. The electron fluxes were decaying from a stream dominated by an interplanetary shock on 18 August, but had bottomed out by ~ 2200 UT, 22 August. During the 0200–1200 UT period of this spectrum the 2–10 keV electron fluxes were very stable and still flowing out away from the Sun, although the anisotropy was small. The energy spectrum is characterized by a power law with spectral slope about − 3, similar to that observed in long-lived streams. Shown for comparison at low energies is the extrapolation of the solar wind halo electron population (Feldman *et al.*, 1975) which can be characterized by a temperature of $kT \sim 60$ eV. Even at this quiet time the observed electron fluxes at > 2 keV energies are many orders of magnitude above the solar wind population. Thus the Sun appears to be a continuous source of non-thermal electrons at energies above ~ 2 keV. The mechanism for producing this non-thermal population is unknown; perhaps these electrons are a byproduct of the coronal heating mechanism or possibly the escaping electrons from hard X-ray micro-flares (Lin *et al.*, 1984).

At energies from ~ 10 keV to ~ 200 keV the electron may also be of solar origin but the measurements are more dominated by instrument background. At energies above a few hundred keV the spectrum appears to have the same spectral slope as observed for the interplanetary electrons from Jupiter. These Jovian electrons are known to dominate the quiet-time interplanetary electron fluxes above a few hundred keV (Teegarden *et al.*, 1974).

4. Solar Type-III Radio Bursts

Table I shows that impulsive electron events are essentially always accompanied by solar type-III radio bursts observed at hectometric and kilometric wavelengths. With the sensitivity of the UCal ISEE-3 particle experiment, the number of impulsive electron events detected is now comparable to the number of discrete type-III radio bursts observed at hectometric and kilometric wavelengths, provided the limited ($\sim 60°$) longitudinal range of the electron events is taken into account.

Type-III radio emission is produced by ~ 2–10 keV electrons impulsively accelerated at the Sun. As these electrons escape from the Sun, the faster ones will run ahead of the slower ones to produce locally a bump-on-tail velocity distribution. Electron plasma (Langmuir) waves are then excited at the local electron plasma frequency via the two-stream instability (Bohm and Gross, 1949). Electron plasma waves with phase velocities corresponding to the positive slope portion of the bump will grow at the expense of the free energy in the bump. If the plasma waves grow rapidly enough, the bump should be flattened out, a process known as 'quasi-linear relaxation' or 'plateau-ing'. For a spatially homogeneous beam-plasma model the fast particle beam is so rapidly diffused in velocity space by the intense plasma oscillations that a coherent stream would only be able to travel a few kilometers from its source before it is plateaued (Sturrock, 1964). For type-III bursts, however, the beam is spatially inhomogeneous, and velocity dispersion will tend to reform the bump that quasi-linear plateauing has flattened. Also, plasma waves produced with a given phase velocity may later be reabsorbed by the electron stream when the bump goes to lower velocities. A number of numerical computations of the quasi-linear evolution of the electron stream, and growth and reabsorption of the plasma waves, have been done (Takakura and Shibahashi, 1976; Magelssen and Smith, 1977; Grognard, 1982), which show that the stream will be able to propagate out at least to distances on the order of an astronomical unit. These numerical models predict electron distributions which are very plateau-like, and plasma wave levels which sometimes exceed the threshold for strong turbulence effects to become important (Nicholson *et al.*, 1978; Goldman, 1983).

Both the plasma waves (Gurnett and Anderson, 1976, 1977) and electrons (Lin, 1970, 1974; Frank and Gurnett, 1972; Lin *et al.*, 1973) associated with type-III bursts have now been detected *in situ* at 1 AU. Figure 14 shows an impulsive electron event with associated type-III radio emission and *in situ* electron plasma waves (Lin *et al.*, 1981). The top panel shows the electric field intensities in four frequency channels from 17.8 to 100 kHz. The smooth intensity variations characteristic of a type-III radio burst, consisting of a rapid rise followed by a slow monotonic decay, are clearly evident in the 56.2 and 100 kHz channels.

The very intense irregular electric field intensity variations in the 31.1 kHz channel from about 20:00 to 21:30 UT are narrow band electron plasma oscillations. Because of the filter overlap, somewhat weaker electron plasma oscillation intensities are also evident in the 17.8 and 56.2 kHz channels.

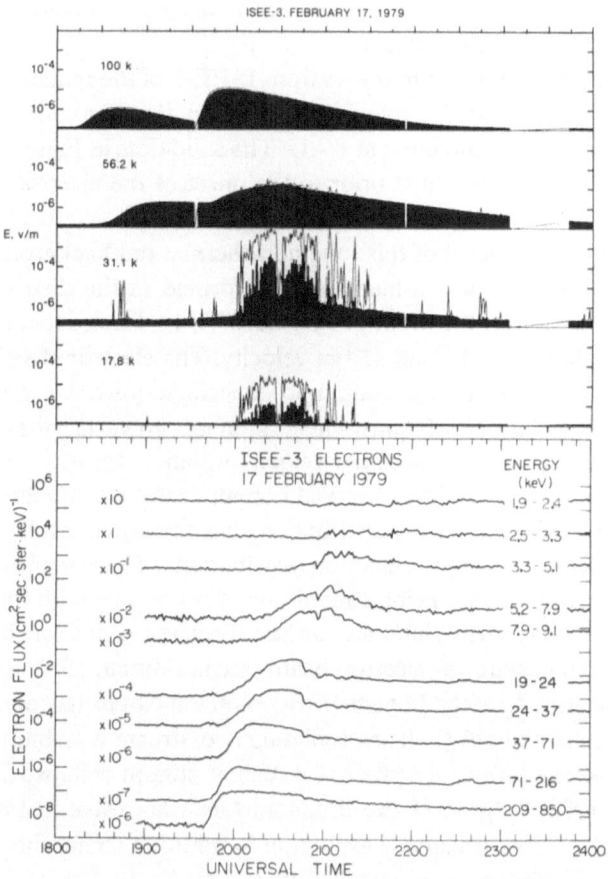

Fig. 14.   The top panel shows the electric field intensity measured from ISEE-3 (from Lin *et al.*, 1981) on 17 February, 1979 in four broad frequency bands for the event of interest. The black areas show the intensity averaged over 64 s. The solid lines give the peak intensity measured every 0.5 s. The smoothly varying profiles in the 100 and 56.2 kHz channels show two solar type-III radio bursts. The second one is of interest here. The intense highly impulsive emissions observed in the 31.1 and 17.8 kHz channels are the plasma waves. The lower panel shows the omni-directional electrons from 2 keV to >200 keV. The velocity dispersion is clearly evident. No significant increase is observed below ∼2.5 keV.

The large difference between the peak and average electric field intensities indicates that the plasma oscillations are very impulsive. The peak amplitudes appear to be limited to a level of a few mV m^{-1}. These very impulsive fluctuations are characteristic of previous observations of electron plasma oscillations associated with type-III radio bursts. Although more intense bursts have been observed closer to the Sun, this is one of the most intense plasma oscillation events observed at 1 AU.

The bottom panel of Figure 14 shows the spin-averaged low-energy electron intensities simultaneously detected by ISEE-3. The solar flare electrons are first detected above 200 keV energy at about 19:40 UT. Electrons are subsequently detected in progressively lower and lower energy channels at later and later times with very clear evidence of velocity dispersion. Simple comparison of the plasma wave and electron intensities suggests that the plasma oscillations are clearly associated with electrons in the energy range of about 3 to 30 keV.

Figure 15 shows the first measurements from ISEE-3 of the reduced (integrated over perpendicular velocities) parallel velocity distribution functions for the fast electrons associated with a type-III radio burst at 1 AU. The solid dots in Figure 15(a) show $f(v_\parallel)$ for the background population just prior to the onset of the electron event. This type of background population of non-thermal electrons is observed at all times, although at varying flux levels. The level of this long non-thermal tail background population is important in determining when a bump will be formed in the distribution. Figure 15 shows $f(v_\parallel)$ every 5 min for the event 19:45–21:55 UT). Each succeeding distribution is shifted to the right by 2×10^9 cm s^{-1} in velocity. The electron distributions develop a well-defined bump-on-tail with the bump progressing to lower velocities with time. In contrast to the expectations of quasi-linear models, however, there is no obvious plateauing of the bump; at a given phase velocity significant positive slopes persist for many minutes. The evolution of the observed Langmuir waves is qualitatively consistent with the variations of the electron distribution, but the wave levels are substantially below the levels predicted from the electron distributions. These measurements indicate that the Langmuir waves were being shifted out of resonance with the electron beam.

Various nonlinear wave-wave interactions have been suggested for shifting the plasma waves out of resonance with the electron beam (see Goldman, 1983, for review). These include induced scattering of the Langmuir waves off ion clouds (Kaplan and Tsytovich, 1968), parametric instabilities such as oscillating two-stream instability (Papadopoulos et al., 1974) and strong turbulence processes such as soliton collapse (Zakharov, 1972; Nicholson et al., 1978). In general, the dominant Langmuir wave transfer processes are those which involve a low-frequency excitation in addition to another Langmuir wave (Tsytovich, 1970). For example, induced scattering, which occurs when $T_e \approx T_i$, can be viewed as the process of decay of the initial or pump Langmuir wave into an acoustic wave which is heavily damped, and another Langmuir wave (Bardwell and Goldman, 1976). For $T_e > T_i$, the ion acoustic wave is more weakly damped and this process becomes the parametric decay of the initial Langmuir wave into an ion acoustic wave and a daughter Langmuir wave.

Density irregularities and ion acoustic waves already present in the solar wind can

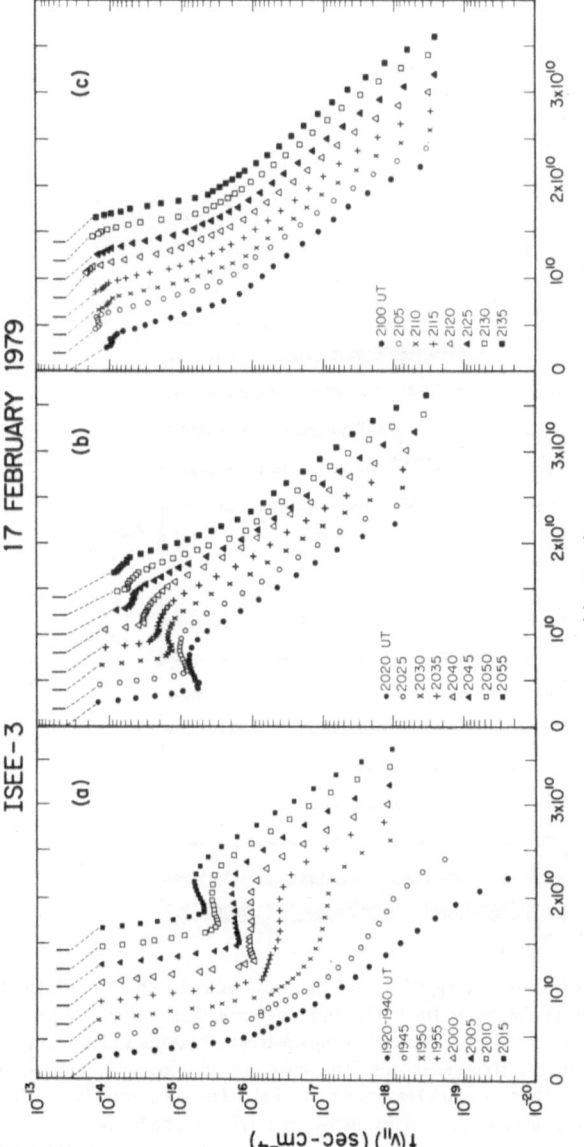

Fig. 15. The comprehensive measurements over energy and angle taken from ISEE-3 on 17 February, 1979 have been used to construct the one-dimensional velocity distribution function of the electrons every 64 s (from Lin *et al.*, 1981). The distribution averaged over 20 min prior to the event onset is indicated by the solid dots in panel (a). The 64 s measurements of the distribution during the event are shown every five minutes. Note the positive slope which develops and moves to lower velocities with time. During the interval ∼ 20:45–20:55 UT the measurements are incomplete because the magnetic field is not contained in the electron detectors' field of view. After 21:00 UT positive slopes are only infrequently observed in the distribution.

also significantly affect the growth and saturation of beam-driven Langmuir waves, to
the point of stabilizing the beam (Escande and de Genouillac, 1978; Goldman and
DuBois, 1982; Russell and Goldman, 1983).

Lin *et al.* (1985) searched for low frequency fluctuations in association with the

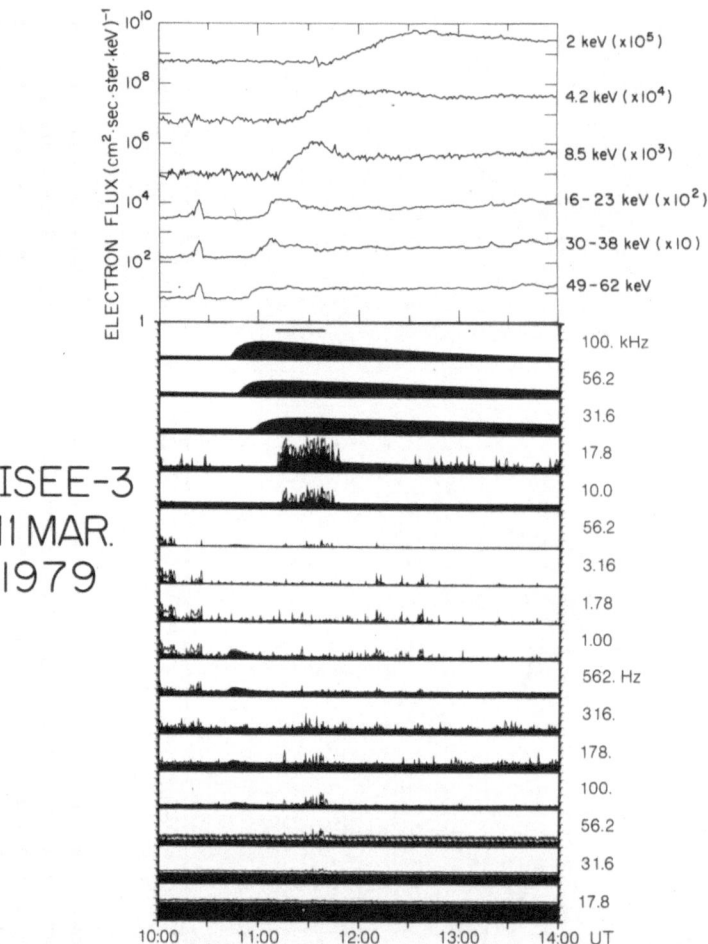

Fig. 16. The top panel shows the spin-averaged flux of electrons from 2 to 62 keV for the type-III solar
radio burst of 11 March, 1979. The spike at 10:20 UT and smaller spikes which occur simultaneously in
several energy channels are due to energetic particles flowing upstream from the Earth's bow shock. Velocity
dispersion is clearly evident for the solar electrons. The lower panel shows the electric field intensity
measured in 16 broadband channels from 100 kHz down to 17.8 Hz. The horizontal bar at the top indicates
times of positive slope in the electron reduced velocity distribution [$f(v_{\parallel})$]. The black areas show the average
intensity over 32 s. The solid lines give the peak intensity. The smoothly varying profiles (100, 56.2, and
31.6 kHz) show the type-III radio burst. The intense, highly impulsive emissions at 17.8 kHz and 10 kHz
are electron plasma waves. The sporadic bursty emissions between 3.16 kHz and ~316 Hz have been
previously identified as short wavelength ion acoustic waves. The impulsive emissions at frequencies from
316 to 31.6 Hz which occur from ~11:15 to 11:45 UT, simultaneous with the electron plasma waves, are
believed to be long wavelength ion acoustic waves (from Lin *et al.*, 1985).

Langmuir waves observed in type-III radio bursts by the ISEE-3 spacecraft. Figure 16 shows a type-III solar radio burst with clearly associated Langmuir waves and energetic electrons, in the format similar to that for Figure 14. For this event $f(v_\parallel)$ exhibits a bump-on-tail from 1109 to 11:42 UT.

The bottom panel of Figure 16 shows the electric field intensities in sixteen frequency channels from 17.8 Hz to 100 kHz. The smooth intensity variations characteristic of a type-III radio burst, consisting of a rapid rise followed by a slow monotonic decay, are clearly evident in the 31.6, 56.2, and 100 kHz channels. The electron plasma oscillations are observed in the 17.8 kHz channel from about 11:10 to 11:50 UT, in good temporal agreement with $f(v_\parallel)$. From ~ 316 Hz down to 31.6 Hz there are highly impulsive bursts at the same time as the most intense bursts of Langmuir waves. In the 56.2 and 31.6 Hz channels, these are the only bursts detectable above the background.

Figure 17 presents plots of the highest time resolution (0.5 s) data of the wave instrument for the 17.8 kHz Langmuir wave channel and the 100 Hz electric field channel for the period 11:36–11:40 UT. The Langmuir waves are extremely impulsive: intensity changes of over two orders of magnitude can occur within 0.5 s and single bursts rarely last more than one or two 0.5 s readouts. Most, but not all, of the Langmuir bursts which exceed ~ 0.1 mV m^{-1} are accompanied by a corresponding burst at 100 Hz. The maximum electric field at ~ 100 Hz is ~ 0.04 mV m^{-1}, almost two orders of magnitude above the background level. The fact that this low frequency noise is not observed in the 100 Hz magnetic channel suggests that it is a low-frequency electrostatic mode, most likely a long wavelength ion-acoustic wave.

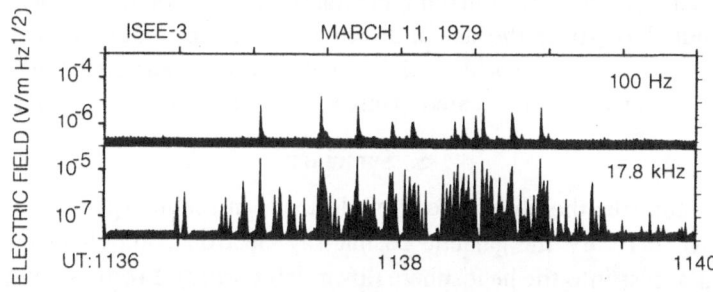

Fig. 17.   High time resolution (0.5 s) plots of the Langmuir wave channel (17.8 kHz) and the 100 Hz long wavelength ion acoustic wave channel for the 11 March, 1979 event (from Lin *et al.*, 1985). Note the close correspondence between the most intense Langmuir wave spikes and the 100 Hz spikes.

The close temporal correlation of the 30–300 Hz low frequency noise bursts with the most intense Langmuir wave bursts suggests that they are the result of some nonlinear wave-wave interaction. Theoretical work and numerical simulations suggest that strong turbulence phenomena such as modulational instability and self focusing of Langmuir waves (soliton collapse) should become important at a threshold of $W = E^2/8\pi nkT_e \approx 10^{-5}$, where E is the Langmuir wave electric field (see Goldman, 1983, for review). Almost all the type-III events observed at 1 AU, however, have maximum values of W substantially below 10^{-5}. Although very intense plasma waves confined to very small spatial regions, such as might be expected for soliton collapse, might be missed by the plasma wave instrument, it appears that, with the possible exception of a few very intense events (such as the 17 February, 1979), it is unlikely that strong turbulence phenomena play a significant role in type-III bursts.

The obvious interpretation of the low frequency, 30–300 Hz, noise observed simultaneously with type-III Langmuir wave bursts is that it is due to long wavelength ion acoustic waves. Because the typical $T_e/T_i \approx 4$, these ion-acoustic waves are not very strongly damped so a parametric decay instability, rather than induced scattering, might be appropriate to the situation for type-III radio bursts observed at 1 AU.

Two three-wave decay processes appear relevant: the decay of a pump Langmuir wave into a daughter Langmuir wave and an ion acoustic wave, $L \rightarrow L' + I$; and into a daughter transverse electromagnetic wave and an ion acoustic wave, $L \rightarrow T + I$. Based on preliminary analysis, it appears that the threshold is exceeded by the observed Langmuir wave levels only for the latter process. The transverse waves from this latter decay process would provide a natural source for type-III radio burst emission at the fundamental, i.e., near f_{p-}.

In the 17 February, 1979 event the radio emission at frequencies near the harmonic, $2f_{p-}$, which was believed to originate near the spacecraft because of its lack of directivity, actually began some 20 min *prior* to the onset of the Langmuir waves which were supposed to produce the radio emission. Lin *et al.* (1981) pointed out that this difficulty could be resolved if what was identified as harmonic radiation produced *in situ* could instead be fundamental radiation produced much closer to the Sun (see Kellogg, 1980). Significant scattering of the fundamental radio emission off density irregularities would be required to explain the lack of directivity. Evidence that fundamental emission dominates from burst onset to peak, and that scattering is important in the interplanetary medium, has recently been presented (Dulk *et al.*, 1985; Steinberg *et al.*, 1985).

5. Summary

Figure 18 summarizes the observations of electrons in the interplanetary medium. At energies above $\sim 10^9$ eV the galactic cosmic-ray electron component has essentially unattenuated access into the heliosphere (lower left corner). From $\sim 3 \times 10^7$ to 10^9 eV galactic cosmic ray electrons are still observed but their fluxes are attenuated and modulated. From a few hundred keV to a few MeV, electrons of Jovian origin dominate at quiet times. Below $\sim 10^2$ keV and down to $\lesssim 2$ keV it appears that the Sun is the dominant source at quiet times. We do not know the production mechanism for the quiet time 2–10^2 keV electrons from the Sun.

The active solar corona is continually accelerating electrons in coronal flares or small flare events which occur, on average, at least several times a day and sometimes, in type-III storms, much more frequently. These events are not known to be accompanied by detectable emissions of $\gtrsim 10$ MeV ions, but ions of lower energies, ~ 1 MeV nucl.$^{-1}$ or less, appear to be commonly accelerated in these small flares and coronal flares. These events preferentially accelerate ^{3}He.

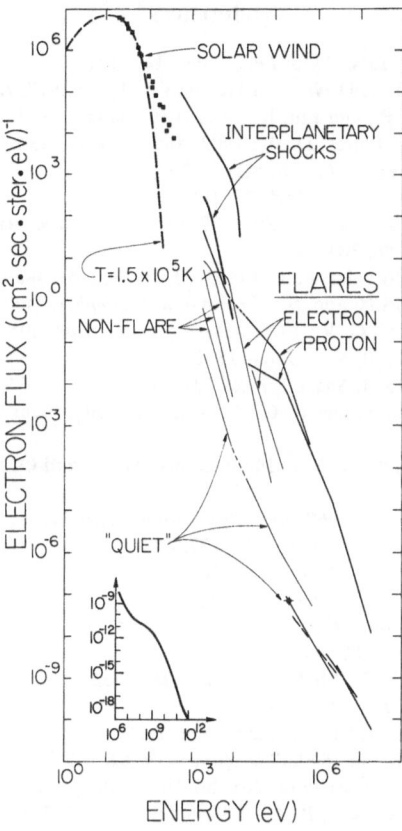

Fig. 18. A summary of the spectra of electrons observed in the interplanetary medium.

Large solar flares are characterized by a much different spectrum of accelerated electrons, which suggests a second acceleration stage occurs in these events. In addition large flares give rise to long-lived sources of low-energy electrons. In the absence of these long-lived streams, which dominate the observed interplanetary electrons in 1978–1979, there appears to be significant emission of 2–10^2 keV electrons by the Sun even at the quietest times.

The ISEE-3 particle and wave data have provided very illuminating measurements of the basic beam-plasma interaction which produce Langmuir waves and in turn, type-III radio emission. Although the basic ideas of Ginzburg and Zheleznyakov (1958) have been confirmed, these processes appear to be far more complicated. More detailed observations are required to provide a quantitative understanding of these plasma processes.

Acknowledgements

This research was supported in part by NASA grants NAG 5–376 and NGL-05–003–017.

References

Anderson, K. A. and Lin, R. P.: 1966, *Phys. Rev. Letters* **16**, 1121.

Anderson, K. A., Lin, R. P., Potter, D. W., and Heetderks, H. D.: 1977, *IEEE Trans.* **GE-16**, 153.

Anderson, K. A., McFadden, J. P., and Lin, R. P.: 1981, *Geophys. Res. Letters* **8**, 831.

Anderson, K. A., Lin, R. P., and Potter, D. W.: 1982, *Space Sci. Rev.* **32**, 169.

Bardwell, S. and Goldman, M. V.: 1976, *Astrophys. J.* **209**, 912.

Bohm, D. and Gross, E. P.: 1949, *Phys. Fluids* **75**, 1851.

Bougeret, J. L., Lin, R. P., Fainberg, J., and Stone, R. G.: 1985, *Astron. Astrophys.* (in press).

Chupp, E. L.: 1983, *Solar Phys.* **86**, 383.

Coplan, M. A., Ogilvie, K. W., Boschler, P., and Geiss, J.: 1983, *Solar Wind Five*, NASA CP-2280, p. 591.

Dulk, G. A., Steinberg, J. L., and Hoang, S.: 1985, *Astron. Astrophys.* (submitted).

Escande, L. A. and de Genouillac, G. V.: 1978, *Astron. Astrophys.* **68**, 405.

Fainberg, J. and Stone, R. G.: 1970, *Solar Phys.* **15**, 222.

Fainberg, J. and Stone, R. G.: 1974, *Space Sci. Rev.* **16**, 145.

Fainberg, J., Bougeret, J.-L., and Stone, R. G.: 1982, in M. Neugebauer (ed.), *Solar Wind Five*, NASA CP-2280, p. 469.

Feldman, W. C., Asbridge, J. R., Bame, S. J., Montgomery, M. D., and Gary, S. P.: 1975, *J. Geophys. Res.* **80**, 4181.

Fichtel, C. E. and McDonald, F. B.: 1967, *Ann. Rev. Astron. Astrophys.* **5**, 351.

Frank, L. A. and Gurnett, D. A.: 1972, *Solar Phys.* **27**, 448.

Frost, K. J. and Dennis, B. R.: 1971, *Astrophys. J.* **165**, 655.

Geiss, J. and Reeves, H.: 1972, *Astron. Astrophys.* **18**, 126.

Ginzburg, V. L. and Zheleznyakov: 1958, *Soviet Astron.* **AJ2**, 653.

Goldman, M. V.: 1983, *Solar Phys.* **80**, 403.

Goldman, M. V. and DuBois, D. G.: 1982, *Phys. Fluids* **25**, 1062.

Grognard, R. J.-M.: 1982, *Solar Phys.* **81**, 173.

Gurnett, D. A. and Anderson, R. R.: 1976, *Science* **194**, 1159.

Gurnett, D. A. and Anderson, R. R.: 1977, *J. Geophys. Res.* **82**, 632.

Hoyng, P.: 1975, Ph.D. Thesis, University of Utrecht, The Netherlands.

Hudson, H. S., Lin, R. P., and Stewart, R. T.: 1982, *Solar Phys.* **75**, 245.

Kaplan, S. A. and Tsytovich, V. N.: 1968, *Soviet Astron.-A.J.* **11**, 834.

Kellogg, P. J.: 1980, *Astrophys. J.* **236**, 696.

Lin, R. P.: 1970, *Solar Phys.* **12**, 266.

Lin, R. P.: 1974, *Space Sci. Rev.* **16**, 189.

Lin, R. P.: 1980, *Solar Phys.* **67**, 393.

Lin, R. P.: 1984, in S. E. Woosley (ed.), *High Energy Transients in Astrophysics*, American Inst. Phys., p. 619.

Lin, R. P., Evans, L. G., and Fainberg, J.: 1973, *Astrophys. Letters* **14**, 191.

Lin, R. P., Lotko, W., Gurnett, D. A., and Scarf, F. L.: 1985, *Astrophys. J.* (submitted).

Lin, R. P., Mewaldt, R. A., and Van Hollebeke, M. A. I.: 1982, *Astrophys. J.* **253**, 949.

Lin, R. P., Potter, D. W., Gurnett, D. A., and Scarf, F. L.: 1981, *Astrophys. J.* **251**, 364.

Lin, R. P., Schwartz, R. A., Kane, S. R., Pelling, R. M., and Hurley, K. C.: 1984, *Astrophys. J.* **283**, 421.

Magelssen, G. R., and Smith, D. G.: 1977, *Solar Phys.* **55**, 211.

Nicholson, D. R., Goldman, M. V., Hoyng, P., and Weatherall, J. C.: 1978, *Astrophys. J.* **225**, 605.

Papadopoulos, K., Goldstein, M. L., and Smith, R. A.: 1974, *Astrophys. J.* **190**, 175.

Pan, L. D., Lin, R. P., and Kane, S. R.: 1984, *Solar Phys.* **91**, 345.

Potter, D. W.: 1981, *J. Geophys. Res.* **86**, 111.

Potter, D. W., Lin, R. P., and Anderson, K. A.: 1980, *Astrophys. J. Letters* **236**, L97.

Ramaty, R., Kozlovsky, B., and Lingenfelter, R. E.: 1975, *Space Sci. Rev.* **18**, 341.

Ramaty, R., *et al.*: 1980, in P. A. Sturrock (ed.), *Solar Flares*, Colorado Associated University Press Boulder, Ch. 4, p. 117.

Reames, D. V., von Rosenvinge, T. T., and Lin, R. P.: 1985, *Astrophys. J.* **292**, 716.

Russell, D. A. and Goldman, M. V.: 1983, *Phys. Fluids* **26**, 2717.

Schwartz, R. A.: 1984, Ph.D. Thesis, University of California, Berkeley.

Simnett, G. M.: 1974, *Space Sci. Rev.* **16**, 257.

Steinberg, J. L., Dulk, G. A., Hoang, S., Lecacheux, A., and Aubier, M. G.: 1985, *Astron. Astrophys.* (submitted).

Sturrock, P. A.: 1964, in W. N. Hess (ed.), *AAS-NASA Symposium on the Physics of Solar Flares*, NASA SP-5, p. 357.

Takakura, T. and Shibahashi, H.: 1976, *Solar Phys.* **46**, 323.

Teegarden, B. J., McDonald, F. B., Trainor, J. H., Webber, W. R., and Roelof, E. C.: 1974, *J. Geophys. Res.* **79**, 3615.

Trubnikov, B. A.: 1965, *Rev. Plasma Phys.* **1**, 105.

Tsurutani, B. and Lin, R. P.: 1985, *J. Geophys. Res.* **90**, 1.

Tsytovich, V. N.: 1972, *Nonlinear Effects in Plasmas*, Plenum, New York.

Van Allen, J. A. and Krimigis, S. M.: 1965, *J. Geophys. Res.* **70**, 5737.

Van Hollebeke, M. A. I., MaSung, L. S., and McDonald, F. B.: 1975, *Solar Phys.* **41**, 189.

Zakharov, V. E.: 1972, *Soviet Phys. JETP* **35**, 908.

IMAGING OF CORONAL MASS EJECTIONS BY THE HELIOS SPACECRAFT

B. V. JACKSON

University of California at San Diego, La Jolla, CA 92093, U.S.A.

Abstract. The zodiacal light photometers on board the HELIOS spacecraft can be used to form images of coronal mass ejections in the interplanetary medium. Several aspects of these data are unique: they trace coronal mass ejections using Thomson scattering techniques to distances from the Sun greater than 0.5 AU; their perspective from the HELIOS orbits allow information to be gained about the three-dimensional shapes of specific mass ejections viewed both by coronagraphs and HELIOS; the global view afforded by the spacecrafts photometers can image the mass ejection from within and thus relate *in situ* measurements to the shape of the whole structure. To date, the HELIOS photometers have been used to study coronagraph-observed mass ejections including those which originate at the Sun on 7 May, 24 May, and 27 November, 1979, and 21 May, 18 June, 29 June, and 6 November, 1980. Masses of ejections determined from these data are generally a few times larger than masses determined from SOLWIND coronagraph images.

1. Introduction

Coronal mass ejections represent a great concentration of mass and energy input into the lower corona. It is estimated (Howard *et al.*, 1985) that more than 5% of the yearly solar wind mass flux in the ecliptic plane can be composed of mass from these discrete ejections of material. As coronal mass ejections propagate outward from the Sun, their effects can be felt to at least 1 AU. Disturbances that reach Earth at approximately solar wind speeds following solar flares were known to exist as early as observations of the geomagnetic field of the Earth by Newton (1943). When *in situ* measurements from Earth-orbiting spacecraft became available, the aftermath of a flare-associated disturbances could be measured (e.g., Hundhausen *et al.*, 1970). Coronal mass ejections were first observed near the solar surface in the light of Hα as material motion away from the solar limb (e.g., Pettit, 1925) and later by Earth-orbiting coronagraphs (MacQueen *et al.*, 1974; Koomen *et al.*, 1975). Spacecraft nearer the Sun not in Earth orbit have given *in situ* evidence of the progress of coronal mass ejections as they move outward from the Sun (e.g., Dryer *et al.*, 1982; Burlaga *et al.*, 1982; Sheeley *et al.*, 1985). Remote sensing techniques from Earth such as interplanetary scintillation measurements (Rickett, 1975; Watanabe and Kakinuma, 1984) or kilometric type-II measurements (e.g., Cane *et al.*, 1982) also detect disturbances at large distances from the Sun. However, it is uncertain what these observations, sensitive to small-scale (~200 km) density changes and shock processes, measure in terms of mass. The three-dimensional shapes and positions of coronal mass ejections are uncertain from coronagraph observations alone. Coronagraph polarization measurements give some information of how distant the material of a mass ejection is from the plane of the sky. However, the shapes of mass ejections are not easily determined from these Earth-based observations. The observations of structure are vague enough to give rise to contro-

Solar Physics **100** (1985) 563–574.   0038–0938/85.15.

versies such as the loop versus bubble form for some mass ejections (e.g., see Wagner, 1983).

The HELIOS spacecraft zodiacal light photometers (Leinert *et al.*, 1975, 1981) measure brightness variations globally around the spacecraft. The photometers are sensitive enough to trace mass ejections to 90° elongation and beyond (Richter *et al.*, 1982) by electron Thomson scattering. Thus, in principal, they can determine the gross features of mass ejections in terms of electron density or mass out to the distance of the spacecraft orbits (0.3–1.0 AU). These techniques have been used successfully to determine the gross features of coronal mass ejections and masses in the interplanetary medium more nearly coincident with *in situ* measurements (Jackson and Leinert, 1985; Jackson, 1985; Jackson *et al.*, 1985). Because the HELIOS spacecraft are not Earth-orbiting, they can view mass ejections from an entirely different perspective than Earth. Thus, the HELIOS spacecraft give information on mass ejection three-dimensional structures using stereoscopic techniques. Polarization data from HELIOS photometers further the ability of HELIOS spacecraft observations to disentangle the components of large mass ejections.

2. HELIOS Observations

The HELIOS zodiacal light photometers were originally intended to measure the distribution of dust in the interplanetary medium between Sun and Earth. However, they can also be used to image the variations of brightness produced by large-scale changes in the interplanetary electron content. The HELIOS zodiacal light sensors (Leinert *et al.*, 1975; 1981) consist of three photometers fixed on the spacecraft and rotating with it on an axis perpendicular to the plane of the ecliptic. The photometers of HELIOS B point 16, 31, and 90° north of the ecliptic plane with the 16 and 31° photometers clocking data into 32 longitude sectors at constant ecliptic latitude around the sky. The sixteen sectors nearest the Sun are lengths of 5°.6 in ecliptic longitude; sector lengths of 11.2 and 22°.4 are formed for sectors at greater solar elongations. The 16, 31, and 90° photometers have apertures of 1, 2, and 3°, respectively. HELIOS A photometers point south of the ecliptic plane and are symmetric to those of HELIOS B. The 90° photometer of the HELIOS A spacecraft viewed the large Magellanic cloud; data for this photometer has not been reduced. The data are compiled over an 8.6-min period from a set of ultraviolet, blue, and visual filters, polaroids and each photometer and refreshed in sequence nominally every 5.2 hr.

Figure 1 presents the time series of selected individual photometer sectors as in Richter *et al.* (1982) or Jackson and Leinert (1985) for May, 1979. The HELIOS photometers are shown to be stable over month-long periods; rapid daily variations from the zodiacal light component are negligible even at times when known meteor streams were in view. Measurement to measurement variations in brightness from sources other than electron Thomson scattering would generally be interpreted as noise in the data. Photometer brightness variations, where the electron density variations are expected to be small at large elongations and at the greatest solar distances, show that

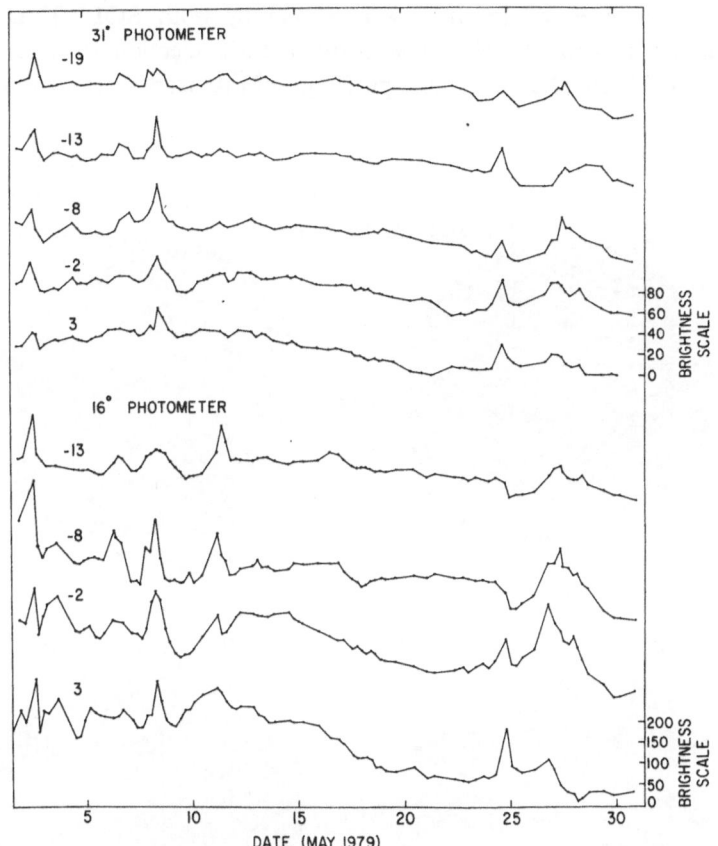

Fig. 1. Time series data for the month of May, 1979 for selected HELIOS B photometer positions presented as in Richter *et al.* (1982) in units of tenth magnitude star brightness normalized to 1 AU for the astronomical *V* (visual) photometric band. The large slowly-varying component from zodiacal light has been largely removed from the data. Sixteen degree photometer sectors with centers at ecliptic longitudes relative to the Sun of 3, − 2, − 8, and − 13° and thirty-one degree photometer sectors at the same relative solar ecliptic longitudes plus an additional sector at − 19° are presented. A brightness scale (number of 10th magnitude stars per square degree) is given for each set of photometers. Evident in the data are the brightness peaks during 8–9 May that are imaged as a mass ejection.

random brightness variations of less than a tenth magnitude star per square degree are typical from one photometer measurement to another. Variations of the signal above this level are taken as significant and assumed due solely to changes in the amount of scattered light from electrons.

The HELIOS photometer data show brightness variations with time that occur later in the 31° photometers than in the 16° photometers and thus are outward propagating disturbances. Richter *et al.* (1982) give velocities for several of these outward-propagating 'plasma clouds' and show that their speeds are consistant with or somewhat higher than solar wind speeds. Jackson *et al.* (1985) and Jackson and Leinert (1985) use these data to produce elongation-time diagrams of the onsets and brightest portions of mass ejections observed in the lower corona by both SOLWIND and Solar Maximum Mission coronagraphs. The coronal mass ejection of 7 May, 1979 as observed by the SOLWIND coronagraph (Figure 2) is used as a demonstration. In

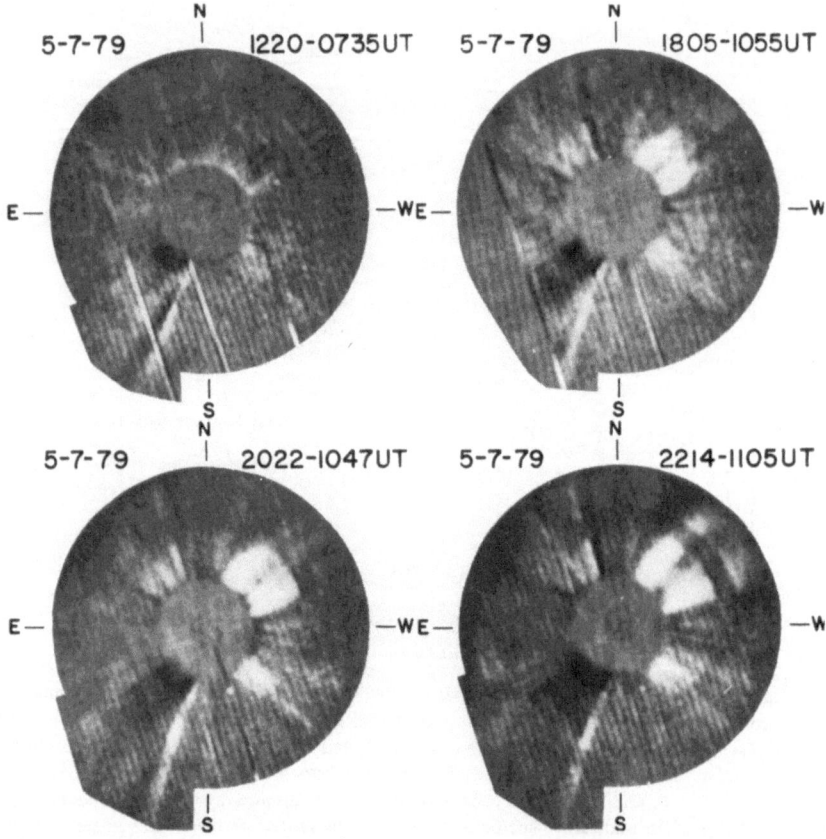

Fig. 2.   SOLWIND coronagraph difference images of the 7 May, 1979 coronal mass ejection. The coronagraph occulting disk at 2.6 R_0 provides the inner mask, while the outer portion of the field of view has been cropped to 8 R_0 at most locations. The 7 May, 1979 coronal mass ejection is observed moving outward to the solar northwest over a period of approximately four hours.

SOLWIND difference images the solar distances of the outer edge of the ejection plotted versus time gives the approximate outward speed (150 km s^{-1}) of the ejection. Figure 3 gives the composite outward angular motion of the 7 May, 1979 mass ejection as observed by the SOLWIND coronagraph and in the HELIOS B spacecraft time series. The plot shows that the outer edge of the mass ejection observed in the lower corona has indeed propagated outward and can be detected in the HELIOS photometers.

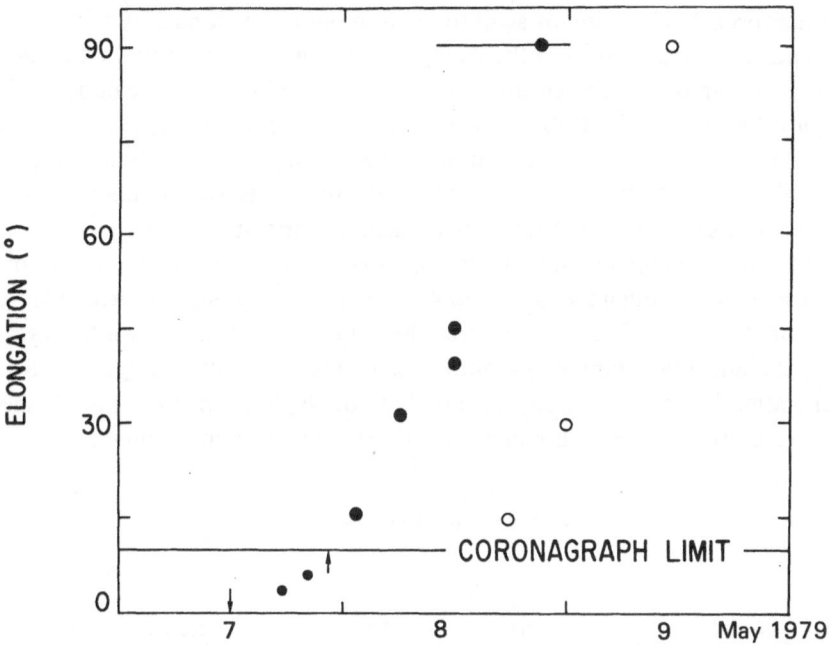

Fig. 3. Elongation-time plot of the 7 May, 1979 ejection. The elongation scale for the coronagraph observations has been expanded to compare with the HELIOS observations. The coronagraph limit of observability is noted. Small dots depict the outermost detected portion of the ejection in coronagraph images; large closed dots depict the same in HELIOS B data. Large open circles give the position of the most dense portion of the ejection.

3. Images of the 7 May 1979 Mass Ejection

To form images of coronal mass ejections from the HELIOS spacecraft photometer data, it is necessary to combine the observations from each sector into a meaningful spatial representation at a single instant in time. To do this, the time series from one photometer position for an eight day period is displayed as in Figure 1. A straight line baseline placed through the time series gives an estimate of the excess brightness at each data transmission time for the period in question. With this procedure, these data are then linearly interpolated to give brightness values at any instant in time. Electron numbers can be related to brightness from Thomson scattering as in Billings (1966). Without knowing the position of the electron condensation along the line-of-sight, an assumption that the electrons are at the point closest the Sun gives a lower limit estimate to the total number of electrons represented by the brightness increase. If brightnesses are interpreted as columnar density, it is possible to combine data from individual photometer sectors much as has been done from coronagraph image pixels (e.g., Hildner *et al.*, 1975; Jackson and Hildner, 1978; Poland *et al.*, 1981). The contoured surface determined in terms of excess density from Helios photometer data is essentially an image of the density distribution around the spacecraft. For the purposes of the contour plots (and masses), the coronal mass ejections are generally assumed at the point of closest approach of the line-of-sight to the Sun out to an elongation of 75°5. At greater elongations, a distance of 0.25 the distance from the spacecraft to the Sun is chosen as more representative of the electron position (see Jackson and Leinert, 1985).

Figure 4 presents HELIOS B images of the ejection of 7 May, 1979 at several stages contoured in quantities of excess mass (see Section 4). The contour plotting program (available at the National Center for Atmospheric Research computing center in Boulder, Colorado, U.S.A.) interpolates and extrapolates from sparsely placed data points by determining directional derivatives of the surface from adjacent positions. In the presentation of Figure 4, all material traveling radially outward from the Sun moves in a straight line away from zero degrees elongation. Note that in these 'fisheye' lens-type views, the antisolar point maps onto a semi-circle at 180° elongation. Because the interpolation is large above 31° ecliptic latitude, high accuracy is not claimed for the contours presented there although they are shown for completeness.

8–9 MAY 1979

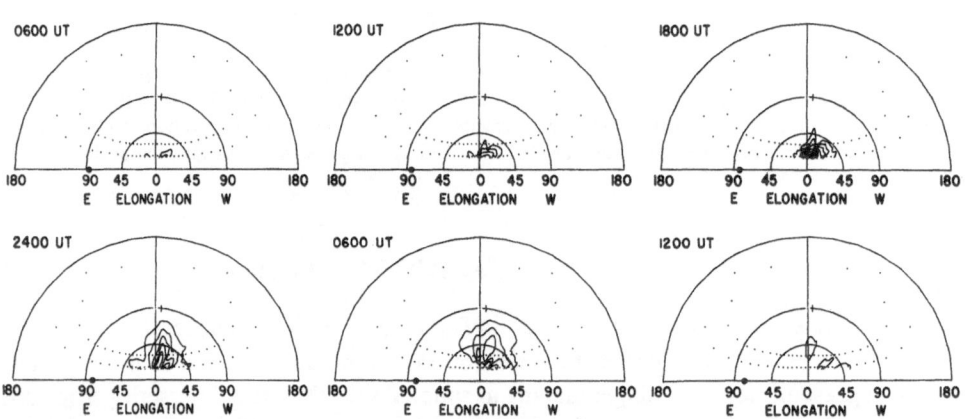

Fig. 4.   HELIOS contour plots for the 7 May, 1979 ejection as it moves outward from the Sun over a period from 06:00 UT on 8 May to 12:00 UT on 9 May. In this presentation, the Sun is centered and various solar elongations labeled on the abcissa form semi-circles above the ecliptic plane (represented by the horizontal line). The vertical line is the great circle to the north of the spacecraft. The position of the Earth is marked as the '⊕' near east 90° and the solar north pole tilt indicated by the short line segment crossing 90° elongation. Positions of the sector centers are marked by dots. Mass is contoured in levels of 3×10^{12} g per square degree beginning at 3×10^{12} g per square degree. The larger elongations are generally the lowest level contoured.

Not used or presented in the contour images of the 7 May ejection is a technique for utilizing the available time series information to form a more complete picture of the event. By knowing elapsed time from a previous observation and material velocity and assuming radial expansion, the projected position of the material at a later time can be determined. With appropriate assumptions about the dispersion of the ejection as it moves outward from the Sun (assumed at constant speed to date and thus a R^{-2} expansion), the extrapolated columnar densities can be determined. This technique has been used successfully in Jackson *et al.* (1985), Jackson (1985), and McCabe *et al.* (1985) to form more complete image representations of HELIOS data and will be demonstrated later. This technique is especially important where data from only a single photometer exists such as for the ejections of 18 and 29 June, 1980. The only firm data points (where contours are exact) are those points (sector centers) shown as dots on the images. No significance should be ascribed to contour variations on scales of less than half the distance between the centers of adjacent photometer sectors.

As viewed in SOLWIND difference images (Figure 2) the ejection of 7 May, 1979 is seen from Earth traveling to the solar northwest, reaching the outer extent of the coronagraph at $10 R_0$ at approximately 22:00 UT 7 May. SOLWIND coronagraph observers describe this ejection as three-pronged and moving at speeds of 150 km s^{-1} (Poland *et al.*, 1981). In SOLWIND data, the coronal mass ejection is confined to approximately 70° of position angle. The ejection on 7 May, 1979 (as in Jackson and Leinert, 1985) depicted in Figure 4 is viewed by HELIOS B nearly 90° from the Sun–Earth line as it travels out to and above the spacecraft situated at 0.3 AU. The HELIOS B spacecraft first observed the outermost portion of the ejection at approximately 01:00 UT 8 May in the innermost photometers ($15 R_0$). As seen from HELIOS B the ejection has two distinct prongs and is confined to 80° of position angle in its initial stages. Further information on the three-dimensional structure of this mass ejection can be derived from the HELIOS polarization information which shows that the portion of the ejection to the north of the Sun is not as highly polarized as the

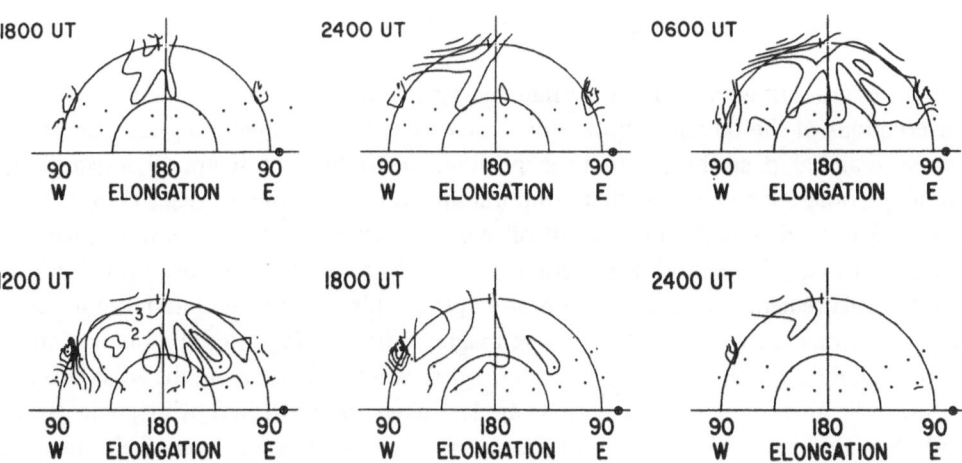

Fig. 4.   HELIOS B contour plot for the 7 May, 1979 ejection. In this presentation the direction 180° opposite the Sun is centered. Electron numbers are contoured in levels of 10^{14} cm^{-2}. At 12:00 UT contour levels are numbered.

material to the solar northwest. This implies that the northern portion of the ejection is heading toward and above the HELIOS B spacecraft. By midday 9 May, all of the spacecraft photometer positions show some evidence of a surrounding increase in electron density.

Figure 5 displays the HELIOS B data of the 7 May, 1979 ejection centered at 180° elongation in contours of electrons cm^{-2}. These columnar density contours again imply that the mass ejection first appeared at large elongations primarily northward of the HELIOS B spacecraft on 24:00 UT 8 May, 1979. The increased density associated with the ejection appeared in essentially all of the photometer sectors at elongations greater than 90° midday on 9 May, 1979 with most of the ejection concentrated towards the west and above the spacecraft. Finally, by the end of 9 May, 1979 the excess mass of the ejection had decreased so that all that could be observed was a small remaining portion to the northwest of HELIOS B.

4. Excess Ejected Mass

Excess masses of the ejected matter are obtained by assuming that each electron is associated with 2×10^{-24} g of material (a 10% helium abundance). The total mass excess of an ejection can be determined in two ways – namely by spatial and temporal integration. In coronagraph observations near the solar surface, observers in the past have simply integrated over the individual area elements of an image assuming all the material to be at the point of closest approach of the line-of-sight to the Sun. This gives a lower limit to the total mass for two reasons. First, it is possible that the material is not at the point of closest approach of the line-of-sight and second; the material of an ejection may continue to flow through the field of view after the outer edge of the ejection has passed the outer limit of the field of view. In the HELIOS data, masses are determined by integrating over the area elements defined by the contour plotting program (a spatial value) when the ejection includes both 16 and 31° photometer sectors.

Masses are also estimated by determining a typical speed for the ejected material from observations of the outward motion of the ejection and by integrating the amount of excess material past the photometer positions with time (a temporal integration). Although others have used similar temporal integration techniques (e.g., Hundhausen et al., 1970) to determine the amount of mass in disturbances from in situ spacecraft measurements, this method of integrating with time to determine mass using corona-graph measurements has seldom been attempted. This is because material motions in the low corona are dominated by strong magnetic fields and material motions can be highly variable. The HELIOS photometers observe material far from the solar surface. At great distances the material motions are less likely to be affected by magnetic fields, and thus the outward motion should be more easily inferred from flight time or brightness change. In situ measurements from HELIOS and other spacecraft within 1 AU (e.g., Burlaga et al., 1982), that supposedly measure velocities within coronal mass ejections, seldom show velocity differentials within ejections greater than a factor of two.

Measurements of mass by the temporal integration technique vary linearly with the speed of the material past the photometer sectors. Therefore, masses by this temporal integration technique are not expected to be in error by more than a factor of two.

Because the calibration of the HELIOS photometers was determined from known stellar brightnesses and checked periodically, the absolute intensity calibration of the instrument is believed to be near 4–5% (Leinert *et al.*, 1981, 1982). This is far better than usually claimed for coronagraph calibrations. However, the problems of determining a baseline limits the accuracy of excess mass estimates; other structures along the line-of-sight can confuse determinations of mass from any given ejection. Because of this and the poor resolution of the imaging technique, in no case is an accuracy greater than 20% claimed for the masses presented in the papers written to date.

5. Results of Previous Work

It is evident from these data that brightness increases associated with coronal mass ejections are observed in HELIOS zodiacal light photometers. These brightness increases can be observed to move outward to heights greater than half the Sun–Earth distance. In general, the gross features of coronal mass ejections are preserved at interplanetary heights. Jackson and Leinert (1985) show that the radial extent of the 7 May, 1979 mass ejection in the interplanetary medium was at least 0.3 AU and took 1.5 days to cross the inner (16°) HELIOS B photometers. Evidently, material was continually being expelled from the inner corona (perhaps within the coronagraph field of view) to produce this ejection. The outer edge of the 7 May, 1979 mass ejection in HELIOS B observations was observed moving outward in elongation consistent with an outward speed of at least 400 km s^{-1}. Since the motion of the outer edge of the ejection in the inner corona was shown to have a speed of approximately 150 km s^{-1} (Poland *et al.*, 1981), this is consistent with an acceleration of the material of the mass ejection above 8 R_0. For this mass ejection Jackson and Leinert (1985) obtain masses of $\sim 2 \times 10^{16}$ g compared to $\sim 10^{16}$ g from Poland *et al.* (1981). The large loop-like coronal mass ejection that began near the Sun on 24 May, 1979 (Sheeley *et al.*, 1981) continued to expel mass through the HELIOS B field of view over a period of many days; masses from the slower-moving persistent features of this ejection ($\sim 5 \times 10^{16}$ g) were greater than those of the early, rapidly-moving portion ($\sim 2 \times 10^{16}$ g).

Jackson *et al.* (1985) specifically use the non-Earth perspective afforded by the HELIOS spacecraft to observe three loop-like coronal mass ejections in an attempt to determine the thickness of the ejections viewed edge-on. The HELIOS views of the three loop ejections give no indication that the material is confined to a very small angular extent when it reaches the HELIOS photometer lines of sight. Jackson *et al.* (1985) thus conclude that loop ejections have substantial non-radial extents along the loop axes when they reach heights available to *in situ* spacecraft measurements. It is also clear from the perspective observations (of the 29 June, 1980 coronal mass ejection) that the small compact loop observed in solar maximum mission coronagraph images is no broader viewed 90° from the Sun–Earth line.

Jackson (1985) uses the HELIOS photometer imaging technique to view a flare-associated coronal mass ejection on 27 November, 1979 that completely circles the Sun in coronagraph images (Howard *et al.*, 1982). The motion of the ejection is traced half way between Sun and Earth. No evidence for the ejection can be observed northward from the HELIOS B spacecraft which is 30° east of the Sun–Earth line, and little evidence of the ejection is observed *in situ*. The observations are consistant with an ejection whose mass is well-collimated along the Sun–Earth line. Some deceleration of the material can be observed as the ejection moves towards Earth. A shock and a mass increase at Earth on 30 November, 1979 are associated with this ejection. Within the limits of the comparison, the excess mass determination from Thomson scattering measurements and *in situ* from Earth-spacecraft are entirely consistent. The position of the shock occurs at the middle of the large mass increase observed at Earth rather than at its onset. This is consistent with the HELIOS observations in that an earlier and more slowly-moving mass ejection can also be observed with an approximate arrival time at Earth slightly ahead of the later 27 November coronal mass ejection.

The non-Earth perspective view from the HELIOS spacecraft can be used to locate the positions of mass ejections and better determine their solar surface origins. In McCabe *et al.* (1985) a massive ejection directed southward of the Sun in a narrow beam

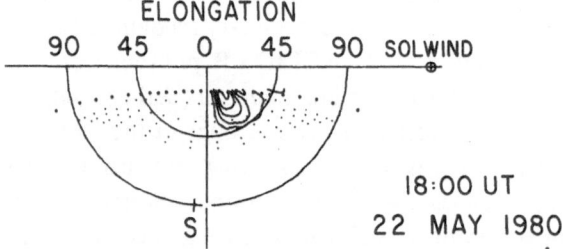

Fig. 6.   HELIOS A contour image of the 21 May, 1980 coronal mass ejection as described in McCabe *et al.* (1985). Electron columnar density is contoured in levels of 3×10^{14} cm^{-2}. The bold dots mark the centers of the 16° photometer sectors. The fainter dots indicate the outward projected positions of these sectors from earlier observations.

was also confined to a small angular dimension from the HELIOS A view and was confirmed to lie primarily along a line 45° below the ecliptic plane. More accurate estimates of the mass of this ejection were possible from both coronagraph (2×10^{16} g) and HELIOS A observations (3.5×10^{16} g) since the true position of the ejected material could be determined. A portion of the ejection is observed as prominence material in Hα at a low southerly latitude and this material must be directed southward of its origin in a highly non-radial fashion at the onset. Figure 6 depicts this ejection; the dots in Figure 6 demonstrate how the image is constructed from only the 16° photometer data.

6. Conclusions

The HELIOS imaging of coronal mass ejections to date has been used primarily to confirm and suppliment information on mass ejections obtained from the two Earth-orbiting coronagraphs. The resolution of the instrument is adequate to show the larger features of coronal mass ejections. Excess mass images are well-calibrated by stars in the photometer fields of view. The difficulty of determining the distance along the line-of-sight to the excess mass is a problem that concerns most measurements by coronagraphs. However, the comparison of the polarization brightness data with Earth perspective views can, as in the case of the 7 May, 1979 ejection, remove the ambiguity in the position of the major mass structures of the ejection.

Other features of the solar corona expected to extend into the interplanetary medium should be possible to image by HELIOS photometers. These features include coronal streamers as well as coronal mass ejections. Perhaps the heliospheric current sheet, thought to be marked in general by increased coronal brightness during solar minimum, can be mapped between Sun and Earth. These data are not unlike interplanetary scintillation measurements obtained from Earth (Gapper *et al.*, 1982) in that they measure enhancements globally. Many of these interplanetary scintillation enhancements seen to engulf Earth could be observed in density by HELIOS to determine whether as structures they were co-rotating or not.

One unique feature of these data is the ability to remotely measure the density of structures that are measured *in situ* at the same time. Thus, three-dimensional shapes of the density increases observed *in situ* may become more clear. These include the density increases behind shocks, the interface regions between high and low speed winds and the enhanced densities associated with the magnetic clouds observed by Klein and Burlaga (1982).

Future spacecraft instruments imploying these Thomson scattering techniques with higher spatial resolution may enable even more complete mapping of the interplanetary medium structures. Operated in conjunction with a wide-field coronagraph, such an 'all sky' imager could trace structures from Sun to Earth. A second all-sky imager in other than Earth orbit – perhaps at the L2 or L3 Lagrangian points could, by stereoscopic techniques, determine even better the dimensions of structures that both observe. The arrival times of shocks nearing Earth might be forecast with precision. In addition, the onsets of high-speed winds, if accompanied by a leading compression region, could also be forecast.

Data from the HELIOS photometers extend over nearly a complete solar cycle. To date, less than 2% of the available data has been used to form images from the HELIOS spacecraft. HELIOS A, launched December, 1974, has operated successfully until the present although various failures and limited power currently decrease the usefulness of the data. HELIOS B, launched January, 1976, operated through the beginning of 1980. The photometer experiment was generally turned-off when the spacecraft were at maximum distance from the Sun; they normally sent photometer data for only four of their six-month orbital periods. Data from the spacecraft zodiacal light photometers is available through the National Space Science Data Center in Goddard, Maryland, U.S.A., or this author on a limited basis.

Acknowledgements

I appreciate many helpful discussions with my colleagues about this review, especially those with Ch. Leinert who has graciously supported my analysis of the HELIOS photometer data from the beginning. Special appreciation is due R. Howard, N. Sheeley, Jr., D. Michels, and M. Koomen of the Naval Research Laboratory, who originally led me to the HELIOS photometer data and have supplied the SOLWIND coronagraph images presented in this review. Particular thanks is due H. Hudson for critically reading this manuscript and his encouraging comments throughout, and UCSD students Steve Hurlbut and Ray Ng whose assistance and analysis of the data have made presentations of the HELIOS images possible. The work of B. V. Jackson is supported by Air Force contract F19628–85–K–0037 and National Science Foundation grant ATM84–06487 with the University of California, San Diego.

References

Billings, D. E.: 1966, *A Guide to the Solar Corona*, Academic Press, New York, p. 150.

Burlaga, L. F., Klein, L., Sheeley, N. R., Jr., Michels, D. J., Howard, R. A., Koomen, M. J., Schwenn, R., and Rosenbauer, H.: 1982, *Geophys. Res. Letters* **9**, 1317.

Cane, H. V., Stone, R. G., Fainberg, J., Steinberg, J. L., and Hoang, S.: 1982, *Solar Phys.* **78**, 187.

Dryer, M., Perez-de-Tejada, H., Taylor, H. A., Jr., Intriligator, D. S., Milhalov, J. D., and Rompolt, B.: 1982, *J. Geophys. Res.* **87**, 9035.

Gapper, G. R., Hewish, A., Purvis, A., and Duffett–Smith, P. J.: 1982, *Nature* **296**, 633.

Hildner, E., Gosling, J. T., MacQueen, R. M., Munro, R. H., Poland, A. I., and Ross, C. L.: 1975, *Solar Phys.* **42**, 163.

Howard, R. A., Michels, D. J., Sheeley, N. R., Jr. and Koomen, M. J.: 1982, *Astrophys. J.* **263**, L101.

Howard, R. A., Sheeley, N. R., Jr., Koomen, M. J., and Michels, D. J.: 1985, *J. Geophys. Res.* **90**, 8173.

Hundausen, A. J., Bame, S. J., and Montgomery, M. D.: 1970, *J. Geophys. Res.* **75**, 4631.

Jackson, B. V.: 1985, *Solar Phys.* **95**, 363.

Jackson, B. V. and Hildner, E.: 1978, *Solar Phys.* **60**, 155.

Jackson, B. V. and Leinert, Ch.: 1985, *J. Geophys. Res.* (in press).

Jackson, B. V., Howard, R. A., Sheeley, N. R., Jr., Michels, D. J., Koomen, M. J., and Illing, R. M. E.: 1985, *J. Geophys. Res.* **90**, 5075.

Klein, L. W. and Burlaga, L. F.: 1982, *J. Geophys. Res.* **87**, 613.

Koomen, M. J., Detwiler, C. R., Bruecker, G. E., Cooper, H. W., and Tousey, R.: 1975, *Applied Optics* **14**, 743.

Leinert, C., Link, H., Pitz, E., Salm, N., and Kluppelberg, D.: 1975, *Raumfahrtforschung* **19**, 264.

Leinert, C., Pitz, E., Link, H., and N. Salm, N.: 1981, *J. Space Sci. Instr.* **5**, 257.

Leinert, C., Richter, I., Pitz, E., and Hanner, M.: 1982, *Astron. Astrophys.* **110**, 355.

McCabe, M. K., Švestka, Z., Howard, R. A., Jackson, B. V., and Sheeley, N. R., Jr.: 1985, *Solar Phys.* (submitted).

MacQueen, R. M., Eddy, J. A., Gosling, J. T., Hildner, E., Munro, R. H., Newkirk, G. A., Jr., Poland, A. I., and Ross, C. L.: 1974, *Astrophys. J.* **187**, L85.

Newton, H. W.: 1943, *Monthly Notices Roy. Astron. Soc.* **103**, 244.

Pettit, E.: 1925, *Publ. Yerkes Obs.* **3**, 205.

Poland, A. I., Howard, R. A., Koomen, M. J., Michels, D. J., and Sheeley, N. R., Jr.: 1981, *Solar Phys.* **69**, 169.

Richter, I., Leinert, C., and Planc, B.: 1982, *Astron. Astrophys.* **110**, 115.

Rickett, B. J.: 1975, *Solar Phys.* **43**, 237.

Sheeley, N. R., Jr., Michels, D. J., Howard, R. A., and Koomen, M. J.: 1981, *EOS* **62**, 153.

Sheeley, N. R., Jr., Howard, R. A., Koomen, M. J., Michels, D. J., Schwenn, R., Mulhauser, K. H., and Rosenbauer, H.: 1985, *J. Geophys. Res.* **90**, 163.

Wagner, W. J.: 1983, *Adv. Space Res.* **2**, 203.

Watanabe, T., and Kakinuma, T.: 1984, *Adv. Space Res.* **4**, 331.

THE SOLAR WIND

W. I. AXFORD

Max Planck Institut für Aeronomie, Katlenburg-Lindau, F.R.G.

Abstract. The current status of our understanding of the nature and origin of the solar wind is briefly reviewed, with emphasis being placed on the need for wave-particle interactions to account for the main energy source as well as details of the particle distribution functions. There has been considerable progress in the theoretical treatment of various aspects of the physics of the solar wind but a complete understanding is not yet in sight. Arguments concerning the ultimate fate of the solar wind are reviewed, in particular those concerning the distance to the shock wave which marks the termination of supersonic flow. This is of particular significance in view of recent observations suggesting that the termination might occur at about 50 AU from the Sun.

1. Introduction

The solar wind as an inference from cometary observations has been known for about 35 years, as a theoretical development, due to Parker, for almost 30 years and as a subject of direct observations by spacecraft for about 25 years. Both in terms of theoretical advances and the wealth of new information made available from experiments carried by dozens of spacecraft in the last two decades, the study of the solar wind has become one of the most elegant and intellectually satisfying branches of space plasma physics. The general concept of such winds has been extended to include not only stellar winds in general, but winds from comets and the outer envelopes of planets at one extreme, to winds from galactic clusters, galaxies and clusters of galaxies at the other. Consequently, there is a convincing case to be made for studying the solar wind in as much detail as possible in all its aspects since it is rare that we have a chance to make detailed observations of a phenomenon of such general astrophysical significance.

However, before getting too carried away with enthusiasm for the solar wind as a phenomenon, it is worth remembering, that, as far as the Sun itself is concerned, the black-body radiation of the photosphere is far more important than the solar wind in terms of mass loss, energy flux and momentum flux. The present solar wind is by comparison only important for the angular momentum loss of the Sun as a whole (see Table I). Furthermore, despite strenuous efforts in recent years to prove the contrary,

TABLE I

Relative importance of solar wind and radiation

	Solar wind	Radiation
Mass loss rate	10^{12} g s^{-1}	4×10^{12} g s^{-1}
Total mass lost	$5 \times 10^{-5} M_*$	$2 \times 10^{-4} M_*$
Energy flux (1 AU)	0.16 erg cm^{-2} s^{-1}	1.4×10^6 erg cm^{-2} s^{-1}
Momentum flux (1AU)	8×10^{-9} g cm^{-1} s^{-1}	5×10^{-5} g cm^{-1} s^{-1}
Despin time	10^{10} yr	10^{12} yr

Solar Physics **100** (1985) 575–586.   0038–0938/85.15.

the solar wind appears to affect the Earth only in the upper atmosphere above about 100 km altitude and in the polar regions, although there are some fairly esoteric arguments that it may have a dominant role in determining the atmospheric abundances of some minor constituents such as ^{3}He.

On the other hand, in the distant past, when the solar system was being formed, the situation may well have been different. The solar wind at the time, is likely to have been sufficiently strong to have carried away most of the angular momentum of the primordial Sun, to have dissipated the disc-shaped nebula which must have formed its extended outer atmosphere and in so doing may well have determined the different characteristics of the inner and outer planets.

2. Why Does the Solar Wind Blow?

During the mid 1950s it was evident from observations of the behaviour of the plasma tails of comets, the almost continual presence of aurorae and associated phenomena in the Earth's polar regions and the apparent absence of a general, extended hydrostatic solar corona, that the solar wind must exist. Parker (1963, 1969; and earlier papers cited therein) pointed out that a continually expanding wind should be a natural consequence of the existence of a hot corona and furthermore that, since the pressure of the interstellar medium is so small, a hydrostatic corona could not possibly be sustained as a consequence of thermal conduction. In fact the latter argument must be altered slightly if one takes into account the existence of a solar magnetic field, but even then some sort of solar wind would have to flow from the Sun's magnetic polar regions.

Chamberlain (1961, 1965) attacked Parker's views in a series of papers which claimed that, with heat conduction taken into account, the solar wind might be nothing more than an innocuous subsonic solar 'breeze'. These arguments soon lost any force they may have had as a result of *in situ* observations by Soviet and US spacecraft which confirmed the existence of the solar wind, as inferred from cometary observations, and in the form of a supersonic radial outflow with an imbedded spiral magnetic field as predicted by Parker. For reviews of these earlier observations see Axford (1969) and Hundhausen (1968).

Now that we know so much about the solar wind one can return to these early arguments and examine them again. In particular it should be noted that solar wind protons observed at 1 AU typically have kinetic energies of the order of 10^3 eV and more. Since the thermal energy available to such protons and their associated electrons when they form part of the lower corona is only about 430 eV and the gravitational binding energy is about 2000 eV it is immediately apparent that the solar wind must be rather more than simply the adiabatic expansion of a hot corona (see Table II). Evidently some means of transferring energy non-adiabatically to the outer corona, such as heat conduction, must play an important role. However, even taking the effects of heat conduction into account in the most generous way one finds that a solar wind would exist (despite Chamberlain's arguments), but its characteristics would not match those

TABLE II

Energy comparison

Energy	Base of corona	Slow stream	Fast stream
Gravitational ($-GMm/r$)	-2000 eV	–	–
Thermal ($5k(T_p + T_e)/2$)	430 eV	30 eV	60 eV
Kinetic ($mV^2/2$)	–	1000 eV	4000 eV
Total[a]	-1570 eV	$+1030$ eV[b]	$+4060$ eV[b]

[a] The average solar wind energy flux is $\sim 8 \times 10^5$ erg cm^{-2} s^{-1} which is about the same as the transition zone UV radiation flux.

[b] The plasma must be given 2600–5630 eV/$e-p$ pair in addition to the energy it has at the base of the hot corona ($\sim 1.5 \times 10^6$ K).

of the observed solar wind, namely in that it would be slow, the ions too cool and the electrons too hot (Hartle and Sturrock, 1968).

In order to account for the basic features of the solar wind, it is necessary to assume as a minimum that there is an energy source which is distributed throughout the corona and solar wind close to the Sun (e.g., Holzer and Axford, 1970) and that the effects of electron heat conduction are suppressed substantially below those predicted by the Spitzer–Harm model for the conductivity (e.g., Forslund, 1970). In these circumstances it is inappropriate to treat the heating of the corona itself as a separate matter; presumably, the energy sources for the heating of the corona, for the radiation from the chromosphere-corona transition region and for the solar wind are ultimately the same and the whole system must be regarded as a single phenomenon. It is in fact interesting to note that for equal areas at the base of the corona, the energy flux associated with high-speed solar wind streams is comparable to the energy flux of the transition region radiation in regions with generally closed solar magnetic field lines where the corona is effectively hydrostatic and the solar wind at most intermittent. This energy flux is of the order of 10^6 ergs cm^{-2} s^{-1}.

The source of the energy required as far as the corona and solar wind are concerned, seems likely to be associated with hydromagnetic waves of one sort or another which are damped as they propagate away from the Sun (e.g., Hollweg, 1983). Whether the wave amplitudes are so large that they should be considered as shocks or they are no more than midly nonlinear is unclear. However, some remnant of the heating process appears to persist out to very large distances from the Sun, perhaps even to the Earth's orbit. The source of these waves is not known though there are several possibilities, each with an adequate reservoir of energy, notably the relatively large-scale motions associated with the solar supergranulation and, at the other extreme, microflare and other small-scale features such as spicules which may be associated with magnetic reconnection and localized heating.

In passing it should be noted that although the solar wind, particularly in high-speed streams, is almost collisionless in the sense that the mean-free path near the Earth's orbit is typically of the order of an astronomical unit or larger, a purely collisionless 'exopheric' model for the solar wind fails to produce its observed characteristics. For example, if it is assumed that the distribution function is collision-dominated and Maxwellian in the moving frame out to a few solar radii and beyond that totally collisionless and free from any wave forces, it is found that the predicted solar wind speed is far too low and the ratio of parallel to perpendicular proton temperatures greatly exceeds unity in contradiction to the observations. Furthermore, the behaviour of heavier species, especially those with large mass-to-charge ratios, predicted by such a model again cannot be made to agree with observations which show that, in high speed streams where the model is most likely to be appropriate, the relative abundances of various species are close to the expected solar abundances. In the case of low speed streams, Coulomb collisions between ions are very important in maintaining distribution functions which are almost isotropic in the solar wind frame and in forcing different species to move together and have comparable temperatures. For these reasons, it is necessary to adopt a multi-fluid approach to solar wind theory while recognising that collisionless processes are important in determining wave damping and in the transfer of energy and momentum from waves to particles (e.g., Marsch *et al.*, 1982).

It is evident from observations of high speed streams that a relatively large distances, the behaviour of the solar wind is determined to a large extent by wave-particle inter-actions involving waves which originate near the Sun, (e.g., Hollweg, 1974; Feldman, 1979; Barnes, 1979; Schwartz, 1980; Marsch, 1985):

(1) The perpendicular temperature is usually larger than the parallel temperature which is the opposite to what would be expected in the absence of preferential absorption of wave energy into perpendicular motion in the solar wind frame;

(2) all ionic species tend to have the same temperature per unit mass; and

(3) heavier species tend to move slightly faster than the protons which is very difficult to explain in terms of a collison-dominated or totally exospheric model.

It seems possible to understand these effects only on the basis of wave-particle interactions, with cyclotron resonance effects being important in some way since this permits variations in the behaviour of species with differing charge/mass ratios as a result of resonance with different parts of the wave spectrum.

In slow streams, which are evidently associated with magnetic sector boundaries (see Neugebauer, 1983), the above factors are absent, as to a large extent is magnetohydro-dynamic turbulence (Denskat and Neubauer, 1983). The minor species have highly varying relative abundances, although their charge states indicate an origin in a reason-ably hot corona (Bame, 1983). It seems likely that this type of solar wind has its origin in regions of the corona which are magnetically open only transiently so that they do not represent an equilibrium state of the solar wind (Axford, 1977).

3. Two-Fluid Models of the Solar Wind

As a first approach to modelling the solar wind, taking into account the various physical processes alluded to above, one can consider the case of sphericially symmetric flow of a proton-electron plasma with an extremely weak spiral magnetic field. A minimum set of equations describing such a situation would be (see Holzer and Axford, 1970):

$$NUr^2 = F = \text{constant},\tag{1}$$

$$NmU\frac{dU}{dr} + \frac{d}{dr}[Nk(T_p^{\parallel} + T_e)] + \frac{2}{r}Nk(T_p^{\parallel} - T_p^{\perp}) + \frac{GM_*}{r^2} = H,\tag{2}$$

$$\frac{dT_p^{\parallel}}{dr} = \frac{T_p^{\parallel}}{N^2}B^2\frac{d}{dr}\left(\frac{N^2}{B^2}\right) + \frac{2v_{pp}}{U}(T_p^{\perp} - T_p^{\parallel}) + \frac{2Q_p^{\parallel}r^2}{kF},\tag{3}$$

$$\frac{dT_p^{\perp}}{dr} = \frac{T_p^{\perp}}{B}\frac{dB}{dr} + \frac{v_{pp}}{U}(T_p^{\parallel} - T_p^{\perp}) + \frac{2Q_p^{\perp}r^2}{kF},\tag{4}$$

$$\frac{dT_e}{dr} = \frac{2}{3}\frac{T_e}{N}\frac{dN}{dr} + \frac{v_{pe}}{U}(T_p - T_e) + \frac{2}{3kF}\frac{d}{dr}\left[Kr^2\cos^2\psi\frac{dT_e}{dr}\right] + \frac{2Q_e r^2}{kF}.\tag{5}$$

Here U is the solar wind speed, N the number density, G the gravitational constant, T the temperature, k Bolzmann's constant, K the electron thermal conductivity, v the collision frequency, B the magnetic field strength, Q a heat source, H a momentum source, and ψ the angle between the magnetic field lines and the radial direction. Subscripts p, e refer to protons and electrons respectively and the superscripts $\parallel$ and $\perp$ refer to the parallel and perpendicular directions relative to the magnetic field.

Hartle and Sturrock (1968) have solved this set of equations with the simplifying assumption that the temperature is known in the corona and that the heat and momentum sources are otherwise absent, that the magnetic field is radial and that the proton temperature is isotropic. Even modifying their results to allow for a spiral magnetic field and supressing the value of the heat conductivity to below the Spitzer-Harm value, it is not possible to obtain a realistic model of the solar wind with reasonable coronal temperatures. It is evidently necessary that the heat be injected in a distributed manner rather than in a narrow region at the base of the corona, and that the proton temperature at least should be allowed to be anisotropic. Leer and Axford (1972) have explored models of this type, assuming $H = 0$, and find that it is possible to achieve more satisfactory results with a suitable assumption for the heat source, but that it is necessary to have $Q_p^{\perp} \gg Q_p^{\parallel}$ in order to keep $T_p^{\perp} > T_p^{\parallel}$ as is observed. With such models it is, in principle, possible to permit a realistic treatment of the chromosphere-corona transition region by assuming a suitable chromospheric temperature and cooling function. However the procedure is unfortunately rather arbitrary in the sense that the energy and momentum transfer required from the presumed field of magnetohydrodynamic waves is postulated on a completely ad hoc basis.

The relevant equations allowing for the presence of Alfvén waves have been developed by Belcher (1971). As a consequence of the nearly incompressible character of these waves a convergent iterative sheme exists for solving the coupled 'background' and 'propagation' equations. Assuming that the waves do not damp but propagate according to WKB theory, one finds that energy and momentum are transferred from the waves to the fluid without dissipation so that

$$H = -\frac{r^2}{F}\frac{\mathrm{d}}{\mathrm{d}r}\left(\frac{b^2}{8\pi}\right), \qquad Q = 0,$$ (6)

where b is the wave amplitude. A possible resonant contribution to H has been neglected. These equations have been solved under certain simplifying assumptions, notably that the electrons and protons behave isothermally (Hollweg, 1978). However such a model is not entirely satisfactory since the relative wave amplitude becomes large in the absence of dissipation, which is not observed, at least if near-Sun amplitudes sufficiently large to drive the solar wind are required. In fact, some form of non-linear damping may limit the wave amplitude such that $b \lesssim B$. In order to satisfactorily deal with this situation it would be necessary to solve the wave propagation equations allowing for dissipation (possibly nonlinear) in which case H and Q are both nonzero. Consequently the waves may drive the wind directly through their radiation pressure gradient and also indirectly by heating the plasma so that it behaves nonadiabatically.

The observation that $T_p^\perp > T_p^\parallel$ in high-speed streams suggests that cyclotron damping is important at some stage of the wave dissipation process. This complicates matters further, however, as the propagation of Alfvén waves is affected by pressure anisotropies and instabilities can occur if the anisotropy becomes too large in either sense. Perhaps the simplest first approach to dealing with these complexities would be to assume that $b \approx B$ everywhere that H is determined by the equivalent wave pressure gradient as before, that the waves remain Alfvénic and that the lost wave energy appears in $Q_p^\perp$ only (e.g., Isenberg and Hollweg, 1982).

4. The Behaviour of Minor Species

It is striking that the minor constituents of the solar wind plasma in high-speed streams have fairly constant relative abundances (similar to those expected from the photosphere), equivalent temperatures which are roughly proportional to ion mass and a tendency to move faster than the proton component by an amount approximately of the order of the Alvén speed. In low-speed streams the effects of Coulomb collisions are apparent since the temperatures and speeds of all ionic species tends to become equal, but the relative abundances are quite variable.

One cannot account for any of the features of high-speed streams if Coulomb collisions play a significant role other than in the lower corona. Instead it is necessary once again to invoke wave-particle interactions in a manner consistent with that described above for a two-fluid model. Provided the waves concerned are Alfvén or ion cyclotron waves, which are incompressible to first order, a quasi-linear approach similar to that outlined above for the two fluid model permits some progress to be made provided the concentrations of minor species are small and their presence does not affect the behaviour of the proton and electron components.

The relevant equations of motion of minor species with atomic mass Am and charge

Ze, to be combined with (1)–(6) for example are:

$$nAmur^2 = F_i = \text{constant} , \tag{7}$$

$$u \frac{du}{dr} - \frac{w^2}{r} = \frac{\Omega u}{U} \frac{d}{dr} (wr) + v_{ip}(U - u) (1 + (\Omega r/U)^2) - \frac{Z}{Amn} \frac{db_e}{dr} -$$

$$-\frac{1}{nm} \frac{d}{dr} (nkT_i) - \frac{GM_*}{r^2} + H_i + H_i(\text{res}) , \tag{8}$$

where

$$w = (U - u) \Omega r/U , \tag{9}$$

and

$$H_i = \frac{d}{dr} \left[\frac{b^2}{2B_r^2} \left((U + V_{Ar})^2 - u^2 \right) \right] , \qquad Q_i = 0 . \tag{10}$$

Thus

$$u \frac{du}{dr} - U \frac{dU}{dr} = \frac{1}{2} \frac{d}{dr} \left[(u^2 - u^2) \left(\frac{b^2}{B_r^2} + \left(\frac{\Omega r}{U} \right)^2 \right) \right] +$$

$$+ v_{ip}(U - u) (1 + (\Omega r/U)^2) + H_i(\text{res}) - H(\text{res}) -$$

$$- \left[\frac{1}{Amn} \frac{d}{dr} (nkT_i) - \frac{1}{Nm} \frac{d}{dr} (Nk(T_p + T_e)) \right] - \frac{Z}{Amn} \frac{d}{dr} (NkT_e) . \tag{11}$$

Here n, $m_i = Am$, $u = (u, v, w)$, T_i are the number density, atomic mass, flow velocity, and temperature of the minor species, respectively, Ω is the angular frequency of solar rotation and H_i (res) is the resonant component of momentum transfer from waves.

On examining these equations one notes first from (10) that the non-resonant wave acceleration of all minor species vanishes if $u = U + V_{Ar}$, as might be expected since the waves are effectively static structures in this frame of reference. However, if one considers the difference between the minor ion and proton accelerations (Equation (11)) it is evident in the absence of pressure gradients and resonance effects, the solution $u = U$, $w = 0$ is an acceptable one and the pressure gradients alone would tend to make $u < U$ unless T_i is quite large. There is evidently not enough physics in this treatment to suggest why the minor species in high-speed streams should have $u \approx U + V_{Ar}$ and $T_i \approx AT_p$. As with the two-fluid model, a WKB account of the wave propagation is unsatisfactory since it permits the waves to become strongly non-linear ($b > B$) in contradiction to the observations. If wave dissipation is included in the analysis it is no longer obvious that the non-resonant and resonant contributions to H, H_i and to Q, Q_i should not change from the values given in (6) and (10).

Since most observed Alfvén waves in high-speed streams are such that $b \approx B$ it would appear that their amplitudes may be controlled by nonlinear damping processes, regardless of the presence of minor species. However at any given distance from the Sun, the wave spectrum affects, and is affected by, various species according to their charge/mass ratios and their concentrations, with protons perhaps being least prone to resonance effects. The observation that $u \approx U + V_{Ar}$ may be the result of resonant acceleration/heating of minor species being otherwise dominant in equations (8) and (11), which should then be approximated by $H_i(res) = 0$ which would imply the required result. As a corollary it should be expected that the most important component of the momentum transfer is associated with the Lorentz force of the waves rather than the 'centrifugal' force (McKenzie *et al.*, 1979, Appendix D). However, the argument does not suggest an explanation for the observation that $T_i \approx A T_p$ in high-speed streams and indeed any approach along the lines outlined must fail since only incompressible (transverse) waves are permitted with the temperatures playing no clear role. Models with compressible modes are required and it has been demonstrated that an equalization of ion thermal speeds can be achieved under suitable conditons (see Marsch, 1983, for a review).

The only attempt at incorporating all these effects into a solar wind model is that of Isenberg and Hollweg (1983) (see also Isenberg, 1983, 1984), who found that there are difficulties in explaining the preferential energization of heavy ions with reasonable assumptions concerning the spectrum of the waves.

Self-consistency is a difficulty in any solar wind model involving minor species, since the concentration of helium ions for example, is about 15–20% of that of protons by mass and this is not negligible. As a result of resonances, even tiny concentrations of minor species can affect wave propagation profoundly at the appropriate frequencies (e.g., Isenberg, 1983, 1984). Furthermore, simple three fluid models of the solar wind, excluding wave effects, suggest that the helium can play an important role in determining the detailed form of the solution even if its general nature is unchanged (e.g., Yeh, 1970; Joselyn and Holzer, 1978). Not surprisingly little progress has been made in carrying out a completely self-consistent treatment of the solar wind problem.

5. Termination of the Supersonic Solar Wind

Since the solar wind is supersonic it must, in order to adapt to the interstellar medium, be terminated by a shock wave beyond which the plasma flow is subsonic. It is possible to construct situations in which cosmic ray pressure gradients and loss of momentum resulting from charge-exchange with interstellar neutral gas can produce a smooth (transonic) transition to subsonic flow (Wallis, 1973). However, conditions must be rather special for this to happen if the overall pressure of the interstellar medium is finite.

If the effects of the solar magnetic field are neglected, the region of subsonic flow between the shock wave and the outer boundary of the heliosphere (or 'heliopause' – the interface between solar and interstellar plasma) would have almost constant pressure. In this case, the position of the shock wave can be determined by equating the dynamic pressure of the solar wind at the shock to the pressure of the interstellar medium just beyond the heliopause (see Axford, 1972, for a review). Thus

$$N_0 \overline{m} U_s^2 / R_s^2 = K P_g, \tag{12}$$

where N_0 is the solar wind number density at 1 AU, $\overline{m} = 2 \times 10^{-24}$ g is the mean mass of solar wind ions, U_s is the solar wind speed just within the shock position, R_s is the distance to the shock in AU, $K = 1.13$ for a strong shock and P_g is the total pressure of interstellar medium just beyond the heliopause. Ignoring to begin with the possible contribution from galactic cosmic rays, the main contributors to the total pressure of the interstellar medium at the heliopause are the pressure of the interstellar magnetic field (B_g) the pressure of the interstellar plasma and the dynamic pressure of the interstellar plasma, thus

$$P_g = \alpha B_g^2 / 8\pi + n_g (2kT_g + \overline{m} V_g^2), \tag{13}$$

where n_g is the number density of the interstellar plasma, V_g is its speed relative to the Sun, and T_g its temperature. We have included a factor α to allow for the possibility that the magnetic pressure is enhanced as a result of being wrapped around the roughly spherical upstream face of the heliopause. For example, if the interstellar magnetic field direction is perpendicular to the relative velocity vector of the interstellar plasma with respect to the Sun it might be appropriate to take $\alpha = 2.25$.

It is not possible to measure the various quantities which determine P_g directly, but measurements of the dispersion and Faraday rotation of pulsar emissions can provide us with average values of n_g and B_g throughout a sphere of about 100 parsecs radius surrounding the Sun. Provided the interstellar medium within the sphere is fairly uniform these are the best values available, although the possibility of local variations cannot be excluded. Assuming that U_s has the same value as observed near the Earth, namely 400 km s^{-1} on average, that the flux is 2×10^8 cm^{-2} s^{-1} at 1 AU and that $\alpha = 2.25$, one can estimate a (minimum) value for R_s. For example, taking $V_g = 20$ km s^{-1} as suggested by observations of the neutral interstellar hydrogen, with $n_g = 0.05$ cm^{-2} and $B_g = 3 \times 10^{-6}$ G as suggested by pulsar observations, and with $T_g = 10^4$ K, we find from (12) and (13) that the minimum distance to the shock termination of the solar wind as a supersonic flow is about 100 AU. It is interesting to note that in this case the three components of P_g contribute in the ratio 81 : 14 : 40 so that the magnetic pressure is most important. Some observations of the relative velocity of the interstellar helium suggest that $V_g = 27$ km s^{-1} and if we were to assume this value, together with n_g and B_g unchanged, it is possible to argue that the shock might occur as close as 90 AU.

Recent observations by Kurth et al. (1984) suggest, on the basis of persistent radio emissions in the 2–3 kHz range observed from the Voyager spacecraft, that the shock termination occurs at a distance of 46 AU. The authors suggest that the radiation is emitted at the second harmonic of the plasma frequency corresponding to the density just behind the shock and that it penetrates in towards the Sun to a point where this is also the local plasma frequency in the solar wind. If this interpretation is correct the conclusion is important because it suggests that the shock will be encountered by Pioneer and Voyager spacecraft in a few years time. For the present, we are in the situation of being able to question the interpretation or of finding alternative explanations.

It is interesting to note that the plasma frequency of the interstellar medium lies in the range 1–3 kHz if the electron number density is 0.01–0.1 cm^{-2}. This suggests that the heliosphere, with the exception of the region around the Sun out to about 30 AU, is essentially a cavity surrounded by a medium in which such low frequency waves cannot easily propagate Accordingly, one would expect that waves originating anywhere within the cavity with frequencies below this cut-off frequency of a few kHz, would simply accumulate until their sources are balanced by losses due to damping or escape into the wake of the heliosphere. The intensity at a given frequency and also the upper cut-off frequency would be fairly constant unless observations are being made in the vicinity of possible sources such as interplanetary shocks. The characteristics of such a radiation field would be essentially independent of the distance to the shock wave terminating the supersonic solar wind and the inference made by Kurth *et al.* concerning its distance from the Sun would have to be abandoned.

Let us assume, however, that the shock termination indeed occurs at a distance of the order of 50 AU and seek possible explanations. This position for the shock wave could be achieved if the electron number density in interstellar space were as large as 0.3–0.4 cm^{-3}, or if the interstellar magnetic field strength were 12×10^{-6} G with $\alpha = 2.25$. However, with the pulsar values for these quantities and $V_g = 20$–30 km s^{-1}, to reduce R_s to less than 50 AU, it is necessary to find additional processes which either change the pressure balance condition or reduce the dynamic pressure of the solar wind (cf. Axford, 1973). It should be noted that we have in any case adopted fairly low values for the average solar wind speed and flux (e.g., Schwenn, 1983).

Although the plasma flow speed in the region between the shock wave and the heliopause is subsonic, it is not necessarily compressible nor must the total pressure be uniform, provided the magnetic field strength can build up appreciably. This will occur quite naturally even if the flow is strictly adiabatic and radial as shown by Axford (1972). However, with $R_s = 50$ AU, the heliopause would have to be at a distance of 100 AU or more for this effect to be appreciable. This distance is drastically reduced if the flow is non-radial so that magnetic flux tubes are evacuated of their plasma, and/or the flow is non-adiabatic as a result of charge-exchange between the initially 10^3 eV solar wind plasma and the 5 eV interplanetary hydrogen which permeates the region. As a consequence of these two effects, the magnetic field strength in the region must build up until at the heliopause the (solar) magnetic field pressure balances the total interstellar pressure, while at the shock wave the total pressure taken up by the solar wind dynamic pressure may be somewhat larger. The net result is a diminution of R by an amount of the order of d/R_s, where d is the thickness of the subsonic region in astronomical units; the diminution is perhaps 10–20%.

A second way of altering the pressure balance condition is to assume there is a so far undetected component of the galactic cosmic radiation which is either excluded from the heliosphere entirely or is unable to penetrate significantly within the shock. If these particles are excluded by the heliopause their pressure must simply be included in the pressure of the interstellar medium. If on the other hand they are able to penetrate the subsonic region but are effectively kept out of the region of supersonic solar wind flow the effect is the same as far as the relation (12) is concerned. Ip and Axford (1985) have estimated the spectrum of galactic cosmic-rays based on current ideas concerning their acceleration and have concluded that as a result of collisional losses and convective escape from the galaxy the spectrum falls off steeply at low energies and there is in fact no significant contribution to the interstellar pressure from (unobserved) low-energy cosmic rays. Most of the cosmic-ray pressure in this case is concentrated in the relativistic range where particles enter the supersonic region of the solar wind fairly freely so that the pressure balance condition is not directly affected. As pointed out most recently by Suess and Dessler (1985), if the (unmodulated) interstellar spectrum of galactic cosmic-rays continues to rise even fairly modestly in the unobserved subrelativistic range it is, ignoring any understanding we might have of the physics involved, possible to produce an interstellar pressure of the order of 6×10^{-12} dyn cm^{-2} necessary to bring the shock termination to the position inferred by Kurth *et al.* Whether or not the pressure of low-energy galactic cosmic-rays is indeed ten times that of relativistic cosmic-rays can only be determined by making the appropriate cosmic-ray measurements beyond 50 AU.

It is possible to reduce the distance to the terminating shock by modifying the ram pressure of the solar wind on the left-hand side of (12) (Axford, 1973). There are three ways in which this can be done. Firstly, the solar wind can be made to slow down by charge-exchange with interstellar neutral hydrogen atoms with penetrate freely to within about 4 AU of the Sun. The number density of the neutral hydrogen is observed to be not more than about 0.1 cm^{-2} and according to the radial flow model developed by Holzer (1972) this is sufficient to reduce the solar wind speed by about 15% at a distance of 50 AU without changing its flux. This translates to a 15% decrease of the solar wind dynamic pressure at the shock so that to bring the shock to this point would still require P_g to be substantially larger than we have considered reasonable. Secondly, the adverse pressure gradient of the (modulated) galactic cosmic-rays also tends to slow the solar wind down, again without changing the flux. However, a simple analysis, assuming that a cosmic-ray gradient of 1–2% AU^{-1} is maintained out to 50 AU, indicates that the diminution of the solar wind speed is only about 10% at the very most and so the requirement for a rather high interstellar pressure remains. The third effect, pointed out by Seuss and Dessler (1985), involves non-radial flow of the solar wind due to the fact that, if there is spherical symmetry near the Sun, the latitudinal pressure gradient of the interplanetary field deflects the solar wind away from equatorial regions at large distances, reducing the flux but maintaining U constant. According to these authors the effect is not more than 20% at $R = 50$ so that again it is not possible to avoid a fairly large value of P_g if the shock is to occur at this distance.

So many factors are unknown at present that it is unreasonable to make any hard and fast pronouncements. If the terminating shock occurs near 50 AU as inferred by Kurth *et al.* and this is confirmed by the Pioneer and Voyager spacecraft, measurements of cosmic-ray and solar wind parameters are likely to give an indication of the cause.

If not, the explanation that the radiation is trapped in the heliosphere cavity may have to be preferred.

References

Axford, W. I.: 1968, *Space Sci. Rev.* **8**, 331.

Axford, W. I.: 1972, *Solar Wind,* NASA Space-308, p 609.

Axford, W. I.: 1973, *Space Sci. Rev.* **14**, 532.

Axford, W. I.: 1977, in *Study of Travelling Interplanteary Phenomena*, D. Reidel Publ. Co., Dordrecht, Holland, p. 145.

Bame, S. J.: 1983, *Solar Wind 5,* NASA/CP2280, p 575.

Barnes, A.: 1979, in *Solar System Plasma Physics,* North-Holland, Amsterdam.

Belcher, J. W.: 1971, *Astrophys. J.,* **168**, 509.

Chamberlain, J. W.: 1961, *Astrophys. J.* **133**, 675.

Chamberlain, J. W.: 1965, *Astrophys. J.* **141**, 320,.

Denskat, K. U. and Neubauer, F. M.: 1983, *Solar Wind 5*, NASA/CP2280, p 81.

Feldman, W. C.: 1979, *Rev. Geophys. Space Phys.* **17**, 1973.

Forslund, D. W.: 1970, *J. Geophys. Res.* **75**, 17.

Hartle, R. E. and Sturrock, P. A.: 1968, *Astrophys. J.* **151**, 1155.

Hollweg, J. V.: 1975, *Rev. Geophys. Space Phys.* **13**, 263.

Hollweg, J. V.: 1978, *Rev. Geophys. Space Phys.* **16**, 689.

Hollweg, J. V.: 1983, *Solar Wind 5,* NASA/CP2280, p 5.

Holzer, T. E. and Axford W. I.: 1970, *Ann. Rev. Astron. Astrophys.* **8**, 31.

Holzer, T. E.: 1972, *J. Geophys. Res.* **77**, 5407.

Hundhausen, A. J.: 1968, *Space Sci. Rev.* **8**, 690.

Ip, W.–H. and Axford, W. I.: 1985, *Astron. Astrophys.* **149**, 7.

Isenberg, P. A.: 1983, *Solar Wind 5,* NASA/CP2280, p 655.

Isenberg, P. A.: 1984, *J. Geophys. Res.* **89**, 2133.

Isenberg, P. A. and Hollweg J. V.: 1982, *J. Geophys. Res.* **87**, 5023.

Isenberg, P. A. and Hollweg J. V.: 1983, *J. Geophys. Res.* **88**, 3923.

Joselyn, J. and Holzer, T. E.: 1978, *J. Geophys. Res.* **83**, 1019.

Kurth, W. S., Gurnett, D. A., Scarf, F. L. and Poynter, R. L.: 1984, *Nature* **312**, 27.

Leer, E. and Axford, W. I.: 1972, *Solar Phys.* **23**, 238.

Marsch, E., Goertz, C. K., and Richter, K.: 1982, *J. Geophys. Res.* **87**, 5030.

Marsch, E.: 1983, *Solar Wind 5,* NASA/CP2280, p. 355.

McKenzie, J. F., Ip, W.–H., and Axford, W. I.: 1979, *Astrophys. Space Sci.* **64**, 183.

Neugebauer, M.: 1983, *Solar Wind 5,* NASA/CP2280, p. 135.

Parker, E. N.: 1963, *Interplanetary Dynamical Processes,* Interscience, New York.

Parker, E. N.: 1969, *Space Sci. Rev.* **9**, 325.

Schwartz, S. T.: 1980, *Rev. Geophys. Space Phys.* **18**, 313.

Schwenn, R.: 1983, *Solar Wind 5,* NASA/CP2280, p 481.

Suess, S. T. and Dessler, A. J.: 1985, unpublished preprint.

Wallis, M.: 1973, *Astrophys. Space Sci.* **20**, 3.

Yeh, T.: 1970, *Planetary Space Sci.* **18**, 199.

MASS LOSS FROM SOLAR-TYPE STARS

L. HARTMANN

Harvard-Smithsonian Center for Astrophysics, Cambridge, MA 02138, U.S.A.

Abstract. Winds are directly detected from solar-type stars only when they are very young. At ages $\sim 10^6$ yr, these stars have mass loss rates $\sim 10^6$ times the mass flux of the present solar wind. Although these young T Tauri stars exhibit ultraviolet transition-region and X-ray coronal emission, the large particle densities of the massive winds lead to efficient radiative cooling, and wind temperatures are only $\sim 10^4$ K. In these circumstances thermal acceleration is unlikely to play an important role in driving the mass loss. Turbulent energy fluxes may be responsible for the observed mass loss, particularly if substantial magnetic fields are present.

The presence of stellar mass loss is indirectly shown by the spindown of low-mass stars as they age. It appears that many solar-mass stars spin up as they contract toward the Main-Sequence, reaching a maximum equatorial velocity of 50 to 100 km s^{-1}. These stars spin down rapidly upon reaching the Main Sequence. Spindown may be enhanced by a decoupling or lag between convective envelope and radiative core. Because this spindown occurs fairly early in a solar-type star's history, the internal structure of old stars like the Sun may not depend upon initial conditions.

1. Introduction

The wind of the present-day Sun is too feeble to be detected at large distances. We presume that the Sun is not unique; we know that many other Main-Sequence stars possess hot coronae, so that physical conditions commonly seem favorable for the production of solar-type winds. But nothing is known about ejection from direct wind measurements of solar analogues, and we can only infer that mass loss must be occurring from measurements of the spindown of stars.

While stellar observations provide few insights into the physics of the present solar wind, such data provide an opportunity to discover the properties of solar mass loss at early and late times in the Sun's evolution, when mass ejection rates are largest. The student of solar activity should have a particular interest in the early behaviour of the Sun, for it is in this stage that magnetic activity is at its greatest intensity. It is thought that a complicated feedback mechanism controls the level of magnetic activity. The interaction of rotation and convection generates magnetic fields; mass loss is produced by magnetic effects; and angular momentum is carried away by the wind at a rate determined by the mass flow and magnetic field properties, slowing the rotation of the star. The complex interaction of these effects over time are responsible for the rather modest magnetic field of the Sun today.

The subject of stellar spindown has received special impetus from recent observations of rotation in young cluster dwarfs. These data suggest that when the Sun first arrived on the Main Sequence, it may have been rotating as much as fifty times faster than today. The rapidly-rotating G dwarfs spin down very quickly on the Main Sequence; it is not clear whether the convective envelope and radiative core spin down together, or whether the core lags behind, producing substantial differential rotation. The new observations

Solar Physics **100** (1985) 587–597. 0038–0938/85.15.

promise to expand our understanding of the connections between rotation, magnetic activity, and mass loss.

In the following sections I shall describe the present picture of mass loss from solar-type (i.e., low-mass) stars, with special emphasis on winds from pre-Main-Sequence stars.

2. Winds from T Tauri Stars

2.1. BASIC PROPERTIES

According to the standard picture, newly-formed, low-mass stars are first optically detected as T Tauri variables. These stars are distinguished by extremely strong emission in lines from the hydrogen Balmer series and low-excitation species such as Ca II, Mg II, Fe II, and Na I. The total energy radiated in these lines and in continuum emission in excess of normal photospheric levels is very large, up to $\sim 10^{-1}$ of the total stellar luminosity (cf. Cohen and Kuhi, 1979; Calvet, 1983).

As T Tauri stars are clearly young, it is tempting to identify the energy source for all this activity with accretion (cf. Walker, 1972; Lynden-Bell and Pringle, 1974; Appenzeller and Wolf, 1977; Wolf et al., 1977; Ulrich, 1976; Uchida and Shibata, 1983). However, the case for radial infall is extremely weak; most objects exhibit Hα P Cygni profiles indicating outflow (cf. Herbig, 1977). Furthermore, it is unlikely that most T Tauri stars are in the gravitational collapse stage, as this should take place over $\sim 10^5$ yr, while the lifetime of the T Tauri phase is between 10^6 and 10^7 yr. It seems more likely that the collapse phase occurs when the star is shrouded from view by its dusty cocoon, so that the optically-observable T Tauri stars represent a more advanced stage of evolution.

Disk accretion remains a possible energy source (cf. Strom, 1983). A variety of observations suggest that circumstellar disks are common (e.g., Cohen, 1983; Beckwith et al., 1985; Appenzeller et al., 1985), but the best direct observational evidence for active disk accretion lies in the rather atypical FU Orionis outbursts (cf. Hartmann and Kenyon, 1985).

Herbig (1970) suggested that the emission envelopes of T Tauri stars could be regarded as a dense chromospheres, pointing out resemblances of the emission spectrum to solar active region and plage spectra. It is presumed that the large amounts of mechanical energy needed to produce the observed emission fluxes are produced in some way by magnetic fields, in analogy with the Sun. Indeed, given the envelope density estimates gives below, it is difficult for acoustic waves to carry the required mechanical energy fluxes. The inference of powerful magnetic activity is strengthened by the observation of rotational modulation of photospheric light observed in several older T Tauris, indicating the presence of large 'spot' areas (Kappelmann and Mauder, 1981; Schaefer, 1983; Rydgren and Vrba, 1983).

Plane-parallel chromospheric models for T Tauri emission have been investigated by several authors (cf. Dumont et al., 1973; Cram, 1979; Calvet et al., 1984). While these models are successful in explaining many properties of the observed emission, they are unable to reproduce the large Balmer decrements observed in many T Tauri stars. An extended component of the chromosphere (i.e., a chromospheric region with a thickness comparable to the stellar radius) is required to explain the observed $H\alpha/H\beta$ ratios (Hartmann *et al.*, 1982). The P Cygni profiles observed in the $H\alpha$ line (cf. Kuhi, 1964; Kuan, 1975; Mundt, 1984) must be produced in an expanding, and probably extended, chromospheric envelope.

From a simple analysis of the $H\alpha$ fluxes observed in T Tauri stars, DeCampli (1981) inferred number densities of 10^{11}–10^{12} cm^{-3} in the emission envelopes, and mass loss rates in the range 3×10^{-9}–1×10^{-6} $M_\odot$ yr^{-1}. However, it is unlikely that the true mass loss rate is near the upper end of this range, based on consideration of other spectral lines. At $M \gtrsim 10^{-7} M_\odot$ yr^{-1}, the radiative losses of gas at temperatures $\gtrsim 2 \times 10^4$ K exceed the stellar luminosity, while a wind with a temperature $\lesssim 1 \times 10^4$ K would produce unobserved damping wings in the blue-shifted absorption of Mg II h and k. Although the expanding envelopes are probably not isothermal, the largest number of constraints from different lines can be roughly accomodated if the envelope properties are $T \sim 10^4$ K, $N_e \sim 10^{11}$ cm^{-3} and, therefore, $M \sim 10^{-8} M_\odot$ yr^{-1}. The radiative losses in such an envelope are $\sim 10^{33}$ erg s^{-1}, or about 10% of the stellar luminosity.

The detection of X-ray emission from some T Tauris (cf. Gahm, 1980; Geigelson and DeCampli, 1981; Walter and Kuhi, 1981) clearly shows that these stars have coronal regions. The detection of uv transition-region emission (Imhoff and Giampapa, 1980; Cram *et al.*, 1980; Brown *et al.*, 1984) also indicates that simple isothermal models cannot be correct. However, the bulk of the envelope emission measure is in low-temperature gas. There is simply too much material in the optically-emitting envelope to be heated to temperatures of 10^5 K or beyond; to balance the radiative losses would require more energy input than the photospheric luminosity of the star. Thus, T Tauri stars can not have predominantly coronal winds. As the bulk of the envelope mass cannot be at the highest temperatures on energetic grounds, ignoring the high-temperature material does not affect the mass loss rate estimated from optical indicators.

2.2. MASS LOSS MECHANISMS

The surface escape velocities of T Tauri stars are on the order of 300 km s^{-1}. A spatially-extended chromosphere of material at 10^4 K, with a thermal velocity ~ 10 km s^{-1}, cannot be maintained by gas pressure alone. The mass loss rate for a thermally-driven wind at this temperature for any reasonable initial density would be ridiculously low. The temperature of a thermally-driven wind must be much higher than 10^4 K, and to have $M \gtrsim 10^{-8} M_\odot$ yr^{-1} would exceed the total radiative losses of the star. Some other mechanism is required to drive the mass ejection.

DeCampli (1981) suggested that 'turbulent' motions were present in the extended envelopes of T Tauri stars, based on the broad bur nearly symmetric emission line profiles generally observed. This result suggests that some type of wave mode might be responsible for extending the chromosphere and driving the outflow.

The most promising modes for driving T Tauri mass loss are magnetic waves. The large radiative energy fluxes observed ($\sim 10^9 \, \mathrm{erg \, cm^{-2} \, s^{-1}}$) cannot be carried by acoustic waves at the inferred densities. Compressive waves generated by turbulent photospheric motions generally have damping lengths that are a small fraction of the stellar radius. While the dissipation of such waves may play a major role in heating the base chromosphere, it is difficult to propagate enough wave energy out to a sufficiently large distance for efficient mass ejection.

Alfvén wave-driven winds have been discussed by several authors (DeCampli, 1981; Hartmann et al., 1982; Lago, 1984), and appear to provide a possible explanation for T Tauri mass loss. The basic properties of the Alfvén wave wind solutions are:

(1) One can achieve mass loss rates of 10^{-9}–$10^{-8} \, M_\odot \, \mathrm{yr^{-1}}$ with wave energy fluxes $\sim 10^{-1} \, L_*$.

(2) In order to carry sufficient energy (i.e., the wave amplitudes $\delta B \ll B$), the surface magnetic field strengths must be a few hundred gauss.

(3) At the critical point, the wave velocity amplitude must be comparable to the escape velocity and to the local flow velocity. This result follows from the equation of motion; for steady flow in spherical geometry,

$$\frac{dv}{dr} = \frac{2v}{r} \frac{\left[a^2 - \dfrac{GM}{2r} + \dfrac{\varepsilon}{4\rho} \dfrac{(3M_A + 1)}{(M_A + 1)} \right]}{\left[v^2 - a^2 - \dfrac{\varepsilon}{4\rho} \dfrac{(3M_A + 1)}{(M_A + 1)} \right]}$$

(cf. Hartmann et al., 1982), where v is the gas velocity, ρ is the gas density, $\varepsilon = \langle \delta B^2 \rangle / 8\pi$ is the time-averaged wave energy density, and M_A is the ratio of the outflow velocity to the Alfvén velocity A. This equation is analogous to the critical-point form of the thermally-driven wind equations, with the turbulent magnetic pressure assuming the role of the gas pressure. As we expect that the thermal pressure is negligible for T Tauri winds, the critical point relations for small Alfvénic mach number reduce to

$$\varepsilon / 4\rho_c \sim GM/2r_c$$

and

$$v_c^2 \sim \varepsilon_c / 4\rho_c.$$

Using the result that $\delta B / B = \delta v / A$, one finds

$$\varepsilon / 4\rho_c = \delta v_c^2 / 8 \sim GM/2r_c.$$

Thus, the wave amplitudes are of the order of the escape velocity at the critical point, and are comparable to the expansion velocity. In the inner envelope, the wave motions are much larger than the expansion velocity. This can be understood by noting that the amplitude of an Alfvén wave tends to vary as $\rho^{-1/4}$, while the expansion velocity depends on the density as ρ^{-1}, so that the wave amplitude relative to expansion grows as one proceeds inwards from the critical point. The net effect is to produce a turbulent inner envelope; this is consistent with observations, as modelling of emission line profiles generally requires mean turbulent velocities on the order of 50 km s^{-1} (cf. DeCampli, 1981; Hartmann *et al.*, 1982). Note that purely acoustic motion at this velocity in a plasma near 10^4 K would imply very strong shocks, while an Alfvén wave with this amplitude can be a *linear* mode if the magnetic field strength is $\gtrsim 100$ G.

Models have been constructed for T Tauri envelopes that are reasonably successful in explaining the observed line emission (cf. Hartmann *et al.*, 1982). However, these models used only the simplest estimate of wave dissipation in order to construct the temperature structure of the outer envelope. It is highly unlikely that the crude parameterization of wave damping provides an adequate description of the envelope properties. A wide variety of wave modes are probably present, with different damping lengths. For example, the wind models with constant damping length are unable to account for the observed Ca II IR triplet emission (cf. Hartmann *et al.*, 1982; Avrett and Hartmann, 1985, in preparation), and one must instead resort to a 'deep chromosphere' (cf. Dumont *et al.*, 1973; Cram, 1979; Calvet *et al.*, 1984), i.e. a high-density chromospheric layer. This probably demands mechanical energy deposition with a short damping length. Although waves with such short damping lengths cannot be effective in driving mass loss, they can change the density scale height near the star quite dramatically, and thus affect the wind acceleration by changing the initial conditions.

Holzer *et al.* (1983) have pointed out the difficulties in producing low-velocity winds using Alfvén waves. These objections do not apply with equal force to T Tauri winds, as the observed terminal velocities are $\sim \frac{1}{2}$ of the surface escape velocity. Variations in magnetic field geometry and the possible presence of other wave modes with short damping lengths can quite easily affect expansion velocities, and observational selection effects will tend to favor the detection of low-velocity flows. These considerations indicate that the observed terminal velocities do not pose any insuperable difficulties for the Alfvén wave mechanism.

3. Angular Momentum Loss

As low-mass stars move toward the Main Sequence, the optical emission and absorption lines characteristic of T Tauri activity disappear. When this happens, we lose all direct indicators of mass loss. The only thing left is to try to make use of observations of stellar spindown.

It is well-known that solar-type stars spin down as they age on the Main Sequence (cf. Kraft, 1967). The standard picture invokes a wind coupled to the star by the magnetic field; as mass is ejected, it gains angular momentum by coupling to the field lines, spinning the star down (Schatzman, 1962). Unfortunately, this mechanism depends not only on the mass loss rate and the magnetic field strength, but it is especially sensitive to the magnetic field geometry (cf. Mestel, 1984) – which, of course, is unknown for stars. Thus, the exact relation between spindown and mass loss is ambiguous.

In recent years a rather dramatic revision has occurred in our understanding of the evolution of rotation among solar-type stars. It began with the discovery by van

Leeuwen and Alphenaar (1982) that some of the K stars in the Pleiades cluster were regular photometric variables with periods less than one day. It quickly became apparent that many K and M stars in the Pleiades cluster rotate at speeds 50–100 km s^{-1} (cf. Soderblom *et al.*, 1983; Stauffer *et al.*, 1984; Figure 1), while similar stars in the older Hyades cluster rotate at $\lesssim 10$ km s^{-1}. It is now accepted that the light variations are due to rotational modulation of the observed photosphere caused by large starspots.

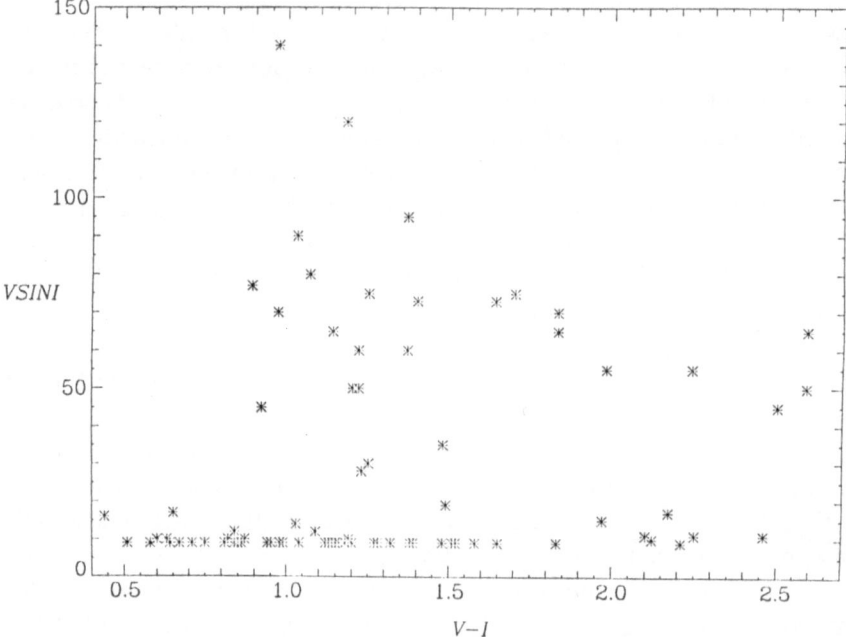

Fig. 1. Distribution of rotational velocities among low mass stars in the Pleiades cluster (Stauffer *et al.*, 1984). The G dwarfs ($0.5 < V - I < 0.9$) rotate slowly, but the K and M dwarfs ($V - I > 0.9$) have a large population of rapid rotators ($v \sin i \sim 50$ km s^{-1}).

As Stauffer *et al.* (1984) pointed out, the large rotational velocities observed for these stars implied very little angular momentum loss during contraction to the Main Sequence. The progenitor T Tauri stars are mostly slowly-rotating (cf. Vogel and Kuhi, 1981), and detailed calculations show that low-mass stars can lose only relatively small amounts of angular momentum as they contract to the Main Sequence.

The angular momentum loss rate due to wind ejection may be written as

$$-\frac{\mathrm{d}J}{\mathrm{d}t} \sim \dot{M}\Omega r_{\mathrm{A}}^{n},$$

where $\dot{M}$ is the mass loss rate, Ω is the angular velocity of the star, and r_{A} is the Alfvén radius. The Alfvén radius in this approximation marks the average radial distance to which the magnetic field enforces corotation. In the simple Weber and Davis (1967) model, in which all the magnetic field lines are open and $B \sim r^{-2}$, the exponent $n = 2$. Adoption of other, more realistic magnetic field geometries yields smaller values of n (cf. Mestel, 1984).

Rowse and Roxburgh (1981) and Roxburgh (1983) have constructed a model for angular momentum loss, adopting a dipolar magnetic field. In this model, the Alfvén radius is given by

$$\frac{r_A}{R} = (B_0^2 R^2 / \dot{M} v_A)^{1/3},$$

(cf. Mestel, 1984, Equation (11)), where B_0 is the surface field strength at the stellar radius R, and v_A is the wind speed at the Alfvén radius. The appropriate exponent of r_A to use in the calculation of the angular momentum loss rate is $n = 1$.

Hartmann et al. (1982) showed that their wind model 5 could adequately represent most of the strong emission-line T Tauri stars. The parameters of this model are $M = 1 M_\odot$, $R = 3 R_\odot$, $B_0 = 300$ G, $\dot{M} = 7 \times 10^{-9} M_\odot$ yr^{-1}, and has a terminal velocity ~ 200 km s^{-1}. This model was calculated assuming $B \sim r^{-2}$, not a dipolar field, but let us use these parameters for the approximate calculation, setting $v_A = 200$ km s^{-1}. Then $r_A/R \sim 7$, and the resulting time-scale for magnetic braking is $\tau \sim 3 \times 10^6$ yr, of the same order as the Kelvin–Helmholtz contraction time for a typical T Tauri star. Thus, Alfvén wave-driven wind models for T Tauri stars are not in conflict with the limits on angular momentum loss, particularly if Mestel (1984) is correct in arguing that the real braking time-scale is less sensitive to the Alfvén radius than given by the Rowse and Roxburgh model (i.e., $0 < n < 1$).

The Pleiades observations (Stauffer et al., 1984) show that the K and M dwarfs are rapid rotators (in general; but see below) but that the G stars are not. Does this mean that $1 M_\odot$ stars do not spin up? Stauffer et al. hypothesized that such was not the case. They suggested that the solar-type stars also spin up during contraction to the Main Sequence, but that they rapidly spin down once they stop contracting. The G stars in the Pleiades have reached the Main Sequence about 3×10^7 yr ago according to standard evolutionary calculations, and must have spun down during this time.

Endal and Sofia (1981) calculated a model of solar rotational evolution which predicted such an effect. Part of the rapid spindown on the Main Sequence was the result of the angular momentum loss formula they used, which increased $\dot{J}$ at the ZAMS. However, much of the rapid spindown effect was caused by a decoupling of the convective envelope and radiative core. The envelope was spun down first essentially as a solid body, but the spindown of the core occurred on a much longer time-scale in this model. As the envelope of a $1 M_\odot$ ZAMS star contains only about 10^{-1} of the moment of inertia of the entire star, the envelope can be spun down rather rapidly.

The hypothesis that $1 M_\odot$ stars are rapid rotators when they first arrive on the Main Sequence has now been confirmed by observations of the younger α Per cluster (Stauffer et al., 1985); many G tars in α Per are rotating at 50 km s^{-1} or more (Figure 2).

The recognition of a phase of rapid rotation in the evolution of low-mass stars has interesting implications for the study of solar activity. Generation of magnetic fields by

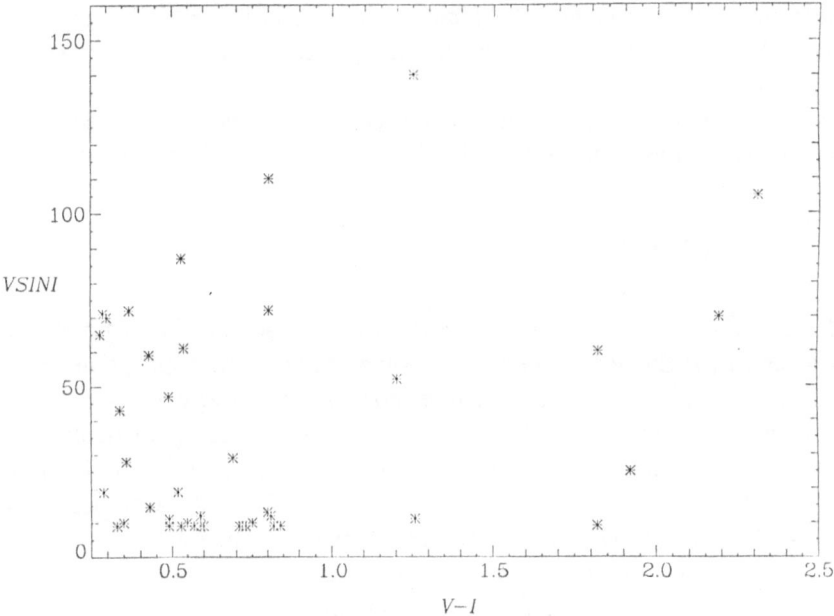

Fig. 2.   Distribution of rotational velocities among low-mass stars in the α Per cluster (Stauffer *et al.*, 1985). In this cluster, the G stars rotate rapidly. Since the age of the α Per cluster given by the upper-Main-Sequence turnoff point is slightly younger than the age of the Pleiades, this result strongly suggests that solar-type stars slow down rapidly between the ages of 5×10^7 and 7×10^7 yr.

dynamo activity should depend on velocity shear fields in the interior of the star. The rapid rotators may have enhanced rotational velocity gradients, and it will be interesting to see what effects this might have on their magnetic activity. Presumably the initial rapid rotation of the core has had time to be damped out at the present age of the Sun, but the remaining differential rotation may still reflect the period of rapid spindown.

Can we say anything about the mass loss of solar-type stars during this period? The combined α Per and Pleiades data suggest that the spindown time of a G dwarf rotating at 50 km s^{-1} is much less than 2×10^7 yr^{-1} – the estimated age difference between the two clusters. If r_A is close to R, or $n \sim 0$, then $\dot{M} \gtrsim 10^{-9} M_\odot$ yr^{-1}, $\gtrsim 10^{-8} M_\odot$ yr^{-1} if the radiative core and convective envelope are strongly coupled. It seems likely that $\dot{M}$ is smaller than this, and that r_A plays an important role in spinning down these stars. Marcy *et al.* (1985) suggest $\dot{M} \approx 10^{-9} M_\odot$ yr^{-1} from observations of the Hα emission of the rapidly-rotating K dwarf H II 1883 in the Pleiades. This suggests that solar-type stars on the Main Sequence may possess substantial winds, although the derived mass loss rate is very uncertain.

Caillault and Helfand (1985) found that the X-ray emission from the rapid rotators in the Pleiades was not very much larger than from the slow rotators. The upper limit to the X-ray luminosity of H II 1883 is 10^{30} erg s^{-1}, implying an emission measure of 10^7 K gas of $\lesssim 10^{53}$ cm^{-3}. This emission measure is comparable to that derived for 10^4 K gas from the Hα emission. It is likely that any thermally-driven wind material will be at a temperature more like 10^6 K, which is not directly observed; but if we assume an emission measure of 10^{53} cm^{-3} for this gas as well, and if we assume an extension of this hot wind material of roughly a stellar radius, we find a mass loss rate of $\sim 3 \times 10^{-10} M_\odot$ yr^{-1}. If the Sun is any guide, we would expect the bulk of the X-ray emitting gas to be in closed magnetic field structures, so that the actual mass loss rate would be considerably lower.

The process of angular momentum loss by winds from solar-type stars may play an important role in the evolution of close binary systems. If two stars are sufficiently close that tidal interactions enforce synchronous rotation, then spin angular momentum loss from one or both stars produces *orbital* angular momentum loss, causing the stars to spiral in towards each other. This process may be responsible for producing the W UMa binaries (Huang, 1966; Mestel, 1968; Okamoto and Sato, 1970; Eggleton, 1976; van 't Veer, 1979; Vilhu, 1982). These are contact binary systems in which each component is very roughly one solar mass. The space density of W UMa's is much larger than one might expect from the distribution of periods among binary systems, and there is some evidence that these systems are older (Guinan, private communication). Angular momentum loss would provide a plausible explanation for these facts. Another possible role for wind angular momentum loss is the production of large accretion rates in certain cataclysmic variable systems (cf. Verbunt and Zwaan, 1981; Spruit and Ritter, 1983; Taam, 1983; Rappaport *et al.*, 1983; Patterson, 1984).

To date, such calculations generally assume some extension of the Skumanich (1972) relation for angular momentum loss as a function of rotation. This relation was originally derived for stars rotating at <20 km s^{-1}, so that its application to dwarfs rotating at 200 km s^{-1} may seem rather questionable, but observations of Pleiades and α Per stars rotating at 50–100 km s^{-1} indicate angular momentum loss at a rate within an order of magnitude of that predicted by the Skumanich law. Further observations of rapidly-rotating young dwarfs will help test the idea that wind angular momentum loss is important in close binary evolution.

One other point concerning stellar angular momentum loss deserves mention. The observations clearly support the idea that wind braking is a feedback mechanism, i.e., stars with larger rotational angular momentum spin down more rapidly. The low-mass stars in the Pleiades and α Per clusters do not all have the same rotational velocities, but low-mass dwarfs in the older Hyades cluster have the same rotation at a given mass (cf. Radick *et al.*, 1983). Thus, as far as surface rotation is concerned, all memory of initial rotation is wiped out by the age of $\sim 6 \times 10^8$ yr. In turn, this result suggests that the Sun should be completely typical in its magnetic activity.

4. Future Research

We can expect further advances in our understanding of the powerful mass loss observed from young stars. Ultraviolet spectra obtainable with Space Telescope should provide better estimates of mass loss rates and a clearer picture of physical conditions in the envelopes of these stars. Additional observations of the rotation of young stars now in progress will refine our understanding of stellar spindown and turn up new aspects of the connection between mass loss, rotation, and magnetic activity.

On the other hand, it will be difficult to interpret much of the stellar data in a useful way without an improved understanding of the physics of the solar wind. While many stellar winds are not driven by thermal gas pressure, their acceleration mechanisms may also play a role in the dynamics of the solar wind. The interaction of magnetic fields and mass ejection in producing angular momentum loss also deserves further study. Stellar observations will demonstrate dependence of mass and angular momentum loss on global parameters, but the physical mechanisms involved can best be resolved through solar research. As we cannot detect the mass ejection from old, slowly-rotating Main-Sequence stars, our understanding of such winds will depend entirely on studies of the Sun.

Acknowledgement

This work was supported in part by NASA grant NAGW-511.

References

Appenzeller, I. and Wolf, B.: 1977, *Astron. Astrophys.* **61**, 21.
Appenzeller, I., Jankovics, I., and Ostreicher, R.: 1985, *Astron. Astrophys.* (in press).
Avrett, E. H. and Hartmann, L.: 1985, in preparation.
Beckwith, S., Zuckerman, B., Skrutskie, M. F., and Dyck, H. M.: 1985, *Astrophys. J.* (in press).
Brown, A., Ferraz, M., and Jordan, C.: 1984, *Monthly Notices Roy. Astron. Soc.* **207**, 831.
Caillault, J.-P. and Helfand, D. J.: 1985, *Astrophys. J.* **289**, 279.
Calvet, N.: 1983, *Rev. Mex. Astron. Astrophys.* **7**, 169.
Calvet, N., Basri, G., and Kuhi, L. V.: 1984, *Astrophys. J.* **277**, 725.
Cohen, M.: 1983, *Astrophys. J.* **270**, L69.
Cohen, M. and Kuhi, L. V.: 1979, *Astrophys. J. Suppl.* **41**, 743.
Cram, L. E.: 1979, *Astrophys. J.* **234**, 949.
Cram, L. E., Giampapa, M. S., and Imhoff, C. L.: 1980, *Astrophys. J.* **238**, 905.
DeCampli, W. M.: 1981, *Astrophys. J.* **244**, 124.
Dumont, S., Heidmann, N., Kuhi, L. V., and Thomas, R. N.: 1973, *Astron. Astrophys.* **29**, 199.
Eggleton, P. P.: 1976, in P. Eggleton, S. Mitton, and J. Whelan (eds.), 'Structure and Evolution of Close Binary Systems', *IAU Symp.* **73**, 209.
Endal, A. S. and Sofia, S.: 1981, *Astrophys. J.* **243**, 625.
Feigelson, E. D. and DeCampli, W. M.: 1981, *Astrophys. J.* **243**, L89.
Gahm, G. F.: 1980, *Astrophys. J.* **242**, L163.
Hartmann, L. and Kenyon, S. J.: 1985, *Astrophys. J.* (in press).
Hartmann, L., Edwards, S., and Avrett, E. H.: 1982, *Astrophys. J.* **261**, 279.
Herbig, G. H.: 1970, *Mem. Roy. Sci. Liège, Ser. 5* **9**, 13.
Herbig, G. H.: 1977, *Astrophys. J.* **214**, 747.
Herbig, G. H. and Soderblom, D. R.: 1980, *Astrophys. J.* **242**, 628.
Holzer, T. E., Flå, T., and Leer, E.: 1983, *Astrophys. J.* **275**, 808.
Huang, S.-S.: 1966, *Astron. Astrophys.* **29**, 331.
Imhoff, C. L. and Giampapa, M. S.: 1980, *Astrophys. J.* **239**, L115.
Kappelmann, N. and Mauder, H.: 1981, *ESO Messenger*, No. 23, p. 18.
Kraft, R. P.: 1967, *Astrophys. J.* **150**, 551.
Kuan, P.: 1975, *Astrophys. J.* **202**, 425.
Kuhi, L. V.: 1964, *Astrophys. J.* **140**, 1409.

Lago, M. T. V. T.: 1984, *Monthly Notices Roy. Astron. Soc.* **210**, 323.

Lynden-Bell, D. and Pringle, J. E.: 1974, *Monthly Notices Roy. Astron. Soc.* **168**, 603.

Marcy, G. W., Duncan, D. K., and Cohen, R. D.: 1985, *Astrophys. J.* **288**, 259.

Mestel, L.: 1984, in S. L. Baliunas and L. Hartmann (eds.), *Cool Stars, Stellar Systems, and the Sun,* Springer-Verlag, Heidelberg, p. 49.

Mundt, R.: 1984, *Astrophys. J.* **280**, 749.

Okamoto, I. and Sato, K.: 1970, *Publ. Astron. Soc. Japan* **22**, 317.

Patterson, J.: 1984, *Astrophys. J. Suppl.* **54**, 443.

Radick, R. R. *et al.*: 1983, *Publ. Astron. Soc. Pacific* **95**, 621.

Rappaport, S., Verbunt, F., and Joss, P. C.: 1983, *Astrophys. J.* **275**, 713.

Rowse, D. P. and Roxburgh, I. W.: 1981, *Solar Phys.* **74**, 165.

Roxburgh, I. W.: 1983, in J. O. Stenflo (ed.), 'Solar and Stellar Magnetic Fields', *IAU Symp.* **102**, 449.

Rydgren, A. E. and Vrba, F. J.: 1983, *Astrophys. J.* **267**, 191.

Schaefer, B.: 1983, *Astrophys. J.* **266**, L45.

Schatzman, E.: 1962, *Ann. Astrophys.* **25**, 18.

Skumanich, A.: 1972, *Astrophys. J.* **171**, 565.

Soderblom, D. R., Jones, B. F., and Walker, M. F.: 1983, *Astrophys. J.* **274**, L37.

Spruit, H. C. and Ritter, H.: 1983, *Astron. Astrophys.* **124**, 267.

Stauffer, J. R., Hartmann, L., Soderblom, D. R., and Burnham, N.: 1984, *Astrophys. J.* **280**, 202.

Stauffer, J. R., Hartmann, L., Burnham, N., and Jones, B.: 1985, *Astrophys. J.* **289**, 247.

Strom, S. E.: 1983, *Rev. Mex. Astron. Astrophys.* **7**, 201.

Taam, R. E.: 1983, *Astrophys. J.* **268**, 361.

Uchida, Y. and Shibata, K.: 1983, *Publ. Astron. Soc. Japan* **36**, 105.

Ulrich, R. K.: 1976, *Astrophys. J.* **210**, 377.

van Leeuwen, F. and Alphenaar, P.: 1982, *ESO Messenger*, No. 28, p. 15.

van 't Veer, F.: 1979, *Astron. Astrophys.* **80**, 267.

Verbunt, F. and Zwaan, C.: 1981, *Astron. Astrophys.* **100**, L7.

Vilhu, O.: 1982, *Astron. Astrophys.* **109**, 17.

Vogel, S. N. and Kuhi, L. V.: 1981, *Astrophys. J.* **245**, 960.

Walker, M. F.: 1972, *Astrophys. J.* **175**, 546.

Walter, F. M. and Kuhi, L. V.: 1981, *Astrophys. J.* **250**, 254.

Weber, E. J. and Davis, L., Jr.: 1967, *Astrophys. J.* **148**, 217.

Wolf, B., Appenzeller, I., and Bertout, C.: 1977, *Astron. Astrophys.* **58**, 163.

THE FUTURE OF SOLAR PHYSICS*

E. N. PARKER

Dept. of Physics, Fermi Institute, and Dept. of Astronomy and Astrophysics, University of Chicago, Chicago, IL 60637, U.S.A.

Abstract. The future of solar physics is founded on the existing fundamental unsolved problems in stellar physics. Thus, for instance, the physics of stellar interiors has been called into serious question by the very low-measured neutrino flux. The ^{71}Ga neutrino detection experiment is the next step in unravelling this mystery. If that experiment should find the expected neutrino flux from the basic p–p reaction in the Sun, then astrophysics is in a difficult situation, because the most likely explanation for the low neutrino flux found in the ^{37}Cl experiment would be an error in our calculation of the opacity or an error in our understanding of the elemental abundances in stellar interiors, with serious implications for present ideas on stellar structure and the age of the galaxy.

The new methods of helioseismology, for probing the interior of the Sun, have already found the primordial rapid rotation of the central core. The forthcoming world-wide helioseismology observing network will permit fuller exploitation of the method, promising to provide the first direct sounding of the interior of a star, hitherto known to us only through theoretical inference and the discrepant neutrino emission.

The activity of all stars involves much the same phenomena as make up the activity of the Sun. The effects are too complex, and too foreign to the familiar dynamics in the terrestrial laboratory, to be deciphered by theoretical effort alone. It has become clear through the observational and theoretical work of the past decade or two that much of the essential dynamics of the activity of the atmosphere takes place on scales of the order of 10^2 km. Thus, an essential step in developing the physics of stellar activity will be the Solar Optical Telescope (presently planned by NASA to be launched early in the next decade) to permit a microscopic examination of the surface of the Sun to study the source of the action. The activity and X-ray emission of other stars depend on much the same effects, so that the study is essential to determining the significance of the X-ray emission from other stars.

1. Introduction

The observed behaviour of the Sun defies theoretical understanding, and therein lies the future of solar physics. Indeed, it is the future of all stellar physics, for a dilemma posed by the Sun is a dilemma for all stars otherwise too distant to be properly studied. The Sun, after several decades of scrutiny, has become as enigmatic as the quasar, but with the advantage that the mysteries can be intimately probed by new techniques and instrumentation. The table of contents of this one hundredth volume of *Solar Physics* provides a list of the current problems and fields of exploration of the Sun, as well as the instrumental means for delving into those problems. The contents, then, provide a tabulation of the problems of all stars.

Space prohibits us from commenting on all the problems, but the individual articles make up for what the present article lacks. For instance, we have slighted the fundamental and exciting futures for X-ray and gamma-ray studies of the Sun, the Ulysses mission into the unexplored high latitudes of the heliosphere (for which the

* This work was supported in part by the National Aeronautics and Space Administration under grant NGL–14–001–001.

reader is referred to the excellent articles by Dennis, Hudson, and Lin), and the distant heliosphere being traversed by Pioneers 10 and 11 and the Voyager spacecrafts (reviewed by Burlaga, 1984; and discussed in the article by Axford). We have also failed to discuss the future prospects for instrumental development, which are really the foundations of the future, referring the reader to the articles by Dunn, Gough, Bahcall, and others.

In view of Bahcall's penetrating article on the problem of solar neutrinos and the clear exposition of helioseismology in the article by Gough, we might have devoted less space to these topics in the present writing. Bahcall and Gough are leaders in their fields, and the present author is indebted to both for his own limited knowledge of their subjects. However, when writing on the future of solar physics, the neutrino emission and the seismololigical probing of the interior of the Sun occupy the center of the stage, along with the high-resolution observations planned for the 'solar microscope' – the Solar Optical Telescope. So we have made the present article self-sufficient in its limited choice of topics (see also the article by Roxburgh on present problems with the solar interior).

We begin with the general remark that the scientific oppurtunities in solar physics are legion, with the progress and achievement over the next two decades limited only by our personal energies and technical resources. Who could have imagined thirty years ago that a placid, pedestrian star could be so complex an engine, presenting so many outstanding puzzles to the inquiring scientist? And, by inference, that the other stars, appearing as faint unresolved points of light, are hiding so much from us. Only the Sun is close enough to be sounded, and even with its proximity the necessary sensitivity and angular spectral resolution are pressing the best of modern technology. Indeed, there was no way to anticipate the complexity of the Sun and the existing theoretical dilemmas, until the observations thrust them upon us. The activity of the Sun involves fluid dynamical effects hitherto unknown and, because of their turbulent nature, not subject to theoretical prediction. As physicists we can proceed only when observations provide instructions, and we can be sure of where we have arrived only when further observations confirm our position.

The physics of the terrestrial laboratory was sufficient to guide us to the concept of the static star, i.e., the gross hydrostatic structure of a star of moderate mass, evolving slowly along the Main Sequence. The theoretical evolution of the quasi-hydrostatic star has been followed for a variety of star masses all the way into the final convulsive stages, where the overall complexity of the phenomena, and unanswered questions on the fundamental nature of matter, become barriers to progress. Similar barriers arise in the theoretical pursuit of the day-by-day small-scale suprathermal activity of the Sun. Consequently, the interrelations of the magnetic fibrils, spicules, sunspots, flares, fast particles, gamma-ray emission, X-ray filaments, radio emission, coronal eruptions, etc., are understood only in fragmentary ways, at best.

The pioneering work of Wilson (1978) and Linsky (1980) studying the chromospheres of other stars, together with the more recent UV, XUV, and X-ray observations, have brought home the fact that essentially all stars are active, for a variety of reasons (deep convection, magnetic fields, close companions, accretion of matter, etc.). The study of chromospheric activity has been greatly expanded in an extensive dedicated observing program in recent years (cf. Linsky, 1980; Baliunas *et al.*, 1983). The activity of most other stars appears to involve the same generic phenomena seen in detail on the Sun, although sometimes on vastly different scales and levels of energy. Hard experience with the Sun shows that such dynamical fluid effects can be properly understood only when guided by close observational scrutiny. There are too many speculative possibilities to do hard science without precise and detailed observational guidance and constraints.

2. The Hydrostatic Sun

We must not get the idea, however, that the future of solar physics depends entirely on the activity of the Sun and other stars, for there are fundamental questions concerning the hydrostatic star itself. Everything that we think we know about the Sun below the visible photosphere is based on a number of 'reasonable' assumptions permitting a simple theoretical extrapolation downward from the surface. The downward extrapolation is straightforward in principle, based on the 'reasonable' assumptions that the conditions are described by (a) the barometric law $\nabla p + \rho \nabla \Phi = 0$; (b) the Newtonian gravitational equation $\nabla^2 \Phi = 4\pi G \rho$; (c) the calculated opacity of the gas as developed from the usual quantum mechanical approximations for the interaction of radiation and matter, including all the known higher-order effects such as dielectronic recombination, etc., as well as the other higher-order effects that have been neglected, knowingly or unknowingly; (d) assuming the 'standard' abundances of the elements, which are deduced by (e) assuming that the terrestrial and meteoritic relative abundances of the heavier elements are the same as for the Sun; (f) assuming that the relative abundance of elements inferred from spectroscopic studies of the solar atmosphere reflect the abundances in the deep interior; and (g) assuming that there is no mixing of material in the radiative core beneath the convective zone, except for a modest amount to account for the depletion of Li and Be, but not B (cf. Bodenheimer, 1965, 1966; Wallerstein and Conti, 1969; Giampapa, 1984), presumably as a result of 'convective overshoot' extending downward from the turbulent convective zone (Strauss *et al.*, 1976; Vauclair *et al.*, 1978a, b) or as a consequence of the thermal shadows of magnetic fields (Parker, 1984a, b). As the reader will soon see, there is a lot more to be said on the subject of mixing. To go on with the inventory of 'reasonable' assumptions it is (h) generally assumed that the Sun has accreted no significant quantity of matter during its life time (although the crust of Earth was reformed about 3.2×10^9 yr ago by causes believed to involve the heavy accretion of matter from space). In treating the energy sources in the central core of the Sun one adopts (i) the nuclear interaction cross-sections at kilovolt energies obtained by extrapolating downward from the 100 keV energies at which the cross-sections can be measured in the laboratory. One also assumes (j) that the neutrino capture cross-sections in terrestrial neutrino detectors are properly known. And one supposes (k) that there are no additional particle interaction effects (such as the catalysis of baryon decay by magnetic monopoles (cf. Sur and Boyd, 1985) outside the established nuclear reactions (see the article by Kleczek). Bahcall (1985) gives a summary of the estimates of these uncertainties. It is also assumed, of course, that (l) the Sun is the same age as Earth (believed to be at the respectable age of 4.5×10^9 yr, from studies of strontium–rubidium ratios and lead isotope ratios in terrestrial rocks.)

The result of this standard approach was a theoretical prediction that the emission of neutrinos from the nuclear reactions in the central core of the Sun is 25 SNU for capture in ^{37}Cl nuclei (1 Solar Neutrino Unit represents 10^{-36} reactions ($\nu + {}^{37}$Cl $\rightarrow {}^{37}$A $+ e^-$) per second per target nucleus ^{37}Cl). Davis (1964) undertook the construction and operation of the huge, but exceedingly delicate, ^{37}Cl experiment, involving the monthly extraction of a dozen or so ^{37}Ar atoms from the 615 tons of C_2Cl_4 in a tank 1.5 km down in the Homestake Gold Mine, near Lead, South Dakota. The accumulated result is a measured neutrino flux that is not in excess of 2.1 ± 0.3 SNU (1σ error) (Bahcall et al., 1982; Bahcall and Davis, 1982; Bahcall, 1985). The neutrinos detected by Davis are attributed mainly to the decay of ^{8}B into ^{8}Be $+ e^+$, providing a maximum neutrino energy of 14.06 MeV. The great flood of neutrinos from the basic p–p reaction has a maximum energy of only 0.42 MeV, well below the 0.81 MeV threshold for the ^{37}Cl reaction. There is only a small contribution (0.2 SNU) from the pep reaction, producing monoenergetic neutrinos at 1.44 MeV. Fortunately, or unfortunately, only two in 10^4 of the overall reactions 4^1H $\rightarrow {}^4$He are calculated to pass through ^{8}B, contributing the high-energy neutrinos detectable in the ^{37}Cl experiment. The rarity of ^{8}B is a consequence of the high Coulomb barrier ($Z = 5$) of this nucleus. Hence, the formation of ^{8}B, resulting from protons far out on the tail of the Maxwellian velocity distribution, is not only unlikely, but highly sensitive to the temperature. Small revisions in the calculated opacity and in the nuclear cross-sections have large effects on the predicted rate of emission of neutrinos. Thus, motivated by the discrepancy between the observed and predicted neutrino emission rate, the nuclear cross-sections have been re-measured and improved, with the result that the best theoretical value is now 6 ± 1 SNU (Bahcall and Davis, 1982; Bahcall et al., 1982). Bahcall (1985) suggests that further refinements and revisions of atomic and nuclear data cannot be expected to close the gap between theory and experiment. But something is wrong! Only a small reduction, of a few percent, of the calculated central temperature would resolve the dilemma. Thus, perhaps we still do not know enough about the complicated quantum mechanical atomic effects that collectively make up the opacity (Filippone and Schramm, 1982). If that is the case, there are errors – not necessarily small – in the theoretical models of the many different classes of stars. Or perhaps the discrepancy arises from the simple fact that the neutrinos have a rest mass (of the order of 10 eV) so that they oscillate between, say, the electron and muon-neutrino states.

Or perhaps we have the wrong idea about the relative abundances of the elements in the core of the Sun. The standard view is that the core contains the primordial abundance of ^{4}He (about 1 in 13 by number, Gough, 1983b) plus the ^{4}He from the basic thermonuclear process 4H $\rightarrow {}^4$He accumulated over the past 4.5×10^9 yr, to give a present abundance of about 1 in 6. The core is stably stratified as a consequence of the efficient radiative transfer (so that there is only a modest outward decline of the temperature) and further stabilized by the assumed increased abundance of ^{4}He toward the center. But what if, as a consequence of some unanticipated mechanism, the core has turned over a couple of times, so that the abundance of ^{4}He is nearer 1 in 13 than 1 in 6? The temperature would be reduced a few percent because the opacity is lower, · and the hydrogen is more abundant so that the basis nuclear reaction $p + p \rightarrow {}^2He + \nu$ proceeds more efficiently. The predicted neutrino flux from ^{8}B would be close to the observed 2 SNU.

Turning over the core a couple of times would somewhat relieve another dilemma, viz., the calculated evolutionary brightening of the Sun by some 30% over the past 3×10^9 yr. It is a geological fact that at least in the last 3.2×10^9 yr Earth has not been completely frozen over. The accumulated strata of sedimentary rocks show that the surface of the planet has been subject to continual erosion by running water. Yet, numerical simulations of terrestrial atmospheric models show that when the luminosity of the Sun is reduced by only 3–5% there is progressive freezing, proceeding over 10^3 yr all the way to the equator as the snow accumulates and the oceans freeze. The climate does not recover from the 'deep freeze' when the solar luminosity is restored to its normal value because the albedo of the snow-covered planet is so high. How then could Earth have survived in an unfrozen condition when the Sun was 30% fainter? It has been suggested that the atmosphere contained enough CO_2, CH_4, H_2O, etc. that a compensating greenhouse effect maintained a temperate climate, the atmosphere evolving with time at precisely the rate to keep from overheating or freezing as the Sun became brighter. Perhaps the Earth was so lucky as to walk this tight rope, falling-off neither to the cold side like Mars nor the hot side like Venus. It is an important factor in the possible survival of life forms, certainly. Or perhaps the core of the Sun mixed enough that the problem is not quite so severe. Perhaps there has been only a 20% brightening. Schatzmann (1977), Genove and Schatzmann (1979), Schatzmann and Maeder (1981), Bienayme et al. (1983), Tassoul and Tassoul (1984a), and others have studied the effect of a modest mixing of the solar interior (see review by Spruit, 1985) showing the profound consequences. Press (1981) and Press and Rybicki (1981) have explored the mixing effects, and the effects on diffusion, caused by possible internal gravity waves in the radiative core. It has not been established that there is an effective means of mixing, nor has the contrary been proved. It is an open question for the future. The question of mixing is not unrelated to the angular velocity and magnetic fields in the core, on which more will be said later when we discuss the exploratory potential of helioseismology.

The immediate problem is the neutrino emission from the supposed thermonuclear reactions in the center of the Sun. Davis's ^{37}Cl experiment is the only check on the present theory of stellar interiors, and the experiment shows a serious discrepancy. Unfortunately the extreme temperature sensitivity of the ^{8}B emission rate leaves too many hypothetical alternatives to explain the discrepancy, as outlined above. In the present dilemma, therefore, it is essential to go ahead with other tests of the conditions in the central core of the Sun. The fundamental test is to measure the flux of neutrinos from the basic p–p reaction, believed to proceed at the rate necessary to produce the observed luminosity of the Sun. The only assumption in calculating the emission rate is that the Sun is operating in a steady state with $p + p \rightarrow d + e^+$ as the basic reaction. The calculated neutrino flux at Earth involves the additional assumption that the neutrinos, once emitted, are stable particles, so that their intensity declines outward as $1/r^2$. It appears that the best means to detect these neutrinos is the ^{71}Ga experiment, employing about 30 tons of Ga (Bahcall *et al.*, 1978; Zatsepin, 1982; Bahcall, 1985). The predicted rate is 80 SNU for the neutrinos from the p–p reaction. An additional 30 SNU is estimated as a result of the reaction $^7Be + e^- \rightarrow {}^7Li + \nu$, which is somewhat dependent on the temperature, but not nearly to sensitive as the production of ^8B. The chemistry for the experiment is based upon the water soluble chloride, and Davis has already operated a pilot experiment to prove the efficacy of the method. The full scale experiment could begin at any time that the astronomical community wishes to divert the necessary funds from other activities. The question would seem to be of central interest to stellar physics. Fortunately the physics community has expressed sufficient interest that plans for the ^{71}Ga experiment are being carried forward by laboratories in the Soviet Union, West Germany, and the United States. Two or more independent determinations of the emission are highly desirable for ultimate verification of the effect.

Suppose, then, that we are some years in the future and the ^{71}Ga experiment has been satisfactorily performed, with a firm result. The consequences of the measurement are far reaching. Suppose, for instance, that the measured neutrino capture rate is in the expected range, i.e. somewhere in the vicinity of 100 SNU. Then we would have direct experimental evidence that the basic ideas about stellar interiors are correct. We would also have direct experimental evidence that neutrinos are relatively stable particles. Turning to the ^{37}Cl experiment, then, we would be forced to interpret it as an indication that our presently calculated temperature of the central core of the Sun is too high by several percent. That would imply that there is mixing in the core, or that we still do not fully understand all the quantum mechanical effects that contribute to the opacity of the standard abundance of elements. If the explanation is the latter, we have a formidable and important task ahead of us, with corrections to be made to the present models of all classes of stars. On the other hand, if the explanation is the former, then the theoretical calculations of the age at which stars leave the Main Sequence are in serious error, with profound consequences for the estimated age of globular clusters and the galaxy. There would also be revisions in the theoretical radial structure of a star in its final epoch. Thus the cosmological consequences of central mixing are nontrivial. But, of course, it would then be necessary to understant the causes of the mixing, on which more will be said in the next section. So far there is no evidence of anomalous abundances from helioseismology (Gough, 1983a, b, 1984). The evidence from the 'standard' depletion of Li in convective stars is that some solar-type stars show no depletion at all, while the Li in the Sun is depleted by a factor of 50, and some more massive stars, where no convective mixing might be expected, are strongly depleted (Duncan, 1981; Duncan and Jones, 1983; Cayrel *et al.*, 1984; Giampapa, 1984). Hence, the standard abundances and the standard anomalies cannot be as simple as conventionally assumed, and the missing neutrino emission from the Sun may be another voice that is crying out this message.

It might be simpler, therefore, if the ^{71}Ga experiment produced an unexpected neutrino flux, say half the expected value of 100 SNU. Then either neutrinos are unstable, proving that they have a rest mass, or the Sun (and, hence, most other stars) is operating in an oscillatory state, or something else equally exotic. Unstable neutrinos would permit the comfortable assumption that the unseen matter in the Universe is 'nothing more' than low-energy neutrinos. Alternatively, we would be faced with a difficult problem in stellar interiors and the oscillatory state of stars on the Main Sequence.

In summary, then, the future ^{71}Ga experiment will have profound implications for the physics of stars and for the present cosmology based on inferences from the standard evolutionary models of stars. So long as the dilemma of the solar neutrinos remains unsolved, the foundations of modern astrophysics remain open to serious question.

The possibility of variation in the luminosity of the Sun is not limited to the neutrino dilemma, of course, but has recently come into its own through speculation on the cause of trends in terrestrial climate (cf. Mitchell, 1976) and through the development of precision absolute radiometers that measure changes in the solar luminosity of one part in 10^3 over periods of months and years (Newkirk, 1984). According to the best existing theory of stellar interiors, the luminosity of the Sun is constant, apart from the slow evolutionary brightening over periods of $10^9 - 10^{10}$ yr, already mentioned. The observed dark sunspot phenomenon, produced in some way by magnetic fields that sporadically bob up and down through the surface in times of days or weeks, has long suggested that there may be transient reductions in the solar luminosity by 2 or 3 parts in 10^3. It was assumed, however, that the sunspots block the flow of heat only at shallow depths, so that the heat soon flows around the spots and emerges elsewhere. Hence, it was a suprise when the precision monitoring of the solar luminosity showed that the heat does not reappear during the life of the spot. The appearance of sunspots was found to cause the expected decline in solar luminosity of as much as 3 parts in 10^3, but with the luminsoity reduced for a week or more (Willson *et al.*, 1981; Hudson *et al.*, 1982; Hoyt *et al.*, 1983; LaBonte *et al.*, 1984). Indeed, it is not known how long it takes for the deficit to be made up, which would give some idea of the depth of the blocking of heat.

There is preliminary evidence of a slow decline of the radiometer measurements of nearly one part in 10^3 per year, which has important implications for both the Sun and Earth if it is confirmed.

There is evidence (Eddy *et al.*, 1982) that the coming and going of sunspots has a small effect on terrestrial climate. One wonders whether it is a variation in solar luminosity responsible for the century long depression of the mean annual temperature at middle northern latitudes (by 0.5 °C) recorded in the 15th and 17th centuries and dubbed the 'Little Ice Age' (Mitchell, 1976; MacNeil, 1976; Eddy, 1976)? Did an enhanced luminosity cause the elevated temperatures of the 12th century? It is a curious fact that these centuries of unusual terrestrial heat and cold coincided with the centuries of greatly enhanced and greatly reduced solar activity, respectively (Eddy, 1976). The EUV and

X-ray luminosities of the Sun vary considerably with the level of activity, but their contribution to the total energy is small, of the order of one part in 10^4. If it is the variation in the UV and X-rays that causes the climatic changes, then it is through some subtle and extraordinarily sensitive aspect of the terrestrial atmosphere. There is the possibility that the close tracking of the terrestrial climate and the century long average of solar activity is pure chance in spite of the number of independent events over which it is observed. One can imagine some complicated conspiracy of winds and ocean currents, atmospheric waves and weather patterns, clouds and snow cover, volcanic pollution of the atmosphere, etc., which just happened to follow the level of solar activity over the last several thousand years. But no one knows, and the connection to solar variations cannot be ruled out. Atmospheric electricity, controlled by the electrical conductivity of the atmosphere, which is created by the cosmic rays and strongly reduced during years of high solar activity, is a prominent and unknown factor in terrestrial climate (Markson, 1978, 1983). In the context of solar physics it is essential to monitor the luminosity of the Sun with absolute precision of one part in 10^3 over at least one sunspot cycle, and perhaps much longer if there is evidence for a long term variation. Indeed the study of long-term variations of solar and stellar luminosity could be greatly accelerated by a dedicated observational program of direct precise comparisons of the luminosities of stars of comparable spectral class, apparent magnitudes, and small angular separation in the sky.

Theoretical studies have established that there may be small variations of the solar luminosity as a result of the cyclic variations of strong magnetic fields beneath the surface of the Sun (Thomas, 1979; Dearborn and Blake, 1980, 1982; Gilliland, 1982; see also Endal and Twigg, 1984). Our general ignorance of the strength of the fields deep in the convective zone of the Sun (they could be as strong as 10^4 G or more) makes the effects difficult to assess, so that a clear observational record of the luminosity variations is an essential step in defining and understanding the problem.

3. The Dynamical Sun

It is evident from the foregoing discussion of the several dilemmas posed by the interior of the Sun that the main problems will not be answered unambiguously unless there are additional means for probing the interior. The neutrino emission is closely tied to the rate of generation of energy in the central core, providing information, therefore, on the gross conditions there. But the neutrinos do not tell much else, so that when it comes to questions of elemental abundances, mixing, and general variability, we must turn to something for guidance. Fortunately, in this hour of need, the subject of solar, or helioseismology has developed to the point that it promises, over the next decade, to probe precisely those basic questions of abundances, angular velocity, meridional circulation, turbulence, and magnetic fields throughout the solar interior. Indeed helioseismology, reviewed in the article by Gough, promises a complete view of the physical conditions throughout the solar interior. Such comprehensive basic information would provide a foundation for discussing the problems and questions concerning the structure of the interior, and particularly those raised by the neutrino emission. The opportunity arises from the excitation (presumably by the jiggling in the convective zone) of the many eigenmodes of oscillation of the Sun. There is an enormous number of basic modes of oscillation (Ulrich, 1970; Leibacher and Stein, 1971) with observational evidence that many thousands of separate modes can ultimately be observed (see the article by Gough). The modes are only weakly excited, fortunately, so that a linear treatment is sufficient (although Press (1981) and Press and Rybicki (1981) have explored the possibility of internal gravity waves of large amplitude in the radiative core, with significant consequences for mixing and heat transfer). The modes are compressional, p-modes, or internal gravity waves, g-modes.

The p-modes are sound waves trapped between the rapidly declining density at the surface and the increasing temperature toward the center. The modes are described by the order l and m of the surface harmonic $P_l^m(\cos \theta) \exp(\pm im\varphi)$ describing their angular distribution. The radial order n describes the mean radial wave number $n\pi/R_\odot$. The l and m can be determined directly from the velocity distribution over the surface. The radial number n can be determined only by comparing the observed period of the oscillation and the associated l and m with theoretical models. The higher p-modes ($l \sim 10^2$–10^3) have periods of the order of 5–10 min. The deeply penetrating p-modes have periods extending up to about an hour, which is essentially the sound propagation time over a solar radius. It is the longer periods, penetrating to the center, that are of primary interest in probing conditions in the core, whereas the higher radial modes reflect the structure of the outher regions of the Sun.

Of particular importance is the line splitting, or fine structure, that arises from the rotation of the Sun. The period of rotation of the equator of the Sun is 25–27 days when viewed from the moving Earth. This corresponds to a surface peripheral velocity v_Ω of the order of 2 km s^{-1}. Hence, the waves of frequency ω propagating in the direction of rotation are seen by a nonrotating observer to have slightly higher frequency than those with retrograde propagation. This Doppler shift is $\Delta\omega = v_\Omega/v_\phi$ where v_ϕ is the phase velocity of the mode at the surface where the waves are observed. The observational determination of $\Delta\omega$ provides a direct measure of the mean v_Ω at the caharcteristic depth of the mode. It is in this way that the angular velocity of the Sun can be determined as a function of radius and latitude from the different $\Delta\omega$, observed for many different values, of l, m and n.

Turning to the g-modes, they exist because the radiative core of the Sun is stable to convective overturning, so that a perturbation introduced into the radiative core propagates as an internal gravity wave. The phase velocity of the waves is of the order of a tenth of the sound speed and the periods are correspondingly longer. The amplitudes of the g-modes are attenuated by the convective zone, which lacks the stability for their propagation, so that the signal at the surface is weak. On the other hand, their deep penetration into the Sun makes them ideal for probing the central regions, and their smaller phase velocity make them more sensitive to internal circulation of the Sun. They nicely complement the p-modes. Their periods range upward from about an hour. Some 32 frequencies attributable to g-modes have been observed (Severny et al., 1984) with periods of 116 to 200 min. Of particular interest is the notorious 160 min oscillation, which was held at arms length for several years because 160 min is precisely one-ninth of the rotation period of Earth, suggesting that the effect originated in the terrestrial atmosphere. The oscillation, whatever its origin, has maintained phase coherence for more than seven years so that an accurate determination of the period has become possible. The period is found to be 160.0102 ± 0.0005 min and not 160.00 min. Hence, its phase drifts a little more than half an hour per year, or about four hours in the seven years of precise observations. The oscillation appears not to be tied to the rotation of Earth. Evidently it is a g-mode in the Sun (Severny et al., 1976, 1984; Kotov et al., 1978; Scherrer et al., 1979; Scherrer and Wilcox, 1983). The ubiquitous period of 160 min is a curiosity in itself, turning up in variety of astrophysical circumstances. The reader is referred to the article by Kotov in this volume for a presentation of the observations.

So far as the future of solar physics is concerned, the crucial fact is that the observations have established that both the p- and g-modes have long coherence times. Indeed, the coherence times of the lower modes are at least as long as the length of any available data sets, including a 5-day continuous observing run from the South Pole (Grec et al., 1980; Claverie et al., 1980). Hence, it is possible to establish the frequencies of each (l, m, n) with great precision. It has become clear that a world-wide network of observing stations is needed (to provide long unbroken data sets) if we are to exploit the full diagnositc potential of the oscillations (see the article by Gough, as well as Hill and Newkirk, 1985). Indeed, it may ultimately be necessary to observe from a spacecraft to eliminate atmosphere noise and obscurity, perhaps from a point 90° ahead or behind the Earth to assist in identification of the individual modes.

Recent reviews (Gough, 1984; Toomre, 1985; Christensen-Dalsgaard et al., 1985) give a concise summary of the potential for exploring conditions in the Sun, based on present observations and theoretical models (Leibacher and Stein, 1981; Deubner and Gough, 1984; Brown et al., 1985; see also the extensive bibliography, Leibacher, 1985; and the articles by Roxburgh and Gough in this volume). To take the sounding of the solar interior one step at a time, note that the frequencies of each mode (l, m, n) follow precisely from the density, temperature, and elemental abundances given as a function of radial distance from the center of the Sun. Hence, accurate observational frequencies for a few hundred identifiable modes over a wide range of (l, m, n) provide a detailed and essentially unique determination of the internal ρ, T, and μ as a function of radius r. Indeed, there are already some curious discrepancies of about one part in 500 between the best adjusted theoretical modes and the observed frequencies. The observed frequencies are believed to be accurate to about one part in 1200. Adjustments of elemental abundances, etc., are not sufficient to bring the theoretical models precisely into line with all the observed frequencies of a large variety of modes. There is no indication of anomalous abundances within the present data, although the results are probably not precise enough yet to settle the question in the central core.

The next step is to examine the fine-structure of modes of all orders, collectively giving the angular velocity Ω as a function of r and θ, and showing to what degree the core may rotate faster than the surface of the Sun (cf. Dicke, 1982). It is essential to know

$\Omega(r, \theta)$ through the convection zone in order to construct an explicit model of the solar dynamo that is presumed responsible for the generation of magnetic fields. The analysis should also determine any strong – presumably primordial – fields in the core. All of these bear directly on the possibility of mixing in the radiative interior, with the profound implications discussed above. What is more, a rapidly rotating core is an interesting phenomenon in itself (Rosner and Weiss, 1985), representing a primordial conditon entirely obscurred by the outer envelope of the Sun. And if the core should prove to be only slowly rotating, it raises questions about how it lost its angular momentum (cf. Tassoul and Tassoul, 1984b). It is interesting to note, then, that analysis of the existing data indicates a slight decrease of angular velocity downward across the convective zone with a more rapidly rotating core at about twice the surface angular velocity (Duvall and Harvey, 1984; Duvall *et al.*, 1984; Christensen-Dalsgaard *et al.*, 1985). The downward decrease of angular velocity across the convective zone is the opposite to what is normally assumed in dynamo models which produce migration of the azimuthal fields from high to low latitudes. If that result holds up with more and better data, we will be obliged to re-examine the theoretical models for the solar dynamo (see discussion in Glatzmaier, 1985). The rapidly rotating core provides no significant contribution to the quadrupole moment J_2 (solar oblateness) in terms of corrections to the precession of the perihelion of Mercury and contradictions of general relatively. The article by Schröter provides a summary of the present knowledge on the rotation of the Sun, whose variations represent the sum of total of our knowledge of the variations of the angular velocity of all stars.

Now the convection and circulation in the convective zone of the Sun is not known from theory. Its existence is inferred and its theoretical structure is represented by a simple mixing length and characteristic velocity. Theoretical modelling of the convective zone, working directly with the hydrodynamic equations is extraordinarily difficult, as a consequence of the large Reynolds number, strong density stratification, and large magnetic Reynolds number, so that extensive theoretical studies (Glatzmaier and Gilman, 1982; Durney, 1983a, b; Gilman, 1983; Tassoul and Tassoul, 1984b; Glatzmaier, 1985) have greatly expanded current knowledge of the dynamics of meridional circulation and large scale convection in stars, without succeeding in deducing a firm and unique result for the meridional flow and distribution of angular velocity. It is particularly exciting, then, to have the ability to infer the properties of the meridional circulation, the giant cells, and the general state of the turbulent convection from the fine structure and frequencies of the *p*- and *g*-modes. Such quantitative guidance can go a long way to directing future theoretical hydrodynamical studies. The inferences from seismology will then provide a clearer idea of the instrumental problems that still remain to be solved before we succeed in direct observation of the meridional circulation and the giant cells at the surface. Of course, the necessary observations of the solar oscillations have not yet been accomplished, but the world wide network of observing stations, presently being set up, should ultimately be able to make those observations. The analysis of the data will be a monumental task, of course, with the necessity of identifying thousands of modes, and then fitting the precise frequencies, line widths, and fine structure, with distributions $\rho(r)$, $\mu(r)$, $\Omega(r, \theta)$, and perhaps $B_\varphi(r, \theta)$. On the one hand we employ the presumption that the proper identification of each observed frequency can be made and that the theoretical frequencies can be brought into line with the observed frequencies. To fail to bring them into line would leave open the question of unknown physical effects beyond the ken of the standard equations for hydrostatic equilibrium, opacity, thermonuclear rates, etc. On the other hand, when exploring for the first time a region so complex and unknown as the interior of a star, we must expect surprises.

Until such time as the mapping and interpretation of the vast spectrum of p- and g-modes is completed, we must be content with studies of the angular velocity at the visible surface (Howard, 1976, 1984; Howard and LaBonte, 1980a; Gilman and Howard, 1984; see also the article by Schröter), the preliminary results already noted (Duvall and Harvay, 1984; Duvall *et al.*, 1984), and the hope that the weak motions of the giant cells may one day be detected in a reproducible manner at the surface (Howard, 1979; Howard and LaBonte, 1980b; Gilman and Glatzmaier, 1980; LaBonte *et al.* 1981; Brown and Gilman, 1984).

In summary, then, the combined neutrino and seismological observations and theory promise a quantitative sounding of the interior of the nearest star. It seems reasonable to expect that if the initiatives are exploited, then by the end of this century we will be in possession of the knowledge of conditions throughout the Sun, hopefully with an understanding of how they are interrelated. We should confirm the evidence of a rapidly spinning core, of primordial origin, and whether that core contains fields of 10^6 G (see discussion in Parker, 1984a, b). We should know whether presently accepted ages of distant stars, based on observed departures from the main sequence, are correct, or whether they must be revised as a consequence of internal mixing. In short, if we exploit the present opportunities, then we shall have unscrambled some of the basic dilemmas of the stellar interior, thereby rescuing the physics of stars from its present fragile position.

4. Solar Microscopy and Stellar Activity

We turn now to the activity of stars. They are, in fact, not static structures at all, but highly active so that their surfaces and outer atmospheres are continually disturbed, sometimes violently, producing X-rays, fast particles, magnetic eruptions, and a variety of other suprathermal effects. The reader is referred to the articles by De Jager and Švestka, Dennis, Lin, and Hudson for a detailed description. The exterior activity is driven by the dynamical convective effects of the interior. As noted in the introduction, none of the activity of the Sun or any other star was anticipated theoretically. It was thrust upon us by observation. The activity involves hitherto unknown magnetohydrodynamic and plasma effects and needs now to be explained by theory. Observations show that the activity has its origins at the small scales (50–100 km) of the individual magnetic fibrils in the Sun. Hence, the observational study of stellar activity begins with the microscopy of the surface of the Sun.

Microwave studies at high resolution (cf. Marsh and Hurford, 1982; Erskine and Kundu, 1982) have begun to map some of the fine sturcture of flares, etc., reviewed in the article by Kundu. Ground-based observations in visible light are used to study all aspects of the activity on scales down to about 500 km, showing that the activity is basically a magnetic effect and that the magnetic field at the visible surface is concentrated into separate, intense, active fibrils of a kilogauss or more. Extensive theoretical studies indicate that the basic magnetohydrodynamical effects occur on scales as small as 500–100 km, producing the fibrils and causing them to behave in the strange collective manner that forms sunspots, initiates coronal eruptions and flares, etc., described in the articles by Muller, Zwaan, Stenflo, and De Jager and Švestka. The next essential step in getting at the physics of stellar activity is the Solar Optical Telescope, which is presently a part of the space science program of the National Aeronautics and Space Administration. The SOT is to operate from orbit, observing in the infra-red, visible, and ultraviolet with a resolution of 75 km or better. The SOT is designed to resolve the individual fibrils, providing direct information on their internal structure, and on their individual motions which drive the activity in the atmosphere above. The activity in the atmosphere can be observed in the radio, UV, X-ray, and gamma-ray emissions. Radio, visible, UV, and X-ray studies of other stars (Worden, 1974; Linsky, 1980; Meyer, 1985a, b; Rosner et al., 1985) establish the general occurrence of the many suprathermal phenomena of stellar activity in essentially all classes of stars. None of the activity in other stars can be resolved, of course, but the gross features, inferred from the integrated light, suggest that it is made up of the same general effects as the activity of the Sun. The exceptions are the extraordinarily active stars, involving accretion of matter from a companion onto spinning compact objects, etc. The Sun, therefore, serves as the laboratory prototype for most X-ray emitters, which otherwise lie outside the reach of hard theoretical physics. The SOT is the essential microscope to provide the physical conditions for understanding and interpreting the results of the International Ultraviolet Explorer and the Einstein X-ray Observatory, in preparation for the Advanced X-ray Astronomical Facility to be flown in the next decade.

The solar-stellar connection, then, begins with high-resolution studies of the Sun to determine the physics of its activity. The pioneering work of Wilson (1978) was the first systematic exploratory extension of activity studies to other stars, with the IUE satellite and the Einstein X-Ray Observatory providing a great leap forward in establishing the existence of chromospheres and coronas for virtually all stars (cf. Linsky, 1980; Ayres et al., 1982a, b; Simon et al., 1982, Linsky et al., 1982; Hartmann et al., 1982; Baliunas and Dupree, 1982; Dupree and Raymond, 1982; Cassinelli et al., 1983; Baliunas et al., 1983; Ayres et al., 1983a, b; Linsky and Gary, 1983; Haisch et al., 1983). Subsequently an extensive cooperative ground based observing program at the Mt. Wilson Observatory has followed the activity of a large number of stars, showing the variety and the variability of the activity and providing precise values for rotation rates, hitherto unavailable (Baliunas et al., 1983; Brown et al., 1984; Walter et al., 1984; Dupree et al., 1984) as well as recording many individual explosions and flare-like outbursts (Blair et al., 1984; Baliunas et al., 1984). The VLA has been used effectively in studying the flare on an M dwarf at 6 cm (Gary *et al.*, 1983) and of course microwave studies of the Sun play a major role in mapping the radio acitivty (Marsh and Hurford, 1982; Erskine and Kundu, 1982). The most intense stellar activity appears to be driven by the tidal effects and vigorous mass transfer between close binary companions. But a star is active whether it has a close companion or not, as a result of its internal convection and the magnetic fields generated by that convection. For it has become clear that the activity of a solitary star like the Sun is a consequence of the dynamical properties of its magnetic fields. The variety of effects are reviewed in the article by Noyes, and the relations between activity and stellar rotation are reviewed in the article by Belvedere in the context of the general principles of the hydromagnetic dynamo. Hence, the discussion of the activity is closely tied to the questions of angular velocity and circulation in the interior, described in the foregoing sections. That is to say, the future results of the microscopic examination of the surface of the Sun with the Solar Optical Telescope, as well as the ongoing high-resolution ground-based observations of the Sun and other stars, are to be interpreted in the light of future results of the solar neutrino and seismology observations.

Now, the magnetic fields that extent outward from the visible surface of the Sun are generated in the convective zone, evidently by the differential rotation and cyclonic convection. The field is primarily in the azimuthal direction as a consequence of the presumed differential rotation $\Omega(r, \theta)$, forming oscillatory dynamo waves that migrate from both sides toward the equator. The 22-year period of the waves provides the 11-year half-cycles associated with each sunspot cycle (Parker, 1955, 1957, 1970, 1979, 1984c; Steenbeck *et al.*, 1966; Leighton, 1969; Moffatt, 1978; Krause and Radler, 1980; Rai Choudhuri, 1984; see also the article by Ruzmaikin). The theoretical models fit most simply to the observations if the angular velocity increases downward through the convective zone. The theoretical dynamo models depend upon an effective turbulent magnetic diffusion coefficient of the order of $\eta \sim 10^{12}$ cm^2 s^{-1}, consistent with the mixing length estimates from standard models of the convective zone (Leighton, 1969; Köhler, 1973; Yoshimura, 1971, 1973, 1975a, b). The resulting dynamo waves present time-dependent patterns resembling the magnetic patterns observed at the surface of the Sun when one assumes that the cyclonic velocities in the convective zone are of the order of 5 cm s^{-1}, with a 10% increase of the angular velocity across the depth of the convective zone. All of these numbers should be kept in mind as the results of solar seimological probing begin to emerge. It should be emphasized that the theoretical dynamo models represent both the magnetic diffusion effects and the cyclonic generation of meridional field in terms of a simple mixing length representation (see also Gilman, 1983). We clearly need quantitative guidance from solar seismology if we are to improve this idealized formulation of the dynamo problem.

For the fact is that all stars seem to have magnetic fields. Some of them may be primordial (Parker, 1979a, pp. 764–769) but with stars like the Sun, where the fields reverse on a regular basis with no consistent north–south asymmetry, the primordial fields that may be trapped in the radiative zone and central core can have little or no presence in the convective zone where the surface fields are generated (Levy and Boyer, 1982; Boyer and Levy, 1984).

On the other hand, there are only weak theoretical limitations on the field in the central core of the Sun, where solar seismology already indicates a high-angular velocity (Duvall *et al.*, 1984). The field could perhaps be as strong as 10^7 G (Parker, 1974, 1975, 1979a, pp. 147–151) in the central core, although the thermal shadows of a primordial field in the radiation zone below the bottom of the convection limit the field strengths there to 5×10^5 G (Parker, 1984b; see Parker, 1984a for a summary of the many ideas to be found in the literature). Levy and Boyer (1982) and Boyer and Levy (1984) show that no more than a few gauss can penetrate into the convective zone without producing magnetic effects at the surface contrary to those observed. It is to be hoped that the analysis of solar oscillations will provide some clarification of the question of the magnetic fields in the deep interior. Once we understand conditions in the Sun, it may be possible to determine something about the situation in the cores of other stars.

The origin of solar magnetic fields is only the beginning of the problem of solar and stellar magnetic activity, of course. The fact is that the magnetic fields behave in the most extraordinary manner when interpreted within our limited knowledge of the theory of magnetohydrodynamics. The convection causes the field to be clumped into widely separated intense flux tubes, with field strengths of the order of 10^3 G and diameters ranging from perhaps 20 km (3×10^{15} Mx) up to 300 km (10^{18} Mx) and larger (Harvey, 1977; Frazier and Stenflo, 1978; Zirin, 1985; Wang *et al.*, 1985). These diameters are well below the usual limits (500 km) of resolution under good seeing conditions, with the result that the structure of the individual magnetic flux tubes, or fibrils, is not known from observation, or is it inferred from elementary theoretical principles.

The magnetic energy of a mean field $\langle B \rangle$ is increased by the factor $B_f / \langle B \rangle$ when it is broken into separate compressed fibrils with fields B_f. The factor is 10 in an active region where $\langle B \rangle = 10^2$ G, and 10^2 in a quiet region where $\langle B \rangle = 10$ G. On the other hand, the compression into widely separated fibrils avoids the endless wrapping and stretching to which a continuum field is subjected by the small-scale convective over-turning, with the result that a fibril field offers less impediment to the convective transport of heat. As a result the total energy of the convective zone, including the thermal energy along with the magnetic energy, is a minimum for fibrils of a few kilogauss (Parker, 1984c).

The intense concentration of field at the surface appears to the result of the reduced opacity and subsequent cooling of the evacuated interior of the individual fibrils (see Deinzer *et al.*, 1984a, b; Hasan, 1985, and references therein). The SOT will resolve the larger fibrils well enough to test the validity of these theoretical modes. Note that this cooling would not occur far below the surface, where the only know effect would be the strong inhibition of convective heat transport within the individual fibril. So it is not clear whether there is a concentrated fibril state of the field far below the surface, whatever may be the total energy reduction if it could be achieved.

To continue with the general discussion, we note that the active regions at the surface of the Sun are bipolar magnetic regions where the powerful azimuthal field has momen-tarily bobbed up through the surface. The ephemeral active regions appear to come up from depths of the order of 10^4 km while the normal active regions come from depths of 10^5 km (Golub *et al.*, 1977, 1981; Parker, 1984d). Following a sojourn of days or weeks at the surface the field appears to pull back again into the Sun with little or no net loss of magnetic flux (Wallenhorst and Topka, 1981; Parker, 1984e; Zirin, 1985). Gaizauskas *et al.* (1983) report a flux emergence at any one time in a large activity complex to be of the order of 10^{23} Mx. Eruptions of new flux of this magnitude occurred every three or four weeks as the older flux disappeared. The feat could not be produced with less than two or three bundles of 10^{23} Mx each, talking turns bobbing through the surface. A much larger total flux of 10^{24} Mx would be required if there were less cooperation. And that might be only a fraction of the total flux in the convective zone beneath the activity complex. Note that 10^{24} Mx over a latitude interval of $10°$ in the lower half (10^5 km) of the convective zone translates into 10^4 G, producing an Alfvén speed of the order of 0.5 km s^{-1}. Such fields might have significant consequences for the extension of the *g*-modes up through the convective zone. We shall have to wait to see what information may derive from the planned observations.

A fundamental question is the rate at which the individual magnetic fibrils are shaken and shuffled about by the convection. Observations (Zirin, 1985; Wang *et al.*, 1985) of individual unresolved fibrils indicate that they sometimes move about with velocities of the order of 0.5 km s^{-1}. There are no observations of transverse or torsional oscillatory motions of the fibrils at the photosphere for the simple reason that the fibrils are not resolved. The SOT will provide the first reliable data on these fundamental questions. Bulk motions of whole regions of fibrils are known from the variation of large-scale flux distributions provided by the standard magnetograms, of course. The observations summarized in the article by Zwaan (1978) have established some of the amazing behavior of the fibrils. For instance, the individual fibrils are observed to clump together in newly emerging flux, forming pores and sunspots (Zwaan, 1978). The clumping is in opposition to the mutual repulsion of the expanded fields above the visible surface and can only be interpreted as forced by unseen hydrodynamic forces below the surface (Meyer *et al.*, 1977; Parker, 1978, 1979a, pp 152–161, 1979b–g; Tsinganos, 1979). The total flux of 10^{22} Mx of a large sunspot, then, is built up from the equivalent of 10^4 individual fibrils of 10^{18} Mx each.

Apart from sunspots, the most spectacular aspects of solar activity are the suprathermal effects, e.g., the solar flare, the coronal mass ejection, and the formation of the chromosphere and corona, described in the articles by Athay, Zirker, Hirayama, and Jackson. It appears that all of these are the result of displacements of the footpoints of the external fields of the Sun, i.e., the individual or collective transport of the individual fibrils. Thus, flares and coronal transients and mass ejections occur when the footpoints are sufficiently displaced by shearing motions at the surface or when one region of flux is rammed into another, producing sufficient distortion of the fields. Low (1977a, b, 1980, 1981, 1985; Low, *et al.*, 1982; Wagner, 1984) has demonstrated how equilibrium ceases to exist when magnetic arches are sheared beyond a certain limit, leading to an outward eruption of the field into space (see the article by Low). Priest (1981) gives an extensive discussion of the possibilities for the formation of flares. Recent observations (Tanaka and Zirin, 1985) indicate that the energy release in a large solar flare may perhaps be as large as 10^{33} ergs, most of it at photon energies of 20 keV and more.

Now, in addition to the large-scale collective displacement of the the fibrils, producing flares and coronal eruptions there are the individual displacements of each fibril. Oscillatory motions with periods of five minutes and less produce Alfvén waves in the field above, while cumulative wandering of individual fibrils among neighboring fibrils wraps and winds the flux tubes about each other in the region above the surface. These motions are opposed by the stresses in the field, so that they represent net work done by the convective motions on the field. The energy goes into the increased strains in the field, whose dissipation evidently produces the heat that is responsible for the existence of both the active X-ray corona (where the fields are rougly 20–100 G) and the coronal holes (where the fields are below about 20 G). The energy input is estimated (Withbroe and Noyes, 1977) to be 10^7 ergs cm^{-2} s^{-1} to maintain the X-ray corona, mostly lost by radiation and thermal conduction from temperatures of 2.5×10^6 K and 3×10^9 atoms cm^{-3}. Rosner *et al.* (1978) make the fundamental observational point that the heating is directly associated with the local magnetic field, on a one to one basis. On the other hand, the energy input is estimated to be 10^6 ergs cm^{-2} s^{-1} to the coronal hole, most of it going into the expansion that produces the solar wind. Indeed, it is this minute fraction of the total solar luminosity, going into the corona, that creates the mass loss and the vast heliosphere, described in the articles by Axford and Hartmann.

It appears at the present time that the coronal holes can only be heated by the dissipation of Alfvén waves propagating outward over distances of 10^{12} cm or more. The X-ray coronal heating requires ten times as much energy input and exhibits the interesting property that the rate of heat input (ergs cm^{-2} s^{-1}), i.e., the X-ray brightness, differs by only a factor of two between the X-ray bright points (ephemeral active regions) with dimensions $L \sim 10^4$ km and the large normal active region with $L \sim 2 \times 10^5$ km. The Alfvén speed is typically 2×10^3 km s^{-1} in the active corona, so that the wave transist time is 10 s for the X-ray bright point and in excess of 10^2 s for the normal active region. Unless there is some vigorous source of 10 s waves, it is not clear how the uniformity in brightness can be achieved over so wide a range of scales and wave transit times. Instead, it appears that the principal source of heating of the X-ray corona may be the cumulative shuffling and resulting nonequilibrium of the fibrils, rather than their periodic agitation. It is clear that the quantitative question of the wave spectrum and the speed with which the individual fibrils are shuffled among each other can be answered only by observations at high resolution, such as will be provided by the Solar Optical Telescope sometime in the nineties. Until then we can develop the theory of wave dissipation and the theory of dynamical nonequilibrium (Syrovatskii, 1971, 1978, 1981; Parker, 1972, 1979a, pp 359–439, 1981a, b, 1982e, 1985; Yu, 1973; Bobrova and Syrovatski, 1979; Priest, 1982; Tsinganos, 1982; Tsinganos *et al.*, 1984; Hollweg, 1984, Hollweg and Sterling, 1984; Sterling and Hollweg, 1984; Fla *et al.*, 1984; Van Ballegooijen, 1985; Vainshtein and Parker, 1984; in particular see the article by Low) but we cannot be sure of coronal heating in the Sun, or in other stars, until the observations show clearly the motions of the fibrils at the photosphere. Hence, we cannot understand the Einstein X-ray results, or the anticipated X-ray survey by AXAF, until the SOT has provided direct quantitative information on the agitation of magnetic fibrils.

The Solar Optical Telescope, then, has many tasks already well defined. It has the potential to get us safely over several chasms of ignorance that presently limit understanding. It will be surprising, too, if the observations carried out with the Solar Optical Telescope do not provide some surprises, just as the ^{71}Ga experiment and helioseismology promise to do.

References

Ayres, T. R., Linsky, J. L., Basri, G. S., Landsman, W., Henry, R., Moos, H. W., and Stencel, R. E.: 1982a, *Astrophys. J.* **256**, 550.

Ayres, T. R., Simon, T., and Linsky, J. L.: 1982b, *Astrophys. J.* **263**, 791.

Ayres, T. R., Linsky, J. L., Simon, T., Jordan, C., and Brown, A.: 1983a, *Astrophys. J.* **274**, 784.

Ayres, T. R., Stencel, R. E., Linsky, J. L., Simon, T., Jordan, C., Brown, A., and Engvold, O.: 1983b, *Astrophys. J.* **274**, 801.

Bahcall, J. N.: 1985, in S. Shapiro and S. Teukolsky (eds.), 'Solar Neutrinos', *Highlights in Astrophysics: Concepts and Controversies,* John Wiley Publ., New York.

Bahcall, J. N. and Davis, R.: 1982, in C. A. Barnes and D. Schramm (eds.), *Essays on Nuclear Astrophysics*, Cambridge University Press, Cambridge, p. 243.

Bahcall, J. N., Cleveland, B. T., Davis, R., Dostrovsky, I., Evans, J. C., Frati, W., Friedlander, G., Lande, K., Rowley, J. K., Stoenner, R. W., and Weneser, J.: 1978, *Phys. Rev. Letters* **40**, 1351.

Bahcall, J. N., Heubner, W. J., Lubow, S. H., Parker, P. D., and Ulrich, R. K.: 1982, *Rev. Mod. Phys.* **54**, 767.

Baliunas, S.L. and Dupree, A. K.: 1982, *Astrophys. J.,* **252** 668.

Baliunas, S. L., Hartmann, L., and Dupree, A. K.: 1983, *Astrophys. J.* **271**, 672.

Baliunas, S. L., Guinan, E. F., and Dupree, A. K.: 1984, *Astrophys. J..* **282**, 733.

Baliunas, S. L., Vaughan, A. H., Hartmann, L., Middelkoop, F., Mihalas, D., Noyers, R. W., Preston, G. W., Frazer, J., and Lanning, H.: 1983, *Astrophys. J.* **275**, 752.

Bienayme, O., Maeder, A., and Schatzman, E.: 1983, *Astron. Astrophys.* **131** 316.

Blair, W. P., Raymond, J. C., Dupree, A. K., Wu, C. C., Holm, A. V., and Swank, J. H.: 1984, *Astrophys. J.,* **278**, 270.

Bobrova, N. A. and Syrovatskii, S. I.: 1979, *Solar Phys.* **61**, 379.

Bodenheimer, P.: 1965, *Astrophys. J.* **142**, 451.

Bodenheimer, P.: 1966, *Astrophys. J.* **144**, 103.

Boyer, D. and Levy, E. H.: 1984, *Astrophys. J.* **277**, 848.

Brown, T. M. and Gilman, P. A.: 1984, *Astrophys. J.* **286**, 804.

Brown, A., Jordan, C., Stencel, R. E., Linsky, J. L., and Ayres, T. R.: 1984 *Astrophys. J.* **283**, 731.

Brown, T. M., Mihalas, B. W., and Rhodes, E. J.: 1985, in P. A. Sturrock, T. E. Holzer, D. Mihalas, and R. K. Ulrich (eds.), *Physics of the Sun*, D. Reidel Publ. Co., Dordrecht, Holland.

Burlaga, L. F.: 1984, *Space Sci. Rev.* **39**, 255.

Cassinelli, J. P., Hartmann, L., Sanders, W. T., Dupree, A. K., and Meyers, R. V.: 1983. *Astrophys. J.* **268**, 205.

Cayrel. R., DeStrobel. G. C., Campbell, B., and Dappen, W.: 1984, *Astrophys. J.* **283**, 205.

Christensen-Dalsgaard, J., Duvall, T. L., Gough, D. O., Harvey, J. W., and Rhodes, E. J.: 1985, *Science* (in press).

Claverie, A., Isaak, G. R., McLeod, C. P., vander Raay, H. B., and Roca Cortes, T.: 1980, *Astron. Astrophys.* **91**, L9.

Cleveland, B. T., Davis, R., and Rowley, J. K.: 1984, in *Proceedings of the Second International Symposium on Resonance Ionization and Its Applications*, (Knoxville, Tenn.), Adam Higler Ltd, Bristol.

Davis, R.: 1964, *Phys. Rev. Letters* **12**, 303.

Dearborn, D. S. P. and Blake, J. B.: 1980, *Astrophys. J.* **237**, 616.

Dearborn, D. S. P. and Blake, J. B.: 1982, *Astrophys. J.* **257**, 896.

Deinzer, W., Hensler, G. , Schussler, M., and Weishaar, E.: 1984a, *Astron. Astrophys.* **139**, 426.

Deinzer, W., Hensler, G. Schussler, M., and Weishaar, E.: 1984b, *Astron. Astrophys.* **139**, 435.

Deubner, F.-L. and Gough, D.: 1984, *Ann. Rev. Astron. Astrophys.* **22**, 593.

Dicke, R. H.: 1982, *Nature* **300**, 693.

Duncan, D. K.: 1981, *Astrophys. J.* **248**, 651.

Dupree, A. K., and Raymond, J. C.: 1982, *Astrophys. J.* **263**, L63.

Dupree, A. K., Hartman, L., and Avrett, E. H.: 1984, *Astrophys. J.* **281**, L37.

Durney, B.: 1983a, *Astrophys. J.* **267**, 822.

Durney, B.: 1983b, *Astrophys. J.* **269**, 671.

Duvall, T. L., Dziembowski, W. A., Goode, P. R., Gough, D. O., Harvey, J. W., and Leibacher, J. W.: 1984, *Nature* **310**, 22.

Eddy, J. A.: 1976, *Science* **192**, 1189.

Eddy, J. A., Gilliland, R. L., and Hoyt, D. V.: 1982, *Nature* **300**, 689.

Endal, A. S. and Twigg, L. W.: 1984, *Astrophys. J.* **260**, 342.

Erskine, F. T. and Kundu, M. R.: 1982, *Solar Phys.* **76**, 221.

Filippone, B. W. and Schramm, D. N.: 1982, *Astrophys. J.* **253**, 293.

Flå, T., Habbal, S. R., Holzer, T. E., and Leer, E.: 1984, *Astrophys. J.* **280**, 382.

Frazier, E. N. and Stenflo, J. O.: 1978, *Astron. Astrophys.* **70**, 789.

Gaizaukas, V., Harvey, K. L., Harvey, J. W., and Zwaan, C.: 1983: *Astrophys. J.* **265**, 1065.

Gary, D. E., Linsky, J. L., and Dulk, G. A.: 1983, *Astrophys. J.* **263**, L79.

Genove, F. and Schatzman, E.: 1979, *Astron. Astrophys.* **78**, 323.

Giampapa, M. S.: 1984, *Astrophys. J.* **277**, 235.

Gilliland, R. L.: 1982, *Astrophys. J.* **253**, 399.

Gilman, P. A.: 1983, *Astrophys. J. Suppl.* **53**, 243.

Gilman, P. A. and Glatzmaier, G. A.: 1980, *Astrophys. J.* **241**, 793.

Gilman, P. A. and Howard, R.: 1984, *Solar Phys.* **93**, 171.

Glatzmaier, G. A.: 1985, *Astrophys. J.* **291**, 300.

Glatzmaier, G. A. and Gilman, P. A.: 1982, *Astrophys. J.* **256**, 316.

Golub, L., Krieger, A. S., Harvey, J. W., and Vaiana, G. S.: 1977, *Solar Phys.* **53**, 111.

Golub, L., Rosner, R., Vaiana, G. S., and Weiss, N. O.: 1981, *Astrophys. J.* **243**, 309.

Gough, D. O.: 1983a, *Phys. Bull.* **34**, 502.

Gough. D. O.: 1983b, in P. Shaver, D. Kunth, and K. Kjar (eds.), *Proc. ESO Workshop on Primordial Helium*, Garching, European Southern Observatory.

Gough, D. O.: 1984, *Adv. Space Res.* **4**, 85.

Grec, G., Fossat, E. and Pomerantz, M.: 1980, *Nature* **288**, 541.

Haisch, B. M., Linsky, J. L., Bormann, P. L., Stencel, R. E., Antiochos, S. K., Golub, L., and Vaiana, G. S.: 1983, *Astrophys. J.* **267**, 280.

Hartmann, L., Dupree, A. K., and Raymond, J. C.: 1982, *Astrophys. J.* **252**, 214.

Harvey, J.: 1977, *Highlights of Astronomy* **4**, part II, 223.

Hasan, S. S.: 1985, *Astron. Astrophys.* (in press).

Hill, F. and Newkirk, G.: 1985, *Solar Phys.* **95**, 201.

Hollweg, J. V.: 1984, *Astrophys. J.* **277**, 392.

Hollweg, J. V. and Sterling, A. C.: 1984, *Astrophys. J.* **282**, L31.

Howard, R.: 1976, *Astrophys. J.* **210**, L159.

Howard, R.: 1979, *Astrophys. J.* **228**, L45.

Howard, R.: 1984, *Ann. Rev. Astron. Astrophys.* **22**, 131.

Howard, R. and LaBonte, B. J.: 1980a, *Astrophys. J.* **239**, L33.

Howard, R. and LaBonte, B. J.: 1980b *Astrophys. J.* **239**, 738.

Hoyt, D. V., Eddy, J. A., and Hudson, H. S.: 1983, *Astrohys. J.* **275**, 878.

Hudson, H. S., Silva, S., Woodward, M., and Willson, R. C.: 1982, *Solar Phys.* **76**, 211.

Köhler, H.: 1973, *Astron. Astrophys.* **25**, 467.

Kotov, V. A., Severny, A. B., and Tsap, T. T.: 1978, *Monthly Notices Roy. Astron. Soc.* **183**, 61.

Krause, F. and Rädler, K. H.: 1980, *Mean-Field Magnetohydrodynamics and Dynamo Theory*, Pergamon Press, Oxford.

LaBonte, B. J., Howard, R., and Gilman, P. A.: 1981, *Astrophys. J.* **250**, 796.

LaBonte, B. J., Chapman, G. A., Hudson, H. S., and Willson, R. C.: 1984, *Solar Irradiance Variations on Active Region Times Scales*, NASA Conf. Publ. 2310.

Leibacher, J. W.: 1985, *A Selected Bibliography of Helioseismology*, Global Oscillation Network Group, Report No. 2, National Solar Observatory, NOAO.

Leibacher, J. W. and Stein, R. F.: 1971, *Astrophys. Letters* **7**, 191.

Leibacher, J. W. and Stein, R. F.: 1981, in S. Jordan (ed.), *The Sun as a Star*, NASA SP-450, p. 263.

Leighton, R. B.: 1969, *Astrophys. J.* **156**, 1.

Levy, E. H. and Boyer, D.: 1982, *Astrophys. J.* **254**, L19.

Linsky, J. L.: 1980, *Ann. Rev. Astron. Astrophys.* **18**, 439.

Linsky, J. L. and Gary, D. E.: 1983, *Astrophys. J.* **274**, 776.

Linsky, J. L., Bormann, P. L., Carpenter, K. G., Wing, R. F., Giampapa, M. S., Worden, S. P., and Hege, E. K.: 1982, *Astrophys. J.* **260**, 670.

Low, B. C.: 1977a, *Astrophys. J.* **212**, 234.

Low, B. C.: 1977b, *Astrophys. J.* **217**, 988.

Low, B. C.: 1980, *Astrphys. J.* **239**, 377.

Low, B. C.: 1981 *Astrophys. J.* **251**, 352.

Low, E.: 1985, *Astrophys. J.* (submitted).

Low., B. C., Murno, R. H., and Fisher, R. R.: 1982, *Astrophys. J.* **254**, 335.

Markson, R.: 1978, *Nature* **273**, 103.

Markson, R.: 1983, *Proc. Second International Symposium on Solar-Terrestrial Influences on Weather and Climate*, 2–6 August, 1982, Boulder, Colo., U.S.A.

Marsch, K. A. and Hurford, G. J.: 1982, *Ann. Rev. Astron. Astrophys.* **20**, 497.

McNeill, W. H.: 1976, *Plagues and Peoples,* Garden City, Anchor Press, New York.

Meyer, J. P.: 1985a, *Astrophys. J. Suppl.* **57**, 151.

Meyer, J. P.: 1985b, *Astrophys. J. Suppl* **57**, 173.

Meyer, F., Schmidt, H. U., Weiss, N. O., and Wilson, P. R.: 1974, *Monthly Notices Roy. Astron. Soc.* **169**, 35.

Meyer, F., Schmidt, H. U., and Weiss, N. O.: 1977, *Monthly Notices Roy. Astron. Soc.* **179**, 741.

Mitchell, J. M.: 1976, *Quatenary Res.* **6**, 181.

Moffatt, H. K.: 1978, *Magnetic Field Generation in Electrically Conducting Fluids*, Cambridge University Press, Cambridge.

Newkirk, G.: 1984, *Ann. Rev. Astron. Astrophys.* **21**, 429.

Parker, E. N.: 1955, *Astrophys.* **122**, 293.

Parker, E. N.: 1957, *Proc. Nat. Acad. Sci.* **43**, 8.

Parker, E. N.: 1970, *Astrophys. J.* **162**, 665.

Parker, E. N.: 1972, *Astrophys. J.* **174**, 499.

Parker, E. N.: 1974, *Astrophys. Space Sci.* **31**, 261.

Parker, E. N.: 1975, *Astrophys. J.* **198**, 205.

Parker, E. N.: 1978, *Astrophys. J.* **222**, 357.

Parker, E. N.: 1979a, *Cosmical Magnetic Fields,* Clarendon Press, Oxford.

Parker, E. N.: 1979b, *Astrophys. J.* **230**, 905.

Parker, E. N.: 1979c, *Astrophys. J.* **230**, 914.

Parker, E. N.: 1979d, *Astrophys. J.* **231**, 250.

Parker, E. N.: 1979e, *Astrophys. J.* **231**, 270.

Parker, E. N.: 1979f, *Astrophys. J.* **232**, 282.

Parker, E. N.: 1979g, *Astrophys. J.* **234**, 333.

Parker, E. N.: 1981a, *Astrophys. J.* **244**, 631.

Parker, E. N.: 1981b, *Astrophys. J.* **244**, 644.

Parker, E. N.: 1982, *Geophys. Astrophys. Fluid Dyn.* **22**, 195.

Parker, E. N.: 1983a, *Astrophys. J.* **264**, 635.

Parker, E. N.: 1983b, *Astrophys. J.* **264**, 642.

Parker, E. N.: 1983c, *Geophys. Astrophys. Fluid Dyn.* **23**, 85.

Parker, E. N.: 1983d, *Geophys. Astrophys. Fluid Dyn.* **24**, 79.

Parker, E. N.: 1983e, *Geophys. Astrophys. Fluid Dyn.* **24**, 245.

Parker, E. N.: 1984a, *Astrophys. J.* **286**, 666.

Parker, E. N.: 1984b, *Astrophys. J.* **286**, 677.

Parker, E. N.: 1984c, *Astrophys. J.* **276**, 341.

Parker, E. N.: 1984d, *Astrophys. J.* **280**, 423.
Parker, E. N.: 1984e, *Astrophys. J.* **281**, 839.
Parker, E. N.: 1985, *Geophys. Astrophys. Fluid. Dyn.* (submitted).
Press, W. H.: 1981, *Astrophys. J.* **245**, 286.
Press, W. H. and Rybicki, G. B.: 1981, *Astrophys. J.* **248**, 751.
Priest, E. R. (ed.): 1981, *Solar Flare Magnetohydrodynamics*, Gordon and Breach, New York.
Priest, E. R.: 1982, *Solar Magnetohydrodynamics*, D. Reidel Publ. Co., Dordrecht, Holland, p. 206.
Rai Choudhuri, A.: 1984, *Astrophys. J.* **281**, 846.
Rosner, R. and Weiss, N. O.: 1985, *Nature* (submitted).
Rosner, R., Tucker, W. H., and Vaiana, G. S.: 1978, *Astrophys. J.* **220**, 643.
Rosner, R., Golub, L., and Vaiana, G. S.: 1985, *Ann. Rev. Astron. Astrophys.* **23** (in press).
Schatzman, E.: 1977, *Astron. Astrophys.* **56**, 211.
Schatzman, E. and Maeder, A.: 1981, *Astron. Astrophys.* **96**, 1.
Scherrer, P. H. and Wilcox, J. M.: 1983, *Solar Phys.* **82**, 37.
Scherrer, P. H., Wilcox, J. M., Kotov, V. A., Severny, A. B., and Tsap, T. T.: 1979, *Nature* **277**, 635.
Severny, A. B., Kotov, V. A., and Tsap, T. T.: 1976, *Nature* **259**, 87.
Severny, A. B., Kotov, V. A., and Tsap, T. T.: 1984, *Nture* **307**, 247.
Simon, T., Linsky, J. L., and Stencel, R. E.: 1982, *Astrophys. J.* **257**, 225.
Spruit, H. C.: 1985, 'Mixing in the Solar Interior', *Hydromagnetics of the Sun*, Proc. Fourth European Meeting on Solar Physics, ESA SP-220, European Space Agency.
Steenbeck, M., Krause, K., and Rädler, K. H.: 1966, *Z. Naturforsch.* **A21**, 369.
Sterling, A. C. and Hollweg, J. V.: 1984, *Astrophys. J.* **285**, 843.
Strauss, J. M., Blake, J. B., and Schramm, D. N.: 1976, *Astrophys. J.* **204**, 481.
Sur, B. and Boyd, S. N.: 1985, *Phys. Rev. Letters* **54**, 85.
Syrovatskii, S. I.: 1971, *Soviet Phys. JETP* **33**, 933.
Syrovatskii, S. I.: 1978, *Solar Phys.* **58**, 89.
Syrovatskii, S. I.: 1981, *Ann. Rev. Astron. Astrophys.* **19**, 163.
Tanaka, K. and Zirin, H.: 1985, *Astrophys. J.* (in press).
Tassoul, M. and Tassoul, J. L.: 1984a, *Astrophys. J.* **279**, 384.
Tassoul, M. and Tassoul, J. L.: 1984b, *Astrophys. J.* **286**, 350.
Thomas, J. H.: 1979, *Nature* **280**, 662.
Toomre, J.: 1985, in R. K. Ulrich (ed.), 'Overview of Solar Seismology', *Solar Seismology from Space*, Jet Propulsion Laboratory, Pasadena, California.
Tsinganos, K. C.: 1979, *Astrophys. J.* **231**, 260.
Tsinganos, K. C.: 1982, *Astrophys. J.* **259**, 832.
Tsinganos, K. C., Distler, J., and Rosner, R.: 1984, *Astrophys. J.* **278**, 409.
Ulrich, R. K.: 1970, *Astrophys. J.* **162**, 993.
Vainshtein, S. I. and Parker, E. N. 1985, *Astrophys. J.* (submitted).
Van Ballegooijen, A. A.: 1985, *Astrophys. J.* (in press).
Vauclair, G., Vauclair, S., and Michaud, G.: 1978a, *Astrophys. J.* **223**, 920.
Vauclair, S., Vauclair, G., Schatzman, E., and Michaud, G.: 1978b, *Astrophys. J.* **223**, 567.
Wagner, W. J.: 1984, *Ann. Rev. Astron. Astrophys.* **22**, 267.
Wallenhorst, S. G. and Topka, K. P.: 1981, *Solar Phys.* **81**, 33.
Wallerstein, G. and Conti, P. S.: 1969, *Ann. Rev. Astron. Astrophys.* **7**, 99.
Walter, F. M., Linsky, J. L., Simon, T., Golub, L., and Vaiana, G. S.: 1984, *Astrophys. J.* **281**, 815.
Wang, J., Zirin, H., and Shi, Z.: 1985, *Solar Phys.* **98**, 241.
Wilson, O. C.: 1978, *Astrophys. J.* **226**, 379.
Wilson, R. C., Gulkis, S., Janssen, M., Hudson, H. S., and Chapman, G. A.: 1981, *Science* **211**, 700.
Withbroe, G. L. and Noyes, R. W.: 1977, *Ann. Rev. Astron. Astrophys.* **15**, 363.
Worden, S. P.: 1974, *Publ. Astron. Soc. Pacific* **86**, 595.
Yoshimura, H.: 1971, *Solar Phys.* **18**, 417.
Yoshimura, H.: 1973, *Solar Phys.* **33**, 131.
Yoshimura, H.: 1975a, *Astrophys. J.* **201**, 740.
Yoshimura, H.: 1975b, *Astrophys. J. Suppl.* **29**, 467.
Yu, G.: 1973, *Astrophys. J.* **181**, 1003.
Zatsepin, G.: 1982, in *Neutrino '82*, Proc. Intern. Conf. Neutrino Phys., Balantonfured, Hungary, p. 53.
Zirin, H.: 1985, in *Proc. Giovanelli Memorial Symposium*, Big Bear Solar Observatory Report No. 0245.
Zwaan, C.: 1978, *Solar Phys.* **60**, 213.

ANNOUNCEMENT

Progress in Solar Physics

Editors: C. de Jager and Z. Švestka

Please note that a hardbound edition of this special issue of *Solar Physics*, Vol. 100, Nos. 1–2 (October 1985), is available from the publishers.

ISBN 90–277–2180–7   Prices: Dfl. 235,– / $99.50 / £65.25

TABLE OF CONTENTS

(Volume 100)

ARTICLES

Unstable Current Systems and Plasma Instabilities in Astrophysics

Proceedings of the 107th Symposium of the International Astronomical Union, held in College Park, Maryland, U.S.A., August 8—11, 1983

Edited by M. R. KUNDU and G. D. HOLMAN

1984, xxii + 566 pp.
Cloth Dfl. 165,—/US$ 62.00   ISBN 90-277-1886-5
Paper Dfl.   80,—/US$ 29.50   ISBN 90-277-1887-3

The new generation of space observations has led to an increasing need for a thorough understanding of processes that occur in magnetized plasmas. The realization that essentially the same plasma processes must be understood for many problems related to astrophysical, space, and man-made plasmas has led to a greater interest in interdisciplinary meetings involving experts from these diverse fields. The Symposium, the Proceedings of which appear in this volume, was the first attempt within the International Astronomical Union to bring together scientists from these disciplines. Included papers cover topics as varied as jets from the nuclei of active galaxies, solar flares, and planetary magnetospheres. Taken as a whole, the Proceedings represent an important step in bringing together in a single volume papers representing recent progress in overlapping disciplines which until now have not interacted strongly.

D. REIDEL PUBLISHING COMPANY

A Member of the Kluwer Academic Publishers Group

P.O. Box 17, 3300 AA Dordrecht, The Netherlands
190 Old Derby Street, Hingham MA 02043, U.S.A.
Falcon House, Queen Square, Lancaster LA1 1RN, U.K.